D0282812

The Changing Carbon Cycle

The Changing Carbon Cycle

A Global Analysis

EDITED BY
JOHN R. TRABALKA AND DAVID E. REICHLE

With 203 Illustrations

Springer-Verlag
New York Berlin Heidelberg
London Paris Tokyo

John R. Trabalka
David E. Reichle
Environmental Sciences Division
Oak Ridge National Laboratory
Oak Ridge, Tennessee 37831, U.S.A.

Library of Congress Cataloging in Publication Data
Main entry under title:
The Changing carbon cycle.
Selected papers from the proceedings of the Sixth
Annual Oak Ridge National Laboratory Life Sciences
Symposium, held Oct. 31–Nov. 2, 1983 in Knoxville, Tenn.
Bibliography: p.
Includes index.
1. Carbon cycle (Biogeochemistry)—Congresses.
2. Atmospheric carbon dioxide—Environmental aspects—
Congresses. I. Trabalka, John R. II. Reichle, David E.
III. Oak Ridge National Laboratory Life Sciences
Symposium (6th : 1983 : Knoxville, Tenn.)
QH344.C46 1986 551.5′2 85-25138

Media conversion by David E. Seham Associates Inc., Metuchen, New Jersey.
Printed and bound by R.R. Donnelley & Sons, Harrisonburg, Virginia.
Printed in the United States of America.

9 8 7 6 5 4 3 2 1

ISBN 0-387-96211-5 Springer-Verlag New York Berlin Heidelberg
ISBN 3-540-96211-5 Springer-Verlag Berlin Heidelberg New York

Foreword

The United States Government, cognizant of its responsibilities to future generations, has been sponsoring research for nine years into the causes, effects, and potential impacts of increased concentrations of carbon dioxide (CO_2) in the atmosphere. Agencies such as the National Science Foundation, National Oceanic and Atmospheric Administration, and the U.S. Department of Energy (DOE) cooperatively spent about $100 million from FY 1978 through FY 1984 directly on the study of CO_2. The DOE, as the lead government agency for coordinating the government's research efforts, has been responsible for about 60% of these research efforts.

William James succinctly defined our purpose when he stated science must be based upon ". . . irreducible and stubborn facts." Scientific knowledge can and will reduce the present significant uncertainty surrounding our understanding of the causes, effects, and potential impacts of increasing atmospheric CO_2. We have come far during the past seven years in resolving some underlyinig doubts and in narrowing the ranges of disagreement. Basic concepts have become less murky.

Yet, much more must be accomplished; more irreducible and stubborn facts are needed to reduce the uncertainties so that we can improve our knowledge base. Uncertainty can never be reduced to zero. However, with a much improved knowledge base, we will be able to learn, understand, and be in a position to make decisions.

U.S. Department of Energy
Frederick A. Koomanoff
Carbon Dioxide Research Division
Washington, D.C. 20545

Preface

The potential importance of the radiatively active gases in the atmosphere in maintaining the heat balance and spectral properties of the earth has been recognized for nearly a century. It is now realized that the earth's atmosphere has evolved over billions of years as a result of natural biogeochemical processes. The influence of the atmosphere on the earth's biological evolution (and vice versa) through geologic time has become a basic scientific premise. We now appreciate that living organisms, and the life processes, have had a marked effect on the development of the earth's atmosphere, producing a closely coupled system whereby the crucial chemical and radiative environment necessary for this planet's habitability is maintained.

Decades ago scientists speculated about the delicate balance of radiative gases in the lower atmosphere and the importance of this balance to contemporary biological systems. More recently, there have been vivid examples of humankind's potential to affect the global environment—the global distribution of xenobiotic chemicals and the extinction of many species are examples. In the 1960s the impetus for scientific investigations to garner a better understanding of global ecology and its importance to human welfare was given by the International Council of Scientific Unions (ICSU). The ICSU initiated the International Biological Program (IBP), which eventually involved scores of nations and lasted well into the 1970s, to examine how species and ecosystems related to their changing environments. Building upon other international scientific programs, such as the International Geophysical Year and the International Hydrologic Decade, research conducted by IBP scientists heightened appreciation of how the linked biogeochemical cycles affect the productivity of biological systems. In the late 1970s, with the United Nations Conference on the Environment, the groundwork for a new era of international awareness and cooperation in addressing global environmental problems was established.

As our sensitivity to the impact of human activities on the environment was developing, industrialization was causing exponential growth in global energy production and consumption. Historically, much of the world's energy production has been derived from the combustion of fossil fuels, and little change in this pattern is foreseen before the end of the 20th century. Farsighted individuals have raised questions about the potential consequences of the annual releases of several billion metric tons of carbon dioxide (CO_2), a radiatively active gas, to the atmosphere from fossil-fuel combustion. A few scattered, imperfect records suggesting a changing CO_2 content of the atmosphere have existed since early in this century, but it was not until the well-documented measurements from the Mauna Loa Observatory, Hawaii (beginning in the late 1950s), became available that a convincing, steady rise in atmospheric CO_2 could be demonstrated. The Mauna Loa record suggested an exponential rise in atmospheric CO_2 which paralleled the accelerating utilization of fossil fuels.

The U.S. Congress enacted the National Climate Program Act of 1978 to establish a comprehensive national policy for dealing with all climate-related issues. This act commissioned several governmental agencies to develop research programs to investigate the various aspects of climate. These included the U.S. Department of Energy (DOE), which serves as the lead coordinating agency for research on atmospheric CO_2. The functions assumed by the DOE Carbon Dioxide Research Program in this role are to coordinate and sponsor research, synthesize our collective knowledge, and communicate findings to scientists and decision makers, both national and international.

While it is simple to state the fundamental policy questions of concern regarding the increase of atmospheric CO_2, it has been more difficult to identify all of the associated scientific and social issues. Our societal concerns relate to factors that increase CO_2 in the atmosphere, time projections of the phenomenon into future decades, understanding of future potential consequences of the "greenhouse effect" on climate, and, ultimately, the effects of any induced climatic changes on both humans and the rest of the biosphere. Obviously, such concerns will not be resolved simply by measuring the injection of fossil fuel CO_2 into the atmosphere. The global biogeochemical cycle of carbon is highly complex, geographically and temporally. Carbon dioxide released to the atmosphere from fossil fuels joins that respired naturally by the biosphere and released through oceanic processes. Both terrestrial vegetation and marine plants absorb CO_2 from the atmosphere in the process of photosynthesis. Deposits of dead organic matter and carbonate precipitates form long-term reservoirs of carbon removed from the atmosphere. Humankind further complicates this cycle by burning fuels, mining limestone, and clearing vast acreages of forested landscape. And so, the scientific issues associated with the carbon cycle alone center on understanding a complex global system, varying spatially and temporally, dynamically regulated by complicated biogeochemical relationships—a cycle which is increasingly being perturbed by human activities. Other aspects of atmospheric CO_2 research, for example, climate effects, are no less complex.

The basic scientific questions concerning atmospheric CO_2 content are how fast is it increasing, what will its future time course be, and what are the contributing sources of CO_2? Important sources of CO_2 include both the combustion of fossil fuels and the clearing of terrestrial vegetation. Knowledge of past histories becomes crucial here if we are to be able to predict that future. The rate of CO_2 uptake by the oceans largely determines the accumulation of excess CO_2 in the atmosphere. However, contemporary estimates of ocean uptake, based upon the difference between present-day combustion releases and observed atmospheric increases, have large uncertainties. Whether the rate of oceanic uptake has (or can be) changed is similarly dependent upon knowledge of all historic atmospheric sources. Whether or not the rate of oceanic uptake will increase in response to increasing releases of CO_2 to the atmosphere is, perhaps, the most singularly important and difficult question to be answered.

As part of its reporting responsibilities, the DOE is scheduled to summarize our scientific understanding of the CO_2-climate change issue, including the global carbon cycle, in a series of technical reviews to be published beginning in 1985. These scientific reviews are to be followed by a synthesis of this knowledge in the context of societal concerns about potential climate change and incumbent policy decisions, should those changes be judged consequential. As a precursor to the review of status and perspectives on gloval carbon cycle research, the Sixth Annual Oak Ridge National Laboratory Life Sciences Symposium was devoted to the topic of "The Global Carbon Cycle: Analysis of the Natural Cycle and Implications of Anthropogenic Alterations for the Next Century." The world's experts from internationally renowned institutions were asked to critically review the key scientific issues. The DOE was joined by the National Science Foundation and the National Oceanic and Atmospheric Administration (NOAA), as well as U.S. industry represented by the Electric Power Research Institute (EPRI) and the Gas Research Institute (GRI), in sponsoring the symposium. From October 31 through November 2, 1983, the authors in this volume, representatives from the above organizations, and an international body of scientists gathered Knoxville, Tennessee, to discuss these issues.

We are indebted to enumerable individuals, without whose assistance this volume would not have been possible. The sponsors and authors have been credited. Particular individuals who helped organize the symposium also deserve acknowledgment. The advisory committee for the symposium provided invaluable advice in selecting topics for presentation and identifying potential speakers; our thanks are extended to Roger C. Dahlman (DOE), John A. Laurmann (GRI); Ralph M. Perhac (EPRI), James T. Peterson (NOAA), and Eric T. Sundquist (USGS). The local ORNL committee of Charles F. Baes, Jr., William R. Emanuel, Bruce B. Hicks, George G. Killough, Jr., Alfred M. Perry, Jr., and Ralph M. Rotty helped organize the program and contacted the speakers. Ms. Joanne S. Sanford, Mrs. Debbie E. Shepherd, and Mrs. Marvel D. Burtis provided superb administrative support and handled most of the organizational burden before, during, and after the symposium. Mrs. Bonnie S. Reesor of the ORNL Conference Office, Mrs. Vivian A. Jacobs of the ORNL Conference Publications Group, and Ms. Donna D. Rhew of the ORNL Environmental Sciences Division Word Processing Center provided professional assistance.

This volume is the result of the presentations at the Sixth Life Sciences Symposium, carefully selected and critically reviewed to represent a summary of our understanding of some of the most important scientific issues relative to the global carbon cycle. While the opinions presented remain those of the authors of individual chapters, and, collectively, of the editors, and not those of the sponsors, it is believed that the contents of the following pages will be useful to the technical and lay community interested in the global carbon cycle.

The contributions to this volume have been deliberately selected and organized, beginning with a statement of the problem from a societal perspective and leading to the scientific issues to be understood before the nature of the problem can be identified or, even more, addressed through rational decision making. The first five chapters discuss the atmospheric concentrations of CO_2, beginning with the most recent patterns of increases and progressing backward in time. This discussion requires examination of the various techniques necessary to piece together the historical atmospheric CO_2 record. Four subsequent chapters focus on a central dilemma of current scientific investigation—that of using sophisticated analytical techniques and interpreting ratios of carbon isotopes in tree rings and geologic media laid down over past centuries to both estimate historic atmospheric concentrations of CO_2 and identify sources both fossil and biological. The use of tree-ring data for such purposes is fraught with controversy at present; our intent was to allow the key (and differing) points of view to be expressed.

The atmosphere is only one of three major global reservoirs of carbon. Understanding the atmospheric exchanges with, and the internal processes of, the biogeochemical cycles of carbon in the biosphere and oceans is central to understanding the dynamics of the global carbon cycle.

An important reservoir of carbon active in the global cycle, terrestrial biota and soils, is discussed in the next four chapters. Clearing land of vegetation, with its resulting release of CO_2 through burning and decomposition, is a major perturbation of the natural gaseous exchanges of photosynthesis and respiration between land and the atmosphere. The first topic addressed is that of estimating historical changes in the carbon pool represented by terrestrial vegetation. This is followed by a chapter on how soil carbon storage changes through time in response to disturbance and recovery of the vegetation cover. Evaluation of the potential of remote sensing to measure contemporary changes in land cover on continental and global scales, with examples from Africa and South America, concludes this section of the volume.

The next five chapters are then devoted to carbon cycles in the ocean. Beginning with a discussion of relatively simple model representations of ocean carbon cycling, subsequent chapters deal with successive, more complex models necessary to capture important biological, chemical, and physical processes and their spatial and temporal variability. A final chapter in this section summarizes recent empirical measurement of CO_2 and alkalinity in ocean waters.

How all of this scientific information can be both synthesized and interpreted and also used to address the central issue of the volume—that of the role of anthropogenic CO_2 emissions to the atmosphere in changing the global carbon cycle—is treated in the penultimate section. Following a chapter which discusses the value and limitations of using geologic analogs in carbon cycle research and modeling, the requirements for a satisfactory model representation of the global carbon cycle are set forth. This is followed by chapters discussing how such models can be calibrated and then validated by observed data, and how the seasonal and geographical patterns of atmospheric CO_2 provide information on the current global carbon balance. The next three chapters address the fossil fuel sources of anthropogenic CO_2 emissions to the atmosphere from the standpoint of recent history, current patterns, and likely future trends, with discussion of the important variables of fossil fuel reserves and future energy production options. Then globally averaged carbon cycle models are used in conjunction with model projections of fuel usage and CO_2 emissions to estimate the likely bounds of future atmospheric CO_2 concentrations. Finally, technological choices which might be made to limit the growth of future atmospheric CO_2 levels and the potential confounding effects of other radiatively active trace gases on such choices are addressed. These chapters raise as many question as they answer but are invaluable in assessing the implications of our present knowledge and in setting the stage for future endeavors.

John R. Trabalka and David E. Reichle
Oak Ridge National Laboratory
Oak Ridge, Tennessee

Contents

Contributors

Charles F. Baes, Jr., Chemistry Division, Oak Ridge National Laboratory, Oak Ridge, Tennessee 37831, U.S.A.

James C. Barnard, Space Sciences and Experimental Methods Section, Geosciences Research and Engineering Department, Battelle-Pacific Northwest Laboratories, Richland, Washington 99352, U.S.A.

Anders Björkström, Department of Meteorology, Arrhenius Laboratory, University of Stockholm, S-106 91 Stockholm, Sweden

Bert Bolin, Department of Meteorology, Arrhenius Laboratory, University of Stockholm, S-106 91 Stockholm, Sweden

A. L. Bradshaw, Department of Chemistry, Woods Hole Oceanographic Institution, Woods Hole, Massachusetts 02543, U.S.A.

Peter G. Brewer, Department of Chemistry, Woods Hole Oceanographic Institution, Woods Hole, Massachusetts 02543, U.S.A.

William D. Dietzman, Energy Information Administration, Dallas, Texas 75242, U.S.A.

James A. Edmonds, Institute for Energy Analysis, Oak Ridge Associated Universities, Washington, D.C. 20036, U.S.A.

William P. Elliott, Air Resources Laboratory, R32, National Oceanic and Atmospheric Administration, U.S. Department of Commerce, Rockville, Maryland 20852, U.S.A.

Ian G. Enting, Commonwealth Scientific and Industrial Research Organization (CSIRO), Division of Atmospheric Research, Mordialloc, Victoria 3195, Australia

Roger J. Francey, Commonwealth Scientific and Industrial Research Organization (CSIRO), Division of Atmospheric Research, Mordialloc, Victoria 3195, Australia

Paul J. Fraser, Commonwealth Scientific and Industrial Research Organization (CSIRO), Division of Atmospheric Research, Mordialloc, Victoria 3195, Australia

Hans D. Freyer, Institut für Chemie, Der Kernforschungsanlage Jülich GmbH, Institut 2, Chemie Der Belasteten Atmosphäre, D-5170 Jülich, Federal Republic of Germany

Inez Y. Fung, NASA Goddard Space Flight Center, Institute for Space Studies, New York, New York 10025, U.S.A.

Richard H. Gammon, PMEL/NOAA, U.S. Department of Commerce, Seattle, Washington 98115, U.S.A.

Robert H. Gardner, Environmental Sciences Division, Oak Ridge National Laboratory, Oak Ridge, Tennessee 37831, U.S.A.

T. E. Goff, NASA Goddard Space Flight Center, Greenbelt, Maryland 20771, U.S.A.

Senator Albert Gore, Jr., U.S. Senate, Washington, D.C. 20510, U.S.A.

Martin Heimann, Geological Research Division, Scripps Institution of Oceanography, University of California, San Diego, La Jolla, California 92093, U.S.A.

B. N. Holben, NASA Goddard Space Flight Center, Greenbelt, Maryland 20771, U.S.A.

Richard A. Houghton, The Ecosystems Center, Marine Biological Laboratory, Woods Hole, Massachusetts 02543, U.S.A.

Charles D. Keeling, Scripps Institution of Oceanography, University of California, San Diego, La Jolla, California 92093, U.S.A.

George G. Killough, Health and Safety Research Division, Oak Ridge National Laboratory, Oak Ridge, Tennessee 37831, U.S.A.

Walter D. Komhyr, Geophysical Monitoring for Climatic Change, Air Resources Laboratory/ERL, U.S. Department of Commerce, National Oceanic and Atmospheric Administration, Boulder, Colorado 80303, U.S.A.

Frederick A. Koomanoff, Carbon Dioxide Research Division, Office of Energy Research, U.S. Department of Energy, Washington, D.C. 20545, U.S.A.

Gregg Marland, Institute for Energy Analysis, Oak Ridge Associated Universities, Oak Ridge, Tennessee 37831, U.S.A.

Charles D. Masters, U.S. Department of the Interioir, U.S. Geological Survey, Reston, Virginia 22092, U.S.A.

Berrien Moore III, Complex Systems Research Center, University of New Hampshire, Durham, New Hampshire 03824, U.S.A.

Hans Oeschger, Physikalisches Institut, Universität Bern, CH-3012 Bern, Switzerland

Archibald B. Park, Natural Resources Consulting Services, Arnold, Maryland, 21012, U.S.A.

Graeme I. Pearman, Commonwealth Scientific and Industrial Research Organization (CSIRO), Division of Atmospheric Research, Mordialloc, Victoria 3195, Australia

Tsung-Hung Peng, Environmental Sciences Division, Oak Ridge National Laboratory, Oak Ridge, Tennessee 37831, U.S.A.

Alfred M. Perry, Energy Division, Oak Ridge National Laboratory, Oak Ridge, Tennessee 37831, U.S.A.

James T. Peterson, Geophysical Monitoring for Climatic Change, Air Resources Laboratory/ERL, U.S. Department of Commerce, National Oceanic and Atmospheric Administration, Boulder, Colorado 80303, U.S.A.

David E. Reichle, Environmental Sciences Division, Oak Ridge National Laboratory, Oak Ridge, Tennessee 37831, U.S.A.

John M. Reilly, Battelle Memorial Institute, Pacific Northwest Laboratories, Washington, D.C. 20036, U.S.A.

Hans-Holger Rogner, Energy Systems Group, International Institute for Applied Systems Analysis (IIASA), A-2361 Laxenburg, Austria

David H. Root, U.S. Department of the Interior, U.S. Geological Survey, Reston, Virginia 22092, U.S.A.

Ralph M. Rotty, Institute for Energy Analysis, Oak Ridge Associated Universities, Oak Ridge, Tennessee 37831, U.S.A.

Jorge L. Sarmiento, Geophysical Fluid Dynamics Program, Princeton University, James Forrestal Campus, Princeton, New Jersey 08542, U.S.A.

William H. Schlesinger, Department of Botany, Duke University, Durham, North Carolina 27706, U.S.A.

B. Stauffer, Physikalisches Institut, Universität Bern, CH-3012 Bern, Switzerland

Gerald M. Stokes, Space Sciences and Experimental Methods Section, Geosciences Research and Engineering Department, Battelle-Pacific Northwest Laboratories, Richland, Washington 99352, U.S.A.

Thomas A. Stone, The Ecosystems Center, Marine Biological Laboratory, Woods Hole, Massachusetts 02543, U.S.A.

Minze Stuiver, Isotope Laboratory, Department of Geological Sciences, Quaternary Research Center, AK-60, University of Washington, Seattle, Washington 98195, U.S.A.

Eric T. Sundquist, Water Resources Division, U.S. Department of the Interior, U.S. Geological Survey, Reston, Virginia 22092, U.S.A.

J. R. G. Townshend, Department of Geography, University of Reading, Reading, Berkshire, United Kingdom

John R. Trabalka, Environmental Sciences Division, Oak Ridge National Laboratory, Oak Ridge, Tennessee 37831, U.S.A.

Compton J. Tucker, Earth Resources Branch, Laboratory for Terrestrial Physics, NASA Goddard Space Flight Center, Greenbelt, Maryland 20771, U.S.A.

L. S. Waterman, Geophysical Monitoring for Climatic Change, Air Resources Laboratory, U.S. Department of Commerce, National Oceanic and Atmospheric Administration, Boulder, Colorado 80303, U.S.A.

R. T. Williams, Scripps Institution of Oceanography, University of California, San Diego, La Jolla, California 92093, U.S.A.

George M. Woodwell, The Woods Hole Research Center, Woods Hole, Massachusetts 02543, U.S.A.

A Congressional Perspective of the Greenhouse Effect

SENATOR ALBERT GORE, JR.

In many ways the greenhouse effect—the projected impact of carbon dioxide (CO_2) buildup on the earth's environment—seems more like a bad science fiction novel than a matter about which public policy is typically concerned. Indeed, it is often difficult to discuss the issue, in view of its almost unthinkable potential consequences. Can we imagine a New York with the climate of Palm Beach, a Kansas that resembles Central Mexico, or 40% of the State of Florida submerged under water? Such projections are not only fantastic but difficult to take seriously. One Tennessee newspaper recently ran a cartoon in which a disembodied voice proclaimed: "Oh boy, who woulda thought that someday 'from sea to shining sea' would mean from Denver to Knoxville?"

But the greenhouse effect is destined to become an increasingly critical issue for those of us in Washington who share some responsibility for designing our nation's scientific policy and priorities. In some ways, it is a model case for examining the nexus between science and politics, and especially the dialogue that exists between scientists and politicians. Oak Ridge knows how important that relationship is, and, in my mind, it will become much more important to the country's social and political well-being in the coming decades—indeed, to the world's social and political well-being.

Until rather recently, it was the perception here in Washington, as well as among the general public, that fears about CO_2 buildup and its projected impacts were only fanciful theories and certainly not real. *The Washington Post* labeled the greenhouse effect as something for the "sandals and granola crowd," and many in the scientific community privately expressed the belief that it was really the pet theory of a few admittedly distinguished but obviously misguided souls.

Although I was initially quite reserved in my opinion of this issue, many of the doubts that I had about the legitimacy of the concern over the greenhouse effect began to be dispelled in July of 1981. At that time, a number of prominent members of the scientific community, including

Roger Revelle and Melvin Calvin, came before my Congressional subcommittee to give their opinion that the greenhouse effect was not a theory any longer but an actual phenomenon. That hearing also began to persuade many of my colleagues in the Congress that this was a matter to be studied much more closely. Consequently, after that hearing, I directed my subcommittee to pursue the matter diligently.

Those efforts led to the second Congressional hearing on the greenhouse effect, in April of last year, when James Hansen and George Kukla from NASA presented to my subcommittee their findings that correlated a rise in the earth's mean temperature, the shrinking of the polar ice caps, and the subsequent rise in the earth's mean sea level. That report awakened some of the more skeptical members of Congress and some in the bureaucracy.

The revelations at last April's hearing represented a significant event for many of us in Congress, because they constituted the initial emergence of a scientific consensus about the buildup of CO_2 and its ramifications. The occurrence of such a consensus as this is always an important point for policymakers, because so often we are presented with divergent, uncertain, and contradictory scientific evidence. We are frequently asked to make policy judgments based on data that are less than clear and less than precise. Public policy judgments are rarely clear-cut, but they are much more difficult when the principal sources of information, such as the scientific community, cannot agree.

Thus, it is common for the congress to monitor developments in science, and when something approaching a scientific consensus with public-policy implications becomes clear, it is common for the Congress to seize on it and to propose action. Once such a consensus exists, whether it comes from the National Academy of Sciences or from more informal mechanisms, the Congress often begins to design some options for action and begins to build a political consensus around the most appropriate and agreed-upon course of action.

In this case, however, the public-policy options presented by the emerging scientific consensus are so vast as to make us blink our eyes and ask the scientists over and over, "Are you sure?" and "How long before we will know more?" Well, those answers will be difficult to obtain. Although it is established that the buildup of CO_2 in the atmosphere is warming the earth's environment, there has been much less of a consensus on either the timeframe within which this warming will occur or when we would feel the more perceptible results. Moreover, uncertainties about the actual temperature sensitivity of the atmosphere, the influence of ocean systems and cloud formation, and nongreenhouse gases also have clouded any conclusion about whether the most dire consequences will be realized. The Congress has come halfway in its understanding of the greenhouse effect, thanks to the scientific consensus that has formed around its basic premise. But the remaining points of uncertainty must be resolved before Congress can reach the point of political consensus.

The recent reports of the National Research Council (NRC, 1983) and the Environmental Protection Agency (EPA) (Seidel and Keyes, 1983) constitute important steps toward helping the Congress reach a political consensus. These important works represent the first major attempts to predict when the impacts of the greenhouse effect will be felt and to delineate some avenues of action that we might take—as a government, as a society, and as a global civilization. Although the reports leave major questions unanswered, in the end, these may be as much political decisions to be made as scientific issues to be resolved.

Let's consider the two studies for a moment.

On the one hand, there is the EPA report, which subscribes to some of the gloomiest and most dire predictions. The report projects a rise in the mean temperature of 2°C by the middle of the next century and a 5°C rise by the year 2100—a rise in temperature sufficient to cause massive disruptions of agricultural patterns, land management, and public works. In short, the report projects a potential cataclysm for our society and a nightmare for those of us concerned about the future of that society.

The EPA report is especially pessimistic regarding potential options to mitigate the cause of CO_2 buildup. The study postulates that measures to drastically reduce or even ban the use of hydrocarbons as an energy source would have no appreciable influence on this warming trend. According to the EPA report, we are basically "stuck," and even the most politically and economically infeasible options such as the extreme measure of banning coal would not mitigate our predicament. Thus, rather than urging strategies to reduce the cause of the greenhouse effect, this report contends that we should follow an "adaptive" strategy and simply prepare ourselves for the effects of the calamity by engineering new crops, relocating cities and public works, and rethinking how we manage our country.

By comparison, but only by comparison, the report of the Carbon Dioxide Assessment Committee at the NRC is a source of some comfort. While it supports the thesis that there will be a doubling of CO_2 levels late in the next century and that there will most likely be a mean temperature increase, the NRC report rejects the nothing-can-be-done position taken by the EPA. Instead, the NRC emphasized what the EPA only briefly mentions: that there are significant uncertainties about the effect on basic physical processes of increased CO_2 levels in the atmosphere. The NRC report points out that there are gaps in our knowledge about the influence of ocean systems, cloud formations, non-CO_2 gases, and deforestation. While it admits that the consequences could be severe, the report concludes that we have a margin of time and that massive shifts in our energy policy are not yet warranted but may eventually be wise. Instead, it urges an aggressive program of research.

Thus, again, as policymakers, members of Congress and others have a dilemma. Do we accept the EPA's doomsday scenarios, and grid ourselves for inevitable massive dislocations in our society? or do we adopt the relative optimism of the NRC report, continue more aggressive research, and wait and see—hoping that our worst nightmares don't materialize? This is certainly not to indict either report; in my view it would be foolish to disregard either one. Certainly the distinguished scientists who developed these reports are aware of the risks and the stakes at hand. But we do have to make some choices fairly soon: This is particularly true because of the timeframes. The greenhouse effect is obviously a long-term dilemma, and its solution will require long-term perspectives. Unfortunately, however, political bodies are too often short-sighted. Looking down the road more than two years is often a tall order for Congress.

I personally believe, however, that the greenhouse effect is one of a number of emerging scientific and technological challenges with which our society will be faced in the coming decades and with which we must absolutely come to grips. There are many other examples. Part of the mission of Congres is to better prepare our society for technological challenges, and this can only be accomplished with a healthy dose of foresight and a whole new habit of looking down the road. C. P. Snow's advice is well taken: "The sense of the future is behind all good policies. Unless we have it, we can give nothing either wise or decent to the world."

It will be far easier to formulate a course of action and to develop a political consensus behind it once there is more agreement about *when* we can expect to see the consequences of the greenhouse effect and about just how bad they are likely to be. Just as science has given us a consensus that the CO_2 buildup is increasing the earth's temperature, it hopefully can render something resembling a consensus on the timetable for its ultimate impacts.

To that end, I would make the following suggestions.

First, I echo the calls of both the EPA and the NRC for an aggressive research program into the effects of increased CO_2 levels on physical processes and on the behavior of the atmosphere. This should include an attempt to alleviate our uncertainties about the roles of oceans and deforestation and the effects of other gases besides CO_2. This is both a scientific and political objective, for unless this element of scientific uncertainty can be removed, a political resolution will be difficult if not impossible. Indeed, it may well turn out to be impossible even with such a resolution.

Second, the appropriate government agencies, as well as the NRC, should begin to suggest and explore possible strategies for addressing the problem of CO_2 buildup. These strategies should proceed along two different avenues.

One option obviously would involve reducing CO_2 emissions into the atmosphere and dealing with the problem at its cause. This, of course, will turn out to be extremely controversial. Moreover, I realize, this position is at odds with the EPA projections; however, I am troubled by their conclusions. Logically, if the greenhouse effect is caused by hydrocarbon emissions, then a reduction in emissions should cause a reduction in those impacts. While I am perfectly willing to accept the EPA's *time* projections, I question both the logic and the spirit of a strategy that is predicated on the tenet that there is nothing we can do, particularly if we, as a civilization, evidently have caused it. Until we increase our body of knowledge in this area, I do not believe that we should merely capitulate and begin to plan a drastic reorganization and relocation of our society.

Because of the possibility—however remote—that our situation might in fact be less hopeful, government and scientific entities should also begin suggesting and exploring strategies for a more adaptive approach: How we could alter our land and water management, city planning, and agricultural techniques to prepare for such impacts before they are upon us.

The research budget for the effect of the CO_2 buildup should be increased, within both DOE and other agencies. It strikes me as curious that precisely when the scientific evidence has given more credence to this potentially disastrous phenomenon the research budget has been cut in half. And to those distinguished scientists from countries around the world, may I ask that we make this an international undertaking, even more so that it presently is.

Finally, I want to make one last observation. I believe that the scientific community in general and the atmospheric disciplines in particular have done an exemplary job of giving much-needed information on the greenhouse effect. That is not always the case in scientific questions with public policy ramifications. As I mentioned previously, the most potent challenges to our society are becoming increasingly technical in nature. Without reasoned and responsible scientific information, policymakers are simply lost in the wilderness. Consequently, a better working relationship between the scientific and political communities is a must if we are going to choose a better future. We need sound scientific advice so that we can keep our end of the bargain and make sound public-policy judgments. I look forward to working with the distinguished group which gathered at the Knoxville Symposium, with atmospheric scientists, and with the larger scientific and technical community on this problem.

In closing, I would like to share one thought. Yesterday an interviewer asked me why I am concerned about the greenhouse effect. My response was that I have children and hope to have grandchildren. I think that I own them the opportunity to live in this country, in a world, and in a society resembling the one that we now have—if not better. We owe it to future generations to reduce the uncertainties that continue to surround the greenhouse effect and to search for solutions if it does turn out to be as real a problem as the scientific consensus now paints it to be.

1
The Global Atmospheric CO_2 Distribution 1968–1983: Interpretation of the Results of the NOAA/GMCC Measurement Program

RICHARD H. GAMMON, WALTER D. KOMHYR, AND
JAMES T. PETERSON

The modern period of precise atmospheric CO_2 measurements began with Keeling's pioneering determinations at Mauna Loa, Hawaii, and the South Pole during the International Geophysical Year. The Mauna Loa record (e.g., Keeling 1983) remains the single most valuable CO_2 time series for carbon cycle model verification. For very recent interpretations of atmospheric CO_2 measurements and carbon cycle relationships, the reader is referred to Cleveland et al. (1983), Keeling (1983), Machta (1983), Mook et al. (1983), Pearman et al. (1983), Keeling et al. (1984), Wong et al. (1984), Bacastow et al. (1985), Komhyr et al. (1985), and Fraser et al. (this volume). In this chapter, the global atmospheric CO_2 records, particularly of the NOAA/GMCC (National Oceanic and Atmospheric Administration/Geophysical Monitoring for Climatic Change) program, are summarized for the period 1968 through 1983, with emphasis on the mean properties of the global carbon cycle as viewed from the atmosphere (i.e., global mean CO_2 concentration, latitude dependence of concentration and seasonal amplitude, airborne fraction, etc.).

Continuous Measurements at GMCC Baseline Observatories (1974–1983)

Background measurements of atmospheric CO_2 have been made continuously since the mid-1970s at four remote NOAA observatories (Barrow, Alaska; Mauna Loa, Hawaii; American Samoa; and the South Pole). The data acquisition and preliminary processing procedures have been previously described (Peterson 1978; Mendonca 1979). The entire record of continuous CO_2 measurements at each station through 1983 has been recently corrected, converted to the World Meteorological Organization (WMO) X81 mole fraction scale (parts per million by volume, ppm, with respect to dry air), and edited to essentially final values. Provisional monthly mean values of the background CO_2 concentration (Table 1.1) at

TABLE 1.1. Provisional monthly mean CO_2 concentrations from continuous measurements at the four GMCC/NOAA baseline observatories in Barrow, Mauna Loa, American Samoa, and the South Pole.

Year	Jan	Feb	Mar	Apr	May	Jun	Jul	Aug	Sep	Oct	Nov	Dec
Barrow observatory												
1973	334.5$^+$	335.5$^+$	336.3$^+$	337.0$^+$	336.6$^+$	334.0$^+$	327.7$^+$	322.0	323.4	327.8	332.7	333.5
1974	336.5	336.0	336.7	337.1	336.9	335.9	328.7	323.5	324.2	328.8	332.0	335.2
1975	336.8	336.6	337.1	337.7	337.9	335.4	330.1	324.4	324.6	329.3	334.3	336.5
1976	337.0	336.7	337.7	338.0	337.8	336.8	331.0	323.5	325.1	329.8	334.1	336.6
1977	336.8	336.9	338.2	339.2	339.4	337.4	330.1	325.4	326.8	332.0	334.7	338.8
1978	339.1	340.2	341.5	341.0	340.8	339.4	332.6	327.7	327.4	332.1	338.8	339.4
1979	340.1	341.3	342.5	342.4	342.9	342.0	333.6	327.7	329.3	333.1	338.1	341.5
1980	341.9	342.4	343.4	343.3	343.3	342.6	337.3	331.8	331.4	337.9	340.5	342.6
1981	344.3	345.5	344.6	345.8	346.8	344.4	338.2	331.8	333.3	338.8	342.2	344.9
1982	345.8	347.2	348.3	348.4	348.0	345.7	339.2	333.2	333.2	337.9	342.1	344.5
1983	347.4*	347.0*	348.2*	348.2*	348.3*	346.8*	339.8*	333.6*	333.8*	339.4*	346.1*	346.9*
Mauna Loa Observatory												
1973	0.0	0.0	0.0	0.0	0.0	0.0	0.0	0.0	0.0	0.0	0.0	0.0
1974	0.0	0.0	0.0	0.0	333.0	332.1	331.0	329.1	327.4	327.3	328.3	329.5
1975	330.7	331.5	332.1	333.3	333.8	333.9	331.8	330.0	328.5	328.3	329.3	330.5
1976	331.7	332.8	333.5	334.9	334.8	334.5	333.0	330.7	329.3	328.7	330.2	331.4
1977	332.0	332.3	334.9	336.2	336.9	336.1	335.0	332.6	331.3	331.3	332.3	333.6
1978	335.1	335.4	336.8	337.8	338.1	338.0	336.5	334.5	332.4	332.4	333.8	334.9
1979	336.3	336.8	338.3	338.9	339.3	339.1	337.6	335.5	333.8	334.2	335.3	336.8
1980	337.8	338.4	340.0	340.9	341.4	341.3	339.5	337.3	336.0	336.1	337.3	338.4
1981	339.2	340.0	341.7	342.4	343.1	342.5	340.7	338.5	337.1	337.3	338.7	339.9
1982	341.2	341.7	342.7	343.8	344.3	343.3	341.8	340.2	337.7	337.9	339.1	340.7
1983	341.6*	342.7*	343.3*	344.9*	345.5*	345.0*	343.1*	341.1*	339.4*	339.9*	341.1*	342.8*

Samoa observatory												
Year	Jan	Feb	Mar	Apr	May	Jun	Jul	Aug	Sep	Oct	Nov	Dec
1973	0.0	0.0	0.0	0.0	0.0	0.0	0.0	0.0	0.0	0.0	0.0	0.0
1974	0.0	0.0	0.0	0.0	0.0	0.0	0.0	0.0	0.0	0.0	0.0	0.0
1975	0.0	0.0	0.0	0.0	0.0	0.0	0.0	0.0	0.0	0.0	0.0	0.0
1976	330.7^{+}	330.6^{+}	330.8^{+}	331.2	330.7	331.2	331.1	331.1	331.3	331.5	331.9	331.8
1977	331.7	332.0	332.0	332.4	332.3	332.6^{+}	332.6^{+}	333.1	332.9	332.9^{+}	333.1^{+}	333.7^{+}
1978		334.2*	334.0^{+}	334.3^{+}	335.0*	335.0*	335.0*	334.3^{+}	334.3^{+}	334.5^{+}	334.4^{+}	334.2^{+}
1979	334.6^{+}	335.4^{+}	335.9^{+}	335.7	335.9	336.0	336.3	336.3	336.3	336.6	336.7	336.8
1980	337.5	337.6	337.9	337.3	337.8	338.0	338.4	338.3	338.2	338.1	338.5	338.9
1981	339.0	339.3	338.8	339.6	338.8	339.3	339.0	339.1	339.2	339.0	339.5	339.8
1982	340.4	340.5	340.4	339.8	339.9	340.1	340.5	340.4	340.1	340.0	340.2	340.2
1983	340.0*	340.5*	340.8*	340.6*	341.1*	341.6*	341.9*	341.9*	341.9*	342.1*	342.4*	342.6*
South Pole Observatory												
1973	0.0	0.0	0.0	0.0	0.0	0.0	0.0	0.0	0.0	0.0	0.0	0.0
1974	0.0	0.0	0.0	0.0	0.0	0.0	0.0	0.0	0.0	0.0	0.0	0.0
1975	329.1^{+}	328.8^{+}	328.5*	328.3*	328.6^{+}	328.6*	329.4*	329.8^{+}	330.1^{+}	330.3^{+}	330.2^{+}	330.3*
1976	330.1	329.8	329.6	329.6	329.7	329.9	330.3	330.7	331.0	331.1	331.1	331.1
1977	330.9	330.3	330.3	330.6	331.0	331.4	331.8	332.2	332.7	333.0	333.0	332.7
1978	332.6	332.4	332.3	332.5	333.0	333.3	333.8	333.4	334.9	335.0	334.9	334.7
1979	334.5^{+}	334.5^{+}	334.6^{+}	334.6^{+}	334.9^{+}	335.4^{+}	335.9^{+}	336.2^{+}	336.4^{+}	336.5^{+}	336.4^{+}	336.2^{+}
1980	336.1	336.0	336.0	336.2	336.6	336.9	337.4	337.7	337.9	338.1	338.0	338.0
1981	337.8	337.8	337.9	337.9	338.1	338.3	338.7	339.0	339.2	339.1	338.9	338.5
1982	338.5	338.6^{+}	338.4	338.6	338.9	338.9	339.2	339.6	339.9	339.8	339.5	339.2
1983	339.1*	339.0*	339.0*	339.4*	340.1*	340.9*	341.1*	341.8*	341.9*	342.0*	342.1*	341.9*

Values have been provisionally selected for background conditions and are expressed in the WMO X81 mole fraction scale (ppm with respect to dry air).

*Less reliable or provisional data.

$^{+}$Substituted flask data.

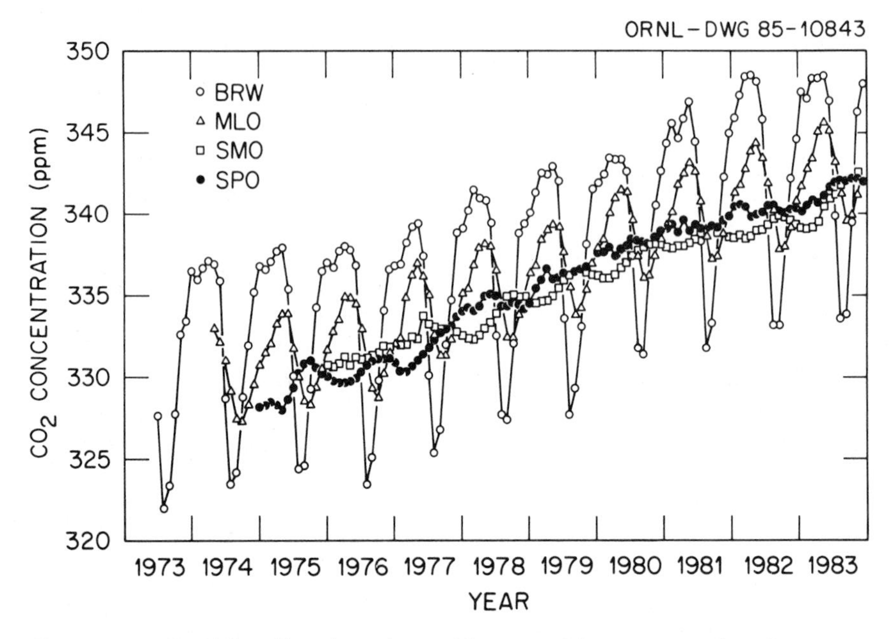

FIGURE 1.1. Provisionally selected monthly mean CO_2 concentrations from continuous measurements at the NOAA/GMCC baseline observatories (Barrow, Alaska [BRW]; Mauna Loa, Hawaii [MLO]; American Samoa [SMO]; South Pole [SPO]). Values are in the WMO X81 mole fraction scale and have been provisionally selected for background conditions. The deseasonalized smooth curve is a 12-month running mean. Refer to Table 1.1 for months with less certain means or for months in which flask means have been substituted for missing continuous means. (Harris and Bodhaine 1983.)

the four observatories are plotted in Fig. 1.1. The major episodes of line contamination and instrumental malfunction have been excluded from these records, and the previous problem of drifting calibration CO_2 standards ($\sim$0.1 ppm • yr^{-1}) with respect to the WMO standards maintained by Keeling at Scripps Institution of Oceanography (SIO) has now been resolved. The NOAA and SIO CO_2 records at Mauna Loa are in excellent agreement (Fig. 1.2).

Each of the four NOAA stations shows a CO_2 increase of very nearly 9.0 ppm for the interval 1976 to 1982, yielding a mean global growth rate for atmospheric CO_2 of 1.50 $\pm$ 0.2 ppm • yr^{-1}. From the data of Rotty and Marland (this volume), the total industrial CO_2 production from fossil fuel combustion and cement manufacture for this same interval (1976 to 1982) may be estimated as $\sim$35.8 x 10^{15}g C or 16.8 ppm atmospheric equivalent. The apparent global average airborne fraction (η) with respect to these cumulative fossil fuel carbon emissions, $\simeq$0.54, for this period is virtually identical to the average (0.53) of the airborne fractions from the CO_2 observations by Keeling at Mauna Loa (MLO) (η = 0.55) and the South

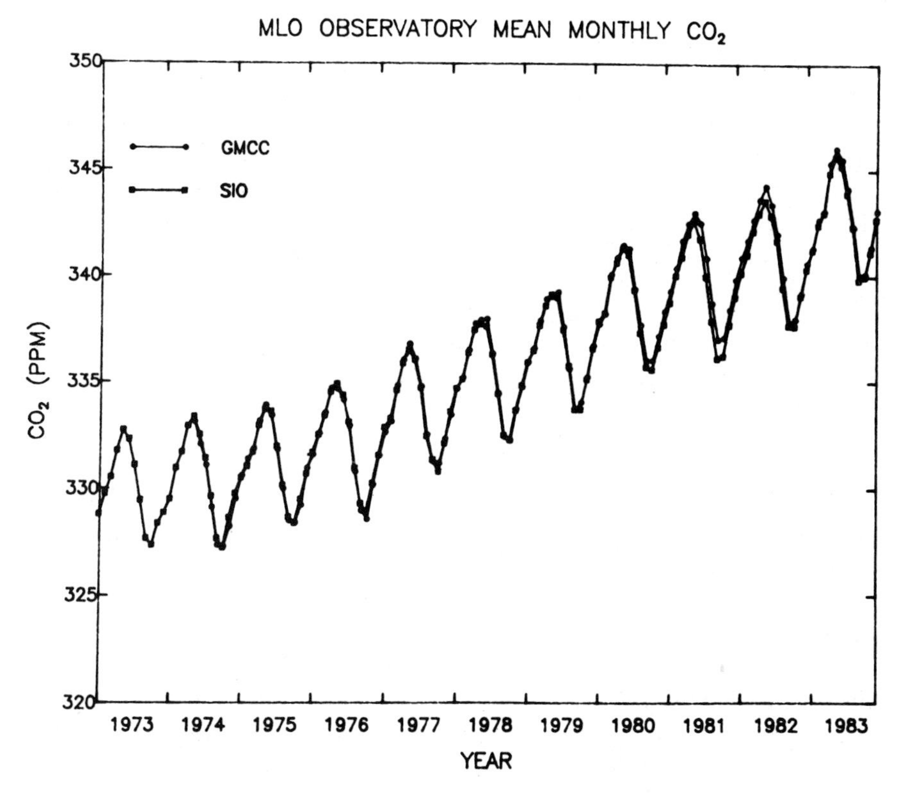

FIGURE 1.2. Comparison of the Keeling (SIO) and NOAA (GMCC) atmospheric CO_2 records at Mauna Loa Observatory for the common period since 1974. Values are edited monthly means from C. D. Keeling (personal communication, 1984) and from GMCC (K. Thoning, personal communication, 1984; very similar to values given in Table 1.1).

Pole (SPO) (η = 0.51) during the previous two decades, 1959 to 1978 (Bacastow and Keeling 1981). Although there is some indication that the latitude dependence of the airborne fraction has weakened in the last decade of slower fossil fuel CO_2 release, a mean value of $\eta \simeq 0.54 \pm 0.02$ still seems appropriate for model projections of future atmospheric increases. This value is only slightly affected by the anomalous growth rates of atmospheric CO_2 in 1982. Extending the averaging period to include 1983, and estimating 1983 fossil fuel use as equal to that in 1982, yields a mean η = 0.55. Because the airborne fraction represents a dynamic response to a particular exponential fossil fuel growth rate, the slower fossil fuel growth since 1975 is expected to result in falling values of the airborne fraction. More recently, Keeling (1983) gives η = 0.57 as the mean value for MLO and SPO over the full period of record 1958 to 1982.

A second conclusion is that the latitude gradient of CO_2 concentration from Barrow (BRW) to SPO has remained near 3.5 ppm during the last decade. The observed interannual variations in this gradient (±0.5 ppm) are likely related to the El Nino/Southern Oscillation (ENSO) events of 1976, 1979, and 1982 (Bacastow et al. 1980; Gammon and Komhyr 1983). The magnitude of the latitude gradient is directly related to the rate of release of fossil fuel CO_2 to the atmosphere of midnorthern latitudes. The fact that this gradient, which increased rapidly from ~1 ppm in 1960 to ~3 ppm in the mid-1970s (Keeling 1983), is now nearly steady is a direct consequence of the abrupt decrease in the rate of global production/consumption of fossil fuels since 1973. (Fossil fuel growth rates have been <2%/year since 1974, with zero or negative growth rates for 1980 to 1982, as compared to the mean growth rate of 4.6%/year for 1950 to 1973; Rotty and Marland, this volume).

Global CO_2 Flask Sampling Network of NOAA/GMCC

In addition to the long time series of continuous CO_2 measurements made at the four GMCC observatories, flask samples of whole air are collected weekly at the approximately 20 cooperating sites of a global CO_2 network and returned to NOAA (Boulder, Colorado) for trace gas analysis (CO_2, CH_4, CO). Although the sampling began before 1970 at a few initial sites, the major expansion of the network to 15 stations took place in 1979 with support from the U.S. Department of Energy. This expanded flask network has allowed the synthesis of a detailed picture of the regional variation of atmospheric CO_2 concentration.

The distribution of cooperating flask sites around the globe (Fig. 1.3; Table 1.2) clearly illustrates the interest of the GMCC program in selecting background, marine locations, and in avoiding the major land biomes. Previously reported details on each site in the flask network (Herbert 1980; DeLuisi 1981) will be updated in a forthcoming WMO special report, and in the research summary of Komhyr et al. (1985). The most recent changes in the network have been the loss of the site in the Falkland Islands in February 1982 (replaced by Christchurch, New Zealand), the start-up of Cape Meares station on the Oregon coast, and the cooperative program with Battelle Laboratories at Kitt Peak, Arizona (Stokes 1982).

The entire flask data set (1968 to 1982) (corrected, converted to the X81 WMO mole fraction scale, and edited for background conditions) has been presented by Komhyr et al. (1985), and is available from the DOE Carbon Dioxide Information Center in Oak Ridge, Tennessee. For both the flask and continuous CO_2 measurement program, GMCC uses CO_2-in-air gas standards directly traceable to the WMO standards maintained by Keeling at SIO. These standards are now being converged with a set of reference materials developed by the U.S. National Bureau of Standards for future use.

As one check on the quality of the flask CO_2 measurements and on the

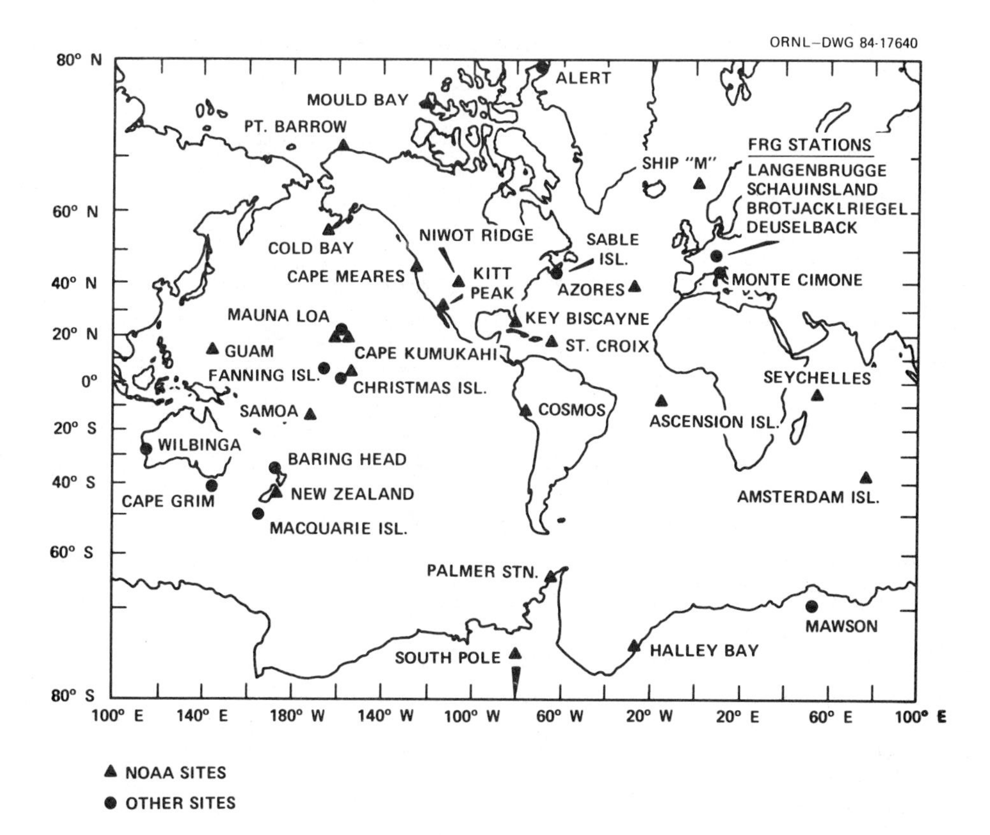

FIGURE 1.3. Geographic location of the cooperating sites of the NOAA/GMCC global CO_2 flask sampling network active in 1983. Refer to Table 1.2 for the coordinates and elevation of each sampling site.

reliability of the in situ CO_2 analyzers at the four GMCC baseline stations, the annual mean concentrations obtained at BRW, MLO, SMO, and SPO from the flask and continuous programs may be compared for the period of common record (1976 to 1983). The difference in the annual mean values is generally <0.2 ppm at each station (Table 1.3). As a second check on the flask results, there was initiated in mid-1981 a flask-flask comparison with Keeling's laboratory at SIO for quasisimultaneously collected flasks at BRW, MLO, KUM, and SMO. The agreement has improved steadily, and the (NOAA–SIO) flask-flask differences now generally fall within the sampling and analysis error of ±0.2 ppm. The annual mean CO_2 concentrations from the GMCC flask network are given in Table 1.4.

With the expansion of the flask network in 1979, the improved latitude coverage permitted the interpolation of the time series at the cooperating sites to create a CO_2 concentration surface representing the global distribution in a zonally averaged sense (Fig. 1.4). The grid points correspond

TABLE 1.2. The cooperating sampling sites of the NOAA/GMCC global CO_2 flask network in 1983.

Code	Site	Latitude	Longitude	Elevation (meters)
MBC	Mould Bay, Canada	76°N	119°W	15
BRW	Barrow, Alaska	71°N	157°W	11
STM	Ocean Station M	66°N	2°E	6
CBA	Cold Bay, Alaska	55°N	163°W	25
CMO	Cape Meares, Oregon	45°N	124°W	30
NWR	Niwot Ridge, Colorado	40°N	106°W	3749
AZR	Terceira Island Azores	39°N	27°W	30
KPA	Kitt Peak, Arizona	32°N	112°W	2095
KEY	Key Biscayne, Florida	26°N	80°W	3
MLO	Mauna Loa, Hawaii	20°N	156°W	3397
KUM	Cape Kumukahi, Hawaii	20°N	155°W	3
AVI	St. Croix, Virgin Islands	18°N	65°W	3
GMI	Marianas Island, Guam	14°N	145°E	2
SEY	Mahe', Seychelles	5°S	55°E	3
ASC	Ascension Island	8°S	14°W	54
COS	Cosmos, Peru	12°S	75°W	4600
SMO	American Samoa	14°S	171°W	30
AMS	Amsterdam Island	38°S	78°E	150
NZL	Kiatorete Spit, New Zealand	44°S	173°E	3
PSA	Palmer Station, Antarctica	65°S	64°W	10
HBA	Halley Bay, Antarctica	75°S	27°W	(10)
SPO	South Pole, Antarctica	90°S	25°W	2810

TABLE 1.3. Differences in annual mean CO_2 concentration determined from comparisons of flask (f) and continuous (c) values from the NOAA and SI0 measurement programs

	NOAA (f)–NOAA (c)				NOAA–SIO	
					(c–c)	(f–f)
Year	BRW	MLO	SMO	SPO	MLO	SPO
1974	(+0.7)*	—	—	—	−0.3	—
1975	(−0.6)*	—	—	0.0	0.0	+0.1
1976	+0.1	0.0	+0.3	−0.1	−0.1	−0.2
1977	−0.2	−0.1	0.0	−0.2	0.0	−0.3
1978	−0.2	−0.3	—	−0.1	+0.2	+0.1
1979	−0.3	0.0	−0.1	—	+0.2	+0.2
1980	+0.3	+0.3	+0.1	−0.4	+0.2	−0.1
1981	−0.3	+0.3	0.0	+0.1	+0.6	+0.4
1982	−0.2	−0.2	+0.2	+0.2	+0.4	+0.2
1983	(+0.2)	(−0.3)	(0.0)	(+0.1)	(−0.1)	—
Mean signed Difference	−0.11 ±0.23	0.00 ±0.23	+0.08 ±0.15	−0.07 ±0.20	+0.13 ±0.27	+0.05 ±0.23

TABLE 1.4. Annual mean CO_2 concentrations from weekly sampling at cooperating sites of the NOAA/GMCC global CO_2 network.

STA	1968	1969	1970	1971	1972	1973	1974	1975
AMS	0	0	0	0	0	0	0	0
ASC	0	0	0	0	0	0	0	0
AVI	0	0	0	0	0	0	0	0
AZR	0	0	0	0	0	0	0	0
BRW	0	0	0	(327.3)	330.1	331.9	333.3	332.8
CBA	0	0	0	0	0	0	0	0
CMO	0	0	0	0	0	0	0	0
FLK	0	0	0	0	0	0	0	0
GMI	0	0	0	0	0	0	0	0
KEY	0	0	0	0	0	(330.7)	0	0
KUM	0	0	0	0	0	0	0	0
MBC	0	0	0	0	0	0	0	0
MKO	0	0	0	0	0	0	0	0
MLO	0		(325.2)	0	0	0	0	0
NWR	(323.1)	324.1	325.6	325.4	326.2	(330.2)	0	0
PSA	0	0	0	0	0	0	0	0
SEY	0	0	0	0	0	0	0	0
SMO	0	0	0	0	0	0	(330.9)	(330.7)
SPO	0	0	0	0	0	0	0	(329.4)
STC	0	324.0	326.4	326.8	(327.5)	0	0	0
STM	0	0	0	0	0	0	0	0

STA	1976	1977	1978	1979	1980	1981	1982
AMS	0	0	0	(335.9)	337.8	339.3	339.4
ASC	0	0	0	0	(338.8)	339.8	340.7
AVI	0	0	0	(337.1)	339.6	340.4	340.9
AZR	0	0	0	0	(338.4)	339.6	341.1
BRW	333.8	334.8	336.5	337.6	340.2	341.4	342.6
CBA	0	0	0	337.7	339.7	341.2	341.8
CMO	0	0	0	0	0	0	341.2
COS	0	0	0	0	(338.7)	0	339.7
FLK	0	0	0	0	0	(339.7)	0
GMI	0	0	0	(337.7)	340.1	341.2	341.0
KEY	(332.7)	335.2	336.7	338.8	(340.1)	342.0	341.5
KUM	(332.3)	334.4	335.7	337.4	339.3	340.4	341.2
MBC	0	0	0	0	(340.3)	341.8	342.4
MKO	0	0	0	337.3	0	0	0
MLO	(332.1)	333.6	335.2	336.8	339.0	340.4	341.0
NWR	332.1	334.8	335.7	336.9	338.0	339.8	341.0
PSA	0	0	(333.9)	335.2	(337.5)	340.1	339.8
SEY	0	0	0	0	339.3	340.1	340.5
SMO	331.5	332.8	334.3	(336.0)	(338.1)	339.2	340.4
SPO	330.2	331.5	333.6	335.0	336.7	338.5	339.3
STC	0	0	0	0	0	0	0
STM	0	0	0	0	0	(341.8)	341.6

Values in ppm relative to dry air, expressed in the WMO X81 mole fraction scale. Values in parentheses represent less certain values and generally contain several months of missing data for which interpolated values have been used to estimate the annual mean. (Komhyr et al. 1985.)

FLK, Falkland Islands; MKO, Mauna Kea, Hawaii; STC, Weathership Charlie (see Komhyr et al. 1985).

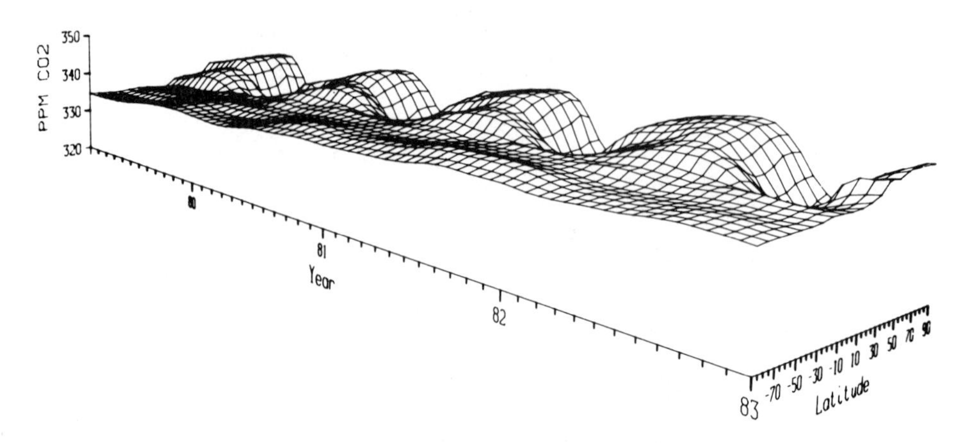

FIGURE 1.4. A three-dimensional perspective of the "pulse-of-the-planet," the variation of the global atmospheric CO_2 concentration in latitude and time based on flask measurements for 1979–1982. This zonally averaged surface has a resolution of 20 days and 10° in latitude and was synthesized from results of ~10,000 individual flask samples returned from the 15 remote sea-level sites shown in Fig. 1.3. (Harris and Bodhaine 1983.)

to a latitude resolution of 10° and a time resolution of 20 days. Subgridscale features of the global CO_2 distribution (e.g., local gradients associated with equatorial ocean upwelling or the InterTropical Convergence Zone) will not be well-represented on this surface. Information about longitudinal gradients (perhaps important at the equator and at midnorthern latitudes) is suppressed in the zonal averaging by latitude bands. In creating these CO_2 surfaces, the flask data were first edited for background conditions. The time series of background values at each station were then fitted to a cubic spline. The set of biweekly values of the spline function at each station then served as the input to the computer graphics program that interpolated the concentration surface. Four years of flask data (1979 to 1982) were fit simultaneously, representing 10,000 separate flask measurements. Except for the South Pole, only island or coastal stations near sea level were included. This means that the seasonal amplitudes represent typical marine locations, and as such are certainly minimum values. Much larger seasonal variations in CO_2 concentration are expected for midcontinental locations, particularly in Siberia and Northern Canada (Fung et al. 1983; see also Fig. 1.5).

The global distribution of the flask sites of the NOAA/GMCC network (Fig. 1.3) permits a precise value of the global average atmospheric CO_2 concentration to be determined (Table 1.5). These annual global means are derived from latitude-weighted, sea level sites, and represent the well-mixed troposphere (Komhyr et al. 1985). An approximation to a global mean, for the years prior to the network expansion in 1979, is provided

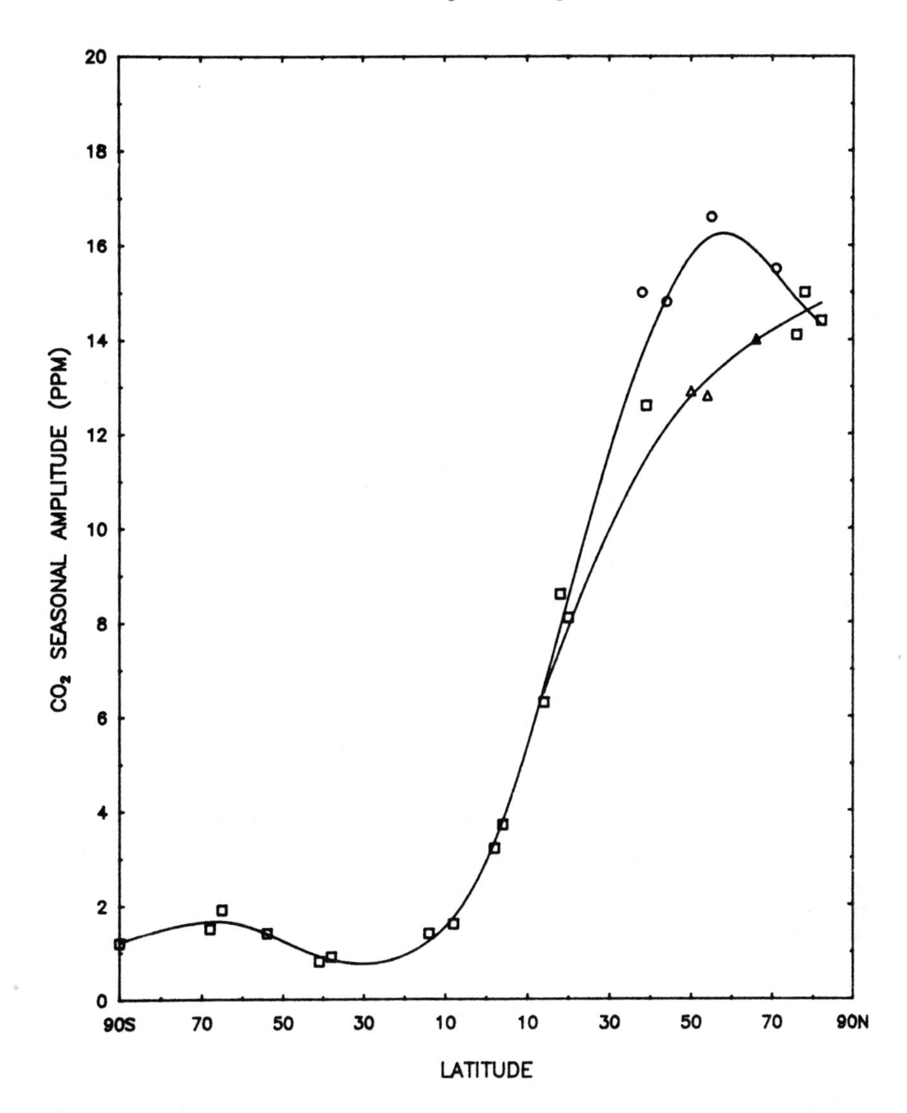

FIGURE 1.5. The latitude dependence of the seasonal amplitude of atmospheric CO_2 concentration observed at clean-air sites of GMCC/NOAA (Harris and Bodhaine 1983; Komhyr et al. 1985; see Fig. 1.3); Ny Alesund 78°N (W. Bischoff, personal communication 1984); Canadian flask stations Alert (82°N), PAPA (50°N), and Sable Island (44°N) (Wong et al. 1984); coastal Japan (~39°N) (Tanaka et al. 1983); Scripps Institution of Oceanography flask stations at Fanning Island (4°N) and Christmas Island (2°N) (Keeling 1983); and CSIRO (Australia) values from Cape Grim (4°S), Mawson (54°S), and Maguarie (68°S) (Fraser et al. 1983). Smooth polynomial fits to the data split at high northern latitude to indicate that midocean stations (▲) Weatherships, P, C, M) show lower CO_2 seasonal amplitudes than do continental and coastal sites (o) (Cape Meares, Oregon; Cold Bay, Alaska; Barrow, Alaska).

TABLE 1.5. Global annual mean tropospheric CO_2 concentration compared with Mauna Loa annual mean from NOAA/GMCC continuous records.

		Global Mean		Difference MLO–Global	
Year	MLO*	Flask†	Continuous‡	Flask	Continuous
1976	332.1	—	331.8	—	+0.3
1977	333.7	—	333.2	—	+0.5
1978	335.5	—	335.1	—	+0.4
1979	336.8	336.6	336.6	+0.2	+0.2
1980	338.7	338.7	338.4	0.0	+0.3
1981	340.1	340.0	339.9	+0.1	+0.2
1982	341.2	340.6	340.8	+0.6	+0.3
1983	342.6	342.3	342.1	+0.3	+0.5
				+0.24	+0.34
1976–1983		Mean difference		±0.23	±0.12

*Mauna Loa annual means from NOAA/GMCC continuous data.

†Global annual mean CO_2 concentration from sea-level sites of the NOAA flask network, latitude-weighted (Komhyr et al. 1985.)

‡Global mean annual CO_2 from average of annual means from continuous data at BRW, MLO, SMO, and SPO. (Harris and Bodhaine 1983.)

—, Missing data.

by the simple average of the annual CO_2 means from the continuous measurements at the four GMCC stations (BRW, MLO, SMO, SPO).

An important assumption often made by carbon cycle modelers is that the MLO record is a good approximation of the global CO_2 level, at least in the annual mean. Data from Mauna Loa typically fall 0.2–0.3 ppm *above* the global annual CO_2 mean (Table 1.5), as determined from either the flask network or the four continuous NOAA stations. With steeper latitude gradients developing in the future from higher global fossil fuel usage, Mauna Loa data will diverge even more from the global mean. For the present, modelers should use concentrations for MLO or MLO–0.2 ppm as being globally representative, and should *not* use the average concentration for MLO and the South Pole, which lies well below the global mean CO_2 level.

Atmospheric CO_2 and Current Questions About the Carbon Cycle

A synthesis of the atmospheric CO_2 measurements from clean-air sites around the world represents the most precise and integrated record available of the workings of the global carbon cycle, of the fluctuating CO_2 exchanges of the atmosphere with both the ocean and the biosphere, and of the response of this complex system to strong perturbations of both natural (ENSO) and manmade (fossil fuel combustion) origins. The global climate change predicted for the coming generations will undoubtedly impress its signature on this CO_2 atmospheric record, which indeed may offer one of the most sensitive possibilities for its early detection.

The current period of precise atmospheric CO_2 records began with Keeling's measurements at Mauna Loa and the South Pole in 1958. These longer records are complemented by the more recently begun, but more geographically complete, CO_2 time series of several national and individual monitoring programs coordinated by WMO. A coordinated interpretation of all such data must be applied to the outstanding questions of the global carbon cycle (e.g., Is the biosphere currently a net source or sink? What will be the future source/sink roles of the biosphere and ocean when the carbon cycle is more strongly perturbed by greater rates of fossil fuel input?). Important clues are certainly imbedded in the complex response of the global carbon cycle to extreme natural fluctuations such as the 1982/1983 ENSO event. Global atmospheric CO_2 monitoring networks must be extended to the major land biomes in both tropical and higher latitudes in close coordination with intensified research on the use of satellite remote sensing to quantify the photosynthetic activity of these biomes.

Atmospheric CO_2 records in the future may well record changes in the apparent airborne fraction of the annual fossil fuel CO_2 input as the dynamic response of the global ocean and biosphere adjusts to changes in the fossil fuel growth rate. Detecting a change in the airborne fraction will in turn require a much improved understanding of the competing signal, the natural interannual variability of the global carbon cycle responding to large-scale perturbations (ENSO, volcanos).

The CO_2-induced climatic change predicted to occur in the coming decades will measurably alter the "breathing of the world" as monitored by the international network of CO_2 sites. The details of the CO_2 response to changing climate will depend on many factors (rate of fossil fuel use, ocean thermal lag, role of other greenhouse gasses, etc.); however, it seems certain that, under altered regional temperature and precipitation regimens, the regional and global seasonality of atmospheric CO_2 will be different than is currently observed. Such CO_2 records now give the most precise and spatially integrated signal of the regional imbalance of respiration and photosynthesis. An evolving biospheric-atmospheric exchange, as seems now to be occurring in the growing CO_2 seasonal amplitude in the northern hemisphere, may be the early signature of the predicted climatic change. For the global biosphere of the future, the relative importance of enhanced photosynthetic uptake (CO_2 fertilization, β-factor response) versus enhanced respiration in the warmer, higher-CO_2 world of the next century cannot yet be predicted with confidence and will probably be determined only from the future CO_2 record itself.

References

Bacastow, R., J. A. Adams, C. D. Keeling, D. J. Moss, T. P. Whorf, and C. S. Wong. 1980. Response of atmospheric carbon dioxide to the weak 1975 El Nino. Science 210:66–68.

Bacastow, R. B. and C. D. Keeling. 1981. Atmospheric carbon dioxide concentration and the observed airborne fraction. In B. Bolin (ed.), Carbon Cycle Modelling, SCOPE 16, pp. 103–112. John Wiley, New York.

Bacastow, R. B., C. D. Keeling, and T. P. Whorf. 1985. Seasonal amplitude increase in atmospheric CO_2 concentration at Mauna Loa, Hawaii, 1959–1982. J. Geophys. Res. (Submitted).

Cleveland, W. S., A. E. Freeny, and T. E. Graedel. 1983. The seasonal component of atmospheric CO_2 information for new approaches to the decomposition of seasonal time series. J. Geophys. Res. 88:10934–10946.

DeLuisi, J. J. (ed.). 1981. Geophysical Monitoring for Climatic Change, No. 9: Summary Report. 1980. NOAA Environmental Research Laboratories, Boulder, Colorado.

Fraser, P. J., G. I. Pearman, and P. Hyson. 1983. The global distribution of atmospheric CO_2: II. A review of provisional background observations 1978–1980. J. Geophys. Res. 88:3591–3598.

Fung, I., K. Prentice, E. Mathews, J. Lerner, and G. Russell. 1983. Three-dimensional tracer model study of atmospheric CO_2: response to seasonal exchanges with the terrestrial biosphere. J. Geophys. Res. 88:1281–1294.

Gammon, R. H. and W. D. Komhyr. 1983. Response of the global atmospheric CO_2 distribution to the atmospheric/oceanic circulation perturbation in 1982. In Hamburg IUGG, Symposium 19, The Oceans and the CO_2 Climate Response, August 1983, paper 19/2, Vol. 2, p. 828.

Harris, J. M. and B. A. Bodhaine. (eds.). 1983. Summary Report 1982, Geophysical Monitoring for Climatic Change. Environmental Research Laboratories/NOAA, U.S. Department of Commerce, Washington, DC.

Herbert, G. A. (ed.). 1980. Geophysical Monitoring for Climatic Change, No. 8: Summary Report 1979. NOAA Environmental Research Laboratories, Boulder, Colorado.

Keeling, C. D. 1983. The global carbon cycle: What we know and could know from atmospheric, biospheric, and oceanic observations. In Proceedings of the CO_2 Research Conference: Carbon Dioxide, Science, and Consensus, DOE CONF-820970, pp. II.3–II.62. NTIS, Springfield, Virginia.

Keeling, C. D., A. F. Carter, and W. G. Mook. 1984. Seasonal, latitudinal, and secular variations in the abundance and isotopic ratios of atmospheric CO_2. J. Geophys. Res. 89:4615–4628.

Komhyr, W. D., R. H. Gammon, J. Harris, L. W. Waterman, T. J. Conway,W. R. Taylor, and K. W. Thoning. 1985. Global atmospheric CO_2 distribution and variations from 1968–1982 NOAA/GMCC flask sample data. J. Geophys. Res. (in press).

Machta, L. 1983. The Atmosphere. In Changing Climate, pp. 242–251. National Academy of Sciences, Washington, DC.

Mendonca, B. G. (ed.). 1979. Geophysical Monitoring for Climate Change, No. 7: Summary Report 1978. NOAA Environmental Research Laboratories, Boulder, Colorado.

Mook, W. M., M. Koopmans, A. F. Carter, and C. D. Keeling. 1983. Seasonal, latitudinal, and secular variations in the abundance and isotopic ratios of atmospheric carbon dioxide (1): results from land stations. J. Geophys. Res. 88:915–933.

Pearman, G. I., P. Hyson, and P. J. Fraser. 1983. The global distribution of atmospheric carbon dioxide: I. Aspects of observation and modelling. J. Geophys. Res. 88:3581–3590.

Peterson, J. T. (ed.). 1978. Geophysical Monitoring for Climatic Change, No. 6: Summary Report 1977. NOAA Environmental Research Laboratories, Boulder, Colorado.

Stokes 1982. Atmospheric carbon dioxide abundance: An archival study of spectroscopic data In W. Clark (ed). Carbon Dioxide Review, Oxford University Press, New York.

Tanaka, M., T. Nakazawa, and S. Aoki. 1983. Concentration of atmospheric carbon dioxide over Japan. J. Geophys. Res. 88:1939.

Wong, C. S., Y. H. Chan, S. Page, R. D. Bellegay, and K. G. Pettit, 1984. Trends of atmospheric CO_2 over Canadian WMO background stations at ocean weather station P, Sable Island, Alert. J. Geophys. Res. 89:9527–9539.

2
Simulating the Atmospheric Carbon Dioxide Distribution with a Three-Dimensional Tracer Model

MARTIN HEIMANN, CHARLES D. KEELING, AND
INEZ Y. FUNG

Atmospheric CO_2 concentrations exhibit spatial and temporal variations reflecting the distribution and time dependence of various natural and anthropogenic CO_2 sources and sinks at the earth's surface. The location, magnitude, and time history of these sources and sinks are of paramount importance for global carbon cycle studies. From data obtained in a global network of suitably located CO_2 monitoring stations, it appears feasible to infer at least some of the characteristics of the sources from measured atmospheric CO_2 concentration data. To do so, the following two main problems have to be dealt with:

1. The atmospheric CO_2 concentration field is the result not only of sources and sinks, but also of the atmospheric circulation. One method of taking into account the effects of the latter is by means of models of atmospheric transport. Depending on the objectives of a study, one- to three-dimensional models have been used in past studies (Bolin and Keeling 1963, Pearman and Hyson 1980, Fung et al. 1983).

2. Several individual sources and sinks contribute to the observed atmospheric CO_2 concentration field. To deal with each source separately, a convenient first step is to decompose the observed concentration record as a function of time into a seasonal component and a slowly varying trend function. Seasonal sources are then principally identified with the seasonal component of the record, whereas stationary or sources varying on time scales longer than a year, e.g., the anthropogenic release of CO_2 from fossil fuel combustion, are identified with the slowly varying trend (Bacastow et al. 1985; Heimann and Keeling, in press). A second means to separate the effects of different sources is provided by the carbon isotope ^{13}C (Keeling et al. 1984). We discuss here mainly the first approach.

In the present study we used the three-dimensional tracer model developed at the Goddard Institute for Space Studies (GISS) in conjunction with the data from the CO_2 station network of the Scripps Institution of Oceanography (SIO).

We address the problem in the forward direction introducing rather simple but plausible source configurations and looking at the concentration patterns generated by the model. No attempt is made to fine-tune the simulated concentration fields to the observations by systematic adjustments of the source terms, although this is intended at a later stage in the project. The present approach, however, permits a detailed investigation of the behavior of the three-dimensional model. At the same time a more comprehensive understanding of the physical processes that determine the observations at the monitoring stations is obtained.

The pioneering study in this field has been conducted by the GISS group (Fung et al. 1983). Since they used only a seasonal biospheric source, the simulated concentration fields could not be compared directly with the seasonal data, because the latter also reflect the effects of other CO_2 sources and sinks. To obtain a fairly complete simulation of the atmospheric concentration field, additional sources were introduced. We considered in particular:

- A seasonal oceanic source driven essentially by the temperature of the surface of the ocean
- A fossil fuel CO_2 source and an associated oceanic sink
- A stationary oceanic source driven by a prescribed mean annual partial pressure distribution of CO_2 in the surface waters of the ocean.

This report summarizes our progress to date. In the following we describe the main features of the GISS tracer model. Thereafter we discuss the assumed sources and sinks and the observational data base. The simulated concentration fields are then compared to the data, thereby distinguishing between the seasonal signal and longer term trend. Finally we discuss some of the effects generated by the model and compare the present findings with results obtained from atmospheric tracers other than CO_2.

The GISS Three-Dimensional Tracer Model

This section describes the three-dimensional tracer model developed at GISS. Since a thorough description is in preparation (G. L. Russell et al. unpublished manuscript) only a brief summary is given of the model version that was used in the present study.

The GISS three-dimensional tracer model numerically solves the continuity equation for a tracer:

$$\frac{\partial}{\partial t}(C\rho) = -\nabla \cdot \mathbf{v}C\rho + \mathrm{CONVEC}(C\rho) + Q \qquad (1)$$

on a three-dimensional grid spanning the whole globe. Here C denotes the concentration of the tracer (expressed as a volume mixing ratio or mole fraction), ρ, the density of air, $\mathbf{v}$, the wind velocity vector, and t, time. CONVEC($C\rho$) represents the change in tracer concentration by vertical convection, discussed below, and Q is a source term that, for the inert CO_2 gas, differs from zero only at the surface of the earth.

The grid has a horizontal spacing of approximately 8° latitude by 10° longitude. There are nine layers in the vertical, of which the topmost two are in the stratosphere (centered at 27 and 103 mb). The lowermost two layers are centered at 959 mb and 894 mb, thus resolving the surface boundary layer with slightly more detail than the rest of the troposphere. The model uses pressure (sigma) coordinates in the vertical (Holton 1979). The tracer is advected by a three-dimensional, time dependent wind field and mixed vertically by convection (Fig. 2.1).

The wind field used in the present investigation was generated by the GISS climate model II (Hansen et al. 1982), and is the same as that used in a previous CO_2 tracer model study (Fung et al. 1983). The numerical scheme that carries out the transports by advection has been described by Russell and Lerner (1981).

In the GISS climate model the vertical convection scheme simulates the action of small-scale vertical motions that are not resolved on the grid of the model. Dry and moist convections are modeled at each time step through air mass exchange between vertical layers that are unstable thermally, moist statically, or both. The monthly average frequencies of vertical convection over each grid column, obtained from the climate model, are used in the tracer model to determine the amount of air that is ex-

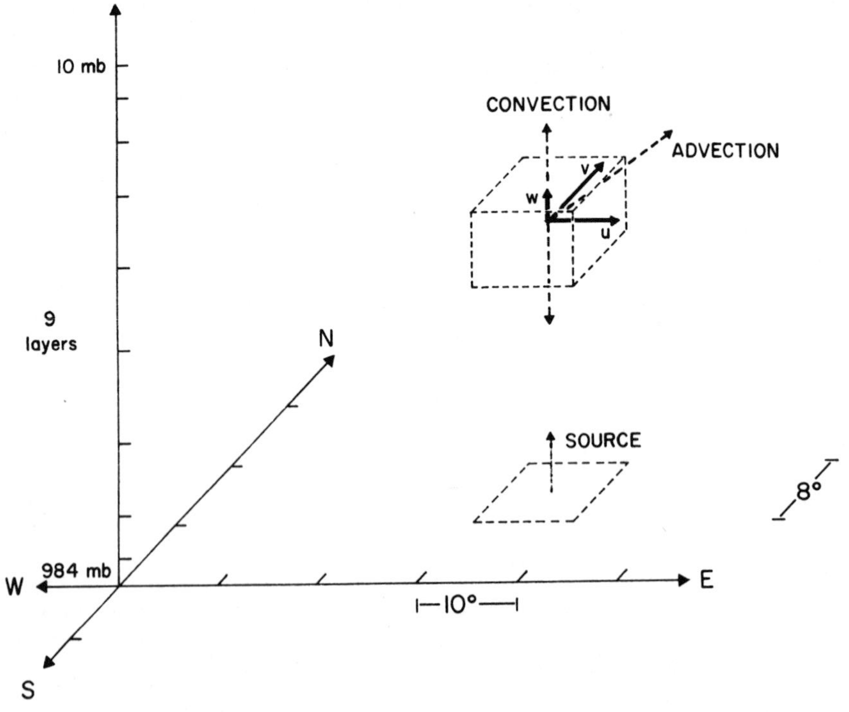

FIGURE 2.1. Schematic illustration of processes represented in the model.

changed in each time step between the various layers (G. L. Russell et al. unpublished manuscript). Tracer experiments with the source at the ground show that almost 50% of the vertical transport is achieved by convection.

The tracer sources and sinks are introduced into the model by specification of the tracer flux at the ground surface in each approximately 8° by 10° gridbox.

The Simulated Tracer Experiments

The basic model Eq. (1) is linear with respect to the source term Q. We assume in the present study that the seasonal and mean annual patterns of the atmospheric CO_2 concentrations can be simulated with a total source, Q_{tot}, consisting of four components:

$$Q_{tot} = Q_{bio} + Q_{ocs} + Q_{foss} + Q_{ocm} \tag{2}$$

Each of the four source terms is described below in greater detail. In the present study the model was run successively with only one of the components present as a source for each run. The results were then combined to obtain the composite solution C_{tot}:

$$C_{tot} = C_{bio} + C_{ocs} + C_{foss} + C_{ocm} \tag{3}$$

where C_i, denotes the model solution corresponding to a source Q_i.

The Terrestrial Biospheric Source (Q_{bio})

The model representation of the seasonal uptake and release of CO_2 by the terrestrial biosphere follows the method used by Fung et al. (1983). They depicted it (in their "experiment 3") as the product of two factors: (1) net primary production (NPP), a latitude- and longitude-dependent function representing the annual uptake of carbon by land plants, and (2) a seasonal vegetation function, describing the time dependency of the net flux of carbon flowing into or out of the biosphere.

The NPP values expressed as a flux of carbon per unit time and area are based on a map compiled by Matthews (1983) and are the same as used in the GISS study.

The seasonal vegetation function originally was constructed by Azevedo (1981) as a simple step function in time. It takes into account the seasonal characteristics of photosynthesis by the land plants and the decay of dead organic matter by heterotrophic respiration. It shows a stronger seasonality in higher latitudes (40 to 70°N and S) than in lower latitudes (10 to 40°N and S). The southern hemisphere fluxes are shifted by 6 months relative to the northern hemisphere fluxes. The equatorial belt between 10°N and 10°S is assumed to have no seasonality in uptake and release, hence the net flux is assumed to be zero.

We replaced the step function for 40° to 70° in the present investigation by two and that for 10° to 40° by one harmonic, thus obtaining a slightly smoother time dependence than Fung et al. (1983) (Fig. 2.2).

The growing season net flux (GSNF), defined as the net uptake of CO_2 during the time interval when the flux is negative, i.e., into the biosphere, is 0.419 times the annual NPP in the higher latitudes and 0.176 times the annual NPP in the lower latitudes. The corresponding fluxes in the GISS study, obtained from Azevedo's (1981) step functions, are 0.500, respectively 0.167 times the annual NPP. These differences are caused by the replacement of Azevedo's step functions with harmonics.

The terrestrial biosphere is assumed to be at steady state. The net flux

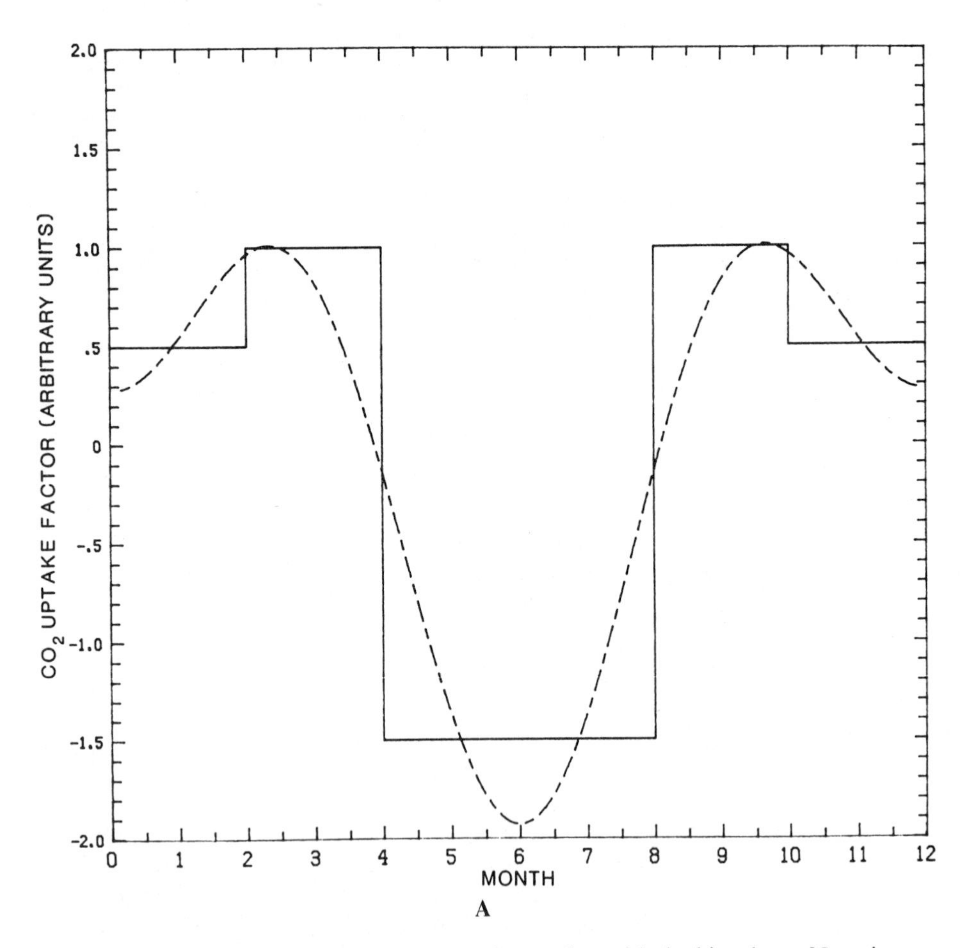

FIGURE 2.2. Time dependence of net exchange flux with the biosphere. Negative values correspond to a net uptake by the biosphere. Step function is seasonality as given by Azevedo (1981); smooth function is representation used in the present study. **A.** 40°–70°N. **B.** 10°–40°N. The functions used in the southern hemisphere are the same but shifted by 6 months.

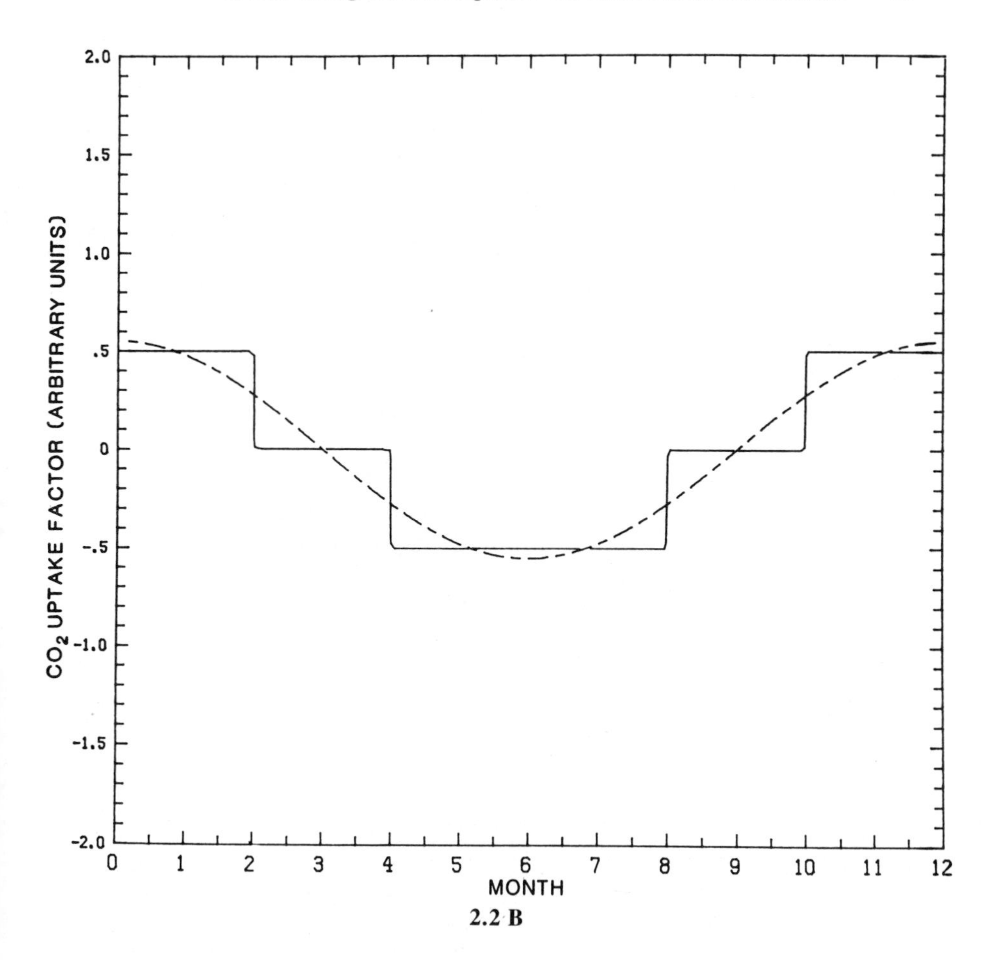

2.2 B

into the atmosphere at any location integrated over 1 year is thus set in the model to be zero. This condition incidentally provides a check on the numerics of the model, since the global CO_2 mass in the atmosphere from the seasonal biosphere should vanish after each simulated year. This was found in our computations to be true within 0.06%.

An additional computer run was performed with only the terrestrial biosphere of the northern hemisphere present. This allowed a separate examination of the signals generated by the land biosphere of each hemisphere.

The Seasonal Oceanic CO_2 Source (Q_{ocs})

In the middle and higher latitudes, substantial seasonal variations of the partial pressure of CO_2 (pCO_2) in the surface waters of the ocean occur. Only part of these variations are caused by the annual changes in temperature that shift the equilibrium of the carbonate chemistry. In addition,

pCO_2 is affected by changes in the rate of photosynthesis of the marine biosphere.

These changes, expressed as differences from the pCO_2 of the overlying air, induce a seasonal flux, F_{sa}, of CO_2 through the air-sea boundary, assumed to be proportional to the CO_2 pressure difference at the interface:

$$F_{sa} = k_{sa} \, (pCO_{2\ sea} - pCO_{2\ air}) \tag{4}$$

We use a constant value for the exchange coefficient k_{sa} of 2.166 kg C day^{-1} m^{-2} atm^{-1} corresponding to a global atmospheric residence time of 7.9 years or a gas exchange rate of 19 mol m^{-2} yr^{-1} (Bolin et al. 1981). This magnitude is known to at most ±25%, thus introducing a considerable uncertainty into the calculation. We neglect a possible latitude dependence of k_{sa}.

The relatively slow exchange of CO_2 between air and sea prevents any substantial depletion of CO_2 in the surface layer of the oceans over the course of one season (Tans 1978, Kamber 1980, Weiss et al. 1982). Therefore, to specify the CO_2 flux, only pCO_2 as a function of time needs to be prescribed. For simplicity we neglect here the second-order effect arising from the seasonal variation of atmospheric CO_2. (The latter amounts to at most 15 ppm, whereas the surface water pCO_2 changes by up to 100 ppm.)

The effect of the carbonate chemistry of seawater has been calculated from the relationship given by Weiss et al. (1982) between pCO_2 and temperature.

In high latitudes, the marine biosphere cancels to a large extent the effect of temperature, since in general photosynthetic activity is highest when light supply is at a maximum (C. S. Wong, personal communication). To allow for this effect, we reduced the temperature-driven variation of pCO_2 linearly to zero from 35° to 50° latitude in each hemisphere. Poleward of 50° in each hemisphere we assumed that no seasonal changes of pCO_2 occur.

The coastal upwelling areas off the western coasts of Africa and South America are treated as special cases. Here the seasonal variations in surface water temperature are caused largely by changes in the rate of upwelling of cooler water. Thus a computation of the oscillation of pCO_2 from the temperature effect alone does not take into account that the upwelling water is enriched in CO_2 relative to surface waters (Broecker and Peng 1982, p. 160). This would indicate a rise in pCO_2 with cooler temperatures, contrary to the aforementioned temperature effect. In addition, the marine biosphere presumably plays an important role in determining the actual seasonal changes in pCO_2. Lacking observational evidence, we reduced the variations of pCO_2 in these regions to the values computed in the subtropical gyres two grid units (20° longitude) further offshore.

Figure 2.3 shows the assumed peak-to-peak variation of sea surface pCO_2, computed by means of the seasonal sea surface temperature data

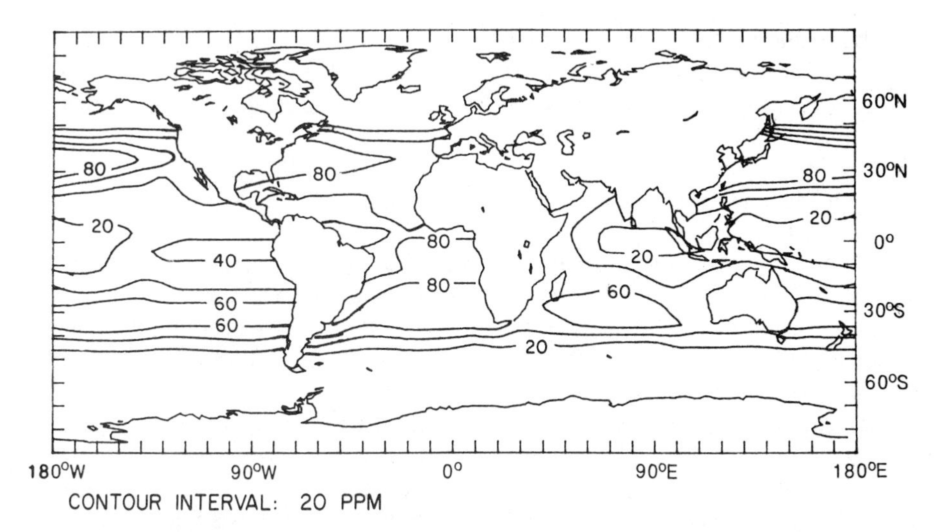

FIGURE 2.3. Peak-to-peak variation of sea surface pCO_2 as used to compute the seasonal oceanic source. Contour lines are labelled in ppm.

of Alexander and Mobley (1974). The phase follows the temperature variation, i.e., with maximum values in late summer of each hemisphere.

Since we are here interested only in the seasonal variation, we adjusted the mean pCO_2 at each location so that the time-integrated flux over 1 year vanishes everywhere.

THE FOSSIL FUEL SOURCE COMPONENT (Q_{foss})

The effect of the CO_2 released by fossil fuel combustion is simulated assuming an annual global CO_2 production rate of 5.29×10^{15} g C yr^{-1}, which represents the average of 1979 and 1980 (Rotty 1983). This source is distributed over the continents according to a map obtained from Rotty (personal communication). According to an estimate by G. H. Kohlmaier et al. (unpublished manuscript) about 10% of the global annual CO_2 release is seasonal, showing a winter maximum in regions where air conditioning is not commonly used, predominantly in Eurasia. We therefore introduced a small seasonality in the release of fossil fuel CO_2 over Eurasia north of 30°N, assuming a winter maximum 18% higher than the annual average.

Not all of the CO_2 released from combustion remains airborne, however. In the 2-year period 1979 to 1980 the atmospheric CO_2 concentration at Mauna Loa rose on average by 1.60 ppm yr^{-1} and at the South Pole by 1.76 ppm yr^{-1}. For the model simulation a global average increase of 1.68 ppm yr^{-1} was assumed, corresponding to 3.53×10^{15} g C yr^{-1}. To balance this mass, a sink of magnitude 1.76×10^{15} g C yr^{-1} was introduced into the model. For simplicity we attribute this sink entirely to the ocean as a constant flux of 5.25 g C yr^{-1} m^{-2} uniformly distributed over the part of the ocean not covered with ice (annual average area = 3.36×10^{14} m^2). With the exchange coefficient k_{sa} prescribed (see earlier discussion), a CO_2 pressure difference between air and sea of 6.6 ppm is required to simulate this flux.

The seasonally changing ice cover (Alexander and Mobley 1974) was found to have a negligible effect on the generated atmospheric concentration patterns.

The simulation of the fossil fuel source assumes an annually averaged constant input, whereas in reality the release of CO_2 has risen almost exponentially over the last decades (Rotty 1983). However the time for an *e*-fold increase is of the order of 25 years, considerably longer than any atmospheric mixing time. If we take 2 years as an upper limit for the interhemispheric exchange time, it is easy to show that the spatial concentration patterns generated by an exponential source differ from those of a constant source by $<5\%$ (Bacastow and Keeling 1981). Thus the fossil-fuel-induced concentration field can be approximated with a source that is constant in time. The solution of Eq. (1) has to be taken at the quasistationary state that emerges after the transients produced by the initial concentration field have died out. We ran the model typically for 4 years, a time sufficient that the annual atmospheric increase at every location very nearly attained the prescribed value of 1.68 ppm yr^{-1}.

The assumption that the nonairborne CO_2 from combustion is taken up by a uniform sink proportional to ocean area requires further justification. First, we notice that this sink corresponds to only 33% of the CO_2 released from fossil fuel use in the 1979-1980 period. Second, we argue that even if the condition of uniformity of the sink is not approximately true, or if the terrestrial biosphere takes up some of this CO_2, it is most likely that the sink is nevertheless widespread and almost balanced between the hemispheres. If this is the case, the resulting atmospheric patterns are caused primarily by the location of the concentrated sources in the few highly industrialized areas of the northern hemisphere.

THE STATIONARY OCEANIC SOURCE COMPONENT (Q_{ocm})

The surface waters of the ocean exhibit large spatial variations of pCO_2 in addition to the seasonal variations referred to in a previous section. These variations drive a flux of CO_2 through the atmosphere that is presumably close to steady state. Interannual variations are inferred to occur (Bacastow 1979). Here we are mainly concerned, however, with the general pattern which is basically the same from year to year (Keeling 1968; Takahashi 1979; Weiss et al. 1982). The general configuration consists of high pCO_2 in the equatorial regions and the major coastal upwelling areas which contrast with areas of relatively low pCO_2 in the subtropical and polar gyres. Unfortunately most of the few observations in higher latitudes represent summer conditions and are therefore biased.

On the basis of the observations reported by Keeling (1968) and from the FGGE expedition in the central Pacific (Weiss et al. 1982) we constructed a fairly crude map of the stationary pCO_2 difference between the ocean and the atmosphere (Fig. 2.4) inferred to exist in preindustrial times. We assumed that the fluxes out of the oceans in regions of high pCO_2 exactly balance the fluxes into the oceans in regions of low pCO_2. To compare this map to contemporary observations perturbed by fossil fuel CO_2, the 6.6 ppm as estimated in the previous section would have to be subtracted from the pCO_2 field shown in Fig. 2.4. This map was then used to simulate a sea-air CO_2 flux according to Eq. (4). With the exchange coefficient defined previously, we obtained a net flux of 1.21×10^{15} g C yr^{-1} into the atmosphere in the region bounded by the equator and 16°S. A similar flux of 0.82×10^{15} g C yr^{-1} was obtained between the equator and 16°N. In order that these fluxes be balanced by corresponding sources in the adjacent oceans south of 16°S, and north of 16°N, a uniform pCO_2 deficit of 11 ppm was postulated in these areas.

As will be discussed below, the comparison of the simulated mean annual meridional concentration gradient with the available data suggest an additional flux of the order of 1.2×10^{15} g C yr^{-1} from the southern into the northern hemisphere. This was tentatively implemented in the model by assuming a pCO_2 excess of 12 ppm in the oceans south of 40°S and by increasing the pCO_2 deficit in the Atlantic north of 24°N to 60 ppm. In postulating this flux we took into account that a flux of this order of magnitude could be generated by the latitudinally asymmetric thermohaline circulation of the oceans. In general, ocean surface waters are depleted in carbon relatively to the deep waters by about 0.25 mol C m^{-3} (Broecker

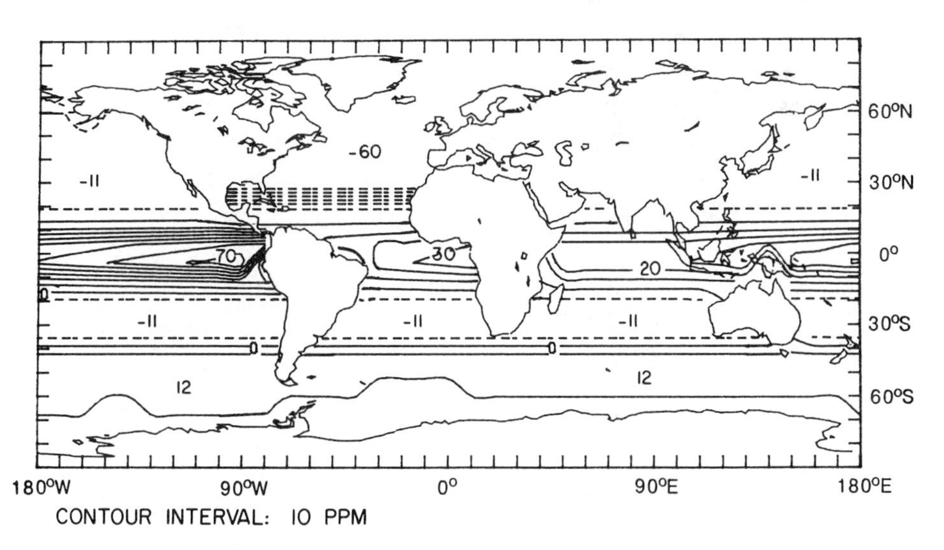

FIGURE 2.4. Mean annual pCO_2 difference (expressed in ppm) between surface ocean and atmosphere as used in the specification of the stationary oceanic source/sink.

and Peng 1982, p. 70) as a result of activity of the marine biosphere. Thus a southward flux of 13×10^6 m^3 s^{-1} (Broecker and Peng 1982, p. 349) in the deep Atlantic Ocean, compensated by an equal but opposite surface backflow, would result in a net southward transport of 1.2×10^{15} g C yr^{-1}.

These changes have been included in Fig. 2.4. Figure 2.5 displays schematically the adopted stationary oceanic CO_2 fluxes as a function of latitude.

The Data Base

Two kinds of atmospheric CO_2 data have been used in the present study: data from air samples taken at land-based stations and data obtained on oceanographic expeditions.

Station Data

The SIO samples air at a network of 10 land stations in the Pacific Ocean region, extending from Alaska to the South Pole. The geographical locations are shown in Fig. 2.6 and listed in Table 2.1. The experimental sampling procedures are as described by Keeling et al. (1968). The time series of recorded CO_2 data from all SIO stations have been analyzed with the method described by Bacastow et al. (submitted). We obtained at each location a representation of the atmospheric CO_2 signal consisting of a seasonal part given by four harmonics, and a trend, depicted by an exponential plus a spline function.

In November 1983, a second set of station data became available from the National Oceanographic and Atmospheric Administration (NOAA). These data, representing a network of 23 stations, cover all three world oceans and include also three high-altitude mountain stations in North

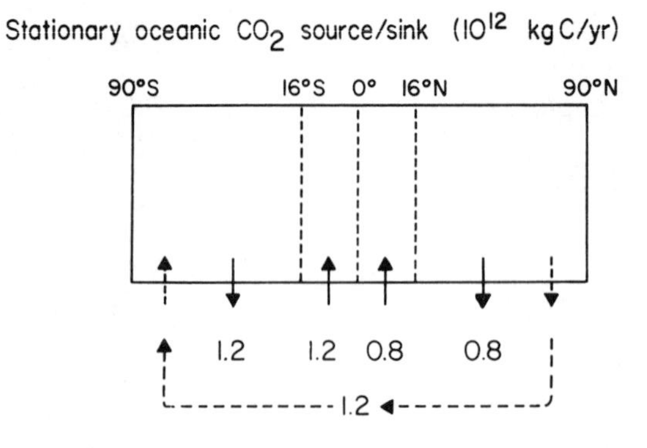

FIGURE 2.5. Schematic representation of latitudinal distribution of the fluxes that constitute the stationary oceanic source Q_{ocm}.

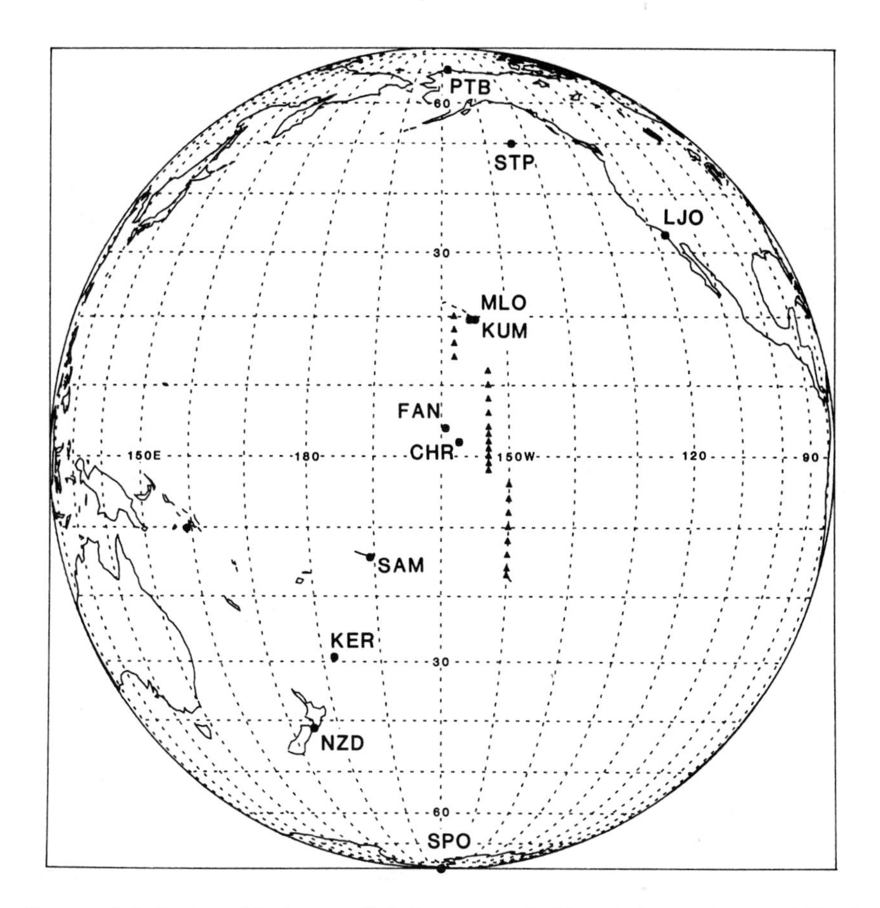

FIGURE 2.6. Geographical map of station network. The station codes are defined in Table 2.1.

and South America. Because most of these stations have been operated regularly only for the last 3 to 5 years, and some stations coincide with SIO stations, only a subset has been included in our investigation. For this report only the data from station SEY (Seychelles, 5°S, 55 °E) have been used. These data have been subjected to the same analysis procedures as the data from the SIO stations.

DATA FROM SHIPS

During 1979 and 1980, air samples taken at sea during the Hawaii to Tahiti shuttle experiment (Wyrtki et al. 1981) were analyzed for their atmospheric CO_2 content at SIO. These data cover the region between 20°N and 17°S in the central Pacific Ocean (160°W) with a spacing of 1-2° latitude and sampling interval of 1–2 months. They have been analyzed with the method described by Heimann and Keeling (in press). The spatial resolution of

TABLE 2.1. Stations where atmospheric CO_2 samples have been collected for analysis at the Scripps Institution of Oceanography.

Station	Code name	Location	Altitude (meters)	Period of record
Pt. Barrow, Alaska	PTB	157°W 71°N	11	1974–present
Ocean Station 'P'	STP	145°W 50°N	10	1969–1981
La Jolla, California	LJO	117°W 33°N	10	1978–present
Mauna Loa, Hawaii	MLO	156°W 20°N	3397	1958–present
Cape Kumukahi, Hawaii	KUM	155°W 20°N	3	1979–present
Fanning Island	FAN	159°W 4°N	2	1972–1983
Christmas Island	CHR	157°W 2°N	2	1975–present
Samoa	SAM	171°W 14°S	30	1981–present
Kermadec Island	KER	178°W 29°S	2	1983–present
Baring Head, New Zealand	NZD	175°E 41°S	85	1973–present
South Pole	SPO	90°S	2810	1957–present

the data set exceeds by far the resolution of the model (8° latitude), thus providing a means to assess subgrid scale variations that are not explicitly treated in the model.

To evaluate fully the performance of a three-dimensional global tracer model, a detailed data base is required. The current network of stations does not suffice in this respect. Large gaps exist over the continents, especially over Eurasia and Africa. The vertical dimension is not covered, except at a few places. The present data base allows, however, a fairly good assessment of the two most prominent features of the atmospheric CO_2 concentration field: the meridional distribution of the seasonal variation and the mean annual gradient between the hemispheres.

Simulation Results

In this section, we present a comparison of the model-generated concentration fields with the available data. We explore first the seasonal cycle and then the mean annual concentration field.

In contrast to our previous modeling efforts (Heimann and Keeling, in press), the time-dependent transport terms [e.g., the wind vectors $\mathbf{v}$ of Eq. (1)] induce a seasonal signal even if the source components themselves are constant in time. To compare the model results to the seasonal signal as determined from observations the complete solution C_{tot} [see Eq. (3)] has to be examined. We decomposed the model-generated concentration field into a seasonal component and a trend, the latter consisting of a uniform linear increase in time (resulting from the fossil fuel source Q_{foss}) and a mean annual field.

To compare the CO_2 time trend at an observing station to the model, the station location with respect to the model grid has to be specified.

Special attention is needed in the case of mountain stations, since a fixed station elevation cannot be unambiguously translated into the model's vertical sigma coordinate. This comes about because the coarse horizontal resolution of the model does not resolve individual mountain ranges. Thus the elevation of a station relative to the model topography has to be determined, which is then transformed into sigma coordinates by means of a standard atmosphere appropriate to the latitude of the station.

An additional complication arises when model results are compared to a station that is located close to local sources. To obtain estimates of the concentration of so-called "background air" an elaborate sampling and data screening scheme is used in the observing programs of SIO and NOAA. However, it is difficult to assess the amount of contamination still present in the data. On the other hand, the model simulates calm, stagnant weather situations, which, in the case of the fossil fuel source, tend to increase the local CO_2 concentration. Thus a dilemma exists between local contamination, which is not resolved in the model, and a possible sampling bias, which underestimates the large-scale average degree of contamination at the site. This problem is especially severe in the case of the mean annual concentration, where gradients of a few tens of ppm imply relatively large fluxes.

The Observed and Simulated Seasonal Cycle

The main features of the seasonal atmospheric CO_2 cycle can be simulated with a terrestrial biospheric source alone as Fung et al. (1983) have shown. The further effect of a CO_2 source owing to the seasonally changing pCO_2 in the ocean surface has been discussed by Fung (this volume). Here we will emphasize the additional seasonal signals induced by the constant sources Q_{foss} and Q_{ocm}. Their contribution to the combined seasonal variation is small, but not negligible, as will be shown below. By way of illustration we give a detailed description of the combined and separate seasonal signals at the South Pole and at Mauna Loa.

As a means to summarize the combined model-generated seasonal field at the surface of the earth, we present amplitude (peak-to-peak) and phase, defined by the date of occurrence of the maximum in Figs. 2.7 and 2.8.

Figure 2.7 illustrates the dominant influence of the land biosphere in the overall seasonal cycle. Maximum variations are found in the center of the continents of the northern hemisphere and, to a lesser extent, of the southern hemisphere. There is a decrease in amplitude going from north to south that is similar to the observations. We note, however, that the model predicts a finite amplitude everywhere in the center of the Pacific, whereas the seasonal signal appears to vanish around 14°S; as observed in the FGGE data (Heimann and Keeling, in press) and at station SAM (American Samoa, 14°S, 170°W).

The phase of the generated variation (see Fig. 2.8) is determined mostly by the prescribed time dependence of the seasonal vegetation function

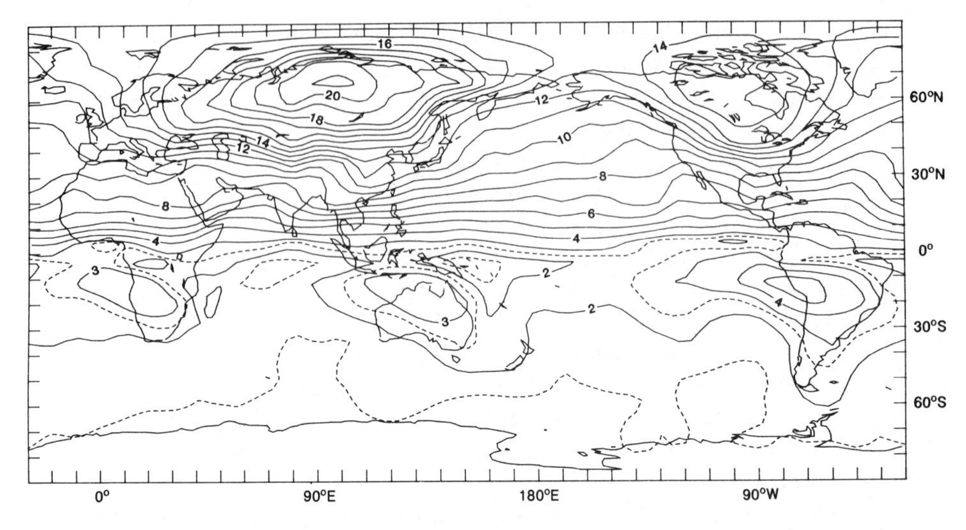

FIGURE 2.7. Amplitude of the seasonal variation (peak-to-peak) as predicted by the model at the surface. Contour interval between solid lines is 1 ppm.

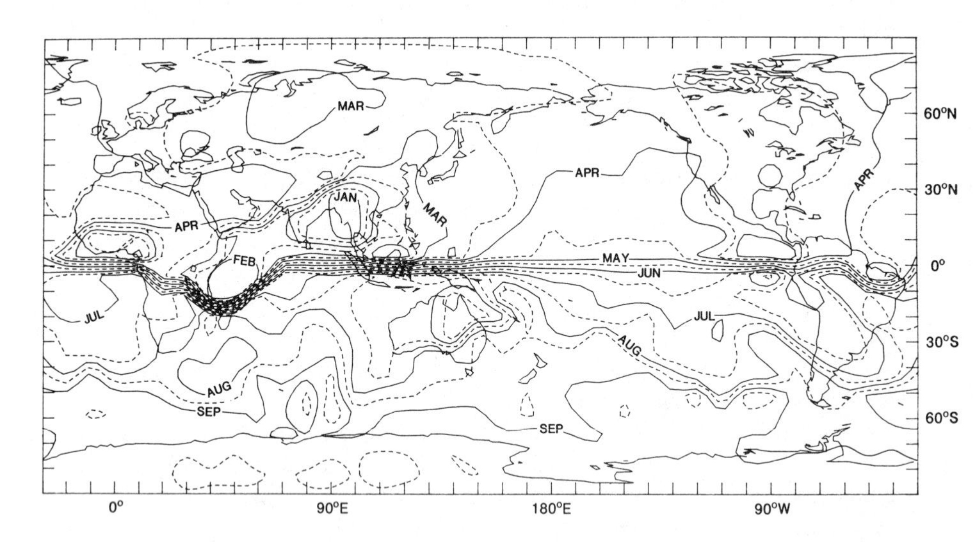

FIGURE 2.8. Phase of the simulated seasonal variation, defined as the date of the arrival of the maximum. Labelled contour lines denote the 15th of the respective month.

(see earlier discussion). In the higher latitudes the minimum of atmospheric CO_2 concentration occurs at the beginning of the growing season of the land biosphere. Along a north-south line in the center of the Pacific the phase in the model is progressively more delayed toward the south, which is confirmed by the observational data.

The most interesting feature in Fig. 2.8 is seen in a region between eastern Africa and Southeast Asia. There the model predicts a maximum signal in January and February. This timing is caused by the lower branch of the winter monsoon which carries CO_2-enriched air from the northern hemisphere toward the south.

The model-predicted timing of the seasonal signal in the monsoon region is corroborated by the data from NOAA station SEY (Seychelles, 5°S, 55°E) as shown in Fig. 2.9. The solid wavy line represents the model-generated daily CO_2 concentrations at that station. Also shown are the smoothed model concentrations using a binomial filter of 60 days width (long dashes) and the observed seasonal signal, represented by four harmonics (short dashes). Figure 2.9a compares the composite model signal (C_{tot}) and Fig. 2.9b compares only the signal from the land biosphere (C_{bio}) to the data. Although the composite model signal still underestimates the magnitude of the actual variation, it is evident that the inclusion of sources other than the land biosphere improved the fit to the data. On the basis of the model calculation, we thus conclude that a sizable fraction of the seasonal variation at station SEY is caused by the stationary sources (Q_{foss} and Q_{ocm}) and the time-varying transport pattern in the monsoon region.

Figure 2.10 shows the various components that constitute the signal at the South Pole. Plotted are the model-computed daily concentration versus time. The bottom panel contains the composite signal (wavy solid line) together with the observational data represented by the sum of four harmonics (Bacastow et al., 1985) (broad dashed line). The panels above show the separate model-derived components of the composite signal. The biospheric signal has been divided into two parts as described earlier: one component arising from the northern (C_{Nbio}) and one from the southern hemisphere land biosphere (C_{Sbio}). To each of the model-generated time series a smooth curve consisting of four harmonics has been fitted (short dashed lines).

Figure 2.11 presents the same components, but now at Mauna Loa. (Notice the different scales in Figs. 2.10 and 2.11.)

The simulated daily values show day-to-day fluctuations about the seasonal variation. This "noise" is associated with simulated air masses of different origin passing by the stations. It is comparable in magnitude to the scatter in the daily values actually observed at the two stations. The root mean square (rms) deviation between the simulated daily values and the smoothed daily values shown in Figs. 2.10 and 2.11 is 0.31 ppm at Mauna Loa and 0.053 ppm at the South Pole. At Mauna Loa, the daily

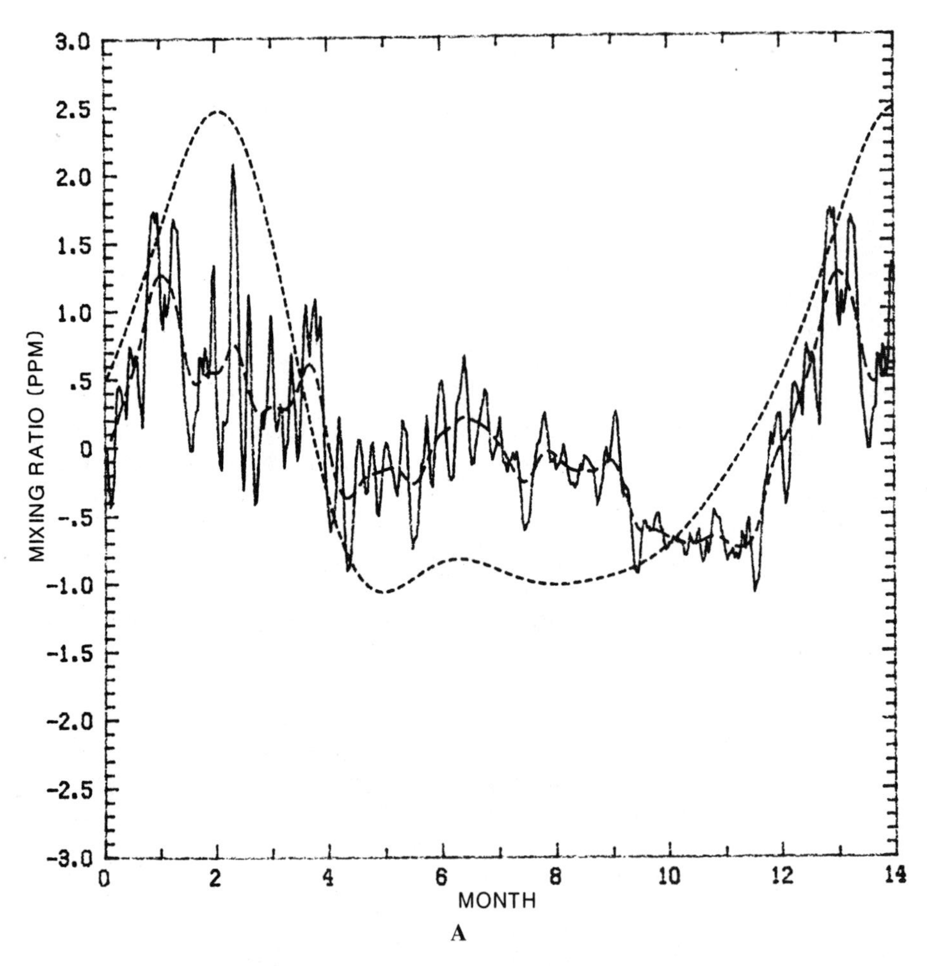

FIGURE 2.9. Model-predicted seasonal daily CO_2 concentration signals (solid lines) and seasonal data (short dashes), represented by four harmonics at station SEY (Seychelles, 5°S, 55°E). The smooth curves (long dashes) have been obtained by filtering the daily model values with a binomial filter of width 60 days. **A.** C_{tot}: Concentration signal was generated from the combined four sources as described in text. **B.** C_{bio}: Signal generated from the land biosphere alone.

averaged observations exhibit a rms deviation of 0.50 ppm about the smooth seasonal trend function (Bacastow et al., 1985). We note, however, that the observational scatter includes also local effects, which are not resolved in the model.

The most important contribution to the seasonal signal at the South Pole is the land biosphere of the southern hemisphere (C_{Sbio}). This con-

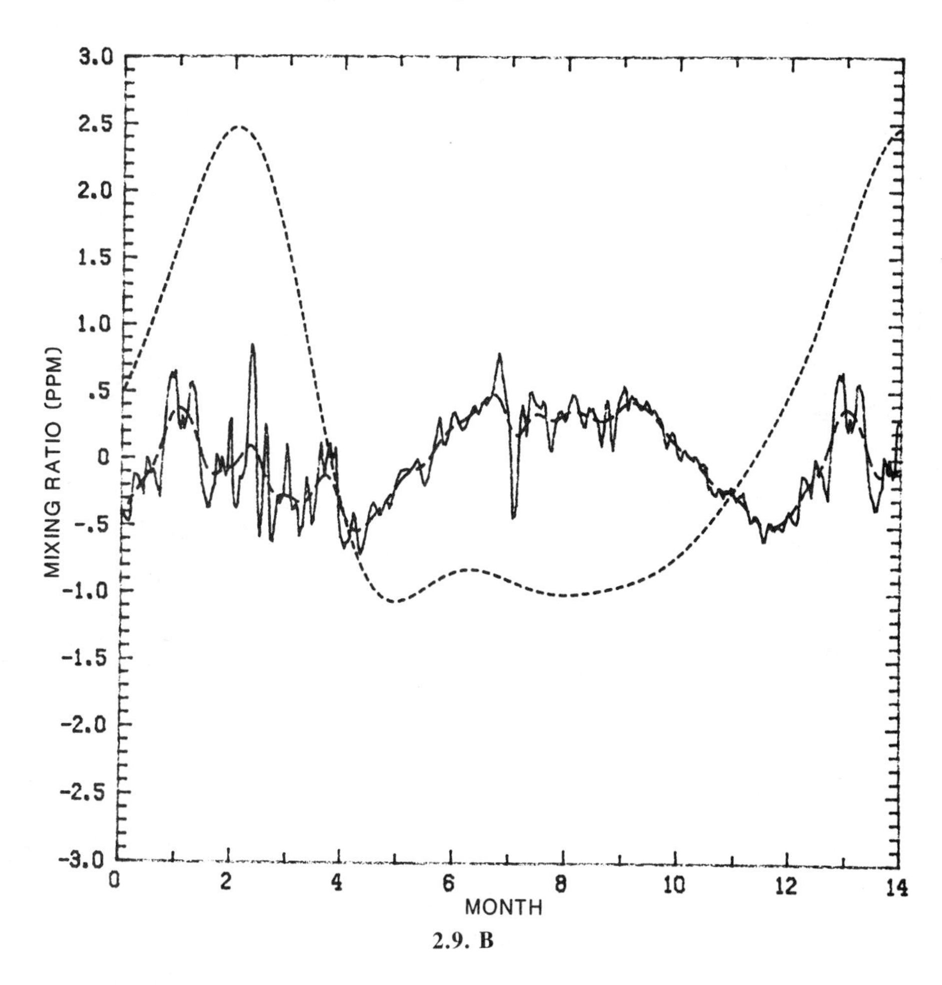

2.9. B

tribution is followed by that from the seasonally changing surface ocean chemistry (C_{ocs}). These two components determine the essential character of the composite solution.

The fact that the influence of the northern hemispheric biosphere at the South Pole is very small is somewhat surprising, since previous investigations found it to be one of the major factors (e.g., Pearman and Hyson 1980). As will be shown later, the GISS model probably underestimates the actual interhemispheric exchange by about 30%. If this defect were corrected for, the signal of the northern hemispheric biosphere would increase by 30% at the South Pole, but would still be small compared to the other components. This depends of course on the assumed ratio of the GSNF in the two hemispheres. However, even an increase of this ratio by a factor of 2 in favor of the northern hemisphere would not reproduce the findings reported by Pearman and Hyson (1980).

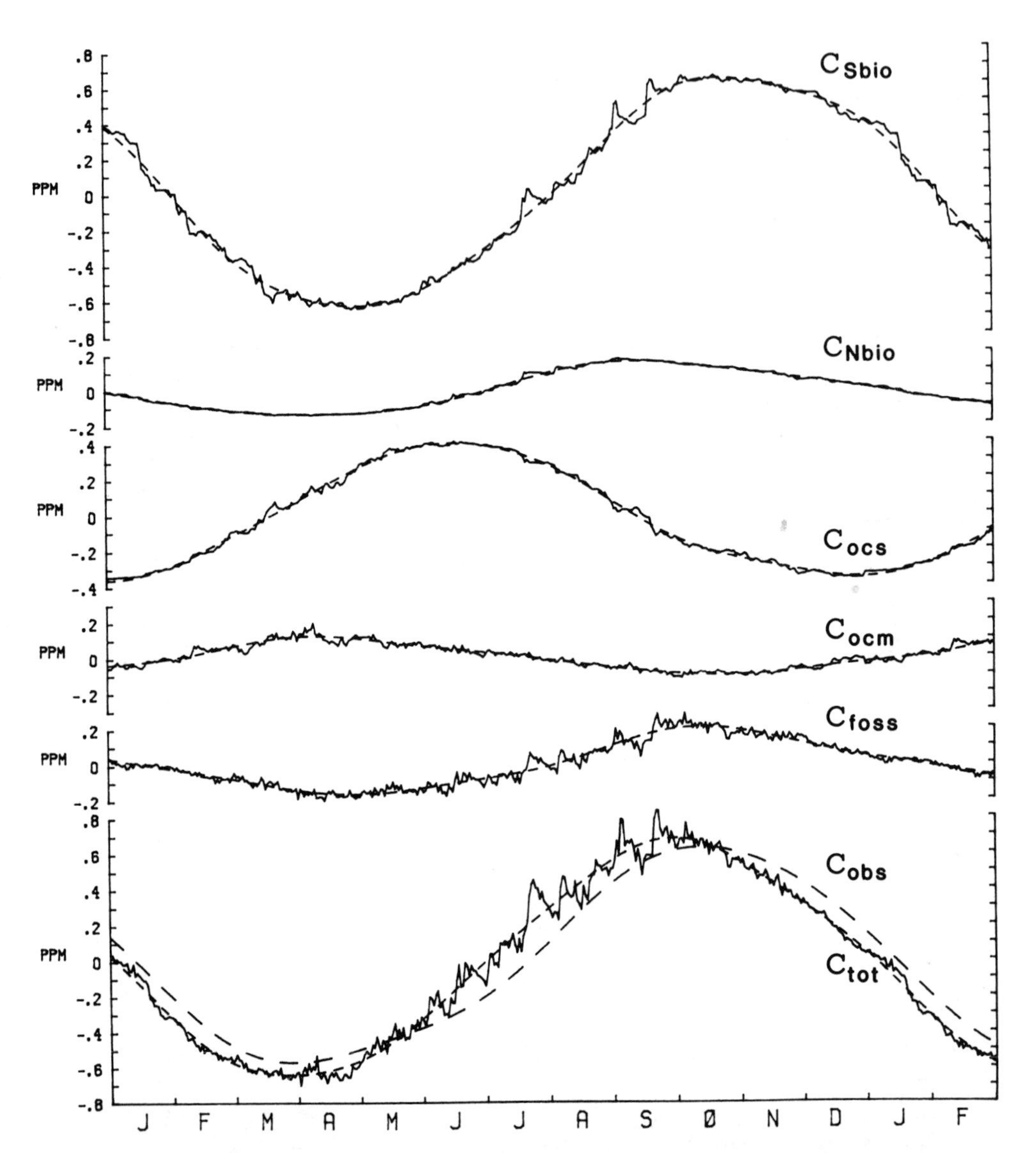

FIGURE 2.10. Simulated daily concentration values (solid curves) at the South Pole. Shown are the various components resulting from the different sources: C_{Sbio}, southern hemisphere land biosphere; C_{Nbio}, northern hemisphere land biosphere; C_{ocs}, seasonal oceanic CO_2 source; C_{ocm}, stationary oceanic source/sink; C_{foss}, fossil fuel source/sink. A four harmonic fit to the simulated daily values is also shown (short dashed curves). The **bottom panel** shows the combined signal C_{tot} together with the South Pole data (Bacastow et al., 1985) (C_{obs}, long dashed curve). Each component has been shifted, so that its annual mean vanishes. The fossil fuel component C_{foss} has been detrended by subtracting a linear increase of 1.68 ppm yr^{-1}.

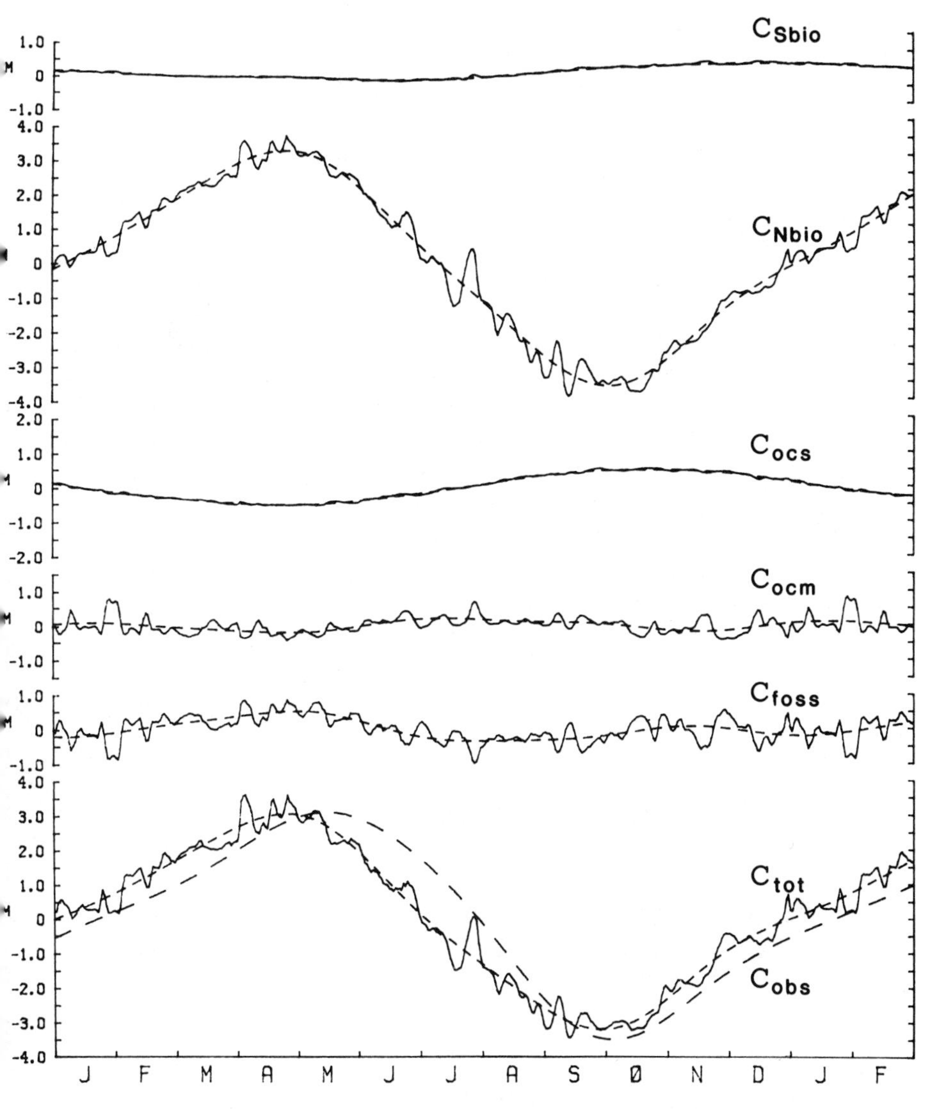

FIGURE 2.11. Same as Fig. 2.10 but for Mauna Loa (20°N, 155°W). Notice the different vertical scale from Fig. 2.10.

THE OBSERVED AND SIMULATED MEAN ANNUAL CONCENTRATION FIELD

The three-dimensional model allows us to simulate the mean annual CO_2 concentration field that results from the four specified sources. We compare here the results to the SIO observational data which are available only near 160°W at the surface. The model not only predicts these but the zonal and vertical mean annual fields as well.

Figure 2.12 shows a cross section at 160°W of the model-predicted mean annual field, C_{foss}, generated from the fossil fuel source alone. Also displayed are the predicted vertically averaged concentration at that longitude, the zonally averaged concentration at the surface, and the combined vertical and zonal average. The baseline of the vertical coordinate in Fig. 2.12 is arbitrary, and has been chosen so that the surface concentration vanishes at the South Pole. The principal feature consists of an interhemispheric concentration gradient, reflecting that >95% of the CO_2 from fossil fuel combustion is released in the northern hemisphere. From the vertical and zonal average concentrations we compute a mean concentration difference of 4.5 ppm between the two hemispheres. To help explain the significance of this number we will use it now to estimate the mean annual interhemispheric exchange time of the model.

Let C_n and C_s denote the average volume mixing ratio and Q_n and Q_s the net sources (expressed in g C yr^{-1}) in the northern and southern hemispheres, respectively. The mass balance in each hemisphere results then in the following equations:

$$\frac{dC_n}{dt} = k(C_s - C_n) + f\frac{Q_n}{M} \tag{5}$$

$$\frac{dC_s}{dt} = k(C_n - C_s) + f\frac{Q_s}{M} \tag{6}$$

where M represents the air mass of one hemisphere (2.53×10^{21} g) and f the molecular weight ratio between dry air and carbon (29/12). The interhemispheric exchange time is denoted by k^{-1}.

Subtracting Eq. (6) from (5), and observing that the time derivatives on the left hand side are equal (this discussion also applies to the quasistationary case described earlier), we obtain:

$$C_n - C_s = (2k)^{-1} f\frac{Q_n - Q_s}{M} \tag{7}$$

Let g denote the fraction of fossil fuel CO_2 released in the northern hemisphere and h the fraction of CO_2 taken up by the oceans north of the equator. Then Eq. 7 can be rewritten as

$$C_n - C_s = (2k)^{-1} f\frac{(2g - 1)Q_f - (2h - 1)S_f}{M} \tag{8}$$

Here Q_f and S_f represent the total release and oceanic uptake of fossil fuel CO_2, respectively. From the figures for these quantities given previously, we obtain:

$$f\frac{Q_f}{M} = 5.05 \text{ ppm yr}^{-1} \tag{9}$$

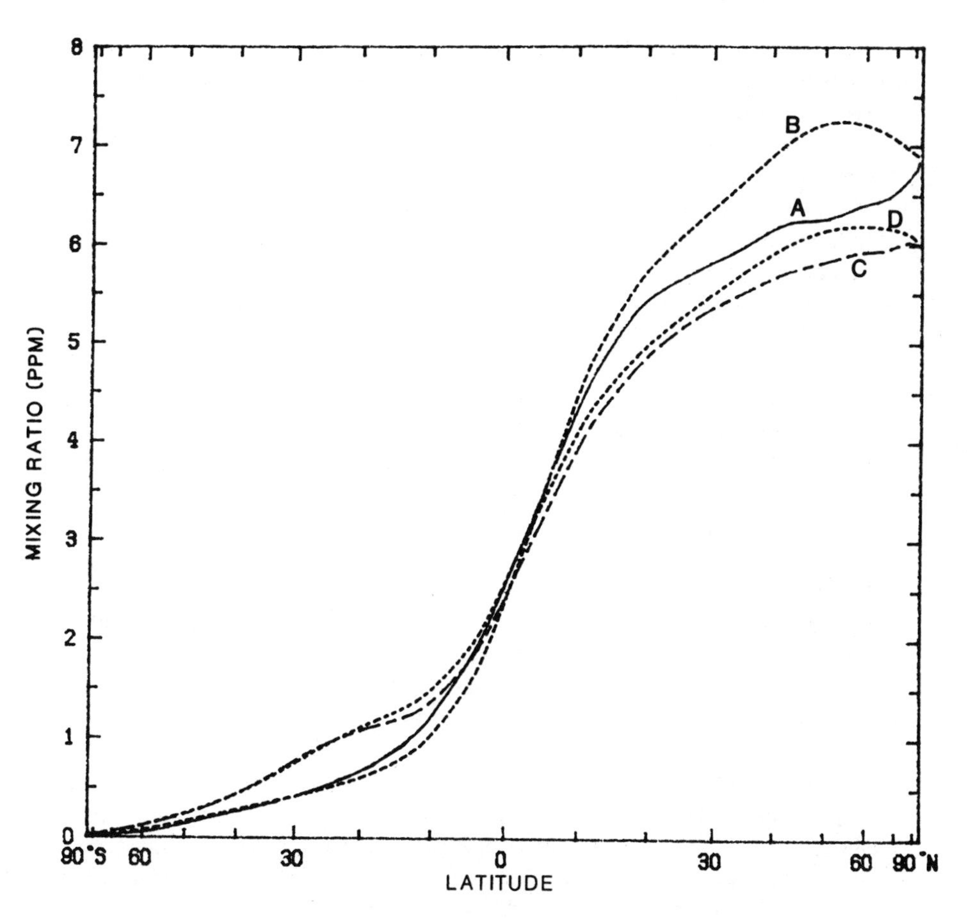

FIGURE 2.12. Mean annual concentration at the surface resulting from the fossil fuel source/sink Q_{foss} alone. The four different curves are: **A**, surface concentration at 160°W; **B**, surface concentration zonally averaged; **C**, vertically averaged concentration at 160°W; **D**, vertically and zonally averaged concentration. The curves have been shifted, so that the concentration at the South Pole vanishes.

and

$$f\frac{S_f}{M} = 1.68 \text{ ppm yr}^{-1} \tag{10}$$

From the map of fossil fuel release we calculate $g = 95.4\%$ and our assumption of the oceanic uptake to be proportional to ocean area results in $h = 43.0\%$.

Together with a vertically averaged concentration difference of 4.5 ppm we obtain

$$k^{-1} = 1.87 \text{ yr} \tag{11}$$

This value compares with an exchange time of 1.40 years which we infer from the observations of $CFCl_3$ during the atmospheric lifetime experiment (Prinn et al. 1983). Similarly, the ^{85}Kr data published by Weiss et al. (1983) indicate an interhemispheric exchange time of comparable magnitude.

If we compare the observed CO_2 gradient in 1980 with corresponding data from 1960 to 1962, an increase of the interhemispheric concentration difference of about 1.5 ppm is found (Keeling and Heimann, in press). If this increase is attributed solely to the increase in CO_2 release from fossil fuels, which approximately doubled during the same period, we obtain an interhemispheric exchange time of 1.2 years.

We thus conclude that the GISS model in its present form underestimates the cross-equatorial exchange by about 30%. It is likely that this deficiency is restricted mainly or entirely to the meridional transport in the tropics since the fluxes in midlatitudes are constrained by the transport of energy, water vapor, and angular momentum, which are quite well represented in the GISS climate model (Hansen et al. 1982).

Figure 2.12 also shows that the model troposphere is fairly well mixed vertically and zonally on a time scale of 1 year. Only around 50°N does the surface zonal mean exceed the surface concentration at 160°W by $>$ 1 ppm. This difference arises because the source is located at the surface over land, whereas the 160°W meridian is located in the middle of the Pacific Ocean, which constitutes a sink for fossil fuel CO_2.

The seasonal terrestrial biospheric source, together with the time-varying transport terms, generates also a mean annual concentration field which is shown in Fig. 2.13 for the surface. Notice that the global average concentration is zero in this case, since the time integral over 1 year of Q_{bio} vanishes everywhere.

Although most variations displayed in Fig. 2.13 are rather small they nevertheless exhibit a consistent pattern that reflects the effects of the various transport processes. There is a concentration peak in a narrow band north of the equator, which is flanked by regions of relatively low concentrations in the regions of the subtropical highs. This pattern can be explained in the terms of the seasonally changing structure of the Hadley cells. In summer and in winter, only one cell is developed and this cell is oriented so that the air in its lower part is flowing toward the summer hemisphere. The winter hemisphere is relatively enriched in CO_2 because of the phase of the seasonal cycle, hence the lower branch of the Hadley cell always transports CO_2-enriched air toward the equatorial region, giving rise to the aforementioned peak.

Also, two regions of high CO_2 concentration show up in Fig. 2.13 over Siberia and Canada. These may be explained by the very different meteorological regime in high northern latitudes during winter and summer. In winter, air tends to sink over the cold continents and there is very little

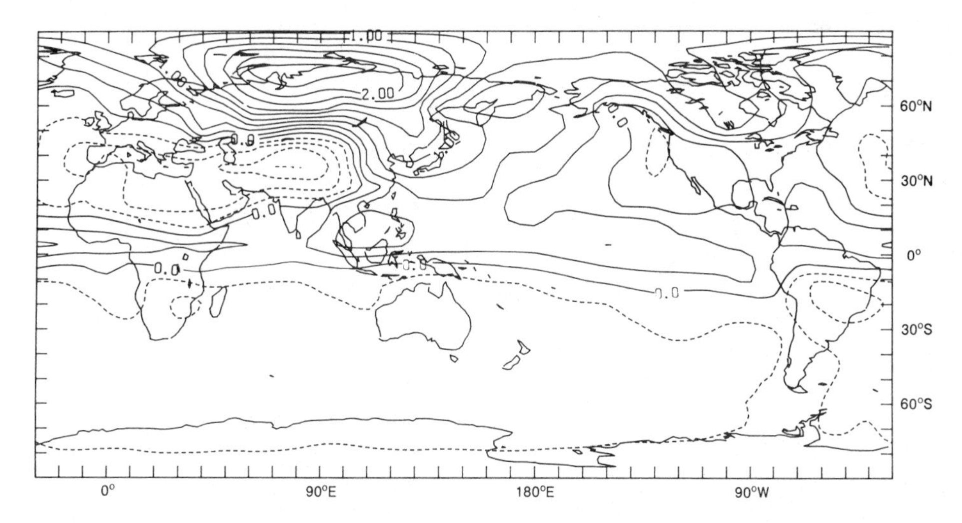

FIGURE 2.13. Mean annual CO_2 concentration at the surface generated by the seasonal terrestrial biospheric source Q_{bio}. Contour interval is 0.25 ppm.

vertical convection. Therefore the CO_2 released from the ground is trapped in the lower troposphere. In summer the continental pressure highs are less marked and there is more vertical convection. Hence the summer low in concentration is more diluted in the vertical, which results in a positive mean annual concentration at the surface.

The foregoing discussion clearly demonstrates that mean annual CO_2 concentration gradients do not necessarily correspond to a net source or sink at the ground. This fact has to be taken into account if flux estimates are to be obtained from atmospheric observations.

The seasonal oceanic source Q_{ocs} was not found to induce any significant mean annual concentration patterns, mainly because it is out of phase with respect to the seasonal transport mechanisms discussed above.

The signal generated by the stationary oceanic source component Q_{ocm} exhibits a strong latitudinal dependence, as displayed in Fig. 2.14. The assumed peak in pCO_2 of the equatorial surface water shows up in the atmospheric concentration profile with a magnitude of about 1.4 ppm. The imposed flux of 1.2×10^{15} g C from the southern hemisphere ocean through the atmosphere and into the North Atlanic results in a south to north gradient of 2.5 ppm. Figure 2.14 shows only the gradient at 160°W in the Pacific Ocean; the corresponding gradient over the Atlantic Ocean would be about 0.8 ppm greater.

Also displayed in Fig. 2.14 are the superimposed gradients discussed above, which result from the seasonal terrestrial biospheric source Q_{bio} and from the fossil fuel CO_2 source component Q_{foss}. Since the latter has

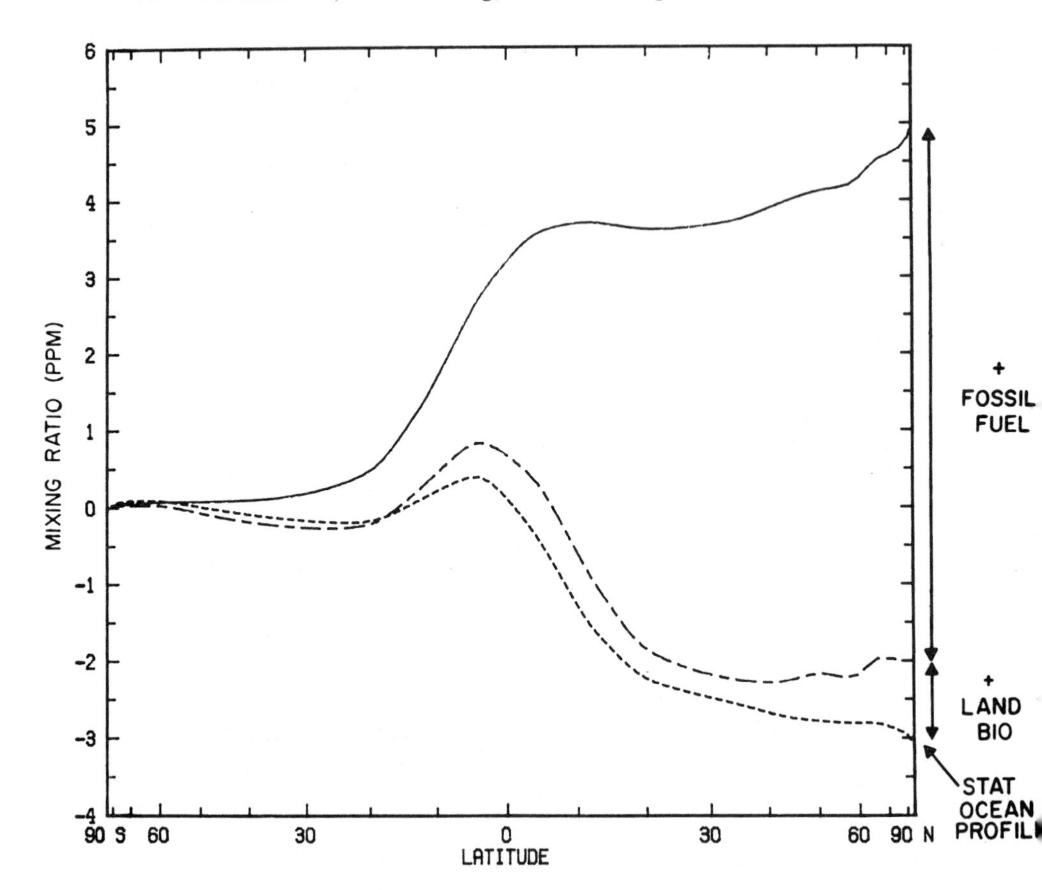

FIGURE 2.14. Superimposed mean annual concentration gradients at the surface at 160°W.

can be reconstructed (Fig. 2.15). In this reconstruction we assumed a time-invariant spatial source and sink distribution, so that the gradients in concentration are directly proportional to the fossil fuel combustion rate. One sees that the contribution from the fossil fuel source increasingly dominates that profile.

Figure 2.16 shows the predicted profile for 1960 together with data from South Pole (SPO), Mauna Loa (MLO), and Pt. Barrow, Alaska (PTB). Additional shipboard data (not shown) obtained in the same period confirm the main features: a peak of about 1.5 ppm at the equator and relatively small interhemispheric concentration difference (Keeling and Heimann, in press). This observational evidence led us to postulate the preindustrial flux of 1.2×10^{15} g C from the southern oceans into the North Atlantic as described earlier. If it were omitted, the simulated interhemispheric gradient would be about 3 ppm (see Fig. 2.15), which clearly violates the data.

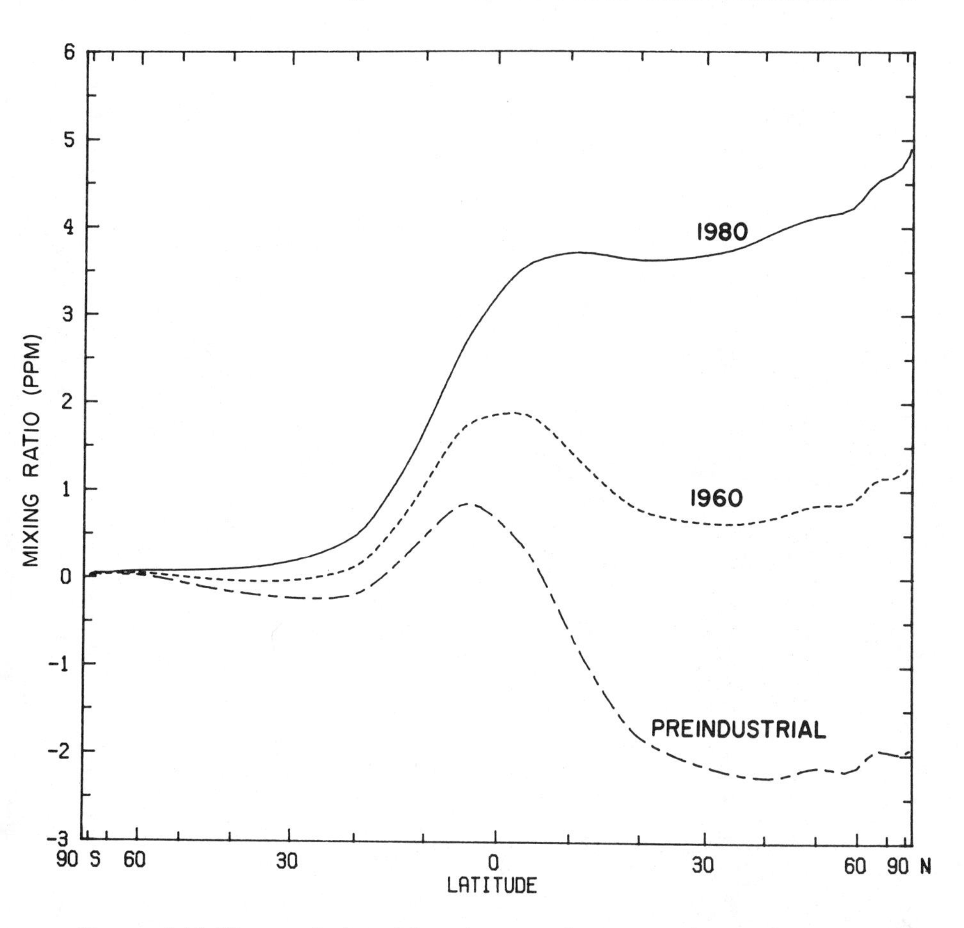

FIGURE 2.15. Time evolution of the mean annual concentration gradient at 160°W.

Figure 2.17 shows the simulated profile at 160°W for the year 1980 together with the data from the SIO station network and from the FGGE expedition. It is seen that the model overestimates the interhemispheric gradient. This could in principle be adjusted by increasing further the preindustrial interhemispheric flux from the stationary oceanic source component. However, this would lead to greater discrepancies between model and data in the 1960 profile. Clearly, the fossil fuel contribution is overestimated in the model, as a result of too small interhemispheric mixing.

Discussion

A conspicuous feature in Figs. 2.10 and 2.11 is that the fossil fuel source and the stationary oceanic source both give rise to seasonal signals in the atmosphere, although both sources have been assumed to be almost con-

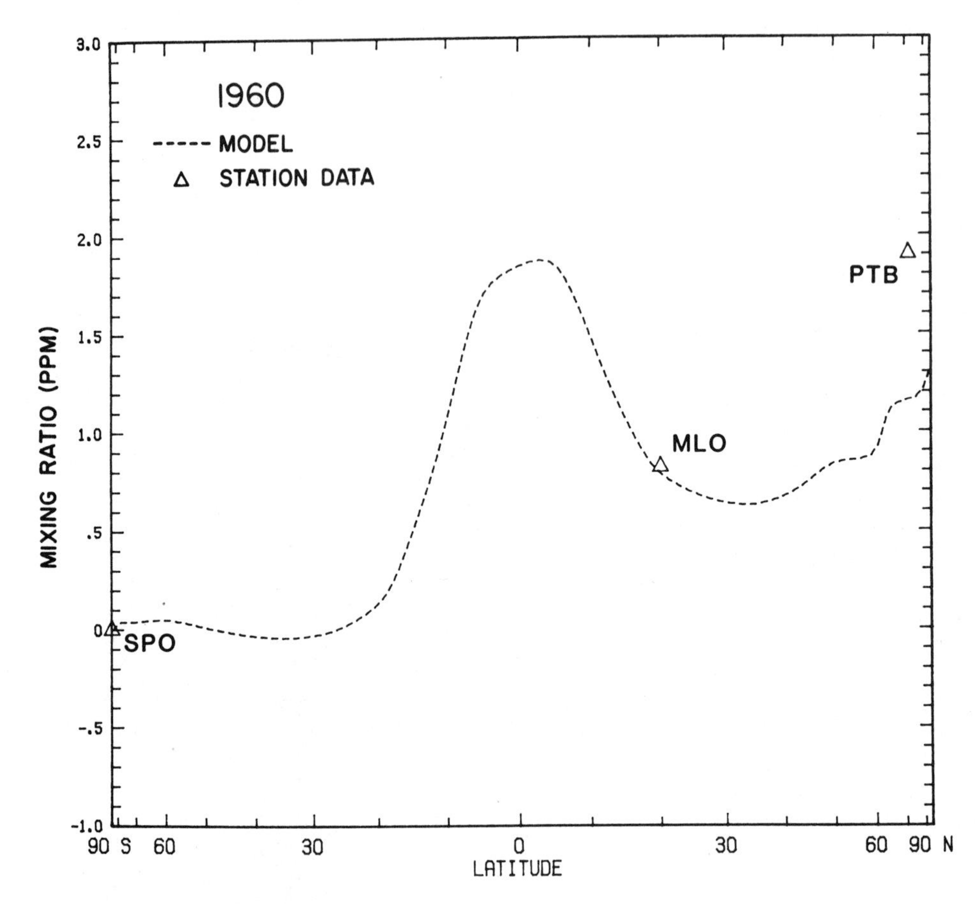

FIGURE 2.16. Simulated mean annual concentration gradient in 1960, compared with available station data.

stant in time.* This effect results from the seasonally changing atmospheric transport patterns as they are represented in the model.

The mean annual concentration fields arising from these two sources exhibit gradients between the hemispheres, as discussed above. These gradients are maintained by the existence of different source/sink configurations in the two hemispheres. The associated interhemispheric flux of CO_2 is then modulated by the changing strength of the Hadley cells and the movement of the intertropical convergence zone, thus giving rise to

*As explained earlier, the fossil fuel source Q_{foss} in the model contains a small seasonal fraction. The comparison with a previous model run that used a constant source in time showed that this seasonal fraction had a noticeable effect only north of about 10°N, with a maximum peak-to-peak variation of <0.3 ppm. Also the seasonally changing extent of sea ice causes some variation, which, however, is even smaller.

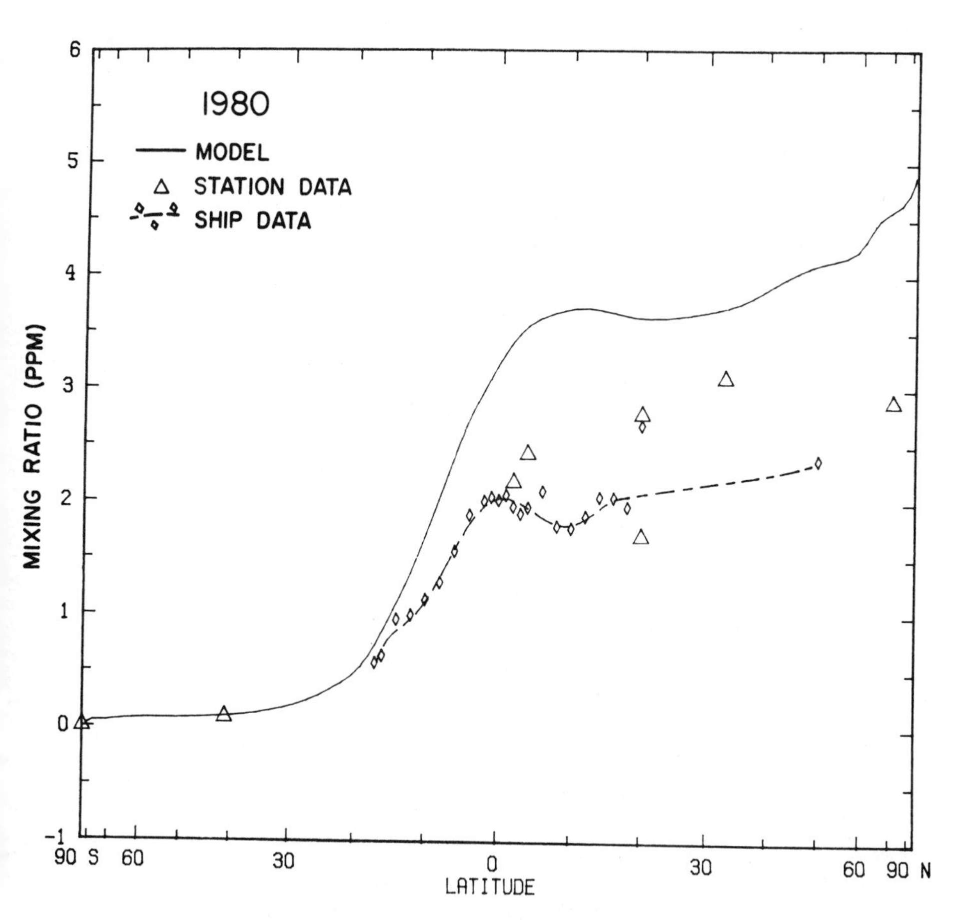

FIGURE 2.17. Simulated mean annual concentration gradient in 1980, compared with station data and shipboard data obtained during the Hawaii to Tahiti Shuttle cruises.

a seasonal variation of atmospheric CO_2 in the individual hemispheres.

The seasonally changing vertical stability of the troposphere also contributes to the fossil-fuel-induced seasonal variation in the northern hemisphere. In summer, the CO_2 released at the ground is diluted vertically into a greater volume, since the enhanced vertical convection over the continents mixes the troposphere more thoroughly. This results in generally lower surface concentrations in summer as compared to winter.

Although small in magnitude, the seasonal signal induced by the fossil fuel source has important consequences. It has been shown recently that the seasonal variation of atmospheric CO_2 increased with time during the last two decades. At Mauna Loa, at 19°N latitude, this increase relative to the variation in 1980 amounted to $0.64 \pm 0.05\%\ yr^{-1}$, computed from the record extending from 1958 to 1982 (Bacastow et al., 1985). At the South Pole the relative increase is much larger, amounting to 1.64 ± 0.46% yr^{-1} (Bacastow et al. 1981). (The greater error is related to the variation at the South Pole being much smaller.) It would appear likely that a substantial fraction of these increases in amplitude are caused by corresponding changes in the seasonal sources, possibly associated with a net growth of the terrestrial biosphere. Our model simulations point to an additional factor. Since the interhemispheric gradient of CO_2 has increased during the last decades (Keeling and Heimann in press), as a result of increasing use of fossil fuel, the seasonal CO_2 signal induced by fossil fuel emissions should have increased also. Thus a fraction of the observed increase of the seasonal variation in CO_2 could be caused by the increasing fossil fuel source.

Quantitatively this effect may be assessed as follows:

Let S_i denote some measure of the seasonal variation of CO_2 signal C_i in the year 1980, where i stands for one of the indices defined in Eq. (2). We shall take for S_i the concentration difference of the signal C_i exhibited between the two time instants, when the composite signal C_{tot} computed for the year 1980 attains a maximum and a minimum. The peak-to-peak variation of C_{tot} in 1980 may then be expressed as

$$S_{tot} = S_{bio} + S_{ocs} + S_{foss} + S_{ocm} \tag{12}$$

If now a linear increase is observed in S_{tot} it may be expressed for time t as S_{tot} $[1 + a(t-t_0)]$, where t_0 represents the year 1980, and a denotes the fractional increase per year. Similarly S_{bio} may be written as S_{bio} $[1 + b(t-t_0)]$, where b denotes the fractional increase of the seasonal uptake and release of the terrestrial biosphere. The effect of the fossil fuel source at any time t is directly proportional to the rate of fossil fuel consumption: $S_{foss}\, Q(t)/Q(t_0)$. Assuming no changes in the oceanic signals S_{ocs} and S_{ocm}, we can now reexpress Eq. (12) for an arbitrary time t. After some algebraic simplifications we obtain:

$$aS_{tot} = bS_{bio} + \frac{Q(t) - Q(t_0)}{(t - t_0)Q(t_0)} S_{foss} \tag{13}$$

which relates the observed fractional increase a to the unknown biospheric fractional increase b.

Table 2.2 lists the magnitudes of the seasonal variations as defined above for Mauna Loa and the South Pole. If we observe that the fossil fuel source increased from 2.27×10^{15} g C yr^{-1} in 1957/58 to 5.29×10^{15} g C yr^{-1} in 1979/1980 (Rotty 1983), we obtain $b = 0.31\%$ yr^{-1} using Eq. (13) and the values listed in Table 2.2 for Mauna Loa. Thus half of the observed increase according to the model calculations is explained by the fossil fuel effect.

TABLE 2.2. Model generated seasonal signals and station data.

	Station MLO (20°N, 155°W)		
	26 April* (ppm)	29 September* (ppm)	Difference (ppm)
S-Biosphere (C_{Sbio})	−0.065	0.065	−0.130
N-Biosphere (C_{Nbio})	3.298	−3.569	6.867
Seas. Ocean (C_{ocs})	−0.511	0.476	−0.987
Stat. Ocean (C_{ocm})	−0.189	0.042	−0.231
Fossil Fuel (C_{foss})	0.524	−0.218	0.742
Composite (C_{tot})	3.058	−3.203	6.261
Observations (C_{obs})**	2.827	−3.461	6.288
	Station SPO (90°S)		
	30 September* (ppm)	28 March* (ppm)	Difference (ppm)
S-Biosphere (C_{Sbio})	0.567	−0.540	1.107
N-Biosphere (C_{Nbio})	0.142	−0.130	0.272
Seas. Ocean (C_{ocs})	−0.124	0.042	−0.166
Stat. Ocean (C_{ocm})	−0.096	0.118	−0.214
Fossil Fuel (C_{foss})	0.191	−0.131	0.322
Composite (C_{tot})	0.679	−0.641	1.320
Observations (C_{obs})**	0.610	−0.569	1.179

*The dates shown indicate the time of the maximum and the minimum of the composite signal C_{tot}.

**Observed seasonal CO_2 concentration signal in 1980.

A similar computation for the South Pole data shows that the fossil fuel effect can explain almost half of the observed increase of the seasonal amplitude, resulting in $b = 0.93\%$ yr^{-1}.

Comparison with Other Tracers

Since the observed CO_2 concentration field is the superposition of the fields generated by several source components, it is difficult to assess the reality of the model-predicted effects caused by a single source component, e.g., the contribution of the fossil fuel source to the seasonal concentration variation discussed above. One way to circumvent this problem is to look at observations of tracers that exhibit source and sink configurations that are similar to one of the CO_2 source components.

We examine briefly the results from observations of the radioactive noble gas ^{85}K and of the halocarbon $CFCl_3$ to investigate to what extent the atmospheric concentrations of these tracers exhibit seasonal variations, which might be caused by atmospheric transport. Both tracers have source configurations that are analogous to the fossil fuel component Q_{foss}. As a quantitative means of comparing the tracer data to the model results obtained for the fossil fuel CO_2 component, we shall use the ratio between the magnitude of the seasonal variations and the interhemispheric concentration difference.

Krypton-85 is a radioactive tracer with a mean lifetime of 15.6 years, which is released in the northern hemisphere as a waste product of nuclear fuel reprocessing. This tracer shows an interhemispheric concentration gradient of about 2 pCi m^{-3} in 1980 (Weiss et al. 1983). Time series of ^{85}Kr obtained in Europe display a seasonality, with highest concentrations in spring and lowest concentrations in late summer. This variation is most probably caused by time varying atmospheric transport, since the ground source of ^{85}Kr should not exhibit any seasonality. The peak-to-peak variation observed in central Europe is about 1 to 2 pCi m^{-3}, or about 50% of the interhemispheric difference. This is considerably more than the model-derived ratio of 16% for the fossil fuel CO_2, obtained from a peak-to-peak variation at Mauna Loa of 0.7 ppm divided by the interhemispheric difference of 4.5 ppm (see earlier discussion). We note, however, that the central Europe observations are fairly close to an ^{85}Kr source region and, as the observers have pointed out, the seasonally varying vertical mixing regime between summer and winter might be the primary cause of the observed variations. This is different at Mauna Loa, which is affected by the seasonally changing tropical circulation. A time series for ^{85}Kr in the southern hemisphere unfortunately does not yet exist.

Other tracers with source and sink configurations similar to the fossil fuel CO_2 are the various halocarbons (Prinn et al. 1983). The data obtained of the longer living species ($CFCl_3$ and CF_2Cl_2) during the first 3 years of the Atmospheric Lifetime Experiment (1978-1981) do not display any prominent seasonal variations. At Cape Grim, Tasmania (41°S, 145°E) the peak-to-peak variation of $CFCl_3$ does not exceed 0.36 ± 0.16 ppt with a maximum occurring in November. Since the interhemispheric difference in $CFCl_3$ is 15 ppt, the variation is only 2.4% of the latter difference. On the other hand, the simulated peak-to-peak variation of 0.3 ppm of the fossil fuel CO_2 signal at the South Pole corresponds to about 6.7% of the interhemispheric difference, i.e., it is appreciably greater. We also note that the halocarbon data obtained at Adrigole, Ireland (52°N, 10°W) do not show a seasonal variation similar to the northern hemisphere ^{85}Kr data, which might be explained with the different continentality of the two observing sites; Adrigole lies at the western edge of Europe and is much less susceptible to the weather patterns with stable vertical air that occur over central Europe in winter.

This discussion is limited, as a thorough, critical appraisal of the tracer data would have to accompany this presentation. Correspondingly, the findings are ambiguous and do not provide strong support for the present model calculations. However they illustrate the potential of the information from tracers other than CO_2 for future research.

Conclusions

The usefulness of the simulation of atmospheric CO_2 with a three-dimensional tracer model has been demonstrated (Fung et al. 1983, Fung this volume). The present study complements these findings with the inclusion of the effects of the time-invariant fossil fuel and stationary ocean CO_2 sources.

Since atmospheric transport processes change with season, it is necessary to consider the full set of CO_2 sources and sinks, to compare model results to the available data. In particular, we have shown that the time-invariant sources contribute to the seasonal signal. This contribution is mostly small, but not insignificant. Similarly, the seasonal CO_2 sources give rise to mean annual fields that have to be taken into account when comparing model and data.

The present assessment of the seasonal cycle depends crucially on the shape and timing of the assumed vegetation function in the specification of the seasonal land biospheric CO_2 source. A significant improvement requires a more realistic vegetation function. With the help of remote sensing data from satellites, which are becoming available from the NOAA weather satellites, a better representation of that function may be obtained in the future.

The GISS model simulation of the mean annual fields predicts a too small interhemispheric exchange. As a next possible step, we will consider the use of observed global wind fields to drive the tracer model, instead of the wind fields generated by the GISS climate model.

We also note that a more detailed representation of the oceanic source/sink terms is needed to obtain a realistic mean annual CO_2 concentration field. This requires either a model of the surface layer of the ocean, which includes the effects of the marine biosphere, or a fairly complete set of observations of pCO_2 as a function of space and time in the surface waters of the oceans.

Acknowledgments

We very much appreciate the efforts of Gary Russell at GISS, who provided us with the first version of the tracer model computer code. The efforts of Sheri Lowe in producing the computer graphics are gratefully acknowledged.

We thank personnel of NOAA for assistance in sampling air at the NOAA environmental monitoring stations at Point Barrow, Mauna Loa, Cape Kumukahi, Samoa, and South Pole; personnel of the Institute of Ocean Sciences of the Government of Canada for sampling at Weathership P; personnel of the Institute of Nuclear Sciences and the Meteorological Service of the Government of New Zealand for sampling at Kermadec

Island and at Baring Head; and Martin Vitousek, Erva Tekaraba, John Bryden, Tioni and Kam Chou for sampling at Christmas Island and Fanning Island.

This work was supported under contract RP2333 with the Electric Power Research Institute, Palo Alto, California. M.H. was also supported by the Swiss National Science Foundation.

References

Alexander, R. C. and R. L. Mobley. 1974. Monthly average sea-surface temperatures and ice-pack limits on a 1° global grid. Rep. 4-1310-ARPA. Rand Corporation, Santa Monica, California.

Azevedo, A. E. G. 1981. Atmospheric distribution of carbon dioxide and its exchange with the biosphere and the oceans, Ph.D. thesis. Columbia University, New York.

Bacastow, R. B. 1979. Dip in the atmospheric CO_2 level in the mid-1960's. J. Geophys. Res. 84:3108–3114.

Bacastow, R. B. and C. D. Keeling. 1981. Hemispheric airborne fractions difference and the hemispheric exchange time. In B. Bolin (ed.), Carbon Cycle Modelling, SCOPE 16, pp. 241–246. John Wiley, New York.

Bacastow, R. B., C. D. Keeling, and T. P. Whorf. 1985 Seasonal amplitude increase in atmospheric CO_2 concentration at Mauna Loa, Hawaii, 1959-1982. J. Geophys. Res. 90:10.529–10.540.

Bolin, B., A. Björkström, C. D. Keeling, R. B. Bacastow and U. Siegenthaler. 1981. Carbon cycle modelling. In B. Bolin (ed.), Carbon Cycle Modelling, SCOPE 16, pp. 1–28. John Wiley, New York.

Bolin, B. and C. D. Keeling. 1963. Large-scale atmospheric mixing as deduced from the seasonal and meridional variations of carbon dioxide. J. Geophys. Res. 68:3899–3920.

Broecker, W. S. and T.-H Peng. 1982. Tracers in the Sea. Eldigo Press, New York.

Fung, I., K. Prentice, E. Matthews, J. Lerner, and G. Russell. 1983. Three-dimensional tracer study of atmospheric CO_2: response to seasonal exchanges with the terrestrial biosphere. J. Geophys. Res. 88:1281–1294.

Hansen, J. G., G. Russell, D. Rind, P. Stone, A. Lacis, S. Lebedeff, R. Ruedy, and L. Travis. 1982. Efficient three dimensional global models for climate studies: Models I and II. Mon. Weather Rev. 111:609–662.

Heimann, M. and C. D. Keeling. Meridional eddy diffusion model of the transport of atmospheric carbon dioxide 1. The seasonal carbon cycle over the tropical pacific ocean. J. Geophys. Res. (in press).

Holton, J. R. 1979. An Introduction to Dynamic Meteorology, p. 199. Academic Press, New York.

Kamber, D. 1980. Modellierung der Variationen von CO_2 and $\delta^{13}C$ in Nord und Sudhemisphare, Masters Thesis. University of Bern, Switzerland.

Keeling, C. D. 1968. Carbon dioxide in surface ocean waters. 4. Global distribution. J. Geophys. Res. 73:4543–4553.

Keeling, C. D., T. B. Harris, and E. M. Wilkins. 1968. Concentration of atmospheric carbon dioxide at 500 and 700 millibars. J. Geophys. Res. 73:4511–4528.

Keeling, C. D., A. F. Carter, and W. G. Mook. 1984. Seasonal, latitudinal, and secular variations in the abundance and isotopic ratios of atmospheric carbon dioxide 2. Results from oceanographic cruises in the tropical Pacific ocean. J. Geophys. Res. 89:4615–4628.

Keeling, C. D. and M. Heimann. Meridional eddy diffusion model of the transport of atmospheric carbon dioxide 2. The mean annual carbon cycle. J. Geophys. Res. (in press).

Matthews, E. 1983. Global vegetation and land use: new high-resolution data bases for climate studies. J. Clim. Appl. Meteorol. 22:474–487.

Pearman, G. I. and P. Hyson. 1980. Activities of the global biosphere as reflected in atmospheric CO_2 records. J. Geophys. Res. 85:4468–4474.

Prinn, R. G., P. G. Simmonds, R. A. Rasmussen, R. D. Rosen, F. N. Alyea, C. A. Cardelino, A. J. Crawford, D. M. Cunnold, P. J. Fraser, and J. E. Lovelock. 1983. The atmospheric lifetime experiment 1. Introduction, instrumentation and overview. J. Geophys. Res. 88:8353-8367.

Rotty, R. M. 1983. Distribution of and changes in industrial carbon dioxide production. J. Geophys. Res. 88:1301–1308.

Russell, G. L. and J. A. Lerner. 1981. A new finite differencing scheme for the tracer transport equation. J. Appl. Meteorol. 20:1483–1498.

Takahashi, T. 1979. Carbon dioxide chemistry in ocean water. In Workshop on the global effects of carbon dioxide from fossil fuels, CONF-770385, pp. 63–71. Carbon Dioxide Effects Research and Assessment Program, U.S. Department of Energy, Washington, D.C.

Tans, P. P. 1978. Carbon 13 and Carbon 14 in trees and the atmospheric CO_2 increase, Ph.D. thesis. University of Gröningen, Netherlands.

Weiss, R. F., R. A. Jahnke, and C. D. Keeling. 1982. Seasonal effects of temperature and salinity on the partial pressure of CO_2 in seawater. Nature 300:511–513.

Weiss, W., A. Sittkus, H. Stockburger, and H. Sartorius. 1983. Large-scale atmospheric mixing derived from meridional profiles of Krypton 85. J. Geophys. Res. 88:8574–8578.

Wyrtki, K., E. Firing, D. Halpern, R. Knox, G. J. McNally, W. C. Patzert, E. D. Stroup, B. A. Taft, and R. Williams. 1981. The Hawaii to Tahiti Shuttle Experiment. Science 211:22–28.

3
Presentation of the 20th Century Atmospheric CO_2 Record in Smithsonian Spectrographic Plates

GERALD M. STOKES AND JAMES C. BARNARD

The history of 20th century scientific measurement programs does not contain many examples of long-running studies in which an important property of the atmosphere is reliably measured by the same person at the same site with precisions of $>1\%$. One outstanding example is C. D. Keeling's Mauna Loa record of CO_2 measurements (Keeling et al. 1982). A program that ran for an even longer period of time in the early and mid-20th century is the Smithsonian Solar Constant Program, directed by Charles G. Abbott. Over the past several years, we have been studying whether a portion of the Smithsonian data set can be used to extend the record of atmospheric CO_2 concentrations back to the early 20th century.

During the course of the Smithsonian Solar Constant Program, a wide variety of measurements were made; the ones of immediate interest are the spectrobolographic measurements. These observations consisted of several measurements of the transparency of the earth's atmosphere at wavelengths between 300 and 2500 nm. Within this wavelength interval are a number of absorption bands of CO_2 and, of course, the strength of these bands is a direct measure of the integrated column of CO_2 between the observer and the sun.

In this chapter we will recount the efforts to use the observed strength of these bands to extract the abundance of CO_2 in the atmosphere over the course of the Smithsonian Solar Constant Program. We will discuss the following topics:

- The relationship between the density of an integrated column of CO_2 and surface measurements
- The sources of error in a measurement of the integrated column
- The details of our analysis of the Smithsonian spectrobolograms, as well as the precision of the results.

Basic Physical Principles

The absorption of solar radiation at a particular wavelength by material in the earth's atmosphere is governed by the Beer-Lambert law:

$$I(\nu) = I_O(\nu)e^{-\tau(\nu)} \quad (1)$$

where $I(\nu)$ is the measured intensity at the instrument and $I_O\nu$ is the intensity of solar radiation at the top of the earth's atmosphere.

The quantity $\tau(\nu)$ is referred to as the optical depth. In most atmospheric applications $\tau(\nu)$ is expressed as the product of $\tau_Z(\nu)$, the optical depth at the zenith, and m, the air mass. The air mass is a geometrical factor that represents the ratio of the actual pathlength that sunlight follows through the atmosphere to the pathlength if the sun were at the zenith. The use of $m\ \tau_z(\nu)$ in place of the actual optical depth is an approximation. The optical depth in Eq. (1) is correctly given by

$$\tau(\nu) = \int \rho(\ell)\alpha(\ell;\nu)\, d\ell \text{ through the atmosphere} \quad (2a)$$

where $\rho(\ell)$ is the number density of the absorber and $\alpha(\ell;\nu)$ is the absorption coefficient of the absorber. This is approximately equal to

$$\simeq m \int \rho(\ell)\alpha(\nu;\ell)\, d\ell \text{ at the zenith}$$

where m is the air mass;

$$\simeq m\ \tau_Z\ (\nu) \quad (2b)$$

Equation (2b) is the approximation discussed above. For well-mixed gases (e.g., relatively constant mixing ratios), the approximation (2b) is fully adequate, and the air mass is relatively straightforward to evaluate. However, as discussed by Young (1974), the problem of air mass determination for any but the simplest cases is extremely difficult.

Returning to Eq. (2a), we can expand the total optical depth in the following way for absorption due to the ith excitation state of a particular molecule:

$$\tau_i(\nu) = \int N_i[T(\ell)]S_i\phi_i[\nu;T(\ell)\ \rho(\ell)]\ d\ell \quad (3)$$

where N_i is the number density of the ith excitation level of the molecule, S_i is the strength of the transition, and ϕ_i is the line profile function.

At this level of detail, we can see some of the problems that may arise in the study of the integrated column density. For example, the absorption at a particular frequency is a function of temperature along the line of sight through the atmosphere. The absorption also depends in detail on the line formation process; note the line profile function ϕ_i, which is a function of atmospheric pressure. Further, Eq. (3) actually refers to the effect of an isolated transition of a particular atmospheric species. In practice, the absorption at any specific wavelength will be represented by a sum of the contributions to the optical depth due to several transitions of a variety of species.

Another complication is that a spectroscopic measurement is made with an instrument that does not have perfect wavelength sensitivity. This means that an instrumental response function must be convolved with the incident intensity to understand the actual spectroscopic measurement. The effect of this is that

$$I'(\nu) = \int_{O}^{\infty} I(\nu')\psi(\nu - \nu')\, d\nu' \tag{4}$$

where I' is the intensity actually measured with the instrument. Because the instrumental sensitivity function, $\psi(\nu - \nu')$, has finite width when compared to actual width of real atmospheric absorption lines, the quantity usually analyzed is the observed equivalent width of a line, w; the width of a hypothetical line that absorbs totally, but has the same area as the actual line.

$$\begin{aligned} \mathrm{w} &= \int_{O}^{\infty} [I_O(\nu) - I(\nu)]\, d\nu \\ &= \int_{O}^{\infty} I_O(\nu_O) \left\{ 1 - \frac{I(\nu)}{I_O(\nu)} \right\} d\nu \end{aligned} \tag{5}$$

Elementary energy conservation considerations can be used to show that the equivalent width of a particular absorption line is independent of instrumental resolution for a suitable range of frequency integration. The relationship between the equivalent width of an absorption line and the number of absorbers required to produce it is commonly called the curve of growth.

Surface Measurements and the Integrated Column

Before turning to the Smithsonian data record, it is extremely important to establish the link between the surface measurement and the integrated column density measurement that is made by spectroscopic means. It is interesting to note that the density of an integrated column of CO_2 is the quantity of interest for climate modeling; the surface measurement is a local measurement from which the column density is inferred. The important point is not so much the relationship between the actual surface value and the actual integrated column; rather, it is whether a measurement of the integrated column can be understood on the basis of the surface measurements that have been made over the last three decades. Only by making this connection can we be sure that the integrated column density measurement is going to make a contribution to the study of the carbon cycle.

The most obvious characteristics that one might hope to see in the integrated column are the seasonal variability and the secular trend. In our search for these features, we have analyzed a modern spectroscopic data set taken at Kitt Peak National Observatory near Tucson, Arizona. Our goal was to understand the detailed variations of the atmospheric CO_2 abundance in the integrated column.

Our analysis of these data is based on a method of spectroscopic analysis (Stokes and Brown 1983; Stokes et al., in preparation) that emphasizes spectral decomposition rather than the more traditional spectral synthesis approach. This latter technique has recently been used to extract CO_2 concentrations from other spectra (Goldman et al. 1983). There are several advantages to the decomposition approach, the most notable of which is that it can be used to limit the importance of a priori information on various stages of the analysis. Of particular concern is the fact that the fundamental spectroscopic parameters of particular transitions might not be accurately known. It is not unusual for absolute line strengths to be in error by 10 to 20% or for the temperature-dependent broadening coefficients of a transition to be in error by a factor of 2.

In Fig. 3.1, we can see the result of an analysis of eight CO_2 transitions near 1.6 μm. Figure 3.1a shows our spectroscopic results and, for reference, the results from the Key Biscayne flask station in the NOAA network. Figure 3.1b shows the same data folded onto a single year. In each of the figures the spectroscopic measurement is a derived surface measurement for which we have assumed an atmospheric profile based on high-altitude measurements by Volz et al. (1981).

From Fig. 3.1, it is clear that we are measuring, in the integrated column, something that we can readily interpret with our knowledge of surface measurements. The seasonal variation is evident, and the long-term secular trend is beginning to emerge. Since Kitt Peak is now an NOAA flask site, we should shortly have a better idea of the precise relationship between the surface measurements and the integrated column measured at the same time.

Figure 3.1 is based on a data set that contains information on a variety of gases other than CO_2, and there are clearly many interesting results concerning these gases to be derived from further analysis of the data.

Approach

When considering possible problems with an integrated column density measurement of an atmospheric species, there are two classes of errors to assess. The first class is associated with the simple problem of making and interpreting an integrated column density measurement, regardless of the observing technique. The second class consists of the errors associated with a particular measurement technique. In our analysis of the Smithsonian data set, we consider both classes of errors and discuss these details in this section.

Returning to Eq. (3), it should be noted that there are several terms in the expression that are often not particularly well-known. Probably the most troublesome of these terms is the absolute line strength. Note that any error in the line strength [S_i in Eq. (3)] translates into a proportional error in the concentration. To circumvent this problem, a spectroscopic analysis is frequently couched in differential terms. That is, rather than computing directly the integrated column density of a particular absorber, it is usual to compare the measurement with some standard measurement

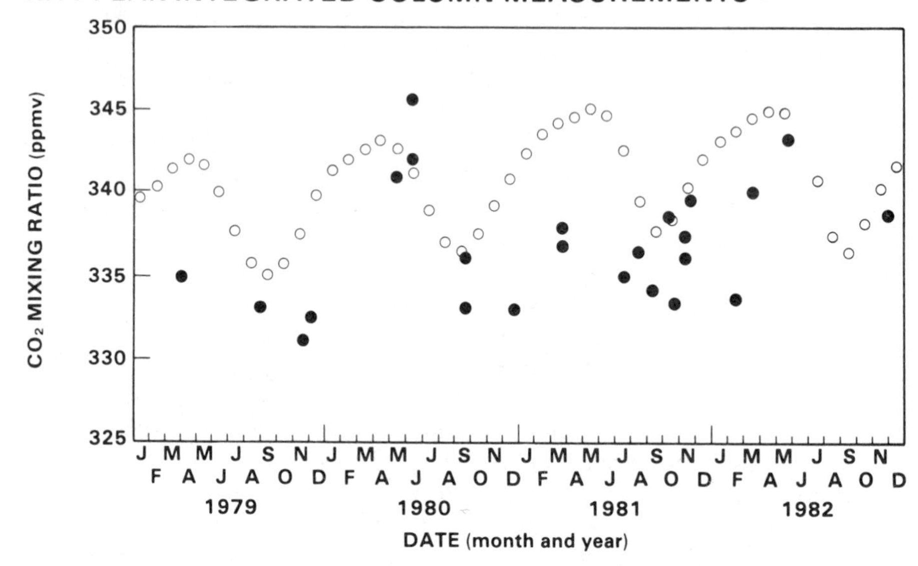

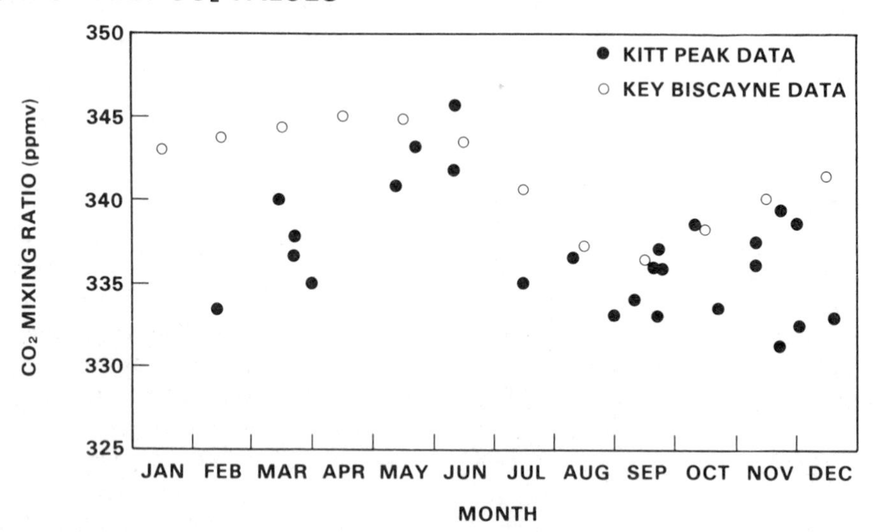

FIGURE 3.1. **A.** The *filled circles* are Kitt Peak integrated column measurements of atmospheric CO_2; the *open circles* are Key Biscayne surface measurements. **B.** Same as A, but the 4 years of data are folded into a single year.

for which the concentration is known. This is the basic principle behind nondispersive infrared spectrometers, in which the measurement of the sample is compared directly with a standard measurement made of a standard gas under precisely the same conditions as the sample (Griffith 1982; Griffith et al. 1982).

We adopt a conceptually similar technique, the differential curve of growth, which is widely used in astrophysical studies (e.g., Mihalas 1970). We have modified the concept somewhat for this study, but it remains a basic differential technique that avoids a number of the problems previously mentioned.

The conventional astrophysical curve of growth technique is usually applied to data taken with the same instrument in an attempt to eliminate the effects of systematic errors produced by specific instrumental characteristics. For the analysis of the Smithsonian spectrobolograms, we have to accommodate the fact that the instruments that took the data no longer exist. For this reason, we have attempted to reconstruct mathematically the performance of the Smithsonian spectrobolometer as an essential step in our differential curve of growth analysis.

Nature of the Data Set

Bolograms were recorded as part of the Smithsonian Solar Constant Program over the period 1902 to 1960. The details of the day-to-day operation of this program are described in volumes I-VII of the *Annals of the Smithsonian Astrophysical Observatory*. The general nature of the program and an excellent review of the actual solar constant results have been described by Hoyt (1979).

The Smithsonian spectrobolograms were taken with the express purpose of measuring the wavelength-dependent attenuation of solar radiation produced by the atmosphere. The results of the spectrobolographic measurements were used to correct total solar flux measurements to the value they would have at the top of the earth's atmosphere. The wavelength coverage was 0.3 to 2.3 μm, and during the course of a day's observations three to seven spectra were recorded over an air mass range from 1.5 to 5.0 air masses. Each set of observations was recoded on a single 8×24 inch glass plate. An example of one day's observations is shown in Fig. 3.2.

These plates are best thought of as glass strip chart recordings of the intensity of solar radiation as a function of wavelength. The device used to create this record was a prism spectrobolometer. The output of the detector, a bolometer, was measured with a galvanometer whose active element was a small needle suspended on a quartz fiber in a magnetic field. A small mirror was attached to the needle, and a beam of sunlight was reflected from this mirror to the photographic plate. As the light struck the detector, the resulting current caused the needle to rotate, moving the light beam across a photographic plate. The spectrum was scanned by

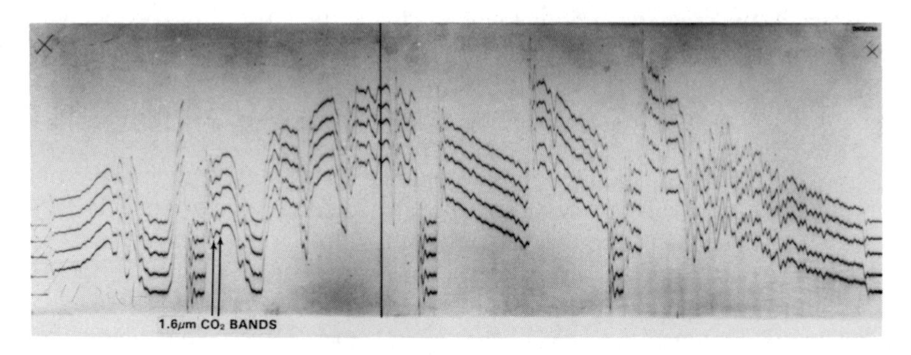

FIGURE 3.2. An 8 × 24 inch glass plate upon which is a Smithsonian spectrobologram. This plate was made on November 11, 1927, as inscribed in the lower left corner. The 1.6 μm CO_2 bands that are used in the analysis are marked with *arrows*.

rotating the prism. Because the prism rotation was directly coupled to the translation of the plate, there was a unique relationship between wavelength and plate position.

The spectrobolographic observations were made from a variety of sites over the course of the program. The sites and the periods over which they were active are summarized in Table 3.1. We have been able to locate the original plate material from only four of these sites: Table Mountain, Mount Harqua Hala, Mount Saint Katherine, and Mount Brukkaros. The largest collection of plates is from the Table Mountain site. The Mount Montezuma site was the site occupied for the longest period, and Butler and Hoyt (1980) have described the day-to-day observations at that location. Unfortunately, the plates from Mount Montezuma were destroyed at the end of occupation of that site.

The original plates are difficult to handle because of their large size and number. To protect the original material and to facilitate our analysis, we photographically reduced all of the plates prior to digitization. Great care was taken during this procedure, and we have demonstrated that the effect on the data quality as a result of this process is negligible. Once reduced, the plates were digitized and the actual record of the spectrum, which we will refer to as a trace, was reconstructed from the digitized image. After the individual traces have been put into a machine-readable form, the process of characterizing the instrument begins.

Removal of Galvanometer Response

The first step in the analysis is to transform the *x-y* coordinates of the trace to detector current as a function of wavelength. There are two steps in this process: a relatively simple baseline correction and the more difficult

TABLE 3.1. Record of spectrobolographic measurements made during the Smithsonian Solar Constant Program.

Site	Latitude	Longitude	Altitude (m)	Period of observation
Mt. Montezuma, Chile	22°40′S	68°56′W	2711	1920–1948
Table Mtn., California	34°22′N	117°41′W	2286	1925–1958
Calama, Chile	22°28′S	68°56′W	2250	1918–1920
Mt. Wilson, California	34°13′N	118° 4′W	1737	1905–1920
Mt. Harqua Hala, Arizona	33°48′N	113°20′W	1721	1920–1925
Mt. Brukkaros, S. W. Africa	25°52′S	17°48′E	1586	1926–1930
Hump Mtn., N. Carolina	36° 8′N	82° 0′W	1500	1917–1918
Washington, D. C.	38°53′N	77° 2′W	10	1902–1907
Mt. St. Catherine, Egypt	28°31′N	33°56′E	2591	1934–1937
Burro Mtn., New Mexico	32°40′N	108°33′W	2440	1940–1945
Miami, Florida	25°N	80°W	10	1948

removal of the galvanometer response function. The galvanometer needle is modeled as an underdamped harmonic oscillator, and the effect of this response on the trace can be seen in Fig. 3.3. The removal is crucial for the study of CO_2, because there is a sharp transition in the vicinity of the CO_2 features we have used in our analysis. The critical nature of this transformation is obvious from the figure, which illustrates the effect of the galvanometer response in the vicinity of the CO_2 band. We analyzed the sensitivity of the final result to errors in the galvanometer response determination, and, over the range of variability of the derived response function, the answer is insensitive to the extraction of the galvanometer response.

It may seem unusual to select a band in a region that may be affected by the galvanometer response analysis; the choice is dictated by far more important concerns, as described in the following section.

BAND SELECTION

The selection of the appropriate CO_2 absorption feature for analysis is an important consideration in the analysis of the Smithsonian data set. There are two overriding concerns governing this selection: sensitivity and uniqueness. Sensitivity refers to the relationship between a change in the absorption in a particular band and the corresponding change in the measured integrated column of CO_2. Uniqueness, on the other hand, deals with the possibility of contamination of the CO_2 band.

Returning to Eq. (1), the Beer-Lambert law implies that the amount of energy absorbed at a particular wavelength is affected not only by the amount of absorber present but by the amount of radiation present as well. Consequently, as the amount of absorber increases, the incremental increase in the energy absorbed (as measured, for example, in the equiv-

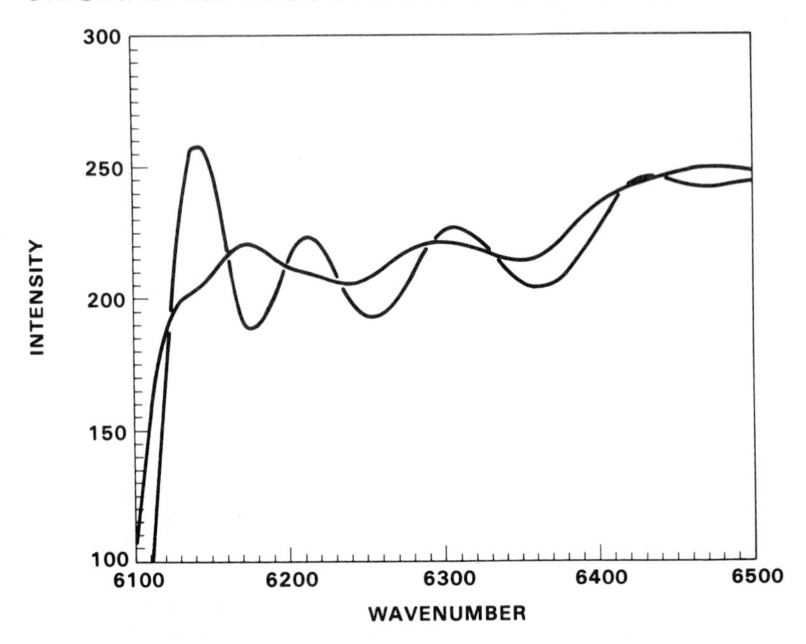

FIGURE 3.3. The original Smithsonian trace near the 1.6-μm CO_2 bands, and the same trace with the galvanometer response removed. The original trace is distinguished by the higher amplitude oscillations.

alent width) decreases. That is, the absorption feature becomes saturated. In selecting the feature to be analyzed, there is an important trade-off between the sensitivity of the absorption to the change in the amount of the absorber and the ability to measure the change in the absorption. The latter is controlled by the signal-to-noise ratio in the original data.

The second issue, that of uniqueness, arises because of spectral contamination. In the earlier discussion of line formation in the earth's atmosphere, we noted that Eq. (3) refers to the effect of a single absorption line. In the real telluric absorption spectrum, it is not unusual to find a large number of spectral lines that partially overlap. Further, these individual lines may very well be caused by a variety of atmospheric constituents.

The entire problem of multiplicity is compounded when the width of the instrumental response function is broad compared with the separation between spectral lines. This is definitely the case for the Smithsonian spectrobolograms, as can be seen in Fig. 3.4. The spectrum in this figure, taken at Kitt Peak National Observatory, was observed with an instru-

---

FIGURE 3.4. A typical slit width of the Smithsonian instrument is indicated by the *solid black line*. The spectral region shown is part of the 1.6-μm CO_2 bands as measured at Kitt Peak.

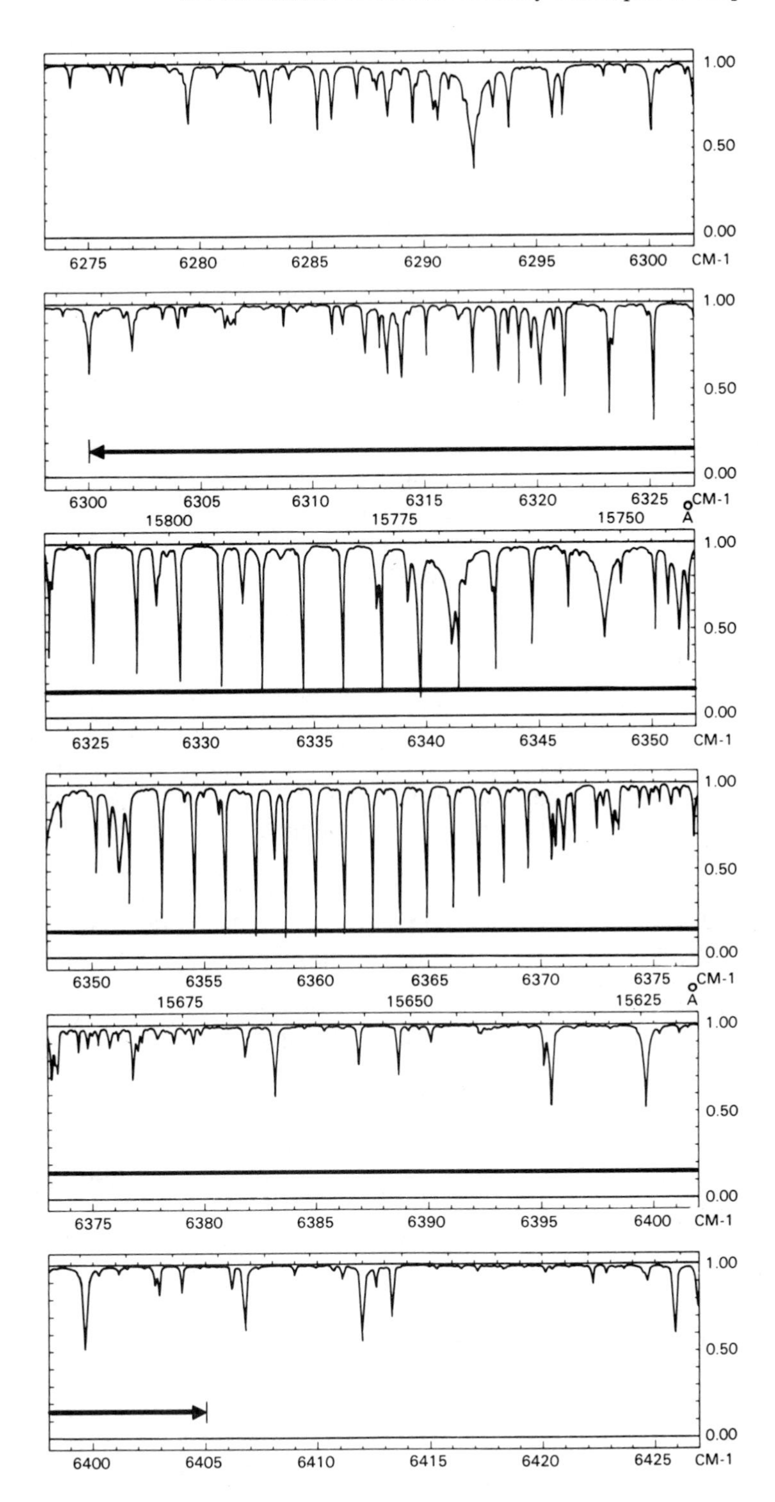
1.00
0.50
0.00
6275
6280
6285
6290
6295
6300
CM-1
6305
6310
6315
6320
6325
15800
15775
15750
Å
6330
6335
6340
6345
6350
6355
6360
6365
6370
6375
15675
15650
15625
6380
6385
6390
6395
6400
6405
6410
6415
6420
6425

mental response function that is negligible compared with the width of the spectral lines. The contrast between that spectrum and the Smithsonian instrumental response function is striking.

The most important problem caused by the breadth of the Smithsonian response function is potential contamination of the measurements by absorption due to other atmospheric gases, particularly water vapor. If the adopted measure of absorption due to CO_2 is contaminated by water vapor absorption, the derived CO_2 concentrations will be similarly contaminated. In view of the large diurnal and seasonal variability in the integrated column of atmospheric water vapor, such contamination is clearly unacceptable, since the variability could not be uniquely associated with changes in the atmospheric CO_2 concentration.

As a result of these considerations, we selected the series of CO_2 bands in the vicinity of 1.6 μm. Other bands, such as the stronger 2.0- and 2.05-μm bands of CO_2, were rejected because of both saturation and water vapor contamination, while the remainder of the CO_2 bands in this region were too weak for reasonable analysis.

Modified Differential Curve of Growth

As was discussed above, our approach is a modified differential curve of growth. It is modified in the sense that, rather than comparing observations made with the same instrument, we attempt to simulate the Smithsonian instrument so that we can use this instrumental model to modify our standard spectra to make them directly comparable with the original Smithsonian data.

At this stage of the analysis the Smithsonian spectra have been transformed to bolometer versus current wavelength. There are three elements that need to be settled to determine the differential curve of growth:

1. Which spectrum will be used as the standard
2. The instrumental sensitivity function (over and above the galvanometer response, which has already been removed)
3. The measure of absorption we will use to make the comparison between the standard and Smithsonian data sets.

There are two choices for the standard spectrum: a synthetic spectrum or another spectroscopic data set. Both are realistic possibilities and should eventually be tried, but the use of another spectroscopic data set has some particularly attractive features. The most important advantage is that the computation of a synthetic spectrum requires a reasonably well-known set of spectroscopic constants. As we have noted above, the required spectroscopic constants may not be known well enough to construct the required set of spectroscopic observations. On the other hand, the use of an actual set of spectroscopic observations is equivalent to letting the atmosphere solve the equations of radiative transfer for us. Within reasonable limits, the problems associated with the kind of analysis we are describing are then concentrated in the characterization of the spectroscopic instrument.

The data set we selected as our standard is the Kitt Peak data set that was described above. In addition to being an exceptionally high-quality data set, as is evident from the analysis that resulted in Fig. 3.1, the detailed analysis that we have performed on the data set gives us particular insight into the sensitivity of the spectra to the properties of the atmosphere. There is, of course, the problem that Kitt Peak is not the same site that the original observations were made from, but, as we shall see, this is relatively easy to accommodate.

The next step is the characterization of the Smithsonian spectrobolograph. Following the removal of the galvanometer response effects, the Smithsonian data set has been transformed to a quantity that is a monotonic function of the incident radiation on the bolometer as a function of wavelength. To transform the Kitt Peak data to a data set directly comparable with the Smithsonian spectra, we need to establish the instrumental sensitivity function, $\psi(\nu - \nu')$ in Eq. (4), for the Smithsonian spectra. In particular, the sensitivity function is directly related to the size of the aperture ahead of the bolometer that determines the region of the spectrum that is actually incident on the detector. For example, the aperture would effectively integrate the energy from the large expanse of spectrum covered by the Smithsonian sensitivity function denoted by the line in Fig. 3.4. In most spectrometers this aperture is usually referred to as the exit slit, and we will refer to this geometrical effect as the slit width.

We have two methods to determine the slit width. The first involves the use of the convolution theorem for Fourier transforms:

$$\psi(\nu) = F^{-1}\left\{\frac{F_S}{F_{KP}}\right\} \tag{6}$$

where $\psi(\nu)$ is the slit width function, F_S is the Fourier transform of the Smithsonian data, F_{KP} is the Fourier transform of the Kitt Peak data, and F^{-1} is the inverse Fourier transform.

This formalism gives us the wavelength-dependent form of the instrumental sensitivity function. It yields a slit function that is reasonable and agrees with the roughly triangular shape that one expects from such an instrument. However, the successful application of this technique requires the use of a rather large portion of the Smithsonian spectrum. As a result, it is very sensitive to noise over the whole region used and usually does not give particularly reliable results.

An alternative approach is to fit several parameters of the instrumental sensitivity function using a minimization scheme such as least squares, to extract the best parameters of that fit. To summarize this procedure, the slit function is assumed to be a triangle. This triangle is convolved with a selected set of Kitt Peak spectra to produce a simulated Smithsonian spectrum. The triangle width is then varied until the least-squares residual between the simulated and actual data is a minimum. The accuracy and internal consistency of this approach appear to be superior to those of the Fourier transform approach.

As a practical matter the slit width is a critical, if not the most critical, parameter in our analysis. Archive research reveals that the slit width was adjusted on a regular basis as part of the observing protocol followed at the various observing sites. Fortunately, it was not adjusted between individual spectra on a given day, and the internal agreement between the traces taken on each day gives us a good check on the internal consistency of the slit width determination.

Finally, the selection of the absorption measure to be used in the analysis is also important. In our choice we were guided by the sensitivity criterion discussed above, as well as the relationship between our absorption measure, the equivalent width of the CO_2 bands, and the air mass. Since air mass is a simple scaling factor that effectively multiplies the total column of CO_2 the sensitivity to air mass is equivalent to the sensitivity in changes in the atmospheric concentration of CO_2. We therefore selected our equivalent width measuring procedure based on the maximization of the sensitivity of the equivalent width to changes in air mass.

Analysis of the Results

The final application of the differential curve of growth was accomplished in the following manner. First the Kitt Peak data were modified using the derived slit function. Approximately 140 spectra were used in this analysis. We then determined the relationship between air mass of the Kitt Peak observations and their corresponding equivalent widths of the CO_2 feature under study. The resulting expression was used to answer the following question: Given an equivalent width measured from the Smithsonian data set, what air mass would be required to produce the same absorption for the Kitt Peak relationship?

The answer to that question yields an equivalent Kitt Peak air mass for each of the Smithsonian observations. This equivalent Kitt Peak air mass and the actual corresponding Smithsonian air mass are related by the following expressions:

$$\begin{aligned} \mathrm{w}_\lambda\ (\mathrm{KP}) &\propto AM_{\mathrm{KP}} N_{\mathrm{KP}} \mathrm{P}_{\mathrm{KP}} \\ \mathrm{w}_\lambda\ (\mathrm{TM}) &\propto AM_{\mathrm{TM}} N_{\mathrm{TM}} P_{\mathrm{TM}} \end{aligned} \tag{7}$$

where N = concentration of CO_2 and P = local pressure.

The modified curve of growth (as practiced here) asks what concentration makes

$$\begin{aligned} \mathrm{w}_\lambda\ (\mathrm{KP}) &= \mathrm{w}_\lambda\ (\mathrm{TM}) \\ N_{\mathrm{TM}} &= N_{\mathrm{KP}} \left(\frac{P_{\mathrm{KP}}}{P_{\mathrm{TM}}} \frac{AM_{\mathrm{KP}}}{\mathrm{AM}_{\mathrm{TM}}} \right) \end{aligned} \tag{8}$$

where N_{TM} is the Smithsonian concentration required to make the Kitt Peak and Table Mountain equivalent widths equal (i.e., the columnar CO_2 abundance for the Smithsonian data). The relative barometric pressures between the two sites reflect the fact that Kitt Peak is at a slightly different altitude from that of the original observations. The value of N_{KP} is arbitrary; for the present analysis, we have chosen 340 ppm.

We have used this technique to determine the CO_2 concentration at the Table Mountain site for 1941. The results are shown in Fig. 3.5. The error bars shown in this figure are the standard error of the mean and an estimate of systematic error due to slit width determination. At this time the major systematic error appears to be slight inaccuracies in the determination of the slit width which would increase the error by a few ppm. It is anticipated that both the systematic and the standard error can be reduced by a factor of 2, resulting in a final precision of ±5 to 6 ppm.

Future Prospects for the Technique

The results shown in Fig. 3.5 are encouraging from two standpoints. First, the precision of the annual averages appears to be reasonable, and there are several things that will improve their precision. Second, the data represented in Fig. 3.5 are only a small fraction of the available data. With the methodology firmly established, the possibility exists that we could

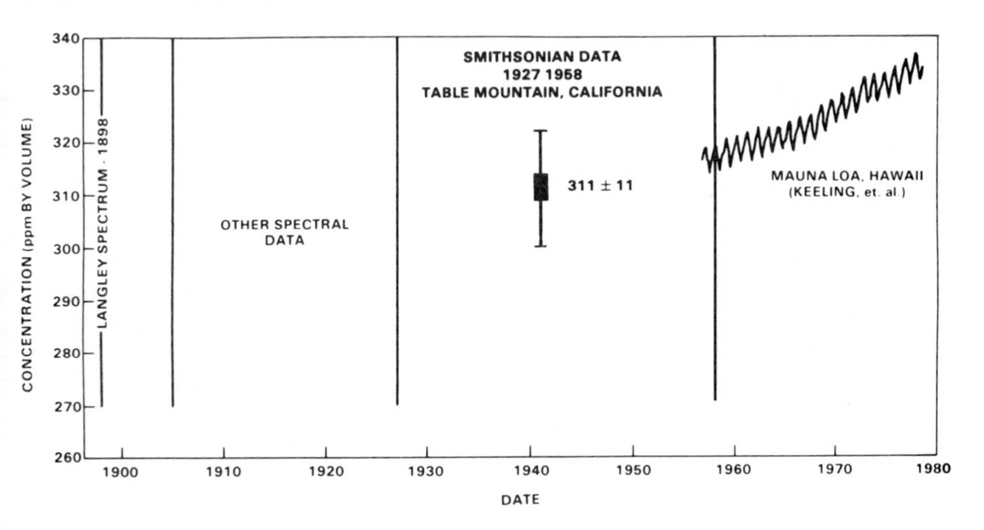

FIGURE 3.5. Results of the preliminary analysis of Smithsonian spectrobolograms for the year 1941. The solid part in the error bar represents the estimated error due to inaccuracies in determining the slit width. Also shown are other data sets suitable for analysis to retrieve historical CO_2 abundances.

construct an annual record of the atmospheric CO_2 concentration throughout the mid-20th century. Such a record, the magnitude of which is shown in Fig. 3.5, would give fresh insight into the natural variability of the atmospheric CO_2 concentration. The analysis would also provide data from the period 1900 to 1950 in which the slope of the anthropogenic CO_2 production was substantially different from the slope today.

Finally, as we can see from Fig. 3.5, the use of the Smithsonian instrumentation extended back to the end of the 19th century. There are peculiar problems associated with the analysis of the older data, but the analysis of the more recent Table Mountain data set has given us considerable experience with those problems, and we are confident that abundances derived from the even older data may soon be available.

Acknowledgments

This research has been supported by the Carbon Dioxide Research Division of the Office of Basic Energy Sciences of the U. S. Department of Energy under Contract No. DE-AC06-76RLO 1830. The authors gratefully acknowledge the assistance of Dr. D. W. Johnson in the preparation of this manuscript. Ms. R. Craig and Mr. N. Larson assisted in large portions of the data reduction. The work itself has benefited from the advice and guidance of Drs. E. W. Pearson, R. A. Stokes, P. A. Ekstrom, N. S. Laulainen, D. W. Johnson, and R. H. Gammon. The preservation of the Smithsonian spectrobolograms has been possible only through the vision of Drs. R. G. Roosen and R. Angione and the hard work and attention to detail on the part of Sarma Modali and Ward Meyer.

Note Added in Proof

Work on the Smithsonian spectrobolographic data has continued. A new method of analysis based on a variant of the minimization technique was developed in 1984 that appears to avoid many of the problems associated with the exact characterization of the instrumental slit width. The details of this technique have been presented recently by *Stokes et al.** The newer results are 297.7 ± 5.6, 298.4 ± 6.3, and 308.3 ± 4.6 ppm for the years 1935, 1941, and 1945, respectively. There may be a residual systematic error of 2 to 3 ppm in these results due to the remaining uncertainties in absolute band strengths and the precise relationship between integrated column density measurements and surface concentration measurements. These results imply an average annual increase of CO_2 through the period of 0.59 ± 0.27 ppm which should not be subject to the systematic errors noted above.

---

*Stokes, G.M., J.C. Barnard, and E.W. Pearson. 1984. Historical carbon dioxide: Abundances derived from the Smithsonian spectrobolograms. DOE/NBB-0063, August 1984.

References

Butler, C. P. and D. V. Hoyt. 1980. Measuring the solar constant at Mt. Montezuma. Sunworld 4:81–86.

Goldman, A., F. G. Fernald, F. J. Murcray, F. H. Murcray, and D. G. Murcray. 1983. Spectral least squares quantification of several atmospheric gases from high resolution infrared solar spectra obtained at the South Pole. JQSRT 29:189–204.

Griffith, D. W. T. 1982. Calculation of carrier gas effects in non-dispersive infrared analysis. I. Theory. Tellus 34:376–384.

Griffith, D. W. T., C. D. Keeling, J. A. Adams, P. R. Girenther, and R. B. Bacastow. 1982. Calculations of carrier gas effects in non-dispersive infrared analyzers. II. Comparisons with equipment. Tellus 34:385–397.

Hoyt, D. V. 1979. The Smithsonial Astrophysical Observatory solar constant program. Rev. Geophys. and Space Physics 17:427–458.

Keeling, C. D., R. B., Bacastow, and T. P. Whorf. 1982. Measurements of the concentration of carbon dioxide at Mauna Loa Observatory, Hawaii. In W. C. Clark (ed), Carbon Dioxide Review 1982, pp. 377–385. Oxford University Press, New York.

Mihalas, P. 1970. Stellar Atmospheres. W. H. Freeman and Company, San Francisco.

Stokes, G. M., J. Brault, M. A. Brown, and D. W. Johnson (in preparation).

Stokes, G. M. and M. A. Brown. 1983. Accurate integrated column density measurements of atmospheric carbon dioxide. In Proceedings of the International Workshop on Atmospheric Spectroscopy, Oxfordshire, U.K., July 19–21, 1983.

Volz, A., U. Schmidt, J. Rudolph, D. H. Ehhalt, F. J. Johnson, and A. Khedim. 1981. Vertical profiles of trace gases at mid-latitudes. Jul-Report No. 1742, Kernforschungsanlage, Julich, Federal Republic of Germany.

Young, A. T. 1974. Atmospheric extinction. In N. P. Carleton (ed.), Methods of Experimental Physics, Vol. 12A: Astrophysics, pp. 123–180. Academic Press, New York.

4
Atmospheric CO_2 Record from Direct Chemical Measurements During the 19th Century

PAUL J. FRASER, WILLIAM P. ELLIOTT, AND
L. S. WATERMAN

In a paper published almost 50 years ago Callendar (1938) suggested that, during the period 1900 to 1935, a 6% increase in atmospheric CO_2 occurred as a result of the combustion of fossil fuels. He calculated that approximately three-quarters of the anthropogenically produced CO_2 remained in the atmosphere, which induced a global warming of 0.1°C and accounted for most of the increase observed in the available long-term temperature records. The calculated atmospheric CO_2 increase was based on a few observations (Brown and Escombe 1905) that, Callendar suggested, indicated a background CO_2 level in 1900 of 274 ± 5 ppmv (parts per 10^6 by volume). As pointed out by Keeling (1978*a*), it is difficult to see how Callendar arrived at such a low concentration from the Brown and Escombe data, and, from a later analysis of additional data sets (Callendar 1940, 1958), he revised his estimate of a late 19th century background CO_2 level to be 288 to 290 (±3) ppmv. From a more rigorous statistical analysis of 19th and 20th century CO_2 data, Slocum (1955) questioned whether concentrations had increased, and Bray (1959) showed that a significant increase is observed only when highly variable data are removed from the observations.

The first clear evidence of an atmospheric CO_2 increase was obtained from data collected in Antarctica (from 1957) and at Mauna Loa (from 1958) using an infrared gas analyzer whose inherent precision was an order of magnitude better than that of the chemical methods (Keeling 1960, 1978*b*; Bolin and Keeling 1963; Pales and Keeling 1965; Brown and Keeling 1965). The Callendar 19th century CO_2 level has been widely used in conjunction with these modern measurements to infer an atmospheric response to the release of fossil fuel CO_2 throughout the 20th century. A back-extrapolation of the Mauna Loa record, assuming a constant airborne fraction of the estimated fossil fuel input (Keeling 1973, Rotty 1983), yields a calculated "preindustrial" value of approximately 295 ppmv. However, geochemical models of the carbon cycle, calibrated with oceanic ^{14}C distributions, predict an additional sink (or an early 20th century source) to reproduce the observed rate of increase in CO_2 at Mauna Loa (Oeschger et al. 1975, Enting and Pearman 1982, Siegenthaler 1983). Either way, these models suggest a lower preindustrial level of about 260 to 280 ppmv (WMO 1983*a*).

The purpose of this chapter is to reexamine these 19th century chemical measurements, in the light of what is now known about the temporal and spatial variability of global atmospheric CO_2 (Fraser et al. 1983), to see if a more reliable estimate of a preindustrial CO_2 concentration can be made. We have restricted our attention to the late 19th century chemical data, since they are considered more reliable (Callendar 1958), and impose more rigorous constraints on carbon cycle models than the 20th century chemical measurements. We have considered only data from original papers available to us at this time, which, fortunately, include all of Callendar's preferred measurements but are $< 50\%$ of the identified sources of late 19th century CO_2 measurements (Callendar 1958).

Modern CO_2 Measurements

All modern long-term measurement programs of background CO_2 concentrations show increases similar to that first observed by Keeling at Mauna Loa (Pearman and Beardsmore 1984). They also show that high-latitude northern hemispheric concentrations are, on an annual average basis, 3 to 4 ppmv higher than southern hemispheric values. Large annual cycles, with spring maxima and late summer minima, are observed in the northern hemisphere, with the peak-to-peak amplitude varying from 14 to 16 ppmv at high latitudes to 0 to 5 ppmv in equatorial regions. The amplitude is virtually constant in the mid to high latitudes of the southern hemisphere, varying from 0.5 to 1.5 ppmv (Bodhaine and Harris 1982; Fraser et al. 1983).

Background Concentrations Show Very Little Diurnal Variation

Since most of the 19th century measurements were made in Europe, it is instructive to examine modern background observations at corresponding latitudes in the Atlantic Ocean. Ocean stations M and C and the Azores (Fig. 4.1) are three North Atlantic locations where background CO_2 observations have been made over several years. The annual cycles observed at these three stations are quite similar (Fig. 4.2), with maxima occurring in April and May and minima in August. The peak-to-peak amplitudes are approximately 12 ppmv. Where possible and appropriate, we have used an average North Atlantic annual cycle [derived from these data (Fig. 4.2)] to adjust 19th century measurements (on a monthly basis) to remove the seasonal variability from the data. For example, for the mid-latitudes of the northern hemisphere, an annual mean based on data collected during the summer months would be approximately 3 ppmv lower than one based on a full year of data. This correction, when applied to the 19th century

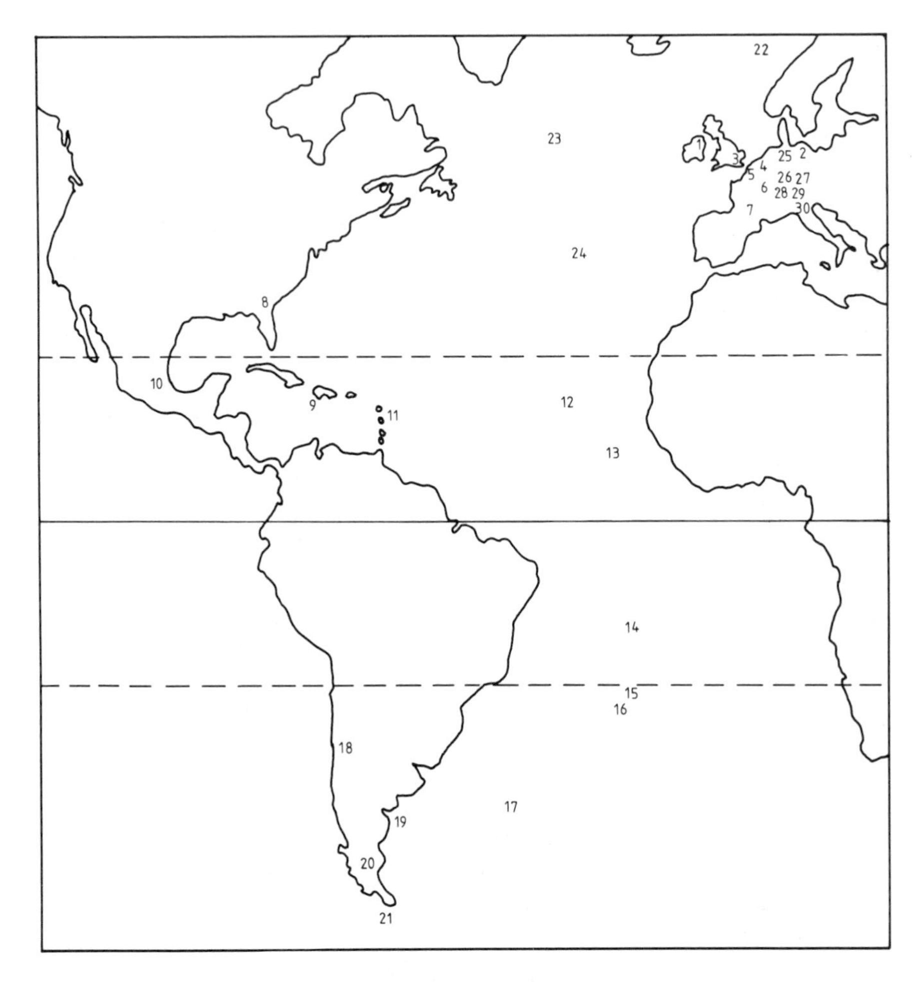

FIGURE 4.1. Location of 19th century sites discussed in this chapter (details in Table 4.2). 1. Belfast; 2. Rostok; 3. Kew; 4. Gembloux; 5. Ecorchboeuf (near Dieppe); 6. Montsouris and Plaine de Vincennes; 7. Pic du Midi; 8. St. Augustine; 9. Petionville; 10. Puebla; 11. Fort-de-France; 12-17. La Romanche Cruise; 18. Cerro-Negro; 19. Chubut; 20. Santa Cruz; 21. Baie Orange. Modern measurements were made at 22. Ocean station M (66°N, 2°E); 23. Ocean station C (54°N, 35°W); 24. Azores (39°N, 27°W); 25. Langenbrugge; 26. Deuselbach; 27. Brotjackriegel; 28. Schauinsland; 29. Wank and Zugspitze; 30. Monte Cimone. More information on continental European stations is listed in Table 4.1.

data, assumes that the annual cycle has not changed markedly over the last 100 years. However, small changes in the annual cycle over this period are quite possible. Pearman and Hyson (1981) and Bacastow et al. (1981*a,b*), and Keeling et al. (1984) have shown that the annual cycle at Mauna Loa may have increased by 0.5 to 1% per year since 1958. Assuming that this change is concentration-dependent it is possible that late

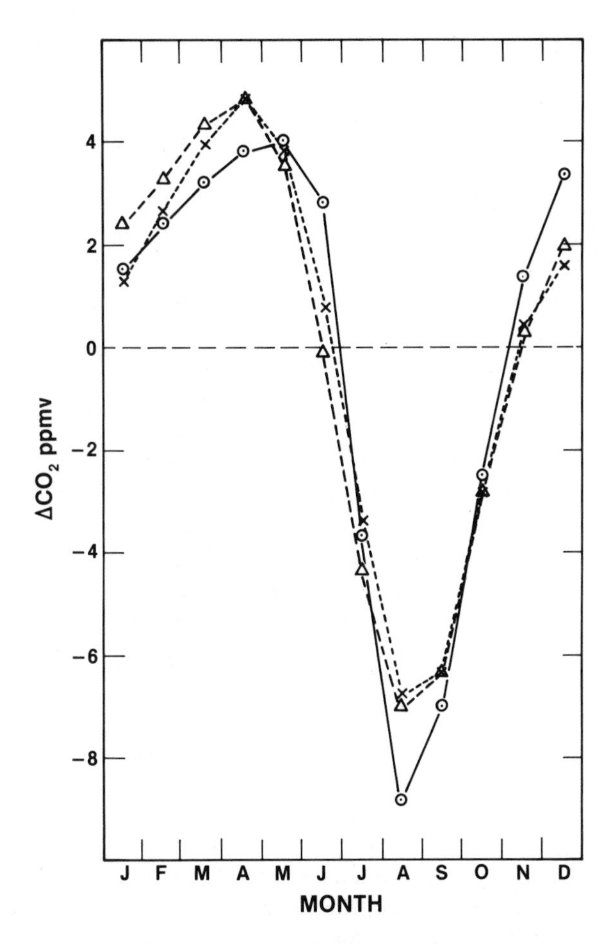

FIGURE 4.2. Annual cycles of CO_2 observed at (⊙), Ocean Station M; (△), Ocean Station C; and (X), Azores. The data are monthly mean standard deviations from linear regressions to 1981 to 1982 data (M), 1969 to 1972 data (C), and 1980 to 1982 data (Azores). The average monthly standard deviations (ppmv) are 1.3 (M), 1.2 (C), and 1.1 (Azores). Data are from the Geophysical Monitoring for Climatic Change (GMCC/NOAA) global CO_2 observational program (Kohmyr et al. 1985).

19th century annual cycles were approximately 35% lower in amplitude than at present. Consequently, northern hemispheric mid-latitude annual cycles may have been 8 ppmv rather than the current 12 ppmv peak-to-peak. If true, the largest error introduced by assuming that the annual cycle has not changed is approximately 2 ppmv (Pic du Midi data, see Table 4.2), which is not considered significant in this analysis.

In attempting to define background CO_2 concentration levels from a particular location, it is important to remove data collected during periods of enhanced atmospheric stability, since that data will be representative of only limited space scales, both horizontal and vertical (Fraser et al.

1983). Since background data should ideally represent global space scales ($>10^6$ m), then as a general rule, only data taken on windy days, preferably those with oceanic fetches whose surface CO_2 fluxes are an order of magnitude smaller than terrestrial fluxes, should be considered. Wherever possible we have attempted to apply these selection criteria to the 19th century measurements.

Measurements of atmospheric CO_2 using in situ infrared gas analyzers are now being made at eight stations (at least that number) in continental Europe (Reiter and Kanter 1982; Ciattaglia 1983; WMO 1983*b*). Unfortunately, the data (Table 4.1) are not corrected for the carrier gas effect, but all stations use the same type of infrared gas analyzer (URAS: Hartman and Braun, Frankfurt/Maintz, Federal Republic of Germany). The record excludes only those observations affected by instrument malfunction. Assuming an average correction for the URAS of 4.6 ppmv (Pearman 1977; Pearman et al. 1983*a*), the five stations located above 1000 m elevation gave a 1981 average CO_2 concentration of 339.9 ($\pm$1.7) ppmv, close to that measured at Azores (339.6 ppmv) and station M (341.9 ppmv) in the North Atlantic, or at Niwot Ridge (339.9 ppmv) in continental North America (Harris and Bodhaine 1983). The three stations below 1000 m gave significantly higher concentrations (Deuselbach, 346.5 ppmv; Langenbrugge, 352.3 ppmv; and Garmisch, 350.7 ppmv), 7 to 12 ppmv above the North Atlantic background. Deuselbach and Langenbrugge are in rural environments; Langenbrugge concentrations are probably affected by pollution sources. Garmisch is in a mountain valley, and the sampling site is close to an urban center.

The data from all stations, except Langenbrugge and Garmisch, show distinct annual cycles, with minima in July and August and broad, poorly defined maxima between December and April (Fig. 4.3). A similar annual cycle of 15 ppmv peak-to-peak amplitude has been reported for Long Island, New York (41°N, 73°W) (Woodwell et al. 1973). By contrast, North Atlantic background observations show a well-defined April maximum (Fig. 4.2), the difference probably being the result of accumulation of CO_2 in stable air masses over Europe in winter. The peak-to-peak amplitudes of the North Atlantic and continental European data are quite similar (12–13 ppmv).

These data suggest that, in continental Europe, CO_2 concentrations near background can be measured above 1000 m, but at lower altitudes, considerably higher concentrations are observed. The occurrence of an annual cycle in the data does not necessarily mean that the observations are background.

Tanaka et al. (1983) have studied variations of atmospheric CO_2 over the Sendai Plain, Japan (38°N, 139°E). They found background concentrations from the surface to the middle troposphere, except during summer, when the midday boundary layer concentration was drawn down by a

TABLE 4.1. Locations and mean concentrations (CO_2) observed at continental European stations in 1981.

Location	Coordinates	Altitude (m)	CO_2 (ppmv)	Analyzer	Reference
Langenbrugge	53°N, 10°E	Sea level	352.3	URAS 2T	WMO (1983*b*)
Deuselbach	50°N, 7°E	480	346.5	URAS 2T	WMO (1983*b*)
Brotjackriegel	49°N, 13°E	1016	341.4	URAS 2T	WMO (1983*b*)
Schauinsland	48°N, 8°E	1205	341.9	URAS 2T	WMO (1983*b*)
Garmisch	48°N, 11°E	700	350.7	URAS 2	Reiter and Kanter (1982)
Wank	47°N, 11°E	1780	338.2	URAS 2	WMO (1983*b*)
Zugspitze	47°N, 11°E	2964	338.4	URAS 2	WMO (1983*b*)
Mt. Cimone	44°N, 11°E	2165	339.7	URAS 2T	Ciattaglia (1983) WMO (1983*b*)

Concentrations are adjusted by 4.6 ppmv to account approximately for the carrier gas effect (see text). The CO_2 value for Garmisch is for 1980.

URAS: infrared gas analyzer, Hartmann and Braun, Frankfurt/Main, Federal Republic of Germany.

maximum of 4 ppmv (August). The data show a clear annual cycle, approximately 14 ppmv in amplitude (peak-to-peak), with a distinct maximum in May and a minimum in August. Winter concentrations are approximately 4 ppmv lower than the May maximum. A similar drawdown (~5 ppmv) of background CO_2 levels over rural locations in southeast Australia has been observed (Garratt and Pearman 1973), and Pearman and Beardsmore (1984) have shown that the magnitude and sign of these terrestrial effects depend on the stability of the lower layers of the atmosphere and the direction of the carbon flux between the surface and the boundary layer.

Previous Assessments of 19th Century Data

The first comprehensive review of 19th century CO_2 data (Letts and Blake 1900), including original observations, provides an extensive bibliography (300 references), experimental summaries, and a synthesis of results from the more important papers. The causes of the variability of atmospheric CO_2 measurements were examined, and significant factors such as the proximity to vegetation or urban centers, latitude, altitude, diurnal and seasonal effects, and the synoptic situation (wind direction, precipitation, etc.) were identified and assessed as to their relative importance. The authors identified the most reliable observers of background CO_2 levels, and based on 10 observers, they found an average northern hemispheric concentration of 286 (±14) ppmv from 14 different locations and an average southern hemispheric concentration of 279 (±29) ppmv from five locations (Fig. 4.4). The highest CO_2 concentration (323 ppmv) in this selected data set was observed at Para, Brazil (1°S, 48°W), at the mouth of the Amazon River. Removing this point from the data gives a southern hemispheric concentration of 267 (±9) ppmv, based entirely on the data of Muntz and Aubin (1886). This concentration is significantly lower than that of the northern hemispheric data and is not consistent with modern observations of the interhemispheric difference (Fig. 4.4). This point will be addressed later when the Muntz and Aubin data are considered in detail.

An annual cycle in background observations was tentatively identified by Letts and Blake (1900), with minimum concentrations occurring in summer and maximum concentrations in winter or spring (or in both seasons). They also identified the two most promising data sets for an analysis of seasonal variations (Reiset 1882, Petermann and Graftiau 1892–1893) (see Fig. 4.5).

Callendar (1938, 1940, 1958) reanalyzed these data using weather maps to identify air mass types. He did not consider data close to strong sources and sinks, or those showing a > 10% difference from the average for a particular time and region. On the basis of the observations of Reiset, Muntz and Aubin, Letts and Blake, Brown and Escombe, and Petermann and Graftiau, a preferred North Atlantic background concentration of 288 to 290 (±3) ppmv was deduced. An annual cycle having a spring maximum-summer minimum and a peak-to-peak amplitude of 7 ppmv was also identified. A comparison of these data with the modern cycle (Fig. 4.5) shows reasonable agreement in phase and amplitude, considering the precision of the chemical measurements (1%). The 3-month running mean of the modern annual cycle shows a peak-to-peak amplitude of 9 to 10 ppmv. These late 19th century data are not inconsistent with the possibility that the annual cycle was smaller than currently observed. The study of the annual cycle from 19th century data supports the proposition that the observed measurements presented the capability for discerning real geophysical phenomena; however, as previously indicated, the identification of an annual cycle does not necessarily prove that the observations are truly background and representative of global space scales.

Reassessment of Preferred 19th Century Data

Reiset

Between 1872 and 1880 Reiset (1882) measured atmospheric CO_2 at Ecorchboeuf (50°N, 1°E) near Dieppe on the coast of France. The exact distance of the observing point from the coast line is not specified. Over a 12-h period, 600 L of air was passed through sulfuric acid to remove water vapor, then through baryta water to remove CO_2, which was later measured by acid titration. Data were collected during the day (7 a.m. to

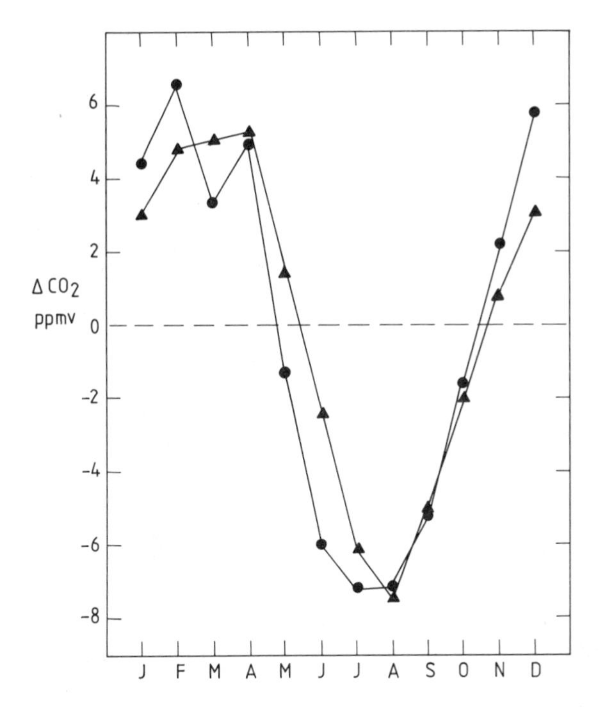

FIGURE 4.3. Annual cycles of CO_2 concentration observed in continental Europe at Mt. Cimone, Italy (1979 to 1983) (▲); and at German stations (●), excluding Langenbrugge and Garmisch (see Table 4.1 for details). The Mt. Cimone data are mean monthly deviations from a linear regression, with an average monthly standard deviation of 1 ppmv; the German data are mean monthly deviations from the annual mean (data detrended by 0.1 ppmv per month, average monthly standard deviation, 2 ppmv).

7 p.m.) and night (7 p.m. to 7 a.m.). Averaged over all wind directions, daytime concentrations were approximately 8 ppmv lower than nighttime values. In an attempt to define background CO_2 levels, we selected only data collected during the day when the wind was between 180° and 270°, a largely maritime sector from which 60% of the winds originated in these experiments. Daytime CO_2 concentrations from this southwest sector were 2 ppmv lower than the daytime mean from all directions. Data were collected from September 1872 to August 1873, from June to November during 1879, and from June to August during 1880. Adjustments were made to the 1879 to 1880 data to account for the missing months, based on the modern North Atlantic annual cycle (Fig. 4.2).

Based on monthly means, the data obtained are listed in Table 4.2. The southwest sector data obtained during 1879 to 1880 are significantly higher than those observed from 1872 to 1873, after adjustment for variability

TABLE 4.2. Average ($[CO_2]$), selected ($[CO_2]$ select), and adjusted ($[CO_2]$ adj) CO_2 concentration measured during the late 19th century.

Date	Location	Coordinates	$[CO_2]$	$[CO_2]_{select}$	$[CO_2]_{adj}$	Selection criteria	Reference
Mar to Jul 1897	Belfast, N. Ireland	55°N, 6°W	292(20)4	281(25)4	278(26)4	SW	Letts and Blake (1900)
Oct 1868 to Jul 1871	Rostok, E. Germany	54°N, 12°E	290(10)32		290(10)32		Schulze (1871)
Jul 1898 to Jul 1901	Kew Gardens, England	51°N, 0°W	300(16)20	285*	286†	D,SW	Brown and Escombe (1905) Callendar (1940)
May 1889 to Apr 1891	Gembloux, Belgium	51°N, 5°E	294(11)21	293(13)21	294(10)21	NW	Petermann and Graftiau (1892–1893)
Sep 1872 to Aug 1873	Ecorchboeuf, France	50°N, 1°E	293(5)12	289(4)12	289(4)12	D,SW	Reiset (1882)
Jun 1879 to Aug 1880	Ecorchboeuf, France	50°N, 1°E	297(7)9	290(5)9	293(4)9	D,SW	Reiset (1882)
May 1881 to Oct 1881	Plaine de Vincennes, France	49°N, 2°E	283(5)5	275(3)5	277(6)5	D,S	Muntz and Aubin (1886)
Aug 1881 to Aug 1883	Pic du Midi, France	43°N, 0°E	279(11)5		286(11)5		Muntz and Aubin (1886)
Nov to Dec 1882	St. Augustine, Florida	30°N, 84°W	291(7)2	290(3)2	291(3)2	D	Muntz and Aubin (1886)
Nov 1882 to Jan 1883	Petionville, Haiti	19°N, 72°W	280(19)3	270(8)2	271(8)2	D	Muntz and Aubin (1886)
Nov 1882 to Dec 1882	Puebla, Mexico	19°N, 98°W	275(8)2	267(21)2	270(21)2	D	Muntz and Aubin (1886)
Oct 1882 to Dec 1882	Fort-de-France, Martinique	15°N, 61°W	280(7)3	274(4)2	276(4)2	D	Muntz and Aubin (1886)

Oct 1883	La Romanche Cruise	15°N, 20°W	270,1		273,1		Muntz and Aubin (1886)
Oct 1883	La Romanche Cruise	7°N, 23°W	249,1		252,1		Muntz and Aubin (1886)
Oct 1883	La Romanche Cruise	17°S, 20°W	270,1		270,1		Muntz and Aubin (1886)
Oct 1883	La Romanche Cruise	24°S, 21°W	272,1		272,1		Muntz and Aubin (1886)
Nov 1883	La Romanche Cruise	26°S, 26°W	277,1		277,1		Muntz and Aubin (1886)
Dec 1882	Cerro-Negro, Chile	33°S, 73°W	270,1	267,1	267,1	D	Muntz and Aubin (1886)
Nov 1883	La Romanche Cruise	43°S, 42°W	274,1		274,1		Muntz and Aubin (1886)
Nov 1882 to Dec 1882	Chubut, Argentina	44°S, 66°N	292(7)2	279,1	279,1	D	Muntz and Aubin (1886)
Oct 1882 to Dec 1882	Santa Cruz, Argentina	50°S, 69°W	270(10)3	264(6)2	264(6)2	D	Muntz and Aubin (1886)
Oct 1882 to Jul 1883	Baie Orange, Chile	56°S, 70°W	259(9)10	259(9)10	259(9)10		Muntz and Aubin (1886)

Averages are means of monthly means; the number in parentheses is one standard deviation; the following number is the number of months of data involved. Selected data are obtained from individual measurements by applying the following selection criteria: D, daytime data only; SW, southwest sector data only; NW, northwest sector data only; S, south sector data only. Adjusted data are selected data adjusted to account for the variability due to the expected annual cycle (see text for detail).

*Callendar's selection based on weather maps and marine air mass from the southwest sector.

†Approximate correction assuming southwest sector data represented in all months for which data were collected.

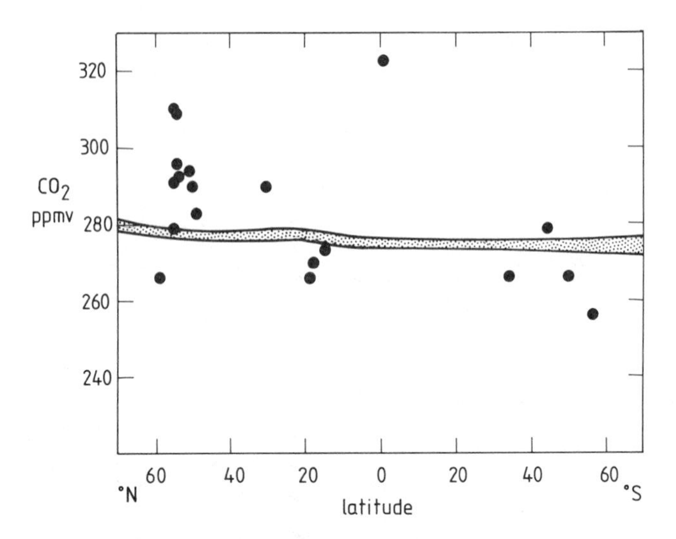

FIGURE 4.4. Selected global background observations of atmospheric CO_2 (Letts and Blake 1900). The *shaded area* is the modern global CO_2 distribution derived from NOAA/GMCC (Harris and Bodhaine 1983) and CSIRO data (Fraser et al. 1983), scaled to 19th century observations.

introduced by the annual cycle. The difference, 3.6 ppmv, is three times greater than the standard error of the difference (1.2 ppmv). Very few long-term CO_2 measurement programs were conducted during the 19th century. The other, at Monsouris, is discussed later. The Reiset data suggest the possibility that during the 1870s, CO_2 concentrations grew at about 0.5 ppmv per year. If true this would imply that significant sources, apart from and significantly greater than fossil fuel combustion, were active during this period, and that CO_2 levels may not have been relatively constant during the latter half of the 19th century. The 12 months of southwest sector data (September 1872 to August 1873) show a minimum CO_2 concentration in August, as do modern North Atlantic and continental data (Figs. 4.2 and 4.3). However, no clear maximum is evident, the data being highly variable from January to June, which is similar to modern continental European data collected below 1000 m altitude. Nevertheless, the difference between maximum and minimum concentrations measured during this period (13 ppmv) agrees with that anticipated from the modern annual cycle (12 to 13 ppmv). The nonadjusted means for the 1872 to 1873 and 1879 to 1880 periods (289 and 290 ppmv, respectively) agree reasonably well with Callendar (1958), who deduced values of 291 and 289 ppmv, respectively, based on selection criteria involving air mass type and a statistical filter.

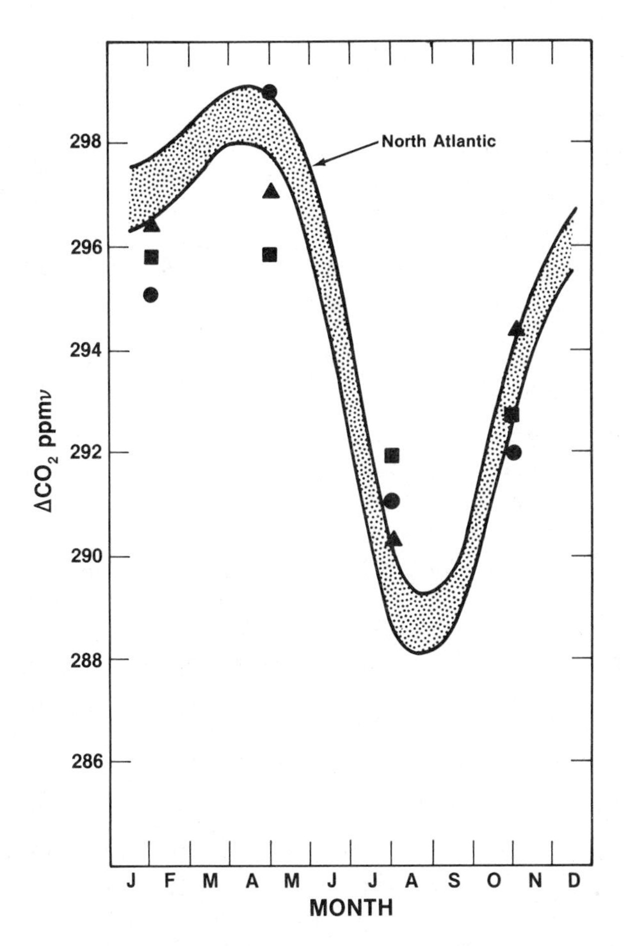

FIGURE 4.5. The annual cycle as observed in seasonally averaged 19th century CO_2 data: (●), Reiset (1882); (■), Petermann and Graftiau (1892–1893); (▲), Callendar (1940). The *shaded area* is the modern North Atlantic annual cycle (from data in Fig. 4.2) presented as a 3-month running mean.

PETERMANN AND GRAFTIAU

Between 1889 and 1891 Petermann and Graftiau (1892–1893), using Reiset's method, measured atmospheric CO_2 levels at Gembloux, Belgium, 120 km southeast of the North Sea. The data are summarized in Table 4.2. Exceptionally high CO_2 concentrations (averaging in excess of 350 ppmv) were observed during the winter of 1890–1891, presumably due to a regional contamination effect or possibly an instrumental problem. Disregarding these data, the average observed CO_2 concentration was 294 ppmv. Meteorological conditions experienced during data collection are available, and, as reported by Petermann and Graftiau (1892–1893) and Letts and Blake (1900), the CO_2 levels do not vary significantly with wind direction. Winds from the northwest sector (from the North Sea?) averaged only 1 ppmv lower than the total data set, and adjustment for the anticipated seasonal variability results in an average concentration of 294 ppmv. By contrast, Callendar (1940), from an analysis of contemporary weather maps, showed significantly lower than average CO_2 concentrations in air masses he characterized as marine (southwest sector, 291 ppmv) and polar-marine (northwest sector, 290 ppmv).

Brown and Escombe

Between 1890 and 1901 Brown and Escombe (1905) measured atmospheric CO_2 concentrations outside the Jodrell Laboratory in the Kew Gardens, 1.4 m above the ground. They used a Reiset (1882) absorption apparatus containing sodium hydroxide that trapped CO_2 from 100 to 300 L of air. The absorbed CO_2 was subsequently determined by a double titration to a stated accuracy of $\pm 1\%$. This method is known as the Pettenkofer process. Data were collected in 1898 (July, August, November, and December), 1899 (January, March, April, and June to August), 1900 (May to September, November, December), and 1901 (January, February, and April to July). The mean and adjusted concentrations (based on the modern annual cycle) are shown in Table 4.2. Brown and Escombe did not publish details of the meteorological conditions prevailing during sample collection. However, Callendar (1940) used weather maps to define marine air masses containing winds from the southwest sector with CO_2 concentrations of, on average, 285 ppmv. All winds with a westerly component averaged 286 ppmv (Callendar 1958). This background level should be increased by approximately 1 ppmv to account for the variability resulting from the expected annual cycle.

Schulze

Between 1868 and 1871 Schulze (1871) measured atmospheric CO_2 concentrations, using the Pettenkofer process, at Rostok, East Germany, on the Baltic Sea. Letts and Blake (1900) and Callendar (1940, 1958) calculate an average CO_2 concentration of 292 ppmv from the data, presumably without using any meteorological selection criteria, since the prevailing synoptic situations were not detailed and the data were presented as monthly means. Average concentrations are listed in Table 4.2. Adjustment for missing months makes little difference to the average concentration of 290 ppmv deduced from these data. Schulze (1871) reports that winds from the southwest (North Sea?) have lower CO_2 concentration than other wind directions.

Letts and Blake

Between March and July of 1897 Letts and Blake (1900) measured CO_2 concentrations in the grounds of Queen's College, Belfast, using a modified Pettenkofer process. The average concentration measured was 292 ppmv. They recorded wind direction and strength for each observation, and, rejecting data collected on calm days, this average was reduced to 288 ppmv. The southwest sector data averaged 281 ppmv. Adjusting these data to account for the expected annual cycle gives 280 ppmv for data collected on windy days and 278 ppmv for southwest sector data.

Muntz and Aubin

Muntz and Aubin (1886) used a different method to determine CO_2 levels in air. From approximately 300 L of air, CO_2 was absorbed over a period of 1 to 2 h in a tube containing pumice impregnated with KOH. The CO_2 was subsequently liberated with acid and its volume measured. Experiments with CO_2-free air and known quantities of Na_2CO_3 showed that the technique was accurate to about $\pm 1\%$. The advantage of this method was that the absorbing tubes could be transported easily to any location, and the trapped CO_2 could be stored indefinitely and then later analyzed in a central laboratory. Subsequently, the method was used by an expedition of the French Academy of Sciences sent to Central and South America to observe the 1882 transit of Venus. All data were examined for correlations between the CO_2 amount, the ambient temperature, and the flow rate of the air sample. Nothing of significance was found. Observations made by the French in various locations around the world (see Table 4.2) are summarized in the following paragraphs.

Plaine de Vincennes

Approximately 35 measurements were made in this rural location southwest of Paris between May and October of 1881. The CO_2 measurements show a dependence on wind direction; the lowest measurements were taken with the wind blowing from southwest to southeast. The few nighttime measurements taken were significantly higher than those taken during the day around the same period.

Pic du Midi

Approximately 40 measurements were made on this mountain (2877 m) in the Pyrenees during August of 1881, 1882, and 1883. In each year the observed CO_2 concentrations did not correlate with wind direction or time of sample collection. Data from the first 2 years averaged around 285 ppmv, whereas the 1883 data were significantly lower (267 ppmv). The causes of the change in measured CO_2 concentrations are not apparent. Since these data were collected in late summer they are subject to the largest correction to remove the seasonal effect (see Table 4.2).

Baie Orange, Chile

Forty measurements were made in this remote location at the southern tip of South America between October 1882 and July 1883. No significant correlation was found between CO_2 level, wind direction, or time of sample collection. No annual cycle was observed, which is as expected, since modern measurements would indicate such a cycle to be quite small (± 1 to 2 ppmv).

Other Locations

Between October 1882 and January 1883 approximately 40 measurements were made at various Caribbean and South American stations set up to observe the transit of Venus. The sample tubes, including those from Baie Orange, were presumably returned to Paris (for analysis) on board the cruise ship La Romanche. Six samples were collected during the return voyage. (See Fig. 4.1 and Table 4.2 for details.)

LEVY

The longest continuous record of atmospheric CO_2 measurements was made between 1877 and 1910 by A. Levy at the Montsouris Observatory on the outskirts of Paris. The method used was similar in principle to that used by Muntz and Aubin (1886) and Letts and Blake (1900), except that measurements were made over a 24-h period. Callendar (1940, 1958) calculated an average concentration of 292 ppmv from these observations for the period 1876 to 1887. He did not explain why he ignored the Montsouris data collected after 1887, and he did not give the observations a "preferred" status, presumably because of the proximity of the Observatory to a large urban center.

Stanhill (1982) analyzed the Montsouris data and concluded that the overall precision of the method was $> 2\%$. He also suggested that the data showed no evidence of significant urban contamination and that the observed CO_2 rise (30 ppmv) from the decade 1880 to 1889 to the following decade (1890 to 1899) was real and indicated that a major, nonfossil fuel source of atmospheric CO_2 was active during this period. An annual cycle was identified in the data, with a minimum in July to August and a maximum in January. The peak-to-peak amplitude was approximately 8 ppmv. The January maximum seems indicative of a site where CO_2 observations are affected by enhanced boundary layer stability in winter.

Waterman (1983) questioned the validity of the Montsouris measurements on the basis of their month-to-month variability and what is known about large-scale atmospheric CO_2 variability from modern measurements. He suggested that the Montsouris results could be affected by the air sampling procedure and the local meteorological conditions prevailing during sampling. Stanhill (1983), in reply to Waterman (1983), gave further details of the Montsouris measurements, including a suggestion that the method of analysis was changed in July 1890, a change, which Starhill (1984) suggests, did not affect the subsequent measured CO_2 concentrations. The 1880 to 1889 decadal mean (283 ± 8 ppmv) is significantly lower than the 1890 to 1899 mean (313 ± 6 ppmv) (Stanhill 1982). During 1890, the January to June average was 282 (±4) ppmv, whereas the August to December average was 324 (±14) ppmv. These data strongly suggest that the change in method in July introduced a sudden increase of 30 to 40 ppmv in the observed CO_2 concentration. It is difficult to believe that this sudden, irreversible change occurred as a result of a real variation in atmospheric CO_2, which is representative of large space scales. Since the pre-July 1890 data (averaging 283 ± 8 ppmv) seem to agree with the Plaine de Vincennes data (277 ± 6 ppmv) collected near Paris (Muntz and Aubin 1886), it is perhaps justifiable to ignore the post-June 1890 data or even to apply a correction for an average bias introduced by the change in method. However, we have chosen not to include the Montsouris data in this analysis because we cannot apply selection criteria based on wind strength and direction or time of day. Modern results suggest that it is difficult to obtain background measurements from an inland, near-sea-level site such as Montsouris; therefore, selection criteria, which consider data collected during the times of maximum atmospheric mixing (around midday and windy), should be applied to ensure that the data are representative of large space scales.

Discussion

The selected and adjusted CO_2 data (Table 4.2) are shown in Fig. 4.6. Northern hemispheric mean CO_2 concentrations of 286 (±7) ppmv from six observers (Reiset, Muntz and Aubin, Letts and Blake, Brown and Escombe, Petermann and Graftiau, Schulze) or 280 (±12) ppmv from 13 different locations can be deduced from the data. Callendar (1958) calculated a preferred value of 289 (±3) ppmv from the same six observers. The difference (3 ppmv) between our northern hemispheric mean and Callendar's, although not significant (standard error of the difference is 3 ppmv), is the result of lower concentrations we deduce from the data of Letts and Blake (278 ppmv compared with 289 ppmv by Callendar) and Muntz and Aubin (277 ppmv compared with 287 ppmv by Callendar). These differences arise largely because Callendar used weather maps to define the days when background observations were to be made, whereas we simply used reported wind directions. The corrections introduced by the annual cycle variability are small (2 to 3 ppmv).

A southern hemispheric mean of 270 (±7) ppmv from eight different locations can be deduced from the data—all from the same laboratory (Muntz and Aubin). When comparing northern and southern hemispheric data, it is perhaps appropriate to consider only the Muntz and Aubin data. The absolute CO_2 concentration may not be correct, but the interhem-

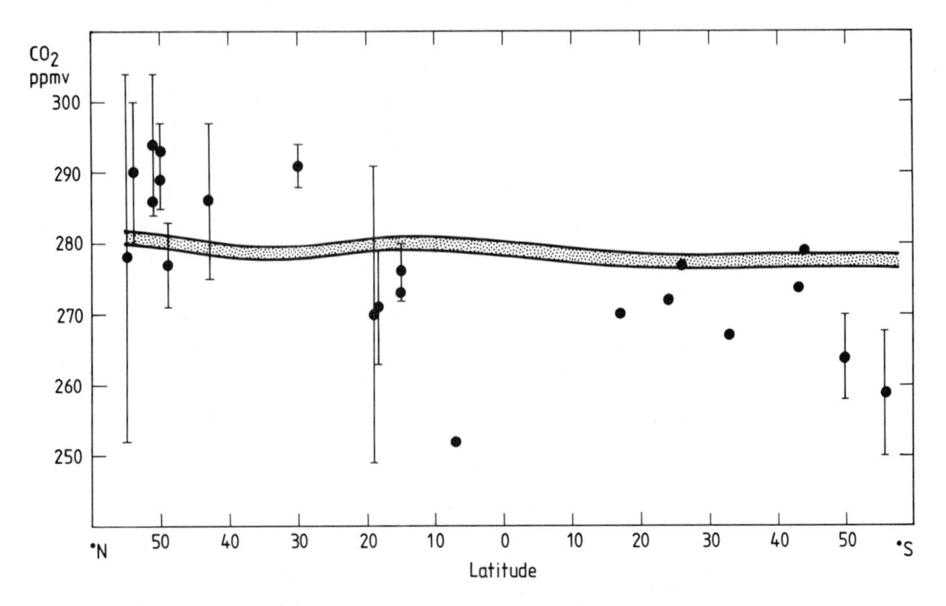

FIGURE 4.6. Global background observations of atmospheric CO_2 based on 19th century data (see Table 4.2). The *shaded area* is the modern global CO_2 distribution (see Fig. 4.4), scaled to 19th century observation.

ispheric difference may be reasonably well-defined. Muntz and Aubin northern hemispheric data show a mean CO_2 concentration of 275 ($\pm$12) ppmv from eight locations. The interhemispheric difference (5 ppmv) is not significant since the standard error of difference is also 5 ppmv.

The Muntz and Aubin data collected from 14 different locations outside France average 271 ($\pm$9) ppmv, whereas from two locations inside France they average 282 ($\pm$6) ppmv. The difference (11 ppmv) is perhaps significant since the standard error of difference is 5 ppmv. Assuming all samples collected outside France were returned to Paris in late 1883 (via La Romanche), the time between sample collection and sample analysis for the land-based stations was at least 12 months. Presumably, the analysis delay for samples collected within France was small by comparison, thus suggesting that some CO_2 loss may have occurred during sample storage. Excluding the data collected inside France, the Muntz and Aubin northern hemispheric mean of 272 ($\pm$12) ppmv from six locations is quite similar to their southern hemispheric mean [270 ($\pm$7) ppmv from eight locations]. Unfortunately, the time between sample collection and analysis was not recorded in the Muntz and Aubin (1886) report; thus, the suggestion of a time-dependent loss of CO_2 from their samples cannot be investigated properly with the current literature. Wigley (1983) has argued that the southern hemispheric data of Muntz and Aubin are probably the most reliable of the 19th century measurements, suggesting, as a consequence, that a representative preindustrial CO_2 level of 260 ppmv is more reasonable than currently accepted values. However, Siegenthaler (1984) has suggested that the data of Muntz and Aubin may be systematically low due to problems associated with the storage of alkaline carbonate solutions in glass containers, which may become more serious as the sample storage time is increased, and also due to interference from water vapor.

As indicated earlier, measurements of CO_2 made at globally distributed stations indicate that, on an annual average basis, northern hemispheric concentrations are approximately 3 ppmv higher than those in the southern hemisphere (Fraser et al. 1983). Model simulations (Pearman et al. 1983*b*) suggest that this difference is the result of the CO_2 from fossil fuels released predominantly in the northern hemisphere, and that around 1920, when fossil fuel CO_2 releases were a factor of 5 smaller than they are now (Keeling 1973, Rotty 1983), this interhemispheric difference was negligible.

Unfortunately, we have to rely solely on Muntz and Aubin data to obtain a picture of interhemispheric CO_2 differences (if any existed) in the late 19th century. Their data are ambiguous. Assuming all data are equally reliable, they possibly suggest that a difference existed on the order of the current difference, implying that there were nonfossil fuel releases in the northern hemisphere on the order of current fossil fuel CO_2 releases. However, because of the variability of the observations, the significance of the suggested interhemispheric difference is minimal, especially if only the data collected outside of France are considered. The late 19th century interhemispheric difference (2 to 5 ppmv), grossly uncertain as it is, is the second item of tenuous evidence supporting the occurrence of nonfossil fuel CO_2 releases in the northern hemisphere during this period, the first being the suggestion from the Reiset (1882) data that CO_2 levels may have increased by approximately 5 ppmv through the 1870 to 1880 decade. This increase does not imply a unique biospheric source strength, even in the context of a specific carbon cycle model. Enting and Pearman (this volume) have emphasized that a determination of the rate of increase of CO_2 concentration (or equivalently a determination of the apparent airborne fraction) at one particular time gives a single number as a constraint on the entire prior biospheric release history. This constraint has been defined by Oeschger and Heimann (1983) although Enting (1985) has pointed out that their analysis can be expressed more concisely to show that the instantaneous rate of change of CO_2 concentration determines a weighted average of the rate of increase of biospheric release.

The biospheric release scenarios, reconstructed by Enting and Pearman (this volume) using a constrained inversion technique for model calibration, give approximately 1 gigaton of biospheric carbon per year around 1880 and increases of about 3 ppmv over the period 1870 to 1880 (see their Figure 4.4; I. Enting, personal communication).

Muntz and Aubin (1886) were the only observers to attempt to calibrate their technique against a prepared CO_2-in-air standard. According to Letts and Blake (1900), they found a CO_2 concentration 1.5% lower than their assigned value. If their prepared standard was accurate, and if this experiment represents the true accuracy of their technique, then it could be argued that their measured CO_2 levels are approximately 4 ppmv too low. Unfortunately, other observers did not independently calibrate their methods in this way.

Analysis of modern CO_2 measurements suggests that the best locations for making background CO_2 observations in Europe are at altitudes higher than 1000 m (and preferably above the tree line) or at the surface around midday, when the air mass concerned has had a lengthy trajectory over the ocean. Once substantial trajectories over land are involved, midday CO_2 levels can be perturbed by up to ± 5 ppmv, depending on the time of the year and the stability of the atmospheric boundary layer. Unfortunately, few 19th century measurements were made that satisfy the above criteria. The candidates are the Reiset (1882) data from coastal France (averaging 289 ppmv when the wind was from the southwest sector) and the Muntz and Aubin (1886) measurements made on Pic du Midi (286 ppmv). The small number and substantial year-to-year variability of the Muntz and Aubin measurements suggest that the Reiset data represent the most important 19th century European measurements which should, perhaps, be investigated more thoroughly. The latter contain a substantial number of observations ($\sim$230) and are the only reliable record that span more than a few years.

Conclusions

At a recent World Meteorological Organization meeting (WMO 1983*a*; Elliott 1984, Pearman 1984), scientists considered "preindustrial" measurements of atmospheric CO_2 made by direct methods such as chemical absorption, solar spectroscopy, and analysis of old air trapped in ice, as well as indirect methods involving ocean chemistry and isotopic tree-ring records. Those in attendance concluded that chemical methods indicate a range of preindustrial CO_2 concentrations from 250 to 290 ppmv, or 260 to 280 ppmv if the Muntz and Aubin (1886) data are considered in isolation. It was suggested that the late 19th century measurements should be reassessed in the light of what is now known about the spatial and temporal variability of global atmospheric CO_2 measurements.

We support the conclusions made at the WMO meeting. Using selection criteria designed to accept data collected during periods of maximum atmospheric mixing, a background CO_2 concentration for the northern hemisphere of 286 (± 7) ppmv was found, based on six independent observers, or 280 (± 12) ppmv, based on 13 different sampling locations. However, it could be argued that these concentrations represent upper limits since, in general, the data still show more variable and possibly elevated concentrations in winter, based on comparison with the modern background annual cycle. A small, barely significant, interhemispheric difference of 2 to 5 ppmv was deduced from the Muntz and Aubin (1886) data. The

Reiset (1882) data suggest that atmospheric CO_2 concentrations may have increased in the 1870s. These latter observations indicate that a significant, nonfossil fuel-derived CO_2 source may have been active in this period. In some of the data sets examined, annual cycles were found that were, in general, similar to the variability seen in modern data, except for the shape and phase of the winter maximum. The observation of annual cycles in the older measurements is encouraging; however, this does not necessarily mean that background concentrations were observed.

The Reiset (1882) data were identified as most likely to benefit from a detailed reanalysis, perhaps with the aid of contemporary weather maps. Measurements were made frequently in locations where long oceanic trajectories are possible and span approximately 7 years. The fundamental problem of calibration remains unresolved; however, the time spent to build a Reiset apparatus and to calibrate it against modern standards should be worthwhile.

Acknowledgments

Helpful discussions concerning the content of this chapter were held with several participants at the Sixth Life Sciences Symposium, especially with Dr. C. D. Keeling. Dr. G. I. Pearman (CSIRO) kindly provided advice and reprints of some of the 19th century literature. P. J. Fraser would particularly like to thank the National Oceanic and Atmospheric Administration, the Cooperative Institute for Research in Environmental Sciences, and Oak Ridge National Laboratory for their financial assistance in this project.

References

Bacastow, R. B., C. D. Keeling, and T. P. Whorf. 1981*a*. Seasonal amplitude in atmospheric CO_2 concentration at Canadian Weather Station P, 1970–1980. In WMO/ICSU/UNEP Scientific Conference on Analysis and Interpretation of CO_2 Data, Bern, Sept. 14–18, 1981, pp. 163–168. World Meteorological Organization, Geneva.

Bacastow, R. B., C. D. Keeling, and T. P. Whorf. 1981*b*. Seasonal amplitude in atmospheric CO_2 concentration at Mauna Loa, Hawaii, 1959–1980. In WMO/ICSU/UNEP Scientific Conference on Analysis and Interpretation of CO_2 Data, Bern, Sept. 14–18, 1981, pp. 169–176. World Meteorological Organization, Geneva.

Bodhaine, B. A. and J. M. Harris. 1982. Geophysical Monitoring for Climatic Change No. 10 Summary Report 1981. Environmental Research Laboratory, U.S. Department of Commerce, Boulder, Colorado.

Bolin, B. and C. D. Keeling. 1963. Large-scale atmospheric mixing as deduced from the seasonal and meridional variations of carbon dioxide. J. Geophys. Res. 68:3899–3920.

Bray, J. R. 1959. An analysis of the possible recent change in atmospheric carbon dioxide concentration. Tellus 11:220–230.

Brown, C. W. and C. D. Keeling. 1965. The concentration of atmospheric carbon dioxide in Antarctica. J. Geophys. Res. 70:6077–6085.

Brown, H. T. and F. Escombe. 1905. On the variations in the amount of carbon dioxide in the air of Kew during the years 1893–1901. Proc. R. Soc. Lond., Biol. 76:118–121.

Callendar, G. S. 1938. The artificial production of carbon dioxide and its influence on temperature. Q. J. R. Meteorol. Soc. 64:223–240.

Callendar, G. S. 1940. Variations of the amount of carbon dioxide in different air currents. Q. J. R. Meteorol. Soc. 66:395–400.

Callendar, G. S. 1958. On the amount of carbon dioxide in the atmosphere. Tellus 10:243–248.

Ciattaglia, L. 1983. Interpretation of atmospheric CO_2 measurements at Mt. Cimone (Italy) related to wind data. J. Geophys. Res. 88:1331–1338.

Elliott, W. P. 1984. The pre-1958 atmospheric concentration of carbon dioxide. Eos 65:416–417.

Enting, I. G. 1985. Green's functions and response functions in geochemical modelling. PAGEOPH. (in press).

Enting, I. G. and G. I. Pearman. 1982. Description of a one-dimensional global carbon cycle model. Division of Atmospheric Physics Technical Paper No. 42, Commonwealth Scientific and Industrial Research Organization, Australia.

Fraser, P. J., G. I. Pearman, and P. Hyson. 1983. The global distribution of atmospheric carbon dioxide 2. A review of provisional background observations, 1978–1980. J. Geophys. Res. 88:3591–3598.

Garratt, J. R. and G. I. Pearman. 1973. CO_2 concentration in the atmospheric boundary layer over southeast Australia. Atmos. Environ. 7:1257–1266.

Harris, J. M. and B. A. Bodhaine. 1983. Geophysical Monitoring for Climatic Change No. 11 Summary Report 1982. Environmental Research Laboratory, U.S. Department of Commerce, Boulder, Colorado.

Keeling, C. D. 1960. The concentration and isotopic abundances of carbon dioxide in the atmosphere. Tellus 12:200–203.

Keeling, C. D. 1973. Industrial production of carbon dioxide from fossil fuels and limestone. Tellus 25:174–198.

Keeling, C. D. 1978*a*. Atmospheric carbon dioxide in the 19th century. Science 202:1109.

Keeling, C. D. 1978*b*. The influence of Mauna Loa Observatory on the development of atmospheric CO_2 research. In Mauna Loa Observatory, A 20th Anniversary Report, pp. 36–54. NOAA Special Report. Environmental Research Laboratory, U.S. Department of Commerce, Boulder, Colorado.

Keeling, C. D., A. F. Carter, and W. G. Mook. 1984. Seasonal, latitudinal, and secular variations in the abundance and iostopic ratios of atmospheric CO_2. J. Geophys. Res. 89:4615–4628.

Komhyr, W.D., R.H. Gammon, T.B. Harris, L.S. Waterman, T.J. Conway, W.R. Taylor, and K.W. Thoning 1985. Global atmospheric CO_2 distributions and variations from 1968–1982 NOAA/GMCC CO_2 flask sample data. J. Geephys. Res., 90:5567–5596.

Letts, E. A. and R. F. Blake. 1900. The carbonic anhydride of the atmosphere. R. Dublin Soc. Rep. 9:107–270.

Muntz, A. and E. Aubin. 1886. Mission scientifique du Cap Horn, 1882–1883. Recherches sur la constitution chimique de l'atmosphere. Tome III. Les Ministeres de la Marine et de l'Instruction publique, Paris.

Oeschger, H. and M. Heimann. 1983. Uncertainties of predictions of future atmospheric CO2 concentrations. J Geophys. Res. 88:1258–1262.

Oeschger, H., U. Siegenthaler, U. Schotterer, and A. Gugelmann. 1975. A box-diffusion model to study the carbon dioxide exchange in nature. Tellus 27:168–198.

Pales, J. C. and C. D. Keeling. 1965. The concentration of atmospheric carbon dioxide in Hawaii. J. Geophys. Res. 70:6053–6076.

Pearman, G. I. 1977. Further studies of the comparability of baseline atmospheric carbon dioxide measurements. Tellus 29:171–181.

Pearman, G. I. 1984. Preindustrial atmospheric carbon dioxide levels: a recent assessment. Search 15:42-45.

Pearman, G. I. and D. J. Beardsmore. 1984. Atmospheric carbon dioxide measurements in the Australian region: ten years of aircraft data. Tellus 36B: 1–24.

Pearman, G. I., D. J. Beardsmore, and R. C. O'Brien. 1983*a*. The CSIRO (Australia) atmospheric carbon dioxide monitoring program: ten years of aircraft data. Division of Atmospheric Physics Technical Paper No. 45. Commonwealth Scientific and Industrial Research Organization, Australia.

Pearman, G, and P. Hyson. 1981. The annual variation of atmospheric CO_2 concentration observed in the northern hemisphere. J. Geophys. Res. 86:9839–9843.

Pearman, G. I., P. Hyson, and P. J. Fraser. 1983*b*. The global distribution of atmospheric carbon dioxide: 1. Aspects of observations and modelling. J. Geophys. Res. 88:3581–3590.

Petermann, A. and J. Graftiau. 1892–1893. Recherches sur la composition de l'atmosphere. Premiere partie. Acide carbonique contenu dans l'air atmospherique. Bruxelles Mem. Couronn. 47:1–79.

Reiset, J. A. 1882. Recherches sur la proportion de l'acide carbonique dans l'air. Ann. Chim. Phys. 26:145–221.

Reiter, R. and H.-J. Kanter. 1982. Time behaviour of CO_2 and O_3 in the lower troposphere based on recordings from neighboring mountain stations between 0.7 and 3.0 km ASL, including the effects of meteorological parameters. Arch. Meteorol. Geophys. Bioklimatol. B. 30:191–225.

Rotty, R. M. 1983. Distribution of and changes in industrial carbon dioxide production. J. Geophys. Res. 88:1301–1308.

Schulze, F. 1871. Tägliche Beobachtungen über den Kohlensäuregehalt der Atmosphäre zu Rostok vom 18 October 1868 bis 31 Juli 1871. Landwirtscha. Vers. Stn. 14:366–388.

Siegenthaler, U. 1983. Uptake of excess CO_2 by an outcrop-diffusion model of the ocean. J. Geophys. Res. 88:3599–3608.

Siegenthaler, U. 1984. 19th century measurements of CO_2-a comment. Climatic Change 6:409–411.

Slocum, G. 1955. Has the amount of carbon dioxide in the atmosphere changes significantly since the beginning of the twentieth century? Mon. Weather Rev. 83:225–231.

Stanhill, G. 1982. The Montsouris series of carbon dioxide concentration measurements, 1877–1910. Climatic Change 4:221–237.

Stanhill, G. 1983. Reply with some additional details on "The Montsouris series of carbon dioxide concentration measurements, 1877–1910". Climatic Change 5:417–419.

Stanhill, G. 1985. A further reply concerning the accuracy of the 'Mont souris series of carbon dioxide concentraion measurements, 1877–1910.' Climatic Change 6:413–415.

Tanaka, M., T. Nakazawa, and S. Aoki. 1983. Concentration of atmospheric carbon dioxide over Japan. J. Geophys. Res. 88:1339–1344.

Waterman, L. S. 1983. Comments on "The Montsouris series of carbon dioxide concentration measurements, 1877–1910" by G. Stanhill. Climatic Change 5:413–415.

Wigley, T. M. L. 1983. The pre-industrial carbon dioxide level. Climatic Change 5:315–320.

Woodwell, G. M., R. A. Houghton, and N. R. Tempel. 1973. Atmospheric CO_2 at Brookhaven, Long Island, New York: patterns of variation up to 125 metres. J. Geophys. Res. 78:932–940.

World Meteorological Organization. 1983*a*. Report of the WMO (CAS) meeting of experts on the CO_2 concentration from preindustrial times to the I.G.Y., Boulder, Colorado, June 1983.

World Meteorological Organization. 1983*b*. Environmental Pollution Monitoring Program No. 15. Provisional daily atmospheric carbon dioxide concentrations as measured at BAPMON sites for the year 1981.

5 Review of the History of Atmospheric CO_2 Recorded in Ice Cores

HANS OESCHGER AND B. STAUFFER

Since the pioneering attempts by Scholander et al. (1961), the observation that the porous spaces in natural ice contain samples of ancient air, the study of ice cores for potential insights into the history of the atmospheric CO_2 concentration has received great attention from scientists interested in the reconstruction of environmental parameters. Progress, however, was made possible only because deep ice cores from Greenland and Antarctica, which are continuous sequences of generally high-quality samples formed during the last 100,000 and 50,000 years, respectively, were available for study. Research has led to new techniques for extracting gases from ice and to recent developments of sensitive and accurate techniques for the analysis of gas. Today, analysis of gas concentrations in air entrapped in natural ice is considered to be the most promising method for reconstructing the history of the atmospheric CO_2 concentration (WMO 1983).

In this chapter, we describe the state of the art of this research. A crucial question relates to the occlusion of air in ice and to possible mechanisms leading to deviations of the gas composition of the trapped air from that of the atmosphere at the time of ice formation and during the long storage time of the air bubbles in the surrounding ice matrix. For ice formed in very cold regions, it can be expected that the CO_2 content of the occluded air indeed closely reflects that of the ancient atmosphere. Therefore, we give an overview of data that we consider to be representative of the atmospheric CO_2 concentration and its variation during the recent preindustrial period, and also during the last glaciation (the Wisconsin) with its termination. The Wisconsin glacial stage was a period with environmental conditions drastically different from those of today and with events of rapid climatic change. The CO_2 concentration during the recent preindustrial period is closely related to questions in carbon-cycle modeling regarding natural and anthropogenic sources and sinks; it is also closely related to the detection of CO_2-induced warming and the sensitivity of climate to rising CO_2. The CO_2 changes observed during periods of rapid climatic change lead to fundamental questions regarding the mechanisms that determine the atmospheric CO_2 concentration and the role of CO_2 in natural climatic change. The history of carbon isotopes ($^{13}C/^{12}C$ and $^{14}C/^{12}C$) in the atmosphere and the ocean as recorded in tree rings, corals, and lake and sea sediments gives indirect information on atmospheric CO_2 changes. The results obtained during the last few years show that the study of information on environmental parameters recorded in natural ice and in other natural archives promises real progress regarding many aspects of the CO_2 issue.

History of the Method

The possibility of reconstructing the history of gas composition of the atmosphere by studying the gases occluded in gas bubbles in polar ice has interested scientists for about 30 years. Let us briefly summarize the pioneering work by Scholander et al. (1961) on glaciers and icebergs of West Greenland. They found that, in general, the gas composition deviated appreciably from that of air, especially concerning the CO_2/air ratio. They attributed this to interaction with liquid water (melting and refreezing) in the accumulation area and at the outlet sites of ice streams. However, ^{14}C and ^{18}O determinations showed that ice of the greatest age and coldest origin had relatively low variability and had CO_2 concentrations close to that of the atmosphere. Scholander et al. therefore concluded that the analysis of ice of the coldest possible origin might still enable the reconstruction of the atmospheric CO_2 concentration, in spite of their discouraging results.

A great breakthrough was achieved by the successful drilling of ice cores through the ice sheets in Greenland, at Camp Century in 1967 (Ueda and Garfield 1968), and in Antarctica, at Byrd Station in 1968 (Ueda and Garfield 1969). These drillings provided ice samples formed during the last 100,000 and 50,000 years at annual temperatures of -24 and -28°C, respectively, for detailed laboratory analysis. In summer 1981 a U.S.-Danish-Swiss deep ice core drilling program was successfully finished with the penetration of the Greenland ice sheet at the radar station Dye 3, South Greenland, where bedrock at a depth of 2037 m was reached. This new ice core enabled especially exciting, detailed studies regarding rapid climatic changes during the Wisconsin glaciation. Other ice cores that have been studied regarding the CO_2 content of the occluded air came from the South Pole and the Antarctic stations Dome C and D 10 (Raynaud and Lebel 1979).

The occluded air is collected either by melting or by crushing the surrounding ice. Both methods have weak and strong points. In the melting procedure, CO_2 that might have been preferentially dissolved in the ice is extracted; on the other hand, contamination by CO_2 from decaying carbonates in the ice or at the surface of the sample is more probable. The dry extraction procedure, based on crushing the samples, avoids such contamination problems, but may not lead to a release of CO_2 proportional to that of the other air gases. Experiments in the Bern and Grenoble laboratories clearly indicated that the dry extraction technique leads to more reliable results (Delmas et al. 1980; Zumbrunn et al. 1982), and we will limit ourselves to the discussion of data obtained by the dry extraction method in both laboratories.

In the more recent studies, gas analysis has been performed by gas chromatography and laser infrared spectroscopy. The experimental errors of the gas analysis itself are of the order of ±1% (≃3 ppm). The overall error is significantly larger, ±3% (≃10 ppm), and it is still controversial whether this is due to a natural variability of the CO_2 content of the air of samples considered as identical or is the result of an additional error during the extraction procedure. To compare different extraction and measuring procedures, the laboratories in Bern and Grenoble studied samples from the same East Antarctic ice core (Dome C) (Barnola et al. 1983). The results agree nicely within the error limits (Fig. 5.1), and suggest that the mean CO_2 level recorded by Antarctic ice for the period 800 to 2500 years before the present (BP) is about 260 ppmv. However, studies on ice cores from other regions in Antarctica and Greenland in the Bern laboratory show higher preindustrial CO_2 concentrations of about 270 ppm.

Recently, a new grinder for larger ice samples (0.5 to a few liters) has been constructed in Bern (Moor and Stauffer 1984). After the samples are ground, the released gases are transferred to a cold trap at 15 K, cooled by a cryocooler. All the air gases, except helium, are frozen out in this

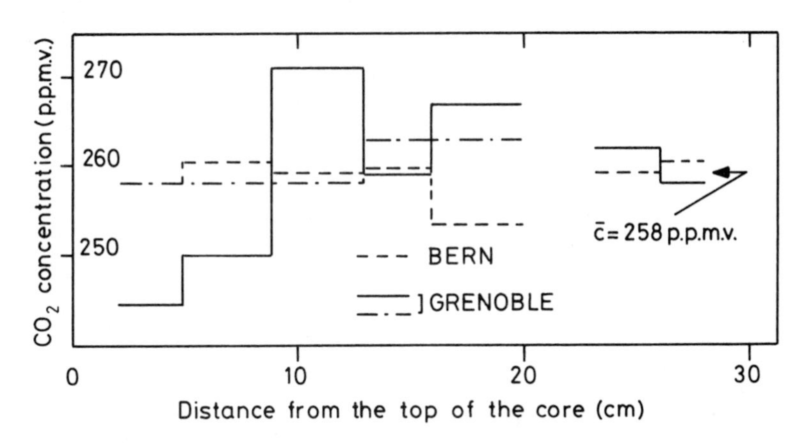

FIGURE 5.1. Concentrations of CO_2 measured in Bern and Grenoble on a core section from Dome C (East Antarctica, depth: 132.9 m below the surface). Experimental precisions are 3% for the Bern measurements; 10% for set 1, measured in Grenoble; and 3% for the 2, also measured in Grenoble. The results obtained in both laboratories agree within the limits of experimental error. The air was trapped by the ice during the time interval 800 to 2500 years BP. The average CO_2 value is 258 ppmv (Barnola et al. 1983. Reprinted by permission from *Nature*, 303:410–413. ©1983 Macmillan Journals Limited).

cold trap. The CO_2 concentration is measured with a gas chromatograph. Only a few samples have been measured as yet with this system, but, interestingly enough, on preindustrial samples from the South Pole, CO_2 concentrations in the range of 280 to 285 ppmv have been observed, i.e., 10 to 15 ppmv higher than obtained with smaller ice samples. These new measurements give rise to questions regarding the *absolute* accuracy of the concentrations obtained with the other methods, in which the air is released through cold traps at $-80°C$ (to trap the water vapor) into the absorption cell for the laser measurement or into the sampling loop of the gas chromatograph. Tests in our laboratory indicate that, during gas transport and diffusion processes, CO_2 shows an affinity for water vapor that might lead to a demixing of CO_2 relative to the main air components N_2, O_2, and argon. Therefore, at present we cannot exclude the possibility that, in spite of the good agreement in the Bern-Grenoble interlaboratory comparison test, such a demixing process leads to absolute CO_2 concentrations that are too low by a few percent.

The most direct way to answer the uncertainty regarding the absolute accuracy of the measured CO_2 concentrations in gases extracted from ice cores would be calibration by the measurement of ice samples that contain air of known composition. Unfortunately, at present such samples are not available.

The problem of demixing of the air components is not the only problem encountered during the measuring procedure. Especially in the case of the measurement of small samples, for which the surface of the walls of the instruments is relatively large compared with the sample volume, an observed release of CO_2 from the surfaces in the presence of water vapor is a disturbing factor, which in the beginning led to CO_2 concentrations that were too high (Zumbrunn et al. 1982). For the measurements presented here, this effect can be kept relatively small and satisfactory corrections may be applied.

Trapping of Air in Natural Ice

Figure 5.2 aids in understanding the process of trapping gases in ice. In the uppermost centimeters of an ice cap or glacier, newly deposited snow flakes degenerate to firn grains of essentially spherical shape. With increasing pressure resulting from consecutive snow deposition, the firn grains sinter together, reducing the porous air-filled space in the firn. At a certain depth (see Fig. 5.2), the channels between the firn grains start to get closed off, and the pore volume becomes separated into isolated bubbles that have no further interaction with the atmosphere (Stauffer 1981).

The analysis of the $N_2/O_2/Ar$ ratios in ice originating from very cold areas with no summer melting shows that, within the error limits, these ratios agree with those in air (Raynaud and Delmas 1977), indicating that

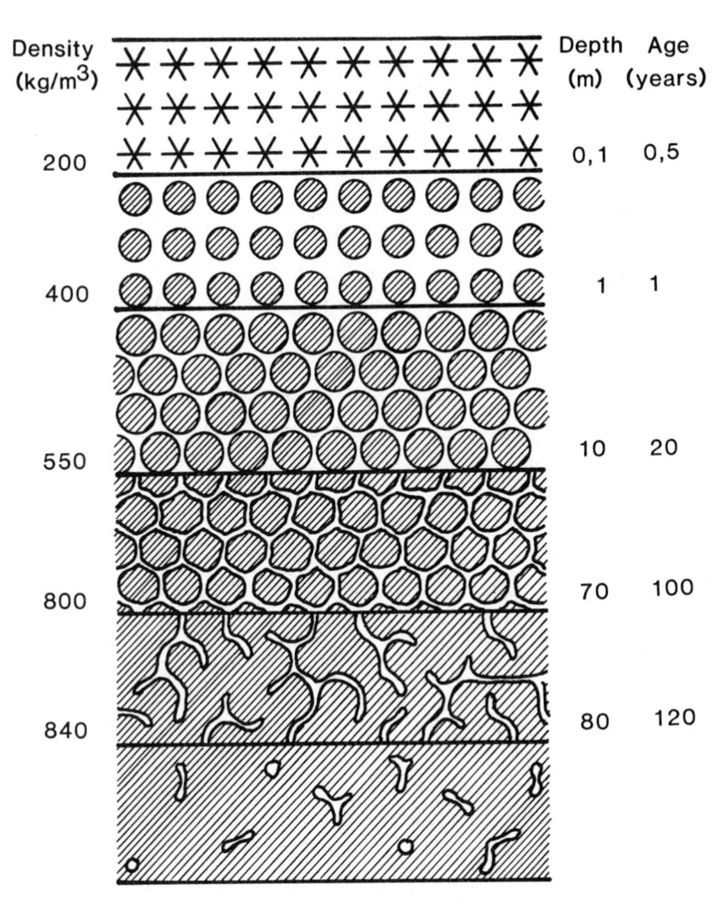

FIGURE 5.2. Metamorphosis of snow to firn and ice: depth-age relationship typical for Greenland ice cores.

the enclosure of air in ice is essentially a physical process of collection of air samples without differentiation of the gas components. However, CO_2 has a much higher solubility than the other air gases and different adsorption properties. Therefore, one cannot automatically conclude that the CO_2 content of the occluded air in the ice corresponds to its atmospheric content. For example, in melt layers CO_2 is significantly enriched; it has also been observed that, in cold, relatively dry accumulation areas, freshly fallen snow in microbubbles contains air with an increased CO_2 content. A CO_2 enrichment might also be caused by adsorption of CO_2 on firn grains. Measurements on snow a few days after its fall, however show a decrease of the CO_2 excess (Stauffer et al. 1981); and estimates indicate that the processes mentioned here for ice from very cold accumulation areas could lead only to relatively insignificant CO_2 enrichments of up to 10 ppmv (Stauffer et al. 1983).

As Fig. 5.2 shows, the air with its specific composition at a certain time is occluded at a greater depth in the ice than the corresponding precipitation. This means that the air composition signal in the ice "leads" the precipitation composition signal by a depth corresponding at Dye 3 in Greenland, for example, to ~100 years. Also, the gas occlusion is a continuous process, taking place during a time interval corresponding to one-fifth to one-third of the age difference between occluded air and ice. We estimated the age difference between the air and precipitation information, assuming that the air is well-mixed down to the depth where the air occlusion mechanism starts. To estimate the time interval for the air occlusion process, we used a model based on the firn/ice density information. The data we obtained for relevant drill sites are listed in Table 5.1. Schwander and Stauffer (1984) observed on samples from Dye 3, Greenland and Siple Station, Antarctica, that gas enclosure in fine-grain winter firn occurs earlier than in the coarse firn originating from summer precipitation. This, as well as other observations, indicates that field studies on newly recovered firn cores from the transition-zone firn-ice might make it possible to improve the time resolution of the CO_2 information in young samples, which is especially important regarding the reconstruction of the atmospheric CO_2 increase since 1850 AD.

Interaction Between the Air in the Bubbles and the Surrounding Ice

As discussed in the preceding section, good evidence exists that, after enclosure, the CO_2 content of the trapped air is close to that of the atmosphere. A further question to be investigated now concerns a possible change of the gas composition of the air in the bubbles during the long storage times involved. The CO_2 concentration might be affected by interaction with the ice itself, but also by interaction with occluded impurities (carbonates).

In regard to the possibility of direct interaction of the air in the bubbles with the ice, density measurements of newly recovered ice indicate that the bubble size is shrinking faster than expected from the increasing hydrostatic pressure. Indeed, at a certain depth the bubbles start to disappear, and it is assumed and confirmed that air hydrates are formed (Miller 1969). After decompression of the ice cores at the surface, air bubbles start to form again; and one might assume that the gas composition of the air in these reformed bubbles differs from the original composition. To investigate a possible enrichment or depletion of the CO_2 content during air hydrate formation and bubble reappearance, we investigated the CO_2 content of the air in ice cores 1 week, 2 months, and 1 year after recovery (Neftel et al. 1983). Within the experimental errors, the same CO_2 concentrations were observed, although in the 1-week-old samples, after the ice was crushed, the air pressure rose significantly more slowly than in

TABLE 5.1. Characteristics of drill sites in Greenland and Antarctica.

Site	Location	Mean annual air temperature (°C)	Annual accumulation [m (w.e.)]	Age interval from beginning to end of air enclosure (years)	Difference between age of ice and mean age of air (years)
Dome C	74°39′S, 124°10′E	−53	0.036	370	1700
South Pole	90°S	−51	0.084	220	950
Byrd Station		−28	0.16	54	240
North Central	74°37′N, 39°36′W	−31.7	0.11	76	350
Crete	71°07′N, 37°19′W	−30	0.265	46	200
Camp Century	77°11′N, 61°09′W	−24	0.34	31	130
Dye 3	65°11′N, 43°50′W	−19.6	0.5	22	90

Schwander and Stauffer (1984).

the older samples, indicating that the ice relaxation and bubble reformation processes had not yet come to an end.

A second possible reason for a shift in the CO_2 concentration of the occluded air is interaction with the chemical impurities contained in the ice. Indeed, the increase of the atmospheric CO_2 concentration at the end of the last glaciation coincides with a significant change of the chemical constituents of the ice. Again, based on experimental observations, we tend to exclude the possibility that this interaction plays an important role, since CO_2 shifts have been observed in both Greenland ice and Antarctic ice from the same time period, which show strongly different impurity contents.

Data on the Preindustrial Atmospheric CO_2 Concentration for the Last 2000 Years

The CO_2 concentration data obtained on air extracted from ice cores and measured in the laboratories in Bern and Grenoble have been discussed and compared at the WMO (CAS) Meeting of Experts on the CO_2 Concentrations from Preindustrial Times to IGY (1958), at Boulder, Colorado, June 22–25, 1983 (WMO 1983). The two laboratories reported data for the last 2000 years from six locations in Greenland and Antarctica. The data from the interlaboratory comparison have been shown in Fig. 5.1. Additional data from Greenland sites, measured at Bern, are shown in Fig. 5.3. On the basis of the data given in Figs. 5.1 and 5.3 one would conclude

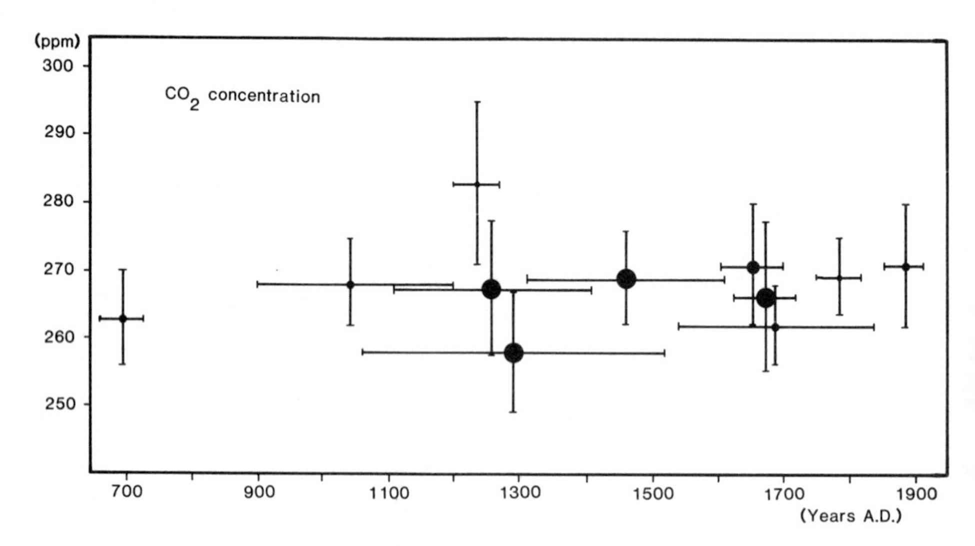

FIGURE 5.3. Concentration of CO_2 in air occluded in ice cores from Greenland and Antarctica. The errors are the standard deviations of the single measurements. Measurements at Bern using the crushing system for 1-cm^3 ice samples (Oeschger et al. 1983).

that the preindustrial concentration lay in the range of 260 to 270 ppmv. As mentioned earlier, however, more recent data, from a procedure using a new gas extraction system, suggest somewhat larger preindustrial values (~280 ppm) (Moor and Stauffer 1984). Additional laboratory tests in Bern, as well as intercomparisons with the Grenoble laboratory, must be performed to determine whether the new results are more correct from an experimental point of view.

In any case, the ice core data suggest that the preindustrial CO_2 concentration was significantly (10 to 30 ppmv) lower than the 295 ppmv obtained by back-extrapolation of the Mauna Loa record assuming an input of fossil CO_2 only and a constant airborne fraction (WMO 1983). The data further show the difficulty of the absolute calibration of the method and the reconstruction of the increase in atmospheric CO_2 content from the beginning of the last century.

As noted before, we have greater confidence in the *relative* accuracy of the CO_2 concentrations than in the *absolute* accuracy. The reason for this is that, by comparison of values measured on the same ice core, uncertainties (such as possible fractionation of the gases during trapping in the ice and also during the measurements) should affect all the data almost equally. The time series of the samples, shown in Fig. 5.3, therefore deserves special attention. Measured with the crushing system for 1 cm^3, the average CO_2 concentrations covering the period 900 to 1900 AD measured on samples from Greenland and Antarctic ice cores lie in the 260 to 270 ppmv concentration band with almost no exceptions. (The one sample set from Byrd Station, covering the period 1180 to 1260 AD, is probably contaminated because of bad core quality; that is, cracks in which later air with higher CO_2 concentration was occluded.) This data set clearly shows that during the last 1000 years, the atmospheric CO_2 concentration (averaged over the period of occlusion of the air in the ice of 30 to 500 years, depending on the accumulation rates) did not show CO_2 excursions as large as that which we observe at present. We consider this to be a strong indication that the presently observed increase is an anthropogenic phenomenon.

A data set obtained by the Grenoble group (Raynaud and Barnola, 1985) seems to indicate variations of the CO_2 concentration of about 10 ppmv over several centuries. On the basis of our data set, as well as on their information, we do not exclude the possibility that during the last 1000 preindustrial years the atmospheric CO_2 concentration might have shown fluctuations of the order of ± 10 ppmv.

Implications of a Low Preindustrial CO_2 Concentration Regarding CO_2 Sources and Sinks and a CO_2-Induced Warming

The general implications of a preindustrial CO_2 concentration significantly lower than that obtained by back-extrapolation of the Mauna Loa data, assuming a "fossil only" CO_2 input, have been discussed in the Report of the WMO (CAS) Meeting of Experts on the CO_2 Concentrations from Preindustrial Times to IGY (WMO 1983). We briefly summarize the major points:

- A low preindustrial CO_2 concentration of 260 to 270 ppmv, or perhaps 280 to 285 ppmv, indicates a major nonfossil, probably biospheric CO_2 source in the 19th century and the first half of the 20th century.
- The possible existence of natural fluctuations of the order of ± 10 ppmv during the last few thousand years could mean that the anthropogenic increase, especially the Mauna Loa data series, is superimposed not on a constant natural level, but on a slightly variable baseline. Fluctuations in atmospheric CO_2 levels that correlate with the El Nino phenomenon indeed are observed, but there might also exist longer term fluctuations on a decadal time scale. Such variations, if occurring during the last 25 years, could add significant uncertainty to the prediction of the future atmospheric CO_2 increase.
- Regarding the detection of a CO_2-induced climatic warming, the implications of a lower preindustrial CO_2 concentration are twofold: First, at present the CO_2-induced change in the global radiation balance may be more advanced than hitherto assumed. Second, the observed temperature trend of the last 100 years would agree better with the estimated temperature increase calculated for a fossil plus biospheric CO_2 increase starting at 265 ppmv around 1820 (Fig. 5.4) (Siegenthaler and Oeschger 1984).

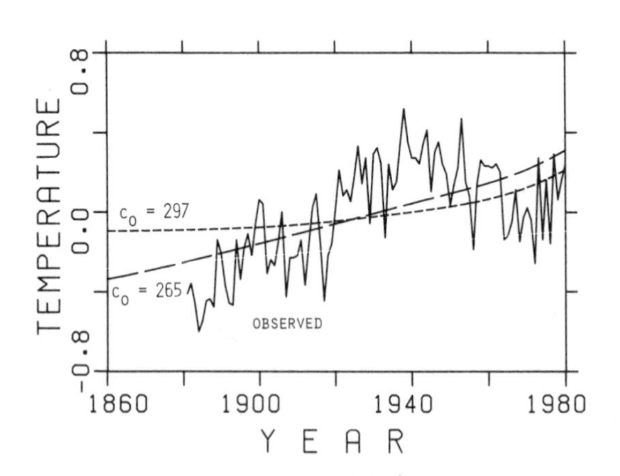

FIGURE 5.4. Comparison of observed global temperature increase with that calculated using different assumed atmospheric CO_2 increases. The short dashed line with indication C_o = 297 ppmv corresponds to the calculated temperature increase based on a "fossil-only" CO_2 scenario. The long dashed line corresponds to the temperature increase calculated for a CO_2 increase starting at an initial CO_2 concentration of C_o = 265 ppmv, as suggested by some of the ice core data (Siegenthaler and Oeschger 1984).

Data on Atmospheric CO_2 Concentration from 50,000 to 1 BC

When measuring the CO_2 concentrations in ice covering the Glacial–Postglacial transition and parts of the Wisconsin glaciation, very surprising results were obtained. In 1979, the laboratories in Bern and Grenoble observed that the CO_2 concentration in several ice cores from both hemispheres was 180 to 200 ppmv at the end of the Wisconsin glaciation and increased at the transition to the Holocene to values in the 260 to 300 ppmv band (Berner et al. 1980, Delmas et al. 1980, Neftel et al. 1982).

The new deep ice core from Dye 3, Greenland, enabled a much more detailed study of the time dependence of the CO_2 increase mentioned above, as well as detailed measurements of sections of the Wisconsin part of the ice core. Based on other information, the Wisconsin part showed significant, rapid climatic variations (Herron and Langway 1985; Dansgaard et al. 1982). The Holocene part of this core shows significant CO_2 enrichments with respect to the actual atmospheric concentrations (Stauffer et al. 1985). However, we believe that during the Wisconsin glaciation, because of the significantly lower temperature and summer insolation, relatively little summer melting occurred and, therefore, the CO_2 concentrations of the occluded air during that time period closely reflect the atmospheric values. The results are discussed in more detail by Stauffer et al. (1985). The next several paragraphs are a summary of the most important observations.

The basic climatic information in an ice core is represented by the parameter $\delta^{18}O$ measured in the water molecules. Periods of relatively high $\delta^{18}O$ values reflect warmer climate whereas relatively low values indicate colder conditions. A dense data set over the entire core has been measured by Dansgaard et al. (1982). The parameter $\delta^{18}O$ is the per mil deviation of the $^{18}O/^{16}O$ ratio in the ice compared with the ratio in standard mean ocean water. In Fig. 5.5, $\delta^{18}O$ values are plotted versus depth for the two Greenland ice cores, from Camp Century, 1967, and Dye 3, 1981. The strong $\delta^{18}O$ shift at about 10,000 BP corresponds to the final warming from the Pleistocene to the Holocene. Both ice cores show an earlier significant $\delta^{18}O$ increase around 13,000 BP. Based on comparison with $\delta^{18}O$ data obtained on lake carbonate from a Swiss lake, indirectly ^{14}C-dated from its pollen composition, we conclude that this first trend to higher $\delta^{18}O$ values corresponds to the transition from the Oldest Dryas cold period to the Bolling-Allerod warm period, whereas the climatic deterioration during the Younger Dryas cold phase is reflected by the low $\delta^{18}O$ values between 11,000 and 10,000 BP (Oeschger et al. 1983).

An additional important observation is the fact that, during the Wisconsin glaciation, rapid climatic transitions reflected in the $\delta^{18}O$ shifts occurred, in contrast to the relatively stable $\delta^{18}O$ values during the last 10,000 years.

In Fig. 5.6, the CO_2 (and ^{10}Be) concentrations and the $\delta^{18}O$ values are plotted versus depth. The comparison of the two profiles shows that the

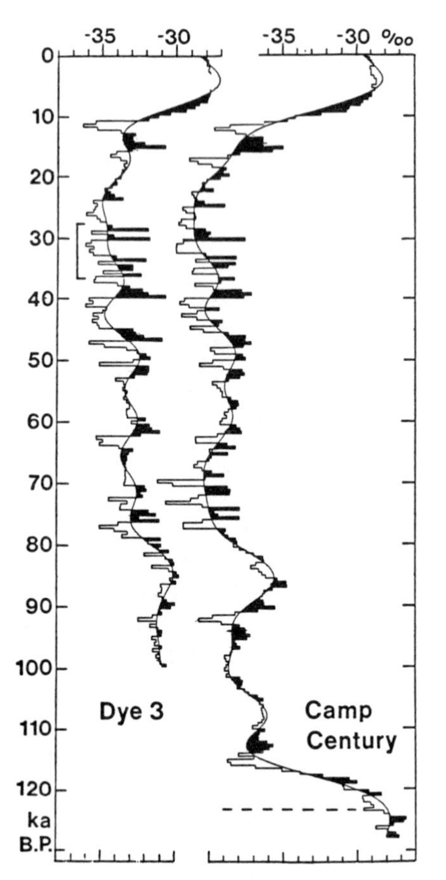

FIGURE 5.5. Profiles of $\delta^{18}O$ measured in Copenhagen along the Dye 3 (0 to 1982 m depth) and the Camp Century (0 to 1370 m depth) ice cores plotted on a common linear time scale based on considerations discussed by Dansgaard et al. (1983). In the time interval 40,000 to 30,000 BP on the left side of the Dye 3 $\delta^{18}O$ profile, the core increments analyzed in more detail regarding CO_2 concentrations in this chapter are indicated.

increase to higher values at the first warming transition, 13,000 BP, occurred within about one century. In view of the fact that the information on the air composition is recorded in ice layers that are ~100 years older than the ice layers containing the $\delta^{18}O$ information, the two transitions occurred almost during the same time period, within ±100 years. Another interesting observation is the relatively high CO_2 concentration value at a depth of 1890 m, corresponding to an estimated age of ~40,000 BP. This high CO_2 value coincides with one of the high $\delta^{18}O$ periods during the Wisconsin glaciation. This observation inspired us to study in detail the relationship between $\delta^{18}O$ and CO_2 in the 30,000- to 40,000-year-old section of the Dye 3 ice core. The results are shown in Figs. 5.7 and 5.8. The surprising result is that all the rapid $\delta^{18}O$ oscillations are accompanied by simultaneous, perfectly correlated CO_2 oscillations (Stauffer et al. 1983). Again, even very detailed measurements did not enable elaboration of a phase shift between the two parameter-time series, and at present we tend to conclude that the CO_2 and the $\delta^{18}O$ changes occurred simultaneously (phase shift $<\pm 1$ century). Besides $\delta^{18}O$ and CO_2, Dansgaard et al. (1982),

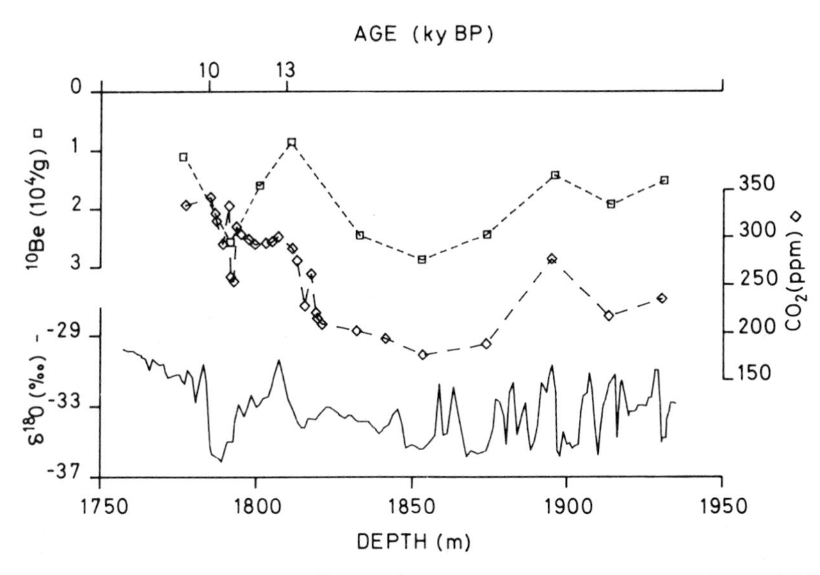

FIGURE 5.6. Concentrations of ^{10}Be (10^4 atm per g of ice), concentrations of CO_2 (ppm), and $\delta^{18}O$ data obtained for the Dye 3 ice core. The tentative time marks are suggested by the comparison with ^{14}C-dated European lake sediments (Oeschger et al. 1983).

as well as Finkel and Langway (1985) and Beer et al. (1984), observed other parameter sets over this time interval, all of which show correspondingly rapid changes, the NO_3^-, Cl^-, SO_4^{2-}, ^{10}Be, and dust concentrations being significantly higher during the cold periods than during the warm periods. Indeed, these parameters indicate the existence of a bistable climate system during the Wisconsin glaciation defined by two sets of the above parameters, one describing the warm and the other the cold climate system state.

Mechanisms Regulating the Atmospheric CO_2 Concentration

Although the observation of rapid atmospheric CO_2 concentration changes during the Wisconsin glaciation needs further confirmation by measurements on other ice cores to exclude occurrence of artifacts due to melt layers or interactions with the impurities in the ice lattice, we currently believe that the observed CO_2 concentration changes of the air occluded in ice indeed represent atmospheric CO_2 changes. (A similar CO_2 transition coinciding with a $\delta^{18}O$ transition has been observed in the Wisconsin part of the Camp Century ice core, originating from a colder accumulation area.)

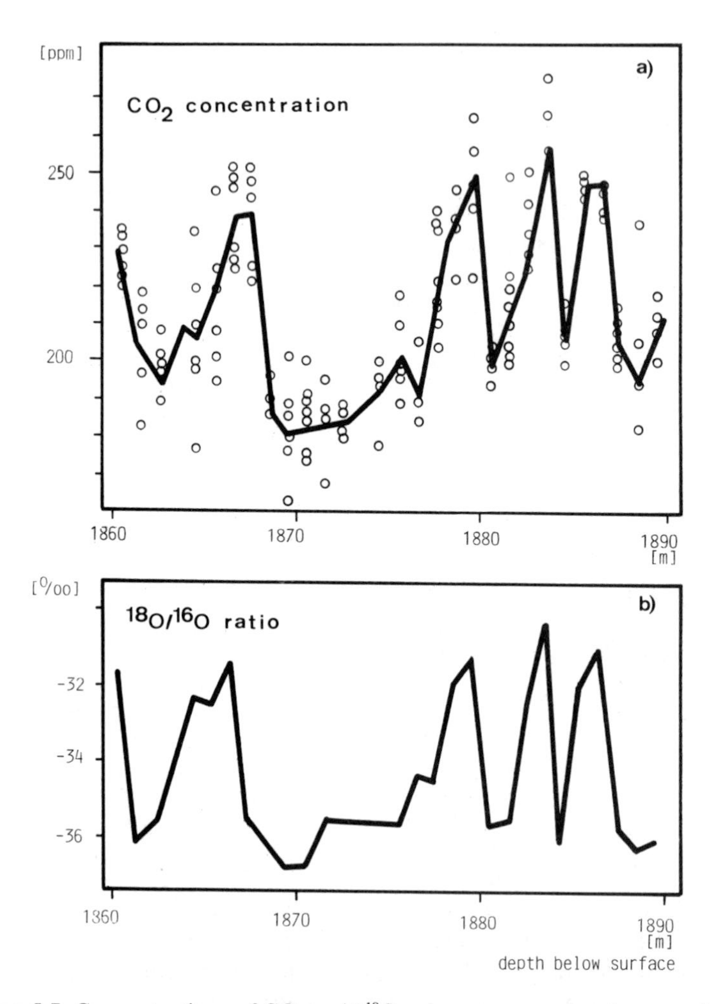

FIGURE 5.7. Concentrations of CO_2 and $\delta^{18}O$ values measured on ice samples from Dye 3 (the 30-m increment corresponds to about 10,000 years; it is indicated in Fig. 5.5). Top: Circles indicate the results of single measurements of the CO_2 concentration of air extracted from ice samples; the solid line connects the mean values for each depth. Bottom: The solid line connects the $\delta^{18}O$ measurements on 0.1-m core increments.

These experimental results have inspired the discussion of mechanisms that might produce atmospheric CO_2 changes of the observed extent. In the early phase, based on fewer data, there was the impression that the CO_2 concentration was low during the entire Wisconsin glaciation and then changed to a higher Holocene concentration; Broecker proposed that this phenomenon could be the result of an alteration in the nutrient element

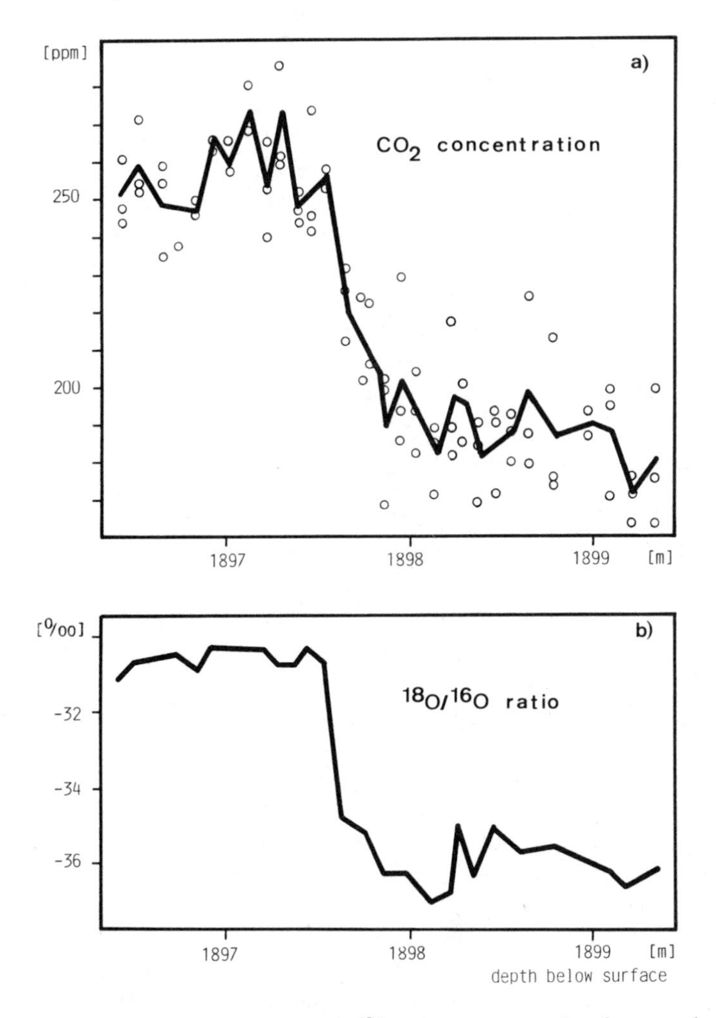

FIGURE 5.8. Concentrations of CO_2 and $\delta^{18}O$ values measured on ice samples from Dye 3 (the 3-m increment corresponds to about 1000 years). Top: Circles indicate the results of single measurements of the CO_2 concentration of air extracted from ice samples; the solid line connects the mean values calculated for increments of about 0.1 m. Bottom: The solid line connects the $\delta^{18}O$ measurements on 0.1-m core increments.

chemistry of the entire ocean, leading to ocean surface chemical conditions corresponding to the CO_2 partial pressures observed during the Glacial and the Postglacial epochs (Broecker 1982, 1983). The recent observation of the rapidity of the CO_2 changes, however, points at changes affecting only the surface ocean chemistry relative to the overall ocean chemistry, due to changing ocean mixing and circulation properties.

The observed CO_2 changes have stimulated the thinking about the mechanisms controlling the atmospheric CO_2 concentration. Before these data existed, it had been generally assumed that natural controlling mechanisms tended to keep the atmospheric CO_2 concentration within a relatively narrow band. The observation of a natural variability of the atmospheric CO_2 concentration beyond that expected from the physicochemical ocean conditions is also of significance for the CO_2 issue: Increasing ocean temperature due to a rising CO_2 concentration and changing wind patterns might induce a change in the ocean circulation and mixing, rapidly affecting the biological and chemical processes in the ocean surface. Such a change could result in either a dampening of the anthropogenically induced CO_2 increase or a more rapid CO_2 increase than expected.

The Climate Impact of Past Atmospheric CO_2 Concentrations

Atmospheric CO_2 concentration variations in the past are ideally suited for sensitivity tests of climate to CO_2 and should attract the attention of climate modelers. The observed CO_2 variations during the Wisconsin glaciation and at the transition to the Holocene, together with other parameters describing the state of the climate system, in principle enable comparisons with climate model calculations of the entire hierarchy of sophistication.

Based on the present information, we think that the CO_2 increase has accompanied the climatic changes reflected in $\delta^{18}O$, probably providing a significant feedback and certainly an almost immediate coupling between the two hemispheres. Since atmospheric CO_2 shifts are global, all the CO_2 shifts observed in the Greenland cores should find their counterparts in the Antarctic ice cores. Their detection in Antarctic ice cores would unambiguously confirm them as atmospheric CO_2 variations. If they were also related to almost simultaneous climatic changes in Antarctica, reflected in $\delta^{18}O$ shifts, the thesis that the CO_2 changes indeed played a major role during the rapid climate oscillations in the Wisconsin glaciation would gain considerable support.

In the following, we compare the probable temperature changes for the rapid Wisconsin fluctuations with model estimates of the global temperature change due to CO_2 changes by a factor of about 1.4. The Wisconsin temperature changes based on the $\delta^{18}O$ shifts might roughly be estimated to half that for the Wisconsin-Holocene transition, for which global values of 4 to 5°C are estimated. If we assume a logarithmic dependence of temperature on the atmospheric CO_2 concentration, we obtain for a CO_2 doubling an upper limit of 4 to 5°C, which corresponds to the upper limit of climate model estimates. However, during the Wisconsin glaciation, the snow and ice cover of the earth was much more pronounced than at present and the sensitivity of climate to a changing radiation balance may have been different from the sensitivity today. In addition, factors other than CO_2, such as changing North Atlantic ice cover and circulation or solar luminosity, may have played an important role during the rapid Wisconsin climatic oscillations (Dansgaard et al. 1983, Oeschger et al. 1983).

Independent (Isotopic) Information on Atmospheric CO_2 Variations

Besides the direct information on the history of the atmospheric CO_2/air ratio from ice cores, independent isotopic information on CO_2 system changes can potentially be derived from the carbon isotopic information ($^{13}C/^{12}C$ and $^{14}C/^{12}C$) in natural records such as tree rings, corals, and sediments.

The carbon in the different CO_2-exchanging reservoirs, atmosphere, biosphere, and ocean shows different isotopic ratios. As an example, due to fossil fuel consumption and deforestation the atmospheric $^{13}C/^{12}C$ ratio has decreased, since these two CO_2 sources have lower $^{13}C/^{12}C$ ratios than the atmosphere (Freyer 1979; Stuiver et al. 1983).

The atmospheric $^{14}C/^{12}C$ ratio as recorded in tree rings shows natural (prenuclear) fluctuations due to a varying ^{14}C production rate (Suess 1968, 1980; Stuiver and Quay 1980), as confirmed by ^{10}Be measurements on Greenland ice cores (Beer et al. 1983).

The atmospheric $^{14}C/^{12}C$ ratio should also change significantly with changes of the atmospheric CO_2 concentration, since the CO_2 partial pressure determines the ^{14}C flux into the oceanic sink (Siegenthaler et al. 1980). Also, since the $^{14}C/^{12}C$ ratio in the total CO_2 in the ocean is lower than the ratio in the atmosphere, an oceanic CO_2 input into the atmosphere should lead to a $^{14}C/^{12}C$ decrease.

The comparison of $^{14}C/^{12}C$ ratios in benthic and planktonic foraminifera of the same age in ocean sediments enables in principle the determination of changes in the oceanic turnover time, which may have been important regarding the chemical and physical role of the ocean in climatic change (Andree et al. 1985).

The difference between the $^{13}C/^{12}C$ ratios of benthic and planktonic foraminifera in ocean sediments gives indications regarding the chemical differences (due to biological activity) between the ocean surface and the deep ocean, which might express themselves in atmospheric CO_2 changes. Indeed, Shackleton et al. (1983) recently reconstructed an atmospheric CO_2 record, based on these data, that was similar to that obtained from the ice cores. Their record extends further back in time and indicates that, during the peak of the last interglacial stage, the CO_2 level was higher than the preanthropogenic Holocene level, and that the level remained high during the early stage of glacial advance.

Future Program and Recommendations

The reconstruction of the atmospheric CO_2 concentrations and the understanding of the CO_2 system and the CO_2-climate interaction based on experimental studies has shown unexpected, exciting results but is still far from being fully exploited. In the following, we list a program of urgent studies for the near future:

- The discrepancy of the CO_2 concentration measurements with the different extraction methods must be resolved.
- The accuracy of the CO_2 concentration measurements on the air occluded in ice cores must be pushed to the limits given by the intrinsic CO_2/air variations in the ice itself.
- Studies in cold firn-ice transition regions need to be performed in an attempt to learn more about the trend of the CO_2 increase since the beginning of the last century. Information on the age distribution of gas in the firn and the young ice might be obtained from ^{85}Kr, CH_4, and Freon measurements.
- Attempts should be made to detect smaller CO_2 variations, of the order of 10 ppm, during the last 10,000 years. Especially valuable would be precise data on the CO_2 concentration during the climatic optimum, which requires new ice core material.
- Special emphasis should be given to the verification, by measurements on Antarctic ice cores, of the fast CO_2 concentration changes during the Wisconsin glaciation.
- CO_2/air measurements on ice dating back to the last interglacial stage (120,000 years BP) should be performed to check whether, during that warm period with higher sea level, the atmospheric CO_2 level indeed was higher than during the Holocene.
- The research on CO_2/air ratios in ice cores should be accompanied by studies of time series of the isotopic ratios $^{13}C/^{12}C$ and $^{14}C/^{12}C$, as summarized in the last section.
- Besides the CO_2/air ratio, a wide spectrum of additional parameters from ice cores and other natural archives should be measured with high time resolution to provide a thorough basis for tests of environmental system processes, such as: the carbon cycle, its dynamics, and its variability; the physical and chemical role of the ocean in climatic changes; the sensitivity of climate to changing concentrations of CO_2 and other gases.

Acknowledgments

The ice cores from Dye 3 were collected in the framework of the International Greenland Ice Sheet Program, which was funded by the U.S. National Science Foundation, the Danish Natural Science Research Council, and the Swiss National Science Foundation.

Laboratory work was supported by the Swiss National Science Foundation, the U.S. Department of Energy, and the University of Bern.

References

Andree, M. and twelve others, 1985. Accelerator radiocarbon ages on foraminifera separated from deep-sea sediments. In: The Carbon Cycle and Atmospheric CO_2: National Variations Archean to Present, pp. 143-153. American Geophysical Union, Washington, D.C.

Barnola, J. M., D. Raynaud, A. Neftel, and H. Oeschger. 1983. Comparison of CO_2 measurements by two laboratories on air from bubbles in polar ice. Nature 302:410–413.

Beer, J., H. Oeschger, M. Andree, G. Bonani, M. Suter, W. Wölfli, and C. C. Langway. 1984. Temporal variations in the ^{10}Be concentration levels found in the Dye 3 ice core, Greenland. Ann. Glaciol. 5:16–17.

Beer, J., U. Siegenthaler, H. Oeschger, M. Andree, G. Bonani, M. Suter, W. Wölfli, R. C. Finkel, and C. C. Langway. 1983. Temporal ^{10}Be variations. Cosmic Ray Conference, Bangalore, August 1983.

Berner, W., H. Oeschger, and B. Stauffer. 1980. Information on the CO_2 cycle from ice core studies. Radiocarbon 22:227–235.

Broecker, W. S. 1982. Ocean chemistry during glacial time. Geochim. Cosmochim. Acta 46:1689–1705.

Broecker, W. S. 1983. The ocean. Sci. Am. September 1983, pp. 100–112.

Dansgaard, W., H. B. Clausen, N. Gundestrup, C. U. Hammer, S. J. Johnsen, P. M. Kristindottir, and N. Reeh. 1982. A new Greenland deep ice core. Science 218:1273–1277.

Dansgaard, W., S. J. Johnsen, H. B. Clausen, D. Dahl-Jensen, N. Gundestrup, C. U. Hammer, and H. Oeschger. 1983. North Atlantic climatic oscillations revealed in deep Greenland ice cores. In J. E. Hansen and T. Takahashi (eds.), Climate Processes and Climate Sensitivity, pp. 288–298. American Geophysical Union, Washington, D.C.

Delmas, R. J., J. M. Ascencio, and M. Legrand. 1980. Polar ice evidence that atmospheric CO_2 20,000 γ BP was 50% of present. Nature 284:155–157.

Finkel, R. C. and C. C. Langway, Jr. 1985. Global and local influences on the chemical composition of snowfall at Dye 3 Greenland: the record between 10 KaBP and 40 KaBP. Earth Planetary Sci. Lett. 73:196–206.

Freyer, H. D. 1979. On the ^{13}C record in tree rings. Part I. Tellus 31:124–137.

Herron, M. M. and C. C. Langway, Jr. 1985. Chloride, nitrate and sulfate in the Dye 3 and Camp Century, Greenland ice cores. In C. C. Langway et. al. (eds.), Greenland Ice Core: Geophysics, Geochemistry and the Environment, pp. 77–84. American Geophysical Union, Washington, D.C.

Miller, S. L. 1969. Clathrate hydrates of air in antarctic ice. Science 165:489–490.

Moor, E. and B. Stauffer. 1984. A new dry extraction system for gases in ice. J. Glaciol. 30 (106): 358–361.

Neftel, A., H. Oeschger, J. Schwander, B. Stauffer, and R. Zumbrunn. 1982. Ice core sample measurements give atmospheric CO_2 content during the past 40,000 y. Nature 295:220–223.

Neftel, A., H. Oeschger, J. Schwander, and B. Stauffer. 1983. CO_2 concentration in bubbles of natural cold ice. J. Phys. Chem. 87:4116–4120.

Oeschger, H., J. Beer, U. Siegenthaler, B. Stauffer, W. Dansgaard, and C. C. Langway. 1983. Late-glacial climate history from ice cores. In J. E. Hansen and T. Takahashi (eds.), Climate Processes and Climate Sensitivity, pp. 299–306. American Geophysical Union, Washington, D.C.

Raynaud, D. and R. Delmas. 1977. Composition des gaz contenu dans la glace polaire. IAHS-Publication 118, pp. 377–381.

Raynaud, D. and B. Lebel. 1979. Total gas content and surface elevation of polar ice sheets. Nature 281:289–291.

Raynaud D. and T. M. Barnola, 1985. An Antarctic ice core reveals atmospheric CO_2 variations over the past few centuries. Nature 315:309–311.

Scholander, P. F., E. A. Hemmingsen, L. K. Coachman, and D. C. Nutt, 1961. Composition of gas bubbles in Greenland icebergs. J. Glaciol. 3:813–822.

Schwander, J. and B. Stauffer. 1984. Age difference between polar ice and the air trapped in its bubbles. Nature 311:45–47.

Shackleton, N. J., M. A. Hall, T. Line, and Cang Shuxi. 1983. Carbon isotope data in core V19-30 confirms reduced carbon dioxide content in ice age atmosphere. Nature 306:319–322.

Siegenthaler, U., M. Heimann, and H. Oeschger. 1980. ^{14}C Variations caused by changes in the global carbon cycle. Radiocarbon 22:177–191.

Siegenthaler, U. and H. Oeschger. 1984. Transient temperature changes due to increasing CO_2 using simple model. Ann. Glaciol. 5:153–159.

Stauffer, B. 1981. Mechanismen des Lufteinschlusses in natürlichem Eis. Z. Gletscherkd. Glazialgeol. 17:17–56.

Stauffer, B., W. Berner, H. Oeschger, and J. Schwander. 1981. Atmospheric CO_2 history from ice core studies. J. Gletscherkd. Glazialgeol. 17:1–16.

Stauffer, B., H. Hofer, H. Oeschger, J. Schwander, and U. Siegenthaler. 1983. Atmospheric CO_2 concentration during the last glaciation. Ann. Glaciol. 5:160–164.

Stauffer, B., A. Neftel, H. Oeschger, and J. Schwander. 1985. CO_2 concentration in air extracted from Greenland ice samples. In C. C. Langway et. al. (eds.), Greenland Ice Core: Geophysics, geochemistry and the environment, pp. 85–89. American Geophysical Union, Washington, D.C.

Stuiver, M., R. L. Burk, and P. D. Quay. 1983. $^{13}C/^{12}C$ ratios and the transfer of biospheric carbon to the atmosphere. J. Geophys. Res. (submitted).

Stuiver, M. and P. D. Quay. 1980. Changes in atmospheric carbon-14 attributed to a variable sun. Science 207:11–19.

Suess, H. E. 1968. Climatic change, solar activity and the cosmic-ray production rate of the natural radiocarbon. Meteorol. Monogr. 8:146–150.

Suess, H. E. 1980. The radiocarbon record in tree rings of the last 8000 years. Radiocarbon 22:200–209.

Ueda, H. T. and D. E. Garfield. 1968. Drilling through the Greenland ice sheet. U.S. Army CRREL Special Report 126.

Ueda, H. T. and D. E. Garfield. 1969. Core drilling through the Antarctic ice sheet. U.S. Army CRREL Technical Report 231.

WMO. 1983. Report of the WMO (CAS) meeting of experts on the CO_2 concentration from preindustrial times to IGY WMO Project on Research and Monitoring of Atmospheric CO_2, Report 10, WCP-53.

Zumbrunn, R., A. Neftel, and H. Oeschger. 1982. CO_2 measurements on 1-cm^3 ice samples with an IR laser spectrometer (IRLS) combined with a new dry extraction device. Earth Planet. Sci. Lett. 60:318–324.

6
Ancient Carbon Cycle Changes Derived from Tree-Ring ^{13}C and ^{14}C

MINZE STUIVER

Current research places major emphasis on atmospheric CO_2 concentration because this variable is important in modeling climatic change. The history of atmospheric CO_2, however, is described not only by concentration change but also by changes in its isotopic composition. Two stable isotopes, ^{12}C and ^{13}C, exist for carbon, whereas ^{14}C, a radioactive isotope, is also present in the natural carbon cycle.

The average $^{13}C/^{12}C$ isotope ratio on earth is $\sim 1.12 \times 10^{-2}$. Appreciable variations are encountered in this ratio because the physicochemical properties of molecules are mass-dependent. A typical example is the CO_2 exchange between atmosphere and oceans that results in a 9-per-mil lower $^{13}C/^{12}C$ ratio of atmospheric CO_2 relative to oceanic bicarbonate.

The ^{14}C isotope, with a half-life of 5730 years, is present in atmospheric CO_2 with a $^{14}C/^{12}C$ ratio of about 10^{-12}. This isotope is continuously produced in the upper atmosphere by the interaction between thermal neutrons and nitrogen $^{14}_{7}N + ^{1}_{0}n \rightarrow ^{14}_{6}C + ^{1}_{1}H$. Although generated by cosmic radiation, the neutrons are not part of the incoming cosmic ray flux. They are the end products of the splitting of atmospheric atoms by highly energetic cosmic ray protons.

The history of the $^{13}C/^{12}C$ isotope ratio of atmospheric carbon is partly related to changes in the size of, and exchange rate between, the various terrestrial carbon reservoirs. The $^{14}C/^{12}C$ history is tied to these aspects as well. However, other major components of the ^{14}C history are the geomagnetic and solar changes that affect the flux of cosmic ray particles arriving at the earth's atmosphere.

The discussions of atmospheric $^{14}C/^{12}C$ presented later include (1) long-term (10^3 to 10^4 years) change, (2) shorter-term (10 to 10^2 years) variability, and (3) the impact of fossil fuel CO_2 (which lacks ^{14}C) on atmospheric $^{14}C/^{12}C$ ratios. The $^{13}C/^{12}C$ analysis focuses on the complexities of this isotope signal in tree rings and on a discussion of the magnitude of biospheric CO_2 release to the atmosphere.

^{14}C and ^{13}C Nomenclature

The $^{13}C/^{12}C$ sample isotope ratios are expressed as $\delta^{13}C$, which is the relative deviation from the Peedee Belemnite (PDB) standard in per mil according to

$$\delta^{13}C = [\frac{(^{13}C/^{12}C)_s - (^{13}C/^{12}C)_{PDB}}{(^{13}C/^{12}C)_{PDB}}]\ 1000‰ \quad (1)$$

Mass-spectrometric techniques are also available for the direct measurement of $^{14}C/^{12}C$ ratios. Most determinations, however, are made by counting the ^{14}C radioactivity of the sample. Using identical amounts (in moles) of sample and standard, the sample activity A_s is compared to 0.95 times the age-corrected National Bureau of Standards' (NBS) oxalic acid standard activity (A_{ox}), resulting in

$$\delta^{14}C = [\frac{A_s - A_{ox}}{A_{ox}}]\ 1000‰ \quad (2)$$

Because counting efficiencies are identical and the amount of ^{12}C is approximately the same for both sample and standard (a difference in $\delta^{13}C$ of 9‰ results in a 0.1‰ difference in the number of ^{12}C molecules), the above expression is for practical purposes identical to

$$\delta^{14}C = [\frac{(^{14}C/^{12}C)s - (^{14}C/^{12}C)ox}{(^{14}C/^{12}C)ox}]\ 1000‰ \quad (1)$$

The sample activity, A_s, in Eq. (2) is the measured activity for modern samples. For older materials of known age, a correction for ^{14}C decay is applied. Here, tree-ring $\delta^{14}C$ values represent the ^{14}C activity at the time of formation—relative to the oxalic acid standard activity.

Trees may have variable discrimination against the heavier ^{13}C and ^{14}C isotopes during photosynthesis. The fractionation for ^{14}C is twice the ^{13}C fractionation, and differences in isotope fractionation can be accounted for by calculating the ^{14}C activity for a fixed $\delta^{13}C$ value of -25 per mil. This results in the $\Delta^{14}C$ term, with

$$\Delta^{14}C = \delta^{14}C - 2\ (\delta^{13}C + 25)\ (1 + \frac{\delta^{14}C}{1000}) \quad (2)$$

For the above normalization (see also Stuiver and Polach 1977), the assumption is made that the differences in tree-ring $\delta^{13}C$ values are all due to isotope fractionation during photosynthesis or sample treatment. However, tree-ring $\delta^{13}C$ values also reflect changes in atmospheric $\delta^{13}C$. These changes should, ideally, not be incorporated in the $\delta^{13}C$ normalization procedure. Unfortunately, the lack of a well-defined atmospheric $\delta^{13}C$ time history prevents the exclusion of these changes.

The errors introduced in $\Delta^{14}C$ values by including atmospheric $\delta^{13}C$ variability during $\delta^{13}C$ normalization are restricted to a few per mil. For comparison, it should be noted that (1) $\Delta^{14}C$ has changed by several percent during the past, and (2) the statistical errors in the ^{14}C radioactivity determination are seldom <1.5 per mil.

For the determination of a radiocarbon age (t), the ratio of the sample activity A_s (normalized on $\delta^{13}C = -25\%$) to 0.95 times the NBS oxalic acid activity A'_{ox} (normalized on $\delta^{13}C = -19\%$) is used according to $t = -8033 \ln A_s/A'_{ox}$. Radiocarbon ages are relative to 1950 AD. This year equals 0 year BP (before present).

Long-Term Natural ^{14}C Variability

The limiting factor in radioactivity measurements is the statistically indeterminate number of counts (n). The relative error ($\sigma = n^{-1/2}$) decreases with increasing n. Precise ^{14}C determinations, therefore, require large samples and long counting times. A typical tree-ring counting precision of 1.5 per mil is obtained at the Quaternary Isotope Laboratory for 6 g carbon samples counted for 4 days.

Most suitable for the determination of past atmospheric ^{14}C changes are dendrochronologically dated tree-ring samples. The cellulose formed annually in each ring ceases to exchange carbon afterwards. Noncellulose compounds, added or exchanged after the year of formation, can be removed by chemical treatment of the sample. Even when this removal is not 100% complete [which is the case for the so-called de Vries method (Cain and Suess 1976)], the resulting errors in natural $\Delta^{14}C$ values can be reduced to a few tenths per mil (Stuiver and Quay 1980).

A radiocarbon age, as defined in the preceding section, is calculated from the ratio of remaining sample activity to oxalic acid standard activity. A $\Delta^{14}C$ value gives the decay-corrected (back to 1950 AD) sample activity relative to the standard activity. Delta values can be calculated only when the dendro-age of a sample is known (dendro-years will equal calendar years for faultless chronologies). Thus, in tree-ring research, plots of radiocarbon age (representing remaining activity) versus dendro-age, as well as plots of $\Delta^{14}C$ (representing original activity) versus dendro-age, are possible.

The longest continuous sequence of dendrochronologically dated wood currently available is an 8600-year-long Bristlecone Pine chronology (Ferguson and Graybill 1983). The absolute dendro-ages of floating chronologies (tree-ring series not corrected to the present) from other regions, that is, those of German (Becker 1983) and Irish oak (Baillie et al. 1983), can be determined by matching the radiocarbon age patterns (Kruse et al. 1980) with the Bristlecone Pine chronology.

The use of varved deposits extends the study of ^{14}C variations beyond dendrochronological limits. The varves are produced in sedimentary environments by annual layers of different composition or texture. The silt and clay couplets in Swedish proglacial lakes are the most extensively studied varve series (Tauber 1970).

A few lakes in temperate zones also contain rhythmites (sedimentary succession of interbedded laminae). The laminated sediments of Lake of the Clouds, Minnesota, contain more than 10,000 couplets (Stuiver 1970*a*). Another "varved" series has been reported from Faulense, Switzerland

(Vogel 1970), whereas marine sediments in Saanich Inlet, British Columbia (In Che Yang and Fairhall, 1972), also contain rhythmites.

A major problem with varve chronologies is the inherent assumption of (1) lack of suppression or duplication of annual deposition of couplets, (2) lack of missing (or added) sections, and (3) constant ^{14}C reservoir deficiency. Agreement with the ^{14}C tree-ring results obtained for the last 8600 years (or even 9200 years, see below) is essential. Only the Lake of the Clouds chronology passes this test, as can be seen in Fig. 6.1 where the Δ values derived from radiocarbon ages of bristlecone pine wood and varved sediment of the Lake of the Clouds are compared. The matching of a floating German oak chronology with bristlecone pine dates (Bruns et al. 1983; Stuiver, unpublished) confirms the Lake of the Clouds results (Fig. 6.1.) back to 9200 years BP.

A comparison of Lake of the Clouds radiocarbon ages of 8600 and 9050 years BP (9740 and 10,070 varve years BP) with corresponding radiocarbon ages of the Swedish varve series (Tauber 1970) shows a shortage of about 1000 varves for the Swedish count by 10,000 varve BP. This proven error in the upper portion of the Swedish chronology, derived from varves studied mainly at a single site (Fromm 1970), is not necessarily also a problem

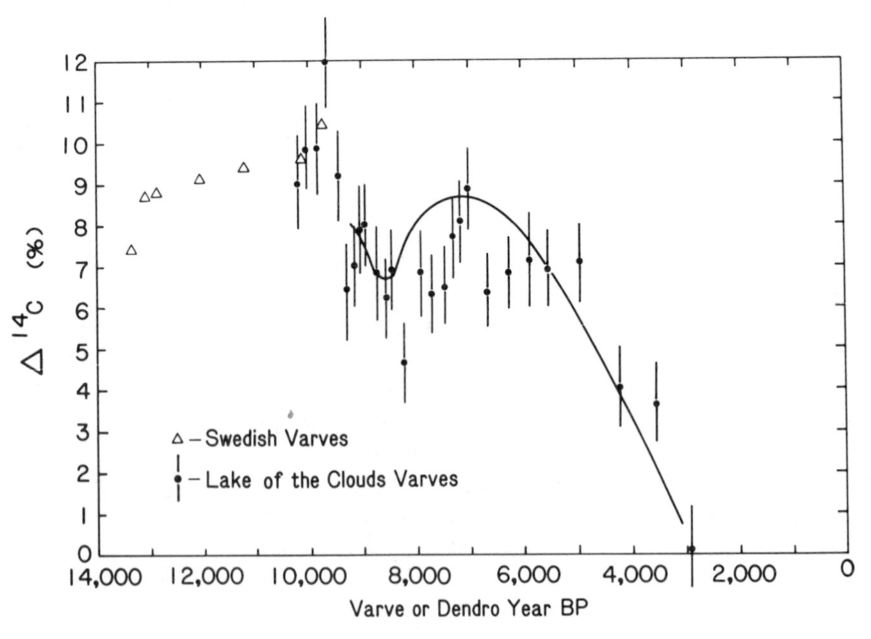

FIGURE 6.1. Long-term trend (*solid line*) of atmospheric $\Delta^{14}C$ values derived from bristlecone pine (Neftel et al. 1981) and German oak (Bruns et al. 1983). The other $\Delta^{14}C$ values were derived for Lake of the Clouds (Stuiver 1970*b*) and Swedish (Tauber 1970) varves.

for the older portion where the chronology has been derived from several closely overlapping varve sequences. Accepting a 1000-year offset in the upper portion, but assuming no further errors, yields the extension of varve-derived $\Delta^{14}C$ values in Fig. 6.1 back to 13,400 years BP.

Atmospheric $\Delta^{14}C$ is less well known beyond 13,000 years BP. The good agreement between ^{14}C and the ionium dates of Searles Lake sediment (Stuiver 1978; Stuiver and Smith 1979; Peng et al. 1978) shows that $\Delta^{14}C$ departures from the oxalic acid baseline are restricted to about 10% between 23,000 and 32,00 years BP. A comparison between ^{14}C and ionium dates of a stalagmite (Vogel 1983) yields a 50% increase in $\Delta^{14}C$ near 18,000 years BP and an approximately normal level near 29,000 years BP. Vogel's analysis suggests ^{14}C levels 1.5 to 2.0 times the present for the interval between 35,000 to 40,000 years BP. The change in ^{14}C age versus depth of Lake Jih Tan sediment suggests a 20% higher $\Delta^{14}C$ value about 20,000 years ago (Stuiver 1970*b*).

Different processes, working in opposite directions, affect the atmospheric ^{14}C and CO_2 levels during glacial episodes. These include (Stuiver 1978; Sarmiento and Toggweiler 1984; Siegenthaler and Wenk 1984) (1) a reduction in ^{14}C atmosphere-ocean exchange caused by a lower sea level and increased ice cover, (2) less downward advection of ^{14}C because of a reduced rate of bottom water formation in the Atlantic and the concomitant reduction in the worldwide rate of oceanic upwelling that brings less ^{14}C-deficient water to the surface, and (3) changes in biological productivity of high-latitude surface waters.

The above processes (1) and (2) would point to a higher atmospheric ^{14}C level. However, downward eddy diffusive transport of ^{14}C over the oceanic thermocline increases during glacials because oceanic thermal gradients are less.

Changes in atmospheric CO_2 content (Barnola et al. 1983) and $\Delta^{14}C$ are interrelated. A 100-ppm decrease in CO_2 content during the previous glacial would result in a 2.5% increase in atmospheric $\Delta^{14}C$ if all other global reservoirs remained constant (Keir 1983). Keir also estimates a 7% $\Delta^{14}C$ reduction in the deep ocean because of a 7% increase in inactive oceanic ^{12}C during glacial conditions. This change in carbon content would occur if the concept of phosphorus extraction on continental shelves, as suggested by Broecker (1982), is correct.

For a cessation of north Atlantic deepwater formation, the model calculations indicate atmospheric $\Delta^{14}C$ increases in the 7 to 10% range, assuming a constant ^{14}C source and a constant transfer rate between atmosphere and ocean (Lal and Venkatavaradan 1970). Increased density stratification related to an increase in freshwater input during the glacial-interglacial transition also would increase $\Delta^{14}C$ values. Keir (1983), when assuming that vertical exchange decreases linearly with the rate of decrease in ice volume, calculates an atmospheric $\Delta^{14}C$ pulse around 10,500 ^{14}C years BP of about +10%.

Long-term changes in the ^{14}C production rate also have to be taken into account. The flux of cosmic ray particles arriving in the atmosphere depends on the strength of the earth's geomagnetic field. The period prior to 12,000 years BP is generally characterized by lower geomagnetic dipole moments than the present one. Barbetti (1980) suggests an average $\Delta^{14}C$ increase of 10% between 12,000 and 25,000 years BP. There is appreciable scatter in the paleomagnetic determinations, and $\Delta^{14}C$ values as high as +50 and as low as −20% are possible (Barbetti 1980).

The fairly recent increase in dipole moment, with a broad maximum between 500 BC and 500 AD, is usually associated with the reduction in the ^{14}C level since 7000 years BP (Fig. 6.1). The high dipole moment of the last 4000 years is an unusual feature of the geomagnetic field that has not occurred during the previous 40,000 years (Barton and Merrill 1983). Current evaluations of the magnetic data set therefore do not support the 8000-year sinusoidal geomagnetic cycle that occasionally is used in carbon cycle modeling (among others, Keir 1983).

The 5000 to 2000 year BP $\Delta^{14}C$ decline can be explained adequately by the earth's geomagnetic field change. However, anomalies exist for the intervals between 0 and 1500 years and 5000 and 8000 years BP, where $\Delta^{14}C$ values are more positive than magnetic field intensity alone would indicate (Sternberg and Damon 1983).

The assignment of the entire long-term ^{14}C increase to geomagnetic causes also appears to conflict with the information derived from the cosmogenically produced ^{10}Be in ice cores. For the Camp Century ice core ^{10}Be does not increase during the first and second millenium BC, suggesting the need for a reexamination of the relationship between geomagnetic variations and radioisotope production (Beer et al. 1984*a*).

The entire $\Delta^{14}C$ change tied to long-term changes in carbon reservoir parameters has to be small if the pre-4000 year BP geomagnetic field reduction indeed causes $\Delta^{14}C$ to be 10% higher during the last glacial. In that case, the 7% reduction in $\Delta^{14}C$ caused by the oceanic ^{12}C concentration change has to be approximately cancelled by the increases from (1) lowering of the atmospheric CO_2 concentration ($\Delta^{14}C$ change about 2.5%) and (2) increased oceanic stratification (plus a reduction in the rates of North Atlantic deepwater formation and upwelling). Note that the Fig. 6.1 interpretation of the varve record does not support a +10% $\Delta^{14}C$ increase centered around 10,500 years BP.

Fine-Structure of $\Delta^{14}C$

The atmospheric ^{14}C variations lasting a few hundred years or less (Fig. 6.2) that are superimposed on the long-term trend are tied to solar (helimagnetic) modulation of the cosmic-ray flux in the upper atmosphere (Stuiver and Quay 1980). The modulation by the sun of the cosmic-ray flux (and its associated atmospheric ^{14}C production) occurs through

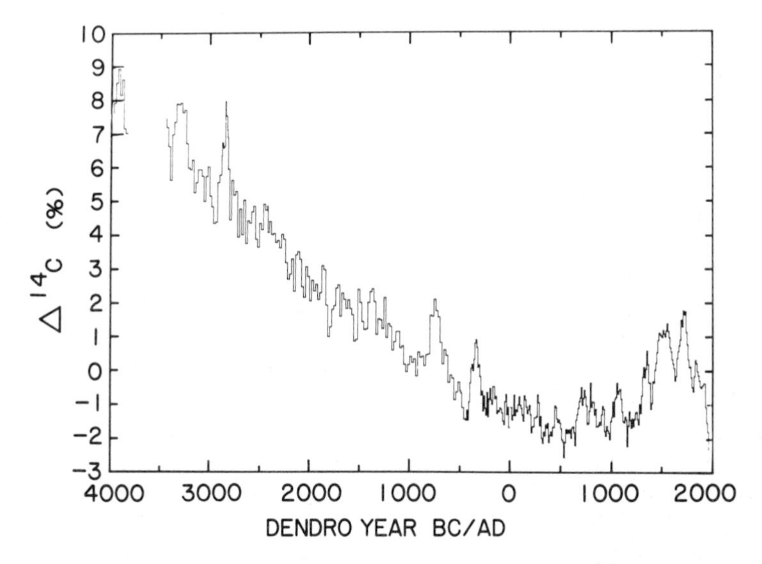

FIGURE 6.2. Detailed $\Delta^{14}C$ record of the last 6000 years. The pre-500 BC data are from Belfast Irish oak measurements (Pearson et al. 1983). The post-500 BC data of the Seattle Quaternary Isotope Laboratory are for trees of the western United States (Stuiver 1982).

changes in the magnetic shielding properties of the solar wind. The solar wind plasma near the earth can be considered an extension of the solar corona, and changes in solar wind properties reflect coronal change. With sunspot numbers being the most obvious indicator of convection changes near the sun's surface, it is no surprise that the observed atmospheric ^{14}C production of the last 34 years correlates very strongly with them (O'Brien 1979). The solar wind magnetic property changes are such that a higher ^{14}C production rate is tied to lower sunspot numbers.

On removal of the long-term trend shown in Fig. 6.2, the past ^{14}C production rate changes, Q, can be calculated through carbon-reservoir modeling. The calculated Q changes of the 1640 to 1860 AD interval are in excellent agreement with the Q changes derived from the historical sunspot record. The high Q values for the 1650 to 1710 AD interval, which is also the Maunder Minimum in sunspot numbers (Eddy 1976), have recently been confirmed by the high levels of the cosmogenically produced ^{10}Be isotope in ice cores (Beer et al. 1984*b*).

A paleorecord of sunspot numbers can be derived from the ^{14}C changes by assuming that solar modulation is also responsible for the pre-AD 1640 fine structure of the ^{14}C record (Stuiver and Quay 1980). The extended record of solar change derived from tree-ring ^{14}C values can also be used for an evaluation of sun-weather relationships (Stuiver 1980; Sonnett and Suess 1984). A spectral analysis of the ^{14}C data yields periodicities near 45,120, and 200 years (Stuiver 1980; Neftel et al. 1981).

Atmospheric ^{14}C and Fossil Fuel Combustion

The natural ^{14}C variability has to be taken into account for an accurate understanding of the atmospheric ^{14}C dilution (Suess effect) because of fossil fuel-produced CO_2. For trees in the state of Washington, the reduction in the atmospheric ^{14}C level was 20.0 ± 1.2‰ between 1860 and 1950 (Stuiver and Quay 1981). This reduction, which is represented by the last steep decline in the ^{14}C level in the curve in Fig. 6.2, is of a magnitude similar to that of the natural variability.

Twentieth century natural ^{14}C levels can be calculated through carbon-reservoir modeling that takes into account solar modulation. The calculated $\Delta^{14}C$ values are reduced when the fossil fuel CO_2 flux is added to the model. This approach was used by Stuiver and Quay (1981), who used a box-diffusion model (Oeschger et al. 1975) with an oceanic eddy diffusion coefficient of 3 $cm^2 \cdot s^{-1}$, a CO_2 atmosphere-ocean gas exchange rate of 21 mol $m^{-2} \cdot y^{-1}$, and a biospheric residence time of 60 years.

The model calculations yield a fossil fuel-induced atmospheric $\Delta^{14}C$ lowering of 15.7‰ between 1860 and 1950. Solar modulation lowers $\Delta^{14}C$ by an additional 4.8‰, whereas a continuation of the long-term trend would increase $\Delta^{14}C$ values by 2.1‰. The total calculated −18.4‰ change is in reasonably good agreement with the −20.0 ± 1.2‰ observed in trees of the Pacific Northwest.

The above measured change represents the change in Northern Pacific Ocean air. This change is not necessarily a global average. An increased Suess effect has been observed in Europe, where by 1940, the $\Delta^{14}C$ values were 2 to 7‰ lower in Dutch and German trees (de Jong and Mook 1982). Southern Hemispheric trees from Chile and Tasmania showed a $\Delta^{14}C$ decline of about 17‰ between 1860 and 1950 (M. Stuiver, unpublished). The true global atmospheric ^{14}C change related to fossil fuel combustion cannot easily be established without a large number of additional measurements.

The $^{13}C/^{12}C$ Signal in Tree Rings

For the previously discussed ^{14}C investigations, the effect of in situ isotope fractionation was minimized by recalculating ^{14}C activities for a fixed $\delta^{13}C$ ratio of −25‰. The rationale of this approach is that $^{14}C/^{12}C$ isotope fractionation is twice the $^{13}C/^{12}C$ fractionation, and that differences in tree-ring $\delta^{13}C$ (δ_t) values are assumed to result entirely from differences in isotope fractionation.

For $^{13}C/^{12}C$ studies, the differences in average level of isotope fractionation are accounted for by considering only the δ_t variability around an average pre-1850 value ($\Delta\delta_t$). The $\Delta\delta_t$ signal is the result of changes in (1) isotope fractionation due to physiological processes, (2) local biospheric contributions to atmospheric CO_2, and (3) global $\delta^{13}C$ of atmospheric CO_2 (δ_a). Although factors 2 and 3 are neglected for ^{14}C normalization, a detailed understanding of factor 1 (isotope fractionation during photosynthesis and transport of photosynthetic products) is needed in ^{13}C research for a derivation of the δ_a signal. Success in the quest for a global δ_a record has been limited because the fractionation response of individual trees to environmental and physiological factors is complicated.

The influences of environmental variables (temperature, precipitation) and physiological change (juvenile effect) on δ_t have been discussed in detail by Francey and Farquhar (1982). When temperature and precipitation records are available, it is possible to investigate their effect on δ_t through partial correlations and regressions of δ_t with climatic variables (Tans and Mook 1980; Leavitt and Long 1983). Ring-area normalization also has been attempted (Stuiver et al. 1984). Although the overall influence of several factors on the δ_t record is fairly well understood (Farquhar 1980), it is much more difficult to predict why a specific tree for a certain time interval deviates by x per mil from the global δ_a record. Historical records of local temperature and precipitation or nutrient supply are either absent or too short. Regional paleoclimatic information is generally derived from tree-ring thickness change and thus is not a quantity determined independently of the growth history of the tree. This limits attempts to correct long-term records for isotope fractionation to δ_t—ring area (or ring-indices) correlations.

For the reconstruction of δ_a from δ_t, it is generally assumed that an atmospheric CO_2 concentration (c_a) change does not affect the tree's isotopic fractionation. The $\delta^{13}C$ values of tomato plants indeed are not very sensitive to c_a change (Park and Epstein 1961; Vogel 1980). However, these results do not exclude possible δ_t changes of a few tenths per mil for c_a changes comparable to those experienced since 1850. Changes of a few tenths per mil are important, because the entire post-1850 δ_a change is <2 per mil.

Large-scale combustion of fossil fuel has appreciably influenced c_a and δ_a since 1850. Biospheric CO_2 release (among others, through deforestation) also played a role during this period. The latter may have influenced, to a limited extent, δ_a values prior to 1850 as well. Although a record entirely free of man's influence may be difficult to obtain, we classify, as a first approximation, the pre-1850 δ_a and δ_t values as "natural." The magnitude of natural δ_t variability is needed for a proper evaluation of the post-1850 anthropogenic influence on δ_t.

Biospheric Fluxes and Atmospheric CO_2 Levels

The $\Delta\delta_t$ records of six Pacific Coast trees, which are described in detail in Stuiver et al. (1984), are presented in Fig. 6.3. Ring-area (r_a) normalization was applied. This technique first investigates whether a significant δ_t to r_a correlation exists, whereupon the δ_t values are recalculated for a constant ring area. Three of the records in Fig. 6.3 [(2), (3), and (5)] were

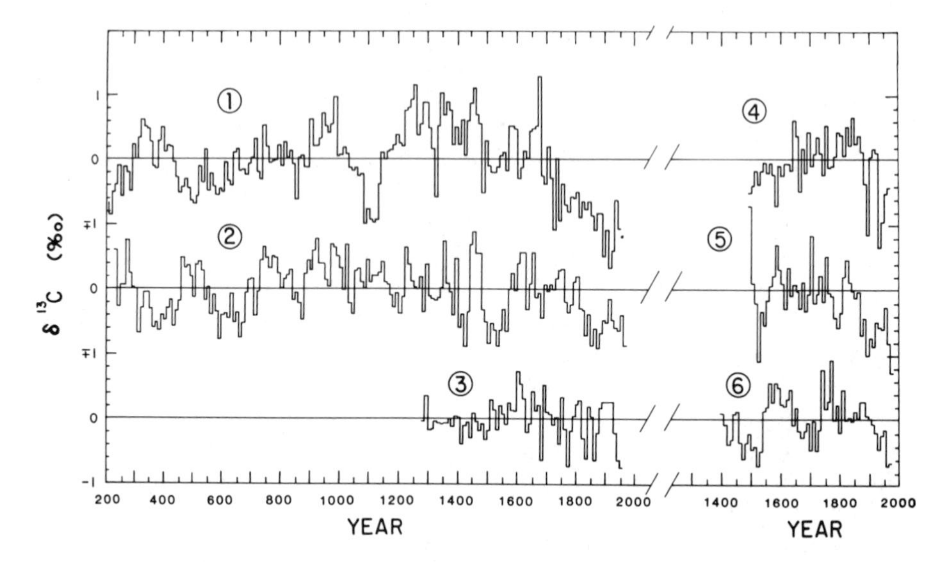

FIGURE 6.3. Records of $\delta^{13}C$ (δ_t relative to the pre-1850 average $\delta^{13}C$ level) of (1) Inyo bristlecone pine (37°N), (2) National park sequoia (37°N), (3) Whittaker sequoia (37°N), (4) Coos Bay fir (43°N), (5) Bjorka spruce (57°N), and (6) Valdivia Alerce (40°S). (Source: Stuiver et al. 1984.)

corrected in this manner. The other δ_t records lacked a significant δ_t to r_a relationship.

Appreciable pre-1850 $\Delta\delta_t$ variability (see Fig. 6.4 average) exists. Thus, a portion of the 19th and 20th century $\Delta\delta_t$ signal certainly could be of natural origin.

The controversial aspects of reconstructing δ_a from $\Delta\delta_t$ involve the separation of locally produced isotope signals associated with individual trees from the global $\delta^{13}C$ record. Although the normalization on constant r_a eliminates some portion of the locally induced δ_t variability, the record (see Fig. 6.4) surely contains a residual component not related to a global δ_a change.

The average time history of the $\Delta\delta_t$ change, when interpreted as a δ_a change and when used in conjunction with a carbon-reservoir model (Oeschger et al. 1975), results in the determination of the integrated biospheric carbon flux to the atmosphere. In these calculations, it is assumed that the δ_a change results from fossil fuel combustion and changes in the size of the biospheric carbon (with $\delta^{13}C = -25‰$) reservoir. Thus, the calculated atmospheric CO_2 concentration change prior to 1850 is entirely of biospheric origin. Changes in the CO_2 level related to climate-induced oceanic variability are not part of the calculation procedure.

For calculating atmospheric CO_2 levels, the initial atmospheric CO_2 concentration at the start of a time series is adjusted in such a manner

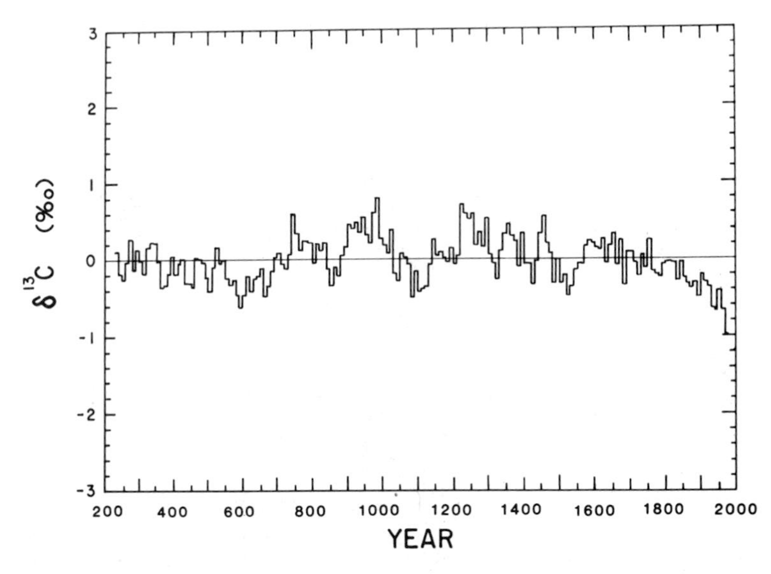

FIGURE 6.4. Average ring-area normalized $\delta^{13}C$ change (relative to the pre-1850 average) of the trees listed in Figure 6.3. (Source: Stuiver et al. 1984.)

that the calculated value for 1965 matches the measured 320-ppm value of Keeling et al. (1982).

The deconvolution of the isotope record (see Fig. 6.4) yields the biospheric carbon fluxes (gigatons per year) and atmospheric CO_2 levels plotted in Fig. 6.5. The calculated preindustrial values (235 to 1850 AD) vary between 240 and 310 ppm and average 276 ppm. The integrated biospheric carbon release (in gigatons) is plotted in Fig. 6.6. The cumulative amount of biospheric carbon of 150 ± 100 Gt released between 1600 and 1975 is similar in magnitude to that released by fossil fuel combustion by 1975 (135 Gt). Houghton et al. (1983) estimated a biospheric release between 1860 and 1975 of 126 to 213 Gt. Deconvolution of the δ_t record yields a net biospheric flux of only 40 ± 30 Gt over the same interval. These conflicting values can be reconciled if CO_2 fertilization has increased yearly rates of carbon transfer to the biosphere (growth factor, β, approximately equal to 0.13).

Freyer and Belacy (1983) arrive at a composite δ_t record that yields an appreciably larger integrated biospheric flux of 345 Gt between 1860 and 1975 (see Fig. 6.6) when using the same calculation procedures. Their 1983 composite δ_t record, with a relatively steep decline during the 19th and 20th centuries, has recently been modified by Freyer (unpublished) by adding δ_t profiles of nine Arizona Pinyon pines. These profiles, measured by A. Long and S. Leavitt of the University of Arizona, show a δ_t decline between 1700 and 1970 similar to the Fig. 6.4 curve and the modified Freyer curve has a reduced 19th and 20th century δ_t decline and

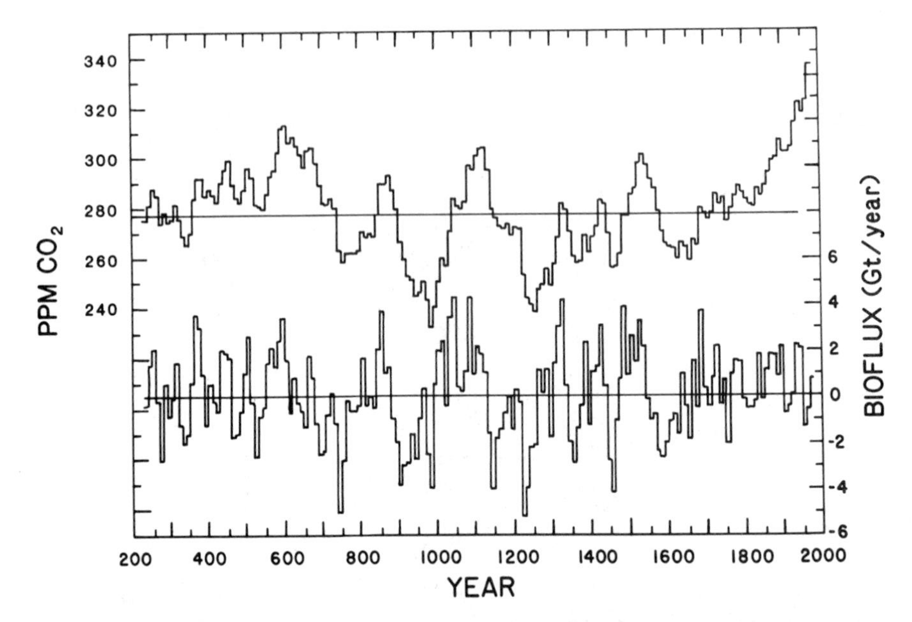

FIGURE 6.5. Biospheric fluxes (gigatons per year) and atmospheric CO_2 levels calculated from the Fig. 6.3 $\Delta\delta_t$ record. The calculation of the atmospheric CO_2 level also includes the addition of CO_2 through fossil fuel combustion. (Source: Stuiver et al. 1984.)

should yield an integrated biospheric flux less than the 345 Gt calculated above.

The calculated pre-1850 atmospheric CO_2 level averages 276 ppm (Fig. 6.5). This value is in excellent agreement with recent determinations of preindustrial CO_2 levels of gases in antarctic ice cores (Friedli et al. 1984, Oeschger and Stauffer, *this volume*).

A lower value of 257 ppm for the year 1860 is derived from the Houghton et al. (1983) biospheric release pattern, after adding fossil fuel CO_2 fluxes to the model calculations. Similarly, a low value of 230 ppm around 1800 results from the Freyer and Belacy (1983) data. Using the same data, Peng et al. (1983) calculated a slightly larger CO_2 concentration of 242 ppm. The difference in calculated values, 230 versus 242 ppm, results from modeling differences, i.e., direct deconvolution of the measured $\delta^{13}C$ record versus curve-fitting, and the use of a changing biospheric reservoir versus a constant reservoir.

The δ_t records presented here were all measured at the Quaternary Isotope Laboratory of the University of Washington. Although reasonable preindustrial atmospheric CO_2 levels were obtained, the δ_t record in Fig. 6.4 is not necessarily fully identical to the global δ_a record. A wide diversity exists among δ_t tree-ring records (see among others, Francey 1981). An unknown portion of the model-calculated atmospheric CO_2 variability

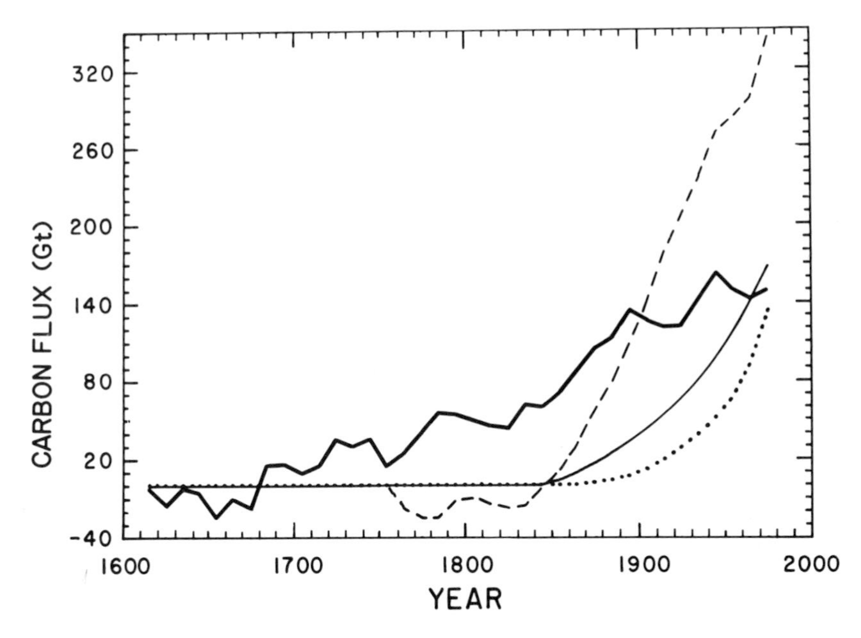

FIGURE 6.6. Integrated biospheric carbon flux calculated for the Fig. 6.4 isotopic record (*heavy solid line*), the Freyer and Belacy (1983) δ_t record (*dashed line*), the Houghton et al. (1983) estimates of biospheric CO_2 release (*light solid line*), and cumulative fossil fuel CO_2 flux only (*dotted line*). (Source: Stuiver et al. 1984.)

(standard deviation 16 ppm for pre-1850 values) should be attributed to variations in the δ_t trend caused by variable isotope fractionation by the trees. Assuming that the entire variability is related to this physiological fractionation effect leads to a range in the calculated pre-1850 atmospheric CO_2 level of 260 to 290 ppm.

By 1980, δ_t (Fig. 6.4) is 1.0‰ below the pre-1850 average value. The $\delta^{13}C$ values of CO_2 separated from air bubbles in Antarctic ice core samples exhibit a similar decline, i.e., air bubbles 400 to 700 years old are 1.1‰ higher in $\delta^{13}C$ than 1980 air (Friedli et al. 1984). The measured CO_2 content of these ice samples (284 ± 5 ppm) also falls within the 260 to 290 ppm range calculated from tree-ring data. Thus, the δ_t isotopic approach applied to Pacific coastal trees yields results fully compatible with those derived from ice core data.

Acknowledgments

The ^{14}C research of the Quaternary Isotope Laboratory reported here was supported by the Climate Dynamics Program of the National Science Foundation under grant ATM 8022240. The ^{13}C research was supported by the U.S. Department of Energy Carbon Dioxide Program. This paper was written during a sabbatical leave at K. O. Munnich's Institut of Umweltphysik at Heidelberg, Germany. The author's leave was made possible by a senior scientist award of the Alexander von Humboldt Foundation.

References

Baillie, M. G. L., J. R. Pilcher, and G. W. Pearson. 1983. Dendrochronology at Belfast as a background to high-prevision calibration. Radiocarbon 25:171–178.

Barbetti, M. 1980. Geomagnetic strength over the last 50,000 years and changes in atmospheric ^{14}C concentration: emerging trends. Radiocarbon 22:192–199.

Barnola, J. M., D. Raynaud, A. Neftel, and H. Oeschger. 1983. Comparison of CO_2 measurements by two laboratories on air from bubbles in polar ice. Nature 303:410–413.

Barton, C. E. and R. T. Merrill. 1983. Archaeo- and paleosecular variation, and long term asymmetries of the geomagnetic field. Rev. Geophys. Space Phys. 21:603–614.

Becker, B. 1983. The long-term radiocarbon trend of the absolute German oak tree-ring chronology, 2800-800 BC. Radiocarbon 25:197–293.

Beer, J., M. Andree, H. Oeschger, G. Bonani, H. J. Hofmann, E. Morenzoni, M. Nessi, M. Suter, H. Wölfli, R. Finkel, and C. Langway, Jr. 1984*a*. Abstract, Third International Symposium on Accelerator Mass Spectrometry, Zurich, Switzerland, April 10–13, 1984.

Beer, J., U. Siegenthaler, H. Oeschger, M. Andree, G. Bonani, M. Suter, W. Wölfli, R. C. Finkel, and C. C. Langway. 1984*b*. Temporal variations in the ^{10}Be concentration levels found in the Dye 3 ice core, Greenland. Ann. Glaciol. 5:16–17.

Broecker, W. S. 1982. Ocean chemistry during glacial time. Geochim. Cosmochim. Acta 46:1689–1750.

Bruns, M., M. Rhein, T. W. Linick, and H. E. Suess. 1983. The atmospheric ^{14}C level in the 7th millenium BC. PACT 8:511–516.

Cain, W. F. and H. E. Suess. 1976. Carbon-14 in tree rings. J. Geophys. Res. 81:3688–3694.

de Jong, A. F. M. and W. G. Mook. 1982. An anomalous Suess effect above Europe. Nature 298:641–644.

Eddy, J. A. 1976. The Maunder Minimum. Science 192:1189–1202.

Farquhar, G. 1980. Carbon isotope discrimination by plants. In G. I. Pearman (ed.), Carbon Dioxide and Climate: Australian Research, pp. 195–110. Australian Academy of Science, Canberra.

Ferguson, C. W. and D. A. Graybill. 1983. Dendrochronology of bristlecone pine: a progress report. Radiocarbon 25:287–288.

Francey, R. J. 1981. Tasmanian tree rings belie suggested anthropogenic $^{13}C/^{12}C$ trends. Nature 290:232–235.

Francey, R. J. and G. D. Farquhar. 1982. An explanation of the $^{13}C/^{12}C$ variations in tree rings. Nature 297:28–31.

Freyer, H. D. and N. Belacy. 1983. $^{13}C/^{12}C$ records in northern hemispheric trees during the past 500 years, anthropogenic impact and climatic superpositions. J. Geophys. Res. 88:6844–6852.

Friedli, H., E. Moor, H. Oeschger, U. Siegenthaler, and B. Stauffer. 1984. $^{13}C/^{12}C$ ratios in CO_2 extracted from Antarctic ice. Geophys. Res. Lett. 11:1145–1148.

Fromm, E. 1970. An estimation of errors in the Swedish varve chronology. In I. U. Olsson (ed.), Radiocarbon Variations and Absolute Chronology. Proceedings of the 12th Nobel Symposium, pp. 163–172. Wiley Interscience, New York.

Houghton, R. A., J. E. Hobbie, J. M. Melillo, B. Moore, B. J. Peterson, G. R. Shaver, and G. M. Woodwell. 1983. Changes in the carbon content of terrestrial biota and soils between 1860 and 1980: a net release of CO_2 to the atmosphere. Ecol. Monogr. 53(3):235–262.

In Che Yang, A. and A. Fairhall. 1972. Variations of natural radiocarbon during the last 11 millenia and geophysical methods for producing them. R. Soc. N. Z. Bull. A:44–57.

Keeling, C. D., R. B. Bacastow, and T. P. Whorf. 1982. Measurements of the concentration of carbon dioxide at Mauna Loa Observatory, Hawaii. In W. C. Clark (ed.), Carbon Dioxide Review: 1982, pp. 377–385. Oxford University Press, New York.

Keir, R. S. 1983. Reduction of the thermohaline circulation during deglaciation: the effect on atmospheric radiocarbon and CO_2. Earth Planet. Sci. Lett. 64:445–456.

Kruse, H. H., T. W. Linick, H. E. Suess, and B. Becker. 1980. Computer-matched radiocarbon dates of floating tree-ring series. Radiocarbon 22:260–266.

Lal, D. and V. S. Venkatavaradan. 1970. Analysis of the causes of ^{14}C variations in the atmosphere. In I. U. Olsson (ed.), Radiocarbon Variations and Absolute Chronology. Proceedings of the 12th Nobel Symposium, pp. 549–569. Wiley Interscience, New York.

Leavitt, S. W. and A. Long. 1983. An atmospheric $^{13}C/^{12}C$ reconstruction generated through removal of climatic effects from tree-ring $^{13}C/^{12}C$ measurements. Tellus 35B:92–102.

Neftel, A., H. Oeschger, and H. E. Suess. 1981. Secular non-random variations of cosmogenic carbon-14 in the terrestrial atmosphere. Earth Planet. Sci. Lett. 56:127–147.

O'Brien, K. 1979. Secular variations in the production of cosmogenic isotopes in the Earth's atmosphere. J. Geophys. Res. 78:423–431.

Oeschger, H., U. Siegenthaler, U. Schotterer, and A. Gugelmann. 1975. A box diffusion model to study the carbon dioxide exchange in nature. Tellus 27:168–192.

Park, R. and S. Epstein. 1961. Metabolic fractionation of ^{13}C and ^{12}C in plants. Plant Physiol. 36:133–138.

Pearson, G. W., J. R. Pilcher, and M. G. L. Baillie. 1983. High-prevision ^{14}C measurements of Irish oaks to show the natural ^{14}C variations of the AD time period. Radiocarbon 25:179–186.

Peng, T. H., W. S. Broecker, H. D. Freyer and S. Trumbore. 1983. A deconvolution of the tree-ring based $\delta^{13}C$ record. J. Geophys. Res. 88:3609–3620.

Peng, T. H., J. C. Goddard, and W. S. Broecker. 1978. A direct comparison of ^{14}C and ^{230}Th ages at Searles Lake, California. Quat. Res. 9:310–320.

Sarmiento, J. L. and J. R. Toggweiler. 1984. A new model for the role of the oceans in determining atmospheric $^{P}CO_2$. Nature 308:621–624.

Siegenthaler, U. and Th. Wenk. 1984. Rapid atmospheric CO_2 variations and ocean circulation. Nature 308:624–626.

Sonnett, C. P. and H. E. Suess. 1984. Correlation of bristlecone pine ring widths with atmospheric ^{14}C variations: a climate-sun relation. Nature 307:141–143.

Sternberg, R. S. and P. E. Damon. 1983. Atmospheric radiocarbon: Implications for the geomagneticdipole moment. Radiocarbon 25:239–248.
Stuiver, M. 1970*a*. Evidence for the variation of atmospheric ^{14}C content in the late Quaternary. In K. K. Turekian (ed.), Late Cenozoic Glacial Ages, pp. 57–70. Yale University Press, New Haven, Connecticut.
Stuiver, M. 1970*b*. Long-term C-14 variations. In I. U. Olsson (ed.), Radiocarbon Variations and Absolute Chronology. Proceedings of the 12th Nobel Symposium, pp. 197–214. Wiley Interscience, New York.
Stuiver, M. 1978. Radiocarbon timescale tested against magnetic and other dating methods. Nature 273:271–274.
Stuiver, M. 1980. Solar variability and climatic change during the current millenium. Nature 286:868–871.
Stuiver, M. 1982. A high-prevision calibration of the AD radiocarbon timescale. Radiocarbon 24:1–26.
Stuiver, M. 1983. Statistics and the AD record of climatic and carbon isotopic change. Radiocarbon 25:219–228.
Stuiver, M., R. L. Burk, and P. D. Quay. 1984. $^{13}C/^{12}C$ ratios in tree rings and the transfer of biospheric carbon to the atmosphere. J. Geophys. Res. (in press).
Stuiver, M. and H. A. Polach. 1981. Discussion: Reporting of ^{14}C data. Radiocarbon 19:355–363.
Stuiver, M. and P. D. Quay. 1980. Changes in atmospheric ^{14}C attributed to variable Sun. Science 207:11–19.
Stuiver, M. and P. D. Quay. 1981. Atmospheric ^{14}C changes resulting from fossil fuel CO_2 release and cosmic ray flux variability. Earth Planet. Sci. Lett. 53:340–362.
Stuiver, M. and G. I. Smith. 1979. Radiocarbon ages of stratigraphic units. USGS Professional Paper 1043, pp. 69–78.
Tans, P. P. and W. G. Mook. 1980. Past atmospheric CO_2 levels and the $^{13}C/^{12}C$ ratios in tree rings. Tellus 32:268–283.
Tauber, H. 1970. The Scandinavian varve chronology and ^{14}C dating. In I. U. Olsson (ed.), Radiocarbon Variations and Absolute Chronology. Proceedings of the 12th Nobel Symposium, pp. 173–196. Wiley Interscience, New York.
Vogel, J. C. 1970. C-14 trends before 6000 B.P. In I. U. Olsson (ed.), Radiocarbon Variations and Absolute Chronology. Proceedings of the 12th Nobel Symposium, pp. 313–325. Wiley Interscience, New York.
Vogel, J. C. 1980. Fractionation of the carbon isotopes during photosynthesis. In Sitzungsberichte der Heidelberger Akademie der Wissenschaften, Mathematisch—Naturwissenschaftliche Klasse, pp. 111–135 . Springer-Verlag, Berlin.
Vogel, J. C. 1983. ^{14}C Variations during the Upper Pleistocene. Radiocarbon 25:213–218.

7
Interpretation of the Northern Hemispheric Record of $^{13}C/^{12}C$ Trends of Atmospheric CO_2 in Tree Rings

HANS D. FREYER

The present long-term increase of atmospheric CO_2 has been accompanied by changes in the isotopic composition of its carbon atoms. Both the ratios of stable $^{13}CO_2/^{12}CO_2$ and radioactive $^{14}CO_2/^{12}CO_2$ have decreased as a result of anthropogenic activity. The decrease of the latter, until the beginning of nuclear testing caused an increase in the mid-1950s, is almost exclusively due to dilution of atmospheric CO_2 resulting from fossil fuel burning, from which the ^{14}C isotope had already decayed. The decrease of the $^{13}C/^{12}C$ ratio is a better measure of total anthropogenic activity because it reflects not only fossil fuel burning but also CO_2 released as a result of deforestation and agricultural manipulation of the soil. The CO_2 from both of these sources is marked by a $\delta^{13}C$ deficit of about 18 to 20‰ (PDB scale; e.g., Freyer 1979*c*).

The present $^{13}C/^{12}C$ ratio of atmospheric CO_2 is about −7.5‰ (with respect to the PDB standard). Measurements by Keeling et al. (1979, 1980) and Mook et al. (1981) show that the ratio has decreased from −6.69‰ in 1956 to −7.34‰ in 1978 and to −7.46‰ in 1980. The preindustrial ratio is not yet available; however, since carbon in wood is assimilated from atmospheric CO_2, a historical record of the ratio could be stored in tree rings.

Existing measurements leave no doubt that the $^{13}C/^{12}C$ ratio in the cellulose fraction of modern wood has indeed decreased significantly. However, an uncertainty exists in the interpretation of decreasing $^{13}C/^{12}C$ tree-ring data, mainly because the atmospheric $^{13}C/^{12}C$ information is often masked by scatter which is thought to depend on local environmental and climatological factors. This chapter summarizes some of these effects and, in addition, compares all published tree-ring data for the Northern Hemisphere. From these data a mean trend has been constructed, which is interpreted on the basis of a model published recently by Peng et al. (1983).

Effects Influencing the Atmospheric $^{13}C/^{12}C$ Signal in Modern Wood

It has been pointed out that the expected $^{13}C/^{12}C$ decrease caused by increasing atmospheric CO_2 concentrations may be superimposed with other effects (Freyer 1980). Some of these effects are described below. An interpretation of $^{13}C/^{12}C$ tree-ring variations with a model has been given recently by Francey and Farquhar (1982).

Choice of the Wood Constituent

The chemical form of the carbon analyzed in wood is important. For example, a network of resin channels permeates the wood tissue and contains soluble organic material that cannot be correlated to the age of the wood structure. Furthermore, the components that form the wood—cellulose, hemicellulose, and lignin—differ in their isotopic content (Freyer 1979*a*; Mazany et al. 1980), and relative abundances of the components may not be constant over the entire age of wood (Nikitin 1966). Therefore, the preparation of one pure component is necessary, and in most tree-ring records, cellulose was analyzed. The chemical preparation of cellulose, if done carefully, does not introduce artificial isotope fractionation, and most authors report a reproducibility of measurements, including sample preparation of $\pm 0.1‰$ ($1\ \sigma$). In a few records, whole wood (Farmer and Baxter 1974; Galimov 1976; Farmer 1979) or acid-alkali-acid pretreated wood (Tans and Mook 1980) was analyzed.

Inner-Ring Variability

The carbon isotope data within the circumference of a single ring closely agree during the juvenile stage of a tree (Freyer and Belacy 1983), whereas those in the older wood show a variability of between 0.1 and 1‰ (Freyer and Wiesberg 1975; Freyer 1979*a*; Harkness and Miller 1980; Mazany et al. 1980; Freyer and Belacy 1983; Long and Leavitt 1983), but sometimes as large as 4 to 5‰ (Tans and Mook 1980). This later differentiation in the circumference of a ring is probably associated with the growth and development of the crown, indicating different environmental conditions around the tree. Some causes of this differentiation have been discussed by Francey and Farquhar (1982). Changes of the carbon isotope content with height, especially in the same wood fiber, seem to be negligible (Freyer and Wiesberg 1975; Tans and Mook 1980), and no consistent longitudinal gradients have been detected (Long and Leavitt 1983). To avoid inner-ring variations, Tans and Mook (1980), Leavitt and Long (1983), and Long and Leavitt (1983) have analyzed wood material from the full circumference of the ring. In the other studies, mainly only one radial cut was analyzed for each tree, because nearly similar relative $^{13}C/^{12}C$ changes in different radial cuts have been found (Freyer 1979*a*; Mazany et al. 1980), which deviate from the mean trend for the tree at most by 20% (Freyer and Belacy 1983). Long and Leavitt (1983) have concluded from their careful studies that measurements of two radial cuts opposite in direction may provide a representative trend of the tree.

Juvenile Stage; Forest Trees

During the juvenile stage of trees, increasing trends in $^{13}C/^{12}C$ data have been observed in several studies, which were explained as a result of the partial uptake of ^{13}C-depleted soil respiration CO_2 (Craig 1954; Freyer 1979*a*). This effect lasts for about 50 years at most (Freyer and Belacy 1983) and disappears at a certain height or age of the tree, indicating that the $^{13}C/^{12}C$ record of an older tree is more representative for the free atmospheric $^{13}C/^{12}C$ signal. The same effect has also been observed in forest trees after clearing of neighboring trees (Freyer 1979*b*). Similar canopy effects could result from the death or regrowth of trees in a forest stand adjoining the sample tree. It has been found by Freyer and Belacy (1983) that sudden jumps and nonsystematic variations of $^{13}C/^{12}C$ data with a set of forest trees occur, but no correlation is observed with respect to time. Therefore, it has to be concluded that forest tree records are not usable for either $^{13}C/^{12}C$ changes of free atmospheric CO_2 or climatic variations. These effects should be minimized by proper selection of free-standing trees. For calculating average tree-ring records, data from the juvenile stage either were not considered or had already been excluded when originally published.

Pollution Effect

Tree-ring studies of $^{13}C/^{12}C$ have shown that heavy air pollution (e.g., from SO_2) lowers the photosynthetic carbon isotope fractionation (Freyer 1979*b*). This effect was explained by a known inhibition of photosynthesis in plants exposed to air pollution; however, the mechanism of this inhibition is still unproven. It could be caused by the reduction of stomatal conductance, which, according to the model of Francey and Farquhar (1982), would cause increasing $^{13}C/^{12}C$ data in the plant material. However, other biochemical damage could also cause the effect. Similar effects have been observed by Leavitt and Long (1983) in trees from Arizona. Trees taken from open sites in rural localities showed a decreasing $^{13}C/^{12}C$ trend during the last 20 years, whereas a tree in the vicinity of a copper smelter showed an increasing trend approximately coincident with the opening of the smelter. Calcining activities releasing heavy CO_2 ($\delta^{13}C \sim 0‰$) could also counteract the decreasing trend. It has been pointed out (Freyer 1979*b*) that trends of $^{13}C/^{12}C$ data in trees during the 19th century similarly decrease, whereas the $^{13}C/^{12}C$ records of individual 20th century trees exhibit larger variations. This may partly reflect the increase in ambient pollution levels or in acidity of rain during this century, which could cause changes even where the trees had not been exposed to local pollution sources.

Climatic Effects

A summary on $^{13}C/^{12}C$ temperature effects in trees and other nonwoody plants has been given recently by Long (1982). In addition, other climatic effects have been reported by Freyer and Belacy (1983) and Leavitt and Long (1983). These effects may disturb the atmospheric $^{13}C/^{12}C$ signal in tree-ring records, especially during the 1920 to 1960 period, in which larger climatic variations have occurred. Freyer and Belacy have observed in trees from Northern Sweden that first differences of $^{13}C/^{12}C$ and climate data correlate significantly (i.e., differences between following 10-year blocks of data, which take into account the sign and magnitude of change in the $\delta^{13}C$ and climate series). A mean temperature effect of $+0.18‰$ $(\Delta\delta^{13}C)/\Delta T(°C)$ for autumn and a mean precipitation effect of $+ 0.019‰$ $(\Delta\delta^{13}C)/\Delta mm$ (mean monthly) for spring was found. Leavitt and Long have observed in trees from Arizona negative responses of $-0.27‰$ $(\delta^{13}C)/T(°C)$ and $-0.04‰$ $(\delta^{13}C)/mm$ for December temperature and precipitation, respectively. It has been shown in both studies that when applying climate corrections on the original $^{13}C/^{12}C$ data, unexplained variations of the tree-ring records were reduced and a smoother trend was obtained.

The response of $^{13}C/^{12}C$ data to temperature and precipitation is complex and even opposite in sign. A change in sign of the temperature effect is not unreasonable for both sites. Data for $^{13}C/^{12}C$ should be maximal at the temperature optimum for photosynthesis, as discussed in the model of Farquhar (1980) and Francey and Farquhar (1982). Therefore, in cooler climates (Northern Sweden), $^{13}C/^{12}C$ data should increase with increasing temperatures favoring optimal conditions for photosynthesis, whereas in warmer climates (Arizona), the reverse effect should result. A similar positive and negative correlation of temperature with growth is also observed in ring-width studies (Fritts 1976).

It is assumed, that to first order, the $^{13}C/^{12}C$ trend obtained from average tree-ring studies represents $^{13}C/^{12}C$ changes of atmospheric CO_2. However, some effects discussed above may be included in the trend. The choice of the wood constituent could not cause large variations, because cellulose was analyzed in most of the studies. Effects of inner-ring variability, although important within a single tree-ring record, should also be negligible, because data from about 65 trees have been used. Data from obvious forest trees were rejected, as well as data for evident juvenile effects in the trees. Some other canopy effects and effects due to ambient air pollution may be included in several records. All of the records, however, include climatic effects, since only uncorrected data have been used or have been published. Climatic effects should influence the $^{13}C/^{12}C$ data, especially during the 1920 to 1960 period.

Regional $\delta^{13}C$ Tree-Ring Trends and Mean Trend of the Northern Hemisphere

Handling of Data

All known published $^{13}C/^{12}C$ records for trees from the Northern Hemisphere were used for calculating a mean trend. The data were first reduced into 10-year block averages [data from Harkness and Miller (1980) were available as numerical values, those from other authors were taken from graphs]. Data for evident juvenile effects in some records were omitted, as described by Freyer and Belacy (1983). A normalization procedure was performed for constructing 13 different regional trends, some of which include measurements from various authors. The long-lasting record during the 1800 to 1979 period was used as a reference. The other records for the same region were included in the regional trend by center-of-gravity coordinates. Data measured before 1800 were added without normalization. The same procedure was performed for constructing the mean trend of the Northern Hemisphere from the regional trends.

All data taken during 1800 through 1979 were tested for linear correlation and regression of individual records on their regional trend and for regional trends on the mean trend of the Northern Hemisphere.

Regional Trends

Figure 7.1 gives the constructed regional trends in relative units and shows the deviation of 10-year block data from the mean trend. The regional trends are described in detail below.

Record A combines $^{13}C/^{12}C$ data obtained on six free-standing Scots pines (*Pinus silvestris*), five of them from Northern Sweden, measured by Freyer and Belacy (1983), and one from Norway, measured by Harkness and Miller (1980). The record starts in 1530. The mean $^{13}C/^{12}C$ data within the preindustrial period over about 300 years do not show any trend. Then, from about 1850 up to the present, the $^{13}C/^{12}C$ data decrease, with a total of about 1.75‰. This exceeds the decrease obtained in the mean trend for the Northern Hemisphere by about 0.4‰. Record A seems to contain small $^{13}C/^{12}C$ fluctuations (especially near 1700 and during the interval of generally decreasing $^{13}C/^{12}C$ data from about 1920 to 1940). These fluctuations have been discussed in terms of climatic changes by Freyer and Belacy.

Record B is composed of $^{13}C/^{12}C$ data on two Bristlecone pines (*Pinus longaeva*) from California, measured by Grinsted et al. (1979) and Wilson (1978), and on two Douglas firs (*Pseudotsuga menziesii*) from Oregon and Washington, measured by Freyer and Belacy (1983) and Stuiver (1978). Data for $^{13}C/^{12}C$ during the preindustrial period from 1530 to 1749 are documented only by one tree measurement (Grinsted et al.), which shows an increase during the first half of the period and a decrease during the se-

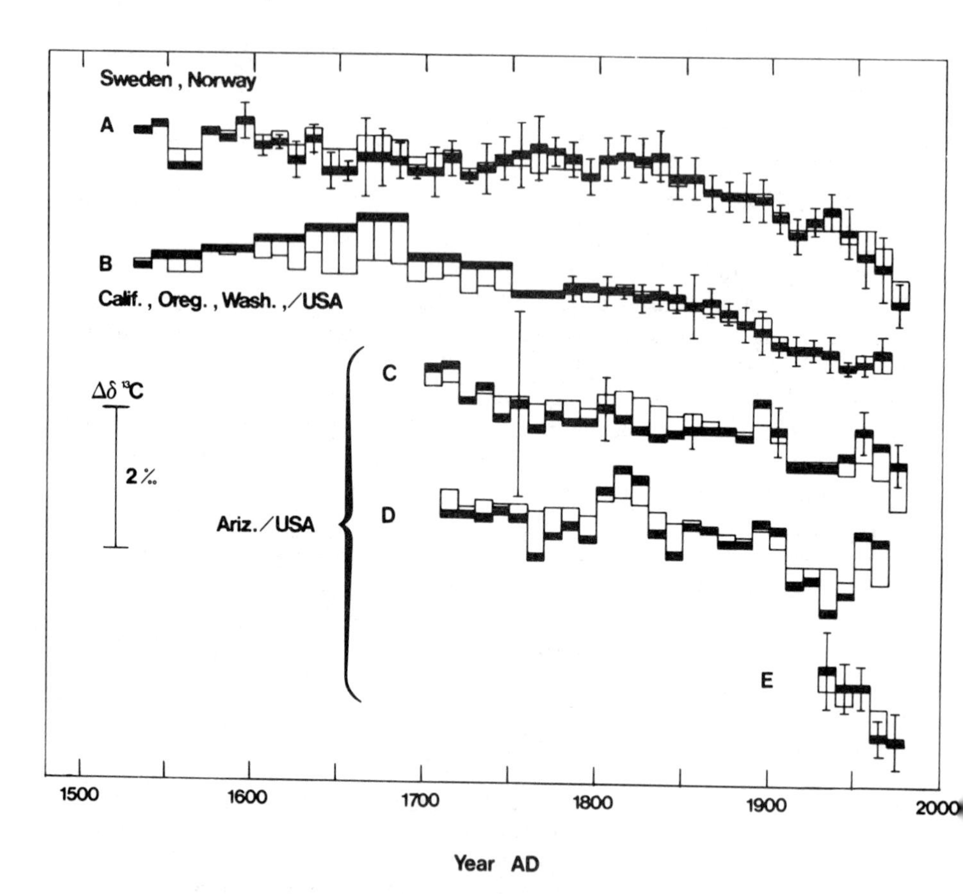

FIGURE 7.1. Regional $\Delta\delta^{13}C$ trends in tree rings at quoted time intervals. *Heavy marks* are mean regional data for each 10-year block. *Thin lines* in the regional records mark the data obtained for the mean trend of the Northern Hemisphere. Error bars represent one standard deviation (1σ) of scatter of individual from mean data. *Hatched marks* in records G and M represent full mean data without omission of those individual records that correlate with the regional trend at levels of significance $P < 0.9$. Record I correlates with the mean trend of the Northern Hemisphere at a level of significance $P < 0.9$. Data in record H without and with omission of individual records deviate by <0.1 ‰.

cond half, with a total variation of $^{13}C/^{12}C$ data of about ±0.5‰. For the period from 1750 to 1850, $^{13}C/^{12}C$ data obtained on three trees are very constant. During the industrial period up to 1950, record B shows decreasing $^{13}C/^{12}C$ data, with a total change of about 0.85‰, which closely agrees with the mean decrease obtained in the trend for the Northern Hemisphere. Finally, $^{13}C/^{12}C$ data in record B, having been completed in 1970, show a slight increase during the last 20 years.

Record C obtained from $^{13}C/^{12}C$ data on nine Pinyon pines from Arizona, measured by Long and Leavitt (1983), covers the period from 1700 up to

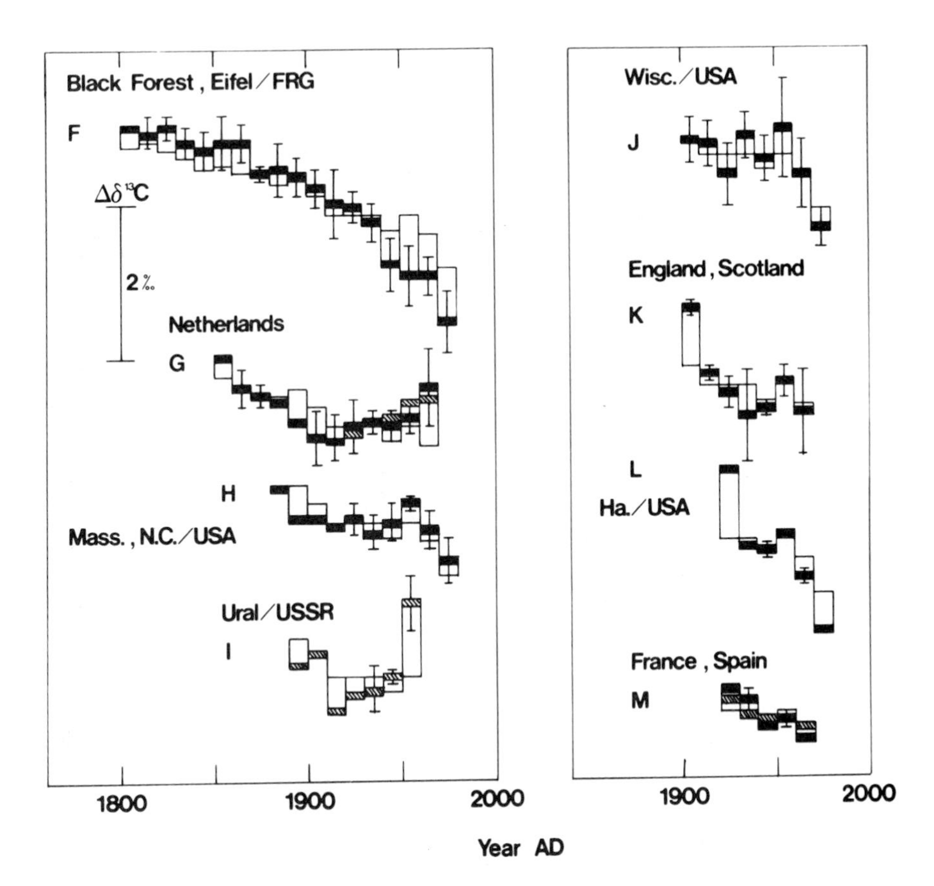

the present. During the 18th century, data for one to three trees are available; the complete data set for the nine trees starts in 1850. Record C data for $^{13}C/^{12}C$ show a decrease from 1700 to 1940 (by about 1.4‰), an increase from 1940 to 1960, and then a decrease. The total decrease from the first half of the 19th century up to present amounts to only about 0.6‰, which is less than half that observed in the mean trend for the Northern Hemisphere. Nevertheless, record C correlates with the mean trend at a level of significance $P > 0.9995$, although the coefficient of regression is as low as $b = 0.42$.

Record D is from $^{13}C/^{12}C$ data measured on one Ponderosa pine (*Pinus ponderosa*) from Arizona by Lerman and Long (1980), which covers the same time period as record C, with data missing for the first and last 10-year periods. Without regard to data from 1800 to 1850, record D looks similar to record C, and the total decrease from 1710 to 1940 amounts to about 1.3‰. Following 1940, record D data for $^{13}C/^{12}C$ increase in a way that no net change during industrial time is observed. Since the decrease from 1800 to 1940 in record D is similar (although steeper) to that in the mean trend for the Northern Hemisphere, and only data for the last 30 years deviate, high correlation between both records with a level of significance $P > 0.9995$ follows.

Record E consists of mean $^{13}C/^{12}C$ data on six Juniper trees from Arizona measured by Leavitt and Long (1983). This record covers the period from 1930 up to the present and obviously deviates from records C and D obtained during that period. The total decrease amounts to about 1.0‰, which exceeds the decrease obtained in the mean trend for the Northern Hemisphere by about 0.3‰. For the periods 1960 to 1969 and 1970 to 1979, a $^{13}C/^{12}C$ flattening in record E is observed, which is not expected for the $^{13}C/^{12}C$ change in atmospheric CO_2 during that period. However, the calculated change for 1956 to 1980 by a least-squares fit of the data agrees with the actual $^{13}C/^{12}C$ decrease observed in atmospheric CO_2.

Record F combines the $^{13}C/^{12}C$ data for one free-standing Scots pine (*Pinus silvestris*), measured in four radial cuts by Freyer and Belacy (1983), and another Scots pine, one chestnut tree (*Aesculus hippocastanum*), four oaks (*Quercus robur*), and two ash trees (*Fraxinus excelsior*), measured by Freyer (1979*a,b*), which had grown in the Black Forest and the Eifel/Germany region. The record starts with almost constant $^{13}C/^{12}C$ data from 1800 to 1870, followed by a steep decrease of about 2.3‰ up to the present. This decrease exceeds that obtained in the mean trend for the Northern Hemisphere by about 1.0‰ for the same period. This fact is reflected by a high coefficient of regression ($b = 1.64$) of record F on the mean trend.

Record G gives the $^{13}C/^{12}C$ data measured by Tans and Mook (1980) on three oaks (two *Quercus robur*, one *Quercus rubra*) and one beech tree (*Fagus sylvatica*), which had grown at a forest border in the Netherlands. The data for the youngest tree (beech tree) in this series correlate with the obtained mean regional record only to a level of significance $P > 0.8$. With omission of these data, record G for the remaining three trees shows a decrease of about 1.0‰ from 1850 to 1920, followed by an increase of about 0.6‰ thereafter up to 1970. The coefficient of regression of record G on the mean trend for the Northern Hemisphere is low ($b = 0.51$), with a lower level of significance for correlation ($P > 0.9$) between record G and the mean trend.

Record H combines the $^{13}C/^{12}C$ data measured by Farmer (1979) on an American elm from Massachusetts and data measured by Freyer (1979*a*) on four oaks (two *Quercus rubra*, two *Quercus alba*), two pines (*Pinus serotina* and *Pinus virginiana*), and one poplar (*Populus alba*) from North Carolina. Data for $^{13}C/^{12}C$ on four of the North Carolina trees do not correlate significantly ($P < 0.9$) with the mean regional trend and were omitted in calculating the final record H. The final record gives a decrease of $^{13}C/^{12}C$ data of about 0.9‰ from 1880 up to the present, which is about 0.2‰ less than that in the mean trend for the Northern Hemisphere.

Record I gives the $^{13}C/^{12}C$ data measured by Galimov (1976) on two spruces from the Ural/USSR. These data show a decrease of about 0.3‰

from 1890 to 1950 and a steep increase of about 1.0‰ during the last measured periods (1940 to 1949 and 1950 to 1959). Record I is not included in the data of the final mean trend for the Northern Hemisphere, as the level of significance for correlation between both records is only $P > 0.7$.

Record J combines the $^{13}C/^{12}C$ data on three Bur oaks (*Quercus macrocarpa*) from Wisconsin measured by Bender and Berge (1982), which covers the time period 1900 up to the present. This record gives a total decrease of $^{13}C/^{12}C$ data of about 1.15‰, which occurs mainly during the last three decades. The decrease exceeds that found in the mean trend for the Northern Hemisphere by about 0.2‰.

Record K gives the $^{13}C/^{12}C$ data measured by Farmer and Baxter (1974) on one oak (*Quercus robur*) and one larch (*Larix decidua*) from England and Scotland, respectively. During the first two decades of the record (1900 to 1909 and 1910 to 1919), a very steep decrease of about 0.8‰ is observed in the data. The decrease thereafter up to 1970 amounts to about 0.5‰, which slightly exceeds the decrease found in the mean trend for the Northern Hemisphere. Because of the first steep decrease of $^{13}C/^{12}C$ data, a high coefficient of regression of record K on the mean trend ($b = 2.45$) follows.

Record L is from the $^{13}C/^{12}C$ data given by Freyer and Belacy (1983) on two False sandalwood trees (*Myoporum sandwicense*) from Hawaii. Data for $^{13}C/^{12}C$ during the first two decades of the record (1920 to 1929 and 1930 to 1939) decrease very steeply by about 1.0‰. The decrease that followed up to the present gives an additional change of about 1.1‰, which exceeds that in the trend for the Northern Hemisphere by about 0.4‰. The coefficient of regression of record L on the mean trend, correspondingly, is high ($b = 2.19$).

Record M consists of $^{13}C/^{12}C$ data measured by Freyer (1979*a*) on four oaks, two *Quercus robur* and two *Quercus lusitanica* (the data for one oak were obtained with three measured radial cuts), and on two pines (*Pinus maritima*) from the French Atlantic coast and the Spanish Pyrenees. This record is the shortest in the regional series and covers a time period (1920 to 1970) for which, in general, larger fluctuations in the $^{13}C/^{12}C$ ratio are observed. The coefficients of correlation between the individual records and the mean regional trend largely disagree, and only two records correlate with the mean regional trend at levels of significance $P > 0.9$. Data for $^{13}C/^{12}C$ in the final record M obtained from these two trees decrease by about 0.65‰, whereas the decrease in the mean trend for the Northern Hemisphere is about 0.3‰ during the same time period.

Summary of Regional Trends

The coefficients for linear correlation and regression of individual tree-ring records on their regional mean trend are given in Table 7.1. Most of the individual records correlate with their regional mean at levels of significance $P > 0.9$. For records beginning before 1800, the level of significance is even better ($P > 0.9995$). Lower significance ($P < 0.9$) or no

TABLE 7.1. Coefficients for linear correlation r with level of significance P, and coefficients of regression b of each individual tree-ring record on the regional $\delta^{13}C$ trend for data during the 1800 to 1979 period.

Location/tree species	References*	Years	Coefficient of regression (b)	Correlation coefficient (r)	Level of significance (P)	With omission of data		
						b'	r'	P'
Sweden, Norway (Record A)								
Pinus silvestris (Pi 7)	(1)	1530–1979	1.10	0.89	>0.9995			
Pinus silvestris (Pi 8)	(1)	1590–1979	1.50	0.98	>0.9995			
Pinus silvestris (Pi 9)	(1)	1720–1979	0.66	0.95	>0.9995			
Pinus silvestris	(2)	1750–1969	1.13	0.85	>0.9995			
Pinus silvestris (Pi 10)	(1)	1780–1979	0.79	0.93	>0.9995			
Pinus silvestris (Pi 11)	(1)	1850–1979	0.76	0.89	>0.9995			
			1.00 ± 0.29†					
California, Oregon, Washington/USA (Record B)								
Pinus longaeva	(3)	1530–1969	1.26	0.98	>0.9995			
Pinus longaeva	(4)	1750–1959	1.18	0.95	>0.9995			
Pseudotsuga menziesii	(1)	1780–1969	0.49	0.71	>0.999			
Pseudotsuga menziesii	(5)	1850–1929	1.36	0.97	>0.9995			
			1.03 ± 0.35					
Arizona/USA (Record C)								
Pinyon pines (nine trees)	(6)	Variable (1700–1981)						
Arizona/USA (Record D)								
Pinus ponderosa	(7)	1710–1969						
Arizona/USA (Record E)								
Juniper trees (six trees)	(8)	1930–1979						

Black Forest, Eifel/FRG (Record F)								
Pinus silvestris (Pi 1)	(9)	1800–1939	1.10	0.97	>0.9995			
Pinus silvestris (Pi 12, rd.1a)	(1)	1810–1949	0.84	0.93	>0.9995			
Pinus silvestris (Pi 12, rd.1b)	(1)	1810–1949	0.96	0.93	>0.9995			
Pinus silvestris (Pi 12, rd.2)	(1)	1810–1949	0.73	0.81	>0.9995			
Pinus silvestris (Pi 12, rd.3)	(1)	1810–1949	1.13	0.89	>0.9995			
Aesculus hippocastanum (Ae 1)	(9, 10)	1890–1949	1.17	0.88	>0.975			
Quercus robur (Qu 14)	(9, 10)	1910–1949	2.44	0.88	>0.9			
Fraxinus excelsior (Fr 2)	(9, 10)	1910–1949	1.58	0.97	>0.975			
Quercus robur (Qu 10)	(9, 10)	1930–1975	1.31	0.86	>0.95			
Quercus robur (Qu 11)	(9, 10)	1930–1975	0.79	0.70	>0.9			
Quercus robur (Qu 12)	(9, 10)	1940–1975	2.00	0.94	>0.95			
Fraxinus excelsior (Fr 1)	(9, 10)	1940–1975	0.68	0.91	>0.95			
			1.09 ± 0.40					
Netherlands (Record G)								
Quercus robur	(11)	1850–1968	0.65	0.72	>0.995	0.69	0.77	>0.995
Quercus robur	(11)	1860–1969	0.81	0.77	>0.995	0.92	0.88	>0.9995
Quercus rubra	(11)	1900–1969	2.38	0.96	>0.9995	2.28	0.94	>0.9995
Fagus sylvatica	(11)	1923–1968	1.60	0.65	>0.8			
			1.18 ± 0.68			1.15 ± 0.64		
Massachusetts, North Carolina/USA (Record H)								
American elm	(12)	1880–1968	0.89	0.71	>0.975	0.88	0.74	>0.975
Pinus serotina (Pi 5)	(9)	1920–1975	0.84	0.75	>0.95	0.72	0.69	>0.9
Quercus rubra (Qu 8)	(9)	1920–1975	1.06	0.84	>0.975	1.05	0.90	>0.99
Quercus alba (Qu 9)	(9)	1930–1975	0.57	0.42	>0.75			
Quercus alba (Qu 7)	(9)	1930–1975	1.56	0.88	>0.975	1.47	0.89	>0.975
Quercus rubra (Qu 6)	(9)	1930–1975	0.32	0.47	>0.75			
Pinus virginiana (Pi 4)	(9)	1940–1975	0.62	0.61	>0.8			
Populus alba (Po 1)	(9)	1950–1975	2.19	0.95	>0.8			
			0.95 ± 0.48			1.00 ± 0.26		

TABLE 7.1. *(Continued)*

Location/tree species	References*	Years	Coefficient of regression (b)	Correlation coefficient (r)	Level of significance (P)	With omission of data		
						b'	r'	P'
Ural/USSR (Record I)								
Spruce	(13)	1890–1959	0.78	0.97	>0.9995			
Spruce	(13)	1930–1959	1.36	1.00	>0.9995			
			0.95 ± 0.28					
Wisconsin/USA (Record J)								
Quercus macrocarpa	(14)	1900–1978	1.47	0.94	>0.9995			
Quercus macrocarpa	(14)	1900–1978	0.78	0.69	>0.95			
Quercus macrocarpa	(14)	1900–1978	0.76	0.69	>0.95			
			1.00 ± 0.34					
England, Scotland (Record K)								
Quercus robur	(15)	1900–1965	1.12	0.91	>0.995			
Larix decidua	(15)	1905–1968	0.88	0.87	>0.99			
			1.00 ± 0.12					
Hawaii/USA (Record L)								
Myoporum sandwicense	(1)	1920–1978	1.03	0.99	>0.9995			
Myoporum sandwicense	(1)	1940–1978	0.91	0.99	>0.995			
			0.98 ± 0.06					

France, Spain (Record M)								
Pinus maritima (Pi 3)	(9)	1920–1969	0.68	0.40	>0.7			
Pinus maritima (Pi 2)	(9)	1920–1969	1.90	0.92	>0.975	0.91	0.88	>0.975
Quercus robur (Qu 1)	(9)	1920–1969	−0.07	−0.03				
Quercus robur (Qu 2)	(9)	1920–1969	−0.23	−0.22	>0.6			
Quercus lusitanica (Qu 4, 3 cuts)	(9)	1930–1969	3.39	0.82	>0.9	1.07	0.98	>0.9995
Quercus lusitanica (Qu 5)	(9)	1930–1969	6.24	0.69	>0.8			
			2.14 ± 2.07			1.02 ± 0.08		

*References: (1) Freyer and Belacy (1983); (2) Harkness and Miller (1980); (3) Grinsted et al. (1979); (4) Wilson (1978); (5) Stuiver (1978); (6) Long and Leavitt (1983); (7) Lerman and Long (1980); (8) Leavitt and Long (1983); (9) Freyer (1979*a*); (10) Freyer (1979*b*); (11) Tans and Mook (1980); (12) Farmer (1979); (13) Galimov (1976); (14) Bender and Berge (1982); (15) Farmer and Baxter (1974).

†All mean data in this chapter are given as weighted means ± 1 SD. The weights consider number of analyzed trees and/or duration of analyzed periods.

correlation is found for the data of only nine of the 65 trees. The nine trees were found in records G, H, and M. Their data, which spanned five decades or less, include parts of the 1920 to 1960 period, during which larger fluctuations in most records were apparent. Because of the internal inconsistency, data of these trees were omitted and correlations and regressions retested. Even with inclusion of these data, the change in the records would be <0.2‰ (see Fig. 7.1). The standard deviation of regression of individual records on their mean trend varies between 10 and 60%, with a weighted mean of about 30%. Tests for correlation and regression were not performed on records C and E. In the case of E, only mean data were available; for C, the individual records were illegible in the graph.

Table 7.2 summarizes mean $^{13}C/^{12}C$ trends found in the regional records.

TABLE 7.2. Mean $\delta^{13}C$ change in the regional trends from preindustrial time up to 1960 (for 1960 to 1980, see Table 7.3).

	Change in Δ^{β} $\delta^{13}C$*			
Record	1550–1800	1800–1850	1850–1920	1920–1960
A	−0.3‰ (Variable, ± 0.3‰)	−0.1‰	−0.7‰	−0.5‰
B	−0.4‰ (Variable, ± 0.5‰)	−0.2‰	−0.6‰	−0.1‰
C	−0.7‰ (1710–1800)	−0.2‰	−0.5‰	+0.4‰
D	(Variable, ± 0.3‰) (1720–1800)	−0.4‰	−0.6‰	+0.6‰
E				−0.5‰ (1940–1960)
F		−0.2‰ (1810–1850)	−0.8‰	−0.9‰
G			−0.8‰ (1860–1920)	+0.4‰
H			−0.3‰ (1890–1920)	+0.1‰
I			−0.6‰ (1900–1920)	+0.8‰ (1920–1950)
J			−0.2‰ (1910–1920)	+0.1‰
K			−0.6‰ (1910–1920)	−0.2‰
L				−0.6‰ 1930–1960)
M				−0.5‰ (1930–1960)
		−0.2 ± 0.1‰	−0.6 ± 0.2‰	−0.1 ± 0.5‰
Mean record	−0.4‰ (variable, ± 0.4‰)	−0.2‰	−0.6‰	−0.1‰

*The changes were calculated from means between following 10-year block data (e.g., data for 1800 are the 1790-1799 and 1800-1809 means).

Whereas no trends or slightly decreasing $^{13}C/^{12}C$ data are observed during the preindustrial period from 1550 to 1800 in three regional records, the data in all records decrease since 1800. The mean decrease from 1800 to 1850 and from 1850 to 1920 amounts to about 0.2 and 0.6‰, respectively. Without exception, all regional records exhibit this decrease. From 1920 to 1960, however, the data fluctuate strongly, as shown in Fig. 7.2.

The fluctuations in record A were checked for possible correlations with climatic factors by Freyer and Belacy (1983). It has been found that first differences of $^{13}C/^{12}C$ and climate data correlate significantly (i.e., differences between following 10-year blocks of data, which take into account the sign and magnitude of change in the $\delta^{13}C$ and climate series). Variations in seasonal temperature of about 2°C and in monthly mean seasonal precipitation of about 15 mm were observed at the site of record A trees from 1920 to 1960. By using a temperature coefficient

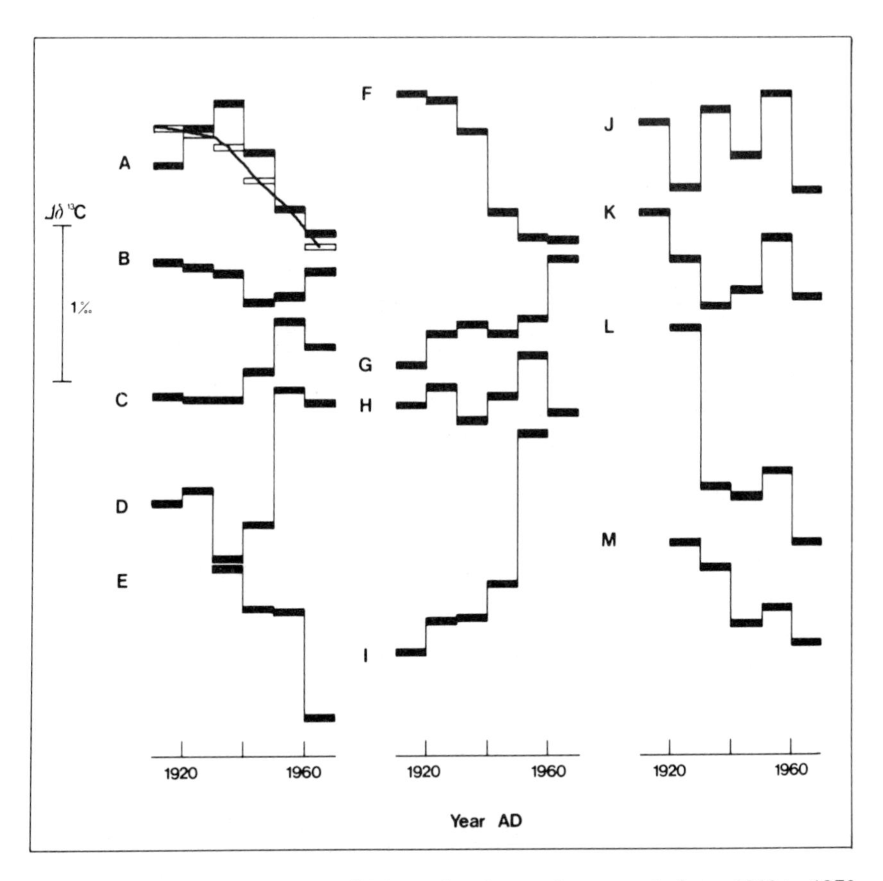

FIGURE 7.2. Fluctuations of $\Delta\delta^{13}C$ in regional tree-ring records from 1910 to 1970. *Heavy marks* in record A represent original data; *open marks* show data corrected for climatic effects.

of $+0.2‰$ $(\Delta\delta^{13}C)/\Delta T$ (°C) and a precipitation coefficient of $+\ 0.02‰$ $(\Delta\delta^{13}C)/\Delta mm$, a total shift of 0.7‰ would be obtained (if temperature and precipitation effect coincide), for which the measured $\delta^{13}C$ data could be corrected. If other tree-ring records were also influenced by climatic effects of this magnitude, much of the disagreement between the regional trends from 1920 to 1960 might be explained. In other periods, less regional and global climate variations were observed.

The trends during the last two decades from 1960 to 1980 are shown in Fig. 7.3 and described in Table 7.3, which gives changes in $\delta^{13}C$ between 1956 and 1978 and also between 1978 and 1980, as calculated by least-square fits of the 10-year block data of the regional records. A decreasing trend is evident in all records covering the 1970 to 1979 period. In addi-

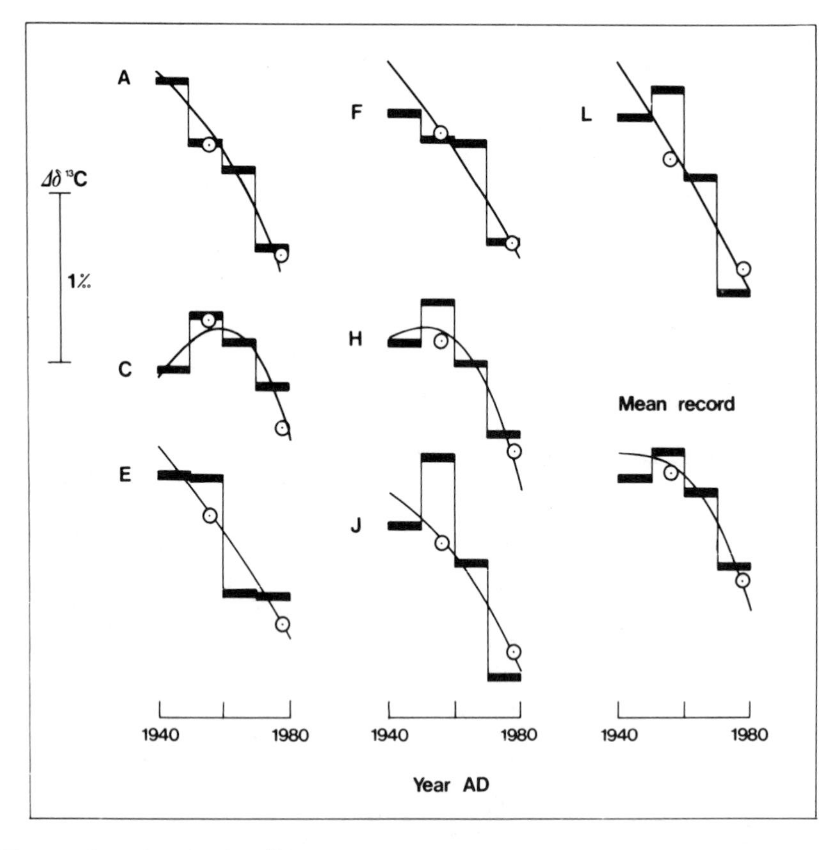

FIGURE 7.3. Trends of $\Delta\delta^{13}C$ in regional tree-ring records from 1940 to 1980 and polynomial fits of the data. Data in records C, H, and in the mean record were fitted by 3rd-degree polynomials; other fits represent 2nd-degree polynomials. *Open circles* give the $\delta^{13}C$ change of $-0.65‰$ in atmospheric CO_2 observed from 1956 to 1978 by Keeling et al. (1979, 1980).

TABLE 7.3. Calculation of 1956–1978 (Δ_1) and 1978–1980 (Δ_2) changes by least-squares fits of the regional $\Delta\delta^{13}C$ trends, using only 20th century data.

	Second degree polynomial				Best or other fit				
Record	Correlation coefficient	Standard error (‰)	Δ1 (‰)	Δ2 (‰)	Degree	Correlation coefficient	Standard error (‰)	Δ1 (‰)	Δ2 (‰)
A	0.95	0.16	−0.83	−0.10	5th	0.98	0.17	−0.74	−0.05
B*‡									
C†					3rd	0.96	0.08	−0.50	−0.14
					5th	0.96	0.11	−0.51	−0.14
D*‡									
E	0.94	0.21	−0.67	−0.07					
F	0.98	0.15	−0.68	−0.07	4th	0.98	0.17	−0.81	−0.15
G*‡									
H	0.64	0.20	−0.34	−0.04	3rd	0.87	0.14	−0.78	−0.15
I*									
J	0.75	0.33	−0.71	−0.09					
K*§					3rd	0.97	0.16	−0.89	−0.19
L	0.89	0.39	−0.76	−0.07					
M*	0.94	0.13	−0.12	0.00	3rd	0.94	0.18	−0.39	−0.06
Mean record	0.89	0.15	−0.38	−0.04	3rd	0.97	0.10	−0.74	−0.13

*1970–1979 data (1960–1979 data for record I) are not available.

†Second-degree polynomial gives positive changes.

‡Extrapolation by 2nd to 5th degree polynomials gives positive changes.

§Extrapolation by 2nd degree polynomial gives positive changes.

tion, the fits of these records closely agree with the $^{13}C/^{12}C$ decrease in atmospheric CO_2 of -0.65 ± 0.13‰ for 1956 to 1978 and of -0.12 ± 0.01‰ for 1978 to 1980, as measured by Keeling et al. (1979, 1980) and Mook et al. (1981). In contrast, some records without 1970 to 1979 data would extrapolate to increasing trends. The agreement of the more detailed record (2-year blocks of data) reported by Freyer (1979*a*), with the atmospheric $^{13}C/^{12}C$ change, has been shown elsewhere (Freyer 1981).

Mean Trend

The mean $\delta^{13}C$ trend of the Northern Hemisphere, as calculated from the regional trends, is shown in Fig. 7.4. The coefficients for linear correlation and regression of regional trends on the mean trend are given in Table 7.4. When all of the regional trends are used, records I and M correlate rather poorly, with levels of significance $P < 0.9$. With omission of record I and further omission of individual records (with $P < 0.9$) in the regional records G, H, and M, all regional trends correlate with the mean trend at levels of significance $P > 0.9$. The standard deviation of regression of regional trends on the mean trend, however, amounts to about 50%. For records beginning in at least 1800, the lowest and highest extreme values of the coefficient of regression are found for record C ($b = 0.42$) and for record F ($b = 1.64$). The time period used for normalization and tests of data is between 1800 and 1979. If this time period is restricted to 1959, coefficients of regression for records C and F will be improved; however, correlations of the 20th century records become worse.

Parameters of least-squares fits of the mean Northern Hemisphere $\delta^{13}C$ trend are given in Table 7.5. The best approximation to the recent $^{13}C/^{12}C$ change in atmospheric CO_2 from 1956 to 1980 is found for a polynomial of the 6th degree. The coefficients of this polynomial and those for the polynomials representing the upper and lower 95% confidence limits of the mean $\delta^{13}C$ tree-ring data are given in Table 7.6. These polynomial fits are shown in Fig. 7.5.

Southern Hemispheric tree-ring records would be expected to give the same $^{13}C/^{12}C$ trend, although with a decrease of some tenths of a mil less than Northern Hemispheric trees (Francey and Farquhar 1982). In fact, measurements by Rebello and Wagener (1976) on a free-standing Brazilian tree agree with the trend for the Northern Hemisphere. Also $^{13}C/^{12}C$ trends

---

FIGURE 7.4. Mean $\Delta\delta^{13}C$ trend for the Northern Hemisphere from 1530 to 1980 obtained from tree-ring records. *Heavy marks* are mean data for each 10-year block. *Hatched marks* from 1910 to 1980 are mean data without any omission of individual records. *Thin lines* show the trend reported previously by Freyer and Belacy (1983). Error bars represent one standard deviation ($1\ \sigma$) of scatter of individual from mean data. The histogram in the lower figure stands for the number of analyzed trees at quoted time intervals. The *dotted line* gives the number of all trees without any omission.

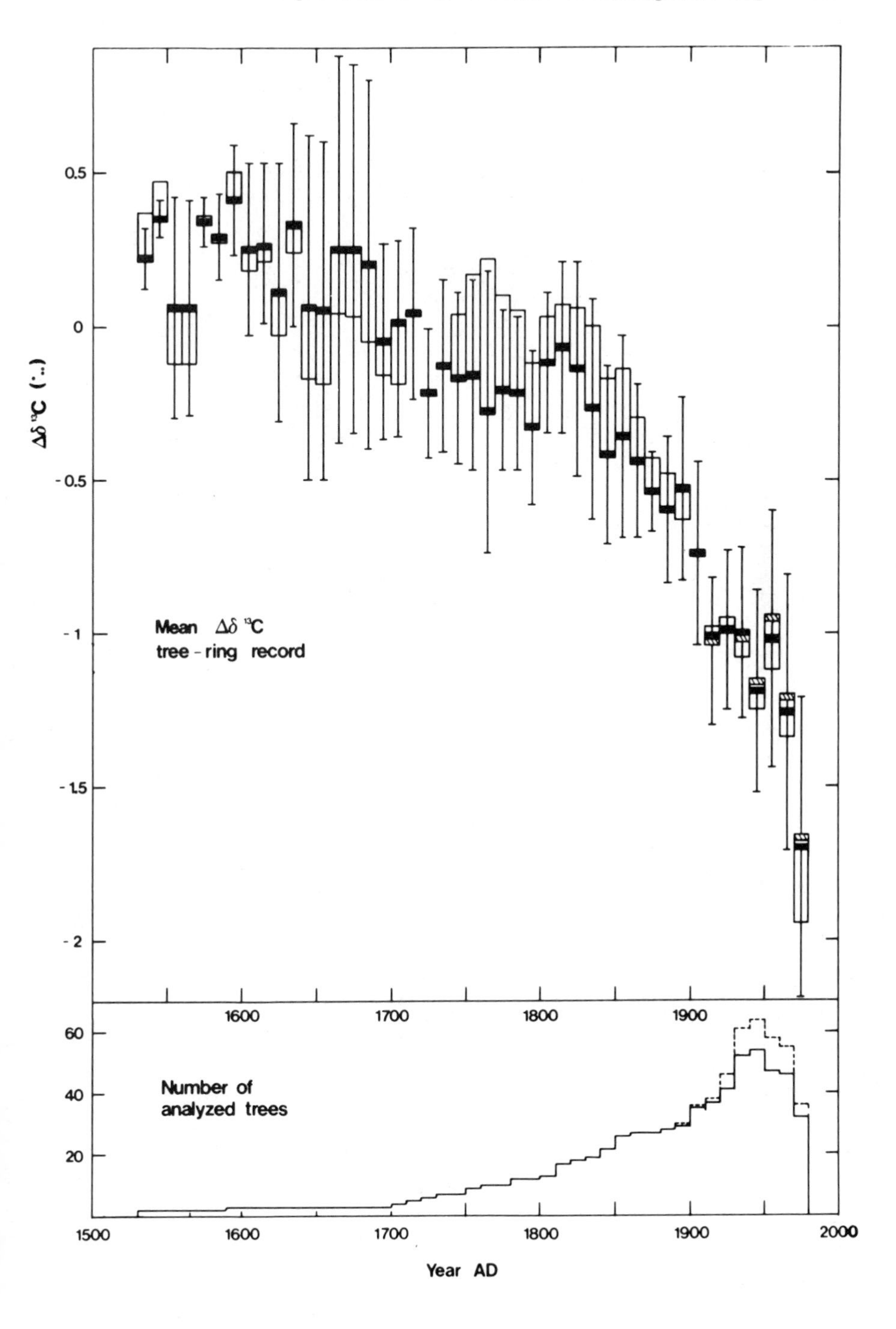
0.5
0
-0.5
-1
-1.5
-2
Δδ 13C (‰)
Mean Δδ 13C
tree-ring record
1600
1700
1800
1900
60
40
20
Number of
analyzed trees
1500
1600
1700
1800
1900
2000
Year AD

TABLE 7.4. Coefficients for linear correlation r with level of significance P, and coefficients of regression b of each regional trend on the mean $\Delta\delta^{13}C$ trend.

	1800–1959 Normalization			1800–1979 Normalization					
	Regression coefficient	Correlation coefficient	Level of significance	Regression coefficient	Correlation coefficient	Level of significance	With omission of data		
Record	b	r	P	b	r	P	b'	r'	P'
A	1.05	0.91	>0.9995	1.22	0.95	>0.9995	1.21	0.96	>0.9995
B	0.99	0.96	>0.9995	0.95	0.95	>0.9995	0.93	0.95	>0.9995
C	0.54	0.74	0.999	0.44	0.73	>0.9995	0.42	0.71	>0.9995
D	1.31	0.86	>0.9995	1.14	0.79	>0.9995	1.08	0.77	>0.9995
E	0.22	0.14		1.24	0.77	>0.9	1.31	0.82	>0.95
F	1.54	0.93	>0.9995	1.66	0.95	>0.9995	1.64	0.96	>0.9995
G	0.71	0.67	>0.975	0.58	0.54	>0.95	0.51	0.49	>0.9
H	0.58	0.73	>0.975	0.59	0.81	>0.995	0.56	0.77	>0.995
I	0.50	0.24	>0.6	0.64	0.30	>0.7			
J	0.39	0.26	>0.6	1.37	0.87	>0.995	1.30	0.86	>0.995
K	3.05	0.91	>0.99	2.74	0.88	>0.995	2.45	0.84	>0.99
L	2.88	0.54	>0.75	2.17	0.83	>0.975	2.19	0.86	>0.975
M	0.91	0.52	>0.75	0.72	0.60	>0.8	1.86	0.88	>0.975
Records A, B, C, D, F	1.01 ± 0.38			1.02 ± 0.48			1.00 ± 0.48		
Records A–M	0.97 ± 0.55			1.01 ± 0.51			1.04 ± 0.53		

obtained on whole wood from three Tasmanian trees, and published by Pearman et al. (1976) and Fraser et al. (1978), are consistent with those of some Northern Hemispheric trees used in this compilation. Francey (1981), however, using the cellulose fraction of wood of the same three Tasmanian trees and including measurements on four additional trees in his record, has reported the absence of any $^{13}C/^{12}C$ trend in tree rings. This is contrary to the findings for Northern Hemispheric trees. However, five of the Tasmanian trees were of forest origin. From measurements on trees within forests, Freyer and Belacy (1983) have observed neither correlation of $^{13}C/^{12}C$ data between individual records nor trends, and they have concluded that forest trees record no $^{13}C/^{12}C$ changes of free atmospheric CO_2.

Interpretation of the Mean $\delta^{13}C$ Trend of the Northern Hemisphere

With inclusion of additional tree-ring data from Arizona (records C, D, and E) and data of one tree from California (in record B), and also slight modifications for normalization of data, a smaller $^{13}C/^{12}C$ decrease from preindustrial time until now was obtained than in the mean $\delta^{13}C$ trend published previously by Freyer and Belacy (1983). The data in Figs. 7.4 and 7.5 are given as deviations from the 1600 to 1800 mean of the mean trend. It is apparent that data from 1750 to 1850, in the present calculation,

TABLE 7.5. Parameters of least-squares fits of the mean Northern Hemisphere $\Delta\delta^{13}C$ trend and calculated 1956–1978 (Δ_1) and 1978–1980 (Δ_2) changes.

Polynomial	Correlation coefficient (*r*)	Standard error of estimate (‰)	Δ1 (‰)	Δ2 (‰)
2nd	0.964	0.115	−0.27	−0.03
3rd	0.965	0.114	−0.24	−0.02
4th	0.966	0.112	−0.32	−0.04
5th	0.975	0.099	−0.52	−0.08
6th	0.977	0.094	−0.64	−0.12
6th (u.)†	0.977	0.099	−0.55	−0.10
6th (l.)†	0.976	0.096	−0.74	−0.13
7th	0.985	0.078	−0.85	−0.21
8th	0.987	0.074	−0.94	−0.28
9th	0.986	0.077	−0.93	−0.28
10th	0.986	0.076	−0.88	−0.23

*Weighted data of the $\delta^{13}C$ trend were used for calculation of the polynomials; the weights consider number of analyzed trees for each 10-year block of data.
†Parameters for polynomials, using the upper (u.) and lower (l.) 95% confidence limits of mean $\delta^{13}C$ data.

TABLE 7.6. Coefficients for fits of mean $\delta^{13}C$ tree-ring data and their 95% confidence limits using polynomials of the 6th degree in the following form: $\Delta\delta^{13}C$ (‰) = $A_0 + A_1X + A_2X^2 + A_3X^3 + A_4X^4 + A_5X^5 + A_6X^6$ (X in units of 10 years, beginning X = 1, 1750–1759).

	Mean $\delta^{13}C$ data	Upper 95% confidence	Lower limits of mean $\delta^{13}C$ data
A_0	−0.3480	+0.0105	−0.7064
A_1	+0.181434	+0.090373	+0.272496
A_2	-7.518462×10^{-2}	-5.499799×10^{-2}	-9.537127×10^{-2}
A_3	$+1.408730 \times 10^{-2}$	$+1.150841 \times 10^{-2}$	$+1.666620 \times 10^{-2}$
A_4	-1.283634×10^{-3}	-1.107256×10^{-3}	-1.460013×10^{-3}
A_5	$+5.364416 \times 10^{-5}$	$+4.754042 \times 10^{-5}$	-5.974790×10^{-5}
A_6	-8.318823×10^{-7}	-7.470853×10^{-7}	-9.166793×10^{-7}

are more negative than those in the previous trend. The fit with the 6th-degree polynomial produces a nearly constant $^{13}C/^{12}C$ level of about −0.2‰ from 1750 to 1829, followed by a decrease during the industrial period of about 1.5‰ in comparison to the decrease of 1.95‰ found in the previous trend.

The previous trend has been adopted by Peng et al. (1983) for calculations of past atmospheric CO_2 levels and evaluation of the net biospheric CO_2 input. According to the model of Peng et al., fossil fuel CO_2 emissions would account for a $^{13}C/^{12}C$ change of about 1.1‰ from 1860 up to 1980. The residual change from the tree-ring data has been attributed to the input of forest and soil CO_2 into the atmosphere. By deconvolution of the residual change using 20-year running means, the integrated forest-soil CO_2 contributions amounted to 22×10^{15} mol. Due to the smaller $^{13}C/^{12}C$ decrease obtained from tree-ring data in the present calculation, these biospheric CO_2 contributions should be reduced considerably and revised estimates are given by Peng and Freyer (*this volume*).

The time history of the estimated biospheric CO_2 contributions depends on the mathematical expression used to represent the tree-ring data. Table 7.5 shows that the 1956 to 1978 and 1978 to 1980 changes produced by a curve fit largely differ with the degree of the polynomial employed, but that there are relatively small variations for correlation coefficients and standard errors of the estimate. A fit of the mean $\delta^{13}C$ trend with a 6th-degree polynomial produces about −0.64‰ for the 1956 to 1978 changes and −0.12‰ for the 1978 to 1980 changes, which agree with the actual $^{13}C/^{12}C$ changes found in atmospheric CO_2 for this period. On the other hand, a fit of the data with a 5th-degree polynomial gave −0.52‰ and −0.08‰ for the 1956 to 1978 and 1978 to 1980 changes, respectively, which underestimate the actual $^{13}C/^{12}C$ decreases.

The residual $\delta^{13}C$ curves obtained from both fits by simply subtracting the fossil fuel $\delta^{13}C$ contributions are compared in Fig. 7.6 with the cor-

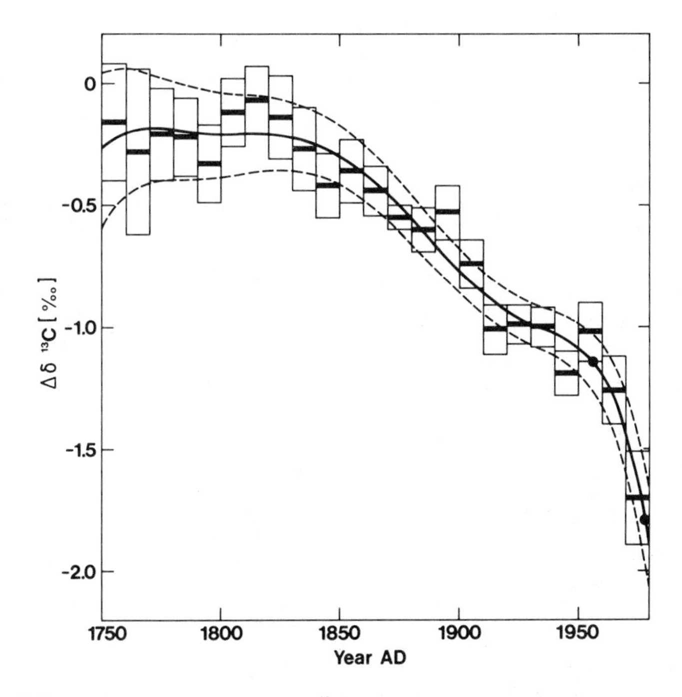

FIGURE 7.5. Mean 10-year block Δ $\delta^{13}C$ data of the Northern Hemisphere tree-ring record from 1750 to 1980 and 6th-degree polynomial fit of the data. The *vertical extension of blocks* represents upper and lower 95% confidence limits of mean data, the *dotted lines* represent their corresponding 6th-degree polynomial fits. The full circles give the $\delta^{13}C$ change of −0.65‰ in atmospheric CO_2 observed from 1956 to 1978 by Keeling et al. (1979, 1980).

responding curve given by Peng and Freyer (based on the technique of Peng et al. 1983). It is observed that the curve obtained from the 6th-degree polynomial fit is similar to that presented by Peng and Freyer (*this volume*). The latter $\delta^{13}C$ change results in a slight biospheric CO_2 source, whereas the curve obtained from the 5th-degree polynomial fit would require a recent biospheric CO_2 sink. It should be emphasized, however, that these distinctions are produced only by the mathematics, and original tree-ring data do not provide a sufficiently detailed record to resolve such differences because of the relatively large residual errors in the mean trend.

Finally, it should be noted that the calculated trend, and thus any conclusions, are based on $^{13}C/^{12}C$ tree-ring data that have not been corrected for climatic influences. These corrections could be especially important for the 1920 to 1960 period, in which larger regional and global climate variations have occurred. Further studies and calculations are required to improve usefulness of the tree-ring record, particularly during this period.

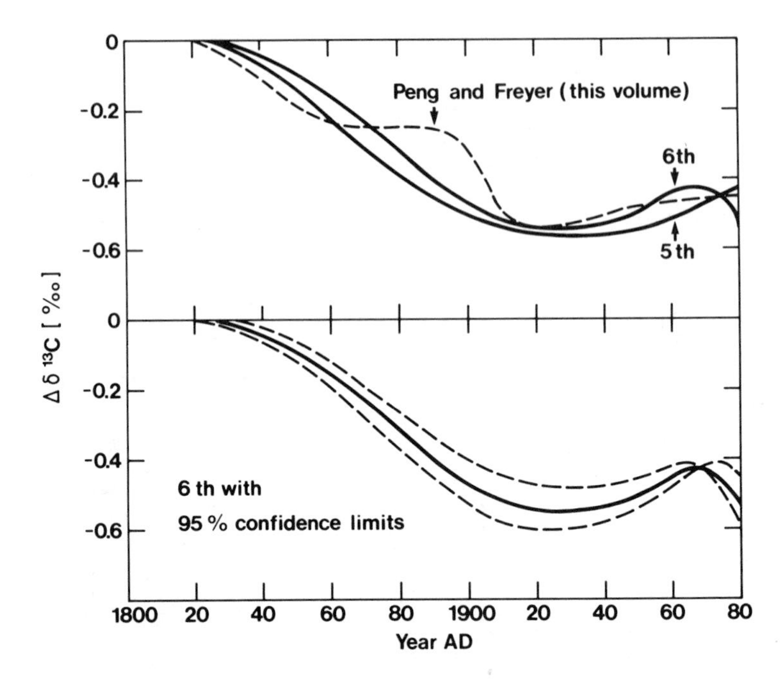

FIGURE 7.6. **Top:** Residual curves representing the biospheric $\delta^{13}C$ contributions from 1820 to 1980 obtained from the 5th- and 6th-degree polynomial fits of mean $^{13}C/^{12}C$ tree-ring data by subtracting the fossil fuel $\delta^{13}C$ contribution compared with the corresponding curve obtained by Peng and Freyer (*this volume*) using the deconvolution procedure. **Bottom:** Residual curves obtained from the 6th-degree polynomial fits of mean $^{13}C/^{12}C$ tree-ring data and their upper and lower 95% confidence limits.

References

Bender, M. M. and A. J. Berge. 1982. Carbon isotope records in Wisconsin trees. Tellus 34:500–504.

Craig, H. 1954. Carbon-13 variations in sequoia rings and the atmosphere. Science 119:141–143.

Farmer, J. G. 1979. Problems in interpreting tree ring ^{13}C records. Nature 279:229–231.

Farmer, J. G. and M. S. Baxter. 1974. Atmospheric carbon dioxide levels as indicated by the stable isotope record in wood. Nature 247:273–275.

Farquhar, G. D. 1980. Carbon isotope discrimination by plants: effects of carbon dioxide concentration and temperature via the ratio of intercellular and atmospheric CO_2 concentrations. In G. I. Pearman (ed.), Carbon Dioxide and Climate: Australian Research, pp. 105–110. Australian Academy of Science, Canberra.

Francey, R. J. 1981. Tasmanian tree rings belie suggested anthropogenic $^{13}C/^{12}C$ trends. Nature 290:232–235.

Francey, R. J., and G. D. Farquhar. 1982. An explanation of $^{13}C/^{12}C$ variations in tree rings. Nature 297:28–31.

Fraser, P. J. B., R. J. Francey, and G. I. Pearman. 1978. Stable carbon isotopes in tree rings as climatic indicators. DSIR Bull. 220:67–73.

Freyer, H. D. 1979*a*. On the ^{13}C record in tree rings. Part 1. ^{13}C variations in northern hemispheric trees during the last 150 years. Tellus 31:124–137.

Freyer, H. D. 1979*b*. On the ^{13}C record in tree rings. Part 2. Registration of microenvironmental CO_2 and anomalous pollution effect. Tellus 31:308–312.

Freyer, H. D. 1979*c*. Variations of the atmospheric CO_2 content. In B. Bolin et al. (eds.), The Global Carbon Cycle, SCOPE 13, pp. 79–99. John Wiley and Sons, New York.

Freyer, H. D. 1980. Factors influencing the ^{13}C record of tree rings in modern wood. WMO Project on Research and Monitoring of Atmospheric CO_2. Rep. 1. App. E. World Meteorological Organization, Geneva.

Freyer, H. D. 1981. Recent $^{13}C/^{12}C$ trends in atmospheric CO_2 and tree rings. Nature 293:679–680.

Freyer, H. D. and N. Belacy. 1983. $^{13}C/^{12}C$ records in northern hemispheric trees during the past 500 years—anthropogenic impact and climatic superpositions. J. Geophys. Res. 88:6844–6852.

Freyer, H. D. and L. Wiesberg. 1975. Anthropogenic carbon-13 decrease in atmospheric carbon dioxide as recorded in modern wood. In Proceedings of the FAO/IAEA Symposium. Isotope Ratios as Pollutant Source and Behaviour Indicators, pp. 49–62. International Atomic Energy Agency, Vienna.

Fritts, H. C. 1976. Tree Rings and Climate. Academic Press, New York.

Galimov, E. M. 1976. Variations of the carbon cycle at present and in the geological past. In J. O. Nriagu (ed.). Environmental Biogeochemistry, Vol. 1, pp. 3–11. Ann Arbor Science Publishers, Ann Arbor, Michigan.

Grinsted, M. J., A. T. Wilson, and C. W. Ferguson. 1979. $^{13}C/^{12}C$ ratio variations in *Pinus longaeva* (bristle-cone pine) cellulose during the last millenium. Earth Planet. Sci. Lett. 42:251–253.

Harkness, D. D. and B. F. Miller. 1980. Possibility of climatically induced variations in the ^{14}C and ^{13}C enrichment patterns as recorded by a 300-year-old Norwegian pine. Radiocarbon 22:291–298.

Keeling, C. D., R. B. Bacastow, and P. P. Tans. 1980. Predicted shift in the $^{13}C/^{12}C$ ratio of atmospheric carbon dioxide. Geophys. Res. Lett. 7:505–508.

Keeling, C. D., W. G. Mook, and P. P. Tans. 1979. Recent trends in the $^{13}C/^{12}C$ ratio of atmospheric carbon dioxide. Nature 277:121–123.

Leavitt, S. W. and A. Long. 1983. An atmospheric $^{13}C/^{12}C$ reconstruction generated through removal of climate effects from tree-ring $^{13}C/^{12}C$ measurements. Tellus 35B:92–102.

Lerman, J. C. and A. Long. 1980. Carbon-13 in tree rings: local or canopy effect? In G. C. Jacoby (ed.). Proceedings of the International Meeting on Stable Isotopes in Tree Ring Research, pp. 22–34. U. S. Department of Energy, Washington, D. C.

Long, A. 1982. Stable isotopes in tree rings. In M. K. Highes, P. M. Kelly, J. R. Pilcher, and V. C. LaMarche (eds.). Climate from Tree Rings, pp. 12–18. Cambridge University Press, Cambridge, Massachusetts.

Long, A. and S. W. Leavitt. 1983. Accurate determination of $^{13}C/^{12}C$ in CO_2 of past atmospheres from $^{13}C/^{12}C$ in tree rings by removal of climatic interferences. Progress Report for Union Carbide Subcontract 19X-22290C.

Mazany, T., J. C. Lerman, and A. Long. 1980. Carbon-13 in tree-ring cellulose as an indicator of past climates. Nature 287:432–435.

Mook W. G., C. D. Keeling and A. Herron. 1981. Seasonal and secular variations in the abundance and $^{13}C/^{12}C$ ratio of atmospheric CO_2. Paper presented at WMO/ICSU/UNEP Scientific Conference, World Meteorological Organization, Geneva.

Nikitin, N. J. 1966. The Chemistry of Cellulose and Wood. Monson, Jerusalem.

Pearman, G. I., R J. Francey, and P. J. B. Fraser. 1976. Climatic implications of stable carbon isotopes in tree rings. Nature 260:771–773.

Peng, T. H, W. S. Broecker, H. D Freyer, and S. Trumbore. 1983. A deconvolution of the tree-ring based ^{13}C record. J. Geophys. Res. 88:3609–3620.

Rebello, A. and K. Wagener. 1976. Evaluation of ^{12}C and ^{13}C data on atmospheric CO_2 on the basis of a diffusion model for oceanic mixing. In J. O. Nriagu (ed.), Environmental Biogeochemistry Vol. 1, pp. 13–23. Ann Arbor Science Publishers, Ann Arbor, Michigan.

Stuiver, M. 1978. Atmospheric carbon dioxide and carbon reservoir changes. Science 199:253–258.

Tans, P. P. and W. G. Mook. 1980. Past atmospheric CO_2 levels and the $^{13}C/^{12}C$ ratios in tree rings. Tellus 32:268–283.

Wilson, A. T. 1978. Pioneer agriculture explosion and CO_2 levels in the atmosphere. Nature 273:40–41.

8
Revised Estimates of Atmospheric CO_2 Variations Based on the Tree-Ring ^{13}C Record

TSUNG-HUNG PENG AND HANS D. FREYER

Since the publication of a paper by Peng et al. (1983) regarding the analysis of the tree-ring-based $\delta^{13}C$ record,* a number of reports giving new measurements of $\delta^{13}C$ in tree rings have emerged (Leavitt and Long 1983; Stuiver et al. 1984; Freyer, this volume; Stuiver, this volume). The results of the original analysis were based on a global $^{13}C/^{12}C$ trend compiled by Freyer and Belacy (1983). Freyer (*this volume*) recalculated the $^{13}C/^{12}C$ trend of the Northern Hemisphere using measurements on 65 trees, including those presented in the publication of Leavitt and Long (1983), and he concluded that no long-term trend exists during the period of a few centuries before 1800 AD. By contrast, a clear trend of decreasing ^{13}C was observed after 1800 AD. Freyer estimated the overall decrease of $\delta^{13}C$ from 1800 to 1980 AD to be about -1.5‰, which is 0.5‰ less than that obtained from the composite ^{13}C trend used by Peng et al. (1983). Therefore, a reevaluation of the terrestrial biosphere contribution to the lowering of $\delta^{13}C$ in the atmospheric CO_2 seems appropriate. This is the main objective of this chapter.

Method

Anthropogenic CO_2 (both from fossil fuel combustion and from deforestation and soil manipulation) has $\delta^{13}C$ values averaging about −26‰ (Schwarz 1970). The value of the preanthropogenic atmosphere was about −6‰ (deduced from Keeling et al. 1980). Therefore, the addition of anthropogenic CO_2 to the atmosphere reduces the atmospheric $\delta^{13}C$ value. The time history of the variation of atmospheric $\delta^{13}C$ is believed to be recorded in tree rings. However, this record is complicated by several factors. The carbon atoms from fossil fuel and from forest plus soil sources will exchange with carbon atoms in the ocean and in the terrestrial bio-

---

$$*\delta^{13}C \text{ per mil} = \frac{{}^{R}\text{sample} - {}^{R}\text{PDB}}{{}^{R}\text{PDB}} \times 1000,$$ where R is the $^{13}C/^{12}C$ isotopic ratio and PDB (Pee Dee belemnite) is the standard used for measuring the $^{13}C/^{12}C$ isotopic ratio.

sphere reservoir. The $\delta^{13}C$ dilution effects caused by this exchange must be properly accounted for if the tree-ring ^{13}C record is to be converted to a record of atmospheric CO_2 history. A global carbon cycle model is needed to unravel the CO_2 history in tree rings.

In addition to atmospheric $\delta^{13}C$ changes, the observed $\delta^{13}C$ variations in tree rings could also be caused by a number of other factors. The local fluctuation in the $^{13}C/^{12}C$ ratio generated by the input of ^{13}C-depleted CO_2 from soil respiration may influence the ^{13}C content of trees. The $^{13}C/^{12}C$ record carried by such trees would not be a reliable representation of $\delta^{13}CO_2$ variations on a global scale. The $^{13}C/^{12}C$ records of oak trees (Freyer and Belacy 1983) from forests (Spessart Mountains in southern Germany) show sudden jumps and nonsystematic variations that are not synchronous with respect to time. The canopy effects of forest trees and partial uptake of soil respiration CO_2 are considered to be the main reason for such $\delta^{13}C$ variations. Therefore, the selection of free-standing trees far from urban areas for $\delta^{13}C$ analysis is necessary to minimize these effects.

Ring-to-ring variations in the chemical makeup of plant material will lead to changes in the $\delta^{13}C$ of whole wood. By analyzing only cellulose, this effect due to variations in the ratio of cellulose to lignin and extractives can be eliminated.

The $\delta^{13}C$ variations in tree rings can also be affected by environmental variability. This is due to the effects of changing irradiance, relative humidity, and growing-season temperature on stomatal geometry, which cause varying isotopic fractionation during photosynthesis. Ideally, free-standing trees that have an adequate supply of moisture and relatively consistent ambient temperatures throughout their growing season should exhibit the least environmental effects. The ^{13}C record from cellulose of such trees would be the best representation of the global atmospheric $\delta^{13}C$ record. However, in selecting trees in Tasmania largely according to the above criteria, Francey (1981) found a ^{13}C change in tree rings smaller even than that expected from fossil fuels alone.

Leavitt and Long (1983) found a correlation between $\delta^{13}C$ of the cellulose fraction of juniper trees in Arizona and the mean December temperature or precipitation at the tree sites. Freyer and Belacy (1983) also demonstrated a correlation between the $\delta^{13}C$ record of the Scots pine trees and autumn temperature and spring precipitation. However, such relationships have not been demonstrated for other trees at different sites and in different climatic zones.

In his recalculation of the global $\delta^{13}C$ trend of tree rings, Freyer (*this volume*) did not correct for environmental effects. Trees that were sampled were mostly free-standing, and cellulose was the material used for most $\delta^{13}C$ analyses. Efforts have thus been made to reduce the local fluctuation and material effects, but it should be kept in mind that the reconstructed global ^{13}C trend of tree rings contains significant environmental "noise." In addition, if the ^{13}C fractionation in trees is correlated with atmospheric pCO_2 (partial pressure of CO_2 gas), because of CO_2 fertilization effects, no amount of averaging will give the correct temporal record for reconstructing atmospheric CO_2 history.

Procedures and results of recalculations for the global ^{13}C trend are given in detail by Freyer (*this volume*). We will use this new composite tree ring ^{13}C record to derive the time history of biospheric CO_2 releases, and we will compare our results with those reported by Stuiver (*this volume*) and Stuiver et al. 1984.

The global carbon cycle model used has been described in detail elsewhere (Peng et al. 1983).

Tans (1981) showed that the $\delta^{13}C$ of fossil fuel CO_2 changes with time because of variations in the proportions of fossil fuels used (coal, oil, natural gas, etc.). This variation in $\delta^{13}C$ is also included in our model computations.

Results

The revised tree-ring-based $\delta^{13}C$ record obtained by Freyer (*this volume*) is shown in Fig. 8.1. For comparison, the Freyer and Belacy (1983) composite $\delta^{13}C$ record and three model-computed curves are also presented in the figure. The curves are obtained by adding the anthropogenic CO_2 into the carbon cycle model atmosphere to simulate human perturbation beginning in 1800 AD. The distribution of ^{13}C in various carbon reservoirs of the model is computed, and the time histories of ^{13}C in the atmosphere resulting from separate CO_2 sources are represented in Fig. 8.1 as a set

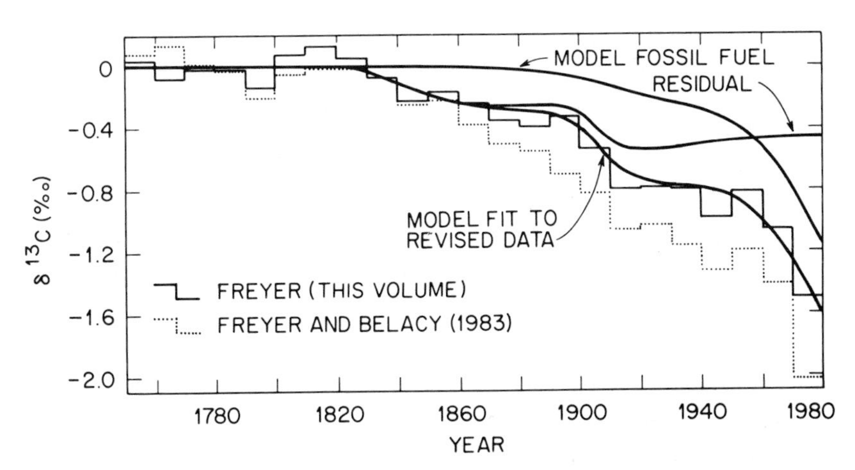

FIGURE 8.1. Terrestrial biospheric CO_2 contribution (*curve* marked "residual") to the decline in atmospheric $^{13}C/^{12}C$ ratio as obtained by subtracting the model-derived contribution of fossil fuel CO_2 (*curve* marked "model fossil fuel") from the model curve that best fits Freyer's (*this volume*) revised ^{13}C record (solid step curve). The *dotted step curve* is the composite ^{13}C record of Freyer and Belacy (1983).

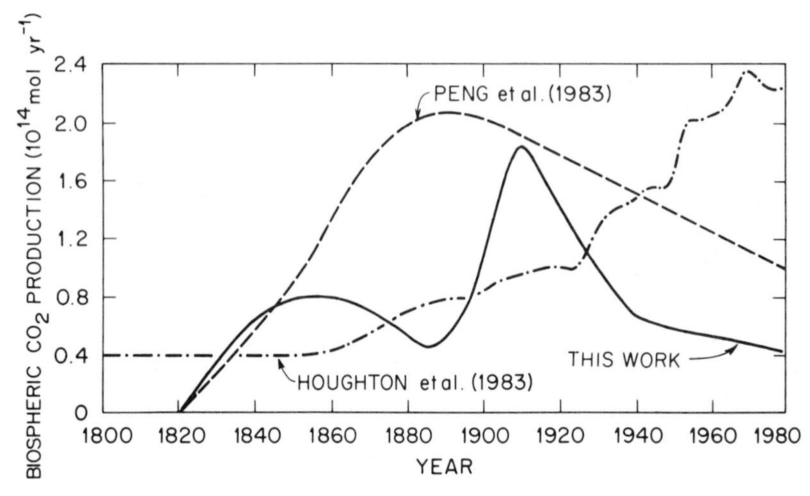

FIGURE 8.2. Comparison of the terrestrial biosphere CO_2 input function derived from this work with that derived by Peng et al. (1983) using the Freyer and Belacy (1983) composite ^{13}C record. The time history of CO_2 release from the terrestrial biosphere, as derived from land-use data (Houghton et al. 1983), is also shown for comparison.

of model-computed curves. The model residual curve represents the ^{13}C changes needed in combination with those caused by the fossil fuel CO_2 inputs to produce the model fit to the data in the revised $^{13}C/^{12}C$ record. This residual is attributed to the input of CO_2 generated by deforestation and soil manipulation.

The estimate of the time history of the forest plus soil CO_2 input needed to produce the residual curve shown in Fig. 8.1 is presented in Fig. 8.2. Also shown in Fig. 8.2 is the time history of the CO_2 input from the same

TABLE 8.1. Results of $\delta^{13}C$ deconvolution.

	Freyer's revised ^{13}C data (this volume)	Freyer and Belacy's ^{13}C data (1983)
Integrated fossil fuel CO_2 input as of 1980	14×10^{15} mol	14×10^{15} mol
Integrated terrestrial biosphere CO_2 input as of 1980	12×10^{15} mol	22×10^{15} mol
ΔpCO_2 from fossil fuel (1958 to 1980)	24×10^{-6} atm	24×10^{-6} atm
ΔpCO_2 from biosphere (1958 to 1980)	1×10^{-6} atm	4×10^{-6} atm
Pre-1850 atmospheric pCO_2	266×10^{-6} atm	243×10^{-6} atm
Ice-core pCO_2 (Barnola et al. 1983; Friedli et al. 1984)	$258–289 \times 10^{-6}$ atm	

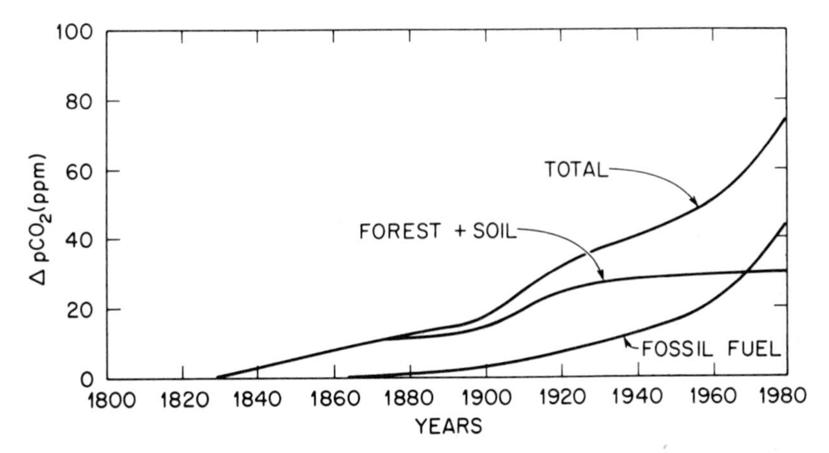

FIGURE 8.3. Increase in atmospheric pCO_2 versus time as obtained from the combined forest plus soil and fossil fuel CO_2 inputs (*curve* marked "total"). Also shown are the individual contributions of the two components. As can be seen over the 20-year period for which we have an atmospheric CO_2 record, the trend is dominated by the fossil fuel CO_2 contribution.

sources obtained from the previous Freyer and Belacy (1983) composite ^{13}C record and from the analysis of Houghton et al. (1983) based on historical land-use data. It is noted that two maxima in CO_2 release are obtained from the revised ^{13}C record; they are centered at around 1860 AD and 1910 AD. The integrated amount of CO_2 releases from this source as of 1980 AD is given in Table 8.1. The previous estimate of 22×10^{15} mol

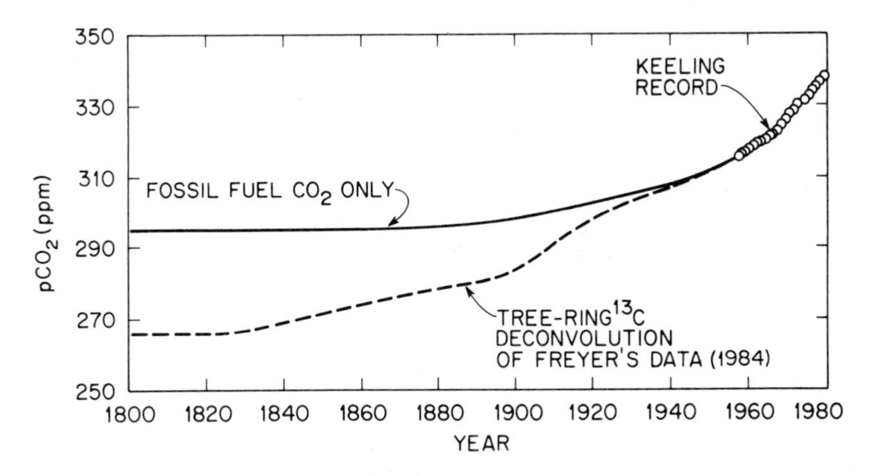

FIGURE 8.4. Time history of atmospheric pCO_2 derived from the deconvolution of tree-ring-based ^{13}C record. The model-based curve for no biospheric input is also shown. The Keeling record is from Keeling et al. (1982); Freyer's data are given in a chapter by Freyer in *this volume*.

(based on Freyer and Belacy 1983 data) is reduced to 12×10^{15} mol using the revised $\delta^{13}C$ record. The latter value for the total terrestrial biosphere CO_2 release is about 90% of the integrated fossil fuel CO_2 release over the same period.

The atmospheric CO_2 increase generated by the terrestrial biosphere scenario alone is shown in Fig. 8.3. Its contribution to increases in atmospheric CO_2 appears to have been negligible since 1940. The actual contribution over the last 20 years (the period of time for which a reliable atmospheric CO_2 record exists) is only 1×10^{-6} atm (see Table 8.1). This is even smaller than the previous estimate (about 4×10^{-6} atm) based on the Freyer and Belacy (1983) record. The fossil fuel contribution over this time interval computed by the same model is about 24×10^{-6} atm. If the revised tree-ring record is an accurate reflection of atmospheric CO_2 history, the atmospheric CO_2 increase over the last 20 years has been dominated by the release of fossil fuel CO_2. These results reinforce our previous conclusion (Peng et al. 1983) that oceanic uptake of CO_2 released from the terrestrial biosphere in the past may have matched the new production from biospheric disturbance during the last 20 years. This is an important consideration, because if relaxation of anomalies generated by releases of terrestrial biosphere CO_2 in the past counterbalance contemporary inputs (e.g., over the last two decades), existing ocean carbon cycle models may provide reasonably accurate simulations of CO_2 uptake.

The preanthropogenic atmospheric CO_2 concentration was calculated by using the time history of forest plus soil CO_2 release derived from the

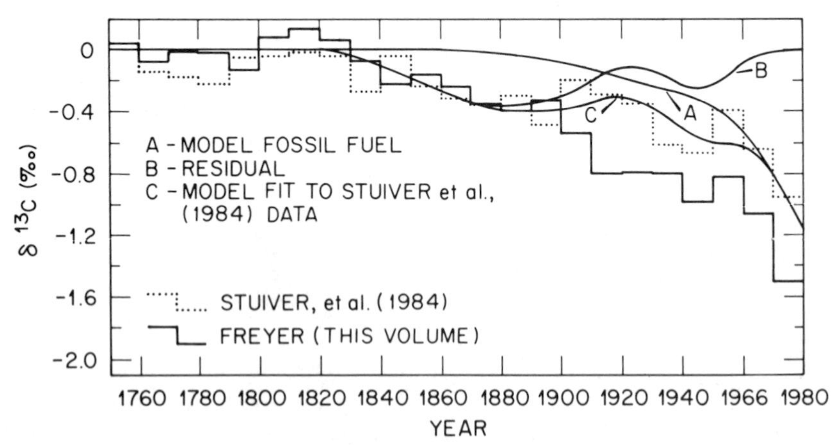

FIGURE 8.5. Model curves to fit the ^{13}C record of Stuiver et al. (1984). Terrestrial biosphere CO_2 contribution (*curve* marked "residual") to the decline in atmospheric $^{13}C/^{12}C$ ratio as obtained by subtracting the model-derived contribution of fossil fuel CO_2 (curve marked "model fossil fuel") from the model *curve* that best fits the Stuiver et al. (1984) ^{13}C record (*dotted step curve*). The *solid step curve* is Freyer's revised ^{13}C record.

tree-ring ^{13}C data. In the model calculation, the atmospheric pCO_2 curve is forced through the observed pCO_2 value of 315.5 ppm for 1958 (Keeling et al. 1982). As shown in Fig. 8.4, the scenario derived from Freyer's revised ^{13}C data yields a pre-1800 pCO_2 of 266×10^{-6} atm. This is to be compared with our previous estimate of 243×10^{-6} atm, using Freyer and Belacy's (1983) ^{13}C data.

Ice-core studies (Barnola et al. 1983; Friedli et al. 1984) suggest that the preanthropogenic atmospheric pCO_2 was in the 258 to 289 ppm range. The results of our analysis of the revised tree-ring ^{13}C record would be consistent with these ice-core data.

Stuiver et al. (1984; see also Stuiver, this volume) reported additional $\delta^{13}C$ records of 11 trees along the Pacific Coast of North and South America. They chose six trees (one from 40°S and five from 37°N to 57°N) to make up a composite $\delta^{13}C$ record for the purpose of analysis. As shown in Figure 8.5, the magnitude of decrease in $\delta^{13}C$ since 1800 AD is about 0.5‰ less than that of Freyer's (this volume) curve. To compare our results more exactly with those of Stuiver et al. (1984), the same model computation processes as described earlier were applied to fit the $\delta^{13}C$ record of Stuiver et al. The resulting model-generated curves are shown in Fig. 8.5. The corresponding time history of biospheric CO_2 release is shown in Fig. 8.6. Comparison of results based on the two methods (Fig. 8.6) shows that the biospheric CO_2 release functions for both methods are essentially the same between 1800 and 1910 AD. Significant deviations are apparent after 1910 AD. Our method using a smooth curve to represent the $\delta^{13}C$ data (C in Fig. 8.5) gives a net integrated biospheric CO_2 release

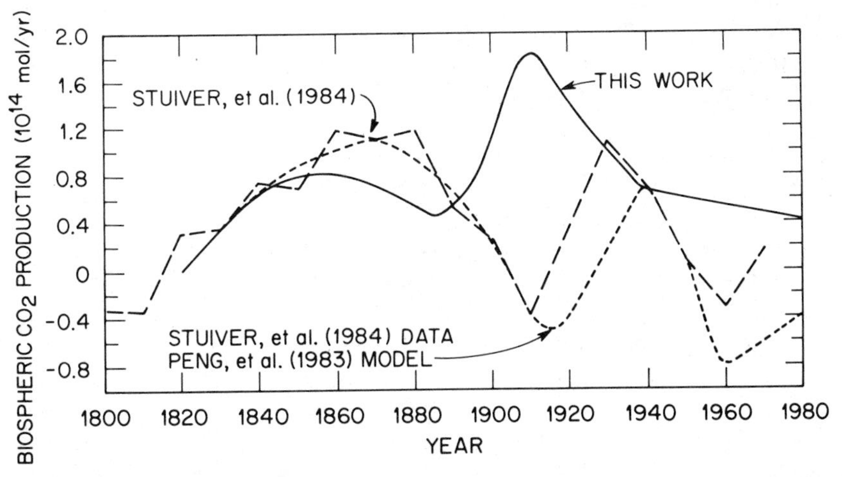

FIGURE 8.6. Comparison of the terrestrial biosphere CO_2 input function derived from this work with that derived using the Stuiver et al. (1984) ^{13}C record. The decadal mean input function as derived by Stuiver et al. using their "direct" deconvolution method is also shown for comparison.

of 2×10^{15} mol for the period between 1860 and 1975, whereas the step function (dotted line in Fig. 8.5) used by Stuiver et al. (1984) gives 2.2×10^{15} mol for the same period. On the basis of the characteristics of this $\delta^{13}C$ record, the land biosphere alternately acts as a sink and source of CO_2 to the atmosphere.

The preanthropogenic atmospheric CO_2 level was calculated by Stuiver et al. (1984) by requiring that the calculated value for the year 1965 match the measured value of 320 ppm (Keeling et al. 1982). The atmospheric pCO_2 levels obtained by Stuiver et al. range from 240 to 310 ppm in the period 235 to 1850 AD, with an average of about 276 ppm. Using our method, the Stuiver et al. (1984) ^{13}C data yield a preindustrial pCO_2 value of 283 ppm, also consistent with the ice-core data.

Conclusions

When more tree-ring ^{13}C data became available, Freyer (this volume) revised his composite ^{13}C record for the Northern Hemisphere. The global carbon model of Peng et al. (1983) was used to reevaluate the impact of biospheric contributions to atmospheric CO_2 history based on this revised ^{13}C record. The following results were obtained:

1. The total release of CO_2 from the terrestrial land biosphere (as of 1980) was estimated to be about 12×10^{15} mol, or 55% of our previous estimate based on the data of Freyer and Belacy (1983). This total release estimate is very similar to that obtained by Stuiver et al. (1984), even though the period of release is much longer in the latter case. The calculated biospheric CO_2 input was about 90% of the fossil fuel CO_2 input.
2. A previous conclusion—that uptake by the ocean of CO_2 released from the terrestrial biosphere in earlier years may have matched the new production by this source during the last 20 years—is reinforced by this study.
3. The atmospheric CO_2 anomaly observed over the last 20 years appears to be dominated by the release of fossil fuel CO_2.
4. The pre-1850 atmospheric pCO_2 is estimated to be 266 ppm. This is somewhat lower than the 276 to 283 ppm range obtained using Stuiver et al. (1984) tree-ring data. However, both results are consistent with data from ice-core studies.

Considering the relatively large changes in the revised global composite ^{13}C record from tree rings produced by addition of newer ^{13}C data, and the corresponding changes in atmospheric fluxes reported here, it is apparent that ^{13}C-based estimates of atmospheric CO_2 variations are highly volatile. Further modifications are likely in the future both as a result of expanded data bases and improved understanding of the process of isotope fractionation in trees.

Acknowledgments

We thank T. J. Blasing, W. S. Broecker, and G. G. Killough for reviewing this manuscript. Research at Oak Ridge National Laboratory was supported jointly by the National Science Foundation's Ecosystem Studies Program under Interagency Agreement BSR 8115316 A03 and by the Carbon Dioxide Research Division, Office of Energy Research, U.S. Department of Energy, under contract DE-AC05-84OR21400 with Martin Marietta Energy Systems, Inc. This is Publication No. 2328, Environmental Sciences Division, ORNL.

References

Barnola, J. M., D. Raynaud, A. Neftel, and H. Oeschger. 1983. Comparison of CO_2 measurements by two laboratories on air from bubbles in polar ice. Nature 303:410–413.

Francey, R. J. 1981. Tasmanian tree rings belie suggested anthropogenic $^{13}C/^{12}C$ trends. Nature 290:232–235.

Freyer, H. D. and N. Belacy. 1983. $^{13}C/^{12}C$ records in Northern Hemisphere trees during the past 500 years: anthropogenic impact and climatic superpositions. J. Geophys. Res. 88:6844–6852.

Friedli, H., E. Moor, H. Oeschger, U. Siegenthaler, and B. Stauffer. 1984. $^{13}C/^{12}C$ ratios in CO_2 extracted from Antarctic ice. Geophys. Res. Lett. 11:1145–1148.

Houghton, R. A., J. E. Hobbie, J. M. Melillo, B. Moore, B. J. Peterson, G. R. Shaver, and G. M. Woodwell. 1983. Changes in the carbon content of terrestrial biota and soils between 1860 and 1980: a net release of CO_2 to the atmosphere. Ecol. Monogr. 53:235–262.

Keeling, C. D., R. B. Bacastow, and P. Tans. 1980. Predicted shift in the $^{13}C/^{12}C$ ratio of atmospheric carbon dioxide. Geophys. Res. Lett. 7:505–508.

Keeling, C. D., R. B. Bacastow, and T. P. Whorf. 1982. Measurements of the concentration of carbon dioxide at Mauna Loa Observatory, Hawaii. In W. C. Clark (ed.), Carbon Dioxide Review: 1982, pp. 377–385. Oxford University Press, New York.

Leavitt, S. W. and A. Long. 1983. An atmospheric $^{13}C/^{12}C$ reconstruction generated through removal of climatic effects from tree-ring $^{13}C/^{12}C$ measurements. Tellus 35:92–102.

Peng, T.-H., W. S. Broecker, H. D. Freyer, and S. Trumbore. 1983. A deconvolution of the tree rings based $\delta^{13}C$ record. J. Geophys. Res. 88:3609–3620.

Schwarz, H. P. 1970. The stable isotopes of carbon. In K. H. Wedepohl (ed.), Handbook of Geochemistry, pp. 1–16. Springer-Verlag, New York.

Stuiver, M., R. L. Burk, and P. D. Quay. 1984. $^{13}C/^{12}C$ ratios and the transfer of biospheric carbon to the atmosphere. J. Geophys. Res. 89:11731–11748.

Tans, P. 1981. $^{13}C/^{12}C$ of industrial CO_2. In B. Bolin (ed.), Carbon Cycle Modeling, SCOPE 16, pp. 127–129. John Wiley, New York.

9
Carbon Isotope Measurements in Baseline Air, Forest Canopy Air, and Plants

ROGER J. FRANCEY

This chapter reviews recent experimental results from a variety of Australian programs, each having potential relevance to the direct or proxy measurement of atmospheric $^{13}C/^{12}C$ variation, in particular the global variation resulting from the combustion of isotopically light carbonaceous material in fossil fuel.

Since 1977, direct measurements of $^{13}C/^{12}C$ in clean air have been made at the Cape Grim Baseline Air Pollution Station. Cape Grim is strategically located in relation to large-scale atmospheric circulation patterns over the southern oceans, and the station collects comprehensive data concerning the history of the air mass sampled. The carbon isotope measurement program has experienced many difficulties; however, a marked reduction in data scatter has occurred in recent years. A preliminary analysis of these data identifies anomalies and permits estimates of the isotopic trends accompanying changing CO_2 concentration in the atmosphere.

One proposed method of extending the record of atmospheric $^{13}C/^{12}C$ over much longer time periods involves the measurement of $^{13}C/^{12}C$ in carbon stored in tree rings. Two of the fundamental assumptions in this method can be stated as follows: (1) the average $^{13}C/^{12}C$ ratio of CO_2 assimilated by the tree from year to year is representative of that in a large volume of "free" atmosphere; and (2) the average $^{13}C/^{12}C$ ratio of cellulose in a ring bears a fixed relationship to the average $^{13}C/^{12}C$ of the CO_2 assimilated by the tree.

By means of techniques developed as part of the Cape Grim program, measurements of CO_2 concentration and isotopic ratio profiles in forest canopy air are reviewed with the aim of commenting on the first of these assumptions. A crude estimate of the average isotopic ratio of air surrounding leaves during times of net CO_2 uptake is presented, and serves to limit the magnitude of tree-ring isotopic variations that can be attributed to regional atmospheric isotope variations.

To comment on the validity of the second assumption requires an understanding of the processes involved in the transport of carbon from the air into the tree ring. Results from a recent major field study on isotopes in tree rings are briefly reviewed to assist in the understanding of these processes. These results demonstrate large systematic variations in carbon isotope variation within a plant and have recently been interpreted by reference to a quantitative model for carbon isotope fractionation during photosynthesis.

An obvious importance of physiological factors to the interpretation of isotopic trends in tree-ring studies is inherent in the model and is also suggested by large differences in measured trends in tree rings. New tree-ring data confirming these differences are presented.

Given the suggested influence of physiological factors on tree-ring isotope records, this chapter explores in a very preliminary fashion isotope records from C_4 plants with a very different metabolism to the C_3 plant material employed hereto.

Carbon-13 in Baseline Air at Cape Grim

Methods

The Cape Grim baseline station is Australia's contribution to the UNEP/WMO Baseline Air Pollution Monitoring Network (BAPMoN). Located at 41°S, 145°E, on the far northwest tip of Tasmania, the station experiences prevailing oceanic winds from the southwest for ~60% of the time.

A description of the Cape Grim facilities and measurement programs can be found in the annual or biannual summary reports, for example in Baseline 1979-1980 (1983) or Baseline 1981-1982 (1984).

The collection of CO_2 for stable carbon isotope measurements began at Cape Grim in 1977 from a temporary caravan. In September 1981, new trapping equipment was installed in the permanent facility, and this provided all results from the beginning of 1982.

The trapping equipment is described elsewhere (Baseline 1981-1982). In brief, air is drawn at ~300 mL/min from the main intake line with its 10-m-high stainless steel mast. An alcohol bath at −50 to −70°C is used to dry the air stream, which is bled into a glass trap immersed in liquid nitrogen and controlled at a pressure of below about 20 to 30 mm Hg. A further liquid nitrogen trap prevents back streaming from the vacuum pump. After 2 h, ample CO_2 is collected in the low pressure trap (~12 mL at STP) for a mass spectrometer analysis. The CO_2 is then transferred to 100-mL glass flasks for transport to a Micromass 602D mass spectrometer at the CSIRO Division of Atmospheric Research in Aspendale.

Measurements of both mass 45/mass 44 and mass 46/mass 44 in the samples are recorded as fractional differences ($\delta^{13}C$, $\delta^{18}O$, respectively) relative to the CSIRO secondary CO_2 isotope standard, which, in turn, has been calibrated against international standards TKL and NBS-19 to an accuracy estimated at ± 0.1‰ (R. J. Francey and H. S. Goodman, private communication). In what follows, $\delta^{18}O$ refers to measured values and $\delta^{13}C$ to values corrected for an O^{17} contribution and converted to the PDB scale (Mook and Grootes 1973). Note that no correction has yet been

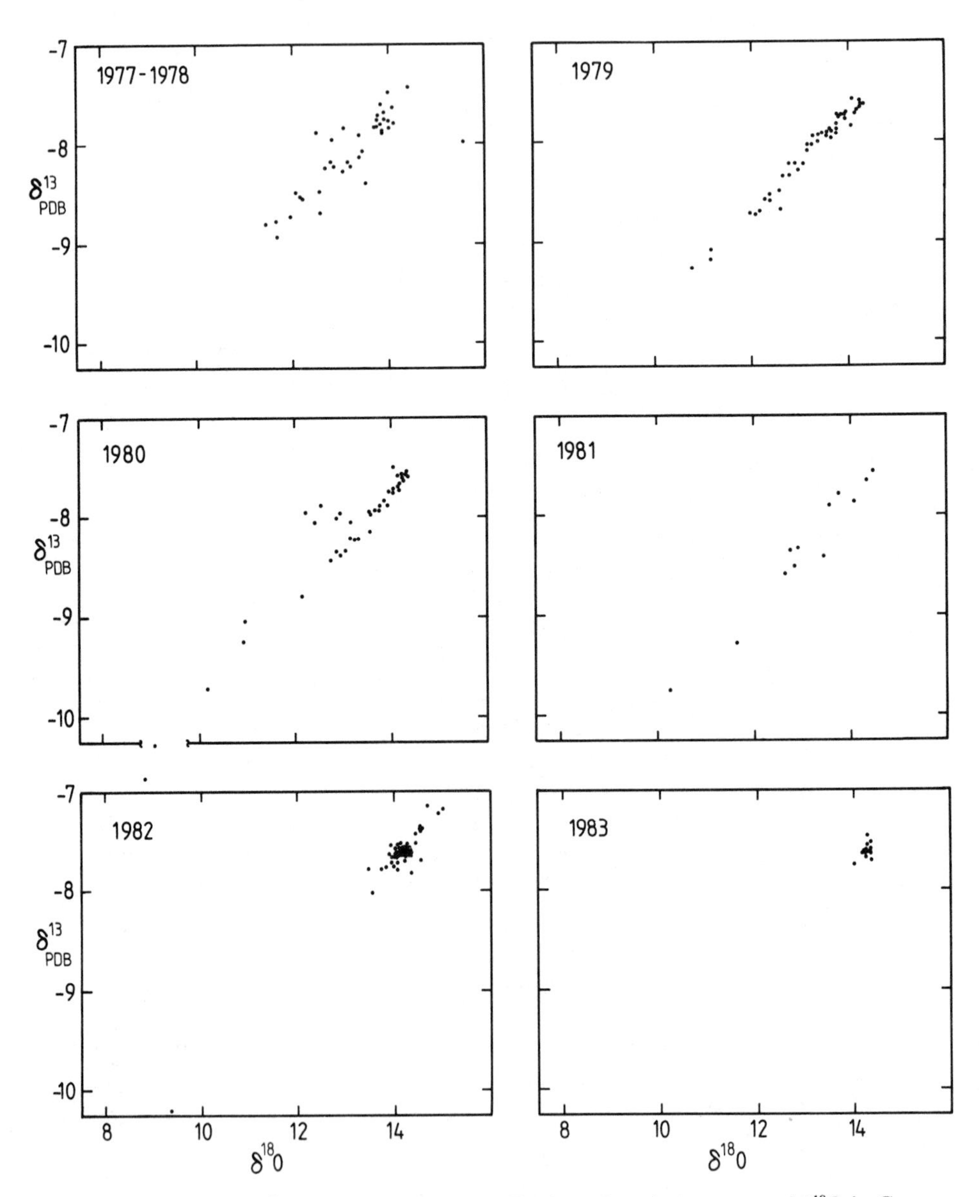

FIGURE 9.1. $\delta_{PDB}{}^{13}C$ (uncorrected for N_2O) plotted against measured $\delta^{18}O$ in Cape Grim air for years between 1977–1978 and 1983.

applied for a nitrous oxide contribution; therefore, approximately 0.3‰ should be added to all $\delta^{13}C$ values for a comparison with Keeling et al. (1979, 1980).

RESULTS

The $\delta^{13}C$ measured in the caravan between October 1977 and November 1981 are characterized by a large scatter in the measured δ values. This can be seen in Fig. 9.1, which plots $\delta^{13}C$ vs $\delta^{18}O$ for different years. In the early years, there is a striking correlation between the isotopes with a mean slope and standard deviation (1978-1982) of

$$0.51 \pm 0.08 \tag{1}$$

which is that expected for processes dependent on the molecular weight of the respective molecules. A general lack of control is evident in the early measurements, as indicated by quite random values for CO_2 yield compared to the more recent samples. The very strong correlation, most evident in the 1979 data, suggests that if the $\delta^{18}O$ value is predictable, then much of the $\delta^{13}C$ information might be recovered. This has been attempted for the pre-1981 data by Goodman (1980) by assuming constant $\delta^{18}O$ for occasions of winds from the southwest (oceanic) sector. This results in a $\delta^{13}C$ trend of $0.03 \pm 0.01‰$ year, similar to that reported by Keeling et al. (1980).

In the 1982 data of Fig. 9.1, the outlying points, which determine the $\delta^{13}C$ vs $\delta^{18}O$ slope of 0.52 ± 0.02, are, in general, identifiable with documented variation or anomaly in technique. For 1982 and 1983, all samples collected in baseline conditions (winds between 190 and 280° and condensation nuclei counts $<600/cm^3$), with no anomaly in technique, show no significant $\delta^{13}C$ or $\delta^{18}O$ correlation (having slopes 0.08 ± 0.06 and -0.06 ± 0.19 respectively).

In this chapter, attention is restricted to the 1982 and 1983 data shown in Fig. 9.2. Both the $\delta^{13}C$ and $\delta^{18}O$ for every sample are shown, connected by a bar and plotted on scales such that $\Delta\delta^{13}C/\Delta\delta^{18}O = 0.50$ [see Eq. (1)].

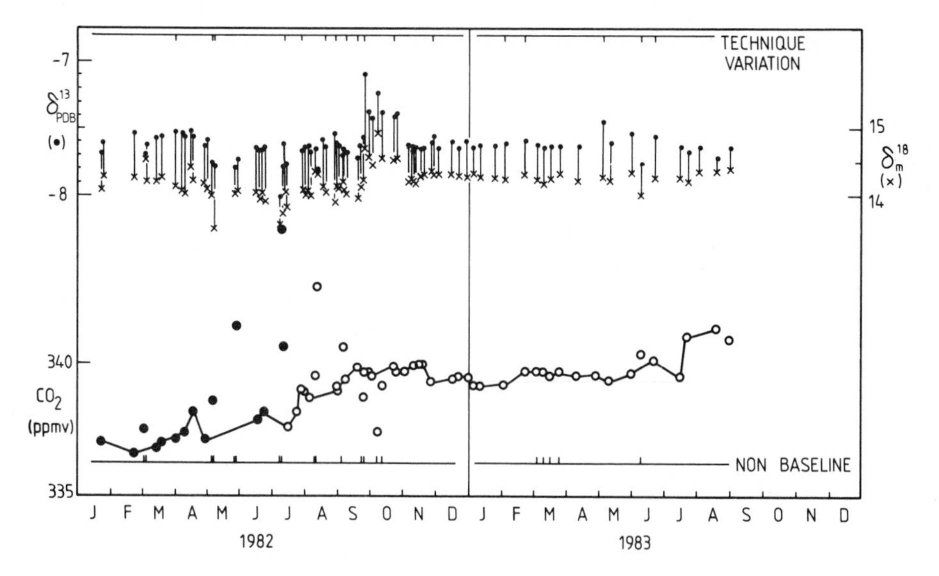

FIGURE 9.2. Cape Grim $\delta_{PDB}{}^{13}C$ (uncorrected for N_2O) measured $\delta^{18}O$ and corresponding preliminary CO_2 concentration data for 1982–1983. Samples with documented technique variation and nonbaseline conditions (see text) are indicated.

Times of technique variation or nonbaseline conditions during sampling are indicated. Technique variation includes deliberate manipulation of flow rates and trapping pressures etc., as well as errors such as pressure anomalies during trapping or transfer, low levels in liquid nitrogen traps, or mass spectrometer malfunctions.

Also shown are CO_2 concentrations measured at Cape Grim during the collection of samples. The CO_2 data depend on limited calibrations of tertiary standards and approximate corrections for carrier gas and nonlinearity effects in the URAS infrared analyzer. Estimated uncertainty is $\pm$ 0.5 ppmv.

The values of both isotopes are quite consistent in time except for well-documented technique variation or nonbaseline conditions, apart from values around September-October 1982 and a single value on May 30, 1983. No clear explanation for these anomalies has yet been discovered, although September-October 1982 was a period of considerable disruption (main intake mast lowered and cleaned, trapping apparatus replumbed, new flowmeter, etc.). On 11 occasions of steady CO_2 conditions, two traps were made per day as an indication of experimental accuracy. The mean difference between the first and second trap is 0.01 $\pm$ 0.04‰ (if one pair is removed, 0.01 $\pm$ 0.02‰).

The $\delta^{13}C$ values, in baseline conditions and for no identifiable technique anomaly, are shown plotted against CO_2 concentration (C in ppmv) in Fig. 9.3. Three points, discussed above, are clearly anomalous.

A linear regression of the remainder of the points is shown and has the form

$$\delta^{13}C = 1.16\ (\pm 1.27) - 0.026\ (\pm 0.004)C\ (n = 36,\ S_{y,x} = 0.03) \quad (2)$$

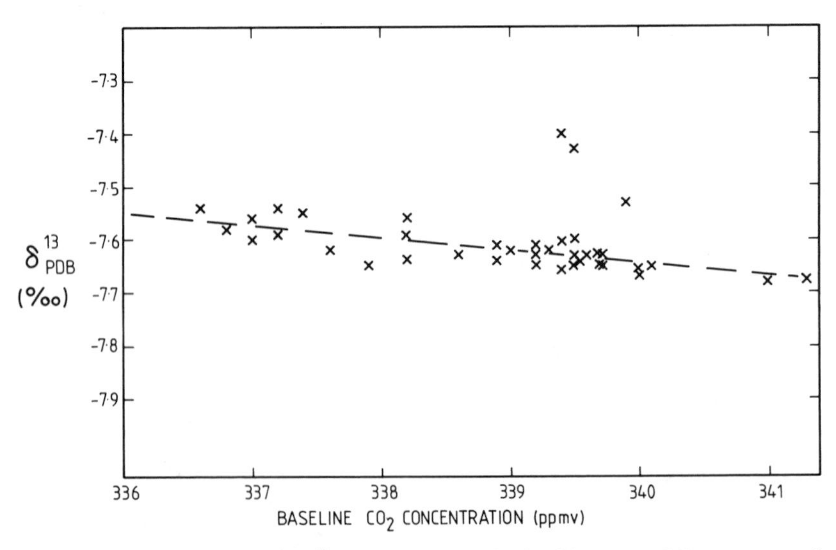

FIGURE 9.3. Atmospheric $\delta_{PDB}{}^{13}C$ (uncorrected for N_2O) versus CO_2 concentration for baseline conditions (oceanic trajectories) at Cape Grim.

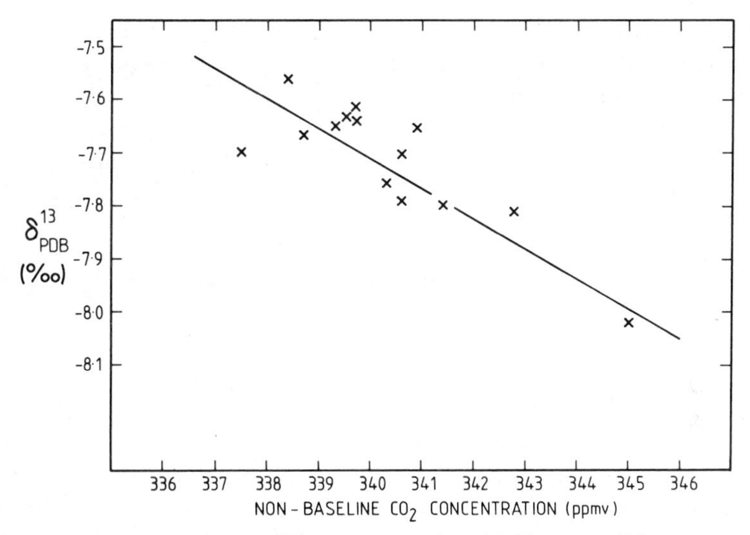

FIGURE 9.4. Atmospheric $\delta_{PDB}{}^{13}C$ (uncorrected for N_2O) versus CO_2 concentration for nonbaseline conditions (continental trajectories) at Cape Grim.

Figure 9.4 shows a similar regression for the nonbaseline values of Fig. 9.2. The regression line is of the form

$$\delta^{13}C = 9.8\ (\pm 3.3) - 0.052\ (\pm 0.010)C \tag{3}$$

DISCUSSION

There is a significant difference in the slopes (‰ per ppmv) of Eqs. (2) and (3). If the data are expressed in terms of a single mixing relationship ($\delta^{13}C \propto C^{-1}$, Keeling 1958), the intercept gives a $\delta^{13}C$ of the component added to the atmosphere; for baseline air

$$\delta^{13}C = -16.4 \pm 1.3‰ \tag{4}$$

and for nonbaseline air (air of recent continental contact)

$$\delta^{13}C = -25.3 \pm 3.3‰ \tag{5}$$

The last value (Eq. 5) is similar to that reported by Keeling et al. (1979) for Northern Hemisphere samples, and is characteristic of the $\delta^{13}C$ of carbon in most vegetation. It is also similar to $\delta^{13}C$ measured in air under forest canopies in Tasmania, as discussed below. The less negative $\delta^{13}C$ for baseline air may be indicative of an oceanic influence on the isotopic ratio of CO_2 at Cape Grim, either as a "dilutant" of a Northern Hemisphere $\delta^{13}C$ signal or as a source of "heavy" oceanic CO_2.

However, the magnitude of the parameters describing the $\delta^{13}C$-CO_2 interrelationship should be treated with some caution. For example, the baseline CO_2 increase from February to September 1982 is in phase with previous increases, but, at ~3 ppmv, the increase is relatively large (D. Beardsmore, private communication). The relationship for these data alone gives a slope of -0.028 ($\pm$ 0.008)‰/ppmv and $\delta^{13}C$ intercept of $-17 \pm$ 2.9‰; it is conceivable that final calibration and corrections of the CO_2 data decrease this change to about 2 ppmv, implying a slope of about -0.04‰/ppmv, which is much closer to the nonbaseline value. After September 1982 the parameters are -0.040 ($\pm$ 0.006)‰/ppmv and -21.1 ($\pm$ 2.3)‰, respectively. Furthermore, the trends in the isotopic data also have apparent discontinuities around September–October 1982. These presumably represent real geochemical phenomena, in which case averaging over the brief period of January 1982 to August 1983 is of limited physical relevance.

Carbon-13 in Air Under Rainforest Canopies

Back-up flask programs are maintained in conjunction with both the in situ CO_2 concentration measurements and the extraction of CO_2 for isotope measurements at Cape Grim. For concentrations, 0.5-L glass flasks are flushed and filled to 100 kPa with magnesium perchlorate-dried air from the main sampling mast, whereas for the isotope measurements, a similar procedure is used to fill 5-L glass flasks. The flasks are returned to CSIRO Aspendale for extraction and measurement.

Similar techniques were employed to measure CO_2 concentrations and isotopic ratios in rain forest air in a major study of tree rings involving expeditions to the Stanley River in Western Tasmania between 1981 and 1982. Full details of the site, methods, and results are given in Francey et al. (1984) and are discussed in a later section.

Approximately one hundred and forty 0.5-L flasks were used to measure CO_2 concentrations at heights between 0.3 and 14 m, as well as on a nearby exposed ridge. To characterize the "growth" environment, sampling was, in the main, restricted to daylight hours (0900 to 1700), with measurements from early spring to late summer at a number of typical sites featuring a closed canopy at a height of 10 to 12 m above ground. Of significance here is the result that, at 14 m, concentrations were always within about 2 ppmv of the annual mean baseline values at Cape Grim. This is also true of the mean 8-m concentration. Above about 5 m in height, the mean values are within about 5 ppmv of the Cape Grim values.

In February 1982, about twenty 5-L flasks were employed to obtain a $^{13}C/^{12}C$ profile throughout the canopy. Expressed as a mixing relationship, the regression of these values against the inverse of CO_2 concentration yields a slope of 0.04 $\pm$ 0.01‰/ppmv, suggesting mixing with CO_2 at -23 $\pm$ 1‰ [see Eq. (2) through (5)].

Several authors have suggested that regional modification of atmospheric $\delta^{13}C$ might explain long-term variations in tree-ring $\delta^{13}C$ of the magnitude of the order of 1‰ or more (e.g., Freyer and Belacy 1983). The density of Tasmanian rainforest, coupled with the fact that these measurements were generally taken under calm summer conditions, suggests a situation of relatively poor atmospheric mixing. Even so, trees with their foliage above about 5 m are exposed to CO_2 with $\delta^{13}C$ on the average being <0.2‰ in difference from that of very clean oceanic air of the southern oceans. This difference is not significantly larger than the reproducibility of $\delta^{13}C$ analyses of cellulose from uniform wood samples. Changes in average $\delta^{13}C$ from year to year in the atmosphere around an assimilating tree, with predominantly canopy top foliage, are likely to be much less than 0.2‰.

Carbon-13 Variations Within a Tree

In the Stanley River field expeditions, emphasis was placed on quantifying local environmental influences on the $\delta^{13}C$ of tree material. In addition to the atmospheric composition measurements described above, and measurements of light levels with the forest canopy, physiological data in the form of stomatal conductances and CO_2 assimilation rates were made on leaves, many of which were harvested for later $\delta^{13}C$ analysis. Further extensive sampling of needle, branch, and tree-ring material for $\delta^{13}C$ analyses was carried out. The methods and results are fully described in Francey et al. (1984).

An interpretation of the combined needle material $\delta^{13}C$, light level, and physiological data was published elsewhere (Francey et al.,1985). In this chapter a brief review is given of the results from both papers, which are perceived to be of significance to the problem of reconstructing past atmospheric $\delta^{13}C$ from tree-ring values.

The Influence of Light Level

Needle-tip material $\delta^{13}C$ near the forest floor was observed to be up to 4‰ more negative than corresponding material at canopy top. This is of a similar magnitude to, and presumably accounts for, the "juvenile effect," or ^{13}C depletion, in the innermost rings of trees, invariably observed from forest locations. Similar leaf observations have been reported (Vogel 1978; Medina and Minchin 1980) and explained in terms of biogenic CO_2 released from soil, an explanation not supported by the direct observations of CO_2 reported in the previous section.

Measurements of fully exposed (northern side) and shaded (southern side) needle tips at 14.3 m, well clear of the canopy, and for which no appreciable atmospheric gradient is conceivable, also exhibited significant $\delta^{13}C$ differences approaching 1‰.

For both the vertical and horizontal gradient situations, the physiological and light-level measurements were used to obtain estimates of p_i/p_a, the ratio of intercellular to atmospheric CO_2 partial pressures in the needles. This parameter appears in the equation

$$\delta_p C \sim \delta_a - a - (b_3 - a)\, p_i/p_a \qquad (6)$$

formulated by Farquhar et al. (1982*b*) to describe carbon isotope fractionation during photosynthesis, where a = 4.4‰, b_3 is thought to be near 27‰ (Farquhar et al. 1982*a*), and subscripts a and p refer to atmosphere and photosynthate, respectively. The application of this equation to tree-ring studies was explored in the paper by Francey and Farquhar (1982). The $\delta_p{}^{13}C$ predicted by Eq. (6), using the measured p_i/p_a (also p_i/p_a inferred from stomatal conductance and light-level measurements), agrees qualitatively with the measured needle tip and branch material $\delta^{13}C$. For both horizontal and vertical gradients, the magnitude of the predicted $\delta_p{}^{13}C$ gradient exceeds the measured differences in both types of material by about a factor of 2.

However, in view of consistent differences in $\delta^{13}C$ observed between the needle tip and adjoining branch, a quantitative assessment of the influence of light requires a decision on which material most closely represents the $\delta^{13}C$ of photosynthate laid down as wood.

Gradients Within a Branch

Strong monotonic $\delta^{13}C$ gradients of 2 to 3‰ were observed in Huon Pine branchlets from the tip to the first woody sections, with the tips more negative than the wood. A similar gradient between the leaves and tree-ring wood was reported by Leavitt and Long (1982) and attributed to additional fractionation of photosynthate.

The $\delta^{13}C$ value of branch wood (typically 200 to 300 mm from the needle tips) in exposed Huon Pine branches was not markedly different from that in the rings. Furthermore, the magnitude of the gradient (with one exception) appears relatively insensitive to position in the forest (e.g., at 14 m above the canopy, typical tip and branch wood values were −27.5 and −24.5‰ respectively, whereas for deeply shaded branches at 1 m, corresponding values were −31.0 and −28.6‰. The one exception was for a very young, deeply shaded individual tree (~1.5 m high), in which the magnitude of the gradient was reduced to 0.4‰.

Cellulose extracted from tips to branch wood of an exposed branch showed an even larger gradient (~4‰), which is consistent with the removal of more isotopically light lignin from the wood than from the tips.

In seeking a single explanation to account for the main features of these tip-to-wood gradients, Francey et al. (1985) consider two broad options:

1. As proposed by Leavitt and Long (1982), further large fractionation occurs after photosynthetic fixation. A mechanism proposed for this fractionation is "maintenance respiration," which is present in all living tissue. An appreciable isotopic fractionation during respiration (resulting in ^{13}C depleted respired CO_2) is implied. Note that the atmospheric $\delta^{13}C$ measurements do not support a significant depletion of ^{13}C relative to tip values of −27‰. Also with this option quantitative agreement with the predicted light effect (previous section) requires a significant sampling error in the measurements of p_i/p_a (e.g., due to the unusually warm dry conditions at the time). No obvious explanation for the small tip-to-wood gradient in the deeply shaded sapling is apparent.

2. Alternatively, the bulk of carbon is fixed at a $\delta^{13}C$ representative of the tree-ring $\delta^{13}C$, that is, approximately −24‰ in this case. The tip tissue, unlike tissue that is older or further "downstream," is predominantly comprised of carbon fixed by immature cells. It is implied that before the cells become fully functional, the p_i/p_a ratio is high and the $\delta_p{}^{13}C$ is correspondingly low. With this explanation good quantitative agreement with the predicted light effect (previous section) can be obtained by invoking translocation of photosynthate from the rapidly photosynthesizing (high light) branches to more shaded branches. The small tip-to-wood gradient in the shaded sapling finds explanation, via the Farquhar model, in the assimilation limit of $p_i/p_a = 1$, i.e., if the low light levels imply p_i/p_a close to 1 then additional influences of immaturity have no effect on $\delta^{13}C$.

On the basis of the evidence summarized here, the second option is preferred.

Discussion

The existence of these strong systematic $\delta^{13}C$ effects, generally much larger than anticipated "signals" resulting from large-scale atmospheric $\delta^{13}C$ changes, and the physical insight provided by the Farquhar et al. (1982*b*) model permit identification of problem areas for the reconstruction of past atmospheric isotope values using tree rings.

Most serious are phenomena that change p_i/p_a in conjunction with changing atmospheric CO_2 (p_a) levels (Francey and Farquhar 1982). Fertilization or poisoning by other industrial pollutants, accompanying fossil fuel combustion, fall into this category.

Other potential causes of long-term p_i/p_a changes that will influence large populations of trees include changes in light level (e.g., changes in shading, particularly in plantations or even-age stands); changes in competition for mineral nutrients, water, etc.; or changes in the average age of photosynthesizing and respiring leaves.

In the Stanley River expeditions, trees have been sampled from forest communities thought to have been stable over thousands of years. Individuals have been selected featuring canopy top foliage only and of considerable age (up to 1100 years old), in the hope of minimizing some of these influences over recent centuries.

Historic Carbon-13 Trends

Francey (1981) published $\delta^{13}C$ data from eight Tasmanian trees from a wide variety of habitats (including an isolated tree on an exposed mountain ridge) showing relatively constant average $\delta^{13}C$ over the last 200 years. This result is in marked contrast to the large trends observed by Freyer and Belacy (1983) and employed in Peng et al. (1983) to reconstruct past atmospheric $\delta^{13}C$ behavior.

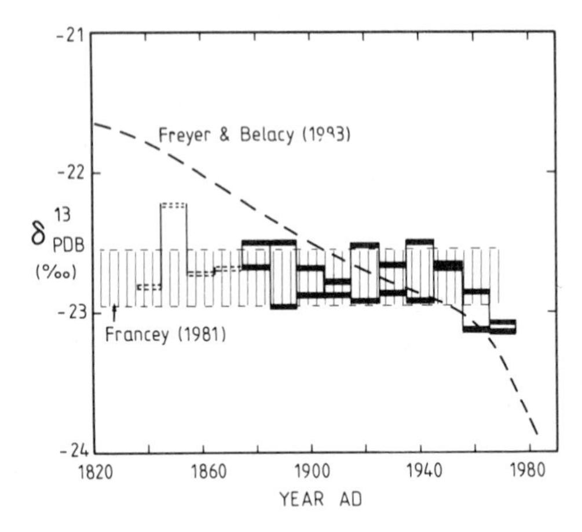

FIGURE 9.5. $\delta_{PDB}{}^{13}C$ values from pooled cores of seven *Phyllocladus aspleniifolius* from Bathurst Harbor. Hatching represents the previous Tasmanian results of Francey (1981), whereas the dashed curve represents Northern Hemisphere results of Freyer and Belacy (1983), normalized to the 1900–1950 Tasmanian data.

NEW TREE RING RESULTS

Further Tasmanian tree ring $\delta^{13}C$ data are shown in Fig. 9.5. A total of 11 cores from seven individual trees were pooled in 10-year blocks for duplicate combustion and analysis.

All trees were *Phyllocladus aspleniifolius* from a small exposed island in Bathurst Harbor on the Tasmanian southwest coast. *Phyllocladus* is characterized by having all its foliage above general canopy height. For example, on the Stanley River expeditions (Francey et al., 1984), 71% of *Phyllocladus* sampled were categorized as having all foliage above the mean canopy height, with another 23% having lowest foliage within 2 to 3 m of canopy top at 12 to 13 m. Before 1870, some trees experienced juvenile effects (obvious in both $\delta^{13}C$ and ring width), and these core sections were excluded.

Superimposed on these data is the Freyer and Belacy (1983) curve and hatching encompassing the Francey (1981) data, both normalized to the Bathurst Harbor data between 1900 and 1950. (For seven of the combined core samples, with sufficient material for both cellulose extractions and whole wood analysis, the respective mean values were -21.5 and -23.1‰, with a correlation coefficient of 0.96; this confirms previous observations of the suitability of whole wood analyses for this purpose.)

Throughout the century (1860 to 1960), the new data support the claim for relatively constant $\delta^{13}C$ in the Tasmanian trees. Subsequently, a decrease in $\delta^{13}C$ very similar to that observed in the atmosphere between 1956 and 1978 (Keeling et al. 1980) is now evident. By comparison with $\delta^{13}C$ data from Freyer and Belacy (1983), used by Peng et al. (1983) to predict a pre-1850 atmospheric concentration of 245 ppm, a higher value

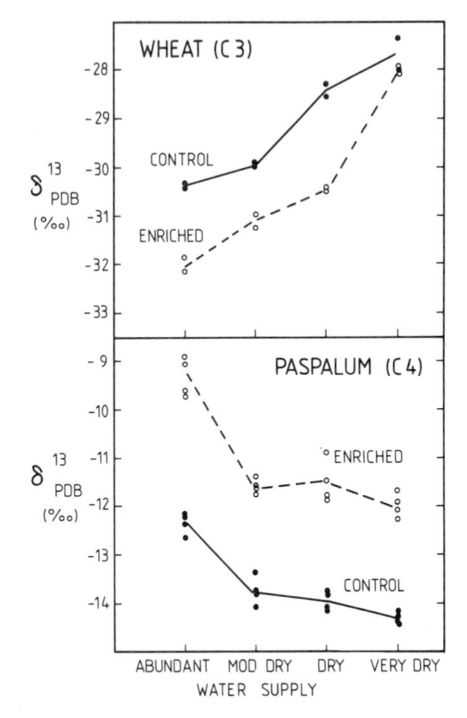

FIGURE 9.6. An illustration of the physiological influences on $\delta^{13}C$ in the different responses of wheat (C_3) and *paspalum* (C_4) grown in different water regimens. Control and enriched samples refer to CO_2 environments of 340 and 590 ppmv, respectively. Note that the *Paspalum*-enriched absolute values are uncertain because of the uncertainty in the $\delta^{13}C$ of the enriching CO_2.

of 280 to 290 ppmv is implied by these new results (if an absence of physiological effects is assumed).

RESULTS FOR C_4 SPECIES

The suggestion above that physiology might influence the $\delta^{13}C$ trends in tree rings has prompted an exploratory analysis of $\delta^{13}C$ behavior in C_4 plants.

In the context of isotopic discrimination there are important differences, both physical and chemical, in the manner in which carbon is fixed in the C_3 and C_4 photosynthetic pathways. A demonstration of these differences has been obtained by an analysis of C_3 plant wheat (*Triticum aestivum* L.) and C_4 plant (*Paspalum*) for four regimens of water availability (Fig. 9.6). The $\delta^{13}C$ trends in the wheat plants for both control and CO_2-enriched treatments are in agreement with predictions by Farquhar et al. (1982*b*), who present a form of Eq. (6) in terms of the water use efficiency of leaves. The opposite trend with water availability (and p_i/p_a) in the *Paspalum* can be explained using a recent model for carbon isotope fractionation in C_4 plants by Farquhar (1983), which in simplified form gives

$$\delta_p{}^{13}C \simeq \delta_a{}^{13}C - a - (b_4 + b_3\,\Phi - a)\,p_i/p_a \tag{7}$$

(see Eq. 6). Here $b_4 = -5.7$ at 25°C, and Φ is the proportion of carbon originally fixed by phospho*enol*pyruvate carboxylation, which subsequently leaks out of the bundle sheath back into the mesophyll cells.

The $\delta^{13}C$ of the control *Paspalum* and the C_4 type are of a value suggesting a Φ of 0.2 to 0.3 (Farquhar 1982), which in Eq. (7) produces a $\delta^{13}C$ dependence of the sign observed. (Note that the absolute $\delta^{13}C$ values of CO_2-enriched *Paspalum* are quite uncertain because of the lack of knowledge of the $\delta^{13}C$ of the enriching CO_2.)

In general, C_4 species are not long-lived; however, modern and historic C_4 material from plants in the form of refined sugar (*Saccharum officinales*) was obtained from four Southern Hemisphere mills; the oldest plant was harvested in 1941. The $\delta^{13}C$ differences between modern and historic C_4 material for the four mills are shown in Table 9.1, compared to values read from the tree-ring composite $\delta^{13}C$ curve of Freyer and Belacy (1983), also from the *Phyllocladus* data of Fig. 9.5. The *Phyllocladus* differences were obtained from a one-dimensional global carbon cycle model match to the data, including both fossil fuel and net historic biospheric release (I.G. Enting, this volume). The results are variable, with the Fiji data showing a $\delta^{13}C$ increase in time. The three Australian mills show decreases in $\delta^{13}C$ which, in general, are significantly greater than that in the Tasmanian trees.

Discussion

There is undoubtedly considerable significance in the different trends of $\delta^{13}C$ observed in plant material, for example, in the trees listed in Table 9.1, in those of Stuiver (1982) and Stuiver (this volume), Leavitt and Long (1983), and the "forest" trees of Freyer and Belacy (1983) and Freyer (this volume), as well as in the Australian sugar plants. However, the existence of such large differences, as well as the emerging theoretical framework for the interpretation of isotopic data, make it impossible to attribute any one set to atmospheric isotope behavior, without much more detailed information on the plant physiology over these time periods.

The processes that most influence the fractionation of carbon, essentially partial pressure gradients, are quite unlikely to be simply related to other gross features such as ring width or area. It is also unlikely that averaging results from a large number of arbitrarily selected trees will discriminate between atmospheric behavior and physiological effects.

TABLE 9.1. Differences of $\delta^{13}C$ in C_4 material between years of historic and modern samples of sugar (*Sacchurum officiales*) from four mills.

		$\Delta\delta^{13}C$(‰)		
Sugar mill	Periods	Sacchurum	Freyer and Belacy	Phyllocladus
Condong*	1963, 1981	-1.3 ± 0.1	-0.8	-0.4
Pyrmont*	1941, 1981	-1.0 ± 0.2	-1.0	-0.5
Macknade*	1969, 1981	-0.4 ± 0.1	-0.6	-0.3
Fiji Islands	1969, 1981	$+0.5 \pm 0.1$	-0.6	-0.3

Also included are corresponding $\delta^{13}C$ data from Freyer and Belacy (1983) and from the *Phyllocladus* of Fig. 9.5 (see text).
*In Australia.

In the immediate future, the flow of information in tree-ring isotope work is likely to be from the geophysical (particularly over recent decades) toward plant physiology.

Acknowledgments

Ian Helmond at Cape Grim and Helen Goodman, David Beardsmore, and Neville Robinson at C.S.I.R.O., Aspendale, assisted with the collection of the Cape Grim data reported here. A NERDDC grant assisted the Stanley River expedition; the many personnel involved are acknowledged in Francey et al. (1984). Dr. Roger Gifford, C.S.I.R.O. Division of Plant Industry, Canberra, provided the wheat and *Paspalum* samples and contributed to the analysis of their isotopic composition, and Dr. John Kering of CSR Limited provided sugar samples. Valuable discussions were held with Dr. Graeme Pearman and Dr. Ian Enting of C.S.I.R.O., Aspendale, and Dr. Graham Farquhar and John Evans of ANU, Canberra.

References

Baseline 1979–1980. 1983. Department of Science and Technology, Australian Government Publishing Service.

Farquhar, G. D. 1983. On the nature of carbon isotope discrimination in C_4 species. Aust. J. Plant Physiol. 10:205–226.

Farquhar, G. D., M. C. Ball, S. Von Caemmerer, and Z. Roksandic. 1982*a*. Effect of salinity and humidity on $\delta^{13}C$ value of halophytes—evidence for diffusional isotope fractionation determined by the ratio of intercellular/atmospheric partial pressure of CO_2 under different environmental conditions. Oecologia 52:121–124.

Farquhar, G. D., M. H. O'Leary, and J. A. Berry. 1982*b*. On the relationship-between carbon isotope discrimination and the intercellular carbon dioxide concentration in leaves. Aust. J. Plant Physiol. 9:121–137.

Francey, R. J. 1981. Tasmanian tree rings belie suggested anthropogenic $^{13}C/^{12}C$ trends. Nature 290:232–235.

Francey, R. J. (ed.). 1984. Baseline 1981–1982. Australian Government Publishing Service, Department of Science and Technology.

Francey, R. J. and G. D. Farquhar. 1982. An explanation of $^{13}C/^{12}C$ variations in tree rings. Nature 297:28–31.

Francey, R. J., M. Barbetti, T. Bird, D. Beardsmore, W. Coupland, J. E. Dolezal, G. D. Farquhar, R. G. Flynn, R. J. Fraser, R. M. Gifford, H. S. Goodman, B. Kunda, S. McPhail, G. Nanson, G. I. Pearman, N. G. Richards, T. D. Sharkey, R. B. Temple, and B. Weir. 1984. Isotopes in tree rings-Stanley River collections 1981/1982. CSIRO Division of Atmospheric Research Technical Paper No. 4.

Francey, R. J., R. M. Gifford, T. D. Sharkey, and B. Weir. 1985. Physiological influences on carbon isotope discrimination in Huon Pines. Oecologia 66:211–218.

Freyer, H. D. and N. Belacy. 1983. $^{13}C/^{12}C$ records in northern hemispheric trees during the past 500 years—anthropogenic impact and climatic superpositions. J. Geophys. Res. 88:6844–6852.

Goodman, H. S. 1980. The $^{13}C/^{12}C$ ratios of atmospheric carbon dioxide at the Australian Baseline Station, Cape Grim. In G. I. Pearman (ed.). Carbon Dioxide and Climate: Australian Research, pp. 111–114. Australian Academy of Science, Canberra.

Keeling, C. D. 1958. The concentration and isotopic abundances of atmospheric carbon dioxide in rural areas. Geochim. Cosmochim. Acta 13:322–334.

Keeling, C. D., R. B. Bacastow, and P. P. Tans. 1980. Predicted shift in the $^{13}C/^{12}C$ ratio of atmospheric carbon dioxide. Geophys. Res. Lett. 7:505–508.

Keeling, C. D., W. G. Mook, and P. P. Tans. 1979. Recent trends in the $^{13}C/^{12}C$ ratio of atmospheric carbon dioxide. Nature 277:121–123.

Leavitt, S. W. and A. Long. 1982. Evidence for $^{13}C/^{12}C$ fractionation between tree leaves and wood. Nature 298:742–744.

Leavitt, S. W. and A. Long. 1983. An atmospheric reconstruction generated through removal of climate effects from tree-ring $^{13}C/^{12}C$ measurements. Tellus 35B:92–102.

Medina, E. and P. Minchin. 1980. Stratification of $\delta^{13}CC$ values of leaves in Amazonian rain forests. Oecologia 45:377–378.

Mook, W. G. and P. M. Grootes. 1973. The measuring procedure and corrections for the high precision mass-spectrometric analysis of isotopic abundance ratios, especially referring to carbon, oxygen, and nitrogen. Int. J. Mass Spectrom. Ion Phys. 12:273–298.

Peng, T. H., W. S. Broecker, H. D. Freyer, and S. Trumbore. 1983. A deconvolution of the tree ring based $\delta^{13}C$ record. J. Geophys. Res. 88:3609–3620.

Stuiver, M. 1982. The history of the atmosphere as recorded by carbon isotopes. In E. D. Goldberg (ed.), Atmospheric Chemistry. Dahlem Konferenzen, pp. 159–179. Springer-Verlag, Berlin, Heidelberg, New York.

Vogel, J. C. 1978. Recycling of carbon in a forest environment. Oecol. Plant. 13:89–94.

10
Estimating Changes in the Carbon Content of Terrestrial Ecosystems from Historical Data

RICHARD A. HOUGHTON

When forests are cleared for agricultural crops, the carbon stored originally in trees is oxidized and released to the atmosphere, either rapidly if the trees are burned or slowly if they are left on the ground to decay. Similarly, the organic matter of soil is reduced through cultivation. Such reductions in the carbon stocks of terrestrial systems occur with the harvest of forests for wood and with the clearing of forests for cropland, pasture, or other uses. On the other hand, the regrowth of forests following harvest, the abandonment of agriculture, or the establishment of plantations increases the storage of carbon on land, both in vegetation and in soils. The balance between the clearing and regrowth of forests is the major factor in determining changes in the net storage of carbon in terrestrial systems. Nonforested systems can also lose or accumulate carbon, such as when grasslands are converted to agriculture; however, the changes in carbon per unit area are much smaller than for forests.

In addition to changes in the stocks of terrestrial carbon brought about by changes in land use, there is the potential for changes in systems not directly disturbed by man. These changes include the possibility of increased or decreased storage of carbon due to increased atmospheric CO_2, changes in climate, or changes in the abundance of nutrients or toxins (Houghton and Woodwell 1983). These environmental changes include both natural cycles and man-caused effects.

Studies based on changes in the use of land (Revelle and Munk 1977; Moore et al. 1981; Houghton et al. 1983; Richards et al. 1983; Woodwell et al. 1983) and on isotopes of carbon in tree rings (Stuiver 1978; Peng et al. 1983; Emanuel et al. 1984) have revealed a net release of carbon from terrestrial ecosystems to the atmosphere in the last century—a release of the same magnitude as the release of carbon from fossil fuels. However, the estimates of the time course of the biotic release differ widely. Most studies based on isotopes show a maximum biotic release in the late 19th or early 20th century with a progressively decreasing release to 1980. In contrast, a study based on changes in land use shows a generally increasing rate of release reaching a maximum between 1950 and the present (Moore et al. 1981; Houghton et al. 1983).

To a large extent, the difference between the two types of analyses is probably due to errors in measurements, assumptions, and understanding of the processes involved; that is, neither method is without valid criticism. On the other hand, the two analyses do not measure the same thing: studies based on isotopes include both fluxes of carbon due to changes in land use and fluxes due to changes in the environment, whereas studies based on changes in land use consider only the former. If both studies could be improved, the difference might define the effect of environmental change on the biota. These effects on the storage of biotic carbon are virtually unknown; arguments for an increased or for a decreased storage of carbon with increasing temperature have both been advanced.

A Global Study by the Marine Biological Laboratory (MBL)

A Model

Empirical evidence of changes in the stocks of carbon in terrestrial systems following disturbance formed the basis of a model (the MBL Terrestrial Carbon Model) developed to calculate the net annual flux of carbon between the land biota and the atmosphere. A description of the model and the data used in previous analyses has been presented in detail elsewhere (Moore et al. 1981; Houghton et al. 1983); therefore only a brief description is given here. In the model, the land surface of the earth was divided into 10 geographical regions, each of which could contain up to 12 types of ecosystems. In actuality, about 60 ecosystems were modeled, including cultivated lands and grazing lands. The types of land-use changes included agricultural clearing and abandonment, clearing of forests for pasture, harvest and regrowth of forests, and afforestation, or the establishment of forests where none had existed in the previous 50 years. The change in carbon stocks of vegetation and soil were described for each kind of ecosystem, in each region, and under each kind of land-use change. In addition, the model accounted for the oxidation of wood products removed from forests. The annual addition to and losses from the vegetation, soil, and wood products of each ecosystem were summed to yield a net change for all systems. The net change is equivalent to the net flux of carbon between terrestrial systems and the atmosphere as a result of changes in the use of land.

Data

The data used in this analysis were of two types: ecological and historical. The ecological data were those that described the stocks of carbon in various ecosystems and the changes in vegetation and soil organic matter following disturbance. The historical data provided the volumes of wood extracted from forests, the areas of arable or cultivated land, and the areas of grazing land. Reliable estimates of forest area worldwide prior to about 1950 were virtually unavailable.

The area in agriculture after 1800 was documented for the nontropics and South and Southeast Asia from the International Institute of Agriculture (1922, 1939), *Production Yearbooks* (FAO 1949–1978), Robertson (1956), Grigg (1974), and the Canadian and U.S. Census Bureaus. For Latin America, tropical Africa, and North Africa-Middle East, the increase in agricultural land was calculated from changes in population (McEvedy and Jones 1978) using a constant per capita ratio for each region prior to 1950 and reducing the ratio by 40% after 1950 to account for increased agricultural yields (USDA 1965, 1970).

The harvest of wood since 1946 was based on the *Yearbook of Forest Products* (FAO, 1946–1979). Earlier estimates for temperate and boreal forests were based on numerous surveys listed in Houghton et al. (1983). The harvest of wood in tropical regions was based on per capita usage for fuel, since up to 80% of the wood harvested in these areas is used for that purpose (Arnold and Jongma 1978).

Changes in the area of pasture produced a net flux of carbon only in Latin America from 1925 to 1980. In other regions, increases in the area of pasture were assumed to have come from grasslands with no change in carbon stocks.

RESULTS

The analysis has shown that changes in the carbon content of terrestrial systems has resulted in a net release to the atmosphere of about 180×10^{15}g C between 1860 and 1980. The net annual flux has increased almost continuously during this interval (Fig. 10.1). The increase from tropical regions has been dramatic in recent decades. In the northern temperate

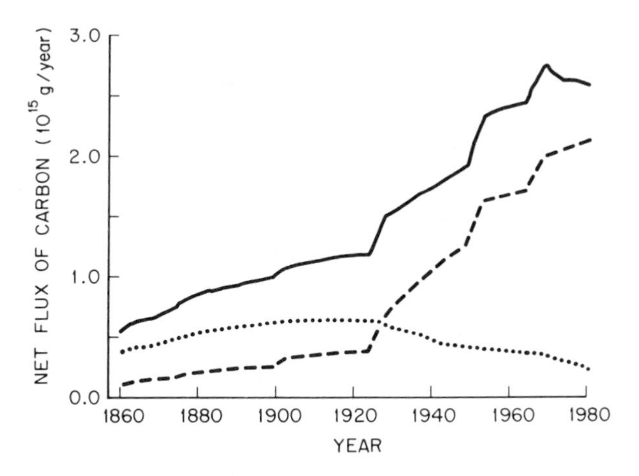

FIGURE 10.1. The net annual flux of carbon from terrestrial systems to the atmosphere for North America, Europe, and the Soviet Union (····), for Latin America, tropical Africa, and South and Southeast Asia (----), and from the entire earth (——). The results are based on the population-based analysis described by Houghton et al. (1983).

and boreal regions, the net flux was highest between 1900 and 1930 and has decreased since then. The reason for the different time courses is largely different rates of clearing for agriculture. In the northern temperate and boreal zones, the area in agriculture has declined slightly since 1960. The cause of a net release of carbon in these zones in 1980 can be attributed to the increasing rate of wood harvest in the Soviet Union. In tropical regions, the area of arable land and permanent cropland continues to increase.

The assumed relationships between population and agricultural area and between population and harvested wood for fuel produced the exponential curve of carbon release from tropical areas. This release was intermediate in 1980 between two other estimates described by Houghton et al. (1983) and Woodwell et al. (1983). These estimates will not be discussed here because they did not affect the time course of the net flux of carbon until after about 1945. The purpose of this chapter is to address the longer time course, that is, from 1860 to 1980.

Other Global Studies of the Long-Term Biotic Release

Two other studies have estimated the reduction of carbon in terrestrial ecosystems since 1860 by estimating the increase in land used for agriculture. In the first study of its kind, Revelle and Munk (1977) calculated the agricultural area in different regions of the world in 1950 and 1970. Using this information in conjunction with world population statistics and per capita agricultural land use, they calculated the agricultural area existing in 1860. Assuming that the types of ecosystems converted to agriculture were in proportion to their original areas, and calculating the changes in carbon stocks involved with such conversion, Revelle and Munk estimated that 72×10^{15} g carbon had been released from the biota and soils as a result of land-use change between 1860 and 1970. They did not indicate the time course of the release.

More recently Richards et al. (1983) calculated the worldwide change in the area of cultivated land between 1860 and 1920 and between 1920 and 1978. They determined the cultivated areas of 176 countries using historical sources and statistical handbooks, and they estimated the kinds of vegetation converted and the amounts of carbon released from vegetation and soil on the basis of historical information and world vegetation maps (Olson et al. 1983). Their estimate of the net loss of carbon from terrestrial ecosystems to the atmosphere was 62.4×10^{15} g, 29.4×10^{15}g between 1860 and 1920 and 33.0×10^{15} g between 1920 and 1978. Finer temporal resolution was not provided.

The differences between the results of the MBL study and the two other studies occur for two major reasons. First, the studies used different estimates of the types of systems converted to agriculture and different estimates of the changes in carbon stocks associated with clearing and cul-

TABLE 10.1. Comparison of three independent studies.

	Period	Release of carbon from changes in land use (10^{15} g)	Change in the area of cultivated land (10^6 ha)
Revelle and Munk 1977	1860–1970	72	853
Richards et al. 1983	1860–1978	62	852
Houghton et al. 1983			
All changes in land use	1860–1980	180	
Changes in agricultural	1860–1980	116	911
area only	1860–1970	100	813

Net release of carbon from terrestrial systems as a result of changes in land use.

tivation. These differences will be discussed below. Second, the MBL analysis considered not only the increase in cultivated land, but the increase in grazing lands and the harvest and regrowth of forests for wood products. When the release of carbon due only to agricultural expansion is identified in the MBL analysis, the three studies are closer in agreement (Table 10.1).

Uncertainties in Historical Data

Table 10.1 and Fig. 10.2 reveal that different estimates of the changes in world agricultural area are apparently in agreement. This is not to say that only one time course of CO_2 release can be generated from the three

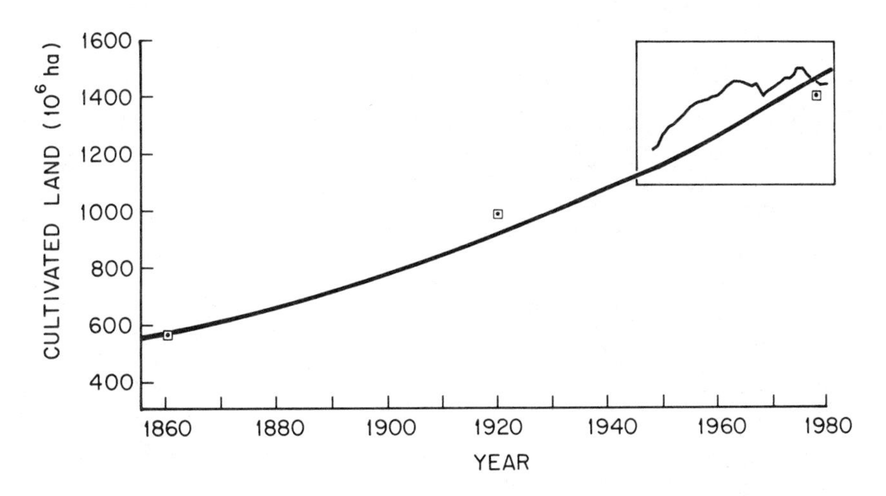

FIGURE 10.2. Estimates of cultivated area worldwide according to the population-based analysis of Houghton et al. (1983) (———), Richards et al. (1983) (⊡) and *Production Yearbooks* of the FAO (1948-1980) (*insert*).

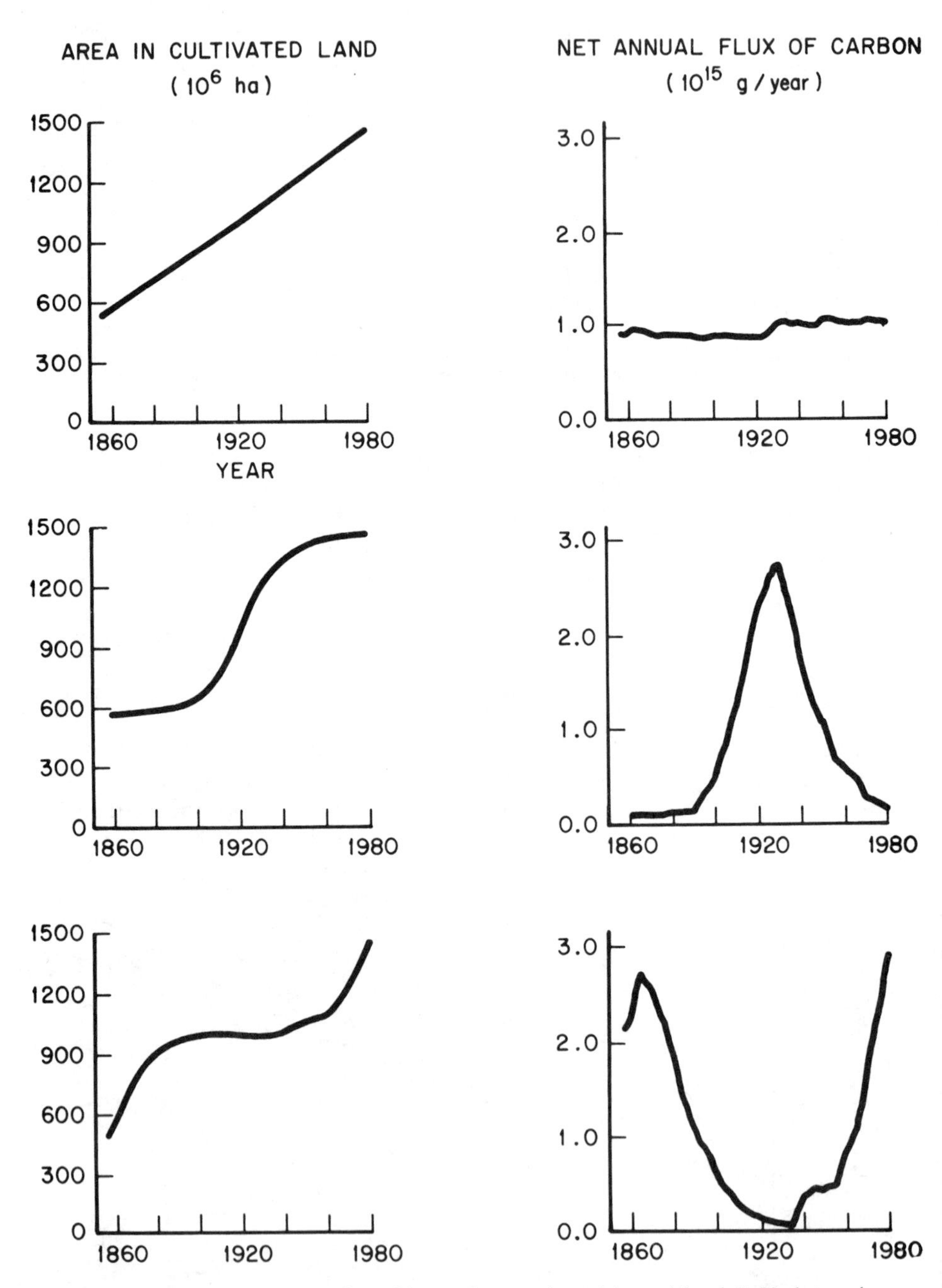

FIGURE 10.3. Three examples of how changes in cultivated land (**left**) determine the time course of carbon released from terrestrial systems (**right**). The areas of cultivated land in 1860, 1920, and 1980 are the same in each sample.

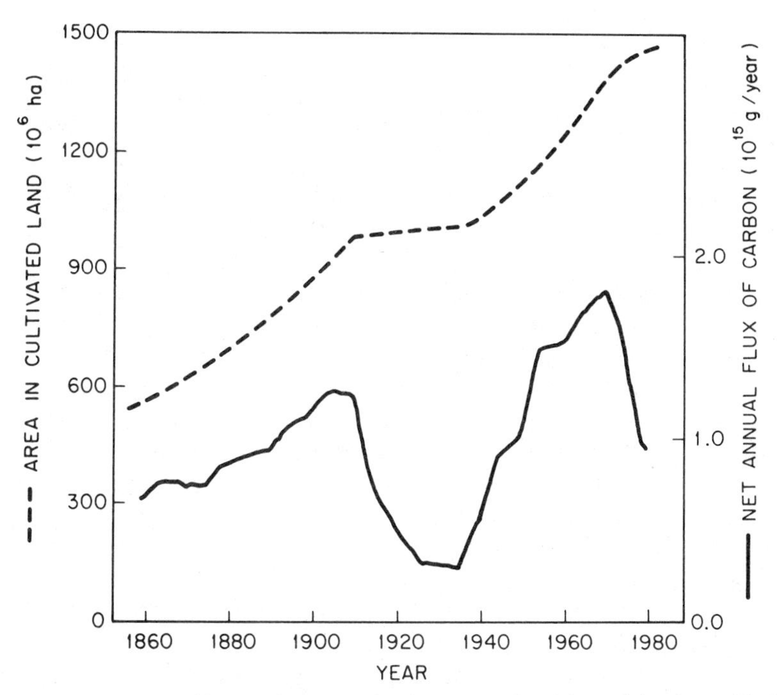

FIGURE 10.4. An estimate of change in the area of cultivated land worldwide (----) (see text) and the net release of carbon to the atmosphere as a result of that change (———).

estimates of agricultural area in 1860, 1920, and 1978. Figure 10.3 shows three different curves, representing agricultural area, that can be drawn through these three estimates of area. The curves to the right show the time course of carbon release calculated from the histories of agriculture shown on the left. The upper left curve assumes that the increase in agricultural area was constant between each date. An increase of about 7×10^6 ha • yr^{-1} resulted in a constant net annual flux of carbon of about 1×10^{15} g C • yr^{-1}, if the kinds of ecosystems converted and their stocks of carbon were the same as in the MBL analysis (Fig. 10.1).

If the most rapid expansion of agriculture (arbitrarily set at 25×10^6 ha • yr^{-1}) occurred around 1920 (middle pair of curves), the shape of the time course of carbon release would be of the type suggested by some records of isotopes in tree rings. At the other extreme, if the most rapid expansion occurred at the start and end of the 1860 to 1980 interval (bottom pair of curves), the net release of carbon would be lowest near 1920 and highest in 1860 and 1980. No one has advanced this world view of agriculture; the curves are included to show how rates of change in agricultural area affect the timing of the release of carbon.

A more realistic estimate of worldwide agricultural area from 1860 to 1980 was made (Fig. 10.4, dashed line) using estimates of area in 1860, 1920 (Richards et al. 1983), and 1980 FAO (1983), and the documentation provided annually since 1948 by the *Production Yearbooks* of the FAO. Between 1860 and the beginning of World War I, an increase in world trade and the growing demand for products stimulated an expansion in cultivated land (Tucker and Richards 1983; Richards et al. 1983). Much of this expansion in the 19th century occurred in colonial tropical regions (under European influence) and in North America and Russia. With the advent of war the expansion of world trade stopped, and the global economy weathered a general depression. Not until after the 1930s did the growing world populations begin clearing new lands for agriculture. The most rapid rise was during the 1950s before the introduction of fertilizers, new varieties of crops, and other products of modern agriculture that increased the yields and, hence, reduced the demand for additional land. To the extent that population growth, world economy, and technology have been the driving forces behind agricultural expansion, the course of the release of carbon from changes in agricultural area alone might have been as it appears in Fig. 10.4 (solid line).

The author knows of no historical data that would support a decreasing rate of agricultural expansion worldwide between 1920 to the present, and thus a decreasing rate of carbon release during this interval. However, the expansion may have slowed and the net release may have decreased over the most recent two to three decades. Such a slowing is uncertain because the best sources of worldwide information, the *Production Yearbooks* of the FAO, have not provided a consistent assessment of arable lands since their first yearbook appeared in 1948. Figure 10.5 shows several estimates of worldwide area in arable land and permanent cropland, all obtained from *Production Yearbooks*. The longest curve (solid) is based on individual yearbooks; that is, each estimate of area is based on the year's publication. The three shorter curves are each based on a single yearbook—1975, 1978, or 1983. Presumably, the information in a single yearbook has been updated and adjusted to meet consistent definitions throughout the period reported. Thus, the three short curves are better estimates of real change than the curve based on successive editions. Unfortunately, revisions have been made only as far back as 1961; therefore, one cannot determine whether the change in the derivative of agricultural area about 1963 is real or an artifact. The revisions made in successive yearbooks, sometimes several times larger than real annual changes, are one of the major reasons for different results among studies relying on *Production Yearbooks* (Table 10.1).

A Detailed Analysis of Southeast Asia

Because data for the MBL global analyses were obtained for most tropical regions from assumptions of per capita use of land and wood, it is important to reexamine in greater detail the histories of land use in the tropics. The tropics are also important because the potential for large releases of carbon is greater there than elsewhere, and because there are sharply contrasting

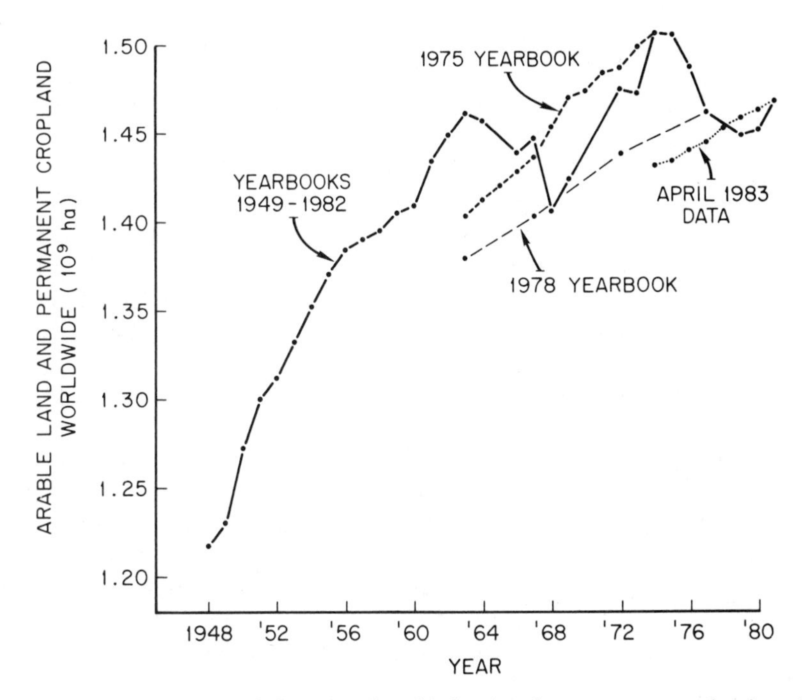

FIGURE 10.5. Estimates of the area of arable land and permanent cropland worldwide according to different *Production Yearbooks* of the FAO. Each point in the *solid line* was obtained from the yearbook published the following year. The three other *lines* are each based on information contained in a single yearbook.

reports concerning the current rates of deforestation in these regions (Houghton et al. 1983). Only for Southeast Asia are the results of an intensified reexamination available to date.

In the earlier analyses (Houghton et al. 1983), the historical information for Southeast Asia was limited, for the most part, to information given by Grigg (1974) for Java and Madura, Cochin China (now part of Viet Nam), and parts of Burma. Changes in agricultural area before 1950 for the remainder of Southeast Asia were assumed to follow population growth, as reported by McEvedy and Jones (1978). After 1950 three different estimates of the release of carbon were calculated from (1) rates of agricultural expansion as reported by the *Production Yearbooks* (FAO 1949 to 1978), (2) assumptions of per capita use of agricultural land, and (3) estimates of deforestation (Myers 1980) adjusted for forest regrowth. Before 1950 the rates of CO_2 flux were the same for each of these analyses. The flux of carbon from Southeast Asia obtained from the analyses based on population and on the *Production Yearbooks* remained similar through 1980 (0.2×10^{15}g C released in 1980); however, the analysis based on Myers' estimate gave an annual net flux in 1980 of 1.5×10^{15}g C.

In the recent reanalysis of Southeast Asia, estimates of the agricultural area in individual countries were obtained from historical references and statistical handbooks (Houghton and Palm, in press). The recent study of tropical forests by FAO/UNEP (1981) was used to provide estimates of biomass and current rates of deforestation. Discussions with Myers revealed that earlier interpretations of his work had been incorrect; when clearing of fallow forests is subtracted from Myers' (1980) total rate of forest conversion for Southeast Asia, his estimate becomes about 1.5×10^6ha • yr^{-1} rather than the 5.7×10^6ha • yr^{-1} assumed by Houghton et al. (1983). The detailed reanalysis also resulted in revisions of the carbon content of vegetation and soils in Southeast Asia and included shifting cultivation as well as permanent agriculture and harvest of forests (Palm et al., in press).

The reanalysis reduced considerably the range of estimates of the biotic release of carbon, largely as a result of the reappraisal of Myers' study. The range of estimates in 1980 was 0.14 to 0.43×10^{15}g C • yr^{-1}, instead of 0.2 to 1.5×10^{15}g. The new range resulted primarily from different estimates of carbon stocks and different estimates of the rate of increase in shifting cultivation. As in the global analyses (Table 10.1), different estimates of change in the area of permanent agriculture contributed <10% to the overall variability; that is, the population-based history of cropland was not very different from the more detailed history derived country by country from numerous sources.

The clearing of land for permanent agriculture was responsible for between 60 and 90% of the net release of carbon between 1860 and 1980, according to the reanalysis. Shifting cultivation, including the degradation of lands that may accompany a reduction in the length of the fallow cycle or other intensification of use, accounted for most of the remainder of the net release. The replacement of traditional shifting cultivation with a permanent form of subsistence agriculture in recent years (Myers 1980; NRC 1982) has resulted in an expansion of these degraded lands. Large areas formerly forested are no longer able to support either subsistence crops or the regrowth of forests. These lands remain as impoverished grasslands. Many investigators believe that all grasslands in Southeast Asia are the result of man's overuse of forests (Grigg 1974; Whyte 1974; Seavoy 1975; Komkris 1978; UNESCO 1978; Kartawinata et al. 1981).

Nonagricultural Uses of Land

One lesson that emerges from these studies is that changes in cultivated land, although they are more extensively and accurately recorded through history than are changes in the area of forests, have been responsible for only a part of the total net flux of carbon from the terrestrial biota and soils to the atmosphere. Worldwide, the harvest (including regrowth) of forests and clearing of forests for pasture accounted for about 35% of the net flux of carbon between 1860 and 1980 (Houghton et al. 1983)(Table 10.1). In Southeast Asia, shifting cultivation, not consistently included in records of cultivated land, was responsible for between 10 and 40% of the 120–year net release. Shifting cultivation is important in Latin America and tropical Africa as well, and increases in the area of pasture may currently exceed the increases in agricultural area in Latin America (Myers 1980; Hecht 1981).

The use of changes in agricultural area to determine the net flux of carbon from terrestrial systems has another shortcoming. The impoverished grasslands of Southeast Asia, which, without investment of fertilizer and energy will support neither forests nor agriculture, are ignored in most statistical accounts; that is, they are not counted as agricultural land. The net conversion of forests to agriculture must, therefore, be greater than the net increase in agricultural area. These impoverished grasslands and shrublands are probably included in the category "other land," which is defined in the *Production Yearbooks* as

> ...unused but potentially productive land, built-on areas, wasteland, parks, ornamental gardens, roads, lanes, barren land, and any other land not specifically included in the other categories [arable land, land under permanent crops, permanent meadows and pastures, forests, and woodland] (FAO 1981).

In Southeast Asia, "other land" is 1.4 times larger than arable land and permanent cropland and has increased 2.2 times more rapidly in the period 1975 to 1980. Worldwide, "other land" is 3 times larger than arable land and permanent cropland and has increased more than twice as rapidly. In fact, worldwide, "other land" is larger in area than any other category in the *Production Yearbook* (FAO 1983). According to this yearbook, the decrease in forest area worldwide is 2.6 times greater than the increase in arable land and permanent cropland. If the carbon stocks of these other lands are as low as the carbon stocks in agriculture and if this inequality in rates of deforestation versus agricultural expansion has existed since 1860, estimates of change in the carbon content of the earth's biota and soils, based on changes in agricultural area alone, might be low by 100%.

Analyses of the history of agricultural expansion must consider not only population growth and cash economies but also the ability of land to support continued use. If the ability were unlimited, one might expect not only the same area but the same land to remain in agriculture under continued use. The ability is not unlimited, however, and new lands must be brought into agriculture to sustain the same yields. If the abandoned lands are degraded to the point where they will not support regrowth of forests, the cycling of land from forests, through agriculture, to other land will release more carbon to the atmosphere than the net change in agricultural area suggests.

Uncertainties in Ecological Data

The effect of different estimates of the carbon stocks of forests on the net flux of carbon is shown in Fig. 10.6. All three curves were based on the same rates of land-use change, that is, the population-based estimate

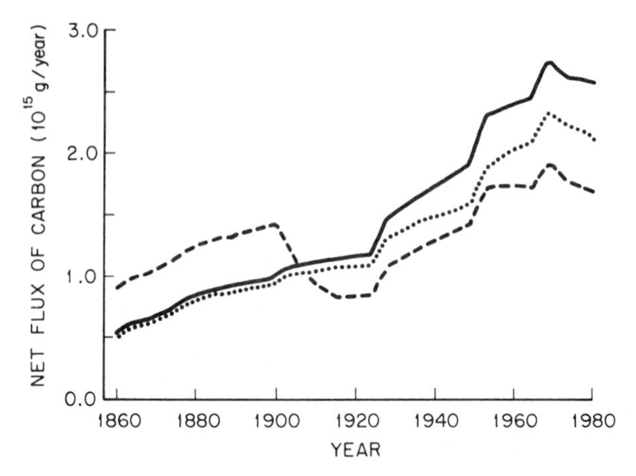

FIGURE 10.6. The net annual flux of carbon from biota and soils worldwide as affected by different estimates of the carbon stocks of forests. The *solid curve* is based on biomass values obtained from ecological studies; the *dotted curve* is based on biomass values derived from the FAO/UNEP study (1981); the *dashed curve* is based on the assumption that forests cleared before 1900 had larger stocks of carbon than forests cleared after 1900.

(Houghton et al. 1983). The solid curve was obtained using average carbon stocks for the world's tropical moist and tropical seasonal forests of 212 and 156 × 10^3kg • ha^{-1}, respectively (Table 10.2). These values were obtained from reviews of the ecological literature (Brown and Lugo 1982, Olson et al. 1983). The dotted curve was based on the recent FAO/UNEP assessment of tropical forests (FAO/UNEP 1981). The conversion of wood volume (as reported by FAO/UNEP) to carbon was made with a factor of 0.54 (2.0 to account for the parts of trees not commercialized and for

TABLE 10.2. Mean biomass of tropical forests (10^3kg C • ha^{-1}).*

	Means from ecological studies		Means from FAO/UNEP study†	
Region	Tropical moist	Tropical seasonal	Tropical moist	Tropical seasonal
Latin America	176 (1)	158 (9)	82	85
Tropical Africa	210 (1)	160 (1)	124	62
Southeast Asia	250 (4)	150 (5)	135	81
Mean	212	156	114	76

*Estimated from a review of ecological studies and as calculated from volumes of wood given by FAO/UNEP (1981); values in parentheses refer to the number of study sites. Tropical moist forests and tropical seasonal forests were differentiated as follows: (1) tropical moist forests, forests with less than 3 consecutive months of precipitation 60 mm; and (2) tropical seasonal forests, forests with 3 or more consecutive months of precipitation < 60 mm (Walter et al. 1975).

†FAO/UNEP values were weighted by the area of forests; both virgin (NHCfluv) and unproductive (NHCf2) forests were included in the mean.

vegetation <10 cm in diameter, 0.6 to convert volume to dry weight, and 0.45 to convert dry weight to carbon) and gave carbon values of 114 and 76 $\times$ 10^3kg • ha^{-1} in the vegetation of tropical moist and seasonal forests, respectively. A similar analysis was carried out independently by Brown and Lugo (1984). The estimate based on ecological studies was almost twice the estimate derived from FAO/UNEP (Table 10.2). It is not clear whether the difference is the result of bias in one or both of the estimates, or whether the conversion factor used for the FAO/UNEP estimate is in error.

The difference in the net flux of carbon resulting from the two estimates of carbon stocks was about 12% over the 120–year period since 1860 and about 20% of the annual flux of carbon in 1980 (Fig. 10.6). The greater difference in more recent years resulted from the increased disturbance of tropical forests relative to other ecosystems. Only the biomass of tropical forests was different in the two analyses.

Errors in ecological data alone will generally not affect the time course of the annual flux of carbon but will change the absolute value. The question of biomass, however, is more complex than the discrepancy between carbon stocks of undisturbed forests. Many of the systems cleared for agriculture, for example, were not forests. And many of the forests converted may have been degraded prior to clearing; that is, their carbon stocks may have been reduced through harvest of fuelwood, for example. In the analysis by Richards et al. (1983) the loss of carbon from vegetation over the period 1860 to 1978 averaged 46 10^3 kg • ha^{-1}. The same loss was 68 10^3 kg • ha^{-1} in the analysis by Houghton et al. (1983). Much of the difference can be explained on the basis of a comparison of the types of systems cleared and the reduction of carbon stocks assumed to have taken place prior to clearing. Richards et al. assumed some clearing of agricultural land from systems previously disturbed, with lower carbon stocks. The average change in the carbon in vegetation was 50 10^3 kg • ha^{-1} in the first 60-year period and 43 10^3 kg • ha^{-1} in the second period. They did not calculate the release of carbon to the atmosphere that would have accompanied this reduction of carbon stocks. In effect they assumed that the reductions had occurred prior to 1860.

Houghton et al. (1983), with one exception, assumed the opposite—that reductions in carbon stocks took place after 1860. The carbon stocks of the systems cleared for agriculture had not been reduced prior to clearing. The exception was in the modeling of forest harvest. In the population-based analysis (Fig. 10.6, solid curve; Houghton et al. 1983), harvested forests regrew to secondary forests with only 75 to 90% of the carbon content of primary forests. Subsequent harvests in the same geopolitical region were taken from secondary forests first; that is, primary forests were logged only when no secondary forests remained. Thus, the harvested forests through time were a mixture of primary and secondary forests, and the secondary forests had lower carbon stocks. The dashed curve in

Fig. 10.6 was based on an analysis using identical harvest and clearing rates; however, in this analysis, only primary systems were affected before 1900, and only secondary forests were affected after 1900. The effect was to reduce the net release of carbon over the 120-year period from 180 to 158×10^{15}g, and to shift more of that release to the earlier period. The time course of the release of carbon was changed not by a different history of land use as much as by an interaction between historical and ecological data. Identification of the systems disturbed and documentation of the extent of degradation would seem to require the close collaboration of historians and ecologists.

The Difference Between Two Methods

None of the estimates of biotic flux based on carbon isotopes in tree rings (Stuiver, Chapter 6, this volume) agrees with the estimates calculated from the history of land use. Differences among the results of isotopic analyses suggest that the ratio of $^{13}C/^{12}C$ is influenced by factors not yet identified, and hence that none of the analyses provides a valid time course of biotic release. If the method were developed to the point where the variation were reduced or understood, a comparison of the results with those obtained independently from analyses of land-use change might reveal similar patterns of biotic flux over the last century.

On the other hand, curves of the biotic release of carbon from the two types of analyses might each be accurate and yet different; for the two methods do not include the same biotic processes. Analyses based on carbon isotopes include the flux of carbon caused by changes in land use, and they include as well any net flux resulting from changes in atmospheric CO_2, temperature, moisture, nutrient availability, and other changes in the environment. Analyses based on the history of land use do not include environmentally induced changes in carbon storage.

The difference between two such curves would be revealing because information about the functioning of natural systems is not currently adequate to predict how carbon storage will change in the short term (1 to 100 years) with the rise in atmospheric concentrations of CO_2, with the expected increased warming, or with any other environmental factor likely to be changed. Many experiments have studied the effects of elevated levels of CO_2 on plant photosynthesis or growth, but neither of these processes is equivalent to storage of carbon in ecosystems.

Using the results of Houghton et al. (1983) and Peng et al. (1983) as examples of the two different measures of biotic flux (Fig. 10.7, top), one can ask whether the difference between them (Fig. 10.7, middle) corresponds to any change in the global environment. Because the flux based on land-use change was subtracted from the total biotic flux (Peng et al. 1983), a positive value indicates a net release of carbon in addition to the

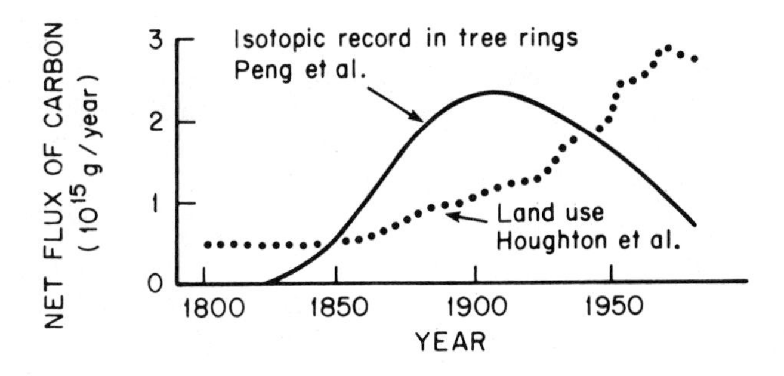

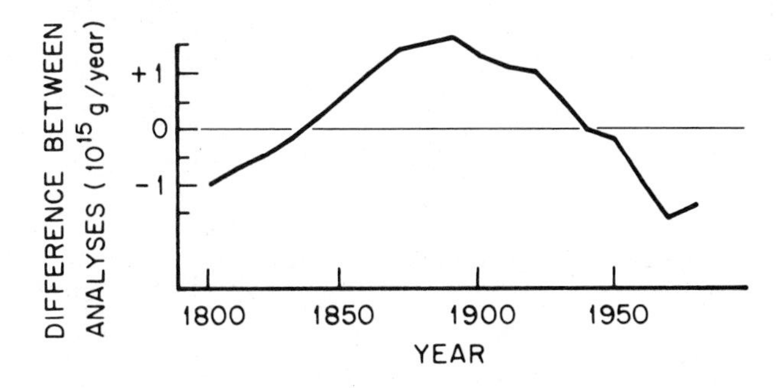

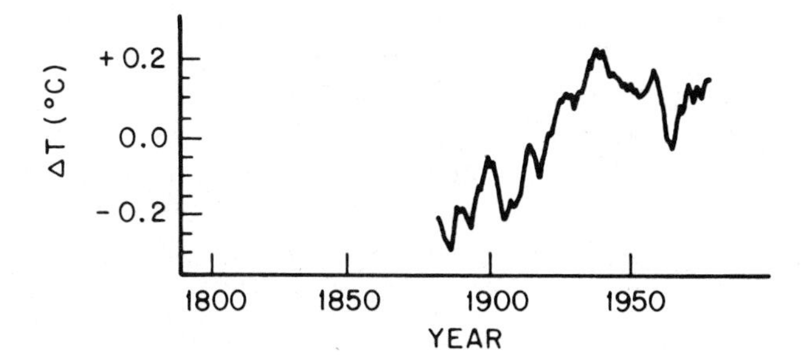

FIGURE 10.7. **Top:** The net annual flux of carbon from terrestrial systems according to analyses based on records of $^{13}C/^{12}C$ ratios in tree rings (Peng et al. 1983) (———) and analyses bases on records of land-use change (Houghton et al. 1983) (....). **Middle:** The difference between the two analyses (**top**), calculated as results of land-use analysis subtracted from results of isotopic analysis. **Bottom:** The change in mean global temperature (from Hansen et al. 1981).

one calculated from changes in land use; a negative value indicates a net storage of carbon countering the release from changes in land use alone. The variability of the difference (middle curve) is not consistent with a CO_2 fertilization effect since the concentration of CO_2 in the atmosphere has increased monotonically since 1800. The curve suggests a possible relationship to mean global temperature (Fig. 10.7, bottom, from Hansen et al. 1981), with a net release of carbon corresponding to a global warming before 1940 and a net storage corresponding to a global cooling from about 1940 to 1965. The correspondence is not perfect and may be fortuitous, but it suggests that a warming of the earth's terrestrial systems will act as a positive feedback on the increasing concentration of CO_2 in the atmosphere. Clearly there are many questions to be resolved before the differences between these two approaches can be expected to provide clues as to the response of the world's terrestrial ecosystems to changes in climate.

Conclusions

Estimates of change in the carbon content of terrestrial systems based on changes in the use of land worldwide since 1860 show a continuous net release of carbon to the atmosphere. The history of agricultural expansion is known well enough to preclude the possibility of a continuously decreasing biotic flux since 1920, although the possibility of a decreasing flux in the decades preceding 1980 cannot be ruled out.

Those factors responsible for the greatest uncertainties in the time course of the net biotic release are the conversion of forests to uses other than permanent agriculture and carbon stocks, not of natural ecosystems necessarily, but of those ecosystems actually degraded or transformed to other uses. The types of analyses required to document and use this detailed information will be geographically complex. They will require geopolitical units smaller than countries and units of vegetation and soil smaller than biomes, and the independent nature of the political and biological units will require the close collaboration of historians and ecologists. There is perhaps no better way to document changes in terrestrial ecosystems in the recent past and future than by observing them directly with remote sensing (Woodwell et al., Chapter 13, and Tucker et al., Chapter 12, this volume).

Finally, changes can occur in the carbon content of terrestrial ecosystems not directly modified by man. Such changes could either augment or cancel the changes associated with land use. The response of natural systems to changes in climate or CO_2 are difficult to measure and difficult to predict. These responses may be tentatively identified by differences between independent analyses, but such methods have been misleading in the past. The best approach toward understanding the cycle would seem to include experiments on the controls of carbon storage in ecosystems and long-term monitoring of carbon stocks or CO_2 flux in undisturbed systems. The longer we wait to begin, the longer we must wait for results.

Acknowledgments

The author is grateful to those who have contributed to this research. Special thanks are extended to R. D. Boone, J. E. Hobbie, J. M. Melillo, B. Moore, C. A. Palm, B. J. Peterson, J. F. Richards, G. R. Shaver, D. L. Skole, and G. M. Woodwell. The research was supported by National Science Foundation Grant DEB-81-10477.

References

Arnold, J. E. M. and J. Jongma. 1978. Fuelwood and charcoal in developing countries. Unasylva 29:2–9.

Brown, S. and A. E. Lugo. 1982. The storage and production of organic matter in tropical forests and their role in the global carbon cycle. Biotropica 14:161–187.

Brown, S. and A. E. Lugo. 1984. Biomass of tropical forests: a new estimate based on volumes. Science 223:1290–1293.

Emanuel, W. R., G. G. Killough, W. M. Post, and H. H. Shugart. 1984. Modeling terrestrial ecosystems in the global carbon cycle with shifts in carbon storage capacity by land-use change. Ecology 65:970–983.

Food and Agriculture Organization (FAO). 1946–1979. Yearbook of Forest Products. FAO, Rome.

Food and Agriculture Organization (FAO). 1949–1978. Production Yearbooks. FAO, Rome.

Food and Agriculture Organization (FAO). 1983. Production Yearbook. FAO, Rome.

Food and Agriculture Organization (FAO). 1981. Production Yearbook. FAO, Rome.

Food and Agriculture Organization United Nations Environment Programme (FAO/UNEP). 1981. Tropical forest resources assessment project. Forest Resources of Tropical America. Forest Resources of Tropical Africa. Forest Resources of Tropical Asia. FAO, Rome.

Grigg, D. B. 1974. The Agricultural Systems of the World: An Evolutionary Approach. Cambridge University Press, Cambridge, England.

Hansen, J., D. Johnson, A. Lacis, S. Lebedeff, P. Lee, D. Rind, and G. Russell. 1981. Climatic impact of increasing atmospheric carbon dioxide. Science 213:957–966.

Hecht, S. 1981. Cattle ranching in the Eastern Amazon: environmental and social implications. In E. F. Moran (ed.), The Dilemma of Amazonian Development, pp. 155–188. Westview Press, Boulder, Colorado.

Houghton, R. A., J. E. Hobbie, J. M. Melillo, B. Moore, B. J. Peterson, G. R. Shaver, and G. M. Woodwell. 1983. Changes in the carbon content of terrestrial biota and soils between 1860 and 1980: a net release of CO_2 to the atmosphere. Ecol. Monogr. 53:235–262.

Houghton, R. A. and C. A. Palm. The reduction of forests in Southeast Asia since

1860: official statistics and assumptions. In J. F. Richards and R. P. Tucker (eds.). The World Economy and World Forests in the Twentieth Century. Duke University Press, Durham, North Carolina (in press).

Houghton, R. A. and G. M. Woodwell. 1983. Effect of increased C, N, P, and S on the global storage of C. In B. Bolin and R. B. Cook (eds.), The Major Biogeochemical Cycles and Their Interactions, pp. 327–343. John Wiley & Sons, New York.

International Institute of Agriculture. 1922. International Yearbook of Agricultural Statistics (1909–1921). IIA, Rome.

International Institute of Agriculture. 1939. The First World Agricultural Consus (1930). IIA, Rome.

Kartawinata, K., S. Adisomarto, S. Riswan, and A. P. Vayda. 1981. The impact of man on a tropical forest in Indonesia. Ambio 10:115–119.

Komkris, T. 1978. Forestry aspects of land use in areas of swidden cultivation. In P. Kunstadter, C. Chapman, and S. Sabhasri (eds.), Farmers in the Forest, pp. 61–70. University of Hawaii Press, Honolulu.

McEvedy, C. and R. Jones. 1978. Atlas of World Population History. Penguin Books, Middlesex, England.

Moore, B., R. D. Boone, J. E. Hobbie, R. A. Houghton, J. M. Melillo, B. J. Peterson, G. R. Shaver, C. J. Vorosmarty, and G. M. Woodwell. 1981. A simple model for analysis of the role of terrestrial ecosystems in the global carbon budget. In B. Bolin (ed.), Modelling the Global Carbon Cycle: SCOPE 16, pp. 365–385. John Wiley & Sons, New York.

Myers, N. 1980. The present status and future prospects of tropical moist forests. Environ. Conserv. 7:101–114.

National Research Council (NRC). 1982. Ecological Aspects of Development in the Humid Tropics. National Academy Press, Washington, D.C.

Olson, J. S., J. A. Watts, and L. J. Allison. 1983. Carbon in live vegetation of major world ecosystems. TROO4. U.S. Department of Energy, Washington, D. C.

Palm, C. A., R. A. Houghton, J. M. Melillo, and D. L. Skole. Atmospheric carbon dioxide from deforestation in Southeast Asia. Biotropica (in press).

Peng, T.-H., W. S. Broecker, H. D. Freyer, and S. Trumbore. 1983. A deconvolution of the tree ring based $\delta^{13}C$ record. J. Geophys. Res. 88:3609–3620.

Revelle, R. and W. Munk. 1977. The carbon dioxide cycle and the biosphere. In Energy and Climate, pp. 140–158. National Academy of Sciences, Washington, D.C.

Richards, J. P., J. S. Olson, and R. M. Rotty. 1983. Development of a data base for carbon dioxide releases resulting from conversion of land to agricultural uses. ORAU/IEA-82-10(M), ORNL/TM-8801, Oak Ridge National Laboratory, Oak Ridge, Tennessee.

Robertson, C. J. 1956. The expansion of the arable area. Scott. Geogr. Mag. 72:1–20.

Seavoy, R. E. 1975. The origin of tropical grasslands in Kalimantan, Indonesia. J. Trop. Geogr. 40:48–52.

Stuiver, M. 1978. Atmospheric carbon dioxide and carbon reservoir changes. Science 199:388–394.

Tucker, R. P. and J. R. Richards. 1983. Global Deforestation and the Nineteenth Century World Economy. Duke University Press, Durham, North Carolina.

United Nations Educational, Scientific, and Cultural Organization (UNESCO).1978. Tropical forest ecosystems, a state-of-knowledge report. Nat. Resour. Res. 14. UNESCO, Paris.

U.S. Department of Agriculture. 1965. Changes in agriculture in 26 developing nations 1948–1963. Foreign Agricultural Economic Report No. 27. Economic Research Service, USDA, Washington, D.C.

Walter, H., E. Harnickell, and D. Mueller-Dumbois. 1975. Climate-Diagram Maps of the Individual Continents and the Ecological Climatic Regions of the Earth. Springer-Verlag, New York.

Whyte, R. O. 1974. Tropical Grazing Lands: Communities and Constituent Species. W. Junk, The Hague, The Netherlands.

Woodwell, G. M., J. E. Hobbie, R. A. Houghton, J. M. Melillo, B. Moore, B. J. Peterson, and G. R. Shaver. 1983. Global deforestation: contributions to atmospheric carbon dioxide. Science 222:1081–1086.

11
Changes in Soil Carbon Storage and Associated Properties with Disturbance and Recovery

WILLIAM H. SCHLESINGER

Organic matter in the world's soils contains about three times as much carbon as the land vegetation. Soil organic matter is labile and is likely to change as a result of human activities. Agricultural clearing, for example, results in a decline in soil organic matter. At the present time, there may be a net release of 0.85×10^{15} g C • yr^{-1} from soils of the world due to agricultural clearing (Houghton et al. 1983; Schlesinger 1984), or about 15% of the annual release from fossil fuels. The release of carbon may have been greater near the turn of the century as a result of more rapid agricultural expansion into virgin areas (Stuiver 1978, Wilson 1978). It is the purpose of this chapter (1) to review briefly the present estimates of the size of the pool of carbon in world soils and (2) to offer a review and analysis of what is known about the effects of agriculture on soil carbon storage.

Soil organic matter exists in many forms, including fresh, incompletely decomposed organic matter on the soil surface (litter or detritus) and humus dispersed through the mineral soil horizons. The turnover time of these fractions varies from <1 year to >1000 years, but it is important to include all of these forms in the soil carbon pool (Schlesinger 1977). Peatland soils present a special problem (Armentano 1979). In many areas, peat accumulates to great depth, forming a continuum with the geological pools of lignite and coal. Most of the discussion in this chapter will focus on carbon stored in soil organic matter; however, in arid and semiarid regions, carbon is also stored as carbonate minerals that precipitate in the lower soil horizons. Changes and turnover in this pool will be considered as well.

The Soil Carbon Pool

There have been several attempts to estimate the storage of soil organic matter and litter in world ecosystems (Table 11.1). Most of these estimates range from 1400 to 1700 $\times$ 10^{15} g C, and they include approaches based on soil groups (Buringh 1984), vegetation groups (Schlesinger 1977, 1984, Bolin et al. 1979, Ajtay et al. 1979), life zone classes (Post et al. 1982),

TABLE 11.1. Estimates of the pool of organic carbon in world soils.

Basis	Reference	Estimate (x 10^{15} g C)
Vegetation groups	Bolin (1970, 1977)	700
	Baes et al. (1977)	1080
	Schlesinger (1977)	1456
	Bolin et al. (1979)	1672
	Ajtay et al. (1979)	2205
		1636
	Schlesinger (1984)	1515
Soil groups	Bazilevich (1974)	1405
	Bohn (1976)	3000
	Buringh (1984)	1477
	Bohn (1982)	2200
Life zone groups	Post et al. (1982)	1395
Modeling	Meentemeyer et al. (1981)	1457

and modeling of plant production and decomposition (Meentemeyer et al. 1981). The most extensive data set is that used by Post et al. (1982), which included values from 2700 measured soil profiles from around the world. Their sampling was least intensive in temperate deserts, but a recent compilation of data from desert soils in Arizona (Schlesinger 1982) substantially confirms the values assigned to desert regions by these workers. Using the data published with the Food and Agriculture Organization's soil maps of the world, Bohn (1976, 1982) has produced consistently higher estimates for the pool of carbon in world soils, but his methods are vaguely described, and the original data do not include measurements of soil bulk density.

Much of the uncertainty in these estimates lies in the treatment of Histosols, the peatland soils of the world. Buringh (1984) suggests that 2.8% of the world total, or 41.5 × 10^{15} g C, resides in Histosols, whereas Bohn (1982) gives a value of 377 × 10^{15} g C. Workers using a vegetation-based approach generally include peatlands in categories such as tundra, boreal forest, and wetlands. These categories contain 489 × 10^{15} g C (or 34%) of Schlesinger's (1977) world estimate of soil carbon. Miller (1981) suggests that field workers in tundra ecosystems have tended to sample more productive sites (e.g., tussock tundra), which are characterized by highly organic soils. These data inflate the mean values assigned to soil carbon in compilations of the literature. Miller (1981) suggests that past estimates of the soil carbon pool in tundra and boreal ecosystems may be too high by a factor of 3, when areas of arctic desert are taken into account. However, the soil carbon values used by Miller (1981) include only the upper horizons, whereas Tarnocai (1972) found that 80% of the soil carbon in spruce forests of Manitoba was located in the permafrost layers.

Tropical soils are of particular interest, inasmuch as tropical regions are likely to undergo major land use changes during the remainder of the century. Brown and Lugo (1982) calculated a weighted mean of 8.6 kg C

• m^{-2} for soil storage in tropical life zones, similar to a value of 8.4 kg C • m^{-2} calculated for tropical soil groups by Sanchez et al. (1982*b*), but 24% lower than the weighted mean of 11.3 kg C • m^{-2} of Post et al. (1982). Nearly one-third of the area included by Brown and Lugo (1982) was assigned to subtropical dry forests with a soil carbon content of 3.9 kg C • m^{-2}. This value seems low for closed forests, but if the category includes sparsely forested savannas, then the low value would agree closely with Schlesinger's (1984) compilation of tropical savanna soils, that is, 4.2 kg C • m^{-2}. Although there is substantial agreement on the carbon content of tropical soils, there is wide disagreement on the land area in each tropical life zone. Brown and Lugo (1982) calculate a total pool of 159 × 10^{15} g C in tropical soils, 14% lower than that of Post et al. (1982), who use a smaller area. Other estimates of the total pool of soil carbon in tropical and subtropical regions differ by a factor of 2.5 (Brown and Lugo 1982), but differences in the area assigned to the tropical region account for nearly half of the range.

The soils of arid and semiarid ecosystems store carbon in inorganic form, primarily as calcium carbonate. This secondary carbonate occurs in a variety of forms, ranging from precipitates in the interstitial spaces of the parent material to almost pure, laminated layers of carbonate (Gile et al. 1981). Highly indurated layers are referred to as calcrete or petrocalcic horizons, and less-cemented layers are known as caliche or calcic horizons (Lattman 1973, Reeves 1976, Dregne 1976). Schlesinger (1982) found that the soils of Arizona contained between 4 and 5.7 times more carbon stored in carbonate than in soil organic matter. Assuming that the Arizona data can be extrapolated to Aridisols and arid Entisols in other regions of the world, the amount of carbon stored in caliche of the world's deserts is 800 × 10^{15} g. Of course, some of these deposits are very old (Reeves 1970, Gardner 1972), but carbonate precipitation is a present-day pedogenic process as well (Buol 1965, Dregne 1976, Jenny 1980). As in the case of peat, caliche spans the realm between the geological and biospheric carbon pools.

Changes in Soil Carbon Storage During the Holocene Period

Prehistoric

When vegetation colonizes a newly available land surface, that is, primary succession, there is often a rapid accumulation of organic carbon in the soil. Vitousek et al. (1983) found accumulation rates of 38 g C • m^{-2} • yr^{-1} in soils derived from volcanic deposits in Hawaii. After many years of soil development, the annual production and decomposition of plant residues nearly balance, and the accumulation of soil organic matter approaches a steady state (Fig. 11.1). Goh et al. (1976) show accumulations of about 6 g C • m^{-2} • yr^{-1} during the first 500 years of soil development on aeolian sands in New Zealand, but after 7000 years of soil development,

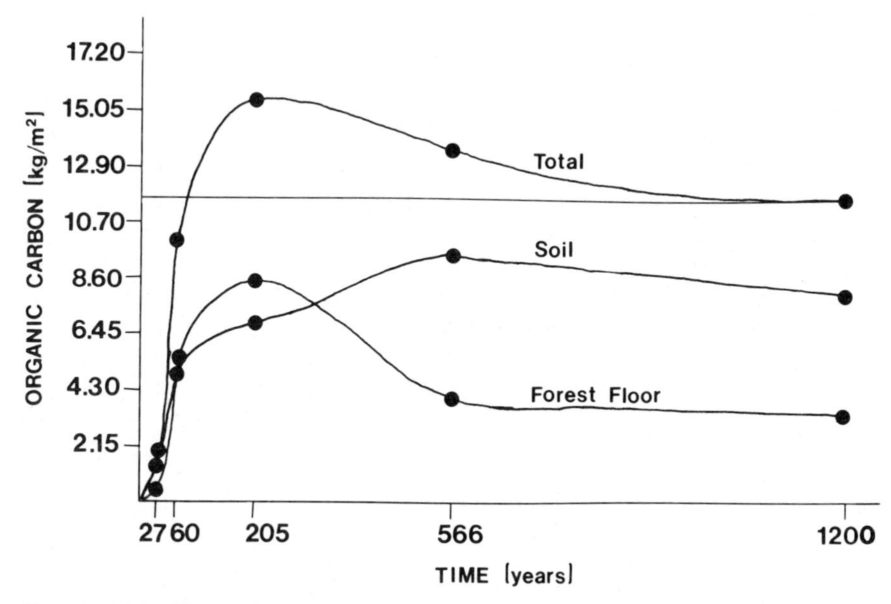

FIGURE 11.1. Change in the content of soil carbon in a chronosequence of ecosystems developed on the volcanic mudflows of Mt. Shasta, California. (From Dickson and Crocker 1953.)

the rate slowed to <0.5 g C • m^{-2} • yr^{-1}. The integrated rate of carbon storage in glaciated soils has been about 1.0 to 2.0 g C • m^{-2} • yr^{-1}, calculated by dividing their current content of 10 to 20 kg C • m^{-2} by the time since the last glacial retreat.

Five hundred years ago, it is likely that the vast majority of the land surface was occupied by vegetation, and changes in soil carbon storage were small. Most upland terrestrial ecosystems probably consisted of a mosaic of sites in various stages of recovery or degradation following natural or human disturbances (Bormann and Likens 1979), yet taken as a whole, the landscape may have approximated a steady state in soil carbon storage. Even following such drastic modern disturbances as clearcutting, there is little change in the total storage of soil carbon, as long as the vegetation is allowed to recover immediately (Harcombe 1977; Bormann and Likens 1979; Gholz and Fisher 1982). Slash and burn agriculture results in little regional change in soil carbon storage; losses in cleared areas are balanced by accumulations of soil carbon in abandoned areas (Nye and Greenland 1960, Aweto 1981).

Before widespread human occupation, net carbon accumulations were probably confined to peatland and wetland soils, where current rates of accumulation range from 5 to 150 g C • m^{-2} • yr^{-1} (Armentano 1979). The area of these ecosystems is small, and the accumulation of carbon in organic soils of the world was only about 0.14×10^{15} g C • yr^{-1} (Armentano 1980). Accumulations of soil carbonate in arid soils, 0.24 g C • m^{-2} • yr,$^{-1}$ would have served as a sink for an additional 0.01×10^{15} g C • yr^{-1} (Schlesinger 1985).

MAN-INDUCED CHANGES

When primitive agricultural practices were abandoned in favor of large-scale and long-term conversions of land to cultivation, there were net changes in soil carbon storage. Cultivation results in a decline in the annual production of plant residues and an increase in decomposition due to increased soil temperature, aeration, and moisture. Erosional losses are also likely to increase. These conditions lead to a decline in the percent organic matter in the surface layers. Most models of the global carbon cycle assume that the loss from the soil pool is released to the atmosphere as CO_2. This assumption will be evaluated in a later section. The total loss from the soil pool is calculated by knowing the area of natural ecosystems that have been converted to agriculture and the percent loss of carbon from undisturbed profiles of these ecosystems. Since most global models agree on the size of the pool of carbon in world soils, the values used for percent loss have the major effect on calculations of net agricultural release from soils (Houghton et al. 1983). In early modeling efforts, Revelle and Munk (1977) assumed a 15% decline in soil organic content, whereas Bohn (1978) used 50%. More recent models include a time course for these losses. Richards et al. (1983) assume that tropical soils lose 20% of their soil carbon in the first 60 years of cultivation and an additional 20% in the next 60 years. For temperate regions, 10% losses were used in each time period. The MBL model (Houghton et al. 1983) includes an initial period of rapid loss followed by a longer period of slower losses (Fig. 11.2). At present, there has been no comprehensive attempt to evaluate the magnitude or the pattern of loss in various world ecosystems.

The compilation here includes two types of studies: (1) those in which paired plots, one cultivated and one under natural vegetation, have been

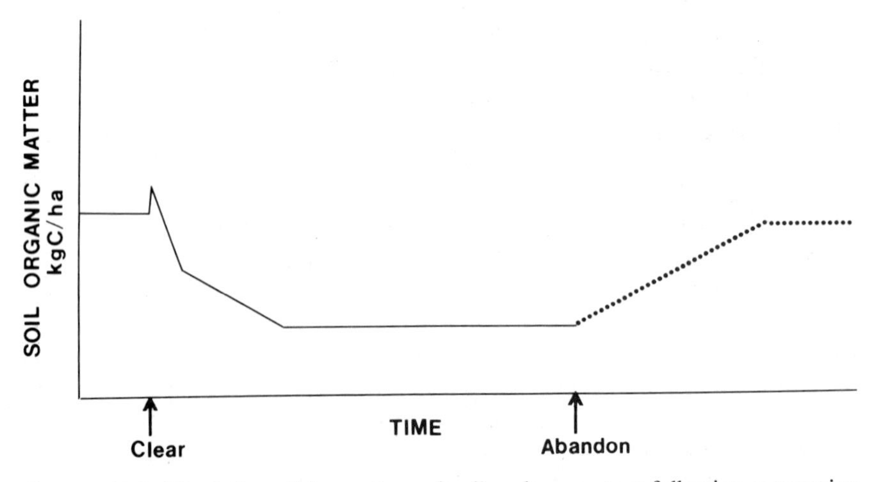

FIGURE 11.2. Simulation of the pattern of soil carbon content following conversion to agriculture. Model of the Marine Biological Laboratory, Woods Hole, Massachusetts. (From Houghton et al. 1983.)

compared after many years of agriculture; and (2) studies in which experimental plots originally under natural vegetation have been observed for many years following conversion to agriculture. Clearly, data of the latter variety are preferable, but they are much less frequent in the scientific literature. In paired-plot studies, it is essential to compare nearby fields that presumably were similar in soil carbon content before land-use conversion.

Time-course studies show that agricultural soils lose soil organic matter most rapidly in the initial years of land-use conversion. The rate of loss declines exponentially, and eventually these soils reach a new steady state, consistent with the input of crop residues and the decomposition characteristics of the field (Jenkinson and Rayner 1977, Lathwell and Bouldin 1981). Thus, studies of paired plots must include some knowledge of the conversion interval.

Most of the agronomic data include measurements of percent organic matter or organic carbon in soil samples. Few studies have calculated the change in the total storage in the soil profile by multiplying measurements of percent organic carbon by field measurements of soil bulk density at various depths. Again, the latter approach is preferable because losses in soil organic matter are often accompanied by increases in bulk density; the change in total storage may not be as great as comparisons of percent soil organic matter indicate (Tiessen et al. 1982).

On agricultural conversion, there are often large changes in the percent soil organic matter in the surface soil layers where bulk density tends to be relatively low. These layers contain easily decomposed plant residues, but only a portion of the total storage of organic matter in the soil profile. Typically, there is less change in the soil organic content in the deeper soil layers where resistant humic materials are predominant. Agricultural clearing of some soils results in increases in soil organic matter in the lower layers, as a result of mixing during cultivation and increased downward transport of soluble organic compounds in percolating waters. The effect of different sampling approaches is clearly shown in Table 11.2; the

TABLE 11.2. Comparisons of the basis of calculating the loss of soil carbon from a Canadian grassland.*

		Loss estimate based on changes in		
Horizon	Soil carbon in virgin condition ($kg\ C \cdot m^{-2}$)	% Organic carbon in each horizon (%)	$kg\ C \cdot m^{-2}$ in each horizon (%)	$kg\ C \cdot m^{-2}$ cumulative† (%)
A	9.41	−54	−57	−57
B	3.80	+ 4	+ 7	−47
C_1	1.62	−23	−21	−37
C_2	0.39	−19	−15	−36

*Data from Voroney et al. (1981); see also Table 11.4.
†Through the horizon indicated.

absolute loss of carbon from the entire profile is 36%, whereas the loss expressed as the change in percent organic carbon in the surface horizon is 54%. Values that are used to assess global changes in the soil carbon pool must be based on studies that have measured changes in the entire profile.

The impact of these factors (experimental design, conversion interval, basis of loss estimate, and depth of sampling) is shown in Tables 11.3 through 11.6, in which values are compiled for the losses of soil carbon during the conversion of undisturbed soils to agricultural use. The mean losses for ecosystem types are first approximations for the purposes of modeling changes in the global carbon cycle. At this point, there is no consideration of the crops grown or of the details of the agricultural practices, such as fertilization or tillage regimens. When agricultural fields are amended with mulch or manure, soil organic matter can be maintained at higher levels than in cultivated fields receiving no treatments (Larson et al. 1972; Jenkinson and Rayner 1977; Hofman and van Ruymbeke 1980). Such factors affect changes in soil carbon, but they differ widely among studies and even among years in most fields that have undergone long-term cultivation.

Seven studies of temperate forest soils show a mean loss of 34% of the original carbon content during long-term conversion to agricultural use (Table 11.3). This loss is equivalent to 4.0 kg C • m^{-2} from a content of 11.8 kg C • m^{-2} in undisturbed temperate forests (Schlesinger 1977). Many of these studies include samples at depths >30 cm in soils that have been cultivated for more than 25 years. Nearly complete removal of virgin forest occurred in both Europe and North America before the establishment of long-term agricultural experiment stations. Thus, data from temperate forest soils are few, and all of the values are from paired-plot studies.

Data from temperate grasslands include 24 studies of more than 370 field plots (Table 11.4). The mean loss was 28.6% (±2.4 SEM), equivalent to 5.5 kg C • m^{-2} from a mean content of 19.2 kg C • m^{-2} in the entire profile of undisturbed grasslands (Schlesinger 1977). This loss is probably an upper limit, because many of the studies in Table 11.4 are based only on changes in percent organic carbon in the surface (<30 cm) soil layers. Several studies show increases in carbon in the lower soil layers (Doughty et al. 1954, Meints and Peterson 1977, Mann, in preparation), and nearly all show greater losses in the upper horizons than lower in the profile (e.g., Hide and Metzger 1939; Newton et al. 1945; Godlin and Son'ko 1970; Voroney et al. 1981; Bauer and Black 1981). Nevertheless, the mean loss found in studies that sampled to depths >30 cm and for periods of 30 years or more was 30.2%. There is a slight trend for decreasing percent loss of soil carbon along a latitudinal transect from Canada to Texas, despite the greater profile storage at higher latitudes (Kononova 1975).

Using data from the Soil Conservation Service, Mann (in preparation) has calculated the soil carbon content for 322 profiles collected in a broad

TABLE 11.3. Changes in soil carbon storage on cultivation of paired-plot temperate forest soils.

Location	Number of field sites	Depth of sampling (cm)	Conversion interval (years)	Loss based on change in		Loss (%)	Reference
				kg C · m^{-2}	% Organic carbon		
USSR							
European region	5	100	100	X		33.0	Rubilin and Dolotov 1967
Dagestan region	?	100	?	X		50.0	Karmanov 1971
USA, Georgia	28	15	25+		X	56.5	Giddens 1957
Canada							
Quebec	3	25–38	50	X		33.0	Martel and MacKenzie 1980
	8	18	75+		X	28.5*	Salisbury and Delong 1940
Ontario	4	30	35	X		3.0	Coote and Ramsey 1983
Australia							
New South Wales	55	10	50		X	31.6	Williams and Lipsett 1961
Mean for temperate forest conversions						33.7	

*Five cultivated soils lost 26.8%; three pastures lost 31.2%.

TABLE 11.4. Changes in soil carbon storage on cultivation of temperate grassland soils.

Location	Number of field sites	Depth of sampling (cm)	Conversion interval (years)	Loss based on		Design of study	Loss (%)	Reference
				kg C · m^{-2}	% Organic carbon			
Canada								
Alberta, Saskatchewan, and Manitoba,	74	30	12–26		X	Paired-plot	19.3	Newton et al. 1945
Saskatchewan	12	15	40–100		X	Paired-plot	47.5	Campbell and Souster 1982
	3	30	14		X	Time-course	22.6	Doughty et al. 1954
	3	243	?		X	Paired-plot	20.4	Doughty et al. 1954
	3	100	70	X		Paired-plot	37.9	Voroney et al. 1981
	4	10	15–60		X	Paired-plot	38.8	Martel and Paul 1974
	3	20–36	65–90	X		Paired-plot	32.3	Tiessen et al. 1982
Alberta	6	Ap*	9–65		X	Paired-plot	46.0	Dormaar 1979
	3	13	64		X	Paired-plot	44.9	Dormaar and Pittman 1980
USA								
North Dakota to Texas	7 to 11	30	36		X	Paired-plot	34.0	Haas et al. 1957
North Dakota	21	45	43		X	Survey	36.5	Haas et al. 1957
	8	30	70	X		Paired-plot	21.5	Bauer and Black 1981

Iowa, Colorado, and Texas	25	15	?		X	Paired-plot	34.0	Thompson et al. 1954
Nebraska	67	30	3–60		X	Paired-plot	22.5	Russel 1929
	4	90–305	50+		X	Paired-plot	34.3	Meints and Peterson 1977
Kansas	20	51	30+		X	Paired-plot	29.2	Hide and Metzger 1939
	20	203	15–45		X	Paired-plot	+2.5	Swanson and Latshaw 1919
Texas and Oklahoma	61	45	Various		X	Paired-plot	14.0	Daniel and Langham 1936
Texas	1	91	60		X	Paired-plot	26.6	Smith et al. 1954
	2	76–107	40–90	X		Paired-plot	21.9	Laws and Evans 1949
New Mexico†	9	Various	19–36		X	Paired-plot	37.5	Chang 1950
Oregon	2	46	17–30		X	Paired-plot	23.0	Bradley 1910
	9‡	15	30–50		X	Paired-plot	32.5	Stephenson and Shuster 1942
USSR—Ukraine region	3§	35	25–100		X	?	12.7	Godlin and Son'ko 1970
Mean for temperate grassland conversions							28.6	

*Plowed layer.
†Semiarid grassland.
‡Orchards.
§Pastures.

survey of cultivated and virgin grassland soils in the loess region of the central United States. The profiles are not paired; however, the mean values suggest that cultivated Alfisols contain 33% less organic carbon than uncultivated Alfisols. Apparent losses from cultivated Mollisols are much less; the data from the Udoll suborder suggest no change in soil carbon content to depths of 150 cm in cultivated profiles. Mann suggests that cultivation may increase the downward transport of organic matter in these soils where it may complex with the high content of calcium in the lower profile.

Growing interest in the potential for intensive agriculture in tropical regions has resulted in a large number of recent studies of soil carbon losses when tropical forests are converted to agricultural use (Table 11.5). Most data are for the changes in the upper soil horizons when second growth forest is converted to short-term agriculture (slash and burn) before abandonment. The losses measured in various studies are highly variable (1.7 to 69.2%). Some of this variation is probably related to a wide variation in initial content. Typically, the highest percentage losses were found in studies that examined only the surface layers. The mean loss of 21.0% (±4.2 SEM) represents a loss of 2.2 kg C • m^{-2} from an undisturbed profile content of 10.4 kg C • m^{-2} (Schlesinger 1977). These losses may be a minimum to be expected if tropical soils are converted to the long-term, mechanized agricultural practices of the temperate zone (Seubert et al. 1977).

The large area of tropical savannas has received little study, despite its extensive use for cultivated and grazing lands in many regions of the world. Jones (1973) noted that 19 cultivated soils in the savannas of Nigeria contained only 56% of the soil carbon that was found in 22 undisturbed soil profiles. This value is based on the upper 15 cm of the profile for soils with an unknown length of land-use conversion. More work is needed on the organic content and changes in organic matter in tropical savanna soils.

This literature search revealed no studies for the boreal forest region, in which only 2% of the land is now cultivated (Revelle and Munk 1977). For purposes of modeling changes in the global carbon cycle, the value for temperate forests can probably be applied to the limited area in the boreal zone with agricultural potential. Losses of carbon from boreal peatlands and from organic soils in various other regions have been carefully reviewed by Armentano (1979) and Armentano et al. (1983, 1984), and at present, there is little potential for an improved analysis with new data. Drainage and cultivation may release 0.03×10^{15} g C • yr^{-1} from organic soils of the world, transforming peatland areas from a sink to a source of atmospheric CO_2. A large uncertainty in this estimate is mainly due to a poor understanding of the areal extent and the pool of carbon in wetland and peatland soils. Mulholland and Elwood (1982) estimate that lake sediments currently sequester 0.02 to 0.2×10^{15} g C • yr^{-1}, a flux that has increased as land areas have been flooded by the construction of reservoirs.

TABLE 11.5. Changes in soil carbon storage on cultivation of tropical forest soils.

Location	Number of field sites	Depth of sampling (cm)	Conversion interval (years)	Loss based on kg C · m^{-2}	Loss based on % Organic carbon	Design of study	Loss (%)	Reference
USA, Puerto Rico	2	38	25		X	Paired-plot	6.2	Smith et al. 1951
Guatemala	6	41	1		X	Paired-plot	31.0	Popenoe 1959
Costa Rica	3	30	15–22		X	Paired-plot	15.4	Krebs 1975
	2	30	1	X		Time-course	4.2	Harcombe 1977
	1	8	<1	X		Time-course	32.4	Ewel et al. 1981
Peru	1*	50	<1		X	Time-course	8.2	Seubert et al. 1977
	1	15	8		X	Time-course	27.2	Sanchez et al. 1982*a*
Columbia	5	50	1	X		Time-course	14.7	de las Salas and Folster 1976
Ghana	3	15	3		X	Time-course	36.3	Cunningham 1963
	5	30	2	X		Time-course	1.7	Nye and Greenland 1964
Sierra Leone	2	18	5		X	Time-course	50.5	Brams 1971
Nigeria	2	10	6–10	X		Paired-plot	69.2	Aina 1979
	2	50	5		X	Paired-plot	4.8	Juo and Lal 1979
	8	15	3		X	Time-course	18.8	Juo and Lal 1977
	5	15	2	X		Paired-plot	10.7	Ayanaba et al. 1976
	1	30	?		X	Paired-plot	34.4	Aweto 1981
India	3	40	1	X		Time-course	10.8	Ramakrishnan and Toky 1981
New Guinea	7	15	1–2		X	Survey	23.5	Wood 1979
Thailand	1	30	?		X	Paired-plot	+1.0	Zinke et al. 1978
Mean for tropical forest conversions							21.0	

*Value is for traditional agriculture; mechanized clearing produced 21.6% loss in the same interval.

When arid and semiarid soils are cultivated, usually through irrigation, soil organic matter may increase (Greaves and Bracken 1946). This response is scarcely universal (Table 11.6); in a broad survey of soils in Arizona, Schlesinger (1982) found that soil organic matter was 14 to 22% lower in agricultural soils. Since arid soils contain small amounts of organic carbon (x = 4.9 kg C • m^{-2}), the losses on cultivation do not add large amounts of CO_2 to the atmosphere. It is surprising that increasing soil moisture by irrigation does not result in greater accumulations of organic matter in arid soils, as observed along natural gradients of increasing moisture in arid regions (Gile 1977).

There is a potential for large changes in the storage of carbon in soil carbonates when arid lands are irrigated. Magaritz and Amiel (1981) found that calcareous soils in Israel lost 50 kg $CaCO_3$ • m^{-2} (= 6.0 kg C • m^{-2}) when irrigated for 40 years, but this change was only 3.1% of the initial pool in the soil (Table 11.6). Similarly, in Arizona, the cultivation of soils derived from limestone parent materials results in losses of carbonate (Schlesinger 1982). Presumably, in these cases, carbonate is lost in runoff waters as HCO^-_3 from the dissolution of $CaCO_3$. When the initial carbonate content is low, cultivation and irrigation result in increases in the content in the soil. The precipitation of $CaCO_3$ occurs when irrigation waters are rich in dissolved salts, including Ca^{2+}. Among cultivated soils derived from noncalcareous parent materials in Arizona, soil carbonate-carbon is 20 to 30 kg C • m^{-2} greater than in undisturbed profiles. Although the world area of cultivated arid lands is not large, it is likely to increase for the remainder of the century, and such large changes in soil carbonate are worthy of further investigation.

The Pattern of Soil Carbon Loss

The decomposition of plant litter often shows an initial period of rapid loss of constituents such as soluble carbohydrates, followed by a long period during which resistant components such as lignin disappear. This pattern is evident in many studies that have followed losses from fresh litter confined in mesh bags on the soil surface (Schlesinger and Hasey 1981, Berg et al. 1982) or the loss of labeled constituents from plant matter grown in a $^{14}CO_2$ atmosphere (Jenkinson 1977). Decomposition losses are often modeled with a negative exponential equation (Olson 1963):

$$\frac{X_t}{X_0} = e^{-kt}$$

where X_t and X_O are the amounts at time t and initially, respectively. Whereas most of the soil respiration of CO_2 results from the decay of labile components of soil organic matter in the surface layers, most of the accumulation of carbon in soils is composed of resistant components that decay very slowly (Minderman 1968). In addition, soil microbes synthesize

TABLE 11.6. Changes in soil carbon storage upon cultivation of arid and semiarid soils.

Location and ecosystem type	Number of field sites	Depth of sampling (cm)	Conversion interval (years)	Loss based on		Loss or gain		Reference
				$kg\ C \cdot m^{-2}$	% C	Organic (%)	Carbonate (%)	
USA, Utah								
Desert scrub	5	60–90	10–30		X	+12.3	+ 3.1	Stewart and Hirst 1914
	21	30	?		X	−19.6		Bracken and Greaves 1941
	8	91	25+	X		+ 1.5		Greaves and Bracken 1946
USA, Arizona								
Desert scrub and grassland								
on limestone	21	100+	?	X		−13.5	−56.1	Schlesinger 1982
on noncalcareous soil	70	100+	?	X		−21.6	+90.3	Schlesinger 1982
USA, New Mexico								
Semi-arid grassland	4	Various	19–36		X	−37.5	+ 5.5	Chang 1950
Israel								
Desert grassland	1	250	40	X			− 3.1	Magaritz and Amiel 1981
Mediterranean woodland	11	90	2–40		X	−24.0		Reinhorn and Avnimelech 1974
USSR								
Grasslands	3	300–400	40	X			− 1.9	Baranovskaya and Azovtsev 1981

resistant humic substances, which are complex molecules containing aromatic rings with organic acid and phenolic groups (Flaig et al. 1975). The decay constant k for fresh litter is much too large to apply to the decay of soil humus or to predict the decline in humus on agricultural conversion.

In natural communities, the change in the mass of soil organic matter is the result of the balance between the rate of production and the rate of disappearance of plant residues in the soil (Olson 1963). The latter is taken as some fraction k of the mass present in the soil. When the mass is not changing, that is, the soil pool is in steady state, then the annual production of detritus equals k multiplied by the soil organic mass. In this case, k is a weighted average value for the labile organic matter in the surface layers and the resistant compounds in the lower soil layers.

Agricultural conversion disrupts the steady-state conditions that may exist in natural communities. Taking an unrealistic case as an extreme, we can consider an agricultural practice that leaves no crop residues as analogous to a litterbag experiment. The soil organic matter decays with no new inputs. Few agricultural practices result in the complete elimination of plant residues. Nevertheless, those studies that have followed changes in the soil organic content at various intervals after conversion of natural ecosystems to agriculture show negative exponential losses of organic carbon (Brams 1971; Martel and Paul 1974; Dormaar 1979) and organic nitrogen (Haas et al. 1957; Hobbs and Thompson 1971; Lathwell and Bouldin 1981). The most rapid losses in temperate grasslands are seen in the first 20 years after conversion (Fig. 11.3); thus, most of the losses found in temperate forests and grasslands (Tables 11.3 and 11.4) are equivalent to the percentage content of labile soil organic matter in undisturbed ecosystems. It is likely that the time to achieve steady-state soil carbon in agriculture is shorter in the tropics (Brams 1971; Krebs 1975; Lathwell and Bouldin 1981).

If soil organic matter could be divided into two functional pools—labile and refractory—then a method of quantifying the storage and turnover in each would allow us to predict the losses on agricultural conversion (Van Veen and Paul 1981). Spycher et al. (1983) used density fractionation to show that the "light fraction" of a forest soil contained 34% of the organic carbon in the profile. They suggest that components of the light fraction comprise the labile fraction of the soil organic matter. Other workers separate soil organic matter into fractions that are soluble in alkaline (humic acids) and acid (fulvic acids) solutions. On the basis of the radiocarbon age in soils, the turnover of fulvic acids is apparently more rapid (Campbell et al. 1967).

Studies of the radiocarbon age and turnover of fractions of soil organic matter are limited by the effect of recent additions of ^{14}C from atomic weapons testing (Goh et al. 1976, 1977, Bottner and Peyronel 1977, O'Brien and Stout 1978). Since ^{14}C decays exponentially, a small amount of fresh plant residue with young ^{14}C age can "contaminate" the weighted mean

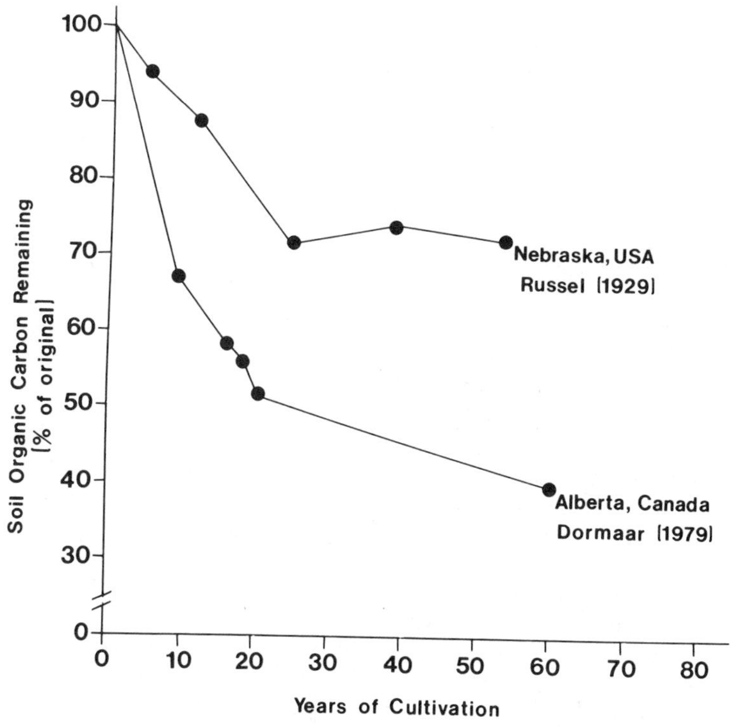

FIGURE 11.3. Loss of carbon from temperate grassland soils as a function of the length of cultivation. Data from both studies are derived from a regional survey of soils in cultivation–not a time course from a single plot (see Table 11.4.)

age for the organic carbon in the soil. O'Brien and Stout (1978) used radiocarbon dating to find that 16% of the organic matter in the soil of a New Zealand pasture was "very old," that is, with a minimum age of 5700 years. Modern carbon was concentrated near the surface and was mostly < 100 years old. When labile fractions are lost on cultivation, the remaining fractions show older ^{14}C ages. Martel and Paul (1974) found a ^{14}C age of 710 years for the organic matter in cultivated Canadian grasslands, whereas the soil organic matter in nearby uncultivated grasslands had an age of 250 years.

The division of soil organic matter into fractions with widely different turnover times is the basis of several recent models of soil carbon dynamics. The model of Van Veen and Paul (1981) varies the proportion of soil carbon that is "protected" by virtue of being complexed with the mineral soil constituents. They suggest that the amount of stable organic matter influences the steady-state levels of soil carbon on cultivation more than does the original decomposition rate of the plant residues (Fig. 11.4). Jenkinson and Rayner (1977) propose a five-fraction model for soil organic matter. When conditions for the production and decomposition of plant residues under agriculture were simulated, the model correctly predicted

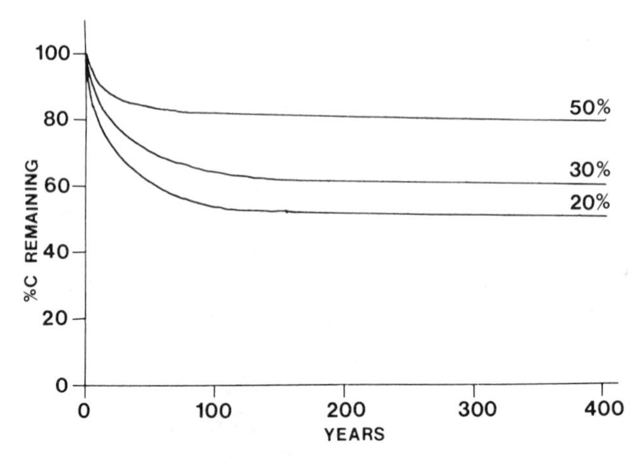

FIGURE 11.4. Effect of changing the proportion of soil carbon that is stabilized in mineral soil components on the loss of soil carbon (0–15 cm) on conversion of a virgin soil to continuous cropping. (From Van Veen and Paul 1981.)

the content of carbon in each fraction and the radiocarbon age of the fraction with the slowest turnover.

Soil organic matter contains high concentrations of nitrogen and other plant nutrients that may be lost on cultivation. Large losses of nitrate-nitrogen in runoff are often observed when terrestrial vegetation is cleared (Likens et al. 1970); the losses are greatest on sites with relatively high initial quantities or high mineralization potentials for soil nitrogen (Porter et al. 1964; Vitousek et al. 1982). Although the changes in streamwater chemistry are spectacular, the loss of total nitrogen from the soil is generally less than the loss of carbon on land-use conversion (Hide and Metzger 1939; Salisbury and Delong 1940; Chang 1950; Haas et al. 1957; Smith and Young 1975); thus, the carbon-to-nitrogen (C:N) ratio of soil organic matter is lower in agricultural fields than in undisturbed areas (Smith and Young 1975; Martel and McKenzie 1980).

The comparative loss of carbon and nitrogen is easily understood in the context of labile and refractory components in soil organic matter. During the decomposition of litter, carbon is mineralized as CO_2 while nitrogen is immobilized in microbial biomass (Staaf and Berg 1982). Thus, the C:N ratio typically declines from the surface litter to the lower layers of the soil (Jorgensen and Wells 1973). The C:N ratio is also higher in the light fraction than in the heavy fraction of soil organic matter (Spycher et al. 1983). The decline in soil nitrogen is less than that of soil carbon on cultivation because the greatest losses of soil organic matter occur among the labile surface materials with a high C:N ratio.

Erosion and Soil Respiration

When natural vegetation is cleared for agriculture, erosion of soil increases. The soil loss is largely derived from the surface layers that are high in organic content. Much of the surface litter is poorly consolidated, low-density material that is easily removed in runoff. The increased loss of soil organic matter occurs in the removal of particulate material; concentrations of dissolved organic carbon may actually be lower in disturbed watersheds (Meyer and Tate 1983). Although the erosional loss is a significant input to stream ecosystems, present evidence suggests that it is not responsible for the rapid decline in the labile pool of soil carbon upon land-use conversion. Losses of 2 to 6 kg C • m^{-2} (see Tables 11.3–11.5) would translate to annual losses of 100 to 300 g C • m^{-2} during a 20-year interval of cultivation. These losses are much larger than actually observed in streams draining small disturbed watersheds (Bishop 1973; Hobbie and Likens 1973; Bormann et al. 1974; Gill et al. 1976; Chichester et al. 1979; Osuji and Babaloa 1983; Tate and Meyer 1983). In a model of soil carbon dynamics in agricultural fields, Voroney et al. (1981) suggest that erosional losses do not become a significant portion of the annual loss of soil carbon on clearing until after the rapid initial losses have attenuated.

If erosional losses of carbon cannot explain the rapid decline on cultivation, then most of the carbon must be released to the atmosphere as CO_2. In cultivated fields, soil temperature and moisture are higher because of the removal of canopy shading and transpiration losses. These factors increase the rate of decomposition. Edwards and Ross-Todd (1983) found that soil respiration from the mineral soil increased from 92 g • m^{-2} • yr^{-1} of CO_2 to 154 g • m^{-2} • yr^{-1} during the first year after harvest of an upland forest watershed. Of course, in fields that have been cultivated for a long period, one would expect that soil respiration rates would be lower than in undisturbed areas (deJong et al. 1974), because there is less plant litter and labile soil organic matter.

Abandonment of Agricultural Lands

Agricultural regions have changed during the last 100 years. Hart (1968) noted that 27 $\times$ 10^6 ha of farmland have been abandoned in the eastern United States, and that 16.7 $\times$ 10^6 ha of this land had returned to forest between 1910 and 1959. Studies of secondary succession are numerous in plant ecology, but few of these have examined changes in soil organic matter. Prince et al. (1938) noted that fields abandoned from agriculture in New Jersey showed increases in soil organic matter ranging from 84% in surface horizons to 63% in the subsoil over a 30-year interval of recovery. Billings (1938) shows linear accumulations of soil organic matter during secondary succession on the Piedmont of North Carolina, with increases of 246% in the surface soil and 149% in the A_2 horizon. A linear recovery of organic matter in the 0- to 15-cm layer is also evident in the data of Hosner and Graney (1970) for abandoned fields in Virginia, and is consistent with the MBL model (Fig. 11.2). Eventually, the production and decomposition of plant residues should again achieve a balance, yielding a steady state in soil organic content. There are few data showing whether the soil carbon content at the end of secondary succession is similar to that found in original, undisturbed vegetation, but any permanent loss of soil fertility might prevent this.

When forests are harvested but allowed to regrow immediately, there is often a decline in the organic content of the forest floor, because production of litter by the regenerating forest does not balance the decomposition of organic matter already in the soil. This period of decline may last for 3 years in tropical forests (Zinke et al. 1978; Aweto 1981) and as long as 15 years in temperate forests (Covington 1981). Although these patterns are striking, they do not imply large changes in the total storage of carbon in the soil, which is dominated by storage in the mineral soil layers (Harcombe 1977; Bormann and Likens 1979; Gholz and Fisher 1982).

Conclusions

The main finding of this chapter is that the loss of carbon from soil profiles on cultivation is about 30% over a 20- to 50-year interval. This value is lower than parameters used in most recent models of anthropogenic changes in the global carbon cycle. Lower values result when field data are expressed in terms of the change in soil carbon storage, rather than the change in the percent organic carbon in soil samples, because losses of carbon are associated with increases in soil bulk density. Losses in the surface layers are often > 30%, but a large amount of soil carbon occurs in the lower portion of the profile and shows little change. More field work is needed in areas of organic soils and in tropical savanna and arid ecosystems.

The pattern of loss of soil carbon on land-use conversion suggests that the soil organic matter can be divided into two fractions: labile and refractory. Changes in ^{14}C age and the C:N ratio with cultivation are consistent with the loss of the more labile fractions comprised of recent plant residues. The major loss of soil carbon occurs in the first 20 years after land-use conversion, and is associated with an increased release of CO_2 from soils to the atmosphere. This rapid decline in soil organic matter should be relatively easy to simulate in "bookkeeping" models of terrestrial carbon storage.

A Note and Acknowledgments

Most of the data in Tables 11.3 through 11.6 were calculated from original measurements that were made and reported in very different ways among studies. In many cases various assumptions and weightings were used to calculate mean loss. The author welcomes inquiries regarding this data set—if future workers are unable to arrive at similar values.

This paper was improved by critical reviews by Dan Binkley, Kate Lajtha, Dan Livingstone, and Peter Vitousek. The investigation was sponsored by the U.S. Department of Energy via a subcontract from Oak Ridge National Laboratory to Duke University.

References

Aina, P. O. 1979. Soil changes resulting from long-term management practices in western Nigeria. Soil Sci. Soc. Am. J. 43:173–177.

Ajtay, G. L., P. Ketner, and P. Duvigneaud. 1979. Terrestrial primary production and phytomass. In B. Bolin, E. T. Degens, S. Kempe, and P. Ketner (eds.), The Global Carbon Cycle, pp. 129–181. John Wiley & Sons, New York.

Armentano, T. V. 1979. The Role of Organic Soils in the World Carbon Cycle. The Institute of Ecology, Indianapolis, Indiana.

Armentano, T. V. 1980. Drainage of organic soils as a factor in the world carbon cycle. BioScience 30:825–830.

Armentano, T. V., A. de la Cruz, M. Duever, O. L. Loucks, W. Meijer, P. J. Mulholland, R. L. Tate, and D. Whigham. 1983. Recent Changes in the Global Carbon Balance of Tropical Organic Soils. Holcomb Research Institute, Indianapolis, Indiana.

Armentano, T. V., E. S. Menges, J. Molofsky, and D. T. Lawler. 1984. Carbon Exchange of Organic Soils Ecosystems of the World. Holcomb Research Institute, Indianapolis, Indiana.

Aweto, A. O. 1981. Organic matter build-up in fallow soil in a part of southwestern Nigeria and its effects on soil properties. J. Biogeogr. 8:67–74.

Ayanaba, A., S. B. Tuckwell, and D. S. Jenkinson. 1976. The effects of clearing and cropping on the organic reserves and biomass of tropical forest soils. Soil Biol. Biochem. 8:519–525.

Baes, C. F., H. E. Goeller, J. S. Olson, and R. M. Rotty. 1977. Carbon dioxide and climate; the uncontrolled experiment. Am. Sci. 65:310–320.

Baranovskaya, V. A. and V. I. Azovtsev. 1981. Effects of irrigation on the migration of carbonates in the soils of the Volga region. Sov. Soil Sci. 13:68–77.

Bauer, A. and A. L. Black. 1981. Soil carbon, nitrogen, and bulk density comparisons in two cropland tillage systems after 25 years and in virgin grassland. Soil Sci. Soc. Am. J. 45:1166–1170.

Bazilevich, N. I. 1974. Geochemical work of the living substance of the Earth and soil formation. In Transactions of the 10th International Congress of Soil Science 6:17–27.

Berg, B., K. Hannus, T. Popoff, and O. Theander. 1982. Changes in organic chemical components of needle litter during decomposition. Long-term decomposition in a Scots pine forest I. Can. J. Bot. 60:1310–1319.

Billings, W. D. 1938. The structure and development of old field shortleaf pine stands and certain associated properties of the soil. Ecol. Monogr. 8:437–499.

Bishop, J. E. 1973. Limnology of a small Malayan River Sungai Gombak. W. Junk, Publishers, The Hague, The Netherlands.

Bohn, H. L. 1976. Estimate of organic carbon in world soils. Soil Sci. Soc. Am. J. 40:468–470.

Bohn, H. L. 1978. On organic soil carbon and CO_2. Tellus 30:472–475.

Bohn, H. L. 1982. Estimate of organic carbon in world soils II. Soil Sci. Soc. Am. J. 46:1118–1119.

Bolin, B. 1970. The carbon cycle. Sci. Am. 223(3):124–132.

Bolin, B. 1977. Changes of land biota and their importance for the carbon cycle. Science 196:613–615.

Bolin, B., E. T. Degens, P. Duvigneaud, and S. Kempe. 1979. The global biogeochemical carbon cycle. In B. Bolin, E. T. Degens, S. Kempe, and P. Ketner. (eds), The Global Carbon Cycle, pp. 1–56. John Wiley & Sons, New York.
Bormann, F. H. and G. E. Likens. 1979. Pattern and Process in a Forested Ecosystem. Springer-Verlag, New York.
Bormann, F. H., G. E. Likens, T. G. Siccama, R. S. Pierce, and J. S. Eaton. 1974. The export of nutrients and recovery of stable conditions following deforestation at Hubbard Brook. Ecol. Monogr. 44:255–277.
Bottner, P. and A. Peyronel. 1977. Dynamique de la matière organique dans deux sols méditerranéens étudiée à partir de techniques de datation par le radiocarbone. Rev. Ecol. Biol. Sol 14:385–393.
Bracken, A. F. and J. E. Greaves. 1941. Losses of nitrogen and organic matter from dry-farm soils. Soil Sci. 51:1–15.
Bradley, C. E. 1910. Nitrogen and carbon in the virgin and fallowed soils of Eastern Oregon. J. Ind. Eng. Chem. 2:138–139.
Brams, E. A. 1971. Continuous cultivation of West African soils: organic matter diminution and effects of applied lime and phosphorus. Plant Soil 35:401–414.
Brown, S. and A. E. Lugo. 1982. The storage and production of organic matter in tropical forests and their role in the global carbon cycle. Biotropica 14:161–187.
Buol, S. W. 1965. Present soil-forming factors and processes in arid and semiarid regions. Soil Sci. 99:45–49.
Buringh, P. 1984. Organic carbon in soils of the world. In G. M. Woodwell (ed.), The Role of Terrestrial Vegetation in the Global Carbon Cycle, SCOPE 23, pp. 91–109. John Wiley & Sons, New York.
Campbell, C. A., E. A. Paul, D. A. Rennie, and K. J. MaCallum. 1967. Applicability of the carbon-dating method of analysis to soil humus studies. Soil Sci. 104:217–224.
Campbell, C. A. and W. Souster. 1982. Loss of organic matter and potentially mineralizable nitrogen from Saskatchewan soils due to cropping. Can. J. Soil Sci. 62:651–656.
Chang, C. W. 1950. Effect of long-time cropping on soil properties in northeastern New Mexico. Soil Sci. 69:359–368.
Chichester, F. W., R. W. Van Keuren, and J. L. McGuinness. 1979. Hydrology and chemical quality of flow from small pastured watersheds. II. Chemical quality. J. Environ. Qual. 8:167–171.
Coote, D. R. and J. F. Ramsey. 1983. Quantification of the effects of over 35 years of intensive cultivation on four soils. Can. J. Soil Sci. 63:1–14.
Covington, W. W. 1981. Changes in forest floor organic matter and nutrient content following clear cutting in northern hardwoods. Ecology 62:41–48.
Cunningham, R. K. 1963. The effect of clearing a tropical forest soil. J. Soil Sci. 14:324–345.
Daniel, H. A. and W. H. Langham. 1936. The effect of wind erosion and cultivation on the total nitrogen and organic matter content of soils in the southern High Plains. J. Am. Soc. Agron. 28:587–596.
DeJong, E., H. J. V. Schappert, and K. B. MacDonald. 1974. Carbon dioxide evolution from virgin and cultivated soil as affected by management practices and climate. Can. J. Soil Sci. 54:299–307.
de las Salas, G. and H. Folster. 1976. Bioelement loss on clearing a tropical rain forest. Turrialba 26:179–186.

Dickson, B. A. and R. L. Crocker. 1953. A chronosequence of soils and vegetation near Mt. Shasta, California II. The development of the forest floors and the carbon and nitrogen profiles of the soils. J. Soil Sci. 4:142–154.
Dormaar J. F. 1979. Organic matter characteristics of undisturbed and cultivated chernozemic and solonetzic A horizons. Can. J. Soil Sci. 59:349-356.
Dormaar, J. F and U. J. Pittman. 1980. Decomposition of organic residues as affected by various dryland spring wheat-fallow rotations. Can. J. Soil Sci. 60:97–106.
Doughty, J. L., F. D. Cook, and F. G. Warder. 1954. Effect of cultivation on the organic matter and nitrogen of brown soils. Can J. Agri. Sci. 34:406–411.
Dregne H. E. 1976. Soils of Arid Regions. Elsevier, Amsterdam.
Edwards, N. T. and B. M. Ross-Todd. 1983. Soil carbon dynamics in a mixed deciduous forest following clear cutting with and without residue removal. Soil Sci. Soc. Am. J. 47:1014–1021.
Ewel, J., C. Berish, B. Brown, N. Price, and J. Raich. 1981. Slash and burn impacts on a Costa Rican wet forest site. Ecology 62:816–829.
Flaig, W., H. Beutelspacher, and E. Rietz. 1975. Chemical composition and physical properties of humic substances. In J. E. Gieseking (ed.), Soil Components, Vol. 1, Organic Components, pp. 1–211. Springer-Verlag, New York.
Gardner, L. R. 1972. Origin of the Mormon Mesa caliche, Clark County, Nevada. Geol. Soc. Am., Bull. 83:143–156.
Gholz, H. L. and R. F. Fisher. 1982. Organic matter production and distribution in slash pine (*Pinus elliottii*) plantations. Ecology 63:1827–1839.
Giddens, J. 1957. Rate of loss of carbon from Georgia soils. Soil Sci. Soc. Am. Proc. 21:513–515.
Gile, L. H. 1977. Holocene soils and soil-geomorphic relations in a semiarid region of southern New Mexico. Quat. Res. 7:112–132.
Gile, L. H., J. W. Hawley, and R. B. Grossman. 1981. Soils and geomorphology in the Basin and Range area of southern New Mexico—guidebook to the desert project. Memoir No. 39. New Mexico Bureau of Mines and Mineral Resources, Socorro.
Gill, A. C., J. R. McHenry, and J. C. Ritchie. 1976. Efficiency of nitrogen, carbon, and phosphorus retention by small agricultural reservoirs. J. Environ. Qual. 5:310–315.
Godlin, M. M. and M. P. Son'ko. 1970. Humus of ordinary steppe chernozems in the Ukraine. Sov. Soil Sci. 1970:8–18.
Goh, K. M., T. A. Rafter, J. D. Stout, and T. W. Walker. 1976. The accumulation of soil organic matter and its carbon isotope content in a chronosequence of soils developed on aeolian sand in New Zealand. J. Soil Sci. 27:89–100.
Goh, K. M., J. D. Stout, and T. A. Rafter. 1977. Radiocarbon enrichment of soil organic matter fractions in New Zealand soils. Soil Sci. 123:385–391.
Greaves, J. E. and A. F. Bracken. 1946. Effect of cropping on the nitrogen, phosphorus, and organic carbon content of a dry-farm soil and on the yield of wheat. Soil Sci. 62:355–364.
Haas, H. J., C. E. Evans, and E. F. Miles. 1957. Nitrogen and carbon changes in Great Plains soils as influenced by cropping and soil treatments. U.S. Department of Agriculture Tech. Bull. 1164.
Harcombe, P. A. 1977. Nutrient accumulation by vegetation during the first year of recovery of a tropical forest ecosystem. In J. Cairns, K. L. Dickson, and E. E. Herricks (eds.), Recovery and Restoration of Damaged Ecosystems, pp. 347–378. University of Virginia Press, Charlottesville.

Hart, J. F. 1968. Loss and abandonment of cleared farmland in the Eastern United States. Annals Assoc. Am. Geogr. 58:417–440.

Hide, J. C. and W. H. Metzger. 1939. The effect of cultivation and erosion on the nitrogen and carbon of some Kansas soils. Agron. J. 31:625–632.

Hobbie, J. E. and G. E. Likens. 1973. Output of phosphorus, dissolved organic carbon, and fine particulate carbon from Hubbard Brook watersheds. Limnol. Oceanogr. 18:734–742.

Hobbs, J. A. and C. A. Thompson. 1971. Effect of cultivation on the nitrogen and organic carbon contents of a Kansas Argiustoll (Chernozem). Agron. J. 63:66–68.

Hofman, G. and M. Van Ruymbeke. 1980. Evolution of soil humus content and calculation of global humification coefficients on different organic matter treatments during a 12-year experiment with Belgian silt soils. Soil Sci. 129:92–94.

Hosner, J. F. and D. L. Graney. 1970. The relative growth of three forest tree species on soils associated with different successional stages in Virginia. Am. Mid. Nat. 84:418–427.

Houghton, R. A., J. E. Hobbie, J. M. Melillo, B. Moore, B. J. Peterson, G. R. Shaver, and G. M. Woodwell. 1983. Changes in the carbon content of terrestrial biota and soils between 1860 and 1980: a net release of CO_2 to the atmosphere. Ecol. Monogr. 53:235–262.

Jenkinson, D. S. 1977. Studies on the decomposition of plant material in soil. V. The effects of plant cover and soil type on the loss of carbon from ^{14}C labelled ryegrass decomposing under field conditions. J. Soil Sci. 28:424–434.

Jenkinson, D. S. and J. H. Rayner. 1977. The turnover of soil organic matter in some of the Rothamsted classical experiments. Soil Sci. 123:298–305.

Jenny, H. 1980. The Soil Resource. Springer-Verlag, New York.

Jones, M. J. 1973. The organic matter content of the savanna soils of west Africa. J. Soil Sci. 24:42–53.

Jorgensen, J. R. and C. G. Wells. 1973. The relationship of respiration in organic and mineral soil layers to soil chemical properties. Plant Soil 39:373–387.

Juo, A. S. R. and R. Lal. 1977. The effect of fallow and continuous cultivation on the chemical and physical properties of an alfisol in western Nigeria. Plant Soil 47:567–584.

Juo, A. S. R., and R. Lal. 1979. Nutrient profile in a tropical alfisol under conventional and no-till systems. Soil Sci. 127:168–173.

Karmanov, I. I. 1971. Cinnamon-brown soils in the foothills of Dagestan. Sov. Soil Sci. 1971:1–14.

Kononova, M. M. 1975. Humus of virgin and cultivated soils. In J. E. Gieseking (ed.), Soil Components, Vol. 1: Organic Components, pp. 475–526. Springer-Verlag, New York.

Krebs, J. E. 1975. A comparison of soils under agriculture and forest in San Carlos, Costa Rica. In F. B. Golley and E. Medina. (eds.), Tropical Ecological Systems, pp. 381–390. Springer-Verlag, New York.

Larson, W. E., C. E. Clapp, W. H. Pierre, and Y. B. Morachan. 1972. Effect of increasing amounts of organic residues on continuous corn. II. Organic carbon, nitrogen, phosphorus, and sulfur. Agron. J. 64:204–208.

Lathwell, D. J. and D. R. Bouldin. 1981. Soil organic matter and soil nitrogen behavior in cropped soils. Trop. Agric. 58:341–348.

Lattman, L H. 1973. Calcium carbonate cementation of alluvial fans in southern Nevada. Geol. Soc. Am. Bull. 84:3013–3028.

Laws, W. D. and D. D. Evans. 1949. The effects of long-time cultivation on some physical and chemical properties of two rendzina soils. Soil Sci. Soc. Am. Proc. 14:15–19.

Likens, G. E., F. H. Bormann, N. M. Johnson, D. W. Fisher, and R. S. Pierce. 1970. Effects of forest cutting and herbicide treatment on nutrient budgets in the Hubbard Brook Watershed-Ecosystem. Ecol. Monogr. 40:23–47.

Magaritz, M. and A. J. Amiel. 1981. Influence of intensive cultivation and irrigation on soil properties in the Jordan Valley, Israel: recrystalization of carbonate minerals. Soil Sci. Soc. Am. J. 45:1201–1205.

Mann, L. K. A regional comparison of soil carbon in cultivated and uncultivated loess derived soils in the central United States (manuscript in preparation).

Martel, Y. A. and A. F. MacKenzie. 1980. Long-term effects of cultivation and land use on soil quality in Quebec. Can. J. Soil Sci. 60:411–420.

Martel, Y. A. and E. A. Paul. 1974. Effects of cultivation on the organic matter of grassland soils as determined by fractionation and radiocarbon dating. Can. J. Soil Sci. 54:419–426.

Meentemeyer, V., E. O. Box, M. Folkoff, and J. Gardner. 1981. Climatic estimation of soil properties; soil pH, litter accumulation and soil organic content. Abstr., Ecol. Soc. Am. Bull. 62(2):104.

Meints, V. W. and G. A. Peterson. 1977. The influence of cultivation on the distribution of nitrogen in soils of the Ustoll suborder. Soil Sci. 124:334–342.

Meyer, J. L. and C. M. Tate. 1983. The effects of watershed disturbance on dissolved organic carbon dynamics of a stream. Ecology 64:33–44.

Miller, P. C. (ed.). 1981. Carbon balance in northern ecosystems and the potential effect of carbon dioxide induced climatic change. U.S. Department of Energy, Washington, D.C.

Minderman, G. 1968. Addition, decomposition and accumulation of organic matter in forests. J. Ecol. 56:355–362.

Mulholland, P. J. and J. W. Elwood. 1982. The role of lake and reservoir sediments as sinks in the perturbed global carbon cycle. Tellus 34:490–499.

Newton, J. D., F. A. Wyatt, and A. L. Brown. 1945. Effects of cultivation and cropping on the chemical composition of some western Canada prairie province soils. Part III. Sci. Agric. 25:718–737.

Nye, P. H. and D. J. Greenland. 1960. The soil under shifting cultivation. Commonwealth Bureau of Soils Technical Communication 51, Harpenden, England.

Nye, P. H., and D. J. Greenland. 1964. Changes in the soil after clearing tropical forest. Plant Soil 21:101–112.

O'Brien, B. J. and J. D. Stout. 1978. Movement and turnover of soil organic matter as indicated by carbon isotope measurements. Soil Biol. Biochem. 10:309–317.

Olson, J. S. 1963. Energy storage and the balance of producers and decomposers in ecological systems. Ecology 44:322–331.

Osuji, G. E. and O. Babalola. 1983. Soil management as it affects nutrient losses from a tropical soil in Nigeria. J. Environ. Manage. 16:109–116.

Popenoe, H. 1959. The influence of the shifting cultivation cycle on soil properties in central America. Proceedings of the Ninth Pacific Science Congress 7:72–77.

Porter, L. K., B. A. Stewart, and H. J. Haas. 1964. Effects of long-time cropping on hydrolyzable organic nitrogen fractions in some Great Plains soils. Soil Sci. Soc. Am. Proc. 28:368–370.

Post, W. M., W. R. Emanuel, P. J. Zinke, and A. G. Stangenberger. 1982. Soil carbon pools and world life zones. Nature 298:156–159.

Prince, A. L., S. J. Toth, and A. W. Blair. 1938. The chemical composition of soil from cultivated land and from land abandoned to grass and weeds. Soil Sci. 46:379–389.

Ramakrishnan, P. S. and O. P. Toky. 1981. Soil nutrient status of Hill agro-ecosystems and recovery pattern after slash and burn agriculture (Jhum) in northeastern India. Plant Soil 60:41–64.

Reeves, C. C. 1970. Origin, classification, and geologic history of caliche on the southern High Plains, Texas and eastern New Mexico. J. Geol. 78:352–362.

Reeves, C. C. 1976. Caliche: Origin, Classification, Morphology and Uses. Estacado Books. Lubbock, Texas.

Reinhorn, T. and Y. Avnimelech. 1974. Nitrogen release associated with the decrease in soil organic matter in newly cultivated soils. J. Environ. Qual. 3:118–121.

Revelle, R. and W. Munk. 1977. The carbon dioxide cycle and the biosphere. In Energy and Climate, pp. 140–158. National Academy of Sciences, Washington, D.C.

Richards, J. F., J. S. Olson, and R. M. Rotty. 1983. Development of a data base fir carbon dioxide releases resulting from conversion of land to agricultural uses. ORAU/IEA-81-10 (M), ORNL/TM-8801. Oak Ridge National Laboratory, Oak Ridge, Tennessee.

Rubilin, Y. V. and V. A. Dolotov. 1967. Effect of cultivation on the amounts and composition of humus in gray forest soils. Sov. Soil Sci. 1967:733–738.

Russel, J. C. 1929. Organic matter problems under dry farming conditions. J. Am. Soc. Agron. 21:960–969.

Salisbury, H. F., and W. A. Delong. 1940. A comparison of the organic matter of uncultivated and cultivated Appalachian upland podsol soils. Sci. Agric. 21:121–132.

Sanchez, P. A., D. E. Bandy, J. H. Villachica, and J. J. Nicholaides. 1982a. Amazon Basin soils: management for continuous crop production. Science 216:821–827.

Sanchez, P. A., M. P. Gichuru, and L. B. Katz. 1982b. Organic matter in major soils of the tropical and temperate regions. 12th International Congress of Soil Science 1:99–114.

Schlesinger, W. H. 1977. Carbon balance in terrestrial detritus. Ann. Rev. Ecol. Syst. 8:51–81.

Schlesinger, W. H. 1982. Carbon storage in the caliche of arid soils: A case study from Arizona. Soil Sci. 133:247–255.

Schlesinger, W. H. 1984. Soil organic matter: A source of atmospheric CO_2. In G. M. Woodwell (ed.). The Role of Terrestrial Vegetation in the Global Carbon Cycle, SCOPE 23, pp. 111–127. John Wiley & Sons, New York.

Schlesinger, W. H. 1985. The formation of caliche in soils of the Mojave Desert, California. Geochim. Cosmochim. Acta 49 57–66.

Schlesinger, W. H. and M. M. Hasey. 1981. Decomposition of chaparral shrub foliage; losses of organic and inorganic constituents from deciduous and evergreen leaves. Ecology 62:762–774.

Seubert, C. E., P. A. Sanchez, and C. Valverde. 1977. Effects of land clearing methods on soil properties of an Ultisol and crop performance in the Amazon jungle of Peru. Trop. Agric. 54:307–321.

Smith, R. M., G. Samuels, and C. F. Cernuda. 1951. Organic matter and nitrogen build-ups in some Puerto Rican soil profiles. Soil Sci. 72:409–427.

Smith, R. M., D. O. Thompson, J. W. Collier, and R. J. Hervey. 1954. Soil organic matter, crop yields, and land use in the Texas Blackland. Soil Sci. 77:377–388.

Smith, S. J. and L. B. Young. 1975. Distribution of nitrogen forms in virgin and cultivated soils. Soil Sci. 120:354–360.

Spycher, G., P. Sollins, and S. Rose. 1983. Carbon and nitrogen in the light fraction of a forest soil: vertical distribution and seasonal patterns. Soil Sci. 135:79–87.

Staaf, H. and B. Berg. 1982. Accumulation and release of plant nutrients in decomposing Scots pine needle litter. Long-term decomposition in a Scots Pine forest II. Can. J. Bot. 60:1561–1568.

Stephenson, R. E. and C. E. Shuster. 1942. Soil properties of tilled orchards compared with untilled areas. Soil Sci. 54:325–334.

Stewart, R. and C. T. Hirst. 1914. Nitrogen and organic matter in dry-farm soils. J. Am. Soc. Agron. 6:49–56.

Stuiver, M. 1978. Atmospheric carbon dioxide and carbon reservoir changes. Science 199:253–258.

Swanson, C. O. and W. L. Latshaw. 1919. Effect of alfalfa on the fertility elements of the soil in comparison with grain crops. Soil Sci. 8:1–39.

Tarnocai, C. 1972. Some characteristics of cryic organic soils in Northern Manitoba. Can. J. Soil Sci. 52:485–496.

Tate, C. M. and J. L. Meyer. 1983. The influence of hydrologic conditions and successional state on dissolved organic carbon export from forested watersheds. Ecology 64:25–32.

Thompson, L. M., C. A. Black, and J. A. Zoellner. 1954. Occurrence and mineralization of organic phosphorus in soils, with particular reference to associations with nitrogen, carbon, and pH. Soil Sci. 77:185–196.

Tiessen, H., J. W. B. Stewart, and J. R. Bettany. 1982. Cultivation effects on the amounts and concentrations of carbon, nitrogen, and phosphorus in grassland soils. Agron. J. 74:831–835.

Van Veen, J. A. and E. A. Paul. 1981. Organic carbon dynamics in grassland soils. I. Background information and computer simulation. Can. J. Soil Sci. 61:185–201.

Vitousek, P. M., J. R. Gosz, C. C. Grier, J. M. Melillo, and W. A. Reiners. 1982. A comparative analysis of potential nitrification and nitrate mobility in forest ecosystems. Ecol. Monogr. 52:155–177.

Vitousek, P. M, K. Van Cleve, N. Balakrishnan, and D. Mueller-Dombois. 1983. Soil development and nitrogen turnover in montane rainforest soils on Hawai'i. Biotropica 15:268–274.

Voroney, R. P., J. A. Van Veen, and E. A. Paul. 1981. Organic C dynamics in grassland soils. 2. Model validation and simulation of the long-term effects of cultivation and rainfall erosion. Can. J. Soil Sci. 61:211–224.

Williams, C. H. and J. Lipsett. 1961. Fertility changes in soils cultivated for wheat in southern New South Wales. Aust. J. Agric. Res. 12:612–629.

Wilson, A. T. 1978. Pioneer agriculture explosion and CO_2 levels in the atmosphere. Nature 273:40–41.

Wood, A. W. 1979. The effects of shifting cultivation on soil properties: An example from the Karimui and Bomai plateaux, Simbu Province, New Guinea. Papua New Guinea Agric. J. 30:1–9.

Zinke, P. J., S. Sabhasri, and P. Kunstadter. 1978. Soil fertility aspects of the Lua forest fallow system of shifting cultivation. In P. Kunstadter, E. C. Chapman, and S. Sabhasri (eds.) Farmers in the Forest, pp. 136–159. University Press of Hawaii, Honolulu.

12

Continental and Global Scale Remote Sensing of Land Cover

COMPTON J. TUCKER, J. R. G. TOWNSHEND,
T. E. GOFF, AND B. N. HOLBEN

In recent years, a number of investigations have indicated that the sensors aboard meteorological satellites have potential for land-cover monitoring at regional, continental, and global scales. The outstanding characteristic of data from such satellites relates to their high temporal resolution; imagery is available for the whole globe on a near-daily basis. Thus the possibilities of obtaining cloud-free imagery are greatly enhanced, and the temporal dynamics of land cover can be observed. The Advanced Very High Resolution Radiometer (AVHRR) of the National Oceanic and Atmospheric Administration (NOAA) series of sun-synchronous, polar-orbiting, operational satellites has been identified as having particular potential in this context (Gray and McCrary 1981, Schneider et al. 1981, Townshend and Tucker 1981, Cicone and Metzler 1982, Ormsby 1982, Schneider and McGinnis 1982, Tucker et al. 1982). This is because the radiometer's first band in the visible-red part of the spectrum and the second band in the near-infrared (Table 12.1) are two bands of particular use in vegetation mapping of green leaf area, green leaf biomass, or the intercepted photosynthetically active radiation (Tucker 1979, Curran 1980, Kumar and Monteith 1982). These bands correspond approximately to bands 5 and 7 of the Multispectral Scanner System (MSS) of the LANDSAT series of satellites.

LANDSAT data have been the principal source of satellite data for land-cover mapping. Their principal limitations have been their relatively low frequency of imaging, which greatly inhibits their use for monitoring—especially in areas of high cloud cover and when a large number of LANDSAT scenes are needed to cover continental-scale areas. In terms of orbital characteristics, the two sets of satellites are similar (Table 12.2). The principal difference between the sensors is in the much larger, instantaneous field of view of the AVHRR, which gives it a much higher temporal resolution than the MSS by virtue of imaging ±56° from nadir. This also leads to spatial resolution that is much coarser: in linear terms, the pixels are approximately 14 times bigger, and areally are almost 200 times larger. It is unrealistic to expect AVHRR data to provide the detail

TABLE 12.1. Spectral bandwidths of AVHRR and the LANDSAT MSS (50% response limits in μm).

TIROS-N AVHRR	NOAA-6, 7, and 8 AVHRR	LANDSAT MSS
Channel 1 0.55–0.90	Channel 1 0.58–0.68	Channel 4 0.5–0.6
Channel 2 0.73–1.10	Channel 2 0.73–1.1	Channel 5 0.6–0.7
Channel 3 3.55–3.93	Channel 3 3.5–3.9	Channel 6 0.7–0.8
Channel 4 10.5–11.5	Channel 4 10.5–11.5	Channel 7 0.8–1.0*
	Channel 5 11.5–12.5†	

*A more accurate representation of the spectral bandwidths than the value of 0.8–1.1 μm which is normally quoted.
†Not on NOAA-6.

of land-cover mapping possible with MSS data. However, for large-area mapping where fine detail is not required, or for situations where the temporal dynamics of vegetation is important, they provide an excellent data source. Apart from their advantage of higher temporal resolution compared with MSS data, the coarser resolution greatly reduces the volume of data and makes data handling and processing significantly easier. In addition,

TABLE 12.2. Comparison of the NOAA AVHRR and LANDSAT MSS.

Characteristic	LANDSAT/MSS*	NOAA/AVHRR
Inclination of orbit	98.91°–99.1°	99.092°
Height above surface	916.6 km	833 km
Number of orbits/day	14	14.2
Times of coverage at equator	09:30	07:30 descending) NOAA-6, -8 19:30 ascending) 02:30 descending)NOAA-7, -9 14:30 ascending)
Orbital period	103.3′	102′
Latitudinal coverage	82°N–82°S	90°N–90°S
Cycle duration	18 days	c. 1 day
Ground coverage	185 km	c. 2700 km
<FOV	±5.78°	±56°
<IFOV	0.086 milliradians	1.39–1.51 milliradians
Ground resolution (nadir)	79 m	1.1 km
Ground resolution (maximum off-nadir)	79.5 m along track 80 m across track	2.4 km along track 6.9 km across track
Samples per IFOV	1.411	1.362
Number of channels (see Table 12.1 for further details)	2 visible 2 near infrared	1 visible 1 near infrared 2 or 3 thermal infrared
Data precision	6 or 7 bit	10 bit

From Kidwell (1979) and General Electric.
*LANDSAT 4 has a lower orbit (725 km) and a cycle duration of 16 days, but overall the MSS has very similar properties to those of previous LANDSATs. It also contains the 7 band, 30 m FOV Thematic Mapper.

operational continuity of the NOAA series of satellites is guaranteed until the mid-1990s. By contrast, there are no firm plans to continue the LANDSAT series of satellites after LANDSAT-5′, which was launched in early 1984.

This paper will review recent research using data from the NOAA-7 AVHRR sensor that is of interest to the role of the biota in the CO_2 cycle in three areas: tropical deforestation monitoring in the Amazon Basin, continental-scale land-cover classification, and global-scale monitoring of green leaf biomass dynamics.

Tropical Deforestation Monitoring

The extent and rate of forest clearing in the Amazon Basin are the subjects of continuing controversy. Estimates of the cumulative area that has been cleared range from about 72,000 km^2 for the total Amazon Basin through 1978 (Tardin et al. 1978, 1979, 1980) to 260,000 km^2 for the rain forest portion of the Amazon through 1976 (Myers 1980, Sioli, 1980). The rate of forest clearing in the total Amazon Basin has been estimated to be as low as 10,000 $km^2 \cdot yr^{-1}$ and as high as 100,000 $km^2 \cdot yr^{-1}$ (Muthoo 1977; Sioli 1980). The lack of any systematic data on the subject has hampered discussion of what many scientists feel to be a major ecological crisis with serious biological, climatic, and political ramifications. One possible means of collecting these data is satellite remote sensing.

Heretofore, the majority of satellite remote-sensing studies of the terrestrial surface have used the LANDSAT satellites (e.g., monitoring tropical forests, Williams and Miller 1979). However, certain features of the LANDSAT MSS are not well suited for accurate assessment of forest clearing and regrowth. There is difficulty, for example, in distinguishing regrowing cleared areas from undisturbed tropical forests. This difficulty results from the spectral configuration of the LANDSAT MSS and the sensitivity of this sensor system to green leaf biomass and not total standing biomass (Tucker et al. 1981). Thus, a regrowing cleared area with a dry leaf biomass of 400 $g \cdot m^{-2}$ and a total dry biomass of 1000 $g \cdot m^{-2}$ can be spectrally identical in the LANDSAT MSS bands to a virgin tropical forest with a dry leaf biomass of 400 $g \cdot m^{-2}$ and a total dry biomass of 40,000 $g \cdot m^{-2}$. This criticism has been raised by Fearnside (1982) in reference to LANDSAT MSS assessments of Amazon clearing by the Brazilian National Institute for Space Research (INPE)(Tardin et al. 1978, 1979, 1980). However, sequential landsat imagery and change detection techniques can be used to overcome this limitation of the LANDSAT MSS (Woodwell et al., Chapter 13, this volume). We now review a study of part of the southern Amazon Basin that has used NOAA-7 AVHRR data to detect large-scale deforestation.

The first two AVHRR channels are spectrally similar to LANDSAT MSS channels 5 and 7, respectively, and both instruments are thus able to monitor the green leaf biomass of plant canopies (Tucker et al. 1981) (Table 12.1). For vegetated targets, channel 1 is sensitive to the in situ chlorophyll density, whereas channel 2 is sensitive to the green leaf density (Tucker 1978). Channel 3 is sensitive to a combination of reflected and emitted radiation, while channels 4 and 5 are sensitive to emitted or thermal radiation from the target (Weinreb and Hill 1980; Matson and Dozier 1981). The outstanding characteristics of data from the NOAA satellites is high frequency of observation; worldwide AVHRR imagery is available daily. In practice, however, more useful data for a given target can be obtained 4 and 5 days out of every nine because of the difficulties in using data with extreme look angles.

We obtained NOAA-7 AVHRR data at 9-day intervals from May to September 1982 over the general area of longitude 50°W to 65°W between latitude 15°S and 5°N, with particular emphasis on the state of Rondonia, Brazil. It has been reported that Rondonia is experiencing the most rapid rate of forest clearing in the Amazon Basin (Table 12.3). One image from July 9, 1982 was largely cloud-free over Rondonia and was processed on the NASA/Goddard Sensor Evaluation Branch's Hewlett-Packard 1000 image-processing computer system. Data from all five AVHRR channels were mapped to a Mercator projection. Inspection of the channel 3 image indicated a series of linear features of higher spectral return radiating from highway BR-364 in Rondonia between the towns of Ariquemes in the north and Pimenta Bueno to the south. Additional linear areas of higher spectral return were apparent south of Vilhena near the Bolivian border (Fig. 12.1). We interpreted these features to be a large-scale, systematic forest disturbance because of their regularity and close association with highway BR-364. Digital count differences between virgin forests and the areas of

TABLE 12.3. Clearing in the Legal Amazon.

State or territory	Total (km^2)	Observed cumulative cleared areas (km^2)		Rate of increase 1975 to 1978 (%)
		1975	1978	
Amapa	139,068	153	171	12
Para	1,227,530	8654	22,445	159
Roraima	243,004	55	144	162
Maranhao	257,451	2905	7334	152
Goias	285,793	3307	10,289	211
Acre	152,589	1166	2465	111
Rondonia	230,104	1217	4185	244
Mato Grosso	881,001	10,124	28,255	179
Amazonas	1,558,987	780	1786	129
TOTAL	4,975,527	28,361	77,074	

From Tardin et al. (1978, 1979, 1980).
States not totally within the Legal Amazon that are included: Goias north of lat. 13°S and Maranhao west of long. 44°W.

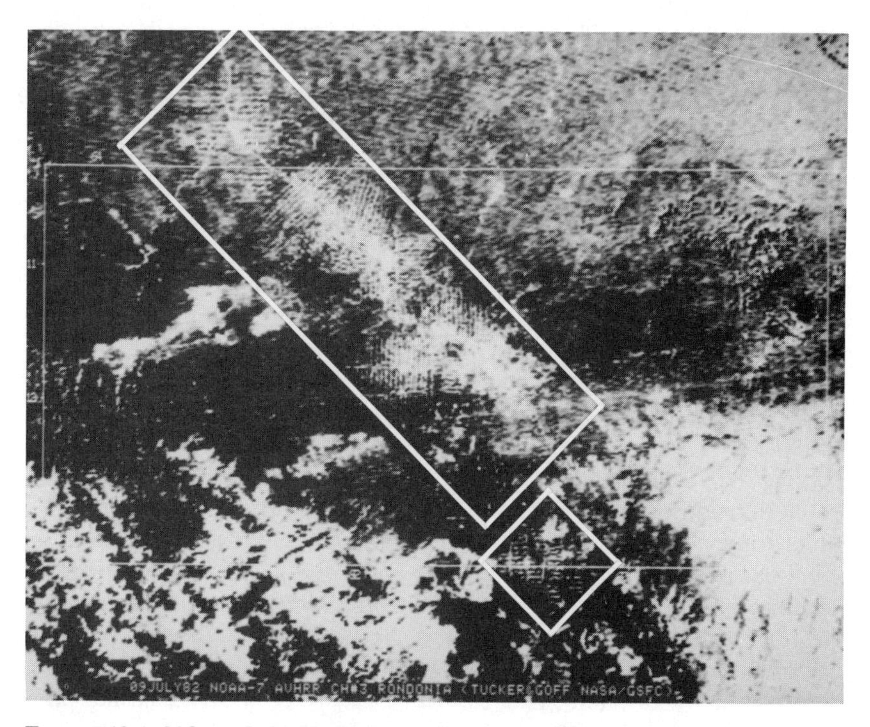

FIGURE 12.1. NOAA-7 AVHRR Channel 3 (3.5–3.9 μm) image obtained at 1430 hours local solar time on July 9, 1982. Note the linear forest clearing roads associated with highway BR-364. See also Fig. 12.2 and Table 12.4.

forest disturbance were found to average 46 counts for channel 3, 20 counts for channel 4, 13 counts for channels 1 and 5, and 3 counts for channel 2 (Table 12.4). The physical reason for the high sensitivity of channel 3 to the forest disturbance is not understood and is possibly related to this sensor's unique response to both emitted and reflected energy. The linear forest-disturbance features were apparent in some areas but not in others

TABLE 12.4. Mean digital counts (possible range: 0–1023) for virgin forest and disturbed areas from July 9, 1982 image from Rondonia.

Spectral channel	Bandwidth	Virgin forest (188 pixels)	Digital counts disturbed areas Ouro Preto (457 pixels)	Ji-Parana (135 pixels)	Colorado (257 pixels)
1	0.55–0.68 μm	28 (0.88)	39 (2.73)	46 (5.94)	37 (3.22)
2	0.73–1.1 μm	91 (1.59)	97 (3.51)	96 (3.87)	88 (1.94)
3	3.5–3.9 μm	820 (2.39)	850 (12.23)	886 (22.58)	861 (23.87)
4	10.5–11.5 μm	811 (1.31)	822 (5.67)	846 (11.33)	826 (10.97)
5	11.5–12.5 μm	805 (1.82)	810 (4.78)	833 (10.11)	812 (9.09)

The means, pixel sample number, and variances (in parentheses) are also given.

when the normalized-difference measure for green leaf density was calculated as [(0.73 to 1.1 μm) – (0.55 to 0.68 μm)]/[0.73 to 1.1 μm) + (0.55 to 0.68 μm)] and displayed. We thus confirmed Fearnside's (1982) observation that green-leaf-density techniques often cannot differentiate between forest and regrowth areas.

A field verification mission was undertaken in September 1982. Site visits were made at numerous locations along highway BR-364 between Porto Velho and Pimenta Bueno, where the linear forest-disturbance features were found to be forest-clearing roads, some more than 80 km long and about 4 km apart (Fig. 12.2). The large-scale forest disturbance apparent in Fig. 12.1 is a government-planned colonization project in which immigrants to Rondonia are sold 100-ha (500 × 2000 m) lots of forest to use for agriculture or pasture. Each lot has about 400 m of road frontage. Clearing on a lot begins nearest the access road, then gradually moves away as the farmer clears more land to maintain or increase agricultural production and to reach soil not yet depleted of nutrients. Currently, the colonization project comprises wide areas of agriculture and pasture adjacent to the access roads, with relatively narrow strips of primary forest between. This pattern accounts for the light and dark parallel lines in the grids of Fig. 12.1.

Economical crop production on a given lot is reportedly limited to 1 to 4 years because of plant diseases, decreasing soil fertility, and the prohibitively high cost of chemical fertilizers. This land obsolescence is com-

FIGURE 12.2. A forest clearing road ~20 km directly south of Ouro Preto, taken in September 1982. See also Fig. 12.1.

pensated for by clearing more of the primary forest. The rate of forest clearing is increasing as new settlers arrive and new colonization projects are begun to accommodate them. The paving of highway BR-364, under funding by the World Bank and scheduled for completion sometime in 1984, is expected to facilitate immigration and enlarge the markets for the agricultural and forests products, thereby increasing the pressure for forest clearing.

Estimates of forest clearing in Rondonia are few. Inspection of the 1972 to 1973 radar imagery for Rondonia at a scale of 1:250,000 showed only minor and scattered forest disturbance along BR-364 (Radam Brazil 1984). More recent (November 11, 1981) radar imagery from the space shuttle Columbia imaging radar indicated an area of forest-clearing roads adjacent to the Rio Guapare at about longitude 62°10′W × latitude 12°30′S. Estimates of primary forest clearing for Rondonia in 1975 and 1978 indicated that 1200 km^2 and 4200 km^2, respectively, had been cleared—an increase of 3000 km^2 in the 3 years (Table 12.3). After a series of classifications, our data of July 9, 1982 indicated that as a most conservative estimate about 9200 km^2 was cleared in the areas adjacent to BR-364. We conclude that previous estimates of the cumulative forest clearing in Rondonia have been conservative and that forest clearing is accelerating along BR-364 and will increase further with the paving of this highway.

This area in Rondonia presents an excellent opportunity for testing remote-sensing techniques for monitoring tropical forest clearing. A combination of coarse-resolution off-nadir viewing satellites, finer spatial-resolution sensor systems for more detailed studies, and radar systems would offer a logical multilevel link to the ground reference data necessary for accurately assessing tropical forest clearing. It is only by collection of these data that the rate of tropical deforestation can be accurately determined. The chapter by Woodwell et al. (*this volume*) reports on the use of multitemporal LANDSAT MSS data for studying the same area of deforestation (see also Tucker et al. 1984).

Continental Land Cover Classification

Traditionally, the principal source of land-cover information was vegetation mapping by ground survey. But for very large areas such ground survey exceeds the capabilities of any single mapping agency, let alone a single individual. For such areas it is inevitable that the observations of many different people and groups have to be synthesized, with all the problems of reconciling disparate observations. The magnitude of the task can be seen in the map and ancillary text on the vegetation of Africa south of the Sahara (Keay 1959). Other vegetation maps of Africa or large parts of it include those of UNESCO (1968), Eyre (1963), the World Atlas of Agriculture Committee (1969), and White (1983).

The use of such traditional methods for the timely production of internally consistent maps and spatial statistics is not feasible for areas on a continental or global scale. The impracticality of such methods is illustrated by the widely varying estimates of surface areas of world terrestrial ecosystems as collated by Ajtay et al. (1979). An alternative is to look to remote-sensing data as a basis for mapping, especially satellite data with a synoptic overview. In numerous studies, such remotely sensed data have been used to map accurately vegetation, crops, and other land-cover types (reviewed in Bauer et al. 1979). The majority of these studies have used LANDSAT data and none of them attempted classification on a continental scale.

The lack of continental-scale vegetation classifications from LANDSAT data stems from their 80-m spatial resolution and 18-day repeat cycle. The 80-m spatial resolution of the LANDSAT MSS and associated 185 × 185 km scene results in a very substantial quantity of data (about 1000 LANDSAT scenes) to cover an area as large as Africa. Thus logistical and financial problems are formidable if the analysis is to be performed in a digital mode.

It is highly desirable, however, to collect several images each year to minimize cloud cover and adequately monitor changes in vegetation over time (Tucker et al. 1981). Multiple imaging within a given year allows classification to be based on differing phenological changes in different vegetation types. For areas such as Africa, such an approach becomes essential if cover types are to be spectrally discriminated, because of the asynchronous responses of vegetation to seasonal rainfall north and south of the equator.

The 18-day LANDSAT repeat cycle, however, is inadequate for acquiring several cloud-free images in a year in many parts of Africa, especially in equatorial rain forests but also in much drier areas. In Sahelian areas, for example, cloud cover averages 40 to 50% throughout the 60- to 80-day wet season, when changes in vegetation are most pronounced.* In addition, substantial parts of the earth's surface are not covered by LANDSAT receiving stations, so MSS data are infrequently available, though with the launching of the two Tracking and Data Relay Satellite Systems this problem may lessen in the future. In summary, LANDSAT data are a potentially useful source for mapping very large areas, but as a practical tool they suffer from high cost, large amounts of data, and inadequate temporal resolution.

As an alternative to LANDSAT we have used AVHRR data from the NOAA meteorological satellites. Such data have already been shown to

---

*The average reported cloud cover was 40 to 50% for LANDSAT images obtained between July 15 to October 15 (i.e., during the growing season) from 1972 to 1982 in northern Senegal, western Mali, and southern Mauritania. For the nine LANDSAT scenes that cover this portion of the Sahel, the total number of images obtained during the wet or growing season ranged from a minimum of 1 to a maximum of 15 over an 11-year period.

have significant potential for assessing and mapping vegetation over relatively modest areas. They have a much coarser resolution (1 and 4 km) and hence a much lower data volume and cost than LANDSAT MSS data. Their temporal resolution is much higher, with 4-km imagery being globally available on a daily basis. We intend to address the questions of whether such data can be satisfactorily managed to provide multitemporal data sets with low cloud frequency and whether these data can provide useful information about vegetation of a complete continent, the one chosen being Africa. Evaluation of the latter question will clearly be a lengthy process, since we are dealing with a very large area. However, by comparison with existing vegetation and land cover maps and with reference to our own field experience, a preliminary evaluation is possible.

It should be stressed that we are attempting to map vegetation formations according to cover criteria rather than using any alternative criteria that have been proposed, such as environmental characteristics, phylogeny, or individual form features (de Laubenfels 1975). Our techniques are based on satellite data collected daily for 1 year from the continent of Africa, and we use the satellite-recorded multitemporal characteristics of land cover for large-scale classification purposes.

Methods

The AVHRR of the NOAA satellite was chosen to provide the data (Table 12.1). Earlier data from this sensor on board the television and infrared observing satellite (TIROS) were unsatisfactory because of the overlap between the visible and near-infrared bands, which made it impossible to produce a spectral green vegetation index (Tucker 1978). We chose data from NOAA-7 because that satellite images in the afternoon, whereas NOAA-6/8 images in the early morning and has significant portions of its broad swath in darkness. A disadvantage of NOAA-7 data compared with NOAA-6/8 data is that the atmosphere tends to be cloudier in the early afternoon than in the early morning, especially in equatorial areas. Global-area coverage (GAC) data with a 4-km resolution were initially chosen rather than the finer resolution local-area coverage (LAC) data with a 1-km resolution, since only the former data are available daily for the whole continent and consequently can provide a geographically comprehensive, cloud-free data set of the continent when composited over several weeks.

The 4-km GAC data are partly resampled 1-km LAC data; the first four 1-km picture elements are averaged; this average is then used to represent a 3 × 5 picture element block (Kidwell 1979). The GAC data are tape-recorded on board the satellite and subsequently transmitted to receiving stations either in Virginia or Alaska.

Three to four NOAA-7 orbits cover Africa each day. The data are processed by NOAA and the channel 1 (C_1) and channel 2 (C_2) GAC values remapped daily into a 1024 × 1024 array for the northern and southern hemispheres (Tarpley et al. 1984). The C_1 and C_2 values are used to generate the normalized-difference vegetation index which has been shown to be strongly correlated with green leaf biomass and intercepted photosynthetically active radiation (IPAR). This index is one of several related linear or ratio combinations of visible and near-infrared spectral data that are used to estimate nondestructively the green leaf area, green leaf biomass, or IPAR of plant canopies (Jordan 1969; Rouse et al. 1973; Tucker 1979; Wiegand et al. 1979; Aase and Siddoway 1980; Bartlett and Klemas 1980; Curran 1980, 1983; Daughtry et al. 1980, 1983; Holben et al. 1980; Brakke et al. 1981; Kimes et al. 1981; Kumar and Monteith 1982; Hatfield et al. 1984; Wiegand and Richardson 1984). Research has also shown that the C_2/C_1 and the $(C_2-C_1)/(C_2+C_1)$ ratios are also influenced by off-nadir viewing, sun angle, canopy morphology, shadowing, soil background, and atmospheric aerosols (Duggin 1977; Kreibel 1978; Dave 1979; Ranson et al. 1981; Slater and Jackson 1982; Kirchner et al. 1982; Kimes 1983). In addition, a shared trait of all spectral green leaf indices is an asymptotic or saturation response to high levels of green leaf area or green leaf biomass.

Every week a normalized-difference vegetation index was selected by NOAA for each picture element in the two 1024 × 1024 arrays, along with the associated pixel values for C_1 and C_2 (Tarpley et al. 1984). We have taken the weekly NOAA AVHRR channel 1 and channel 2, formed the C_2/C_1 and $(C_2-C_1)/(C_2+C_1)$ ratios, and selected the highest value over a 3-week period for both spectral ratios for each grid cell location for Africa. Selecting the respective maximum spectral ratio for a 3-week period simultaneously minimizes the degrading effects of pathlength, directional reflectance, aerosols, clouds, and shadowing, all of which only decrease the spectral ratios (Holben and Fraser 1984). This results in the selection of data that tend to occur near-nadir, are largely cloud-free, and were obtained under low atmospheric aerosol conditions.

The weekly composited data have a grid-cell size of 15 × 15 km at the equator and approximately 30 × 30 km at 90° north or south latitude. The satellite data used to prepare this grid cell could represent an area as small as 1.1 × 4.4 km at nadir or about 8 × 8 km at a ±56° scanning angle. Thus we see that the AVHRR GAC data used for each grid cell are undersampled by a factor of 12 to 100 relative to the actual area and geographic location of each composite grid cell.

Weekly composite AVHRR data were obtained from NOAA from mid-April 1982 to April 1983. Inspection showed that for most weeks the compositing process produced images with over 15% cloud cover, especially in equatorial regions. Consequently, further temporal compositing was used to combine data obtained over 3 weeks to reduce cloud cover to <5%. It proved possible by this process to create 17 separate composite images with cloud cover <5% for the period April 1982 to April 1983. The importance of combining data over periods of time to minimize clouds, sun angles, directional reflectance, and atmospheric effects while main-

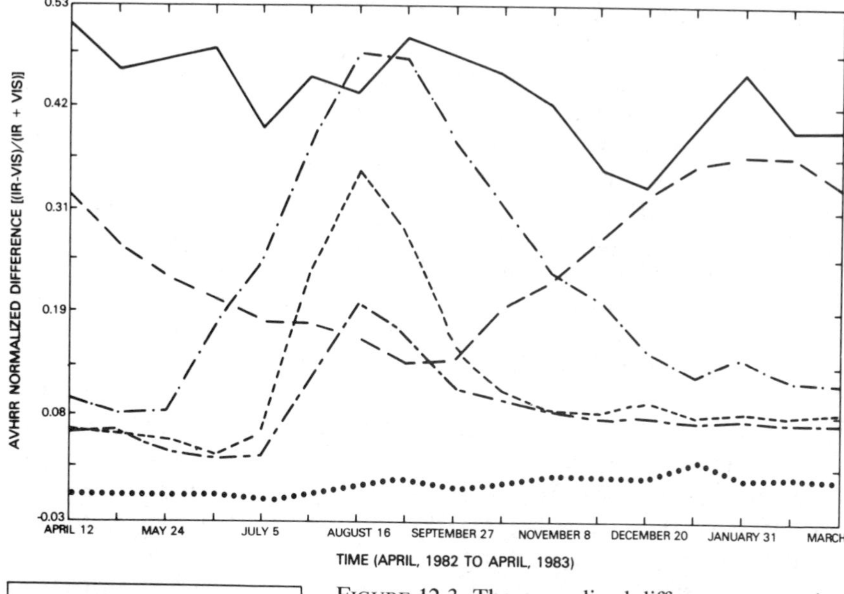

FIGURE 12.3. The normalized difference vegetation index from April 1982 to April 1983 at six African locations. One data value is plotted per location for each 21-day period. The first day of each 21-day period is indicated on the *x*-axis (i.e., April 12 is actually April 12, 1982 to May 2, 1982, etc.). Each plotted value represents the average of nine picture elements.

taining the ability to record the temporal dynamics of vegetation must be stressed; this method is fundamental to the work we describe herein.

The ability to composite successfully daily satellite data on a continental scale provided the means to overcome the problems of extensive cloud cover in equatorial and other areas and the degrading effects caused by variations in sun angle, the atmosphere, and the satellite-target viewing geometry. Once cloud-free data were available at frequent intervals, and the various degrading influences minimized, the large-scale multitemporal dynamics of green leaf biomass could be recorded (Fig. 12.3). Almost identical results were found for the C_2/C_1 ratio and $(C_2 - C_1)/(C_2 + C_1)$ normalized difference vegetation indices. Accordingly, we will present only results from the normalized difference series of analyses and henceforth refer to the normalized difference as the spectral vegetation index.

Land Cover Classification

From the 17 cloud-free geographically referenced data sets, two (April 12 to May 5 and August 16 to September 5) were chosen for land-cover classification. These were chosen because their wide temporal separation provided an opportunity to use phenological changes to classify the vegetation cover. The variable chosen to define the feature space was the spectral vegetation index, for two reasons. First, for a given vegetation type this index is related to variables such as the IPAR and green leaf biomass. As discussed above, the recorded spectral response can also be a function of other variables including shadow and soils. However, this is not necessarily a disadvantage for classification if the spectral responses of the vegetation from the two dates remain different. Second, use of the spectral vegetation index means that differences in radiance received at the sensor as a result of differences in solar elevation should be largely eliminated. Where cloud cover remains particularly persistent, we would expect misclassifications to occur because of the resultant depression of the spectral vegetation index.

The classification procedure used was interactive, relying on the analyst to select and label areas in the feature space belonging to each class. Picture elements of the whole 640 × 512 scene were represented in the feature space on the monitor of the image-processing system. Interactively, the boundaries or areas belonging to each class were selected from the training set data and then applied to the original two data sets. Locating these areas could have been made simply on the basis of a priori knowledge of green-vegetation dynamics. The principles used are illustrated in Figs. 12.3 and 12.4. Figure 12.3 was constructed for locations centrally within the indicated vegetation types. It shows quite clearly that at any one date confusion between classes is likely, but that when values for April and

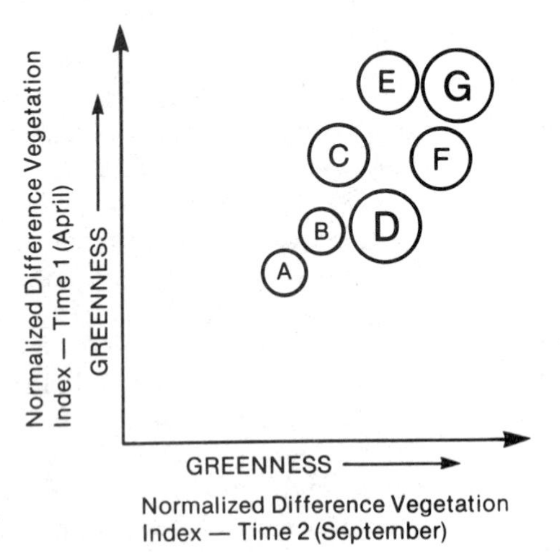

FIGURE 12.4. The two-dimensional representative of the various clusters formed by the normalized-difference vegetation index for April 12 to May 2, 1982, and August 16 to September 5, 1982. (**A**) Water, (**B**), desert and semidesert, (**C**) southern hemisphere savanna, (**D**) northern hemisphere savanna, (**E**) southern hemisphere seasonal forest and grassland, (**F**) northern hemisphere seasonal forest and grassland, and (**G**) closed canopy forest. See also Fig. 12.3.

September are compared, unique characterization can be achieved. Thus tropical rain forest will have high values of the vegetation index for both dates while savanna areal will have higher values for one date and lower for the other. Diagrammatically we would expect the principal cover types to be arranged in the feature space as shown in Fig. 12.4. On the basis of this pattern, boundaries to the classes were located and the classified scene was then produced. Some obvious errors were apparent, so minor changes were made to the feature space boundaries, and a final image was produced with two further iterations (Fig. 12.5). It is clear from Fig. 12.5 that the same class lay in different locations within the feature space according to whether the area was north or south of the equator. Thus, after the initial classification, the corresponding pairs of classes for moist savannas, dry savannas, and wooded grasslands were combined.

Several qualifications should be made concerning the cover map shown in Fig. 12.5. We did not primarily attempt definitive or even quasi-definitive land-cover classifications of Africa but instead attempted to evaluate the feasibility of using NOAA-7 AVHRR data for this purpose. In addition,

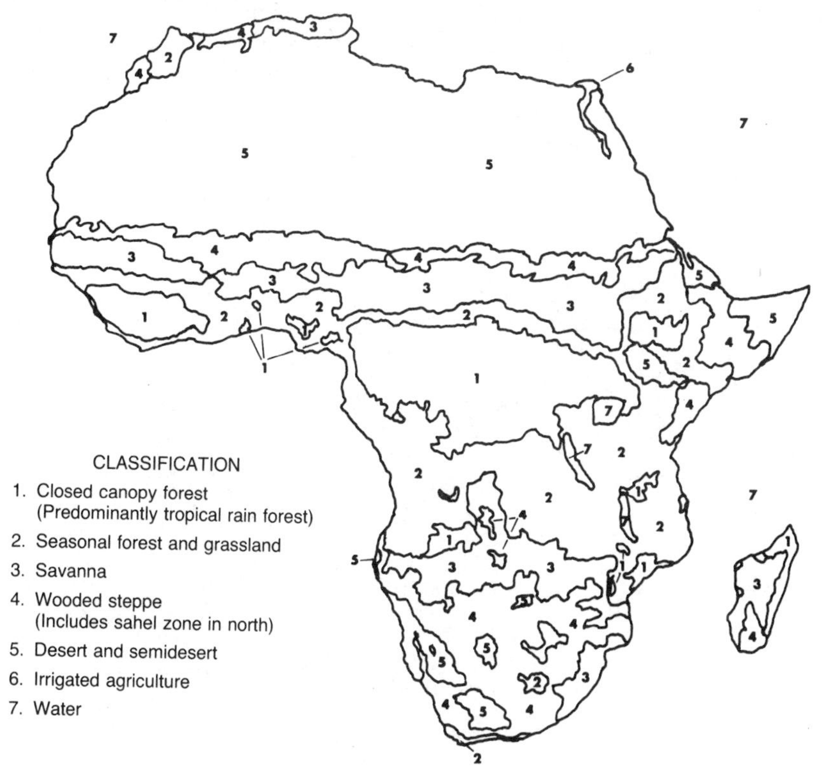

FIGURE 12.5. Preliminary land-cover classification of Africa produced from two 21-day composited images, one April 12 to May 2, 1982, and one August 16 to September 5, 1982. See also Figs. 12.3 and 12.4.

no formal performance evaluation has been carried out. Our evaluation has been hindered by the substantial lack of agreement between existing vegetation maps for many points of Africa. Therefore, we have no objective basis for comparison as yet. Field checking will clearly be extremely time-consuming; a satisfactory sample of test sites may take several years to collect. Nevertheless, our intention is to pursue such a field program in the future. At present, we have to rely on our African field experience and that of our colleagues, along with a qualitative comparison with existing vegetation maps such as the one shown in Fig. 12.6. Comparisons with such maps suggests a considerable measure of agreement, although certain errors can be noted. The classification of desert north of the equator includes both deserts and other semiarid zones with precipitation up to about 300 mm • yr^{-1} (Fig. 12.5). As such this includes a portion of the Sahelian zone (200 to 500 mm • yr^{-1}). The extent of closed-canopy forest in eastern Africa is probably overestimated. In coastal Gabon in western Africa, the amount of rain forest may well have been somewhat underestimated: the remarkably high frequency of clouds and haze in the early afternoon in this area lowered the spectral vegetation index derived from the satellite sensor below what it should have been. Immediately south of the southern limits of the rain forest in Zaire, a small wooded steppe

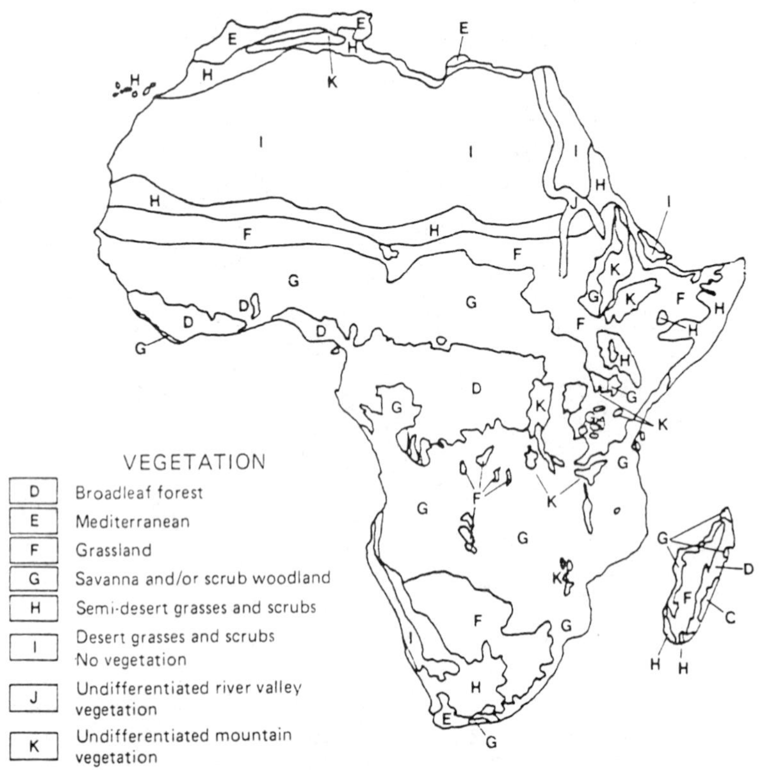

FIGURE 12.6. A generalized vegetation map of Africa.

is incorrectly shown; an explanation for this anomaly is not apparent at present. The classification of vegetation on the Mediterranean coast is described in the same terms as that in the rest of the map. More detailed mapping at a later date will be accompanied by labeling on the basis of the known differences in the assemblage of vegetation information in this area. These several qualifications should not detract from the very large measure of correct classification displayed by our satellite-derived map and the promise of continental-scale land cover mapping by the NOAA series of operational satellites. Additional research in this area has been reported by Tucker et al. (1985).

Global-Scale Remote Sensing of Green Leaf Biomass

The previously discussed continental land cover classification technique used daily AVHRR data from Africa that were composited into two 21-day periods to record the seasonal dynamics of aggregated green leaf biomass while simultaneously minimizing the degrading effects of sun angle, off-nadir viewing, atmospheric pathlength and aerosols, clouds, and shadowing. The same technique can be applied to global-scale data to record the seasonal dynamics of aggregated green leaf biomass or IPAR.

The global data have a grid-cell size of approximately 50 km and are derived from the same data source as the continental land-cover data. Four global green leaf biomass images are shown (Figs. 12.7 and 12.8). In Fig. 12.7 (map A) we see the growing season is just beginning in the northern hemisphere and that South America, Africa, and Southeast Asia contain the greatest amount of green leaf biomass for April 1982. For June 1982 (map B) we observe that the seasonal pulse of green leaf biomass activity has increased in the northern hemisphere over April 1982 and that South America, Africa, and tropical Southeast Asia are at about the same level as 2 months earlier. August 1982 (Fig. 12.8, map C) represents the maximum green leaf biomass distribution with high values throughout the Northern Hemisphere. By October 1982 (map D) the seasonal pulse of green leaf biomass has almost ended in the northern hemisphere and the greatest distributions of green leaf biomass are again in equatorial areas and the Southern Hemisphere.

Figures 12.7 and 12.8 demonstrate that the satellite land-cover monitoring can routinely be undertaken for continental and global-scale vegetation studies using data from the NOAA series of operational spacecraft. These global data have recently been related to atmospheric CO_2 draw-down for a 31-month period by Tucker et al. (1986).

Conclusions

Satellite remote sensing of land cover at regional, continental, and global scales is possible using NOAA-satellite AVHRR data. The daily acquisition of imagery for the entire planetary surface provides the means to overcome

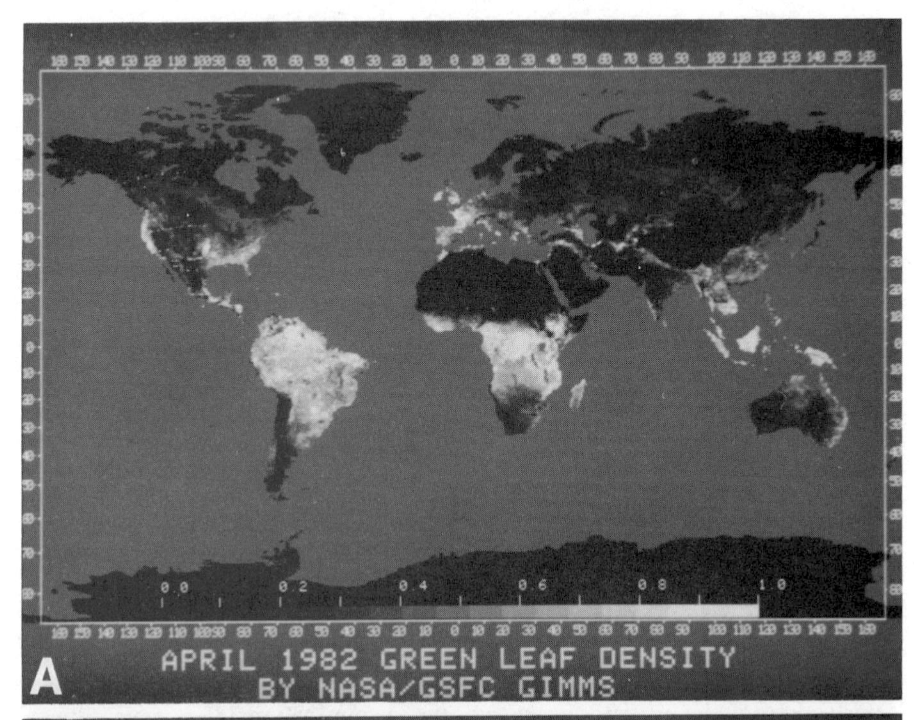

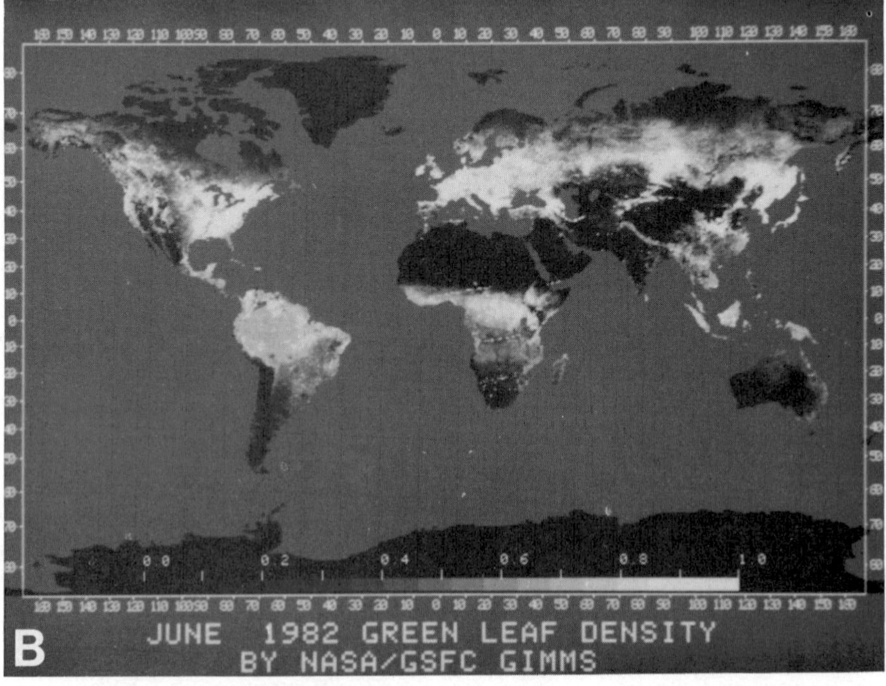

FIGURE 12.7. The aggregated normalized difference green leaf biomass or "greenness" from four individual months of daily NOAA-7 AVHRR imagery. (**A**) April

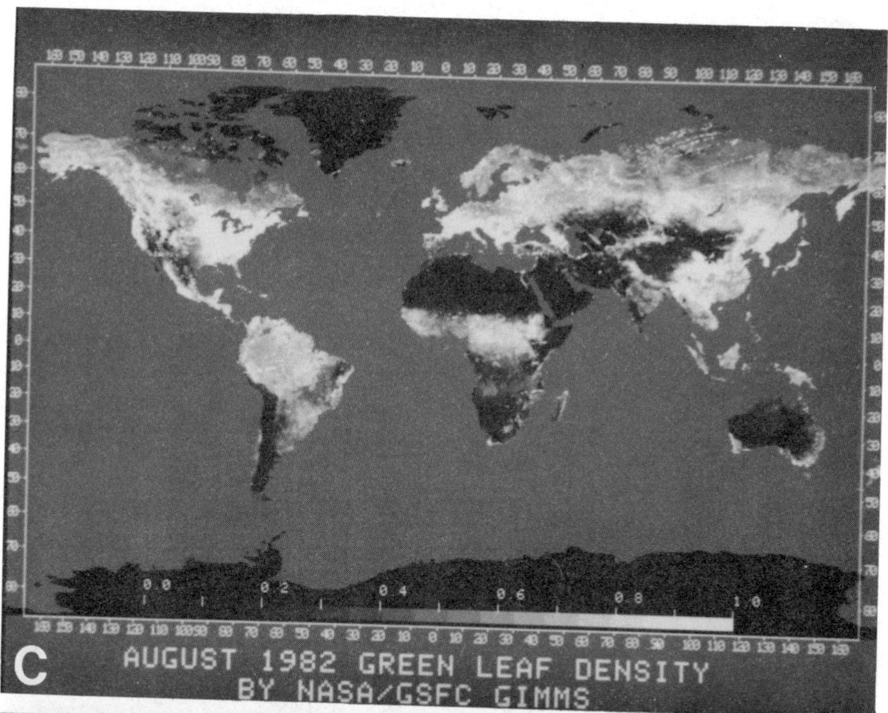

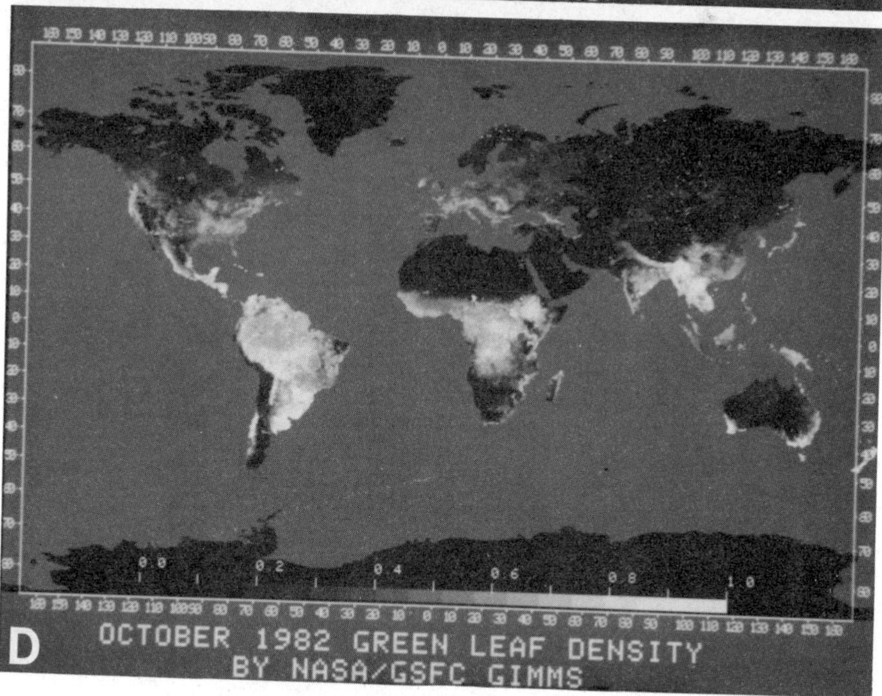

1982, (**B**) June 1982, (**C**) August 1982, and (**D**) October 1982. Note seasonal pulse of photosynthetic activity in the Northern Hemisphere.

cloud cover, atmospheric effects, and other degrading influences while recording the aggregated response of green leaf biomass at scales of 1 km and larger. One-kilometer data have been used to identify areas of tropical deforestation, while remapped and sampled 4-km data have been used to classify the land cover of Africa and document the seasonal dynamics of global green leaf biomass. These data provide the means for monitoring the terrestrial biota and advancing our understanding of the role of the terrestrial biota in the carbon cycle.

References

Aase, J. K. and F. H. Siddoway. 1980. Spring wheat yield estimates from spectral reflectance measurements. Agron. J. 72:149–154.

Ajtay, G. L., P. Ketner, and P. Duvigneaud. 1979. Terrestrial primary production and phytomass. In B. Bolin et al. (eds.), The Global Carbon Cycle SCOPE 13, pp. 129–181. John Wiley, New York.

Bartlett, D. S. and V. Klemas 1980. Quantitative assessment of tidal wetlands using remote sensing. Environ. Manag. 4:337–345.

Bauer, M. C., J. E. Cipra, D. E. Anuta, and J. E. Etheridge. 1979. Identification and area estimates of agricultural crops by computer classification of LANDSAT multispectral scanner data. Remote Sensing Environ. 8:77–92.

Brakke, T. W., E. T. Kanemasu, J. L. Steiner, F. T. Ulaby, and E. Wilson. 1981. Microwave radar response to canopy moisture, leaf-area index, and dry weight of wheat, corn, and sorghum. Remote Sensing Environ. 11:207–220.

Cicone, R. C. and M. D. Metzler. 1982. Comparison of LANDSAT MSS, Nimbus 7 CZCS, and NOAA 6/7 AVHRR sensors for land use analysis. In Proceedings of the Fifth Symposium for Machine Processing of Remotely Sensed Data, pp. 291–297. Purdue University, Indiana.

Curran, P. J. 1980. Multispectral remote sensing of vegetation amount. Prog. Phys. Geogr. 4:315–341.

Curran, P. J. 1983. Multispectral remote sensing for the estimation of green leaf area index. Phil Trans. R. Soc. Lond. 309:257–270.

Daughtry, C. S. T., M. E. Bauer, D. W. Crecelius, and M. M. Hirson. 1980. Effects of management practices in reflectance of spring wheat canopies. Agron. J. 27:1055–1060.

Daughtry, C. S. T., K. P. Galio, and M. E. Bauer. 1983. Spectral estimates of solar radiation intercepted by corn canopies. Agron. J. 75:527–531.

Dave, J. V. 1979. Extensive data sets of the diffuse radiation in realistic atmospheric models with aerosols and common absorbing gases. Solar Energy 21:361–369.

de Laubenfels, D. J. 1975. Mapping the World's Vegetation. Syracuse University Press.

Duggin, M. J. 1977. Likely effects of solar elevation on the quantification of changes in vegetation with maturity using sequential LANDSAT imagery. Appl. Opt. 16:521–523.

Eyre, S. R. 1963. Vegetation and soil, a world picture. Arnold, London.

Fearnside, P. M. 1982. Deforestation in the Brazilian Amazon: how fast is it occurring? Interciencia 7:82–88.

General Electric Co. LANDSAT 3 Reference Manual. Valley Forge Space Center, General Electric, Philadelphia.

Gray, T. I. and D. G. McCrary. 1981. Meteorological satellite data—a tool to describe the health of the world's agriculture. AgRISTARS Report EW-N1-04042, NASA/JSC. Houston, Texas. 7 pp.

Hatfield, T. L., G. Asrar, and E. T. Kanemasu. 1984. Intercepted photosynthetically active radiation in wheat canopies estimated by spectral reflectance. Remote Sens. Environ. 14:65–76.

Holben, B. N. and R. S. Fraser. 1984. Red and near-infrared sensor response to off-nadir viewing. Int. J. Remote Sensing 5:145–160.

Holben, B. N., C. J. Tucker, and C. J. Fan. 1980. Assessing soybean leaf area and leaf biomass with spectral data. Photogramm. Eng. Remote Sensing 46:651–656.

Jordan, C. F. 1969. Derivation of leaf area index from quality of light on the forest floor. Ecology 50:663–666.

Keay, R. W. J. 1959. Vegetation Map of Africa. Clarendon Press, Oxford.

Kidwell, K. A. 1979. NOAA Polar Orbiter Data Users Guide. Department of Commerce, Washington, D.C.

Kimes, D. S. 1983. Dynamics of directional reflectance factor distributions for vegetation canopies. Appl. Opt. 22:1364–1372.

Kimes, D. S., B. L. Markham, C. J. Tucker, and J. E. McMurtrey. 1981. Temporal relationships between spectral response and agronomic values of a corn canopy. Remote Sensing Environ. 11:401–411.

Kirchner, J. A., D. S. Kimes, and J. E. McMurtrey. 1982. Variation of directional reflectance factors with structural changes of a developing alfalfa canopy. Appl. Opt. 21:3766–3774.

Kreibel, K. T. 1978. Measured spectral bidirectional reflection properties of four vegetated surfaces. Appl. Opt. 17:253–259.

Kumar, M. and J. L. Monteith. 1982. Remote sensing of plant growth. In H. Smith (ed.), Plants and the Daylight Spectrum, pp. 133–144. Academic, London.

Matson, M. and J. Dozier. 1981. Identification of subresolution high temperature sources using a thermal IR sensor. Photogramm. Eng. Remote Sensing 47:1311–1318.

Muthoo, M. K. 1977. Perspective tendecias do setor florestal Brasideiro, 1975 a 2000. Institute Brasileiro de Desenvolvimento Florestal, Brasila.

Myers, N. 1980. Conversion of tropical moist forests. National Academy of Sciences, Washington, D.C.

Ormsby, J. P. 1982. Classification of simulated and actual NOAA-6 AVHRR data for hydrological land-surface feature definition. IEEE Trans. Geosci. Remote Sensing 20:262–268.

Projecto Radan Brazil. 1978. Folha Sc. 20 Porta Velho. Ministerio das Minas e Energia. Departmento Nacional de Producao Mineral, Rio de Janeiro.

Ranson, K. J., V. C. Vanderbilt, L. L. Biehl, B. F. Robinson, and M. E. Bauer. 1981. Soybean canopy reflectance as a function of view and illumination geometry. pp. 852–866. In Proceedings of the Fifteenth International Symposium for Remote Sensing Environment.

Rouse, J. W., R. H. Haas, J. A. Schnell, and D. W. Deering. 1973. Monitoring vegetation systems in the great plains with ERT. In Third Symposium on Significant Results Obtained with ERTS-1, NASA SP-351, pp. 309–317. Washington, D.C.

Schneider, S. R. and D. F. McGinnis. 1982. The NOAA/AVHRR: A new satellite sensor for monitoring crop growth. In Proceedings of the 1982 Machine Processing of Remotely Sensed Data, pp. 281–290. Purdue University, West Lafayette, Indiana.

Schneider, S. R., D. F. McGinnis, and J. A. Gatlin. 1981. Use of NOAA AVHRR visible and near infrared data for land remote sensing. NOAA Tech. Report 84. U.S. Department of Commerce, Washington, D.C.

Sioli, H. 1980. Foreseeable consequences of actual development schemes and alternate ideas. In F. Barbira-Scazzocchio (ed.), Land, People, and Planning in Contemporary Amazonia, pp. 257–268. Cambridge University, U.K.

Slater, P. N. and R. D. Jackson. 1982. Atmospheric effect on radiation reflected from soil and vegetation as measured by orbiting sensors using various scanning directions. Appl. Opt. 21:3923-3931.

Tardin, A. T., A. P. dos Santos, E.M.I. Morais Nov., and F. L. Toledo. 1978. Projectos agropecuarios de Amazonia; Desmatemento e fiscalizacaoraltorio. A Amazonia Brasil em foco 12:7–45.

Tardin, A. T., A. P. dos Santos et al. 1979. Levantamento de areas de desmatamento na Amazonia legal atrayes de imagens de satellite LANDSAT. INPE-COM3/NTE, C. D. U. 621.38SR. San Jose dos Campos, Brazil.

Tardin, A. T., A. P. dos Santos et al. 1980. Subprojecto sesmatamento, convenio. IBDF/CNP-INPE 1979, Relatoria IBDF-1 CNPq-INDE. San Jose dos Campos, Brazil.

Tarpley, J. D., S. R. Schneider, and R. L. Money. 1984. Global vegetation indices from the NOAA-7 meteorology satellite. J. Climate Appl. Meteorol. 23:491–494.

Townshend, J. R. G. and C. J. Tucker. 1981. Utility of AVHRR of NOAA 6 and 7 for vegetation mapping. In Matching Remote Sensing Technologies and Their Applications, pp. 97–109. Remote Sensing Society, London.

Tucker, C. J. 1978. A comparison of satellite sensors for monitoring vegetation. Photogramm. Eng. Remote Sens. 44:1369–1380.

Tucker, C. J. 1979. Red and infrared linear combinations for monitoring vegetation. Remote Sensing Environ. 8:127–150.

Tucker, C. J., I. Y. Fung, C. D. Keeling, and R. H. Gammon. 1986. A relationship is found between atmospheric CO_2 variations and a satellite-derived vegetation index. Nature 319 (in press).

Tucker, C. J., J. R. G. Townshend, and T. E. Goff. 1985. African land cover classification using satellite data. Science 227:369–374.

Tucker, C. J., J. A. Gatlin, S. R. Schneider, and M. A. Kuchinos. 1982. Monitoring large-scale vegetation dynamics in the Nile Delta and River Valley from NOAA AVHRR data. In Proceedings Conference on Remote Sensing of Arid and Semi-Arid Lands, pp. 973–987. Cairo, Egypt.

Tucker, C. J., B. N. Holben, J. H. Elgin, and J. E. McMurtrey. 1981. Remote sensing of total dry matter accumulation in winter wheat. Remote Sensing Environ. 11:171–189.

Tucker, C. J., B. N. Holben, and T. E. Goff. 1984. Intensive forest clearing in Rondonia, Brazil, as detected by satellite remote sensing. Remote Sensing Environ. 15:255–261.

United Nations Education, Scientific, and Cultural Organization. 1968. Vegetation Map of the Mediterranean Region, Paris.

Weinreb, M. P. and M. L. Hill. 1980. Calculation of atmospheric radiances and brightness temperatures in infrared window channels of satellite radiometers, NOAA Tech. Report NESS 80. Washington, D.C.

White, F. 1983. The vegetation of Africa, UNESCO, Paris.

Wiegand, C. L. and A. J. Richardson. 1984. Leaf area, light interception, and yield estimates from spectral components analysis. Agron. J. (in press).

Wiegand, C. L., A. J. Richardson, and E. T. Kanemasu. 1979. Leaf area index estimates for wheat from LANDSAT and their implications for evapotranspiration and crop modeling. Agron. J. 71:336–342.

Williams, D. L. and L. D. Miller. 1979. Monitoring forest canopy alteration around the world with digital analysis of LANDSAT imagery. NASA, Washington, D.C., 48 pp.

World Atlas of Agriculture Committee. 1969. World Atlas of Agriculture, Agosteri-Novare, Italy.

13 Changes in the Area of Forests in Rondonia, Amazon Basin, Measured by Satellite Imagery

George M. Woodwell, Richard A. Houghton, Thomas A. Stone, and Archibald B. Park

The transformation of tropical forests to pasture, crops, and barren land, now rapidly underway wherever tropical forests occur (Myers 1980; Fearnside 1982; Lanly 1982), contributes to the increase in CO_2 in the atmosphere (Woodwell and Houghton 1977; Woodwell et al. 1978, 1983*a,b*; Houghton et al. 1983). The greatest promise in measurement of the rate of change in area of forests lies in use of satellite imagery, available since 1972 in the LANDSAT series with a resolution of about 80 m [NASA 1983, Woodwell et al. 1983*a*, Klemas and Hardisky 1983, Woodwell 1984]. We have developed a special technique using LANDSAT imagery from different times to make direct measurements of changes in the area of forests. The technique was developed for Maine and tested in Washington (Woodwell et al. 1983*a*). We report here an application of the technique in the tropical moist forests of the state of Rondonia in the Brazilian Amazon (Fig. 13.1).

We have also explored the possibility of using the lower-resolution (1.1 km) Advanced Very High Resolution Radiometer (AVHRR) imagery from the National Oceanic and Atmospheric Administration (NOAA7) satellite, in combination with LANDSAT imagery to detect areas of extensive deforestation and reduce the cost of obtaining and processing the much more expensive LANDSAT imagery.

The objective of the study was to test the change-detection technique (Woodwell et al. 1983*a*) in tropical moist forests using imagery obtained from a receiving station operated by another country. We chose the Brazilian state of Rondonia in the southwestern Amazon Basin.

Rondonia is being developed rapidly for agriculture under an elaborate governmental plan for resettling people from southern Brazil (Tardin et al. 1980; Fearnside 1982; Tucker et al. 1983). The state has an area of 243,000 km^2. Prior to 1960 Rondonia supported about 10,000 Amerindians and a few rubber and mineral prospectors (World Bank 1981). In the 1970s, large sections of central Rondonia were designated by the Brazilian National Institute for Colonization and Agrarian Reform (INCRA) as colonization areas for immigrants from the increasingly mechanized agricultural regions of the southeast. During the 1970s the population of Rondonia increased at the rate of 15% per year (World Bank 1981). In 1984 Rondonia had a population of about 500,000.

Clearing of forests in the INCRA colonization areas is along an extensive system of parallel roads where the majority of colonists have been assigned 100-ha plots. Most of the clearing is done by hand and the land goes directly into pasture. Other areas are planted in upland rice, manioc, banana, coffee, cacao, or rubber, or annual crops such as beans and corn.

Estimates of the rates of deforestation in Rondonia were made first by Tardin et al. (1980). They used manual interpretation of LANDSAT imagery at 1:500,000 and found that deforestation in Rondonia increased 244% between 1975 and 1978 to cover a total of 4200 km^2. By 1980 7600 km^2 had been cleared (fide Fearnside 1984). The digital information available from each 59 × 79 m picture element (pixel) from LANDSAT was not used. The approach has been criticized (Fearnside 1982) for omitting small clearings and for failing to distinguish secondary from primary forest. Fearnside (1982) also interpreted LANDSAT photographs manually and predicted on the basis of his own measurements that if the rate of deforestation observed between 1975 and 1978 continued, Rondonia would be completely cleared by 1988.

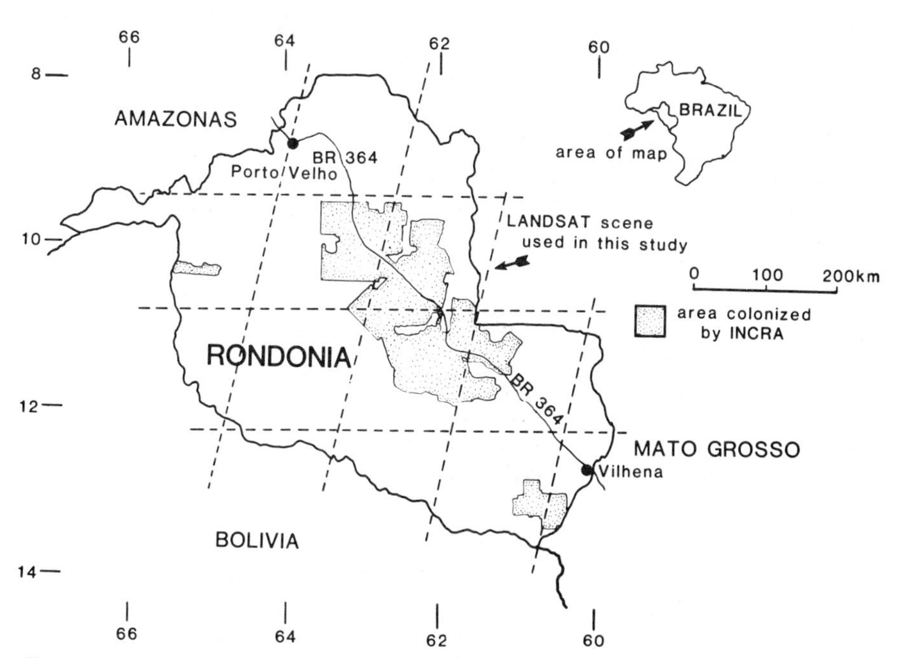

FIGURE 13.1. The Brazilian state of Rondonia and the area covered by this study. Dashed lines indicate approximate boundaries of LANDSAT scenes.

More recent work by Tucker et al. (1983), based on classification of NOAA7 AVHRR data, has suggested that between 9200 and 11,000 km^2 of forest in central Rondonia had been cleared by July 9, 1982.

Tape Acquisition

Computer searches of LANDSAT data were conducted both at the EROS Data Center in Sioux Falls, South Dakota, and at the Instituto de Pesquisas Espaciais (INPE) in Cachoeira Paulista, Sao Paulo, Brazil. An area in central Rondonia, where deforestation was known from the NOAA7 imagery to be extensive, was selected for the work. One LANDSAT tape was ordered from EROS and six tapes were ordered from INPE. The seven data tapes of LANDSAT path-row coordinates 248-67, along with 1:500,000-scale black and white imagery, covered the period from 1973 to 1981. Clouds in the 1973 scene prevented use of these data and a series of three cloud-free LANDSAT scenes from 1976, 1978, and 1981 were selected for the change-detection analysis.

A NOAA7 AVHRR data tape of Amazonia for July 1982 was obtained from C. J. Tucker of the NASA Goddard Space Flight Center to examine the differences between the deforested area determined from the LANDSAT data and the deforested area determined from the NOAA7 data.

Methods

LANDSAT Analysis

The seven LANDSAT data tapes were analyzed by the Earth Satellite Corporation (EARTHSAT) to determine the location of cloud cover and the amount of and location of noise in the data. The three best scenes (Figs. 13.2 through 13.4) were:

INPE Scene ID	*Date*	*Center Point*
176173–125857	June 21, 1976	10° 06′S : 62° 15′W
378216–133336	August 4, 1978	10° 05′S : 62° 08′W
281137–133031	May 17, 1981	10° 08′S : 62° 04′W

Because revegetation in the tropics can occur almost immediately, and detection of deforestation may be difficult using LANDSAT imagery, it was necessary to obtain data on shorter time intervals in the tropics than in the previous work in Maine and Washington (Woodwell et al. 1983*a*). Estimates were made of forest clearing and regrowth from 1976 to 1978, from 1978 to 1981, and from 1976 to 1981.

Each of the three LANDSAT scenes was geometrically and radiometrically corrected and lines in the data caused by LANDSAT sensor miscalibration were removed. Other differences in the data between two dates

FIGURE 13.2. INPE LANDSAT scene 176173-125857 for June 21, 1976 (see Fig. 13.1 for location). Grid patterns are roads and cleared forests within INCRA colonization areas. This scene is about 185 × 185 km.

FIGURE 13.3. INPE LANDSAT scene 378216-133336 for August 4, 1978, covering the same area as Fig. 13.2. Increased forest clearing is evident when compared to Fig. 13.2.

are functions of changes in the atmosphere, changes in sun angle, and other factors such as rainfall. Definition and removal of each of these potential sources of error can be difficult. Our emphasis on changes from forest to nonforest and nonforest to forest meant that we could work with coarse thresholds of change (Woodwell et al. 1983*a*). Nevertheless, EARTHSAT used a regression relationship between paired subscenes (or test sites) to correct for any constant bias, such as sun angle, between the two subscenes. This type of regression correction is commonly applied in the detection of changes (Burns 1983).

Pairs of LANDSAT scenes (185 × 185 km) were superimposed on one another (registered) in the computer to provide periods of 2 years, 3 years, and 5 years for analysis. Within the LANDSAT scene, five test sites or subscenes, each 512 × 512 pixels (30 × 30 km), were chosen to represent different levels of deforestation. Figures 13.5 through 13.7 show the 512 × 512 pixel (30 × 30 km) test site 1 in 1976, 1978, and 1981.

Following geometric and radiometric corrections and noise removal, the change-detection algorithm was applied to each of the five subscenes or test sites as described previously by Woodwell et al. (1983a). In this approach reflected radiance data from LANDSAT bands 5 (visible red) and 7 (near infrared) for each pixel from time 1 was subtracted from the reflected radiance data of the same pixel location from time 2. Areas that showed no spectral change between times 1 and 2 were assumed not to have changed and were not considered further.

The data for an entire subscene were plotted on a graph whose two axes were changes in reflectivity recorded by the scanner's bands 5 and 7. The point where the two axes crossed was defined as no change. The

FIGURE 13.4. INPE LANDSAT scene 281137-133031 for May 17, 1981, covering the same area as Figs. 13.2 and 13.3. Extensive new forest clearing is evident by comparison with Figs. 13.2 and 13.3.

FIGURE 13.5. Test site 1 LANDSAT scene detail for June 21, 1976, showing deforested areas along and perpendicular to BR-364. This 30 × 30 km area is a portion of the Padre Adolpho Rohl Settlement established in 1976.

area immediately around the point of no change was an area of "change" within the threshold. This change was attributed to uncontrolled factors such as changes in forest density. Similar plots of changed pixels were developed for all five test sites and for all three time intervals. From the display of these plots on the computer monitor, an analyst identified parallelepipeds to define various types of changed pixels. Changed pixels within each parallelepiped were color-coded, highlighted on the terminal, and then summed to provide test site statistics.

FIGURE 13.6. Test site for August 4, 1978; LANDSAT scene detail. Same scale as Fig. 13.5.

FIGURE 13.7. Test site 1 for May 17, 1981; LANDSAT scene detail. Extensive increase in clearing is evident by comparison with two earlier images. Same scale as Figs. 13.5 and 13.6.

Because healthy vegetation absorbs red light (LANDSAT band 5) and reflects near infrared radiation (LANDSAT band 7), pixels showing a change from forest to bare ground show a decrease in brightness in band 7 and an increase in brightness in band 5. Such pixels were negative in the band 7 scale and positive in the band 5 scale. The bare ground-to-agriculture or -pasture change showed the opposite effect: that is, pixels were positive in the band 7 scale and negative in band 5. No change from nonforest to forest was found in any of the test sites.

Application of the algorithm to the five test sites (or subscenes) provided information on rates of change within the test sites but not for the entire LANDSAT scene. One way to determine clearing rates for the entire LANDSAT scene would have been a traditional random sampling designed to yield a rate of deforestation for each period. We chose instead to use a direct enumerative approach in which we examined all the pixels in a LANDSAT scene for significant spectral change. In using this approach we first determined which set of parallelepipeds for each period was most successful in separating change categories. The parallelepipeds from each time period that provided the greatest separation of the various classes in the test sites were used to enumerate the pixels of each change class from the entire LANDSAT scene.

NOAA7 AVHRR Analysis

To determine the amount of land cleared up to 1982 in the entire state of Rondonia for which we did not have all the LANDSAT scenes, we used NOAA7 AVHRR data from July 1982. These data have much lower resolution and single scenes cover continental-scale features. Before analysis, the AVHRR data were rectified for display (Fig. 13.8), for the location of the boundaries of the state, and for the location of 1.2-km^2 pixels that represented the four corners of the LANDSAT scene and subscenes. The first stage in the analysis was to define deforested areas by using the thermal infrared band (3.5 to 3.9 μm). This approach takes advantage of the fact that deforested areas are significantly warmer than areas with an intact forest canopy. The IDIMS image processing system at the Woods Hole Oceanographic Institution in Woods Hole, Massachusetts was used. Visual inspection showed that cleared areas were warmer (had a higher number) than the surrounding forest. By assigning colors to several ranges of digital values of the thermal data we determined that all cleared areas had digital values greater than one standard deviation from the mean thermal value for the whole image. To remove the possibility of confusion of cleared forest with areas of similar thermal response, natural savannahs were excluded from the analysis based on their location and shape.

The second stage was to use the 1981 LANDSAT data within the 1982 NOAA7 AVHRR scene to develop calibration factors between the low-resolution NOAA7 data and the higher-resolution LANDSAT data. Because the LANDSAT and AVHRR data were not from the same year a direct comparison was inappropriate. However, a range of estimates of the amount of forest cleared by 1982 could be made in the LANDSAT scene area by extrapolating from 1981 assuming an exponential rate of

FIGURE 13.8. July 1982 NOAA7 thermal image. The grids in the center clearly show some of the colonization areas of central Rondonia. Deforested regions show as warmer areas, probably due to canopy loss, lower rates of evapotranspiration, and soil warming. They contrast strongly with the cooler forested areas.

increase of forest clearing or that the 1978 to 1981 rate prevailed. The calibration factors were determined by a comparison of the area of deforestation measured in the same location by the two satellites. The potential advantages of this technique over a technique based solely on LANDSAT would seem to be substantial.

Results

LANDSAT Analysis

Having used the five subscenes to determine the best classifications of change and the amount of change within those classifications (Table 13.1), we derived rates of change for the entire LANDSAT scene by applying the change detection algorithm to all pixels. These rates were as follows:

1976 to 1978 = 26,900 ha • yr^{-1}
1978 to 1981 = 55,200 ha • yr^{-1}
Total deforestation = 26,900 ha • yr^{-1} × 2 years + 55,200 ha • yr^{-1} × 3 years = 219,400 ha.
1976 to 1981 = 44,300 ha • yr^{-1}
Total 1976 to 1981 deforestation = 44,300 ha • yr^{-1} × 5 years = 221,500 ha.

Rates of deforestation were not linear but increased during the 1978 to 1981 period. The area of total deforestation measured on LANDSAT imagery between 1976 and 1978 and again between 1978 and 1981 was, when combined, within 1% of the rates determined independently for the entire period, 1976 to 1981.

The analysis of the 5-year (1976 to 1981) interval determined the total amount of forest to nonforest change, but was not sufficient to determine by year the annual rates of deforestation. The five-year interval was also insufficient for determining other types of transformations such as bare ground to agriculture or pasture (Table 13.1). A nonforest to forest transformation was not seen in Rondonia, suggesting that all forests cleared are kept in agriculture or pasture.

An enumerative approach for determining changes in land use can succeed for very large areas only if complete LANDSAT coverage is available and used. LANDSAT provides coverage of about 34,000 km^2 per scene. For an area as large as Amazonia, approximately 200 LANDSAT scenes would be required for complete coverage. Use of the change-detection algorithm with LANDSAT data for all of Amazonia would require 400 or more LANDSAT scenes (two dates for each scene) and the expenditure of $260,000 for data alone (1983). A more efficient and economical approach would integrate synoptic NOAA7 AVHRR data and some LANDSAT data by calibration factors that take advantage both of LANDSAT's higher resolution and NOAA7's synoptic coverage.

TABLE 13.1. Results from the application of the change-detection technique* to the five LANDSAT subscene test sites.

Subscenes†	1976–1978	1978–1981	1976–1981
Site 1			
Class 1	6,793‡	14,536	21,948
Class 2	194	9,336	736
Class 3	78,184	61,298	62,486
Site 2			
Class 1	5,895	15,244	20,259
Class 2	122	10,557	1,097
Class 3	84,515	59,370	63,815
Site 3			
Class 1	1,201	8,253	9,778
Class 2	1	1,137	0
Class 3	84,780	75,780	75,393
Site 4			
Class 1	3,439	11,105	17,013
Class 2	79	4,034	266
Class 3	84,028	70,031	67,891
Site 5			
Class 1	3,218	12,465	16,010
Class 2	184	3,184	241
Class 3	84,065	69,522	68,920

*Woodwell et al. 1983*a*.
†Class 1, forest to nonforest; class 2, bare ground to agriculture (pasture class); class 3, no change.
‡Measurements in hectares.

NOAA7 AVHRR ANALYSIS

Analysis of the NOAA7 AVHRR scene was confined to the determination of the deforested and forested areas first in the area bounded by the LANDSAT scene and then for the entire state of Rondonia. Within the area of the LANDSAT scene, we found, using only the NOAA7 AVHRR data, 3677 km^2 of nonforest. Two adjustments were made before the LANDSAT and AVHRR results could be compared. First, the LANDSAT data were adjusted to define not only the change in area of forest but the total area of nonforest in 1981. Then this area of cleared forest was corrected to 1982, the date of the AVHRR data. The area in the LANDSAT scene cleared prior to 1976 was estimated from more recent work in an adjacent LANDSAT scene with a similar area of cleared forest and rates of forest clearing. Classification of the adjacent scene determined that 1519 km^2 of forest had been cleared up to 1976 (unpublished data). A reasonable minimum for the amount of forest cleared prior to 1976 for this scene was 1000 km^2. The total area cleared for this scene was estimated as 2200 plus 1000 km^2 or 3200 km^2, about 10% of the entire LANDSAT scene.

Because the date of the last LANDSAT image was 1981 and the AVHRR image was from 1982, we adjusted the 1981 LANDSAT data to 1982 using both a linear rate of increase and a continuation of the exponential rate of increase observed. Using the linear assumption, we estimated that 552 km^2 were cleared between 1981 and 1982. With the exponential rate the amount cleared between 1981 and 1982 would have been 1120 km^2. Using these two estimates and adjustments for clearing before 1976, the range of estimates for total forest cleared within the LANDSAT scene up to 1982 was 3752 to 4320 km^2 or between 12 and 14% of the entire LANDSAT scene. These estimates were 2–18% higher than the area of deforestation determined with AVHRR data alone. Because of the adjustments to the LANDSAT data, however, a calibration of AVHRR data with the finer resolution LANDSAT data is tenuous. The results suggest the calibration factor would be close to 1.0.

We used this experience to estimate total deforestation in the entire state of Rondonia. The AVHRR data were used to define the area of cleared forest as those pixels with digital numbers (DN) greater than one standard deviation from the mean thermal value. On that basis 11,400 km^2 of forest had been cleared within the areas of Rondonia that were being colonized along the central highway (BR-364). This may be an underestimate of the total area cleared as of 1982 because about 5% of southern Rondonia was cloud covered on the AVHRR image (Fig. 13.8). We could assume either that the distribution of forest to nonforest was the same under the cloud-covered areas or that deforestation occurred only where there were roads. Recent maps show that there are few roads in the cloud-covered region, and that a large portion of the cloud-covered region is contained in Pedras Negras Forest Reserve (rescinded in 1981) and in the Guapore Biological Reserve (World Bank 1981). The large area of reserves suggest that the cloud-covered region is uncut forest. We assumed no deforestation in the cloud-covered region.

How accurate is this estimate of 11,400 km^2? There is no definitive answer. Earlier estimates are available for 1975 and 1978 (Tardin et al. 1980) and for 1982 (Tucker et al. 1983). Fearnside (1982) used the rates of deforestation of Tardin et al. (1980) to predict the complete deforestation of the entire state of Rondonia by 1988 if trends he showed continued (see Fig. 13.9). Tucker et al. (1983), using classifications of NOAA7 AVHRR data, estimated between 9200 and 11,000 km^2 of total deforestation from areas near BR-364 which runs through central Rondonia. Using official Brazilian estimates of cleared forest in Rondonia for 1975 (1217 km^2), for 1978 (4185 km^2) (Tardin et al. 1980), and 1980 (7579 km^2) (fide Fearnside 1984), the exponential rate of increase would predict 13,517 km^2 of forest cleared by 1982. Using simple photogrammetric methods with the AVHRR imagery, we estimated 12,390 km^2 of forest (plus or minus 6%) had been cleared by 1982 in Rondonia. All estimates, the years for which they are made, and Fearnside's (1982) extrapolation from Tardin et al.'s data are shown in Fig. 13.9. At least 5.5% of the forested area of Rondonia seems to have been cleared by 1982.

Although there are difficulties with the availability and use of NOAA7 AVHRR imagery, a more definitive experiment should be done using more than one set of NOAA7 AVHRR data coupled with the LANDSAT data and ground level observations. A three-level multistage approach with the AVHRR and LANDSAT has the potential for improving wide area coverage, of lowering the cost of collecting data, and of establishing a complete data set rather than a sample. With NOAA7 imagery it is possible to provide data in digital map form without the inherent inaccuracies that would accompany interpolated values from a LANDSAT sampling scheme (Woodwell et al. 1983*a*).

CO_2 Flux from the Area of LANDSAT Scene (Rondonia)

The amount of CO_2 released to the atmosphere as a result of deforestation in the area of the LANDSAT scene was determined with a regional version of the Marine Biological Laboratory Terrestrial Carbon Model (MBL/TCM) (Moore et al. 1981, Houghton et al. 1983). A clearing rate of zero was assumed before 1960 because the population of Rondonia prior to the mid-1960s was sparse (World Bank 1981). For the period 1960 to 1970, we assumed that 5000 ha • yr^{-1} of deforestation occurred, a value intermediate between later clearing rates and the zero clearing rate for 1960

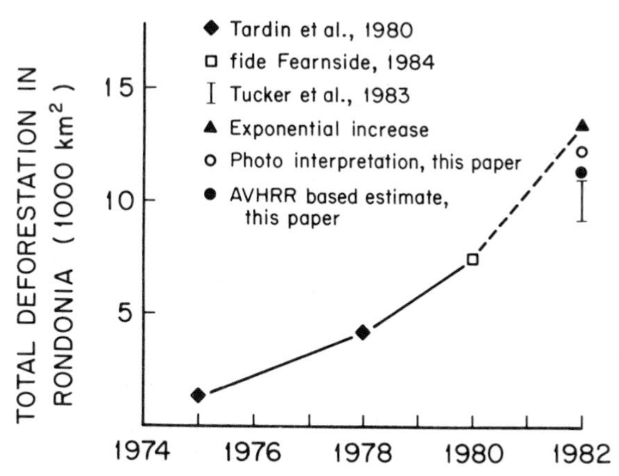

FIGURE 13.9. Total deforestation in Rondonia by several estimates. Units are 1000 km^2. The curve (Fearnside 1982) is an extrapolation from the data of Tardin et al. (1980) using rates of increase in forest clearing calculated for central Rondonia.

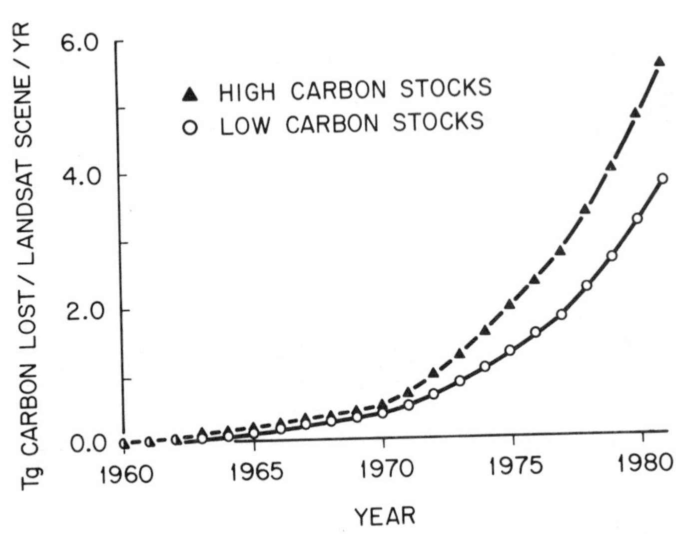

FIGURE 13.10. Carbon (10^{12} g) lost to the atmosphere because of deforestation in the area of the LANDSAT scene studied. Release rates were calculated with the MBL/TCM using the area deforested (as determined by LANDSAT) and two different estimates of carbon stocks. The lower estimates (0) of initial carbon stocks in above-ground biomass were based on the data in DNPM 1978 and the higher estimates ( ) were based on the data of Brown and Lugo 1982.

and earlier. For the 1976 to 1978 and the 1978 to 1981 periods, we used the clearing rates determined from LANDSAT: 26,900 ha • yr^{-1} and 55,200 ha • yr^{-1}, respectively.

Additional inputs to the MBL/TCM included the initial carbon stocks of vegetation and soil. Initial soil carbon data were from Schlesinger (1977). Two sets of data for the initial carbon content of vegetation were used. Data from the RADAM project (DNPM 1978) for areas in central Rondonia yielded somewhat lower values for biomass and carbon content than did the more generalized data from Brown and Lugo (1982). Two separate analyses with the model were carried out, one with each estimate of initial carbon stocks.

The results from the application of the regional version of the MBL/TCM with deforestation rates supplied from our analysis of the Rondonia LANDSAT data are shown in Fig. 13.10. Using the low estimates of carbon stocks data from RADAM (DNPM 1978), we estimated for 1981 that 3.9×10^{12} g C were released to the atmosphere from the area of the LANDSAT scene. Using the high estimates of carbon stocks from Brown and Lugo (1982), we estimated that 5.5×10^{12} g C were released in 1981.

Similar analyses in other regions of the world, using both LANDSAT and NOAA7 AVHRR, would reduce current uncertainties about cutting rates and the biotic release of CO_2 to the atmosphere.

Conclusions

1. Rates and areas of deforestation can be determined from both LANDSAT and NOAA7 satellite data. Information from the analysis of satellite data can be used to lessen current uncertainty about rates of deforestation in tropical forests.
2. LANDSAT tapes of the Amazon Basin are available and easily obtained from both Brazilian and U.S. sources.
3. LANDSAT data tapes from three different years were used to determine the amount of forest cleared in one LANDSAT scene in the Brazilian Amazon Basin state of Rondonia, a state that is rapidly being colonized and probably has the highest rate of forest clearing in the Brazilian Amazon Basin.
4. Rates of deforestation in the area of the LANDSAT scene (34,000 km^2) doubled over the 1976 to 1981 period examined.
5. Rates of deforestation can be determined using a 5-year interval; this interval is too long to determine accurately other types of transformations such as nonforest to agriculture or pasture.
6. No nonforest-to-forest transformation was found in Rondonia, presumably because all areas cleared are kept either in pasture or crops and are not allowed to return to forest.
7. Information acquired from the analysis of LANDSAT data can be used with data from the lower-resolution NOAA7 AVHRR series of satellites to cover even larger areas.
8. With data from LANDSAT and NOAA7 we have determined that at least 11,400 km^2 of Rondonia has been deforested up to 1982. This estimate tends to confirm the higher estimates from other sources.
9. The amount of carbon lost to the atmosphere due to forest clearing within the area of the Rondonia LANDSAT scene was determined using a regional version of the MBL/TCM, with clearing rates determined from the LANDSAT change-detection analysis. Between 3.9×10^{12} and 5.5×10^{12} g C were released in 1981.

Acknowledgments

This research was supported by U. S. Department of Energy Grant DOE P80000184, by the Ecosystems Center of the Marine Biological Laboratory, and the Woods Hole Research Center, Woods Hole, Massachusetts.

References

Brown, S. and A. E. Lugo. 1982. The storage and production of organic matter in tropical forests. Biotropica 14:161–187.

Burns, G. S. 1983. Land cover change monitoring within the east central Louisiana study site—A case for large area surveys with LANDSAT multispectral scanner data. NASA Technical Report DC-Y3-04418, NSTL/ERL-221.

Departamento Nacional Da Producao Mineral. 1978. Folha SC 20 Porto Velho. Projecto Radambrasil, Levantamento De Recursos Naturais, Vol. 16, Ministerio das Minas e Energia, DNPM, Rio De Janeiro.

Fearnside, P. M. 1982. Deforestation in the Brazilian Amazon: how fast is it occurring? Interciencia 7(2):82–88.

Fearnside, P. M. 1984. A Floresta Vai Acabar? Ciencia Hoje 2(10):42–52.

Houghton, R. A., J. E. Hobbie, J. M. Melillo, B. Moore, B. J. Peterson, G. R. Shaver, and G. M. Woodwell. 1983. Changes in the carbon content of terrestrial biota and soils between 1860 and 1980: a net release of CO_2 to the atmosphere. Ecol. Monogr. 53:235–262.

Klemas, V. and M. Hardisky. 1983. The use of remote sensing in global biosystem studies. Adv. Space Res. 3(9):115–122.

Lanly, J. 1982. Tropical forest resources. FAO Forestry Paper No. 30. Food and Agriculture Organization, Rome.

Moore, B., R. D. Boone, J. E. Hobbie, R. A. Houghton, J. M. Melillo, B. J. Peterson, G. R. Shaver, C. J. Vorosmarty, and G. M. Woodwell. 1981. A simple model for analysis of the role of terrestrial ecosystems in the global carbon budget. In B. Bolin (ed.), Carbon Cycle Modelling. SCOPE 16, pp. 365–385. John Wiley & Sons, New York.

Myers, N. 1980. Conversion of Tropical Moist Forests. National Research Council, Washington, DC.

NASA. 1983. Land-related global habitability science issues, NASA Technical Memorandum 85841.

Rambler, M. (ed.). 1983. Global biology research program, program plan, NASA Technical Memorandum 85629.

Schlesinger, W. H. 1977. Carbon balance in the terrestrial detritus. Annu. Rev. Ecol. Syst. 8:51–81.

Tardin, A. T., D. C. L. Lee, R. J. R. Santos, O. R. deAssis, M. P. Barbosa, M. Moreira, M. T. Pereira, and C. P. Filho. 1980. Subprojecto Desmatamento Convenio IBDF/CNP-INPE. Instituto De Pesquisas Espaciais, Sao Jose dos Campos, Brazil.

Tucker, C. J., B. N. Holben, and T. E. Goff. 1983. Forest clearing in Rondonia, Brazil as detected by NOAA7 AVHRR data. NASA Technical Memorandum 85018.

Woodwell, G. M. (ed.). 1984. The Role of Terrestrial Vegetation in the Global Carbon Cycle: Measurement by Remote Sensing. SCOPE 23. John Wiley & Sons, New York.

Woodwell, G. M., J. E. Hobbie, R. A. Houghton, J. M. Melillo, B. Moore, A. B. Park, B. J. Peterson, G. R. Shaver, and T. A. Stone. 1983*a*. Deforestation measured by LANDSAT: steps toward a method. Technical report to the U. S. Department of Energy, TR005. Washington, DC.

Woodwell, G. M., J. E. Hobbie, R. A. Houghton, J. M. Melillo, B. Moore, B. J. Peterson, and G. R. Shaver. 1983*b*. Global deforestation: contribution to atmospheric carbon dioxide. Science 222:1081–1086.

Woodwell, G. M. and R. A. Houghton. 1977. Biotic influences on the world carbon budget. In W. Stumm (ed.), Global Chemical Cycles and Their Alterations by Man, pp. 61–72. Abakon Verlagsgesellschaft, Berlin.

Woodwell, G. M., R. H. Whittaker, W. A. Reiners, G. E. Likens, C. C. Delwiche, D. B. Botkin. 1978. The biota and the world carbon budget. Science 199:141–146.

World Bank. 1981. Brazil, Integrated Development of the Northwest Frontier, Latin America and the Caribbean Regional Office, The World Bank, Washington, DC.

14
One-Dimensional and Two-Dimensional Ocean Models for Predicting the Distribution of CO_2 Between the Ocean and the Atmosphere

ANDERS BJÖRKSTRÖM

To assert anything about what the concentration of CO_2 in the atmosphere will be in the future requires that one understand the oceanic-atmospheric distribution of carbon and the way it may change when excess CO_2 is introduced into the atmosphere. The span of time during which the emission occurs is short compared to the slow oceanic circulation; therefore, it becomes natural to separate the mixing process into two parts: (1) the transport from the atmosphere into a thin surface layer of the ocean and (2) the exchange of water between the surface layer and deeper down.

This chapter is a review of the approaches that have been pursued in constructing mathematical models of the circulation of carbon between the atmosphere and the ocean.

Data for Testing Carbon Cycle Models

The pool of information about the ocean, especially with regard to the distribution of natural and man-made tracers in it, has grown much since the first carbon cycle models were developed. The most useful facts for checking oceanic carbon models belong to the following categories:

1. Data on the natural distribution of the radioactive isotope ^{14}C in various parts of the ocean. Produced by cosmic radiation in the atmosphere, this isotope is less abundant (relative to stable carbon) in the ocean than in air. The depletion is (partly) due to radioactive decay during the time it takes for water to move from the surface to deep regions in the sea. The symbol ρ will be used here to denote this effect. For the ocean as a whole, the average value of ρ is close to 0.84; that is, the abundance of ^{14}C relative to stable carbon is 84% of what it is in the atmosphere. Since one generally wants to model the ocean as in a stationary state (uninfluenced by man), there is some uncertainty about the correct value of ρ. On the basis of the data of Bolin et al. (1981), we will use the preferred

value 0.84 to judge the realism of carbon models. Variations of ρ from place to place will be discussed subsequently.

2. Data on atmospheric CO_2 contents since accurate observations began in the late 1950s. The ratio between the increase in atmospheric content and the total emission of man-made CO_2 during a period is often called the "airborne fraction" for that period. The airborne fraction for the whole period 1959 to 1977 is between 50 and 55% (Bacastow and Keeling 1981). These estimates assume that no significant contributions to the man-made emissions come from sources other than fossil fuel consumption. Since forest clearings in the tropics have taken place at an accelerated rate in this same period, the total airborne fraction is perhaps not more than 40%.

3. Data on the so-called Suess effect. Carbon in fossils has been isolated from the atmosphere for so long that all atoms of the isotope ^{14}C have decayed. The admixture of the present atmosphere with CO_2 from fossil fuels, therefore, means a dilution of the isotope ^{14}C relative to all carbon in the air.

4. Radiocarbon and tritium (^3H) produced at the atmospheric bomb tests have spread from the atmosphere into the sea, particularly at high latitudes. A model of the carbon cycle should be able to reproduce the total amounts taken up and, to the extent model resolution permits, also be correct relative to its geographical location.

5. Biological tracers. The differences in concentration of carbon between different parts of the sea are the result of a competition between, on one hand, circulation and mixing (which tend to homogenize all distributions) and, on the other hand, life processes, whereby carbon is continuously removed from the mixed layer and returned at greater depths where dead biogenic debris redissolves. Both types of processes affect several compounds in addition to carbon (e.g., oxygen, phosphorus, and alkalinity). Therefore, any carbon model that involves the biological fluxes also implies something about the concentration of these other compounds, the implications of which can be checked against actual element distributions.

6. Oceanographic constraints. A carbon cycle model's picture of how the water transports carbon must not contradict what dynamic oceanography tells us about vertical or horizontal motions in the sea. As carbon models gradually develop more geographical refinement, this kind of comparison becomes more and more important.

A principal aspect about forecasting any system deserves to be mentioned in connection with the global carbon cycle: For a model to be credible, the number of assumptions involved in it must be smaller than the number of data the model explains. A well-known example is when trying to adjust a smooth curve to a number of observations along an axis. If permitting polynomials of the same degree as the number of points to be approximated, one can always find a curve that intersects with all the points. However, one does not usually feel confident about this polynomial as a tool for forecasting future observations. A model gains interest only if one can find a simple expression that approximates the points reasonably well. When constructing a model of how the ocean takes up carbon from the atmosphere and distributes it throughout its volume, one faces a much more complex and heterogeneous set of data to be explained, as has been shown. Despite this, the principal requirement of simplicity remains. For a carbon cycle model to be sound, it must reproduce all facts whereby it can be checked and yet permit overview of its basic features.

The Two-Box Model

Probably the simplest way to capture both the air-sea exchange and the slowness of oceanic mixing is to visualize the ocean as two interconnected boxes, one on top of the other. The uppermost box represents the wind-mixed layer, a body of water immediately below the surface, which is completely mixed in the vertical direction within days or weeks. The lower box represents the remainder of the ocean. The wind-mixed layer is assumed, in the simplest model, to cut off all direct connections between the atmosphere and the deep sea.

Thinking of the atmosphere as a third box, one can envision carbon atoms moving between three compartments in a chain. The flux of carbon per unit time from any box A to a neighboring box B is assumed, with one exception, to be proportional to the amount of carbon in box A. This assumption is called first-order kinetics. It is a justified assumption if box A becomes mixed (by internal processes) considerably faster than the expected length of time a carbon atom will spend in A. This is not the case for the deep sea as a whole, and most carbon cycle modeling since the two-box model has been directed toward realistic ways to describe the deep sea.

The exception (from first-order kinetics) referred to above concerns the carbon flux from the mixed layer to the atmosphere. It can be shown that for an x percent increase of carbon content in the mixed layer, the flux to the atmosphere would increase to about 10 times x percent. First-order kinetics would imply an x percent flux increase only; consequently, some other dependence must be assumed. A chemical model first described by Keeling (1973) is frequently used.

Assuming first-order kinetics to govern the exchange of carbon between the two ocean boxes, and using some chemical equation to model the flux between the top box and the atmosphere, one can derive a set of ordinary differential equations that describe how the amount of carbon in each box and in the atmosphere will increase with time, if fossil carbon is injected into the atmosphere. One can also compute what the distribution of ^{14}C between the two ocean boxes and the atmosphere will be. Experiments to this end have been made by Keeling (1973) and Bacastow and Björkström (1981). These experiments demonstrate the inadequacy of the two-box model.

In summary, Fig. 14.1 shows that if the model is to reproduce an airborne fraction <60% and also have a realistic average abundance of ^{14}C in the ocean, then the boundary between the mixed layer and the deep sea has to be located at least 500 m below sea level. Even though rapid vertical mixing may go down to more than 100 m in certain high-latitude areas, the global average depth ought not be more than about 75 m (Bathen 1972).

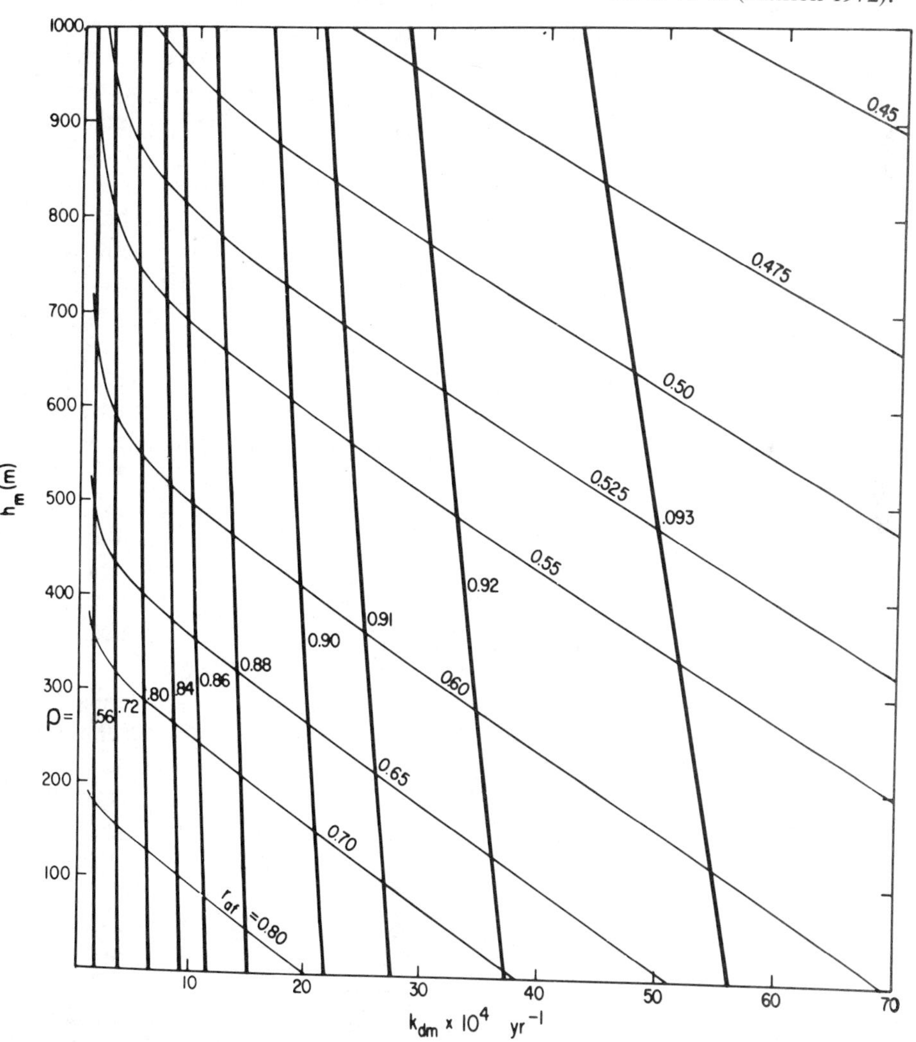

FIGURE 14.1. Analysis of the two-box model. For a variety of values of mixed-layer depth (h_m) and deep-sea turnover time (k^{-1}_{dm}), the airborne fraction (r_{af} and ρ value (see text) have been computed. (From Bacastow and Björkström 1981. Reprinted with permission of John Wiley & Sons.)

Diffusion Models

Inhomogeneities in tracer distributions demonstrate that the deep sea is not mixed infinitely rapidly. The concentration of carbon in the sea varies with location as well as with time:

$$C = C(x,y,z,t)$$

where C denotes concentration of carbon, x, y, and z are coordinates; and t denotes time. Attempts to analyze the distribution of oceanic tracers by formal application of the continuity equation including velocity field (advection) and the diffusive effect of small-scale motions are not new. This "advection-diffusion model" was used by Wyrtki (1962) to describe the oxygen distribution. Munk (1966) applied it also to distributions of temperature, salinity, and radiocarbon. As far as the carbon cycle is concerned, vertical inhomogeneities are more important than horizontal. Therefore, when wishing to devise something better than the two-box model, we need not move in one step to a full three-dimensional simulation of the concentration field. A suitable first step is to consider a model where the carbon concentration is a function of the vertical coordinate (and of time) only:

$$C = C(z,t)$$

The most explicit application of this one-dimensional diffusion-model technique to a model of the global oceanic carbon cycle was made by Oeschger and co-workers. In 1975, their group presented a one-dimensional model that retained the box model of the mixed layer but replaced the deep-sea box by a horizontally homogeneous medium, through which carbon was assumed to spread according to the diffusion equation. No advective motion was included in the model, and the diffusion coefficient was assumed to be the same at all depths.

Given this set of assumptions, one can (as with the two-box model) derive a set of differential equations that describe how the amount of carbon in the atmosphere and the ocean increases when fossil carbon is injected into the atmosphere. A plot, analogous to Fig. 14.1, can be constructed where instead of the deep-sea turnover time we now get the diffusion coefficient as the parameter that describes the deep ocean. This plot is shown in Fig. 14.2.

The result, as shown in Fig. 14.2, is that it is not possible to calibrate the box-diffusion model so that the airborne fraction and the average ^{14}C age of the ocean emerge correct, unless an unphysical value is assigned to the mixed-layer depth. If the ^{14}C age is to be right, the wind-mixed layer has to be about 500 m deep, if the airborne fraction is not to be $<55\%$.

However, although the box-diffusion model is unsatisfactory, it is nearer reality than the two-box model. This can be shown by taking the mixed-layer depth to be 100 m and calibrating each model so that the average ρ

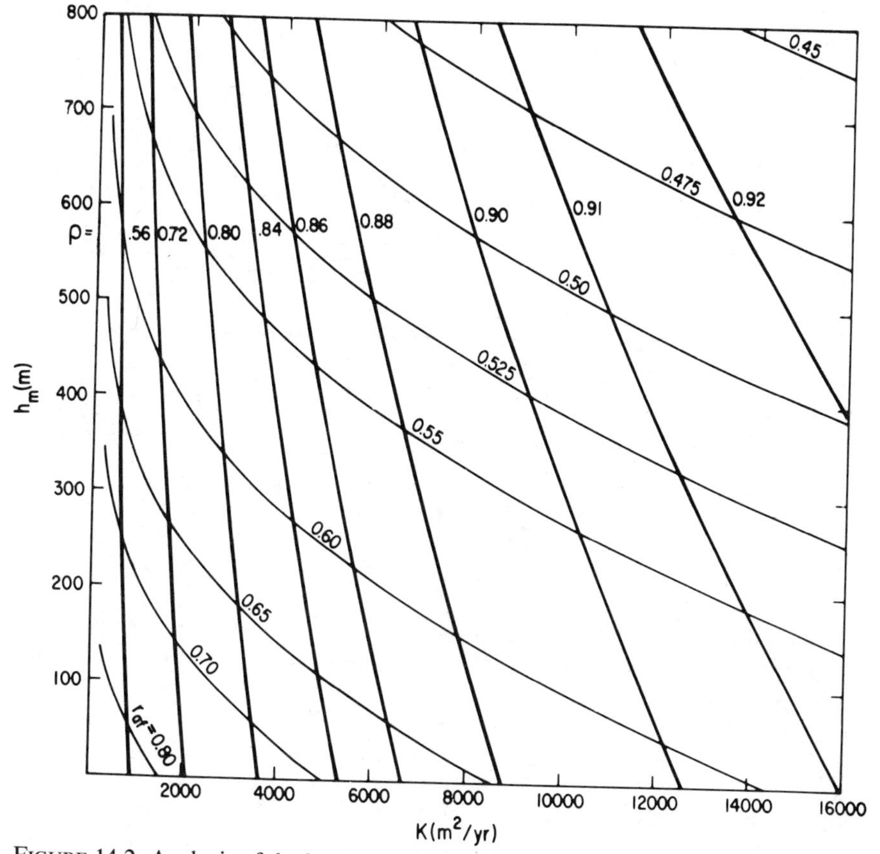

FIGURE 14.2. Analysis of the box-diffusion model. For a variety of values of mixed-layer depth (h_m) and diffusion coefficient (K), the airborne fraction and ρ value (see text) have been computed. (From Bacastow and Björkström 1981. Reprinted with permission of John Wiley & Sons.)

value (see earlier discussion) for the ocean is 84%. Whereas the two-box model then gives an airborne fraction of 80%, the box-diffusion model absorbs almost twice as much, and the airborne fraction becomes about 65%. The box-diffusion method can be regarded as a successful first approach to model the variations in ^{14}C age that prevail between different parts of the global ocean. Significant regions of the sea have, in fact, ρ values in water below 75 m well above 84%, which indicates that water there is in contact with the atmosphere more rapidly than the two-box model can simulate.

The vertical diffusion model appears to offer the possibility to simulate the uptake capacity of these water volumes. Accordingly, when all other factors are the same, the box-diffusion model (ocean) can take up substantially more carbon than the two-box model. The improvement, however, is not large enough for the box-diffusion model to meet the demands

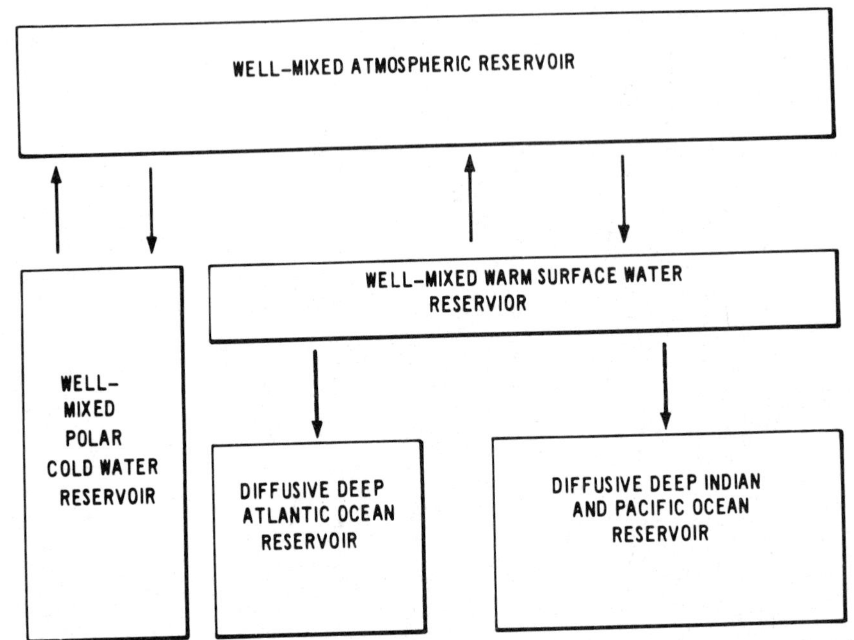

FIGURE 14.3. Five-component model of the ocean. (From Crane 1982. Reprinted with permission of Tellus.)

of reasonable airborne fraction, reasonable ^{14}C inventory, and reasonable mixed-layer depth simultaneously.

It is well established that the Atlantic Ocean has a higher average ρ value than the Pacific and Indian oceans. Differences in ρ also occur within each of these oceans and form a significant pattern, displaying the regions of deep water formation and how the newly formed deep water penetrates into different volumes of the world ocean. Even if the box-diffusion model is an improvement over the two-box model, much of the information on the actual distribution of ^{14}C remains unused. Crane (1982) therefore expanded on the original box-diffusion model (Oeschger et al. 1975) by constructing a five-component system of the sea (Fig. 14.3), where the Atlantic and Pacific oceans are each modeled as diffusive systems joined at the top by a common-surface layer box. It becomes possible thereby to use different turbulent exchange coefficients in the two ocean basins.

Crane's experiments show that although this does reduce the computed airborne fraction, the result still is that <65% of the anthropogenic emissions end up in the atmosphere. The lesson to be learned from this investigation is that diffusion alone cannot realistically explain the penetration of CO_2 into the ocean as it has proceeded until now. The oceanic circulation must be modeled with still greater realism.

Models of Polar Outcropping

In high latitudes, the vertical profile of seawater temperature is different from elsewhere in the ocean because of the low surface temperatures. Often, no level with a pronounced change in temperature and density corresponding to the bottom of a mixed layer is present.

Mixing within a water mass where the density is approximately constant proceeds more rapidly than between masses with different densities. Vertical mixing, therefore, is more rapid at high latitudes. This is evident from data on bomb-produced tritium and radiocarbon in the ocean. These two tracers have spread over especially large volumes in rather well-defined areas around Antarctica and the North Atlantic. The special importance of these areas in taking up carbon from the atmosphere and transporting it into the ocean can be represented in models. For example, one can divide the deep sea into several boxes, where each box is defined to contain water with density between two given limits. Physically, the water in one of these boxes would be located in a "slice" between two almost horizontal surfaces in the sea.

Figure 14.4 [from Defant (1961)] gives an impression of the distribution of density in the Western Atlantic Ocean. The important feature is that since the density increases toward the poles, the slices bend upward, and several of them will be in contact with the air. Because of this, an excess of CO_2 in the model atmosphere will be transferred to the interior of the model ocean more quickly than if it must first pass through the mixed layer and the upper parts of the deep sea.

Early models based on this idea have been developed by Craig (1963) and by Broecker et al. (1971). Siegenthaler (1983), in a further development of the box-diffusion model, has combined polar outcropping with vertical diffusion. Figure 14.5 illustrates his model. Each vertical increment is assumed to have an area facing the atmosphere through which CO_2 can be taken up. Letting the deep sea outcrop to the atmosphere over 10% of the ocean surface, Siegenthaler finds that the airborne fraction decreases from 66.7 to 60.5%. Considering that parts of the polar ocean are ice-covered, 20% of the total ocean surface corresponds roughly to the Arctic and Antarctic ocean areas. If outcropping occurs over 20% of the ocean surface, the airborne fraction is further reduced to 58.5%.

In his previously described geographical refinement of the box-diffusion model, Crane (1982) also carried out an experiment to explore the effect of polar outcropping. His approach was to introduce a separate well-mixed box, with a volume representative of the polar ocean, connected directly to the atmosphere. Even though this box could not pass excess carbon onward to the interior of the other oceans, Crane finds important reductions in the atmospheric CO_2 increase. Airborne fractions were computed to be 65.8%, 64.3%, and 60.5%, respectively, when the polar box covered 0%, 10%, and 20% of the global ocean area. These are larger effects than

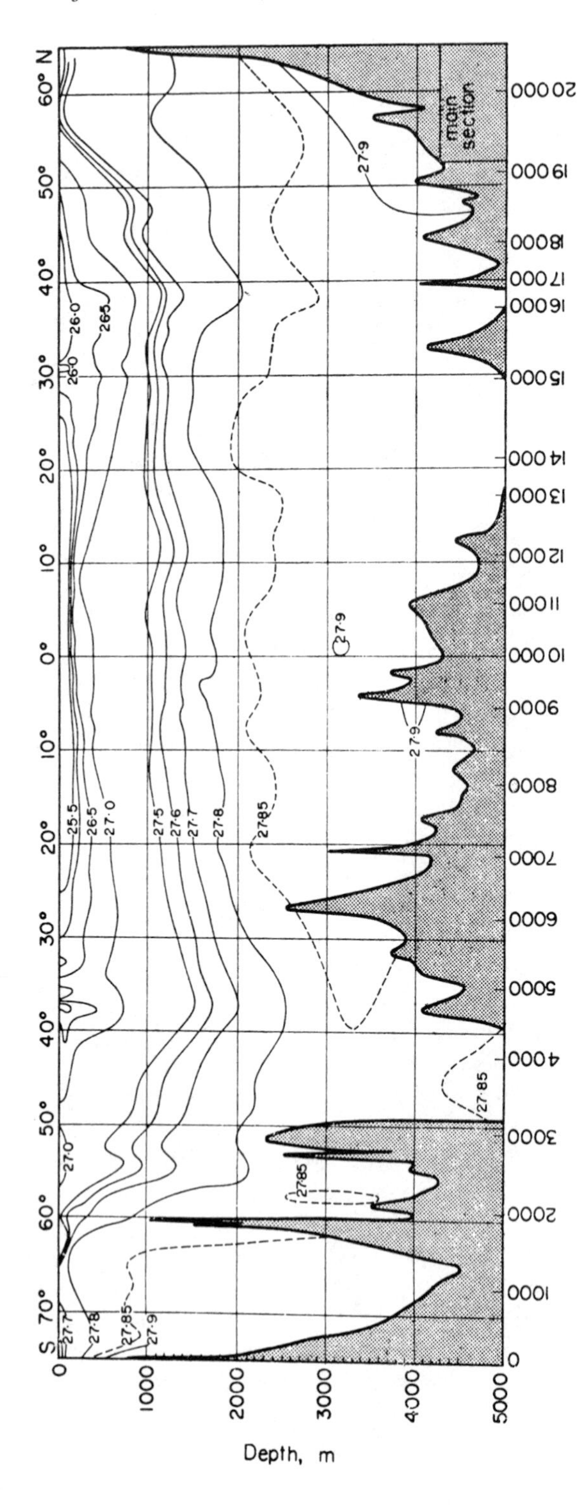

FIGURE 14.4. Longitudinal density section along the Western trough of the Atlantic. (Reprinted with permission from Defant A. Physical Oceanography, Vol. 1, Copyright 1961, Pergamon Press.)

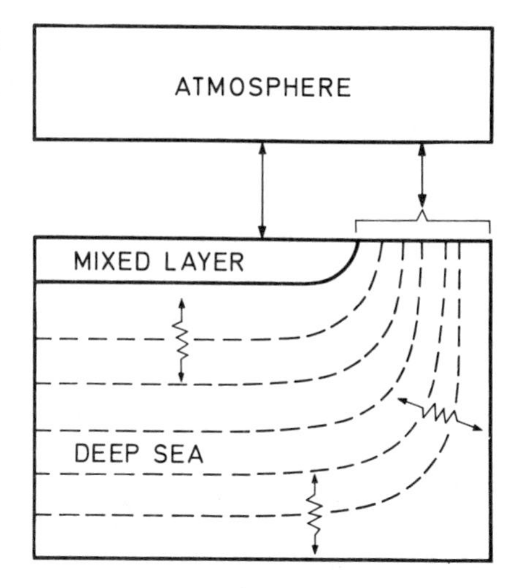

FIGURE 14.5. Outcrop-diffusion model of the ocean. (From Siegenthaler 1983).

those obtained with the same model when increasing the diffusion coefficient for the Atlantic Ocean (as mentioned above).

Although some unrealistic simplifications were made in the two series of experiments described here (for example, Siegenthaler assumes each vertical increment to be internally mixed without any delay), it is safe to say in summary that the polar regions and the main oceanic thermocline (down to about 1000 m below sea level) are potentially more efficient stores for excess CO_2 than purely diffusive models with constant diffusion coefficients suggest. This observation makes it natural to proceed to explore whether there are other aspects of the oceanic circulation patterns that should also be given more consideration.

Models of Circulation Systems

The outcropping described in the previous section affects primarily the water down to about 1000 m below sea level. This water mass, the main oceanic thermocline, or intermediate ocean, can be regarded as a transition zone between the wind-mixed layer and the cold abyssal water below.

The process whereby water is brought into the abyssal parts of the sea is a form of convection that takes place in those regions where surface water and intermediate water are cooled to sufficient density. The convection is sporadic and confined to limited areas in the Weddell Sea and in the North Atlantic. The total amount of water in the abyssal ocean remains constant because the downwelling is compensated by a slow upward motion that goes on in other parts of the ocean, driven partly by heat from the interior of the earth. The total rate of this thermohaline circulation is such that the abyssal ocean turnover time is many hundreds of years.

The amount of excess CO_2 that has actually been removed by this process is as yet small. Nevertheless, the failure to account for thermohaline circulation was recognized early as a principal shortcoming with the box-diffusion model, and attempts have been made to have this motion pattern represented.

Bacastow and Björkström (1981) describe a modification of the box-diffusion model, where a constant flux of water leads from the surface layer direct to the bottom of the diffusive reservoir. The effect of this on the model's performance is only slight. With all other factors unchanged, an upward velocity of 1 m • yr^{-1} reduces the airborne fraction (expressed in percent) by approximately one unit.

Björkström (1979) presented another method to represent downwelling and upwelling (see Fig. 14.6). In this model, separate boxes were used for the surface water in warm and cold regions, respectively. Outcropping from intermediate levels was represented by a two-way exchange between the cold surface water and the two boxes that represent the region above 1000 m. Downwelling was assumed to transfer water from the cold surface reservoir to the eight boxes that represent the ocean below 1000 m.

The Björkström (1979) model has been subjected to parameter calibration tests similar to those discussed above for the two-box and box-diffusion models. The model was parameterized by the aid of two coefficients: one described the outcropping between the intermediate ocean and the cold surface water; the other determined the amount of water circulating through the deep-sea system. The parameter combination required to make the airborne fraction 60% and the ρ value for ^{14}C equal to 0.84 corresponds to a situation with very intensive ventilation of the intermediate ocean. The radiocarbon content in the intermediate ocean emerges higher than it is in reality; however, this is counterbalanced by low ^{14}C content in the abyssal ocean. The total inventory of ^{14}C in the ocean is corrected, but its distribution with depth is obviously out of balance. Thus, this model is incapable of reproducing two of the most important pieces of information simultaneously, just as the box-diffusion model and the simpler box models have proven to be inadequate.

Viecelli et al. (1981) have depicted the global meridional circulation in a model with much improved spatial resolution. These authors construct a two-dimensional mesh with vertical and north-south extensions. They resolve 22 vertical layers and 10 positions along the meridional axis. In addition to admitting a more detailed picture of large-scale circulation than the Bacastow and Björkström (1981) model, it also admits vertical turbulent diffusion. One of the results obtained with this model is that the rise in atmospheric CO_2, as computed until 1980, is very insensitive to the upwelling velocity.

Toward a Realistic Geographical Resolution

The number of parameters in a box model increases with the number of surfaces between boxes and, therefore, grows as fast as the geographical

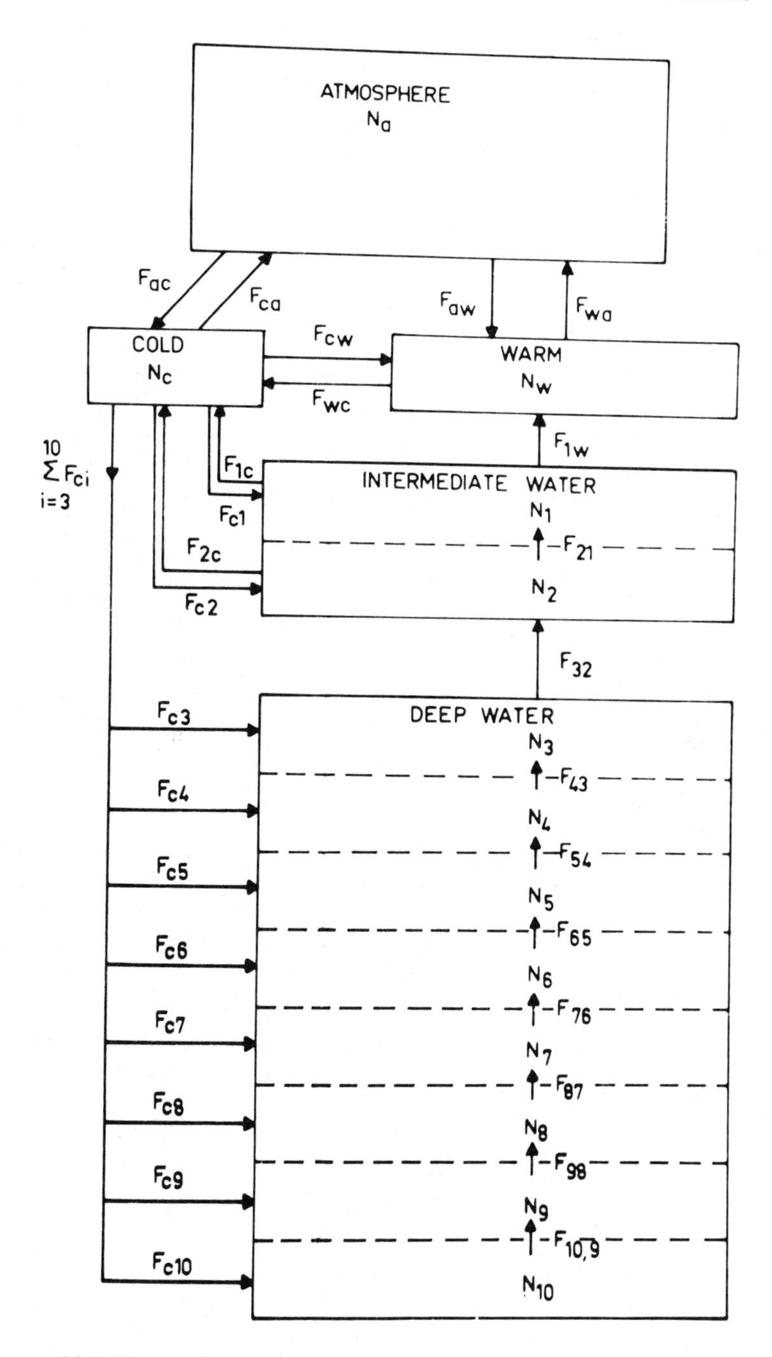

FIGURE 14.6. Vertical box model of the ocean. (From Björkström 1979. Reprinted with permission of John Wiley & Sons.)

resolution is refined. Recalling the principle (discussed earlier) that the number of parameters should not exceed the amount of data explained by the model, we realize that the need for data for model calibration and validation will increase rapidly when the number of boxes increases.

A source of data that has been used sparingly in connection with the simple models previously discussed is the natural distribution of compounds that are necessary for marine life. Keeling and Bolin (1968) made use of several tracers of this kind simultaneously; however, they worked with only a three-box model designed to be a zero-order approximation of the South Pacific Ocean. The availability of observations on oceanic tracers has improved since that time. Based on data from the GEOSECS expedition, Bolin et al. (1983) have formulated a model where the Atlantic Ocean and the Pacific/Indian oceans are represented separately, as is the Arctic Ocean and the circumpolar ocean around Antarctica. Each of these four regions is subdivided vertically, as shown in Fig. 14.7.

The parameters of this model, like those of other box models, are primarily the fluxes of water between the reservoirs. Bolin et al. (1983) used a finite-difference approximation of the continuity equation for tracers and considered the average advection and turbulent exchange through the interconnecting surfaces as unknown. The assumption that a given tracer in a given box is at steady state could then be translated into a linear relation between some of the unknowns. It proved suitable to regard the biological sink and source terms as unknown also, since these same unknowns can be used in the continuity equations for all tracers, with the aid of so-called Redfield ratios.

The Bolin et al. (1983) model contains 64 free parameters. The steady-state hypothesis and other obvious relations provide 80 possibilities to check the model. It turns out that there does not exist any set of parameter values that satisfies all 80 conditions simultaneously. Rather than considering the model to be outruled by this finding, the authors reason as follows: trying to match 80 equations by 64 free parameters is like searching for an exact solution to an overdetermined system; that is, as a rule, it is impossible. The problem could be traced to the lack of representativeness in the data used; therefore, a least-squares solution to the equation system is computed. The credibility of the least-squares solution is not damaged primarily by a certain misfit in the steady-state equations. A more serious shortcoming is that the turbulent transports, at some points, are in the direction toward higher concentration, and that the detritus flux, in some cases, is upward. The solution was therefore adjusted, based on minimizing the errors given that all turbulent fluxes must be down the gradient, that photosynthesis must be greater than decomposition in the top box of each region, and that no photosynthesis must take place except in the surface reservoirs. The solution obtained is shown in Fig. 14.7.

In its overall features, the solution is oceanographically reasonable; therefore, it is of interest to try to validate it by its power to describe the

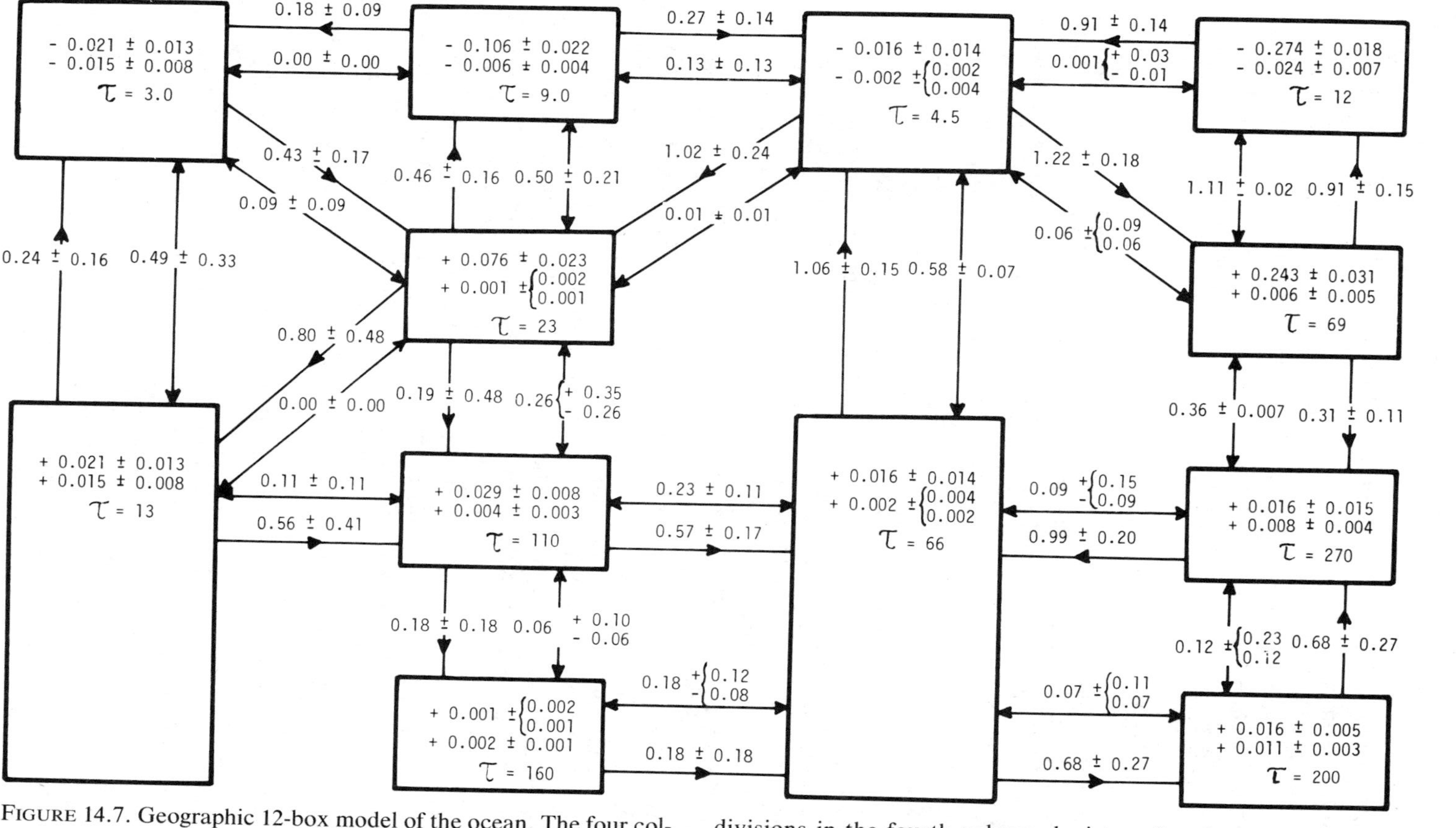

FIGURE 14.7. Geographic 12-box model of the ocean. The four columns symbolize, from left to right: 1, The Arctic basin; 2, The Atlantic Ocean; 3, The Antarctic Circumpolar Ocean. The vertical divisions in the fourth column depict surface layer, main thermocline, deep water, and bottom water, respectively. (From Bolin et al., 1983. Reprinted with permission of Tellus.)

CO_2 increase. However, the result is that the ocean, like earlier models, is too slow in taking up carbon from the atmosphere. The airborne fraction for the solution in Fig. 14.7 is about 75%. The spread of bomb-produced ^{14}C reveals similar behavior. It was concluded, therefore, that the resolution shown in Fig. 14.7 is not adequate for the simultaneous treatment of steady-state and transient properties of the carbon cycle.

Future Model Improvements

No simple model has yet been proven capable of explaining the global carbon cycle satisfactorily. The most important cause of failure seems to be that models cannot simultaneously treat the wide spectrum of time scales that occur in oceanic circulation. It is interesting to note, for example, that when Viecelli et al. (1981) determined the vertical diffusion coefficient from data on tritium uptake in the thermocline, they obtained a value of 16,000 $m^2 \bullet yr^{-1}$—four times more than the value obtained by Oeschger et al. (1975) on the basis of the average radiocarbon profile for the whole ocean. Viecelli's results are influenced by an extrapolation of Northern hemisphere data to the whole ocean. This has certainly made the diffusion coefficient larger than realistic, but we note that the results of testing the simpler models show that the circulation one has to assume in order to get a correct steady-state ^{14}C balance is always "slower" (in the sense defined by the parameters of each particular model) than the circulation that is required to satisfy the transient CO_2 and ^{14}C signals, not vice versa. This is, qualitatively, just what could be expected if in the real ocean the diffusion was more vigorous in the main thermocline than below, while modellers kept trying to interpret the situation using the same diffusion coefficient from top to bottom. A box-diffusion model, adjusted so that K can vary with depth in a simple way, is an improvement over the original box-diffusion model, but, as should be clear from the above, one-dimensional models ought not to be used when calculating the future state of the carbon cycle. Some transient tracers by which one can calibrate models have been injected into the sea in ways that vary importantly with latitude. The input patterns for different tracers are also dissimilar from each other. All the information in these differences and gradients has to be smoothed away when working with a one-dimensional model.

As for models with greater geographical realism, the following can be said about lines of future improvement. Since it is impossible to capture all time scales within the same model, it becomes essential to focus development efforts on those time scales for which improved prediction is of greatest importance. Considering other aspects of the CO_2 issue, what is most needed at the present time seems to be the acquisition of a more certain understanding of how the atmospheric CO_2 concentration will develop during the coming century. For such work, the Atlantic Ocean is more important than the Pacific or Indian ocean, and the thermocline region is more important than the abyssal ocean as a whole. The decision to be made thus concerns the type of models best suited for gathering specific information.

Oceanographers have constructed detailed numerical models of the oceanic circulation on scales down to a resolution of a few degrees latitude and longitude and with a vertical resolution of at least five levels. Some of these models have been equipped with the option to follow tracers being transported with the water and, at least in one case (Maier-Reimer et al. 1984), to also have the power to describe the air-sea exchange of CO_2.

One may then ask if there is any future for box models. The most detailed box models resolve the Atlantic Ocean in less than 100 boxes. Extrapolating this level of resolution to the whole ocean, one would get a box number that seems modest indeed compared to the 15,000 to 20,000 grid points in an ocean general circulation model.

There is still need for box models, however, because they are better suited than general circulation models to answer certain questions about the ocean. The distribution of a tracer compound reflects the average intensity of turbulence, average circulation pattern, and average biological conditions over a long period of time. Even if it is theoretically possible to compute average properties by integrating a general circulation model for a long time, it may not be practically feasible to do so. To exploit tracer distributions is a much better way to approach questions such as: How long does it take, on average, for North Atlantic deep water to move from the Northern to the Southern hemisphere? or What percent of the water in the Arctic Ocean enters the Atlantic in less than 10 years?

The result of these investigations can preferably be explained in statistical terms (e.g., characteristic times or transfer probability per unit time between well-defined regions and water types). This framework leads naturally to the concept of box models.

To calibrate a box model reliably significant differences in tracer concentrations must exist between the boxes. At the present time, data inadequacy limits meaningful resolution. When trying to estimate tracer gradients, one is most often attempting to ascertain a small difference between two large numbers. If data from more than one oceanographic expedition are used, great care must be taken to confirm that they are intercomparable. Accepting that the GEOSECS and TTO data bases have been compiled by similar enough techniques, Bolin et al. (1984) use these two data sets to construct a model of the Arctic and Atlantic Oceans. The model is sketched in some detail by Moore and Björkström (*this volume*). As demonstrated in their Fig. 16.2, the model consists of two meridional box systems, separated at the Mid-Atlantic ridge.

Each system is divided into about five regions, following oceanographic

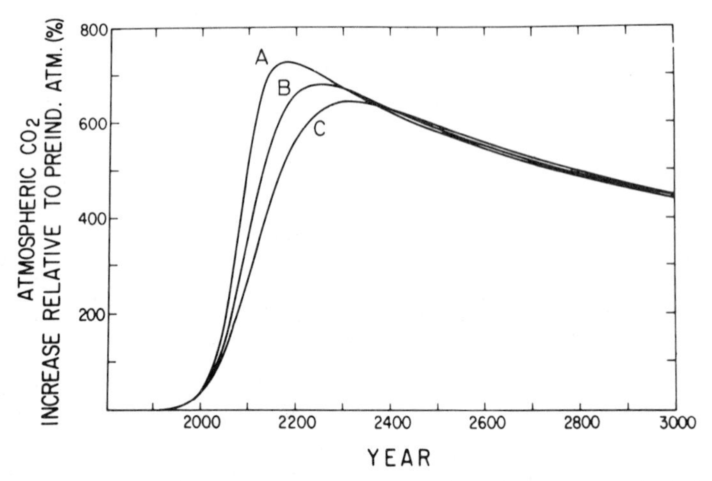

FIGURE 14.8. Atmostpheric response to combustion of approximately 6000 Gt carbon. Computations have been done using the box-diffusion model. Three different scenarios for combustion rate as a function of time were used. The peak annual emission is about 10 times at present in (A) and between four and five times in (C) (From Bacastow and Björkström 1981. Reprinted with permission of John Wiley & Sons.)

or geographical criteria, so that, in all, 12 regions are discerned. Up to eight levels are used for vertical resolution, separated only by surfaces of constant density. About 90 boxes are developed as the result of these subdivisions.

Using the same parameterization method as with the 12-box model described in the previous section, there would be, at each surface that connects two boxes, one advective-velocity and one turbulent-diffusion coefficient (in need of calibration), in addition to two biological parameters for each box. In a large, rigid, orthogonal, three-dimensional grid, this parameterization method requires eight parameters per box. With the geometry as shown in Fig. 14.8, and using the Atlantic ridge to disconnect some neighboring boxes, we obtain a parameter-to-box ratio of between 6:1 and 7:1. If using data on alkalinity, phosphorus, and oxygen (in addition to the obvious carbon and radiocarbon), and if supplementing these biochemical tracers with temperature and salinity data, one obtains seven times as many continuity equations as the number of boxes (eight, if also including water continuity equations). Thus, with about 90 boxes, there is space for about 600 unknowns before the system becomes underdetermined. This reasoning is admittedly a bit mechanical and approximate; however, it should give an idea of the size and detail of a box model that can be developed today. Work on a model of this type is in progress (Moore and Björkström, this volume; Bolin et al. 1984), but it has not been determined how it will respond to injections of CO_2 or other transient tracers.

Predictions of Future CO_2 Levels

Oceanic tracer profiles are the result, on the one hand, of life in the sea, precipitation, evaporation, and other processes that sharpen the concentration gradients, and on the other hand, of water mixing, which tends to smooth out the gradients. The usual way to interpret tracer data is to solve for a time-constant pattern of processes, which, if prevailing for long time, would produce the observed gradients (profiles). Predictions have been made on the theory that the average pattern would be equal to the prevailing pattern and continue to remain so in the future, which must be kept in mind when interpreting the computations.

An analogy with weather prediction may be useful. Meteorological forecasts are based on an analysis of present wind and weather data. Physical principles are used to deduce the state of the atmosphere at a particular later time. A purely statistical (climatological) forecast of the weather for the next few days has little credibility if it disagrees with the meteorological forecast. The situation is different if one wants to predict what the weather will be several seasons ahead. Meteorological forecasting is then impossible; however, statistical information is not completely void of interest and can be used to predict weather conditions for the overall period.

Pursuing this comparison, we must realize that tracer data are much like "climatological statistics," i.e., they reflect average conditions over a long period. Exactly how long is not known, since average conditions vary significantly in parts of the ocean. The abundance of ^{14}C in parts of the North Pacific Ocean corresponds to an age over 1000 years. It is unlikely that the large-scale circulation has been the same for that long a time. Climatic changes, for example, the "Little Ice Age" 1450 to 1850 A.D., may have had an impact on the intensity of abyssal water formation. The concept of a steady-state ocean is thus to an extent artificial. It is, however, the only means we have to say anything about the future uptake of CO_2 by the ocean. There is at present no way to foresee how or if the circulation will change.

Bacastow and Björkström (1981) compared a number of ocean models and obtained approximately the following conclusion: Given that mankind uses up a total of 6000 Gt of carbon in the next two centuries, we will increase the atmospheric CO_2 concentration more than five times, and the excess will go down only slowly through the following millenium, as shown in Fig. 14.8. Their value for fossil fuel use (6000 Gt) was based on a modification of an estimate by Perry and Landsberg (1977). Although it is not an unrealistic figure for the total amount of coal present on earth, its relevance may be questioned on other grounds. Estimates of future energy consumption have, generally, been lower in the last few years than in the 1970s. Nordhaus and Yohe (1983) put the 50th percentile for the likely emissions in 2025 at 10 Gt • yr^{-1}. This is almost half of the smallest of three alternatives used by Bacastow and Björkström in 1981.

Several predictions of atmospheric CO_2 have been made, for example, by Keeling and Bacastow 1977, Revelle and Munk 1977, Siegenthaler and Oeschger 1978, and Viecelli et al. 1981. In their overall characteristics, all extrapolations give similar results, and it is not meaningful to say that one model is more credible than another as far as small differences are concerned.

The only realistic way whereby the decline in CO_2 concentration can be essentially more rapid is to assume that the dissolution of carbonate-containing sediments occurs in response to the lowering concentration of carbonate ion. Keeling and Bacastow (1977) compute that, under certain assumptions, the amount of excess CO_2 in the atmosphere in the year 3000 could be less than half of what it would have been if no dissolution occurred (also see Sundquist, *this volume*). Except for this, any fundamentally different extrapolation builds on some certainly overly optimistic assumption, for example, about the capacity of the terrestrial vegetation as a sink for CO_2.

The similarity between extrapolations with different models on the time scale of 1000 years indicates that the uncertainty that comes from ignorance of which model is best, is smaller than the uncertainties caused by several other factors. Important sources of uncertainty are that we do not know how much CO_2 will be released totally, and, looking at the CO_2 problem as a whole, that the effects on climate of a given CO_2 increase cannot be predicted accurately. It is also impossible to state whether a warming climate would produce changes in the oceanic carbon cycle, for example, by reducing the annual formation of deep water or by affecting marine biological productivity.

Extrapolations concerning the next century or two should preferably be made using models tuned to phenomena on that time scale. Unfortunately, suitable data are lacking. Bomb-produced tracers have been present for too short a time to be fully informative. The build-up of CO_2 has been going on since long before 1958 (but no reliable data exist), and increasing concentrations of carbon in the ocean are not yet quantifiable. In principle, it is also unfortunate that data on the atmospheric CO_2 increase should be used to determine the parameters of the model. Calibration ought rather to be based on data of other kinds, and the ability to reproduce the observed increase should be used to validate the model. Despite this, one may, of course, calibrate a model to these short-term data and use it for predictions to the year 2100 or so. One is then tacitly assuming that the diffusion and mixing parameters in the upper and intermediate ocean will have the same averages for the whole prediction period as they have had for the last 25 years, and that dissolution of calcium carbonates will not modify CO_2 uptake considerably during the period. At present, our inability to foresee changes in these conditions precludes predicting anything except gross features of the rise in atmospheric CO_2.

References

Bacastow, R. and A. Björkström. 1981. Comparison of ocean models for the carbon cycle. In B. Bolin (ed.), Carbon Cycle Modelling, Scope 16, pp. 29–79. John Wiley & Sons, New York.

Bacastow, R. B. and C. D. Keeling. 1981. Atmospheric carbon dioxide concentration and the observed airborne fraction. In B. Bolin (ed.), Carbon Cycle Modelling, Scope 16, pp. 103–112. John Wiley & Sons, New York.

Bathen, K. H. 1972. On the seasonal changes in depth of the mixed layer in the North Pacific Ocean. J. Geophys. Res. 77:7138–7150.

Björkström, A. 1979. A model of CO_2 interaction between atmosphere, oceans, and land biota. In B. Bolin, E. T. Degens, S. Kempe, and P. Ketner (eds.), The Global Carbon Cycle, Scope 13, pp. 403–457. John Wiley & Sons, New York.

Bolin, B., A. Björkström, U. Cederlöf, K. Holmën, and B. Moore. 1984. The analysis of the general circulation of the ocean by the simultaneous use of physical, chemical, and biological data (in press).

Bolin, B., A. Björkström, K. Holmen, and B. Moore. 1983. The simultaneous use of tracers for ocean circulation studies. Tellus 35B:206–236.

Bolin, B., C. D. Keeling, R. B. Bacastow, A. Björkström, and U. Siegenthaler. 1981. Carbon cycle modelling. In B. Bolin (ed.) Carbon Cycle Modelling, Scope 16, pp. 1–28. John Wiley & Sons, New York.

Broecker, W. S., Yuan-Hui Li, and Tsung-Hung Peng. 1971. Carbon dioxide—man's unseen artifact. In D. W. Hood (ed.), Impingement of Man on the Oceans, pp. 287–324. Wiley-Interscience New York.

Craig, H. 1963. The natural distribution of radiocarbon: mixing rates in the sea and residence times of carbon and water. In J. Geiss and E. D. Goldberg (eds.), Earth Science and Meteoritics, pp. 103–114. Noth-Holland, Amsterdam.

Crane, A. J. 1982. The partitioning of excess CO_2 in a five-reservoir atmosphere-ocean model. Tellus 34:398–405.

Defant, A. 1961. Physical Oceanography. Vol. I. Pergamon Press, New York.

Keeling, C. D. 1973. The carbon dioxide cycle. Reservoir models to depict the exchange of atmospheric carbon dioxide with the oceans and land plants. In S. Rasool (ed.), Chemistry of the Lower Atmosphere, pp. 251–329. Plenum Press, New York.

Keeling, C. D., and R. B. Bacastow. 1977. Impact of industrial gases on climate. In Energy and Climate, pp. 72–95. National Academy of Sciences, Washington, D.C.

Keeling, C. D. and B. Bolin. 1968. The simultaneous use of chemical tracers in oceanic studies. Part II. Tellus 20:17–54.

Maier-Reimer, E., K. Hasselmann, D. Müller, and J. Willebrand. 1984. An ocean circulation model for climate variability studies (in press).

Munk, W. 1966. Abyssal recipies. Deep-Sea Res. 13:707–730.

Nordhaus, W. D. and G. W. Yohe. 1983. Future paths of energy and carbon dioxide emissions. In Changing Climate, pp. 87–153. National Academy of Sciences, Washington, D. C.

Oeschger, H., U. Siegenthaler, U. Schotterer, and A. Gugelmann. 1975. A box diffusion model to study the carbon dioxide exchange in nature. Tellus 27:168–192.

Perry, H. and H. H. Landsberg. 1977. Projected world energy consumption. In Energy and Climate, pp. 35–50. National Academy of Sciences, Washington, D.C.

Revelle, R. and W. Munk. 1977. The carbon dioxide cycle and the biosphere. In Energy and Climate, pp. 140–158. National Academy of Sciences, Washington, D. C.

Siegenthaler, U. 1983. Uptake of excess CO_2 by an outcrop-diffusion model of the ocean. J. Geophys. Res. 88(C6):3599–3608.

Siegenthaler, U. and H. Oeschger. 1978. Predicting future atmospheric carbon dioxide levels. Science 199:388–395.

Viecelli, J. A., H. W. Ellsaesser, and J. E. Burl. 1981. A carbon cycle model with latitude dependence. Climatic Change 3:281–302.

Wyrtki, K. 1962. The oxygen minima in relation to ocean circulation. Deep-Sea Res. 9:11–23.

15 Three-Dimensional Ocean Models for Predicting the Distribution of CO_2 Between the Ocean and Atmosphere

JORGE L. SARMIENTO

The basic ingredients necessary to predict fossil fuel CO_2 uptake with a three-dimensional model of the oceans are threefold.

First, a specification of the boundary conditions at the air–sea interface for the entire period of the prediction is needed. The boundary conditions needed include wind stress, heat and water fluxes, and gas exchange rates. A considerable amount of historical data on wind stress (Hellerman and Rosenstein 1983) and on surface temperatures and salinities (Levitus 1982) are available. These data have been used to obtain what appears to be reasonable estimates of climatically averaged yearly and monthly values for many regions of the oceans. The surface temperatures and salinities have been used with some success to provide boundary conditions for heat and evaporation minus precipitation (Bryan and Lewis 1979). Ultimately, however, it will be necessary to provide boundary conditions that evolve with the climate changes induced by the increasing atmospheric carbon dioxide. This will require a coupled ocean-atmosphere model, a subject that will not be considered in this chapter. Gas exchange should be parameterized as a function of wind speed, a problem that continues to be an area of active research (Peng et al. 1979; Kromer 1979).

Second, a model of the interior physical processes of advection and mixing is necessary. There are basically two types of three-dimensional models of physical processes in the oceans that are in active use at present: (1) the primitive-equation approach first introduced by Bryan (1969) and (2) the quasi-geostrophic approach developed for use in the oceans by Holland (1978) and others. The main differences between the present versions of the two types of models that are of primary interest here are that quasi-geostrophic models take the oceanic density field as given and allow no vertical exchange between layers. Primitive-equation models, on the other hand, solve for the temperature and salinity fields and allow for advective, mixing, and convective exchange between layers. The disadvantage of primitive-equation models, in comparison to quasi-geostrophic models, is that they are slower to run. Primitive equation models that resolve mesoscale eddies stretch the outer limits of even the largest of modern computers.

Vertical exchange processes and thermohaline circulation are, of course, essential in any treatment of CO_2 uptake by the oceans. For this reason, it is difficult to use quasi-geostrophic models in their present form to study the uptake of tracers from the surface of the ocean. Our research into the CO_2 problem at Princeton has thus focused on primitive-equation models. The Max Planck Institut für Meteorologie in Hamburg is, perhaps, the only other group working on the CO_2 uptake problem with three-dimensional ocean circulation models. Hasselmann (1982) and his group there are in the process of developing an ocean model that has many of the same characteristics as the models we are working with at Princeton.

One can also use velocity and mixing fields estimated directly from data by a variety of inverse techniques such as that of Wunsch (1978). No attempts have been made to model CO_2 uptake with velocity fields calculated in this manner. My experience with a different type of diagnostic technique, using temperature and salinity fields, was not very successful in predicting vertical velocities (Sarmiento and Bryan 1982). Given that vertical velocities are estimated from the small divergence of large horizontal velocities, it will be important to include data that are sensitive to vertical processes, such as tritium, in performing inverse calculations to obtain velocity fields for use in CO_2 uptake calculations.

Third, a model of the biological and chemical processes affecting the carbon cycle is needed to determine the feedback effects of CO_2 perturbation (and its related climate change) on the carbon cycle. The parameterization of biological and chemical processes is a topic that has recently begun to receive considerable attention in connection with the work with sediment traps and GEOSECS and TTO measurements.

I believe it will be several years before three-dimensional ocean models will be realistic enough that one would have greater confidence in their predictions than in those of the simpler models that have been used for such studies up to now. The reason is basically that primitive-equation models of the oceans continue to have serious deficiencies that will be resolved only when we have large enough computers to develop eddy-resolving primitive-equation models, and (or) when new numerical techniques and solutions to the equations are developed.

In the meantime, I think that the major application of three-dimensional models should be as a form of laboratory where one can test analogue modeling approaches and study the effects of chemical and biological processes on CO_2 uptake. I will give a more specific idea of the present capabilities of ocean circulation modeling by discussing some results from our research at Princeton. The first results are from a tritium calibration study of a model of the North Atlantic (Sarmiento 1983*a*). These results will show that models are already capable of quite realistic simulations of the ocean circulation, although there are problems related to the parameterization of subgrid-scale processes. A significant finding from this work is the great importance of wintertime convection in tracer penetration into the ocean.

The next results are from a series of tests of the analogue modeling approach using a one-dimensional model of the type developed by Oeschger et al. (1975). These tests raise a number of serious questions concerning the assumption of a constant vertical diffusivity with depth and time. They also raise a serious concern relating to the use of tritium (which has an input strongly biased toward high latitudes) to calibrate models that will be used for predicting CO_2 uptake, which has an input with a much smoother spatial distribution.

Our present research is focussed on the exploration of various ways of parameterizing subgrid-scale processes and on the development and testing of predictive equations for biological and chemical processes. The work with biological and chemical processes was initiated quite recently, but has already yielded an extremely important result concerning the importance of high-latitude deep-water formation processes in controlling atmospheric CO_2 (Sarmiento and Toggweiler 1984). This finding may have a dramatic impact on the fossil fuel CO_2 problem, which is poorly understood at present. These results will also be discussed.

Tritium Calibration Study and Test of the Analogue Modeling Approach

Tritium Model

Sarmiento and Bryan (1982) made use of the Bryan (1969) primitive-equation model and temperature and salinity observations [as analyzed by Levitus (1982)] to obtain a solution for the oceanic velocity field in the North Atlantic. The velocity field obtained was validated by simulating the distribution of tritium for the period 1952 to 1972 (Sarmiento 1983*a*). An excellent tritium data set for 1972 is available (Sarmiento et al. 1982). This data set was used for comparison with model predictions.

The version of Sarmiento and Bryan's models used for tritium simulation is the one referred to as the free thermocline model, which has 12 levels in the vertical and 1° resolution in the horizontal. The equations solved are the full equations of motion, continuity, state, and conservation of heat and salt. Subgrid processes are parameterized by an anisotropic viscosity and diffusivity with a constant value in the horizontal and a constant but smaller value in the vertical. The values of the parameters used in the model are shown in Table 15.1.

The model is driven by a combination [Leetmaa and Bunker (1978), Hellerman (1967)] of winds at the surface. The temperature and salinity conservation equations include an extra term that restores the model predictions toward the observations with a Newtonian damping of the form (γ $(T^* - T)$]. The symbol T^* represents data (Levitus 1982), and γ is an inverse time constant. The restoring is done only at the surface and below the main thermocline. At the surface, the restoring provides the boundary

TABLE 15.1. Parameters used in Sarmiento and Bryan (1982) model.

Vertical viscosity	100	$cm^2 \cdot s^{-1}$
Horizontal viscosity	5×10^7	
Vertical diffusion	0.5	
Horizontal diffusion	5×10^6	
γ at surface	1/50	days
γ below 1000 m	1/250	

condition. The restoring is retained below 1000 m to speed convergence to a solution. The values used for γ are shown in Table 15.1.

The tritium simulation is run separately from the ocean circulation model using the steady-state velocity field obtained by the model. The input function for the tritium is that given by Weiss et al. (1979). The diffusivities used in the tritium simulation are the same as those given in Table 15.1.

The ocean circulation is shown in Figs. 15.1 and 15.2, and some of the tritium modeling results are shown in Figs. 15.3 through 15.5. The model reproduces most of the major features observed in the data. The low penetration of tritium in the equatorial region (Fig. 15.3) is a consequence of equatorial upwelling, which one can see in Fig. 15.2. The tritium "front" at about 15°N (Fig. 15.3) is seen to occur within the North Equatorial Current (Fig. 15.1b). This front moves to higher latitudes with increasing depth, coinciding with the decrease in the size of the subtropical gyre (Fig. 15.1c). The northeast to southwest trend of the North Equatorial Current gives rise to the downward dip toward the west of the tritium data in the 14°N section (Fig. 15.5). The intermediate maximum at about 100 to 200 m between 10°N and 30°N (Figs. 15.3 and 15.5) is seen to be a consequence of northward-moving low-tritium equatorial waters at the surface (Fig. 15.1a) underlaid by southward-moving Subtropical Underwater (Fig. 15.1b) (also see discussion of data by Sarmiento et al. 1982).

The main failure of the model is the insufficient vertical penetration of tritium between 30°N and 40°N that is evident in Figs. 15.3 and 15.4. Related to this is the loss of tritium from this same region to higher latitudes, which can be noted in standing-crop maps not shown here. The model thermocline is also somewhat too deep in lower latitudes, giving rise to excessive penetration of tritium toward the equator (Fig. 15.3). We believe that a major reason for these problems is the manner in which vertical processes are parameterized in the model. A new series of experiments will examine this parameterization.

The simplest comparison that can be made between the model and data is to calculate a vertical penetration, $\int T\,dz/T_{\mathrm{surface}}$. The model gives a penetration of 540 m compared with a penetration of 590 m obtained from the data for the same region of the North Atlantic. The comparison is excellent. The discrepancy arises primarily from the low penetration in the 30° to 40°N region mentioned above.

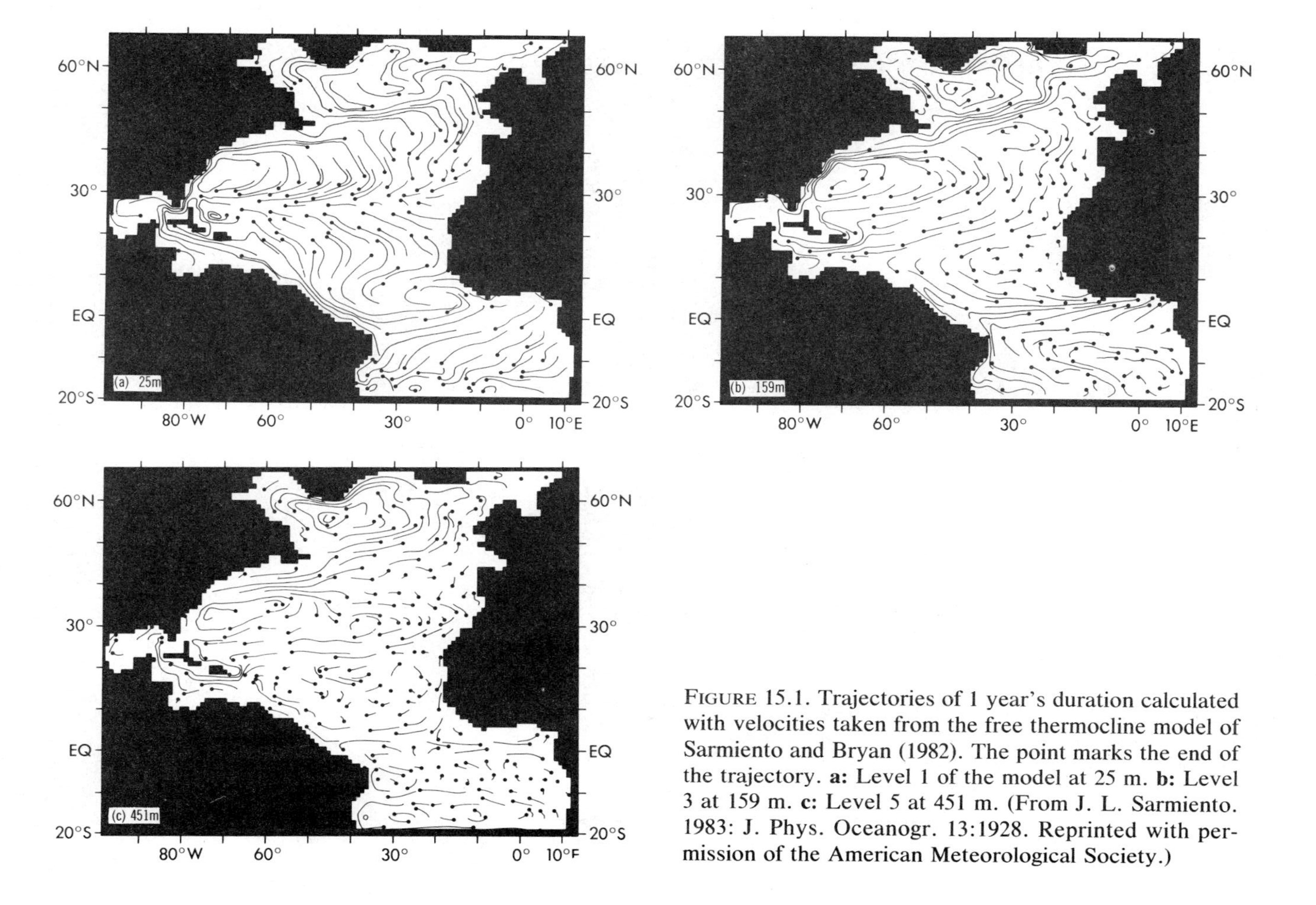

FIGURE 15.1. Trajectories of 1 year's duration calculated with velocities taken from the free thermocline model of Sarmiento and Bryan (1982). The point marks the end of the trajectory. **a:** Level 1 of the model at 25 m. **b:** Level 3 at 159 m. **c:** Level 5 at 451 m. (From J. L. Sarmiento. 1983: J. Phys. Oceanogr. 13:1928. Reprinted with permission of the American Meteorological Society.)

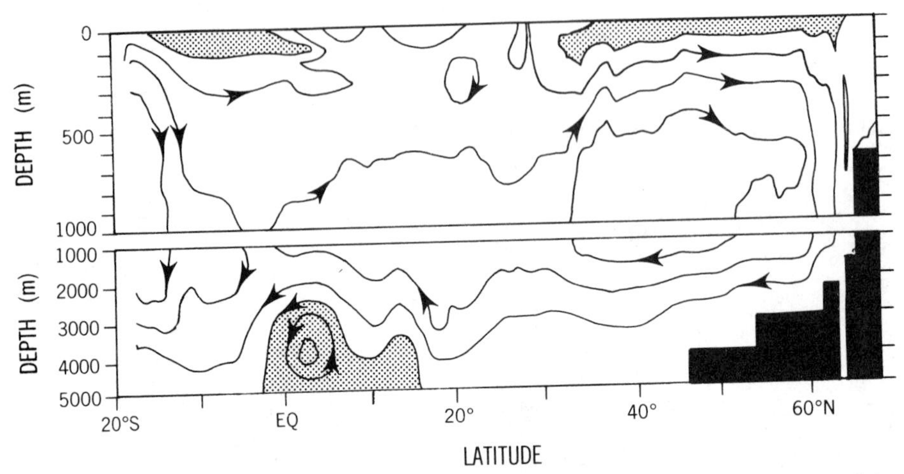

FIGURE 15.2. Zonally integrated meridional transport in the free thermocline model of Sarmiento and Bryan (1982). Contour interval is 5 Sv. Negative values are shaded. (From J. L. Sarmiento 1983*b:* J. Phys. Oceanogr. 13:1928. Reprinted with permission of the American Meteorological Society.)

Study of Analogue Modeling Approach

Perhaps the most authoritative study of ocean uptake of fossil fuel CO_2 to date is that of Oeschger et al. (1975). They used a one-dimensional box-diffusion model in which the vertical eddy diffusivity was assumed to be constant with depth and time. They calibrated their model with the preindustrial ^{14}C signal and the Suess effect. Broecker et al. (1980) sub-

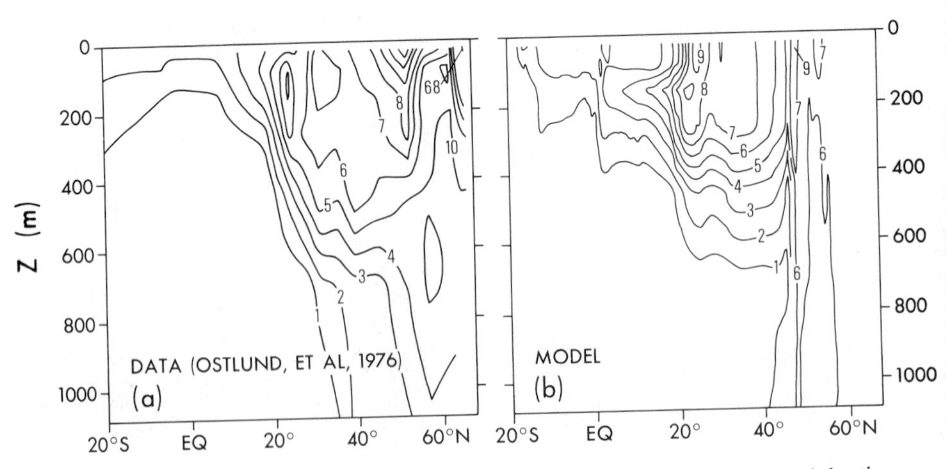

FIGURE 15.3. Tritium on the north-south GEOSECS section in the western Atlantic. **a:** Measurements made in 1972 (Ostlund et al. 1976). **b:** Model prediction. (From J. L. Sarmiento. 1983: J. Phys. Oceanogr. 13:1929. Reprinted with permission of the American Meteorological Society.)

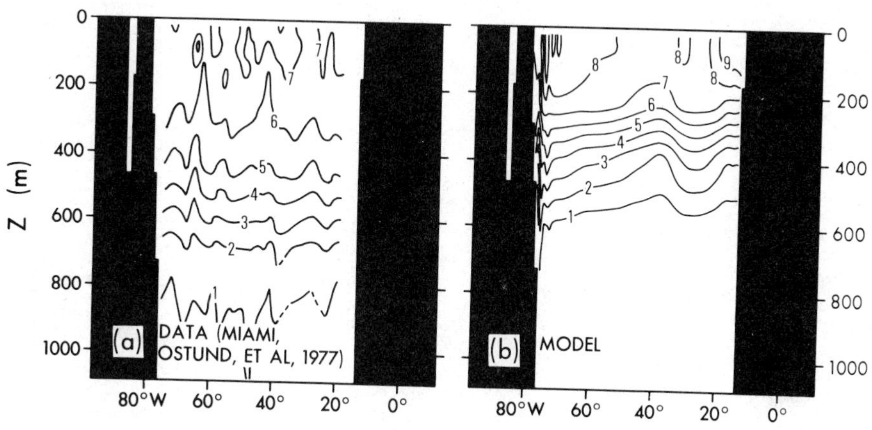

FIGURE 15.4. Tritium on an east-west section at ~29°N in the North Atlantic. **a:** Measurements made in 1972 (Ostlund et al. 1977). **b:** Model prediction. (From J. L. Sarmiento. 1983: J. Phys. Oceanogr. 13:1929: Reprinted with permission of the American Meteorological Society.)

sequently used tritium and bomb ^{14}C to calibrate the same model. There are significant differences obtained with each of the tracers. Table 15.2 summarizes the results of a recent study by Siegenthaler (1983) that shows the discrepancies. His study also includes calculations done with a new model he refers to as the outcrop-diffusion model, which allows direct ventilation of intermediate- and deep-ocean waters. The results in Table 15.2 show a range of >40% in predicted oceanic fossil fuel uptake.

The tritium simulation discussed above can be used to gain some insight into the variation shown in Table 15.2. One can obtain from the three-

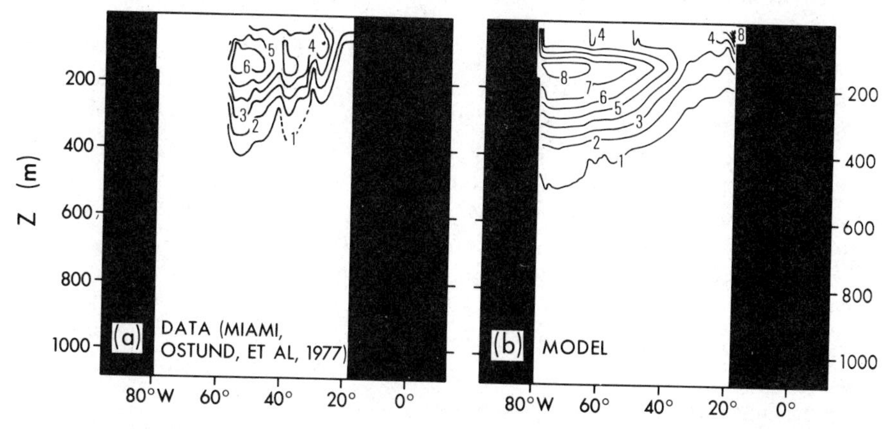

FIGURE 15.5. Tritium on an east-west section at ~14°N in the North Atlantic. **a:** Measurements made in 1972 (Ostlund et al. 1977). **b:** Model prediction. (From J. L. Sarmiento. 1983: J. Phys. Oceanogr. 13:1930. Reprinted with permission of the American Meteorological Society.)

dimensional model an equation equivalent to that solved in the Oeschger et al. (1975) model by taking a horizontal average of the equation solved in the three-dimensional model:

$$\frac{\partial \overline{T}^h}{\partial t} = \frac{\partial}{\partial z}\left(K_{\text{model}}\frac{\partial \overline{T}^h}{\partial z}\right) - \frac{\partial}{\partial z}\overline{\text{w}T}^h + \overline{\text{convection}}^h \qquad (1)$$

All horizontal terms drop out in the averaging procedure. T is tritium, K_{model} is the background vertical mixing provided in the model (0.5 $\text{cm}^2 \bullet \text{s}^{-1}$), and w is vertical advection. The convection term arises from the mixing of tritium to the depth of the observed wintertime mixed layer, as discussed by Sarmiento (1983*a*).

Oeschger et al. solve the equation

$$\frac{\partial \overline{T}^h}{\partial t} = \frac{\partial}{\partial z}\left(K_{\text{Oeschger}}\frac{\partial \overline{T}^h}{\partial z}\right) \qquad (2)$$

The K_{Oeschger} includes three model processes: model mixing, vertical advection, and wintertime convection. Each of these processes can be represented by a vertical diffusivity simply by requiring each of the terms in Eq. (1) to take the form of the right-hand side of Eq. (2). One thus has

$$K_{\text{Oeschger}} = K_{\text{model}} + K_{\text{advection}} + K_{\text{convection}}, \text{ where}$$

$$K_{\text{advection}} = \overline{\text{wT}}^h/(\partial \overline{T}^h/\partial z) \text{ and } K_{\text{convection}} = \int \overline{\text{convection}}^h\, dz/(\partial \overline{T}^h/\partial z) \qquad (3)$$

Figure 15.6 shows $K_{\text{advection}}$ and $K_{\text{convection}}$ and $\overline{T}^h$ versus time. K_{model} is essentially constant with time and depth with a value of 0.5 $\text{cm}^2 \bullet \text{s}^{-1}$. One can see that advective and convective processes are much larger, on the average, than the model diffusivity. The most important point in Fig. 15.6 is that the mixing is not constant either with time or with depth. The basic assumption concerning the nature of mixing in the box-diffusion model is not supported by these results.

The complex pattern of the curves, discussed in detail by Sarmiento (1983*a*), has basically two major features: (1) Convective overturning, which is larger than the advective penetration for most of the time shown, is largest near the time of maximum tritium input to the oceans in 1964. It decreases with time. (2) The advective penetration in the upper part of the ocean behaves the opposite way, in that it is at a minimum at the time of maximum tritium entry and increases with time.

The advective penetration depends on the negative correlation between T and w, which begins small and becomes more negative with time as regions of upwelling, such as the equator, sweep tritium along the surface toward the downwelling regions in the subtropics. Convection, on the other hand, acts instantaneously to smooth out T vertically. Once that has been done, its effectiveness depends on the further addition of tracer from the atmosphere or by horizontal mixing and advection. Thus, it is largest at the times of maximum atmospheric input.

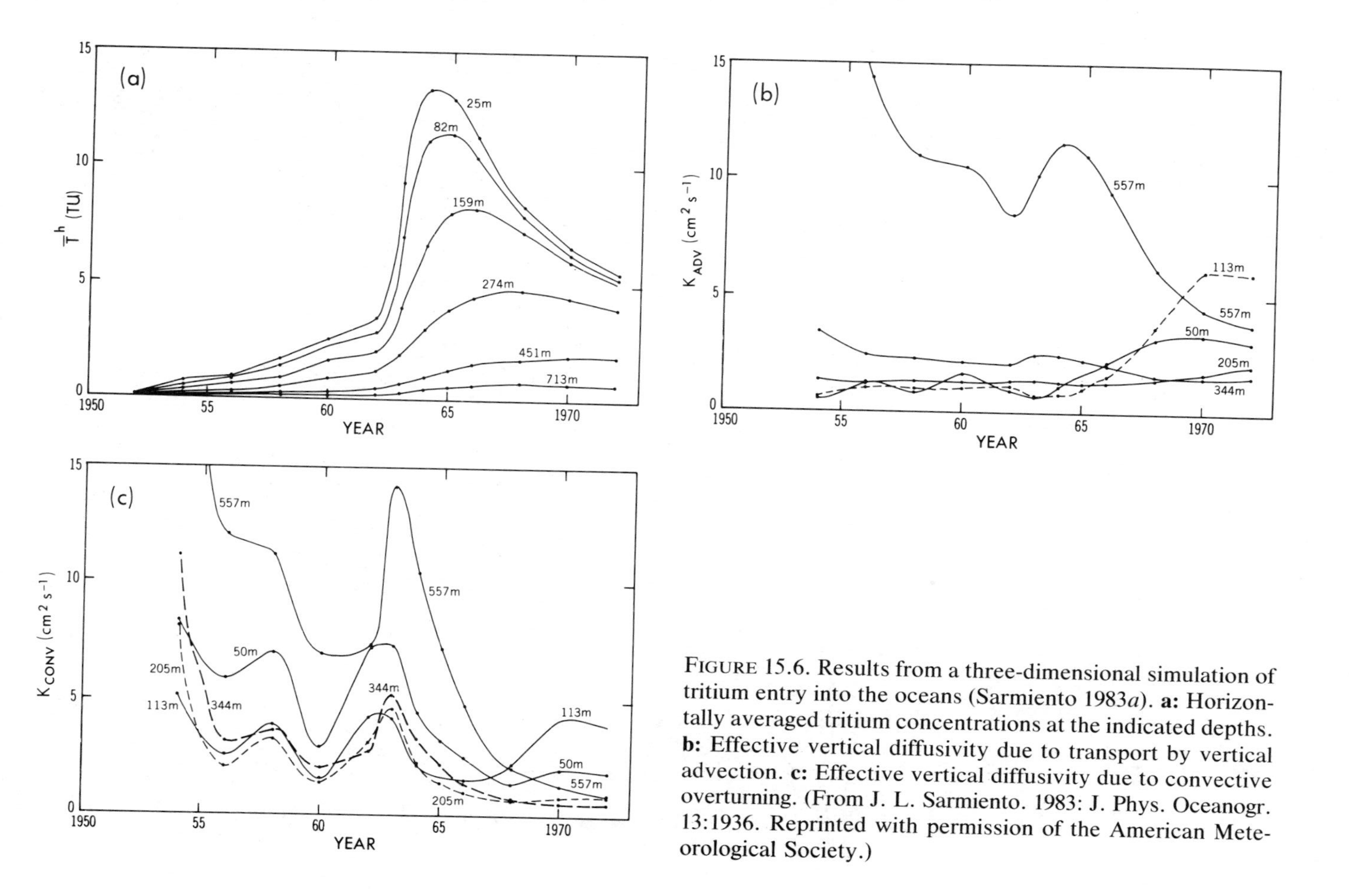

FIGURE 15.6. Results from a three-dimensional simulation of tritium entry into the oceans (Sarmiento 1983*a*). **a:** Horizontally averaged tritium concentrations at the indicated depths. **b:** Effective vertical diffusivity due to transport by vertical advection. **c:** Effective vertical diffusivity due to convective overturning. (From J. L. Sarmiento. 1983: J. Phys. Oceanogr. 13:1936. Reprinted with permission of the American Meteorological Society.)

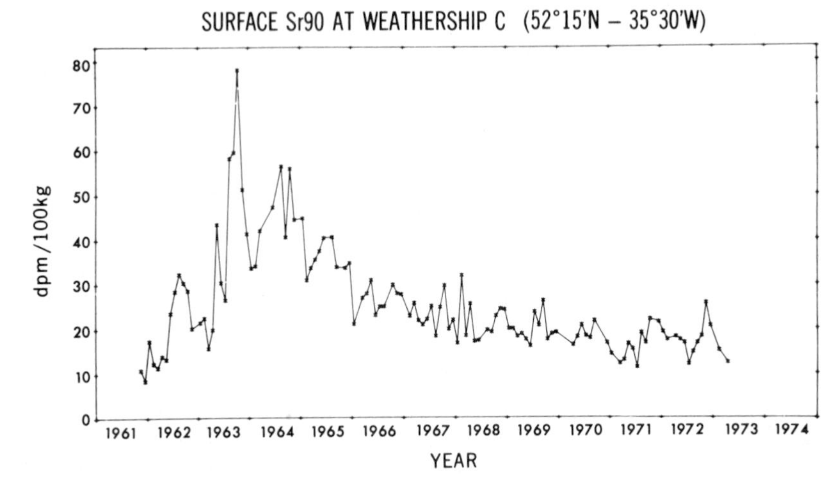

FIGURE 15.7. Strontium-90 concentration at the surface of the ocean as a function of time at Weathership Station C (52°N, 35°W). These measurements were made by V. Bowen at WHOI and provided by S. Kupferman.

The sensitivity of oceanic tracer uptake to time can be further illustrated with a remarkable set of observations of ^{90}Sr made by Bowen on Weathership C during the decade of the 1960s (Fig. 15.7). These show a large seasonal signal resulting from wintertime convection during the time of maximum ^{90}Sr input to the oceans in 1963 and 1964. The seasonal signal becomes smaller with time, because convection has very little effect once the tracer has had time to become more uniform in its distribution.

A remarkable aspect of the ^{90}Sr observations is the apparent lack of significant decrease of surface concentrations after ~1968. Plots from a large number of locations in the North Atlantic show a similar slowing of tracer loss from the surface. One possible explanation is that the isopycnal surfaces in the thermocline become saturated with tracer fairly rapidly (see discussion of isopycnal ventilation by Sarmiento 1983*b*), and then the loss of tracer from the upper ocean slows dramatically. Another possible explanation is horizontal input of ^{90}Sr, although the widespread nature of the effect argues against this. It is interesting that the surface tritium concentrations from the three-dimensional simulation (Fig. 15.6) do not show the same effect.

TABLE 15.2. Siegenthaler (1983) model results: Oceanic fossil fuel CO_2 uptake.

Model	Tracer used for calibration	
	Preindustrial ^{14}C	Bomb ^{14}C
Box-diffusion model	1.00	1.17
Outcrop-diffusion model (outcrop = 10% of surface area)	1.19	1.42

Units are relative. 1.00 is equivalent to an uptake by the oceans of 0.333 of the fossil fuel production.

The differences in the results shown in Table 15.2 are a consequence of the fact that the penetration of tracers into the oceans is highly variable in time. The penetration mechanisms, and thus the types of simplifications that are appropriate, vary according to the time scale being considered. The analogue-modeling approach would be less uncertain if there existed a tracer that entered the oceans on exactly the same time scale as fossil fuel CO_2. In approximately 10 years bomb ^{14}C and tritium will have been present in the oceans on a time scale equal to that of the mean age (30 years) of fossil fuel CO_2 molecules (Broecker et al. 1980). However, the order of magnitude difference in atmosphere–ocean gas equilibration rates between $^{14}CO_2$ and $^{12}CO_2$ will probably give rise to additional complications that we have not yet explored in our models.

Tritium has an additional complication as a tracer of CO_2 because its input is strongly biased toward the high-latitude Northern Hemisphere. $^{14}CO_2$ has such a slow removal time scale from the atmosphere that it has enough time to equilibrate within and between the hemispheres before entering the ocean in substantial amounts. Our tritium experiment had a penetration depth of 540 m in 1972. An experiment was also performed with a smoothed tritium input that more closely approximated the spatial distribution of the input of $^{14}CO_2$. This gave a penetration depth of only 395 m in 1972 because a higher fraction of the tritium entered into low-latitude low-penetration regions such as the equator. Great care must be exercised in using simplified models calibrated with tritium for CO_2 predictions.

On the Role of the Oceanic Biosphere in Fossil Fuel Uptake

The role of oceanic organisms in determining atmospheric CO_2 is understood in broad outline (e.g., Broecker and Peng 1982). Models of fossil fuel CO_2 uptake have generally oversimplified biological processes, because it has always been thought that they operate on such long time scales as to obviate any noticeable impact on fossil fuel CO_2 on the time scales of interest to mankind. However, Stauffer et al. (1984) recently discussed some studies of trapped air bubbles in ice cores and reported that variations on the order of 50 to 100 ppm in atmospheric CO_2 may have occurred on time scales as short as 100 years during the last ice age. Such changes, should they occur today, would have a dramatic effect on the rate of increase of fossil fuel CO_2 in the atmosphere. They suggest extremely rapid uptake or removal of enormous amounts of carbon from some reservoir, most likely the oceans. What are the mechanisms that give rise to such rapid changes, and is it possible that they may operate today?

The first explanation offered for glacial-interglacial CO_2 changes was that of Broecker (1982). His mechanism, which calls for erosion or deposition of large amounts of phosphate from or to the continental shelves, operates on a time scale too long to explain the recent observations. A

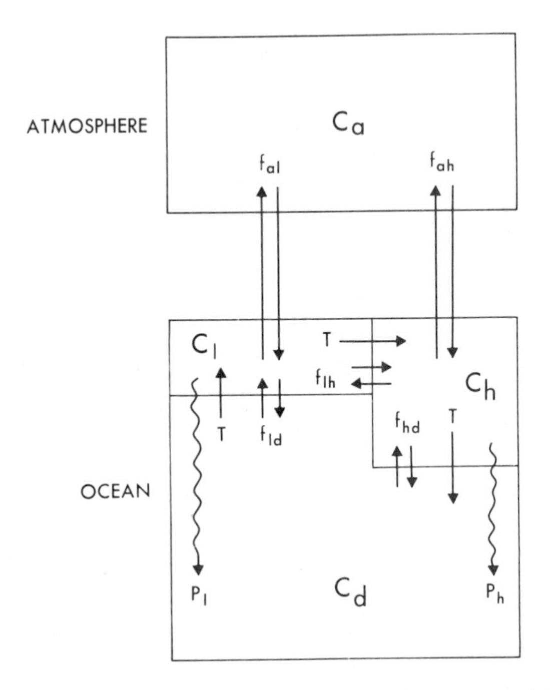

FIGURE 15.8. Schematic of the four-box model of Sarmiento and Toggweiler (1984) used to predict atmospheric CO_2. The f's are all exchange rates in units of $m^3 \cdot s^{-1}$. The subscript a refers to the atmosphere, l to low-latitude nutrient-depleted surface water, h to high-latitude deep-water formation regions, and d to deep ocean. Sarmiento and Toggweiler assume that f_{ld} and f_{lh} are negligible. In addition to exchange rates, the model allows for a shallow thermohaline overturning, T ($m^3 \cdot s^{-1}$), which goes through the low-latitude surface box to the high-latitude surface box before sinking into the deep box. $P \cdot (M \cdot C \cdot s^{-1})$ is the total particulate flux of organic and carbonate carbon. We assume that none is lost to sediments. We visualize the low-latitude surface box as being 100 m thick and occupying 85% of the ocean's surface area. The high-latitude surface box, which represents essentially the subpolar gyres, is 250 m thick and occupies the remaining 15% of the surface ocean. (Reprinted by permission from Nature, Vol. 308, No. 5960, p. 623. Copyright © 1984 Macmillan Journals Limited.)

number of other studies have followed Broecker's (e.g., McElroy 1982; Keir and Berger 1983), all of which suffer from the same difficulty of time scale.

Recently, a new model was developed that appears capable of explaining the observations (Fig. 15.8). Very similar versions of the model were developed simultaneously by three different groups (Sarmiento and Toggweiler 1984; Siegenthaler and Wenk 1984, Knox and McElroy 1984) and explains the CO_2 changes as resulting from variations in the nutrient level of high-latitude surface waters. The variations can be caused by either changes in productivity (P) or changes in nutrient supply. Changes in nutrient supply are caused primarily by variations in high-latitude overturning

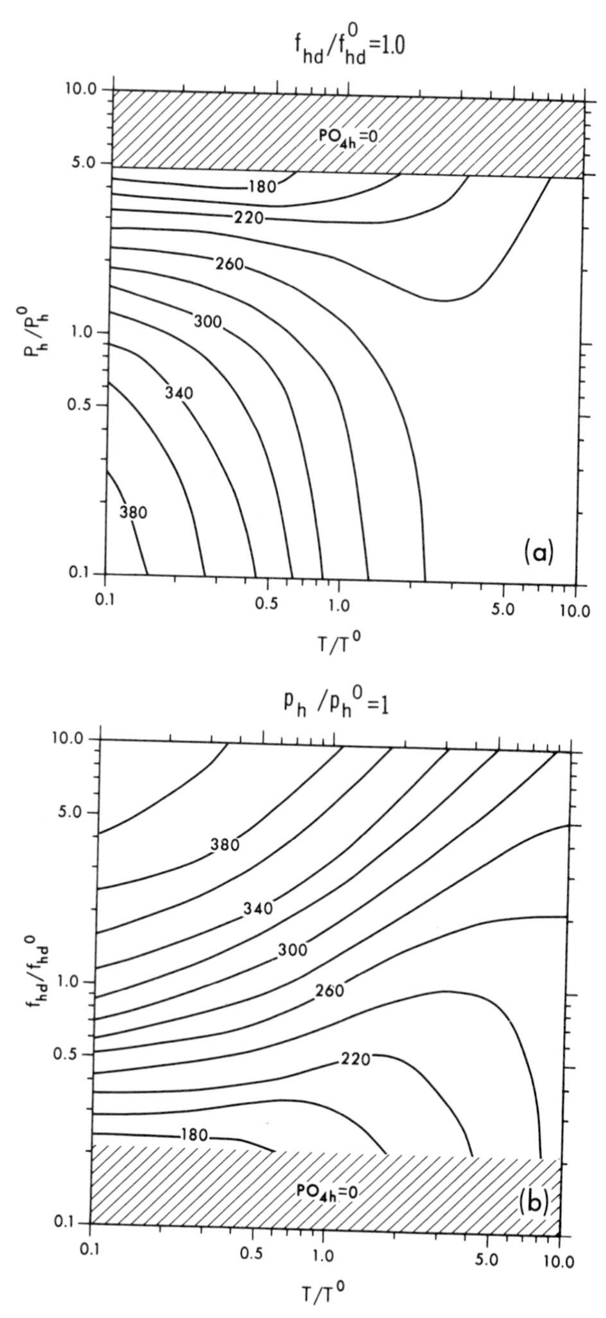

FIGURE 15.9. Atmospheric CO_2 in ppm as predicted by the model of Sarmiento and Toggweiler for various values of the parameters f_{hd}, P_h, and T. $f^\circ_{hd} = 28.1 \times 10^6\ m^3 \bullet s^{-1}$, $P^\circ_h = 2.31 \times 10^6$ mol C $\bullet\ s^{-1}$, and $T^\circ = 25.4 \times 10^6\ m^3 \bullet s^{-1}$ are the best values for each of these parameters as predicted for present conditions. (Reprinted by permission from Nature, Vol. 308, No. 5960, p. 624. Copyright © 1984 Macmillan Journals Limited.)

(f_{hd}), which is directly linked to rates of deep-water formation. Figure 15.9 shows the sensitivity of CO_2 to P and f_{hd}. The shallow thermohaline overturning, T, determines the sensitivity of CO_2 to P and f_{hd}. When T is very large, the model behaves essentially like Broecker's (1982) two-box model of the ocean because of the rapid circulation through the low-latitude nutrient-depleted surface water. The atmospheric CO_2 becomes relatively insensitive to variations in P and f_{hd}. Siegenthaler and Wenk believe that changes in high-latitude nutrient content and in atmospheric CO_2 can occur fast enough to alter the atmospheric CO_2 by as much as 50 to 100 ppm on the observed time scales of 100 years.

It is quite conceivable, indeed probable, that the climate changes accompanying a CO_2 increase will cause some response in the overturning rate in high latitudes. This is an effect that has been virtually unexplored up to now. The complexity of the processes, which involve changes in oceanic overturning, point toward the need for a careful three-dimensional study.

Conclusions

In conclusion, I would like to emphasize three points. The first concerns the feasibility and desirability of three-dimensional ocean models for simulating CO_2 distribution. Fairly realistic three-dimensional models are already in existence and in the process of being improved. However, these models are not likely to be useful as predictors of the absolute amount of CO_2 uptake for many years. Their best use for the present is as laboratories for improving our understanding of physical processes and the role of chemical and biological processes in the carbon cycle, and for testing the simpler models currently being used to make CO_2 predictions.

The second point I would emphasize is the extent to which a three-dimensional model is likely to behave differently than simpler models. I have given a graphic answer to that in discussing one-dimensional models. The difficulty with simpler models is that they must represent a large range of physical processes, such as wintertime convection and vertical advection, with much simpler parameters, such as a constant vertical diffusivity. This problem becomes less important as the models become more realistic; however, one must always retain a healthy skepticism. Unfortunately, we have no tracer that is a perfect analogue for the CO_2 transient. Even if we did, one would still have the very important problem of being unable to predict the changes in ocean circulation that accompany climate changes.

The third important point is the extent to which three-dimensional models could make better use of the available tracer data. Unfortunately, the only place where we have adequate tracer and nutrient data is in the North Atlantic because of the TTO work. In other regions, particularly around the critical deep and intermediate water formation areas of the

Antarctic, information is very sparse. It is for this reason that we at Princeton have concentrated our initial effects in the North Atlantic. Thus far, we have made heavy use of hydrographic and tracer data in our model studies. We are also now making heavy use of the TTO nutrient and carbon system data in developing our model of chemical and biological processes. We will eventually need data of this quality from all the ocean basins if we are to expand our efforts to the world oceans.

Acknowledgments

I thank Marty Jackson, Johann Callan, Evan Romer, and Phil Tunison for help in preparing the manuscript and figures. R. J. Toggweiler made some helpful comments on the manuscript. This work was supported by ARL/NOAA grant NA83RAC00052 and NSF grant OCE8110155.

References

Broecker, W. S. 1982. Ocean chemistry during glacial time. Geochim. Cosmochim. Acta 46:1689–1705.

Broecker, W. S. and T. -H. Peng, 1982. Tracers in the Sea. Eldigio Press, Palisades, New York.

Broecker, W. S., T. -H. Peng, and R. Engh. 1980. Modeling the carbon system. Radiocarbon 22:565–598.

Bryan, K. 1969. A numerical method for the study of the circulation of the world ocean. J. Comput. Phys. 4:347–376.

Bryan, K. and L. J. Lewis. 1979. A water mass model of the world ocean. J. Geophys. Res. 84:2503–2517.

Hasselmann, K. 1982. An ocean model of climate variability studies. Prog. Oceanogr. 11:69–92.

Hellerman, S. 1967. An updated estimate of the wind stress on the world ocean. Mon. Weather Rev. 95:607–626; also Corrigendum 1968. 96:63–74.

Hellerman, S. and M. Rosenstein. 1983. Normal monthly wind stress over the world ocean with error estimates. J. Phys. Oceanogr. 13:1093–1104.

Holland, W. R. 1978. The role of mesoscale eddies in the general circulation of the ocean—numerical experiments using a wind-driven quasi-geostrophic model. J. Phys. Oceanogr. 8:363–392.

Keir, R. S. and W. H. Berger. 1983. Atmospheric CO_2 content in the last 120,000 years: the phosphate extraction model. J. Geophys. Res. 88:6027–6038.

Knox, F. and M. B. McElroy. 1984. Changes in atmospheric CO_2: influence of the marine biota at high latitude. J. Geophys. Res. 89:4629–4637.

Kromer, B. 1979. Gasaustausch zwischen Atmosphäre und Ozean—Feldmessungen mittels der Radonmethode. Inaugural dissertation, Ruprecht, Karl Universität, Heidelberg, West Germany.

Leetmaa, A. and A. F. Bunker. 1978. Updated charts of the mean annual wind stress, convergences in the Ekman layers and Sverdrup transports in the North Atlantic. J. Mar. Res. 36:311–322.

Levitus, S. 1982. Climatological Atlas of the World Ocean. National Oceanic and Atmospheric Administration Professional Paper No. 13.

McElroy, M. B. 1982. Marine biology: controls on atmospheric CO_2 and climate. Nature 302:328–329.

Oeschger, H., U. Siegenthaler, U. Schotterer, and A. Gugelman. 1975. A box diffusion model to study the carbon dioxide exchange in nature. Tellus 2:168–192.

Ostlund, H. G., H. G. Dorsey, and R. Brescher. 1976. GEOSECS Atlantic radiocarbon and tritium results. Data Rep. No. 5. Rosentiel School of Marine and Atmospheric Science, University of Miami, Florida.

Ostlund, H. G., H. G. Dorsey, R. Brescher, and W. H. Peterson. 1977. Oceanic tritium profiles, NAGS cruises 1972-73. Data Rep. No. 6. Rosenstiel School of Marine and Atmospheric Sciences, University of Miami, Florida.

Peng, T. -H., W. S. Broecker, G. G. Mathieu, Y. -H. Li, and A. E. Bainbridge. 1979. Radon evasion rates in the Atlantic and Pacific Oceans as determined during the GEOSECS program. J. Geophys. Res. 84:2471–2486.

Sarmiento, J. L. 1983*a*. A tritium box model of the North Atlantic thermocline. J. Phys. Oceanogr. 13:1269–1274.

Sarmiento, J. L. 1983*b*. A simulation of bomb tritium entry into the North Atlantic Ocean. J. Phys. Oceanogr. 13:1924–1939.

Sarmiento, J. L. and K. Bryan. 1982. An ocean transport model for the North Atlantic. J. Geophys. Res. 87:394–408.

Sarmiento, J. L., C. G. H. Rooth, and W. Roether. 1982. The North Atlantic tritium distribution in 1972. J. Geophys. Res. 87:8047–8056.

Sarmiento, J. L. and J. R. Toggweiler. 1984. A new model for the role of the oceans in determining atmospheric pCO_2. Nature 308:621–624.

Siegenthaler, U. 1983. Uptake of excess CO_2 by an outcrop-diffusion model of the ocean. J. Geophys. Res. 88:3599–3608.

Siegenthaler, U. and T. Wenk. 1984. Rapid atmospheric CO_2 variations and ocean circulation. Nature 308:624–626.

Stauffer, B., H. Hofer, H. Oeschger, J. Schwander, and U. Siegenthaler. 1984. Ann. of Glaciol. 5:160–164.

Weiss, W., W. Roether, and E. Dreisigacker. 1979. Tritium in the North Atlantic Ocean. Behavior of Tritium in the Environment, pp. 315–336. International Atomic Energy Agency (IAEA), Vienna.

Wunsch, C. 1978. The North Atlantic general circulation west of 50°W determined by inverse methods. Rev. Geophys. Space Phys. 16:583–620.

16
Calibrating Ocean Models by the Constrained Inverse Method

BERRIEN MOORE III AND ANDERS BJÖRKSTRÖM

From the results of several investigations (Keeling 1973; Björkström 1980; Killough and Emanuel 1981; Björkström, *this volume*; Bolin 1983; Bolin et al. 1983, p. 231; Fiadeiro 1983; Bolin, *this volume)* the approach of using highly aggregated box models (less than 15 boxes for the world oceans) appears to be inadequate for the task of estimating accurately the current rate at which the ocean is absorbing excess atmospheric CO_2. However, general circulation models of the entire ocean are not at hand; therefore, what is needed is a new generation of models that can serve usefully in the interim to study the important question of the current rate of oceanic CO_2 uptake. Further, such models may have other applications, not the least of which may be their use as diagnostic tools for general circulation models of the ocean (see also Bryan et al. 1975; Sarmiento, this volume).

The construction of more detailed box models should be accomplished within a set of general requirements:

1. Reservoirs should reflect actual oceanic regions.
2. Data that are used to develop the parameters for the model must be used consistently and kept distinct from data used for validation or for testing the model.
3. The assumptions on which the model is built must be shown explicitly.
4. The available data must provide for a calibration that is narrow enough to permit nontrivial conclusions.

We have attempted to honor these guidelines by employing a step-by-step approach to constructing an 84-box model of the Atlantic Ocean; and, when technical details are mastered, we shall apply this approach to the construction of a detailed box model of the world oceans. The results of this work will appear in a series of papers, the first of which is in preparation (Bolin et al. in preparation).

There are six steps in the construction of the 84-box model of the Atlantic:

1. Establish broad oceanographic regions, 12 in the case of the Atlantic; within regions, define a series of density layers.

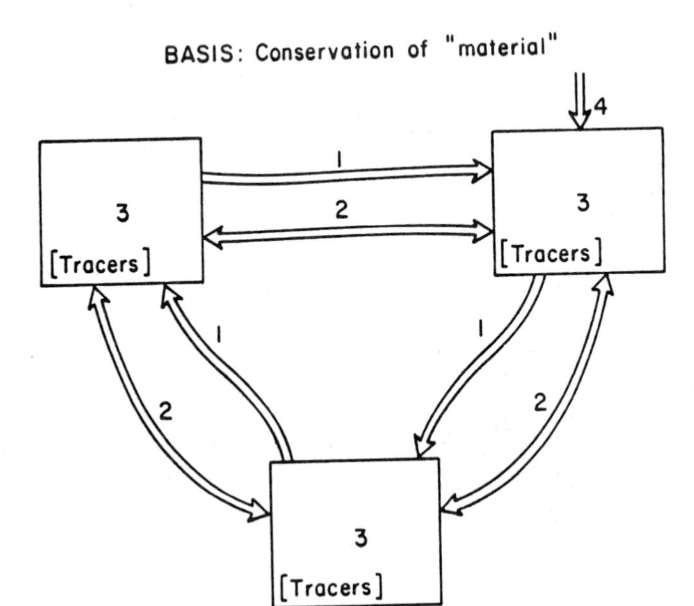

FIGURE 16.1. A general scheme for inversion problems: What is known and what is unknown.

2. Define the topology of the model (i.e., which boxes are connected to which). In the case of the Atlantic Ocean model, eight layers are used, and hence there is the possibility of representing the Atlantic with 96 boxes; however, since not every region contained water of each density type, this number is reduced to 84.
3. Compute, for each box, average concentrations of the tracers that are used to deduce the rates that essentially define the model.
4. Establish the boundary conditions: quantify the fluxes across external boundaries and rates of internal decay.
5. Define the constraints the solution must meet.
6. Perform a constrained inversion of the tracer field against the boundary conditions and thereby calculate a set of rates of exchange: advection, turbulence, biotic formation and decomposition of organic tissue and $CaCO_3$, which best describes the tracer field and still meets the constraints (Fig. 16.1).

FIGURE 16.2. Regionalization of the Atlantic Ocean.

Methodology

DEFINING THE BOXES

The Atlantic Ocean was divided into 12 major regions (Fig. 16.2) that reflect a subjective balance between oceanographic features and the existence of GEOSECS stations, GEOSECS having been selected as the prime oceanographic data base for the construction of not only this model but the world ocean model as well (Bainbridge 1981*a,b;* Craig et al. 1981; Broecker et al. 1982; Spencer et al. 1982). An extensive compilation of

TABLE 16.1. Density limits ($kg \cdot m^{-3}$) used to define layers.

Layer	Top interface	Bottom interface
1		1026.0
2	1026.0	1026.5
3	1026.5	1027.0
4	1027.0	1027.4
5	1027.4	1027.7
6	1027.7	1027.8
7	1027.8	1027.88
8	1027.88	

TABLE 16.2. Box volumes (10^{15} m^3).

Western section							
	Region						
Layer	I	II	IV	VI	VIII	X	XII
1			0.96	1.70	1.22		
2			1.37	0.61	1.12	0.34	
3		0.25	3.04	2.16	2.35	1.29	
4	0.07	0.51	2.57	4.81	4.24	3.92	1.03
5	0.24	1.69	3.53	4.49	4.47	4.49	3.03
6	0.16	3.15	5.18	7.82	3.48	4.24	4.59
7	0.20	2.84	13.29	11.05	16.27	10.73	32.19
8	3.56	0.80	14.09	5.56	1.64	0.02	0.05

Eastern section							
	Region						
Layer	I	III	V	VII	IX	XI	XII
1			0.52	0.57	0.83		
			0.65	0.51	1.09	0.21	
3		0.27	2.45	1.72	2.39	0.92	
4	0.07	1.87	3.62	3.58	4.42	2.28	1.03
5	0.24	1.71	3.79	3.43	5.00	2.83	3.03
6	0.16	2.29	2.96	2.86	4.18	2.47	4.59
7	0.20	3.22	14.64	12.74	13.94	6.99	32.19
8	3.56	1.64	9.61	3.97	5.15	0.09	0.05

temperature and salinity measurements (Levitus 1982*a,b*) was used to define the position, volume, and surface for eight density layers (or less, as applicable) within each of the 12 regions. The density limits used to define the layers are shown in Table 16.1. The resulting volumes and surface area expressions are summarized in Tables 16.2 and 16.3 and Fig. 16.3.

DEFINING THE INTERFACES

The establishment of density volumes essentially defines the topology for the model (Fig. 16.4). In the main, each box has two almost horizontal boundary surfaces, where it borders lighter or heavier water (density types 1 and 8 are obvious exceptions). Each box has also a number of interfaces bordering water of the same density in neighboring regions. To a first approximation, these latter surfaces are just as suggested by Fig. 16.2. However, a few complications arise when the horizontal and quasi-vertical divisions are combined:

1. Because of the mid-Atlantic ridge, there is no east-west connection in layer 8. This eliminates five interfaces (Fig. 16.4).
2. Since constant-density surfaces are not perfectly horizontal, cross-isopycnal flow cannot be identified with vertical motion. In Fig. 16.5, part

TABLE 16.3. Box surface areas (density computed at 50 m depth from Levitus 1982*b*) in contact with the atmosphere (10^{12} m^2).

Western section							
	Region						
Layer	I	II	IV	VI	VIII	X	XII
1			8.17	11.65	8.61		
2			1.69			3.39	
3		2.08	0.63			3.09	
4	0.81	1.19					5.47
5	1.46	1.21					4.96
6	0.68						0.77
7	0.71						
8	0.62						

Eastern section							
	Region						
Layer	I	III	V	VII	IX	XI	XII
1			5.49	7.06	8.17		
2			3.00	0.26	0.99	1.71	
3		2.38	0.59			2.01	
4	0.81	1.56					5.47
5	1.46	0.09					4.96
6	0.68						0.77
7	0.71						
8	0.62						

of the flow from box B to box A is horizontal (F_1) and part is vertical (F_2). At present, the structure of the model prevents separation of these components and this phenomenon makes the interpretation of the orientation of certain advections ambiguous.

3. Some boxes are located like box B in Fig. 16.5 and have no GEOSECS station in the area in contact with the atmosphere. Data are available only from positions where box B is overlaid by water of lower density. Therefore, important conditions governing the exchange between the atmosphere and box B remain unknown. The lack of coverage by GEOSECS prevents a more detailed regionalization which would avoid difficulties raised in points 2 and 3, and this suggests the need to use other existing data sets as well as the importance of developing new data sets.
4. In some places, regional boundaries were defined to follow isopycnals along the sea surface (Fig. 16.3). Situations as shown in Fig. 16.6 were created, where density type 1 in region A borders density type 2 in region B along a line. Because of wind-driven turbulence, the "line" is rather a surface with a depth of some 50 m. Water penetrating here will pass immediately from A1 to B2. In arranging all boxes in a rectangular scheme by region and by density, these fluxes are represented

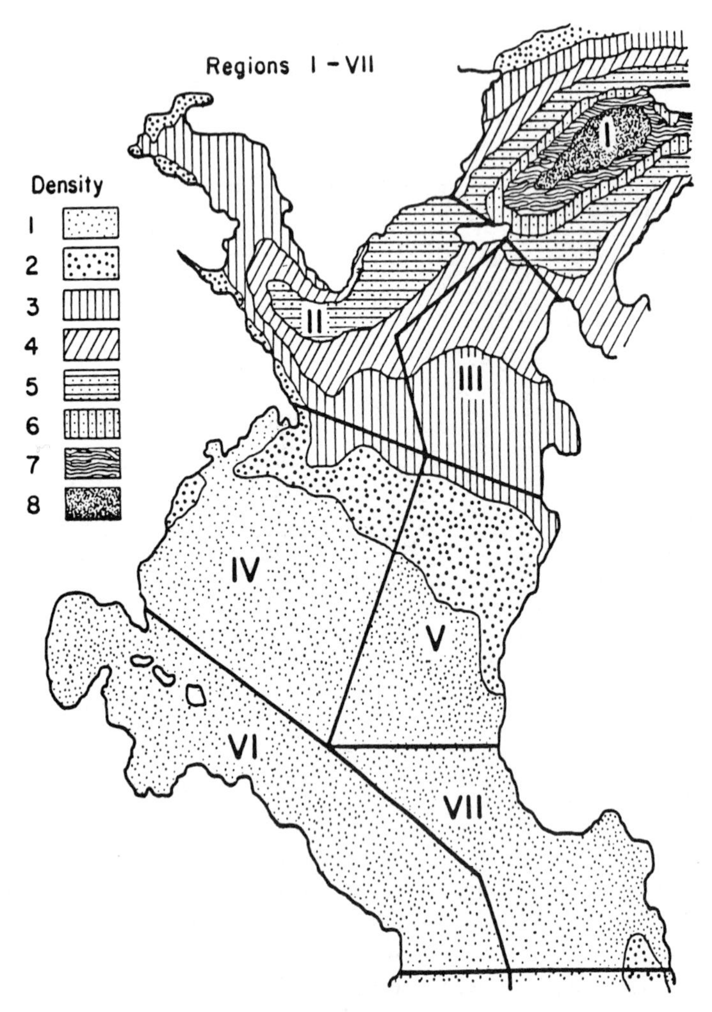

FIGURE 16.3. Surface expression of density layers and regionalization (density computed at 50 m depth, from Levitus 1982*b*). **A:** Northern section.

by "slant" lines (see Fig. 16.4). There are five connections of this type.

5. Because of the form and extent of the density layers in the deep sea, the regional division sometimes cuts off "tiny" volumes like A8 in Fig. 16.7. At present, some very small boxes are still in the model (see Table 16.2), although these may be eliminated at a later stage.

The regionalization and density configuration are discussed further in the section entitled Research Issues and Summary.

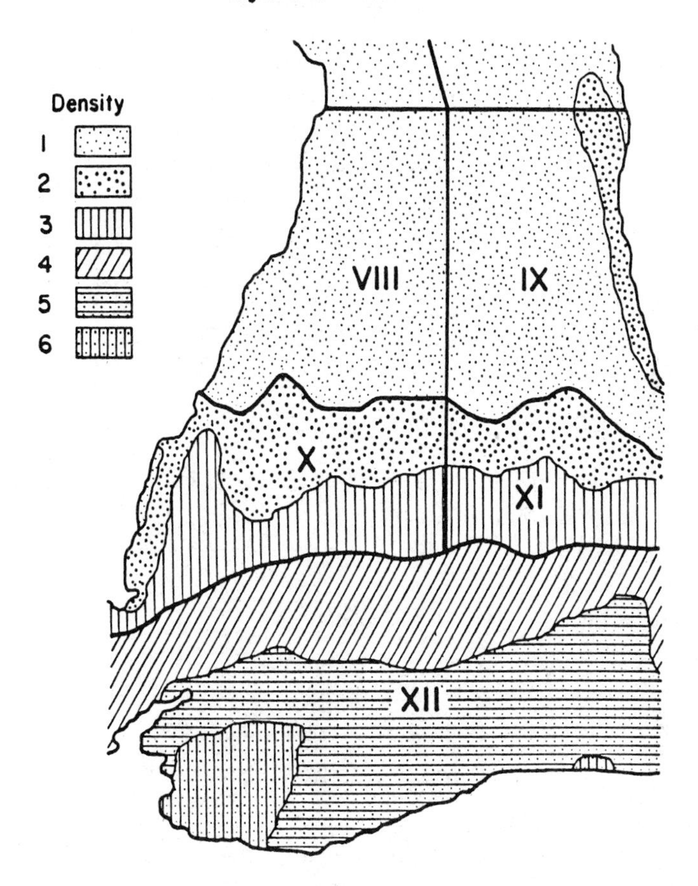

FIGURE 16.3. **B:** Southern section

ESTABLISHING THE TRACER PROFILES

Each of the Atlantic GEOSECS stations is assigned to one of the 12 regions (Fig. 16.8). For each station, depth-defined concentration data $d_k(z)$, k = 1, ..., 7, is extracted, where

$d_1(z)$ = Total carbon, mol • m^{-3};
$d_2(z)$ = Carbon-14, pmol • m^{-3};
$d_3(z)$ = Alkalinity, meq • m^{-3};
$d_4(z)$ = Phosphorus, μmol • m^{-3};
$d_5(z)$ = oxygen, mol • m^{-3};
$d_6(z)$ = salinity, kg • m^{-3};
$d_7(z)$ = heat, Mcal • m^{-3}.

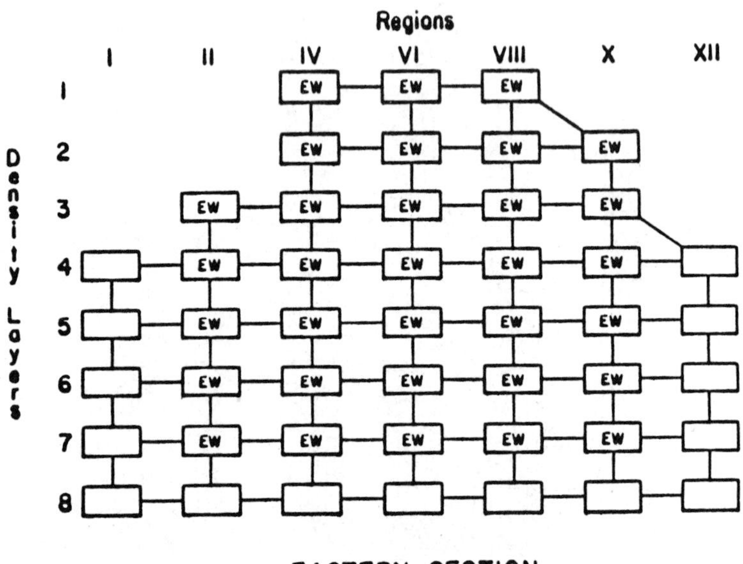

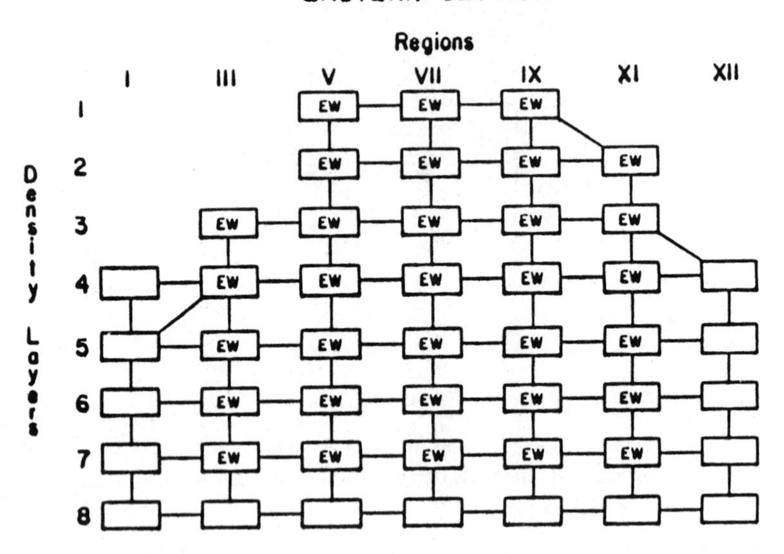

FIGURE 16.4. Box topology of the model. Note: Boxes other than those first in a column also may have an exchange with the atmosphere; see Table 16.3.

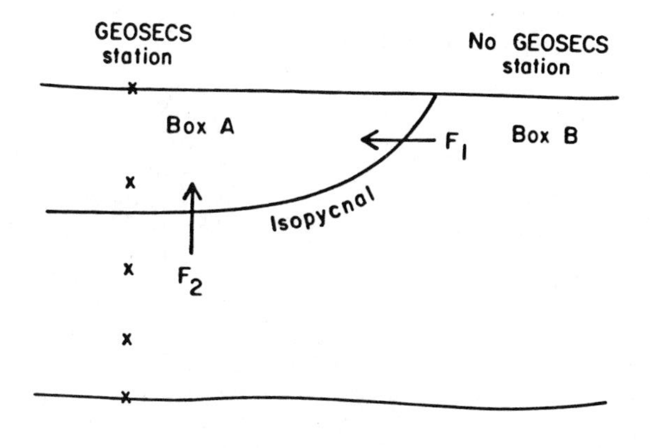

FIGURE 16.5. The occurrence of more than one surface density layer in a region makes the separation of vertical and horizontal flow ambiguous; however, the lack of coverage by GEOSECS makes this difficulty unavoidable. In this vein, a more serious difficulty is the lack of information, in the data from stations such as shown in the figure, regarding important surface dynamics for box B.

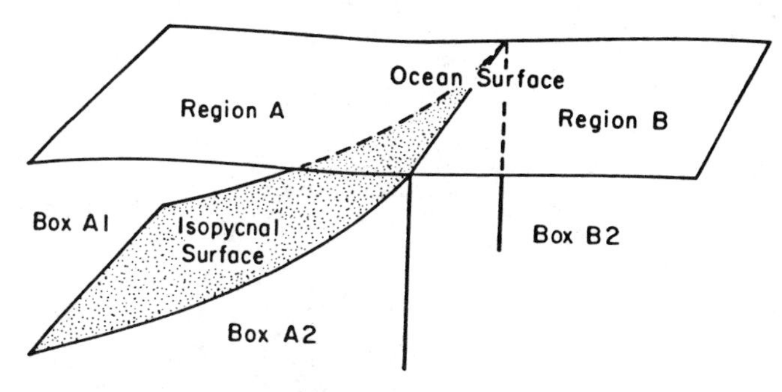

FIGURE 16.6. Example of cross-isopycnal surface flow, such as between region VIII, layer 1 and region X, layer 2 (see Fig. 16.3B).

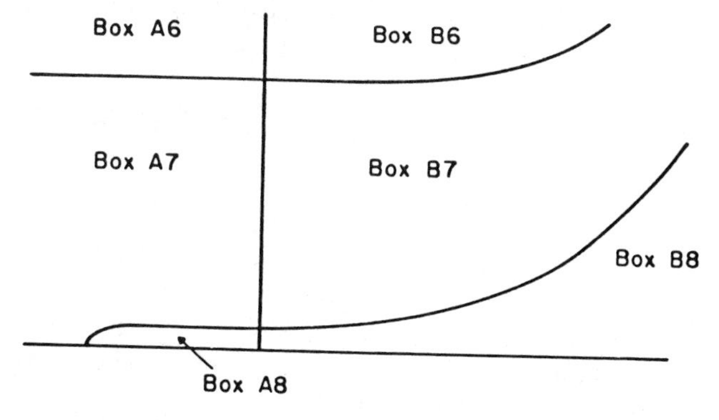

FIGURE 16.7. The model still contains boxes with very small volumes; these likely will be removed in the future.

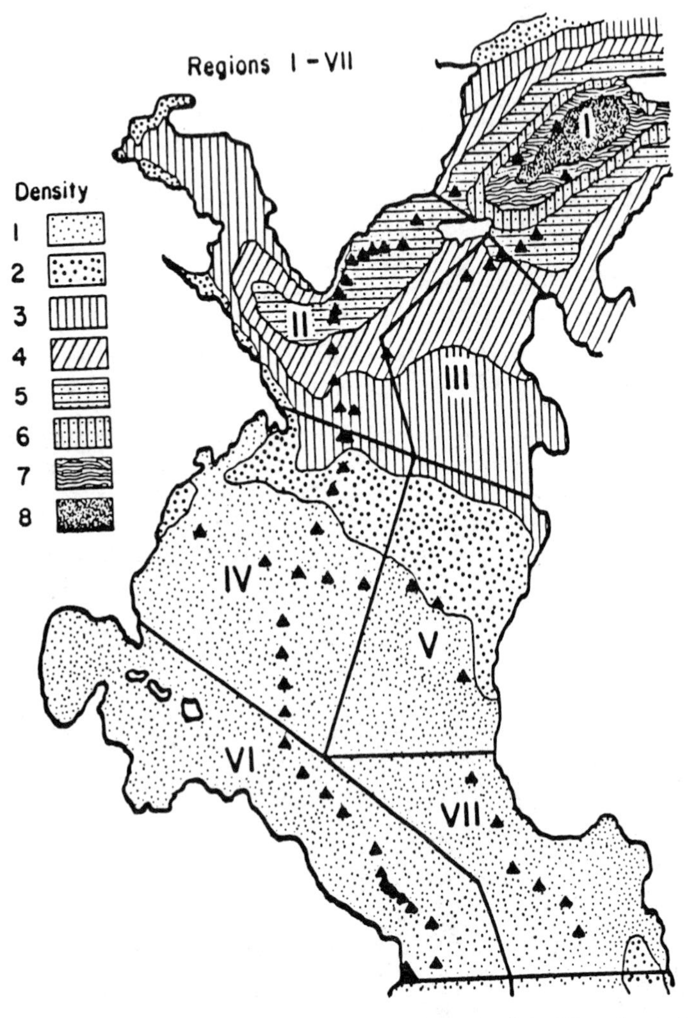

FIGURE 16.8. Location of GEOSECS stations in each of the 12 regions. **A:** Northern section.

For each depth z, $d_6(z)$ and $d_7(z)$ are used to calculate the location in density space of the data vector

$$D(z) = [d_1(z), \ldots, d_7(z)]$$

and to assign this vector to one of the 84 boxes. The location in density space is used to establish a weight for the specific concentration data in $D(z)$ that is proportional to the distance in density space to the boundary of the nearest adjacent density layer (Fig. 16.9). The average value for each tracer for a particular box (a density layer in a region) is found by using this weighting technique and computing the weighted average across depth for all values $d_1(z), \ldots, d_7(z)$ in the given density layer and across all GEOSECS stations in the given region.

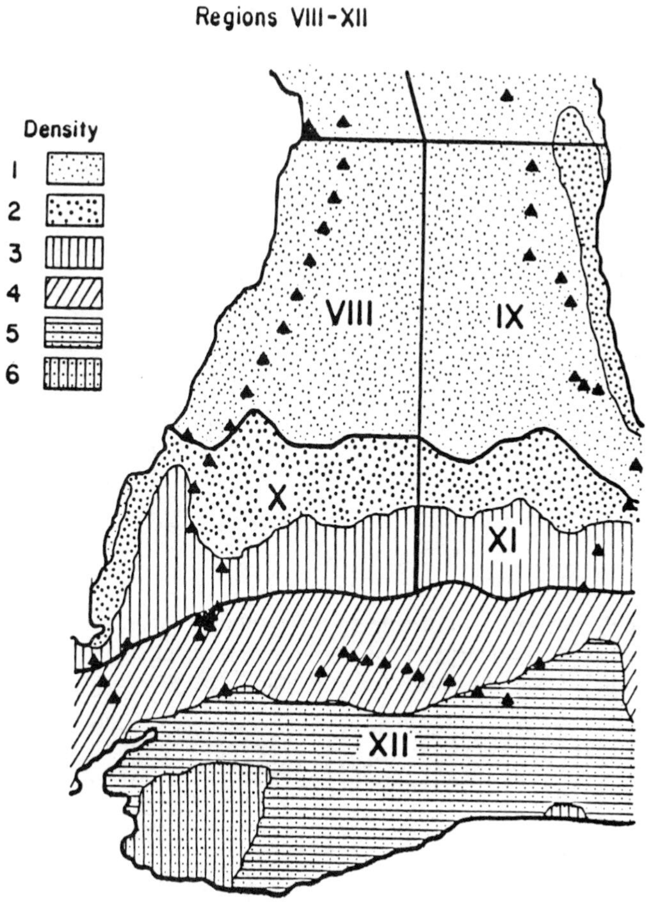

Regions VIII-XII

Density
1
2
3
4
5
6

VIII IX X XI XII

FIGURE 16.8. **B:** Southern section.

The data profiles for the 84-box model that were in use at the time of this report are given in Tables 16.4 to 16.10. (Tables 16.9 and 16.10 present the data in more traditional terms, salinity in parts per thousand and temperature in degrees Celsius). The reader is cautioned that this investigation is in progress, and that therefore any numerical information, including calibration data, must be viewed as preliminary. In particular, carbon data for high northern latitudes were extracted from more recent data sets and are under reappraisal at this time.

Establishing the Boundary Conditions

There are five external media exchanging material with the model region: the atmosphere, the Arctic Ocean, the Pacific Ocean, the Indian Ocean, and the Mediterranean Sea. For boxes with a surface toward any of these,

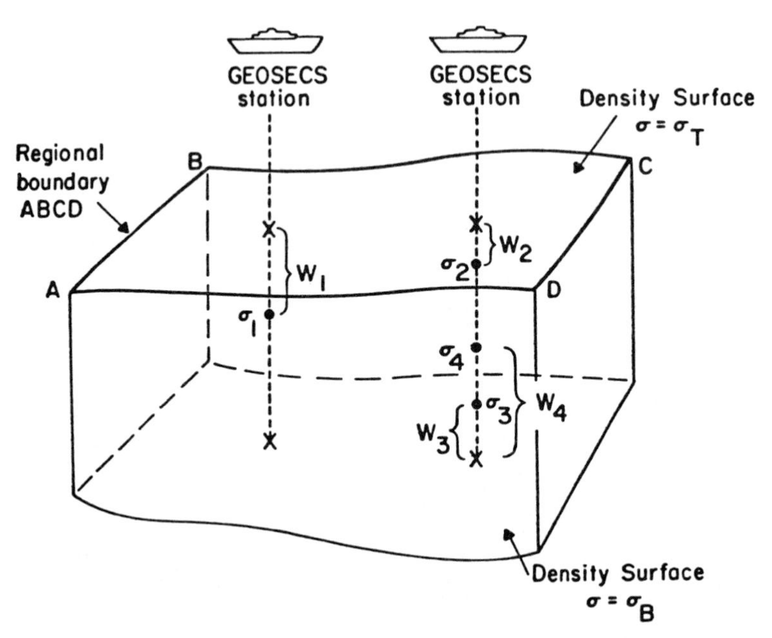

FIGURE 16.9. Illustration of principle for weighting samples. Weights are proportional either to sample density minus top density or to bottom density minus sample density, whichever is the smaller. For example, $W_1 = \sigma_1 - \sigma_T$; $W_3 = \sigma_B - \sigma_3$.

TABLE 16.4. Total carbon ($mol \cdot m^{-3}$).

Western section

Layer	Region I	II	IV	VI	VIII	X	XII
1			2.061	2.027	2.060		
2			2.079	2.160	2.069	2.041	
3		2.060	2.120	2.167	2.117	2.074	
4	2.122	2.108	2.181	2.215	2.172	2.141	2.135
5	2.138	2.156	2.214	2.222	2.218	2.225	2.222
6	2.137	2.200	2.194	2.193	2.199	2.234	2.246
7	2.136	2.196	2.184	2.176	2.181	2.219	2.256
8	2.159	2.195	2.198	2.199	2.208	2.250	2.256

Eastern section

Layer	Region I	III	V	VII	IX	XI	XII
1			2.010	1.980	2.041		
2			2.077	2.134	2.069	2.018	
3		2.103	2.111	2.187	2.161	2.095	
4	2.122	2.139	2.169	2.217	2.212	2.150	2.135
5	2.138	2.176	2.200	2.222	2.233	2.236	2.222
6	2.137	2.191	2.197	2.200	2.223	2.241	2.246
7	2.136	2.183	2.187	2.175	2.202	2.235	2.256
8	2.159	2.186	2.192	2.197	2.226	2.251	2.256

TABLE 16.5. Estimated steady-state distribution of ^{14}C in the oceans.

Western section

Layer	Region I	II	IV	VI	VIII	X	XII
1			0.950	0.942	0.954		
2			0.935	0.935	0.942	0.930	
3		0.945	0.925	0.915	0.910	0.905	
4	0.950	0.940	0.910	0.895	0.880	0.882	0.885
5	0.945	0.935	0.910	0.895	0.875	0.870	0.860
6	0.940	0.930	0.910	0.905	0.885	0.865	0.845
7	0.936	0.925	0.910	0.910	0.895	0.850	0.840
8	0.932	0.920	0.900	0.890	0.860	0.850	0.840

Eastern section

Layer	Region I	III	V	VII	IX	XI	XII
1			0.947	0.935	0.947		
2			0.935	0.930	0.942	0.915	
3		0.945	0.925	0.915	0.910	0.905	
4	0.950	0.937	0.900	0.895	0.890	0.888	0.885
5	0.945	0.928	0.895	0.890	0.875	0.860	0.860
6	0.940	0.920	0.895	0.905	0.880	0.855	0.845
7	0.936	0.913	0.895	0.905	0.899	0.850	0.840
8	0.932	0.907	0.885	0.885	0.865	0.850	0.840

Numbers show ratios of carbon-14 atoms to carbon atoms in the boxes, expressed as fractions of the corresponding ratio in the preindustrial atmosphere. Multiplication by total carbon in mol · m^{-3} (Table 16.4) times 1.176 (Bolin 1981, p. 91) gives concentrations in pmol · m^{-3} necessary for transport calculations.

tracer balance equations must include a term for loss or gain through that surface. In the interest of keeping down the number of unknowns, it is useful to establish estimates for these fluxes whenever possible.

ATMOSPHERE

The atmospheric exchanges of CO_2 and ^{14}C were estimated as follows. For each box that has contact with the atmosphere, a CO_2 pressure was computed, mainly on the basis of GEOSECS data from the uppermost 25 m in the region. As is well known, the CO_2 pressure data in GEOSECS for high northern latitudes are not available. Because of this and the general importance of surface CO_2 data, additional data sources (e.g., Brewer, personal communication) were used to supplement the GEOSECS data set in these latitudes and elsewhere. This will be discussed in detail in subsequent publications. The CO_2 pressure distribution that was obtained was assumed to be in equilibrium with the atmosphere in the sense that no net transfer of CO_2 takes place across the Atlantic surface as a whole. Then gross fluxes of carbon and ^{14}C into and out of each box were de-

TABLE 16.6. Alkalinity (meq · m^{-3}).

Western section

Layer	I	II	IV	VI	VIII	X	XII
1			2.396	2.382	2.419		
2			2.404	2.360	2.362	2.338	
3		2.309	2.378	2.340	2.325	2.320	
4	2.309	2.310	2.344	2.315	2.304	2.287	2.297
5	2.311	2.314	2.335	2.323	2.328	2.326	2.329
6	2.315	2.314	2.334	2.329	2.333	2.345	2.349
7	2.308	2.314	2.328	2.330	2.333	2.347	2.362
8	2.304	2.313	2.343	2.349	2.354	2.363	2.364

Eastern section

Layer	I	III	V	VII	IX	XI	XII
1			2.357	2.336	2.373		
2			2.404	2.345	2.337	2.300	
3		2.296	2.390	2.324	2.315	2.299	
4	2.309	2.295	2.337	2.308	2.305	2.296	2.297
5	2.311	2.308	2.336	2.317	2.325	2.335	2.329
6	2.315	2.309	2.339	2.322	2.339	2.351	2.349
7	2.308	2.308	2.343	2.323	2.335	2.357	2.362
8	2.304	2.309	2.346	2.345	2.362	2.369	2.364

TABLE 16.7. Phosphate (mol · m^{-3}).

Western section

Layer	I	II	IV	VI	VIII	X	XII
1			0.055	0.161	0.150		
2			0.101	1.057	0.391	0.320	
3		0.388	0.418	1.276	1.008	0.768	
4	0.180	0.751	1.181	2.057	1.913	1.805	1.758
5	0.634	1.091	1.583	2.096	2.162	2.300	2.313
6	0.823	1.102	1.308	1.568	1.765	2.093	2.275
7	0.783	1.097	1.207	1.300	1.485	1.871	2.209
8	0.948	1.045	1.356	1.551	1.766	2.210	2.228

Eastern section

Layer	I	III	V	VII	IX	XI	XII
1			0.040	0.171	0.211	0.202	
2			0.082	1.167	0.536	0.513	
3		0.315	0.416	1.600	1.372	1.102	
4	0.180	0.716	0.122	2.221	2.209	1.877	1.758
5	0.634	1.067	1.622	2.160	2.198	2.367	2.313
6	0.823	1.158	1.424	1.658	1.861	2.170	2.275
7	0.783	1.059	1.358	1.411	1.602	1.978	2.209
8	0.948	1.050	1.400	1.460	1.657	2.088	2.228

TABLE 16.8. Oxygen ($mol \cdot m^{-3}$).

	Eastern section						
	Region						
Layer	I	II	IV	VI	VIII	X	XII
1			0.215	0.199	0.214		
2			0.218	0.124	0.217	0.248	
3		0.288	0.203	0.133	0.195	0.251	
4	0.306	0.271	0.163	0.128	0.206	0.264	0.329
5	0.314	0.258	0.172	0.163	0.190	0.194	0.223
6	0.300	0.277	0.227	0.211	0.209	0.191	0.199
7	0.311	0.279	0.255	0.246	0.235	0.214	0.198
8	0.311	0.285	0.262	0.249	0.239	0.223	0.226

	Eastern section						
	Region						
Layer	I	III	V	VII	IX	XI	XII
1			0.215	0.201	0.223		
2			0.230	0.118	0.216	0.257	
3		0.278	0.201	0.097	0.146	0.271	
4	0.306	0.268	0.151	0.100	0.131	0.271	0.329
5	0.314	0.251	0.155	0.150	0.172	0.192	0.223
6	0.300	0.249	0.189	0.200	0.200	0.188	0.199
7	0.311	0.276	0.210	0.236	0.223	0.207	0.198
8	0.311	0.277	0.247	0.246	0.233	0.217	0.226

TABLE 16.9. Salinity (parts per thousand).

	Western section						
	Region						
Layer	I	II	IV	VI	VIII	X	XII
1			36.371	36.221	36.368		
2			36.593	35.822	35.573	35.330	
3		35.125	36.159	35.279	35.087	34.982	
4	34.980	34.926	35.442	34.656	34.399	34.217	34.012
5	35.035	34.919	35.047	34.671	34.532	34.462	34.350
6	34.970	34.915	35.067	34.914	34.811	34.705	34.597
7	34.849	34.935	35.029	34.973	34.900	34.804	34.696
8	34.917	34.928	34.905	34.860	34.790	34.692	34.671

	Eastern section						
	Region						
Layer	I	III	V	VII	IX	XI	XII
1			35.780	35.670	35.779		
2			36.706	35.637	35.433	34.889	
3		34.976	36.385	35.232	35.021	34.298	
4	34.980	35.092	35.514	34.715	34.492	34.123	34.012
5	35.035	35.043	35.212	34.689	34.557	34.455	34.350
6	34.970	35.004	35.310	34.891	34.757	34.688	34.597
7	34.849	34.942	35.263	34.955	34.859	34.765	34.696
8	34.917	34.992	34.962	34.893	34.848	34.730	34.671

TABLE 16.10. Temperature (°C).

Western section							
	Region						
Layer	I	II	IV	VI	VIII	X	XII
1			25.089	25.473	22.013		
2			19.066	16.863	15.759	15.109	
3		11.072	15.868	12.442	11.880	11.410	
4	8.882	8.348	10.967	7.001	5.431	4.318	1.862
5	6.372	5.785	6.852	4.509	3.331	2.647	1.331
6	4.330	4.293	5.434	4.424	3.558	2.638	1.341
7	2.230	3.696	4.431	4.016	3.424	2.518	1.293
8	−0.395	2.374	2.533	2.141	1.483	0.580	0.138

Eastern region							
	Region						
Layer	I	III	V	VII	IX	XI	XII
1			24.240	26.077	21.429		
2			19.316	16.135	15.448	13.526	
3		11.067	16.608	12.508	11.505	7.517	
4	8.882	9.067	11.251	7.339	5.989	3.436	1.862
5	6.372	6.745	7.790	4.725	3.541	2.515	1.331
6	4.330	4.981	6.966	4.261	3.121	2.488	1.341
7	2.230	3.834	6.020	3.860	3.063	2.105	1.293
8	−0.395	3.008	3.051	2.518	2.211	1.193	0.138

termined from the assumptions that (1) the net influx of ^{14}C through the air-sea interface balances the radioactive decay in the Atlantic and (2) since ^{13}C-corrected data of ^{14}C were used, isotopic fractionation at the surface could be neglected. The fluxes used in the current reference case are given in Table 16.11.

For oxygen, no attempt was made to determine the net of absorption and outgassing for individual boxes. These were treated as unknown, and this was accomplished, technically, by downweighting the oxygen balances for surface boxes, i.e., equations governing conservation of oxygen in surface boxes were multiplied by a small positive number.

Estimates of the heat exchange with the atmosphere were derived from Bunker (1980) and Ebensen and Kushmir (1981); estimates of evaporation and precipitation were derived from Baumgartner and Reichel (1975). The values are given in Tables 16.12 and 16.13.

External Oceans and Seas

To a first approximation, the circumpolar current around Antarctica brings the same amount of water into the Atlantic region (south of South America) as it takes out of the region (south of Africa), but it has been estimated that the outflow is slightly dominant. Estimates of the water exchange with the Arctic Ocean, primarily via the Labrador current, and of the flow

TABLE 16.11a. Reference case influxes of carbon pmol · yr^{-1} from the atmosphere.

Western section

Layer	Region I	II	IV	VI	VIII	X	XII
1			0.1007	0.1435	0.1061		
2			0.0208			0.0418	
3		0.0256	0.0078			0.0381	
4	0.0100	0.0147					0.0674
5	0.0180	0.0149					0.0611
6	0.0084						0.0095
7	0.0087						
8	0.0076						

Eastern section

Layer	Region I	III	V	VII	IX	XI	XII
1			0.0676	0.0870	0.1007		
2			0.0370	0.0032	0.0122	0.0211	
3		0.0293	0.0073			0.0248	
4	0.0100	0.0192					0.0674
5	0.0180	0.0011					0.0611
6	0.0084						0.0095
7	0.0087						
8	0.0076						

TABLE 16.11b. Reference case outfluxes of carbon (pmol · yr^{-1}) to the atmosphere.

Western section

Layer	Region I	II	IV	VI	VIII	X	XII
1			0.1141	0.1492	0.0966		
2			0.0192			0.0336	
3		0.0212	0.0083			0.0327	
4	0.0106	0.0139					0.0606
5	0.0188	0.0167					0.0809
6	0.0078						0.0131
7	0.0078						
8	0.0073						

Eastern section

Layer	Region I	III	V	VII	IX	XI	XII
1			0.0665	0.0870	0.0949		
2			0.0341	0.0043	0.0115	0.0162	
3		0.0329	0.0073			0.0219	
4	0.0106	0.0247					0.0606
5	0.0188	0.0015					0.0809
6	0.0078						0.0131
7	0.0078						
8	0.0073						

TABLE 16.11c. Reference case influxes of ^{14}C ($kmol \cdot yr^{-1}$) from the atmosphere.

Western section

Layer	Region I	II	IV	VI	VIII	X	XII
1			0.1184	0.1688	0.1248		
2			0.0245			0.0492	
3		0.0301	0.0092			0.0448	
4	0.0118	0.0173					0.0793
5	0.0212	0.0175					0.0718
6	0.0099						0.0112
7	0.0102						
8	0.0089						

Eastern section

Layer	Region I	III	V	VII	IX	XI	XII
1			0.0795	0.1023	0.1184		
2			0.0435	0.0038	0.0144	0.0248	
3		0.0345	0.0086			0.0292	
4	0.0118	0.0226					0.0793
5	0.0212	0.0013					0.0718
6	0.0099						0.0112
7	0.0102						
8	0.0089						

TABLE 16.11d. Reference case outfluxes of ^{14}C ($kmol \cdot yr^{-1}$) to the atmosphere.

Western section

Layer	Region I	II	IV	VI	VIII	X	XII
1			0.1275	0.1652	0.1084		
2			0.0210			0.0367	
3		0.0236	0.0089			0.0348	
4	0.0119	0.0154					0.0632
5	0.0208	0.0184					0.0817
6	0.0087						0.0130
7	0.0086						
8	0.0080						

Eastern section

Layer	Region I	III	V	VII	IX	XI	XII
1			0.0740	0.0957	0.1057		
2			0.0375	0.0047	0.0127	0.0174	
3		0.0365	0.0079			0.0233	
4	0.0119	0.0273					0.0632
5	0.0208	0.0016					0.0817
6	0.0087						0.0130
7	0.0086						
8	0.0080						

TABLE 16.12. Heat exchange with the atmosphere (TW).

Western section

Layer	Region I	II	IV	VI	VIII	X	XII
1			−327	243	182		
2			−84			−68	
3		−166	−50			31	
4	−32	−71					55
5	−73	−73					−50
6	−34						−31
7	−32						
8	−25						

Eastern section

Layer	Region I	III	V	VII	IX	XI	XII
1			192	373	156		
2			105	13	59	−34	
3		−60	6			40	
4	−32	−94					55
5	−73	−5					−50
6	−34						−31
7	−32						
8	−25						

Negative sign denotes net flux is to atmosphere.

TABLE 16.13. Exchange of water with the atmosphere in $10^9\ m^3 \cdot s^{-1}$.

Western section

Layer	Region I	II	IV	VI	VIII	X	XII
1			−0.09	0.18	−0.11		
2			0.00			0.04	
3		0.02	0.00			0.04	
4	0.03	0.01					0.07
5	0.02	0.01					0.06
6	0.01						0.01
7	0.01						
8	0.01						

Eastern section

Layer	Region I	III	V	VII	IX	XI	XII
1			−0.15	0.05	−0.24		
2			−0.04	0.00	−0.03	0.02	
3		0.03				0.02	
4	0.03	0.02					0.07
5	0.02						0.06
6	0.01						0.01
7	0.01						
8	0.01						

Negative sign denotes net flux is to the atmosphere.

TABLE 16.14. Advective flux of water (10^{12} $m^3 \cdot s^{-1}$) from surrounding non-Atlantic boxes.

To region II (Davis Strait)	
Layer 3	2.03
To region V (Gibraltar Strait)	
Layer 5	0.22
Layer 6	0.57
Layer 7	0.57
Layer 8	0.32
To region XII (Drake Passage)	
Layer 4	9.00
Layer 5	32.00
Layer 6	35.00
Layer 7	48.00

through Gibraltar Strait at different depths were considered together with the loss to the Indian Ocean, and were adjusted so as to give water balance for the Atlantic Ocean as a whole, including precipitation and evaporation. The data are shown in Tables 16.14 and 16.15; all of these data are preliminary, and this is particularly true with regard to regions X, XI, and XII. Loss of a tracer compound with water leaving the model region was estimated as the water outflow times the tracer concentration in the box losing water. For incoming water, tracer data from surrounding regions was estimated to compensate locally for the loss to external regions, in-

TABLE 16.15. Advective flux of water (10^{12} $m^3 \cdot s^{-1}$) to surrounding non-Atlantic boxes.

From region I (Greenland, Barents, and North Seas)	
Layer 5	0.14
Layer 6	0.14
Layer 7	0.14
From region V (Gibraltar Strait)	
Layer 3	0.96
Layer 4	0.79
From region XI (Agulhas Basin)	
Layer 2	1.00
Layer 3	8.00
Layer 4	16.00
Layer 5	12.00
Layer 6	4.54
Layer 7	20.00
From region XII (Atlantic-Indian Basin)	
Layer 4	1.00
Layer 7	63.00

cluding the atmosphere (estimates based, in part, on Sverdrup et al. 1942; Gordon 1971; Stigebrand 1981). Details regarding these important boundary conditions will be addressed more completely in subsequent publications.

Defining the Constraints

The general theme in this method is to determine the rates of exchange between boxes, due to advection and turbulent mixing, and the rates of internal sources and sinks, due to the formation and dissolution of organic and inorganic compounds, that will best reproduce a set of steady-state concentrations (Tables 16.4 to 16.10) given a set of boundary conditions (Tables 16.11 to 16.15) and constraints. In this initial investigation, the constraints referred only to the signs of certain unknowns, as described below.

The matricial expression for this problem

$$Ax = b$$

is constructed as described in Sect. 2 of the paper by Bolin et al. (1983), except with a slight change in the manner of scaling the rows in the matrix. The method employed for the results reported here had two aspects. First, the rows associated with the equations of continuity for a particular tracer were divided by the maximum variation in the data field for that tracer (i.e., tracer became unitless). The second aspect treated the fact that the problem under investigation was overdetermined (incompatible) and, therefore, not all continuity equations for all the tracers could be satisfied exactly. Consequently, subjective weights are used to stress the role of some tracers at the expense of others. Specifically, the equations governing carbon, radiocarbon, and alkalinity were multiplied by 25; the continuity equations for water and detrital material were multiplied by 10^5 so that these are satisfied exactly.

Evidence that the problem is overdetermined can be found in the following "counting exercise." There are 84 boxes and seven tracers; consequently, one can establish 588 continuity equations: one for each tracer and each box. Water continuity yields 84 more equations. Detrital continuity provides an additional 24: two equations (organic and inorganic) for each of the 12 regions (see Eqs. 2.8 and 2.12 in Bolin et al. 1983). Thus there are a total of 696 relations.

A careful counting of the connections in Fig. 16.4 reveals that there are 186 interfaces between boxes including five slant connections (note again that the mid-Atlantic ridge is assumed essentially to prevent east-west flux in layer 8 in regions II to XI); hence, there are 186 advective unknowns and 186 turbulent unknowns. Similarly there are 84 rates associated with formation or decomposition of organic material and 84 rates for inorganic. Thus there are a total of 540 unknowns. In fact, numerical analysis confirms that the columns of the matrix A are linearly independent, so this counting exercise actually describes the relevant dimensions associated with the system $Ax = b$.

The constraints employed in this report were simple sign conditions, as mentioned previously. All 186 turbulent mixing terms are required to be nonnegative. As for the biological terms, we distinguished between "white," "gray," and "black" boxes. In boxes that do not reach the surface (black ones), net dissolution of biogenic material must be nonnegative. There are, at present, 56 black boxes with a total of 112 biological unknowns. White boxes are those with no part covered by the other boxes. Since there can be no gravitational flux into white boxes, net dissolution here must be nonpositive. There are 12 white boxes, with a total of 24 biological unknowns. For gray boxes (i.e., boxes partly sunlit, partly covered) we cannot say beforehand whether biogenic dissolution is to be positive or negative. The total number of sign-constrained variables is thus $186 + 112 + 24 = 322$. It is, of course, perfectly possible to constrain the sign of some advective terms also and, for example, require downwelling at a certain location. Some experiments with constrained advections were made but will not be reported here. The fact that only sign constraints were used greatly simplifies the calculations; more importantly, such constraints avoid the numerical instability that can be associated with the conversion from a general least-squares inequality (LSI) problem to a least-distance programming (LDP) problem when the dimension is large (Lawson and Hanson 1974; Björck 1981; Haskell and Hanson 1981). On the other hand, more complex constraints can arise very naturally (Wunsch, 1984). For example, the biogenic dissolution in a gray box can, at most, be equal to the total net production in the white and gray boxes above it. This has not always been the case in our results, but it could easily be formulated as an inequality constraint on a linear combination of unknowns.

Performing Constrained Inversions

Because of the importance of the inequality constraints, we shall highlight some of the numerical issues associated with overdetermined constrained inverse problems. The work of Lawson and Hanson (1974) is particularly germane to these discussions (see also Veronis 1975; Wunsch 1978; Fiadeiro and Craig 1978; Wunsch 1980; Fiadeiro 1982; Fiadeiro and Veronis 1982; Wunsch and Grant 1982; Wunsch and Minster 1982; Wunsch, 1984; and the Appendix in Bolin et al. 1983).

The general configuration for least-squares problems with linear inequality constraints can be formulated as follows:

Let A be an $m \times n$ matrix, b an m-dimensional vector, G a $k \times n$ matrix, and h a k-dimensional vector. The LSI problem is to minimize the norm of $(Ax-b)$ subject to $Gx \geqslant h$.

There are two important special cases of the LSI problem:

1. *The least-distance programming problem (LDP):* Given a k-dimensional vector h and an $n \times k$ matrix F, minimize the norm of w subject to $Fw \geq h$.
2. *The nonnegative least-squares problem (NNLS):* Given an $m \times n$ matrix A and an m-dimensional vector b, minimize the norm of $(Ax - b)$ subject to $x \geq 0$.

A general LSI problem is solved by transforming it first to an LDP problem and then further to an NNLS problem. The conversion of an LSI problem to an LDP problem can lead to numerical difficulties. Without going into the details (see Björck 1981) of these difficulties, let us simply state that they can arise if the dimension of the system is large (>100) and if one makes the conversion from LSI to LDP via the pseudo-inverse of the matrix A (see Lawson and Hanson 1974, pp. 167–168, and Bolin et al. 1983, p. 223). If the dimension is <100 and if the problem is reasonably well-conditioned, we have found that the conversion is then generally stable. Fortunately, there is now an alternative method for transforming an LSI problem to an LDP problem that has been developed recently by Haskell and Hanson (1979, 1981). This method is numerically stable and commercially available (see Hanson and Haskell 1982).

The constrained solution presented in the next section was not affected by numerical instabilities, since, as mentioned, only sign constraints were employed, and the calculated result is essentially the same whether the method of calculation is NNLS (Lawson and Hanson 1974) or the new algorithm WNNLS (Hanson and Haskell 1982). However, initial explorations with more complex constraints in the Atlantic Ocean model have shown that these more general linear considerations will require the recent LSEI/WNNLS algorithms (Hanson and Haskell 1982) and that the earlier LDP/NNLS algorithms (Lawson and Hanson 1974) are unsatisfactory.

Results

In this section, we present a "steady-state" solution x that best represents the tracer field (Tables 16.4 to 16.10) where the boundary conditions are as given in Tables 16.11 to 16.15. The subjective weighting that was employed in this reference case stressed satisfying the continuity equations for water and detrital decay (both organic and inorganic), and matching the carbon ^{14}C, and alkalinity profiles at the expense of matching the profiles of oxygen, phosphorus, salinity, and temperature.

It must be stressed again that these results are tentative and that the research team is preparing a far more extensive paper (Bolin et al. in preparation; see also Bolin, this volume), which considers a variety of constrained solutions and the corresponding transient responses of the

```
                              WESTERN SECTION ADVECTIONS
1                   2                   4                   6                   8                   10                  12

                                        S********           S********           S********
                                        * 0.090 * 0.059     * 0.088 * 0.114     * 0.060 *
                                        *        **********          **********        *
                                        *  WEST  * <<<<<    *  EAST  * >>>>>    *  WEST  *
                                        *********           *********           **********
                                          0.143               0.263               0.092    *0.262
                                          DOWN                 UP                  UP       DOWN
                                        S********           *********           *********          *S********
                                        * 0.455 * 0.051     * 0.343 * 0.030     * 0.257 * 0.569    * 0.450 *
                                        *        **********          **********          **********        *
                                        *  EAST  * >>>>>    *  WEST  * <<<<<    *  EAST  * <<<<<    *  WEST  *
                                        *********           *********           *********           *********
                                          0.366               0.162               0.190               0.145
                                           UP                 DOWN                DOWN                DOWN
                    S*******E           S********           *********           *********           S********
                    * 0.027 * 0.428     * 0.177 * 0.113     * 0.021 * 0.039     * 0.155 * 0.017     * 0.478 *
                    *        **********          **********          **********          **********        *
                    *  EAST  * >>>>>    *  WEST  * >>>>>    *  EAST  * <<<<<    *  EAST  * >>>>>    *  EAST  *
                    *********           *********           *********           *********           **********
                      0.376               0.124               0.293               0.020               0.074    *0.243
                       UP                 DOWN                DOWN                 UP                  UP       UP*
S********           S********           *********           *********           *********           *********          *S*******E
*        * 0.000    * 0.326 * 0.013     * 0.414 * 0.706     * 0.205 * 0.359     * 0.228 * 0.018     * 1.096 * 1.176    *        *
*         **********         **********          **********          **********          **********          **********        *
*        *          *  WEST  * <<<<<    *  EAST  * <<<<<    *  WEST  * <<<<<    *  WEST  * <<<<<    *  EAST  * <<<<<    *        *
*********           *********           *********           *********           *********           *********           *********
  0.093               0.034               0.402               0.150               0.132               0.011               0.472
  DOWN                 UP                 DOWN                DOWN                 UP                  UP                  UP
S*******E           S********           *********           *********           *********           *********           S*******E
*        * 0.326    * 0.362 * 0.071     * 0.038 * 0.345     * 0.314 * 0.036     * 0.238 * 0.111     * 0.025 * 0.027     *        *
*         **********         **********          **********          **********          **********          **********        *
*        * <<<<<    *  EAST  * >>>>>    *  EAST  * <<<<<    *  EAST  * <<<<<    *  WEST  * >>>>>    *  EAST  * >>>>>    *        *
*********           *********           *********           *********           *********           *********           *********
  0.359               0.795               0.779               0.472               0.042               0.048               0.307
  DOWN                 UP                 DOWN                 UP                  UP                 DOWN                DOWN
S*******E           *********           *********           *********           *********           *********           S*******E
*        * 0.517    * 0.052 * 0.232     * 0.150 * 0.536     * 0.221 * 0.320     * 0.840 * 0.461     * 1.705 * 1.467     *        *
*         **********         **********          **********          **********          **********          **********        *
*        * >>>>>    *  EAST  * <<<<<    *  EAST  * >>>>>    *  EAST  * <<<<<    *  WEST  * >>>>>    *  EAST  * <<<<<    *        *
*********           *********           *********           *********           *********           *********           *********
  0.042               0.097               0.139               0.164               0.018               0.270               0.563
  DOWN                 UP                  UP                 DOWN                DOWN                DOWN                DOWN
S*******E           *********           *********           *********           *********           *********           ********E
*        * 0.049    * 0.159 * 0.147     * 0.069 * 0.036     * 0.136 * 0.088     * 0.032 * 0.090     * 0.669 * 0.546     *        *
*         **********         **********          **********          **********          **********          **********        *
*        * >>>>>    *  WEST  * >>>>>    *  WEST  * >>>>>    *  EAST  * >>>>>    *  EAST  * >>>>>    *  EAST  * <<<<<    *        *
*********           *********           *********           *********           *********           *********           *********
  0.028               0.038               0.042               0.023               0.016               0.238               0.453
  DOWN                 UP                 DOWN                 UP                  UP                 DOWN                 UP
S********           *********           *********           *********           *********           *********           *********
*        * 0.049    * 0.000 * 0.011     * 0.000 * 0.052     * 0.000 * 0.029     * 0.000 * 0.013     * 0.000 * 0.251     *        *
*         **********         **********          **********          **********          **********          **********        *
*        * >>>>>    *        * >>>>>    *        * >>>>>    *        * >>>>>    *        * >>>>>    *        * >>>>>    *        *
*********           *********           *********           *********           *********           *********           *********
```

FIGURE 16.10. A reference case study steady-state solution for advection, in 10^{15} $m^3 \cdot yr^{-1}$. **A:** Western section.

associated atmosphere–ocean systems to perturbations such as fossil fuel CO_2 release or bomb ^{14}C; the latter represents a valuable, independent validation experiment.

Even at this early stage in the investigation, certain oceanic features do emerge in the solution (see Fig. 16.10 to 16.12). Deep water formation is apparent in regions I and III. In low latitudes we observe patterns of equatorial upwelling, though perhaps further south than expected. The preliminary nature of these findings is reflected by other features, such as the advective upwelling (although this may be southward movement of water) in region II (Fig. 16.10A), the turbulent pattern associated with Mediterranean exchanges in layer 3 between regions IV and V (Fig. 16.11A), and the biological productivity that is reasonable overall (though perhaps high; net primary production of 3.0×10^{15} g C per year) may be too low in regions I, III, and V and too high in region XII (Fig. 16.12B). Even in light of these difficulties, if one considers the preliminary status of the results, it would appear that the methodology holds promise; this,

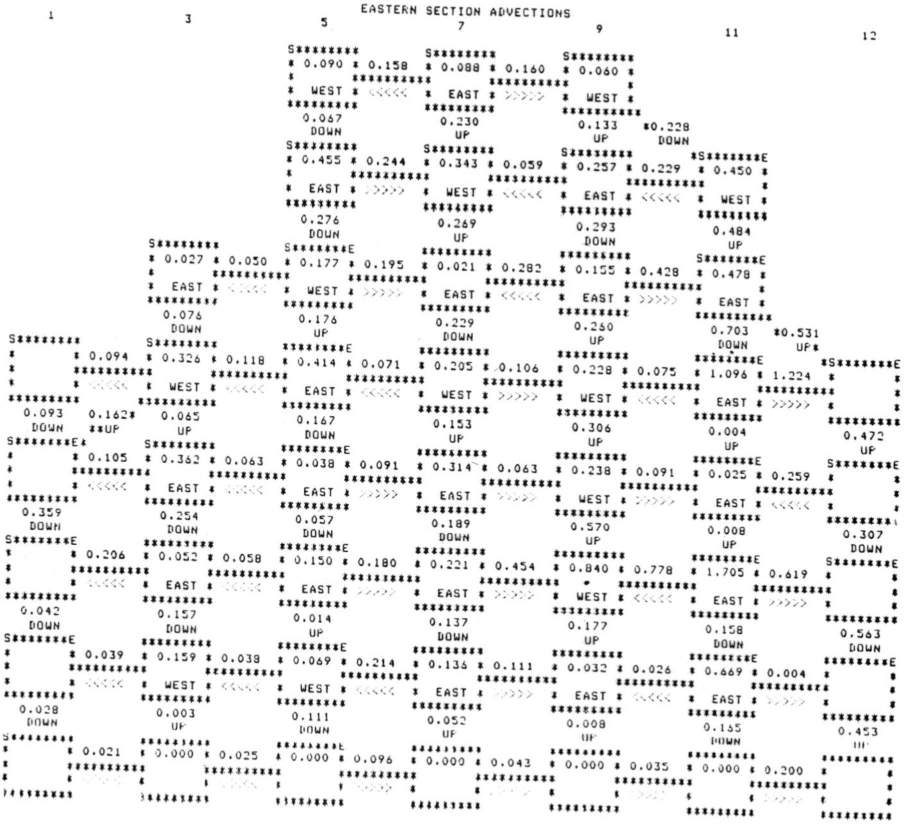

FIGURE 16.10. **B:** Eastern section.

in part, we hope to confirm with the paper in preparation (Bolin et al., in preparation).

Research Issues

At least four major research issues are posed by the inverse methodology as we are applying it. First, since density defines boxes, should not it also define the regionalization? Second, given the density field, and in particular the density profile at the interfaces, can one include in the formulation of matrix *A* the requirements on the advective flows posed by geostrophic considerations. Third, how should one treat "steady-state" tracers that are, in fact, no longer in "steady state"? Finally, the conflict between the real limitations in the geographical scope of currently available tracer data and the needed increases in the geographical resolution of ocean models could lead to a shift from overdetermined models to underdetermined ones; what difficulties might this present?

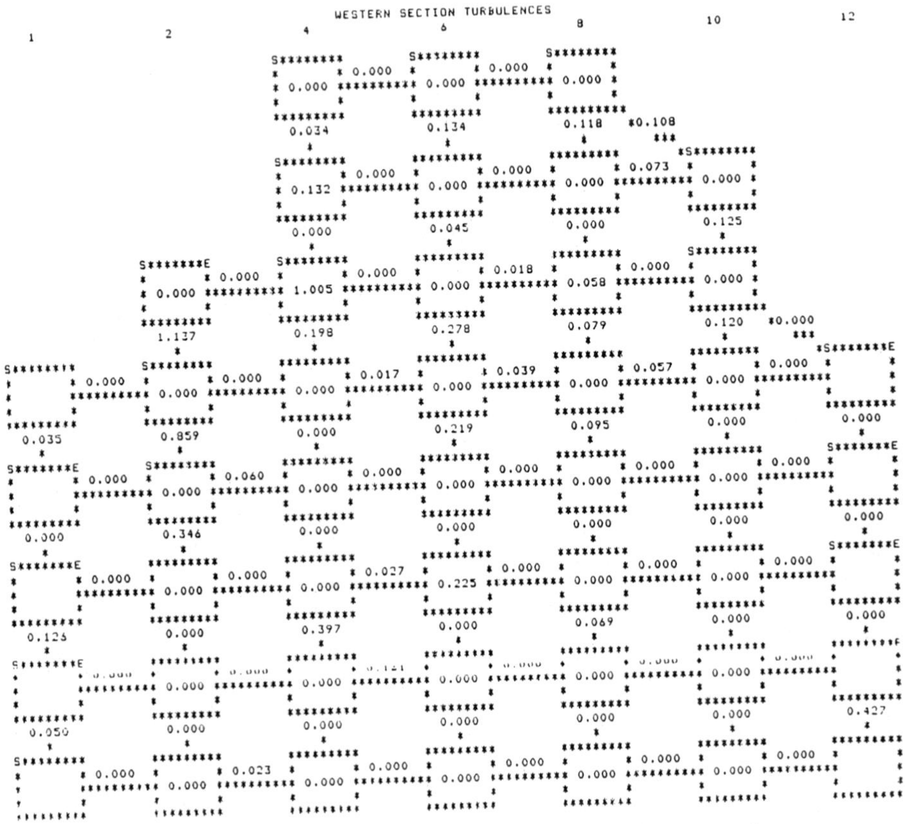

WESTERN SECTION TURBULENCES

1 2 4 6 8 10 12

FIGURE 16.11. A reference case steady-state solution for turbulence, in 10^{15} m^3 • yr^{-1}. **A:** Western section.

DENSITY AND REGIONALIZATION

From the point of view of predicting future atmospheric concentrations of CO_2, the most important purpose of regionalization is to separate more clearly water types with different ventilation times. Although density is a key parameter here, it should not be the only basis for a division of the sea. Water with the same density has indeed different radiocarbon ages in different places, which proves that the isopycnal layers are not well-mixed boxes (e.g., Broecker 1979; Stuiver 1978; Stuiver et al. 1983). Regionalization is certainly needed in the broad oceanic areas of the mid-latitudes, and to separate eastern basin waters from western. In other cases external regionalization may not be needed. For instance, given the detailed density structure north of 30°N, one could consider removing the regional structure (except for east–west separation) altogether. Actually, rather than removing the regional structure, the regionalization should be defined by density layers according to the rule that no region should have more than one density type for its surface water. For instance,

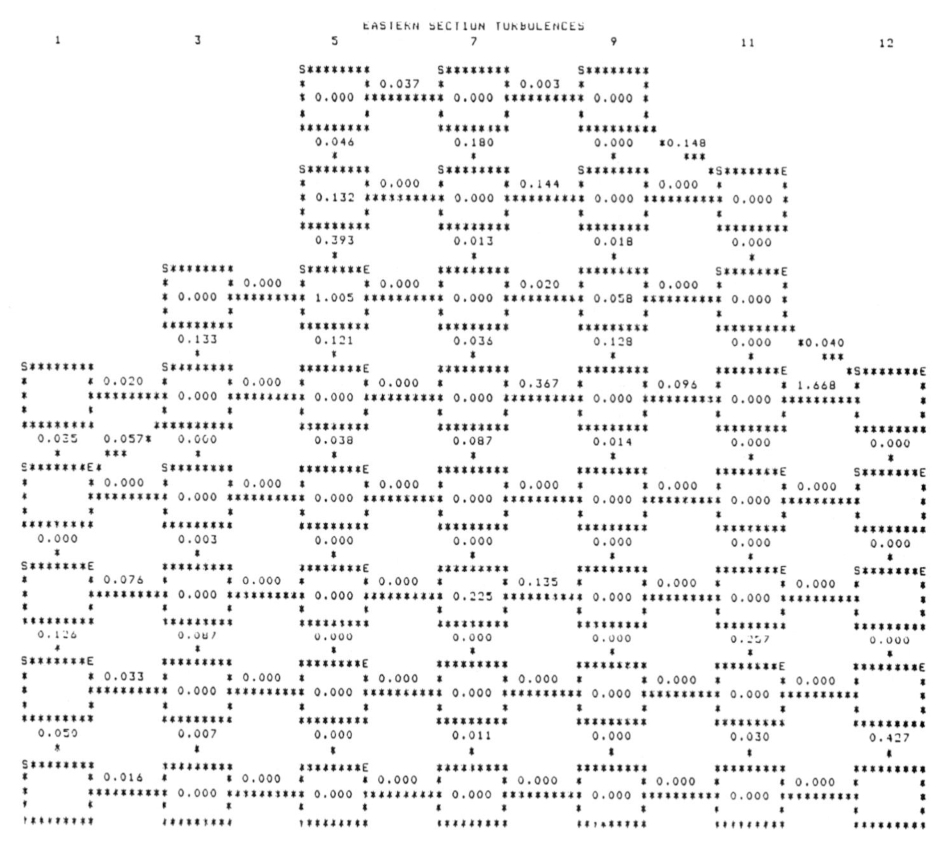

FIGURE 16.11. **B:** Eastern section.

region III would be divided into two regions: one with density layer 3 water at the surface, the other with density layer 4 water at the surface (cf. again Fig. 16.3). A consequence of this procedure is that the total number of boxes increases, as Fig. 16.13 illustrates.

The principal advantage is that this would (1) avoid the ambiguous definition of flow (cf. point 2 in the subsection Defining the Interfaces of the preceding section entitled Methodology) and (2) avoid averaging data for surface water with data for subsurface water (cf. point 3 in the subsection just referred to; it would eliminate the need for "gray" boxes). An example of the ambiguity is apparent in our brief discussion of the model solution. Does the flow in region II from layer 5 to layer 4 represent water upwelling or a surface flow of water southward? Extreme examples of problems in data averaging can be found in region IX. In region IX, there is no GEOSECS station where layer 2 water meets the surface (Fig. 16.8b); therefore, the data for layer 2 depends only on information from water that lies under layer 1 water. (Region IX, layer 2 is thus an example of "box B" in Fig.

WESTERN SECTION DETRITUS FLUXES

1	2	4	6	8	10	12
		S 0.000 / 0.000	S -0.345 / 0.000	S -0.176 / 0.000		
		S -0.099 / -0.006	0.075 / 0.002	0.175 / 0.000	S 0.000 / -0.002	
	S…E -0.200 / 0.000	S 0.018 / 0.013	0.000 / 0.000	0.000 / 0.000	S -0.124 / -0.002	
S 0.000 / 0.000	S -0.001 / 0.030	0.066 / 0.000	0.150 / 0.000	0.012 / 0.000	0.135 / 0.000	S…E -0.666 / -0.012
S…E -0.027 / -0.002	S 0.065 / -0.016	0.018 / 0.002	0.060 / 0.005	0.000 / 0.005	0.000 / 0.000	S…E 0.193 / 0.002
S…E 0.051 / 0.014	0.150 / 0.000	0.000 / 0.000	0.071 / 0.002	0.000 / 0.000	0.000 / 0.010	S…E 0.251 / 0.005
S…E -0.024 / 0.000	0.000 / 0.000	0.000 / 0.000	0.000 / 0.000	0.000 / 0.004	0.000 / 0.000	…E 0.232 / 0.014
S 0.011 / -0.003	0.000 / 0.000	0.066 / 0.002	0.000 / 0.000	0.000 / 0.000	0.000 / 0.005	0.000 / 0.000

FIGURE 16.12. A reference case steady-state solution for biological production and decomposition. The rates of change of calcium carbonate concentration are given in 10^{16} mol C per year, where production is shown as negative (sink) and decomposition is positive (source). **A:** Western section.

16.5.) As a consequence, layer 2 water in region IX shows negative (given as a positive number since it represents a source of carbon) net biological (new) production (Fig. 16.12b), even though layer 2 surface water in region IX is an area of intense upwelling and consequently high biological productivity. This example actually points out another difficulty with a more detailed regionalization which would reflect density differences; namely that the data are not available in GEOSECS for this better regionalization. On balance, it is likely to be more advantageous to have a more detailed regionalization and to use additional data sets or perhaps treat some of the concentrations as unknowns. In fact, additional data are already being used for high northern latitudes, and certainly it would be wrong to ignore the potentially rich data set that the Transient Tracers in the Ocean (TTO) program can provide (e.g., Jenkins, this volume; Brewer 1983, personal

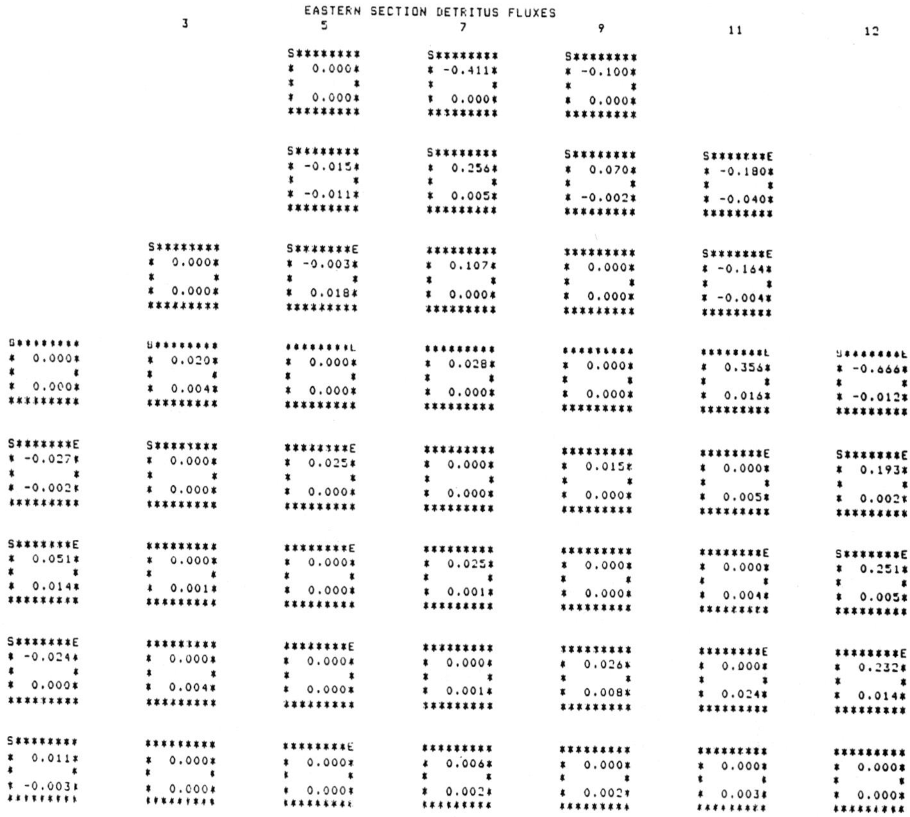

EASTERN SECTION DETRITUS FLUXES

	3	5	7	9	11	12
		(S) 0.000 / 0.000	(S) -0.411 / 0.000	(S) -0.100 / 0.000		
		(S) -0.015 / -0.011	(S) 0.256 / 0.005	(S) 0.070 / -0.002	(S, E) -0.180 / -0.040	
	(S) 0.000 / 0.000	(S, E) -0.003 / 0.018	0.107 / 0.000	0.000 / 0.000	(S, E) -0.164 / -0.004	
(S) 0.000 / 0.000	0.020 / 0.004	(L) 0.000 / 0.000	0.028 / 0.000	0.000 / 0.000	(L) 0.356 / 0.016	(L) -0.666 / -0.012
(S, E) -0.027 / -0.002	(S) 0.000 / 0.000	(E) 0.025 / 0.000	0.000 / 0.000	0.015 / 0.000	(E) 0.000 / 0.005	(S, E) 0.193 / 0.002
(S, E) 0.051 / 0.014	0.000 / 0.001	(E) 0.000 / 0.000	0.025 / 0.001	0.000 / 0.000	(E) 0.000 / 0.004	(S, E) 0.251 / 0.005
(S, E) -0.024 / 0.000	0.000 / 0.004	(E) 0.000 / 0.000	0.000 / 0.001	0.026 / 0.008	(E) 0.000 / 0.024	(E) 0.232 / 0.014
(S) 0.011 / -0.003	0.000 / 0.000	(E) 0.000 / 0.000	0.006 / 0.002	0.000 / 0.002	0.000 / 0.003	0.000 / 0.000

FIGURE 16.12. **B:** Eastern section.

communication). This issue of regionalization or geographical resolution is an important subject for further investigation.

GEOSTROPHIC RELATIONS

The data set (Levitus 1982*a,b*), which is used to determine the box volumes, surface area expressions, and potential interfaces, can be used also to establish geostrophic relationships between the vertical interfaces. When included in the matrix A, these expressions further overdetermine the system and, more importantly, capture critical oceanographic features that the tracers alone do not. In particular, the appropriate orientation and flux of major currents (e.g., Gulf Stream) are achieved in the solution when geostrophic relations are used (Bolin et al., to be published), whereas they are not accurately represented when only chemical tracers are used (Fig. 16.10). In an analogous way, it should be possible also to use surface wind stress data and exploit Ekman pumping.

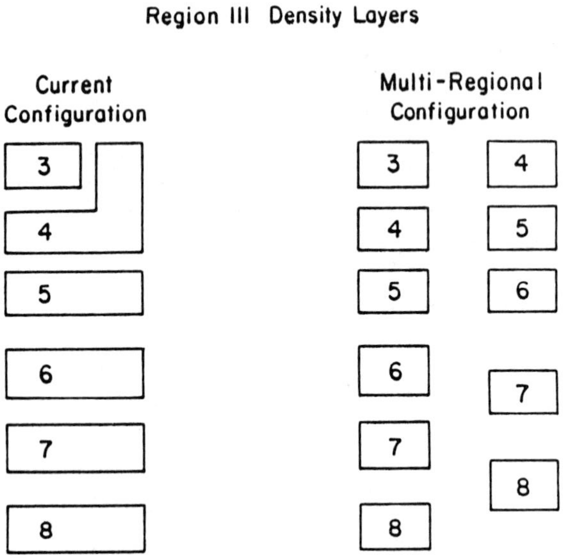

Region III Density Layers

Current Configuration

3 4 5 6 7 8

Multi-Regional Configuration

3 4 4 5 5 6 6 7 7 8 8

FIGURE 16.13. An example of the effect of a density-based regional pattern using the rule of one surface density type per region.

The simultaneous use of the properties of geophysical fluid dynamics with chemical tracers greatly enhances the general inverse methodology and will be explored further (see also Hoffert et al. 1984; Wunsch 1984).

PERTURBED TRACERS

To use contemporary chemical tracer data as if it were steady state is an assumption that is obviously not without risk. This is particularly true when one purpose of the resulting model is to examine the response of the atmosphere–ocean system to perturbations in the carbon forcing function, either CO_2 or ^{14}C. Therefore, almost by definition, the GEOSECS profiles of total carbon and ^{14}C are not in steady state. We have attempted to address this difficulty in two different ways. The ^{14}C data are treated by subjective adjustment based on earlier investigations in which the effects of nuclear testing could be removed or discounted (e.g., Broecker et al. 1960). Total carbon concentration, however, must be treated differently. One strategy that is being explored is first to invert the system as if GEOSECS data represented steady state, and then to time-integrate the model with a constant preindustrial atmospheric CO_2 concentration so that the ocean outgasses CO_2 until a new steady state is achieved. During the integration, it may be important to hold all flows to or from deep ocean boxes equal to zero (under the assumption that in the timeframe of 110 years carbon concentrations in these boxes have not been changed). Finally, the resulting new distribution of carbon instead of the GEOSECS total carbon data is used and the system is inverted again to obtain a new "steady-state" solution. Alternatively, one can attempt a back integration in which the atmospheric CO_2 declines to a preindustrial value over the interval 1980 to 1860 and the ocean model runs "backward" (all rates change sign). This is difficult to accomplish, because of numerical instabilities. Another approach that we are investigating is to capture in the b vector the non-steady-state character of GEOSECS data. This might be done by estimating the time rate of change of surface and mid-depth boxes and incorporating this in the right-hand side. The resulting system could be solved and either integrated forward to a steady-state preindustrial atmosphere or integrated backward, assuming that the technical difficulties can be resolved. The latter is an unlikely approach.

The general subject of the preindustrial condition is an important area of investigation that touches on the work of other investigators (e.g., Brewer 1978), and further work clearly needs to be done.

Overdetermined versus Underdetermined Solution

While the approach we are pursuing uses overdetermined systems, the issue of the limitations in existing data sets suggests that the character of underdetermined systems needs to be explored. In analyzing the sensitivity of an underdetermined 12-box model, we noted a tendency that small perturbations in the data could lead to discontinuous changes in the solution. Closer analysis of the phenomenon suggested (Björkström, in preparation) that this feature will occur often when minimizing the norm of $(Ax - b)$ in a situation where the homogeneous equation $Ax = 0$ has nonzero solutions. The instability arises because the least-norm minimum-error solution x_u usually does not meet the inequality constraints, and therefore some linear combination $\Sigma^i C_i x_{hi}$, where x_{hi} are solutions to the homogeneous equation, must be added to x_u to come to an acceptable solution

$$x_c = x_u + \sum_i C_i x_{hi}$$

without increasing the norm of $(Ax - b)$. If the elements of matrix A are slightly perturbed, the solution x_u will shift and, more important, the vectors x_h will also change. This affects the directions along which one can move using the homogeneous solution to go from x_u to the constrained solution, x_c. As a consequence, the resulting x_c can change greatly from one location in the permissible region to another one.

The stability of the solution to changes in the matrix A as a result of changes in the tracer profiles or changes in the boundary conditions needs further investigation in both underdetermined and overdetermined systems (see Wunsch and Minster 1982).

Summary

To understand more thoroughly the global carbon cycle as a whole and the oceanic sink in particular, it is increasingly evident that the heterogeneity of the various components must be considered. Past descriptions of the major subsystems within the cycle (e.g., oceans, terrestrial biota) using a few boxes linked by simple kinetics have been wrung dry of information. Future progress will require the explicit consideration of spatial heterogeneity and hence more spatially detailed models.

Oceanic models whose components represent actual oceanic regions are clearly necessary. This will require methodologies that are appropriate for systems with hundreds of boxes as well as techniques to extract consistent information from diverse global data sets. The existence of data and the use of data are fundamental to future progress; higher-resolution oceanic models must be built upon extensive sets of consistent oceanographic data. There are no shortcuts possible with regard to the requirement of data.

It would appear that the methodology of constrained inversion that we have presented is particularly well-suited to develop the spatially detailed global ocean models through logical exploitation of the important information contained in GEOSECS and Levitus (1982), the developing data set that TTO hopefully will provide, and region-specific oceanic data sets that exist around the world. Such geographically detailed ocean models should allow more accurate estimates of the ocean uptake of fossil fuel CO_2 over the last 100 years and, hence, indirectly an independent estimate of net exchange of CO_2 between the atmosphere and the biosphere. Obviously, such estimates are important benchmarks in our understanding of the carbon cycle generally and, hence, part of the foundation for any predictions about the future.

Acknowledgments

This research was supported by the National Science Foundation under contract DEB-81-10477 and by a grant from the National Aeronautics and Space Administration under contract NAGW-453 (B. Moore III) and by the Swedish Natural Science Research Council (NFR) under contract E-EG 223-111 (A. Björkström).

References

Bainbridge, A. E. 1981*a*. GEOSECS, Atlantic Expedition, Vol. 1, Hydrographic Data. U.S. Government Printing Office, Washington, DC.

Bainbridge, A. E. 1981*b*. GEOSECS, Atlantic Expedition, Vol. 2, Sections and Profiles. U.S. Government Printing Office, Washington, DC.

Baumgartner, A. and E. Reichel. 1975. The World Water Balance. Elsevier–North Holland, Amsterdam and New York.

Björck, Å. 1981. Least squares methods in physics and engineering. CERN—Organization Europeenne pour la Recherche Nucleaire. CERN [Rep.] 81:16, Geneva, Switzerland.

Björkström, A. 1980. On the inadequacy of one-dimensional ocean models for the global carbon cycle. Report CM-51. Department of Meteorology, University of Stockholm, Stockholm, Sweden.

Björkström, A. On numerical instability in underdetermined systems. (in preparation).

Bolin, B. 1983. Changing global biogeochemistry. In P. Brewer, ed., The Future of Oceanography, 50th Anniversary Volume, pp. 305–326. Woods Hole Oceanographic Institution. Springer-Verlag, New York.

Bolin, B., A. Björkström, K. Holmen, and B. Moore. 1983. The simultaneous use of tracers for ocean circulation studies. Tellus 35B:206–236.

Bolin, B., A. Björkström, K. Holmen, and B. Moore. The analysis of the general circulation of the ocean by the simultaneous use of physical, chemical, and biological data (in preparation).

Brewer, P. G. 1978. Direct observation of the oceanic CO_2 increase. Geophys. Res. Lett. 12:997–1000.

Broecker, W. S. 1979. A revised estimate for the radiocarbon age of North Atlantic deep water. J. Geophys. Res. 84:3218–3226.

Broecker, W. S., R. Gerard, M. Ewing, and B. C. Heezen. 1960. Natural radiocarbon in the Atlantic Ocean. J. Geophys. Res. 65:2903–2931.

Broecker, W. S., D. W. Spencer, and H. Craig. 1982. GEOSECS, Pacific Expedition, Vol. 3, Hydrographic Data. U.S. Government Printing Office, Washington, DC.

Bunker, A. F. 1980. Trends of variables and energy fluxes over the Atlantic Ocean from 1948 to 1972. Mon. Weather Rev. 108:720–732.

Bryan, K., S. Manabe, and R. C. Pacanowski. 1975. A global ocean-atmosphere climate model. Part II. The oceanic circulation. J. Phys. Oceanogr. 5:30–46.

Craig, H., W. S. Broecker, and D. W. Spencer. 1981. GEOSECS, Pacific Expedition, Vol. 4, Sections and Profiles. U.S. Government Printing Office, Washington, DC.

Ebensen, S. K. and Y. Kushmir. 1981. The heat budget of the global ocean: an atlas based on estimates from surface marine observations. Climate Research Institute Report No. 29, Oregon State University, Corvallis, Oregon.

Fiadeiro, M. E. 1982. Three-dimensional modeling of tracers in the deep Pacific Ocean: II. Radiocarbon and the circulation. J. Mar. Res.

Fiadeiro, M. E. 1983. Physical-chemical processes in the open ocean. In B. Bolin and R. B. Cook, eds., The Major Biogeochemical Cycles and Their Interactions, pp. 461–476. John Wiley & Sons, Chichester and New York.

Fiadeiro, M. E. and H. Craig. 1978. Three-dimensional modeling of tracers in the deep Pacific Ocean: I. Salinity and oxygen. J. Mar. Res. 36:323–355.

Fiadeiro, M. E. and G. Veronis. 1982. On the determination of absolute velocities in the ocean. J. Mar. Res. 40 (Suppl.):159–182.

Gordon, A. L. 1971. Oceanography at Antarctic waters. Antarctic Research Series 15, American Geophysical Union, pp. 169–203. National Academy of Sciences, Washington, DC.

Hanson, R. J. and K. H. Haskell. 1982. Two algorithms for the linearly constrained least-squares problem. ACM-Trans. Math. Software 8(3):323–333.

Haskell, K. H., and R. J. Hanson. 1979. Selected algorithms for the linearly constrained least-squares problem—a user's guide. Sand 78-1290. U.S. Department of Commerce, Springfield, Virginia.

Haskell, K. H. and R. J. Hanson. 1981. An algorithm for linear least-squares problems with equality and nonnegativity constraints. Math. Programming 21(1):98–118.

Hoffert, M. I., T. Volk, and C. T. Hsieh. 1984. A two-dimensional ocean model for climate and tracer studies. Report prepared for Department of Energy, NYU/DAS 83-114. New York University, New York.

Keeling, C. D. 1973. The carbon dioxide cycle: Reservoir models to depict the exchange of atmospheric carbon dioxide with the oceans and land plants. In S. I. Rasool, ed., Chemistry of the Lower Atmosphere, pp. 251–329. Plenum Press, New York.

Killough, G. G. and W. R. Emanuel. 1981. A comparison of several models of carbon turnover in the ocean with respect to their distributions of transit time and age, and response to atmospheric CO_2 and ^{14}C. Tellus 33:274–290.

Lawson, C. L. and R. J. Hanson. 1974. Solving Least-Squares Problems. Prentice Hall, Englewood Cliffs, New Jersey.

Levitus, S. 1982*a*. Climatological atlas of world ocean. National Oceanic and Atmospheric Administration Professional Paper 13:83–1480. Washington, DC.

Levitus, S. 1982*b*. Data tape on temperature salinity and oxygen available through Princeton University, Princeton, New Jersey.

Spencer, D., W. S. Broecker, H. Craig, and R. F. Weiss. 1982. GEOSECS, Indian Ocean Expedition, Vol. 6, Sections and Profiles. U.S. Government Printing Office, Washington, DC.

Stigebrand, A. 1981. A model for the thickness and salinity of the upper layer in the Arctic Ocean and the relationship between ice thickness and some external parameters. J. Phys. Oceanogr. 11:1407–1422.

Stuiver, M. 1978. Atmospheric carbon dioxide and carbon reservoir changes. Science 199:253–258.

Stuiver, M., P. D. Quay, and H. G. Ostlund. 1983. Abyssal water carbon-14 distribution and the age of the world ocean. Science 219:849–851.

Sverdrup, H. U., M. W. Johnson, and R. H. Fleming. 1942. Oceanography. Prentice Hall, Englewood Cliffs, New Jersey.

Veronis, G. 1975. The role of models in tracer studies. In Numerical Models of Ocean Circulation, pp. 133–145. Ocean Science Committee, National Academy of Sciences, Washington, DC.

Wunsch, C. 1978. The general circulation of the North Atlantic west of 50°W determined from inverse methods. Rev. Geophys. Space Phys. 16:583–620.

Wunsch, C. 1980. Meridional heat flux of the North Atlantic Ocean. Proc. Natl. Acad. Sci. USA 77:5043–5047.

Wunsch, C. 1984. An eclectic model of the Atlantic circulation, 1, model—part I. The meridional flux of heat. J. Phys. Oceanogr. 14:1712-1733.

Wunsch, C. and B. Grant. 1982. Towards the general circulation of the North Atlantic Ocean. Prog. Oceanogr. 11:1–59.

Wunsch, C. and J. F. Minster. 1982. Methods for box models and ocean circulation tracers: mathematical programming and non-linear inverse theory. J. Geophys. Res. 87:5647–5662.

17
Chemical and Biological Processes in CO_2-Ocean Models

CHARLES F. BAES, JR. AND GEORGE G. KILLOUGH

The amount of CO_2 that will ultimately be released to the atmosphere by mankind's use of fossil fuels will doubtless be several times the amount of CO_2 contained in the preindustrial atmosphere and equivalent to many times the amount of carbon that can be taken up by the terrestrial biosphere. Thus the oceans will become the sink for most of the excess carbon that does not remain in the atmosphere. To avoid too rapid or too great a rise in the atmospheric concentrations of CO_2 during this massive transfer of fossil carbon to the oceans, it will be necessary to predict reliably the rate at which the oceans will take up excess CO_2 from the atmosphere.

This prediction will require models of the ocean that are realistic enough to simulate the relevant processes that occur therein, yet simple enough to be verifiable in terms of the available information about the ocean. These processes include water circulation, the chemical and biological transformations of carbon, and the consequent movement of chemical and radiochemical tracers in the oceans. Most of the models developed thus far have been simple, either one-dimensional (depth) or two-dimensional (latitude $\times$ depth) representations of the oceans. However, none have been able to account adequately for the available data on the history of CO_2 in the atmosphere and the radiotracer data in the oceans (Björkström, Chapter 14, this volume). Moreover, few models have attempted to include in much detail the chemical and biochemical processes that are largely responsible for the nonuniform distribution of carbon and alkalinity in the oceans and, as a consequence, strongly affect the distribution of CO_2 between the ocean and the atmosphere (Baes 1982).

In this chapter, we present some steady-state results from what we consider to be the simplest CO_2-ocean model that can adequately represent the chemical, biochemical, and circulation processes of the oceans. The model has been described in more detail elsewhere (Baes and Killough 1985), including an account of how the various parameters involved were selected to obtain the results depicted in Figs. 17.1 and 17.2. In the present calculations we will explore separately the effects of photosynthesis, nutrient concentration, and increased atmospheric CO_2 on the distribution

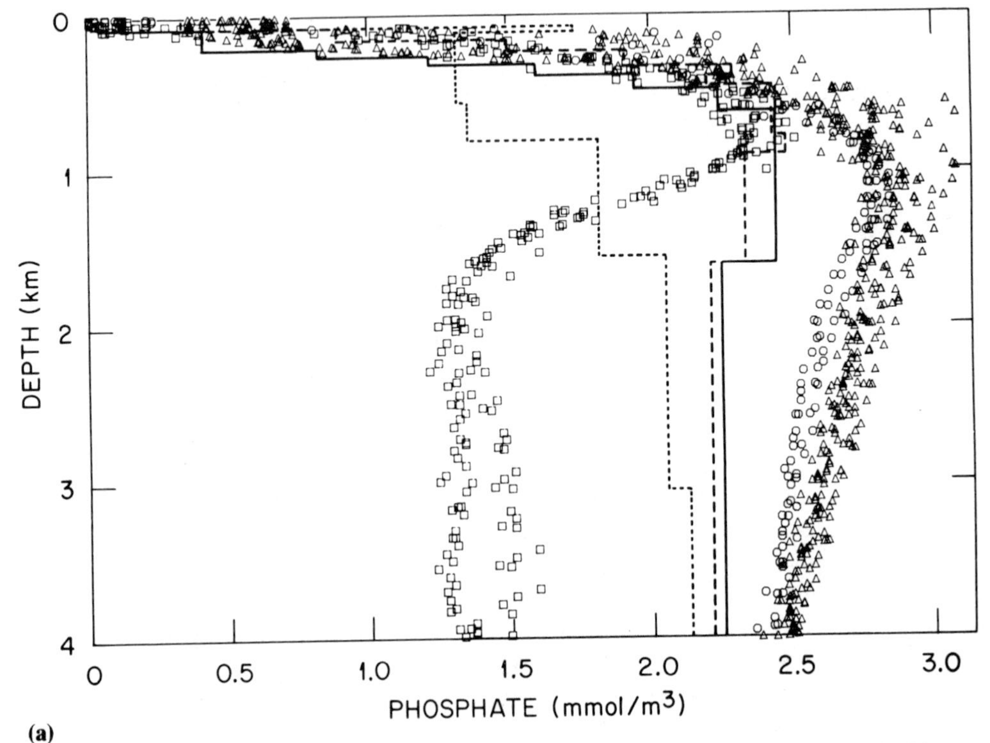

FIGURE 17.1. Comparison of observed depth profiles of (**a**) nutrient, (**b**) *TC*, (**c**) *TA*, and (**d**) oxygen with profiles generated by the model from the set of parameters in Table 17.3. The observed data are from GEOSECS stations within 5° of the equator in the Atlantic (*squares*), the Indian (*circles*), and the Pacific (*triangles*). They have been normalized to a salinity of 35 g • kg^{-1} and converted from the original weight concentrations to volume concentrations using a density of 1.027 g • cm^{-3}. The calculated profiles come from sections through the model at 0° (*solid*), 30° (*dashed*), and 50° (*dotted*) latitude.

of carbon between ocean and atmosphere. The results should make it clear that it is essential to include these processes in models used to represent the time dependence of the response of the oceans to increasing atmospheric CO_2.

The Model

The model (Fig. 17.3) consists of an array of boxes arranged in isothermal regions intended to simulate approximately the form of isotherms in a meridional section of the oceans. Each region "outcrops" at surface boxes that exchange heat and CO_2 with the atmosphere. Water flow is permitted in both directions between all adjacent boxes, with the exception of the

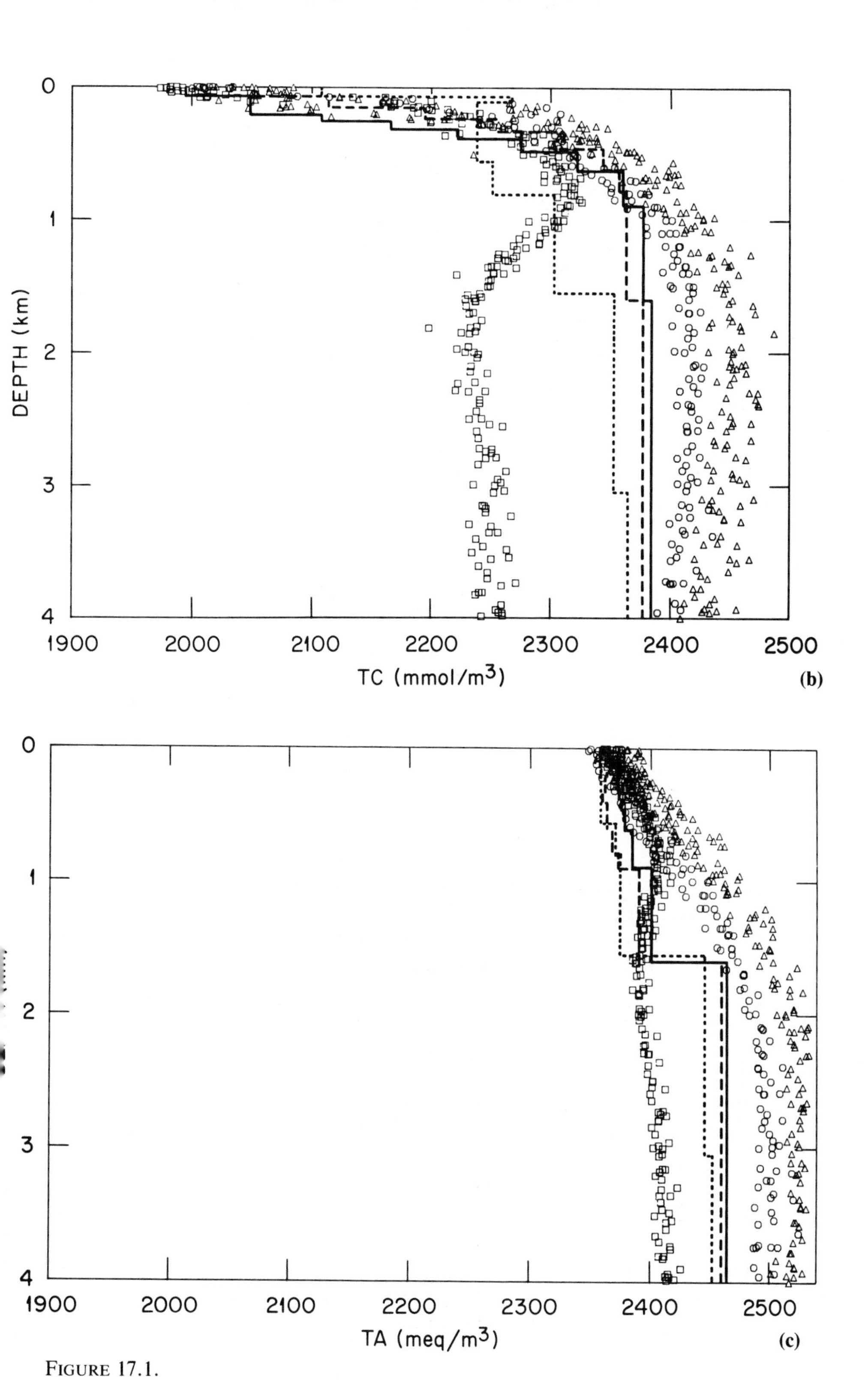
0
1
2
3
4
DEPTH (km)
1900
2000
2100
2200
2300
2400
2500
TC (mmol/m3)
(b)
0
1
2
3
4
1900
2000
2100
2200
2300
2400
2500
TA (meq/m3)
(c)

FIGURE 17.1.

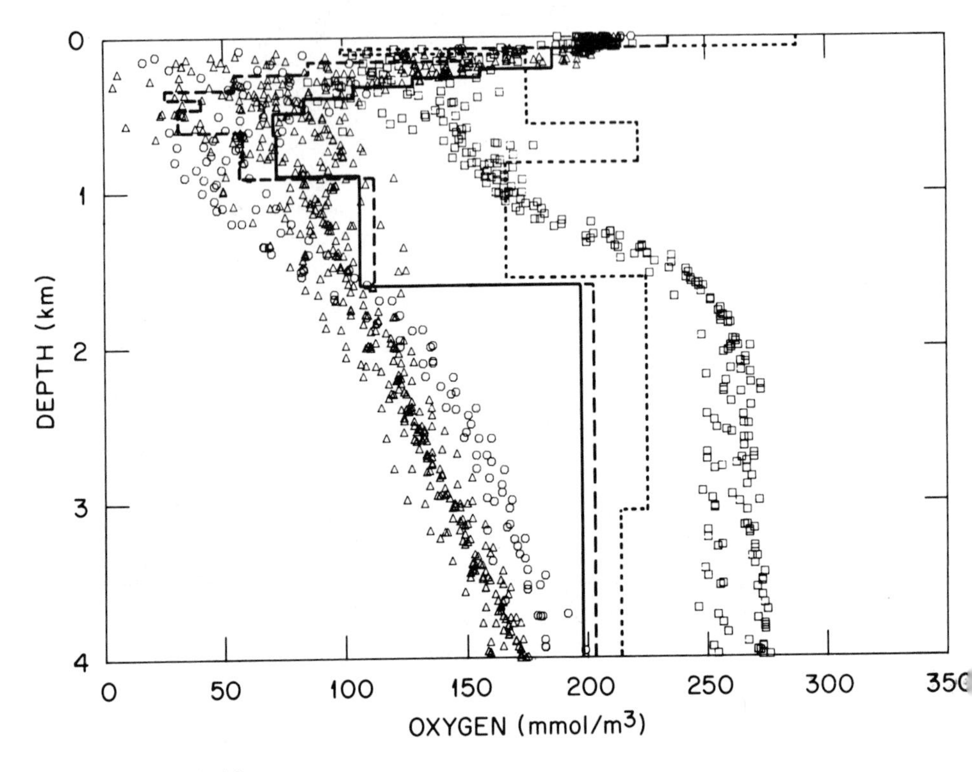

FIGURE 17.1. (d)

coldest boxes. Here, flow from warmer boxes would violate the heat balance that is maintained about each subsurface box.

The many flows in the model are set by means of flux parameters, K ($m^2 \bullet s^{-1}$), which, to the extent possible, are like those of Oeschger et al. (1975), namely, proportionality constants between the flow rate, F_{ij} ($m^3 \bullet s^{-1}$), and the ratio of the interfacial area to the center-to-center distance for adjacent boxes i and j:

$$F_{ij} = K \bullet A_{ij} / D_{ij} \tag{1}$$

Three flux parameters are used, one (K_d) for diathermal flows between boxes, one (K_a) for isothermal flows, and one (K_s) for flows between surface boxes. The heat balance about each box requires a modification of Eq. (1) for flows between boxes of different temperature. In the case of isothermal (or advective) flows, the interfacial areas A_{ij} are taken to be the vertical projections of the nearly horizontal interfaces. To the equal and opposite isothermal flows generated by use of Eq. (1), a residual flow is added in one direction to preserve the water balance about each box.

Within each box of the model, the concentration of seven quantities is calculated:

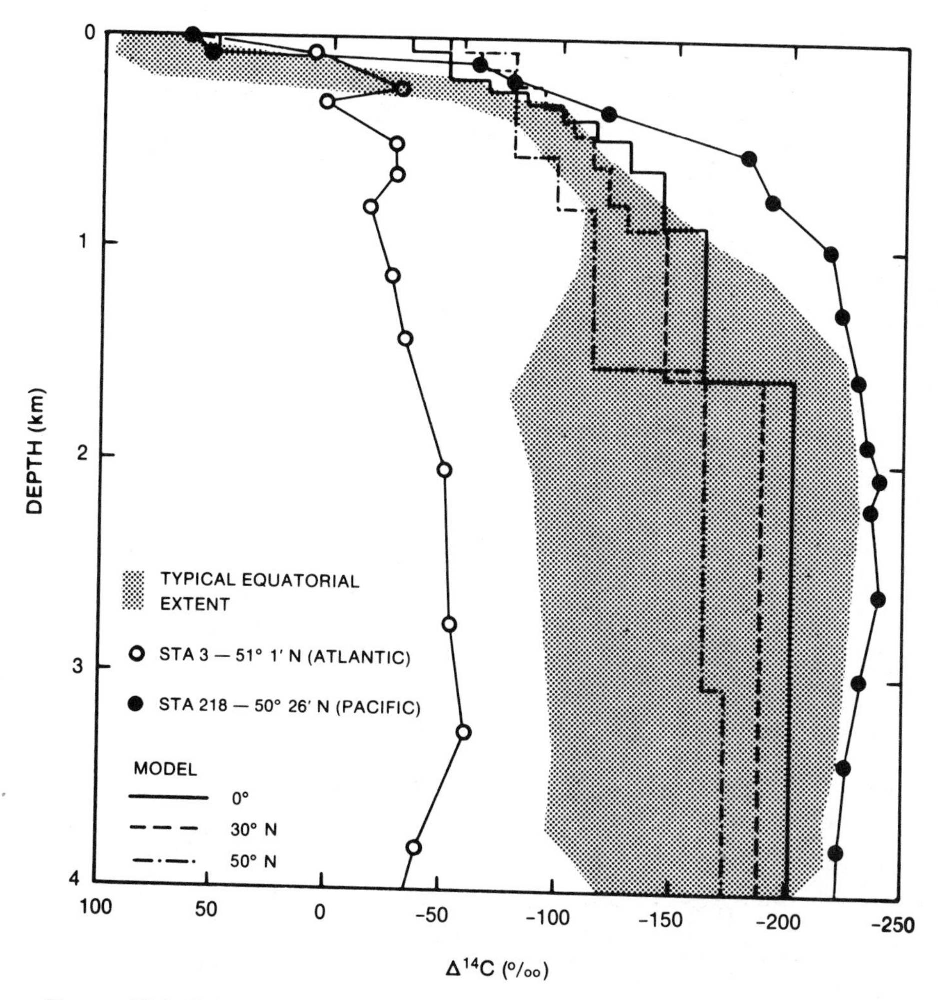

FIGURE 17.2. Comparison of observed depth profiles of ^{14}C with profiles generated by the model from the set of parameters in Table 17.3. The observations (*shaded area and points*) are from the GEOSECS data. The values of $\Delta^{14}C$ in surface waters before nuclear testing began averaged about −50‰ (Oeschger et al. 1975).

CO, suspended organic carbon (mmol • m^{-3})
CI, suspended inorganic carbon (mmol • m^{-3})
CN, dissolved nutrient (mmol • m^{-3})
TC, dissolved inorganic carbon (mmol • m^{-3})
TA, total alkalinity (meq • m^{-3})
O_2, dissolved oxygen (mmol • m^{-3})
^{14}C, the deficit of ^{14}C (‰)

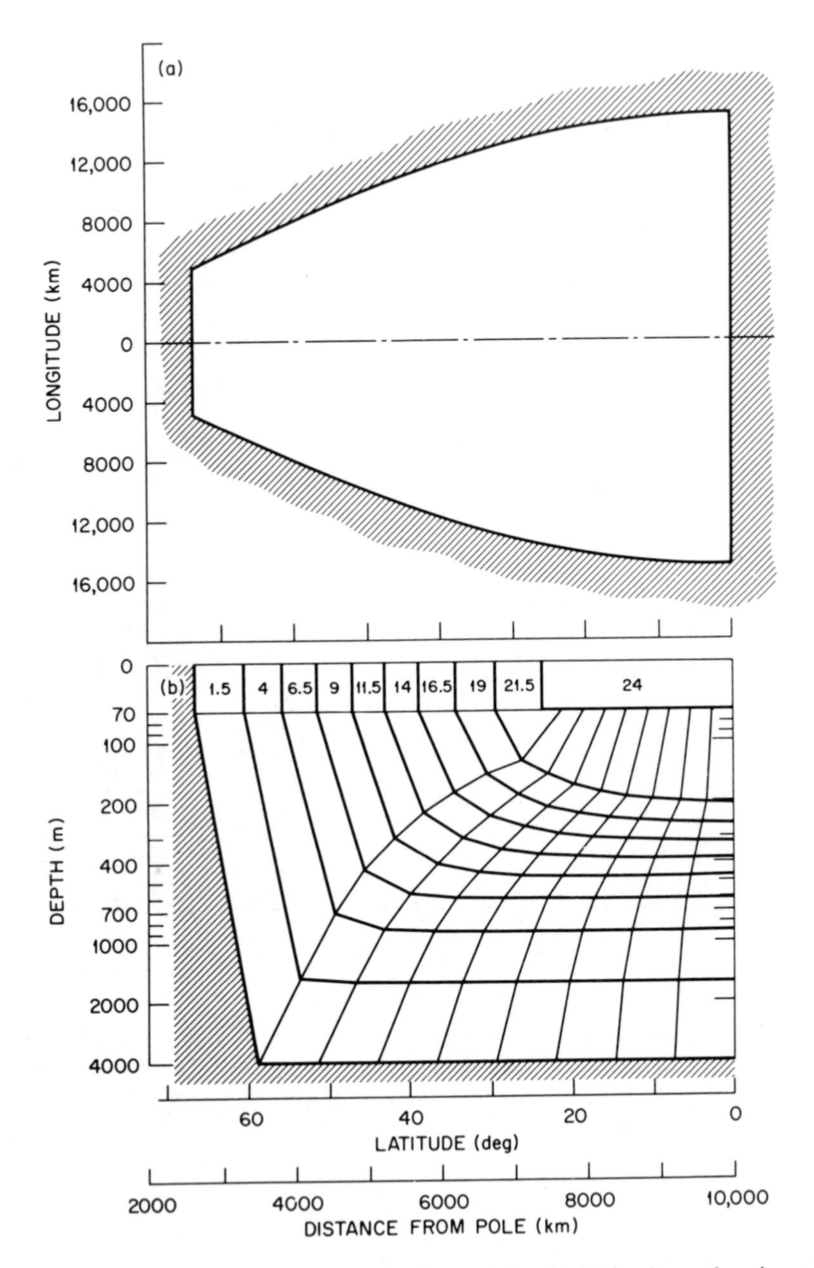

FIGURE 17.3. Arrangement of boxes in the model, with 10 isothermal regions, 10 boxes in each region, a low temperature of 1.5°C, and a temperature increment of 2.5°C. The warmest region is consolidated into a single box.

The values of these quantities are affected by one or more of the following processes (Tables 17.1 and 17.2):

1. Exchange of CO_2 between the atmosphere and surface boxes. The rate of transfer in either direction is assumed proportional to the interfacial area and the partial pressure of CO_2 in the region from which flow originates.
2. Net photosynthesis. Production of organic matter occurs only in surface boxes according to the reaction

$$106\ CO_2 + 16\ NO_3^- + H_2PO_4^- + 17\ H^+ + 122\ H_2O \rightleftharpoons (CH_2O)_{106}\ (NH_3)_{16}\ (H_3PO_4) + 138\ O_2 \tag{2}$$

   based on the ratios of C:N:P = 106:16:1 in marine organic matter reported by Redfield et al. (1963). The rate is assumed proportional to concentration of the limiting nutrient CN. This could be either nitrate or phosphate. In the model it is taken to be nitrate.
3. Production of $CaCO_3$. This is assumed proportional to the production of organic carbon.
4. Fallout of detritus. The suspended organic and inorganic carbon produced in the surface boxes falls at a specified velocity into lower-lying boxes.
5. Oxidation (decay) of organic carbon. This occurs in subsurface boxes and is assumed proportional to the amount of organic carbon.
6. Dissolution of $CaCO_3$. This occurs only in the deepest boxes and the rate is assumed proportional to the depth below a specified compensation depth.
7. Exchange and decay of ^{14}C. Radiocarbon enters each surface box from the atmosphere and is lost by the reverse path. All the processes described above are applied to ^{14}C with the same rate constants (no isotope effect is assumed), with the addition of a decay term for each form in each box.
8. Transport by water flow. The flow of water between boxes affects all seven of the quantities listed above by supplying and removing material from each box. The rates are given by the product of water flow and the concentration in the box from which the flow originates.

The various rate expressions used are listed in Table 17.1, and the specific rate constants are listed in Table 17.2. All of the rate terms were made linear in the unknown concentrations to be determined. In the case of CO_2 transfer between ocean and atmosphere, this condition was met by expressing the difference in the CO_2 partial pressures as an empirical function linear in TC and TA. The approximate values obtained were usually in error by $<10\%$.

For the steady-state calculations performed thus far with this model, the following procedure was used:

TABLE 17.1. Chemical Processes in the Model.[a]

Quantity (mmol · m^{-3})	Surface boxes: Production	Surface boxes: Consumption	Subsurface boxes: Production*	Subsurface boxes: Consumption†
CO	*GP•CN*	*DOI•CO* *DOJ•CO*	$DOI_1 \bullet CO_1$ $DOJ_1 \bullet CO_1$	*DOI•CO* *DOJ•CO* *SOD•CO*
CI	*RI•GP•CN*	*DII•CI* *DIJ•CI*	$DII_1 \bullet CI_1$ $DIJ_1 \bullet CI_1$	*DII•CI* *DIJ•CI* *SID•CI*
CN		(16/106)•*GP•CN*	(16/106)•*SOD•CO*	
TC	*AP*‡ *BP•TC*‡ *CP•TA*‡	*GP•CN* *RI•GP•CN*	*SOD•CO* *SID•CI*	
TA	(17/106)•*GP•CN*	2•*RI•GP•CN*	2•*SID•CI*	(17/106)•*SOD•CO*
O_2	(138/106)•*GP•CN*			(138/106)•*SOD•CO*

[a]The expressions for *TA* are in kiloequivalents per second. See Tables 17.2 and 17.3 for definition of rate coefficients.
*The subscript 1 refers to overlying boxes.
†Detrital fallout terms were zero for the deepest boxes.
‡These terms give the rate of takeup of CO_2 from the atmosphere (Baes and Killough 1985, Table 3).

TABLE 17.2. Formulation of Rate Coefficients (10^6 $m^3 \cdot s^{-1}$).

Coefficient	Description	Definition	Rate constant
GP	Gross photosynthetic productivity	Rate constant (s^{-1}) times box volume (10^6 m^3)	*GK* (I)
	Detrital fallout	Velocity constant ($m \cdot s^{-1}$) times horizontal area of interface (km^2)	
DOI	Organic to adjacent colder box		*VO*
DOJ	Organic to adjacent isothermal box		*VO*
DII	Inorganic to adjacent colder box		*VI*
DIJ	Inorganic to adjacent isothermal box		*VI*
SOD	Oxidation of organic detritus	Rate constant (s^{-1}) times box volume (10^6 m^3)	*SO*
SID	Dissolution of inorganic detritus	Rate constant ($s^{-1} \cdot m^{-1}$) times box volume (10^6 m^3) times depth below compensation depth (m)	*SI*

TABLE 17.3. Parameter Values.

Parameter	Description	Value	Units
pCO_2	Partial pressure of CO_2 in atmosphere	300	μatm
TA (1,1)	Total alkalinity in deepest subequatorial box	2464.8	$meq \cdot m^{-3}$
CN (1,1)	Dissolved nitrate in deepest subequatorial box	36	$mmol \cdot m^{-3}$
CD	Compensation depth	1500	m
KP	Rate constant for CO_2 exchange	2×10^{-6}	$mmol \cdot m^{-2}$ $s^{-1} \cdot \mu atm^{-1}$
K_d	Eddy diffusion coefficient for diffusive flows	4×10^{-5}	$m^2 \cdot s^{-1}$
K_a	Eddy diffusion coefficient for advective flows	1400	$m^2 \cdot s^{-1}$
K_s	Eddy diffusion coefficient for surface box	1400	$m^2 \cdot s^{-1}$
GK	Rate constant for photosynthesis	0.4×10^{-6}	s^{-1}
RI	Inorganic/organic carbon production rate	0.05	s^{-1}
SO	Rate constant for oxidation of organic carbon	8×10^{-7}	s^{-1}
		10^{-6}	
SI	Rate constant for $CaCO_3$ dissolution	4×10^{-9}	$s^{-1} \cdot m^{-1}$
VO	Organic detritus fallout velocity	4×10^{-4}	$m \cdot s^{-1}$
VI	Inorganic detritus fallout velocity	4×10^{-4}	$m \cdot s^{-1}$

1. Values were selected for all the parameters listed in Table 17.3, including *CN* and *TA* in the deepest subequatorial box.
2. The remaining values of *CO, CI, TC,* and *TA* were determined in all the boxes. With 10 isothermal regions and 10 boxes in all but one of them (Fig. 17.3), there were $(5 \times 91) - 2 = 453$ equations in a like number of unknowns to be solved. The resulting 453×453 matrix was a banded one, however, and could be converted to a 151×453 matrix for solution.
3. The values of O_2 in each box were next determined. Air-saturated values were specified in the surface boxes, and these were combined with *CN* values to calculate the conservative quantity

$$CNO = CN + (16/106) \bullet O_2$$

for each surface box. *CNO* is unaffected by photosynthesis and decay, because the accompanying changes in the concentrations of nutrient and oxygen are in opposite directions and are related by a stoichiometry factor (from Eq. 2). Since *CN* and O_2 in subsurface boxes are not affected by any of the other processes, the quantity *CNO* is altered only by water flow. Consequently, it could be determined in each subsurface box by use of a much smaller matrix than previously. From the results and the previously known values of *CN*, the remaining values of O_2 were obtained.

4. The deficit of ^{14}C in each box was determined. Since ^{14}C is associated only with the quantities *CO, CI,* and *TC,* the total amount in each box could be calculated from a set of $3 \times 91 = 273$ simultaneous linear equations involving a like number of unknowns. The coefficients involved the previously determined values of *CO, CI, CN, TC,* and *TA* and the specified rate parameters.

Ocean Simulations

The profiles versus depth shown for *CN, TC, TA,* O_2, and ^{14}C in Figs. 17.1 and 17.2 were generated by the model with the parameter values listed in Table 17.3. As can be seen, they are approximate averages of the observed data for the world's oceans. The systematic increases in *CN, TC,* and *TA* and the decrease in O_2 and ^{14}C in deep water as one proceeds from the Atlantic to the Indian to the Pacific Oceans are caused by a deep circulation in this direction and cannot be represented in a two-dimensional model such as this.

The parameters controlling circulation between surface boxes (K_s) and the fallout of inorganic detritus (*VI*) had no appreciable effect on these profiles, and hence they were placed equal to more important parameters (K_a and *VO,* respectively). Moreover, the ratio of two parameters *(VO/SO)* which corresponds to the *e*-folding depth over which organic detritus is oxidized was found to be far more important than either one. Conse-

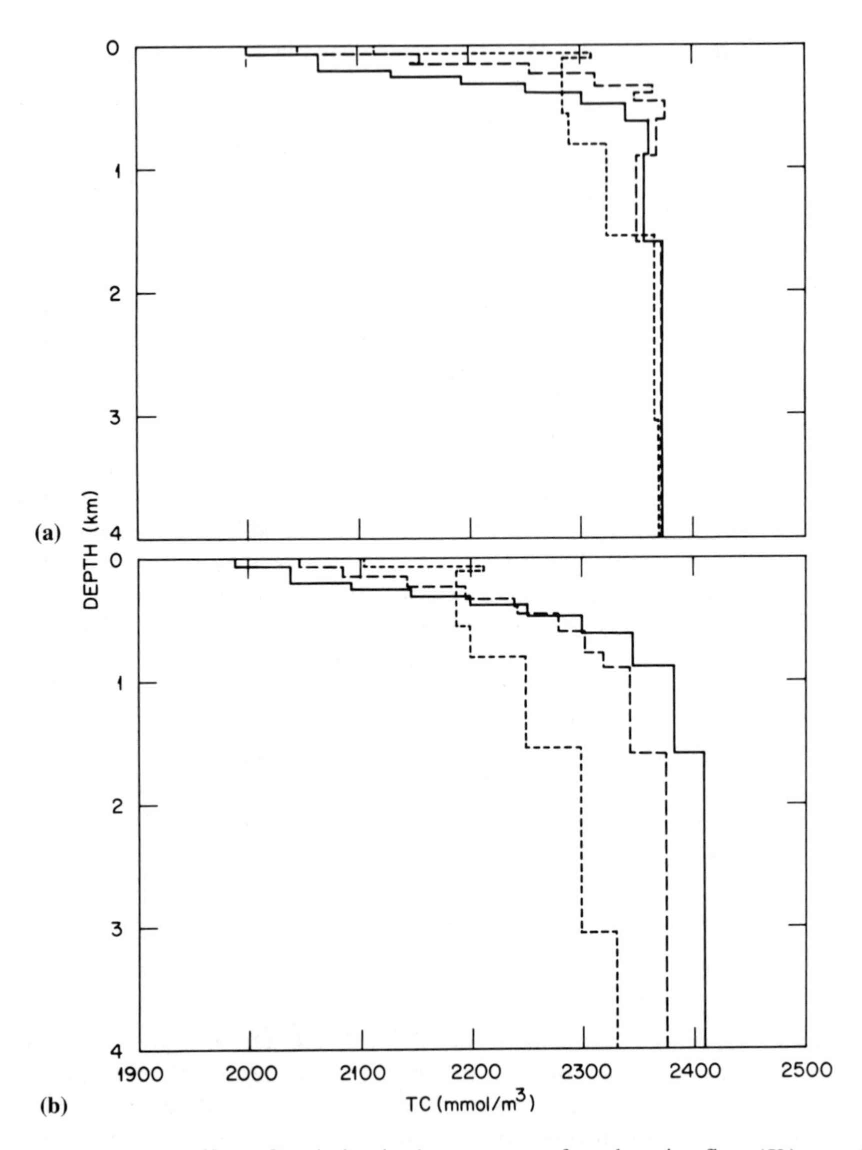

FIGURE 17.4. Effect of variation in the parameter for advective flow (K_a) on calculated TC and O_2 profiles. **A, C**: K_a = 3300 m^2 • s^{-1}. **B, D**: K_a = 500 m^2 • s^{-1}. Other parameters were adjusted (Table 17.4) to preserve a reasonable fit to the data in

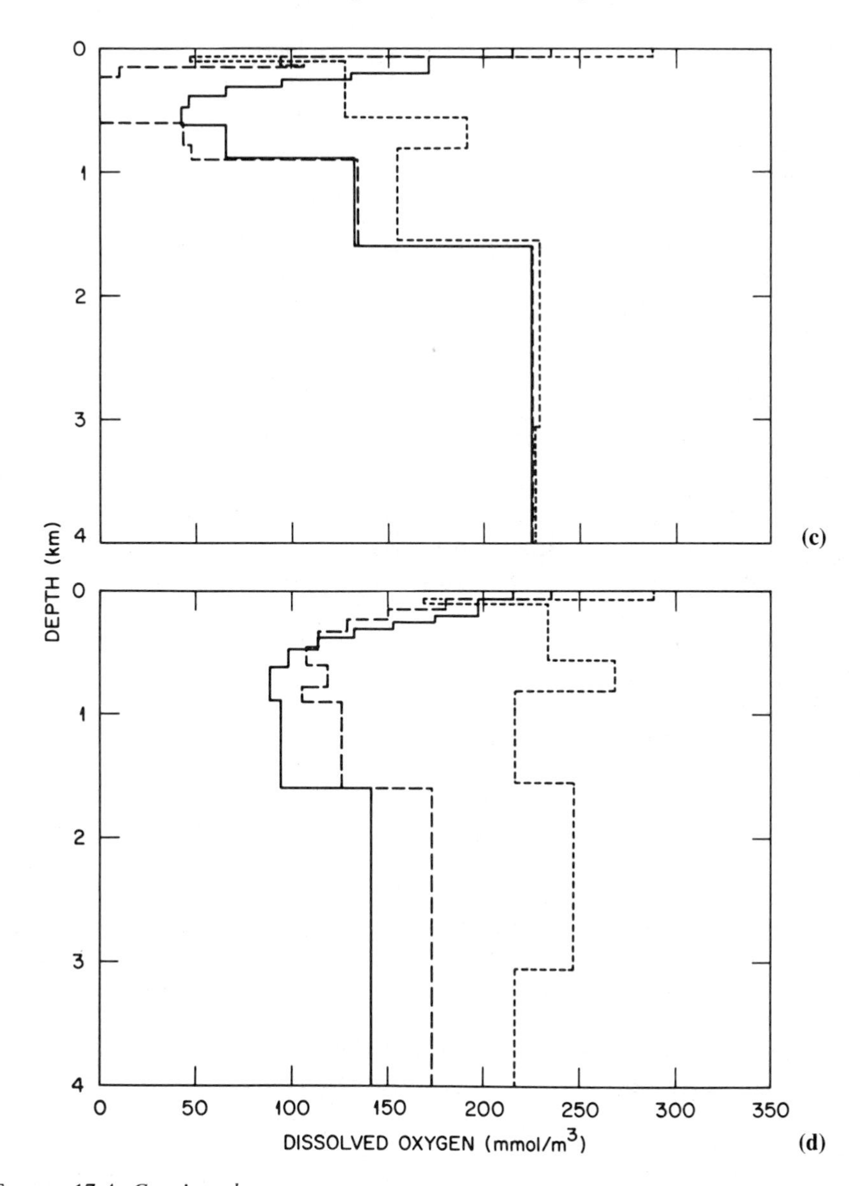

FIGURE 17.4. *Continued*

quently, only the following six parameters were adjusted to obtain this fit: *PC*, K_d, K_a, *GK, RI,* and *VO/SO*. The first, the rate constant for CO_2 transfer between ocean and atmosphere, was determined almost entirely from the reported ^{14}C content of surface waters before nuclear testing began (Oeschger et al. 1975). Because of a scaling effect among all these parameters except *RI,* only four could be independently varied to adjust the profiles of *CN, TC, TA,* and O_2. The scaling factor was determined almost entirely by the deep-water content of ^{14}C.

TABLE 17.4. The Effect of Varying the Advective Flow Parameter.

K_a ($m^2 \cdot s^{-1}$)	*RI*	*VO/SO* (m)	*GK* (s^{-1})
500	0.1	1000	0.1×10^{-6}
1400*	0.05	500	0.4×10^{-6}
3300	0.025	250	1.0×10^{-6}

As K_a varied, the parameters *RI* and *GK* and the ratio *VO/SO* were adjusted as shown to preserve a reasonable fit with observed oceanographic data (Figs. 17.1 and 17.4). The parameter K_s was made equal to K_a. All other parameters remained at the reference values in Table 17.3.
*Reference case.

Advective versus Diffusive Flow

It has often been argued that most of the effects important in the takeup of CO_2 by a model ocean can be simulated by diffusive flow in one (depth) dimension. Although this may prove true, the magnitude of the diffusivity required will probably be determined by use of two- or three-dimensional models. In any case, the present model permits us to examine how the distributions of nutrients, carbon, alkalinity, and oxygen in a two-dimensional model ocean are affected when the ocean is made more or less advective.

The results of two such calculations are shown in Fig. 17.4. The magnitude of K_a was varied by a substantial amount in either direction, and the other parameters were adjusted to produce about the same averages for each set of profiles (Table 17.4). As can be seen, when the amount of advection is increased, the latitude dependence of each quantity is reduced (Fig. 17.4a versus Fig. 17.1b) and the oxygen minimum at about 400 m is strengthened (Fig. 17.4c versus Fig. 17.1d). The two extremes shown here represent the approximate limits within which acceptable average chemical profiles could be obtained. If a model such as the present one were applied to a single ocean, narrower limits could be placed on the range of K_a. This exercise demonstrates the value of the observed chemical profiles in determining the nature of ocean circulation.

A Dead Ocean

A series of runs were made in which the processes of photosynthesis and precipitation of calcium carbonate were turned off (by setting *GK* equal to 0). With all other parameters in the model at the reference values, the partial pressure of CO_2 in the atmosphere and both *CN* and *TA* in the deepest subequatorial box (three boundary conditions) were adjusted in successive trials to give the same total amounts of nutrient and alkalinity in the ocean and the same total amount of carbon in the ocean plus the atmosphere as in the reference calculation (Table 17.5). The resulting partial pressure of CO_2 in the atmosphere was 534 μatm, about 80% higher

TABLE 17.5. Other Steady-State Simulations.

Condition	GK (s^{-1})	K_d ($m^2 \cdot s^{-1}$)	K_a ($m^2 \cdot s^{-1}$)	Average nutrient ($mmol \cdot m^{-3}$)	Average alkalinity ($meq \cdot m^{-3}$)	Average carbon ($mmol \cdot m^{-3}$)	pCO_2 (μatm)
Reference case	4×10^{-7}	4×10^{-5}	1400	33.5 (36.0)	2428.1 (2464.8)	2343.2	300
Dead ocean	0	4×10^{-5}	1400	33.5 (33.5)	2428.1 (2428.1)	2310.7	534
Increased nutrient	4×10^{-7}	4×10^{-5}	1400	67.0 (72.0)	2428.1 (2501.5)	2363.3	156
Decreased circulation	4×10^{-7}	2×10^{-5}	700	33.5 (36.3)	2428.1 (2466.8)	2347.8	267

The value of K_s in each case was equal to that assigned to K_a. Other parameters not listed were set at the reference values in Table 17.3. The boundary conditions CN(1,1), TA(1,1), and pCO_2 were adjusted to produce the indicated average concentrations of nutrient, alkalinity, and carbon. These averages include totals of all forms of each divided by the volume of the model ocean (6.55×10^{17} m^3). The values of $CN(1,1)$ and $TA(1,1)$ are shown in parentheses. The total amount of carbon in the half-ocean plus that in half the atmosphere (containing 9.10×10^{20} mol of all gases) is the same in all cases: $19{,}084 \times 10^{15}$ g of carbon (10^{15} g = 1 Gt).

than the reference value. This is consistent with recent more qualitative estimates by Baes (1982) and Broecker (1983) of the effect of a dead ocean on atmospheric CO_2.

The resulting profiles of *TC* and *TA* with depth are shown in Fig. 17.5a and b. There is no variation of *TA* with depth or latitude, because there is no longer any process in the model that alters *TA*. There is still a depth and latitude dependence of *TC*, because of the temperature dependence of the distribution of CO_2 between the surface waters and the atmosphere, but the variation is much less than in the reference case (Fig. 17.1). The resulting elevated *TA* values in the surface boxes tend to lower the partial pressure of CO_2, but the elevated *TC* values more than offset this effect to produce a net loss of CO_2 to the atmosphere.

The elevated partial pressure of CO_2 thus calculated for a dead ocean may be regarded as a minimum estimate, since inorganic precipitation of calcium carbonate, which would release additional CO_2 to the atmosphere, was not provided for in the model. It would be highly desirable to include this process, but this will not be easy to do credibly.

Increased Nutrient

In another series of runs, the total nutrient (nitrate) in the model ocean was doubled. Again with all other parameters in the model at the reference values, the remaining boundary conditions (the partial pressure of CO_2 in the atmosphere and the alkalinity) were adjusted until the total amounts of carbon and alkalinity were the same as in the reference case. The resulting partial pressure of CO_2 was 156 μatm, about half the reference value. This supports the argument of Broecker (1982) that an increase in limiting nutrient concentration in the ocean would lower the partial pressure of CO_2 in the atmosphere.

The resulting profiles of *TC* and *TA* (Figs. 17.5c and d) show effects opposite to those in the previous calculation. Compared with the reference values, *TA* is lower in surface boxes, tending to increase pCO_2, but *TC* is lowered even more, offsetting the effect of TA and producing a reduced partial pressure of CO_2 in the surface boxes. The minima in the dissolved oxygen profiles are more extreme; in fact all the oxygen in portions of the profiles at 0 and 30° latitude was consumed, indicating that these regions of the ocean would become anoxic.

Decreased Circulation

In another series of runs, the parameters K_d, K_a, and K_s were halved while all other parameters were held at the reference values and the boundary conditions were adjusted to bring the total amounts of carbon, nutrient, and alkalinity to the reference values. The partial pressure of CO_2 in the atmosphere thus obtained was 267 μatm, down about 11% from the reference value.

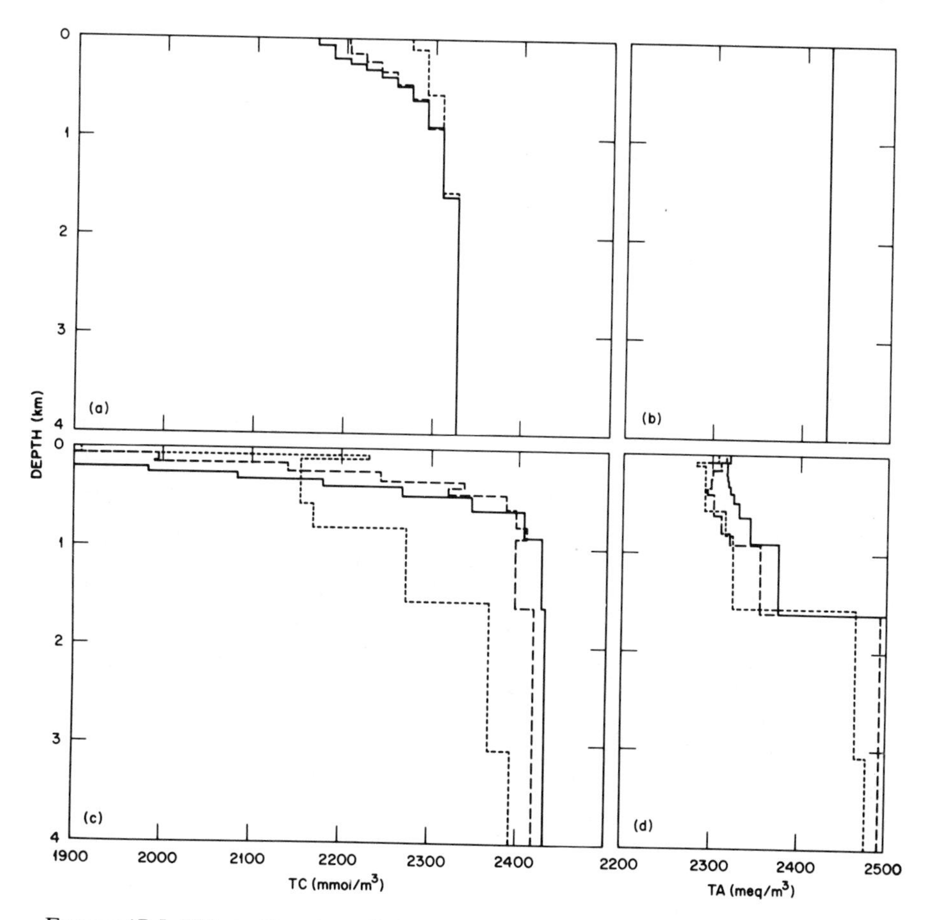

FIGURE 17.5. Effect of turning off photosynthesis (**a, b**) and of doubling the nutrient concentration (**c, d**) on profiles of *TC* and *TA* (Table 17.5).

The resulting profiles are similar to the reference case, the most important effect being a slight lowering of *TC* in the upper ocean. While the decreased circulation reduces biological productivity, it also lowers the rate of replacement of inorganic carbon consumed by photosynthesis and calcium carbonate precipitation.

Since a decrease in ocean circulation would presumably be caused by a general warming of world climate and a decrease in the meridional temperature gradient, this calculation should have included changes in the temperature distribution in the model ocean. No attempt was made to do this, however, since the relationship between temperature gradient and circulation rate in this primitive model is not evident. It can be predicted that a general warming would have the effect of increasing the partial pressure of CO_2 in the atmosphere. Hence, the present result may be regarded as a lower limit. The feedback effect of reduced circulation will probably be a small one.

TABLE 17.6. Steady States at Elevated Levels of CO_2.

pCO_2 (μatm)	Total carbon (Gt)*		Excess carbon (Gt)*			Airborne fraction
	Atmosphere	Ocean	Atmosphere	Ocean	Total	
300	656	36,860	0	0	0	
450	983	37,950	327	1090	1420	0.23
600	1311	38,690	656	1830	2490	0.26
900	1967	39,690	1311	2830	4140	0.32

All parameters other than pCO_2 are those listed in Table 3 for the reference case.
*1 Gt = 10^{15} g.

Increasing Atmospheric CO_2

In a final series of steady-state runs the partial pressure of CO_2 in the atmosphere was increased and the resulting carbon content of the oceans was determined. Again, all other parameters, total nutrient, and total alkalinity were held at the reference values (Table 17.3).

The results of these runs (Table 17.6) suggest that if some 4000 Gt (1 Gt = 10^{15} g) of fossil carbon are ultimately released, a relatively conservative estimate (see Masters this volume), then after a new steady state is established between the atmosphere and the ocean, the atmospheric content of CO_2 will have tripled. Of course, a long time (probably of the order of a thousand years) would be required to approach this new steady state. In the interim, much higher atmospheric levels of CO_2 will doubtless have been reached. It is this transient, of course, which should be the principal concern of those wishing to estimate the response of the ocean to increasing atmospheric CO_2.

Summary

The steady-state calculations reported here (Tables 17.4 to 17.6) were performed to investigate the behavior of our two-dimensional ocean model (Fig. 17.3) when a number of parameters were varied by substantial amounts from the reference values (Table 17.3) established by comparison with oceanographic data (Figs. 17.1 and 17.2). From these tests we may make the following observations:

1. When the parameter controlling advective flow (K_a) was varied and other parameters adjusted to retain a reasonable fit to the oceanographic data (Table 17.4), there were substantial effects in the profiles obtained (Fig. 17.4). Increased advection reduced the latitude dependence of the depth profiles and also strengthened the oxygen minima in the upper ocean.

2. A dead ocean at steady state produces a substantially higher (almost doubled) partial pressure of CO_2 in the atmosphere. If inorganic precipitation of $CaCO_3$ were included in the model, the increase would be even greater.

3. A doubling of the nutrient concentration in the ocean halved the partial pressure of CO_2 in the atmosphere.

4. Halving the water circulation rate lowered the atmospheric CO_2 content by only a small amount (11%). If the reduced temperature gradient implied by such a lowered circulation had been introduced, the reduction in atmospheric CO_2 would have been even less, or perhaps replaced by a small increase.

5. Large increases in atmospheric CO_2 (Table 17.6) suggest steady-state conditions in which 20 to 32% of the excess carbon in the terrestrial carbon cycle would remain in the atmosphere, depending on the amount of the excess. The latter "airborne fraction" corresponds to a release of about 4000 Gt of fossil carbon and a tripled concentration of CO_2 in the atmosphere.

While the model used for these simulations is still a primitive representation of the oceans, the results indeed serve to show the importance of the biological processes on the distribution of CO_2 between the oceans and the atmosphere.

Acknowledgments

The authors wish to thank members of the Carbon Dioxide Information Center, ORNL, for their help in accessing the GEOSECS data tapes, which were obtained from R. T. Williams, Physical and Chemical Oceanographic Data Facility, Scripps Institute of Oceanography, La Jolla, California. This research was supported by the Carbon Dioxide Division, U.S. Department of Energy, under Contract No. DE-AC05-84OR21400 with Martin Marietta Energy Systems, Inc.

References

Baes, Jr., C. F. 1982. Effects of Ocean Chemistry and Biology on Atmospheric Carbon Dioxide. In W. C. Clark, ed., Carbon Dioxide Review 1982, pp. 187–204. Oxford University Press, New York.

Baes, Jr., C. F., and G. G. Killough. 1985. A two-dimensional CO_2-ocean model including the biological processes. U.S. Department of Energy report DOE/NBB-0070 TRO 21.

Broecker, W. S. 1982. Ocean Chemistry During Glacial Time. Geochimica et Cosmochimica Acta 46:1689–1705.

Broecker, W. S. 1983. The Oceans. Sci. Am. 249:146–160.

Oeschger, H., S. Siegenthaler, U. Schotterer, and A. Gugelmann. 1975. A box diffusion model to study the carbon dioxide exchange in nature. Tellus 27:168–192.

Redfield, A. C., B. H. Ketchum, and F. A. Richards. 1963. The Influence of Organisms on the Composition of Seawater. In N. M. Hill, ed., The Sea, Vol. 2, pp. 26–77. Wiley Interscience, New York.

18
Measurements of Total Carbon Dioxide and Alkalinity in the North Atlantic Ocean in 1981

PETER G. BREWER, A. L. BRADSHAW, AND
R. T. WILLIAMS

The ocean uptake of fossil fuel CO_2 has long been recognized as the principal modulator of the rising atmospheric CO_2 level. If we are to observe and understand this effect, then an essential step is the accurate measurement of the CO_2 properties of the ocean. Historically, this has been quite difficult to achieve. Although measurements of some kind date back to the late 19th century, complete, documented, and verifiable measurements are scarce indeed. This chapter describes and documents the series of total CO_2 and alkalinity measurements of seawater made on the North Atlantic Ocean during the Transient Tracers in the Ocean (TTO) expedition in 1981, and presents briefly the signals these data reveal.

The basic features of the problem are well known. Brewer (1983*a*) has reviewed the basis for (and early progress of) the TTO experiment. The recent National Academy of Sciences report (NAS 1983) contains a detailed assessment of the oceanic chemistry involved (Brewer 1983*b*).

The TTO experiment had its antecedents in the GEOSECS program. Here, for the first time, a global survey of the oceanic water column was conducted with the essential components of hydrographic and nutrient data, ocean CO_2 system properties, and radiotracer measurements (^{14}C, ^{3}H). The results of this program established, *inter-alia,* fundamental constraints on oceanic mixing times (Stuiver et al. 1983) and on oceanic uptake of fossil fuel CO_2 (Broecker et al. 1979).

The GEOSECS CO_2 system data consisted principally of potentiometric titrations of seawater for the determination of alkalinity and total CO_2 (Dyrssen and Sillen 1967) and independent measurements of the partial pressure of CO_2 gas (pCO_2) by Takahashi and co-workers (Takahashi et al. 1976). Since any two of the four system properties (pH, pCO_2, alkalinity, and total CO_2) may be combined with the appropriate thermodynamic relationships to compute the remaining properties (Skirrow 1975), then the redundancy implied by three measured variables permits checks on the internal consistency and validity of the data. Takahashi (1977) pointed out that several inconsistencies in the GEOSECS measurements were apparent, leading to the suggestion that the titrimetric total CO_2 data had a small but significant error. The estimated error was $+14$ μmol CO_2 • kg^{-1}; since the surface ocean water CO_2 concentration has probably changed by about $+40$ μmol • kg^{-1} since the industrial revolution, and subsurface changes are very much smaller, then errors of this magnitude were unacceptable for documenting the chemical evolution of the ocean CO_2 system. The source of the error was discovered and corrected by Bradshaw et al. (1981).

The precision of the measurements achieved during the GEOSECS program, [$\pm 0.4\%$ in alkalinity (± 9 μeq • kg^{-1}) and $\pm 0.5\%$ in total CO_2 (± 10.5 μmol • kg^{-1})] was overall remarkably good; however, results significantly better than this were achieved on individual stations or cruise segments. We believed that some improvements in precision and accuracy could be made, and a further goal of the TTO expedition was to reduce the leg-to-leg variability in these measurements.

The Woods Hole group was responsible for the overall direction, calibration, and theory of the potentiometric titrations. The responsibility for the construction, operation, and maintenance of the experimental apparatus during the arduous work at sea was accepted by the Scripps group. Particular recognition must go to the technical staff for their painstaking efforts in tending the instruments and running the thousands of analyses.

Instrumentation

A diagram of the titration system is shown in Fig. 18.1. An earlier description appeared in Bos and Williams (1982). The titration cell, with a final volume of about 110 mL, is constructed of Lucite. It is cylindrical and surrounded by an outer jacket through which water is circulated to maintain constant temperature. The lower part of the cell has two values for filling and draining with the sample. An overflow valve at top center completes the sample entry and containment system. Four ports in the cell top are provided for the glass and reference electrodes, a thermometer, and a volume expansion plunger (Edmond 1970). An additional feed-through for a capillary tube supplies acid for the titration. The cell is stirred with a magnetic stir bar.

The acid, 0.25 *N* HCl fortified to the approximate ionic strength of seawater with 0.45 *N* KCl, is added from a 2.5-mL Gilmont microburet driven by a stepping motor. An optical shaft encoder logs the delivery. The entire ensemble is microprocessor-controlled.

The glass electrodes (Beckman model 41263) are of lithium glass; the reference electrodes are of the calomel type.

The apparatus was contained in a temperature-controlled van mounted outside the interior spaces of the ship, and efforts were made to create a low electrical noise environment. Two titrators were used simultaneously for the expedition, and samples were run randomly on either. Acid was added stepwise in 0.053-mL increments throughout the titration sequence.

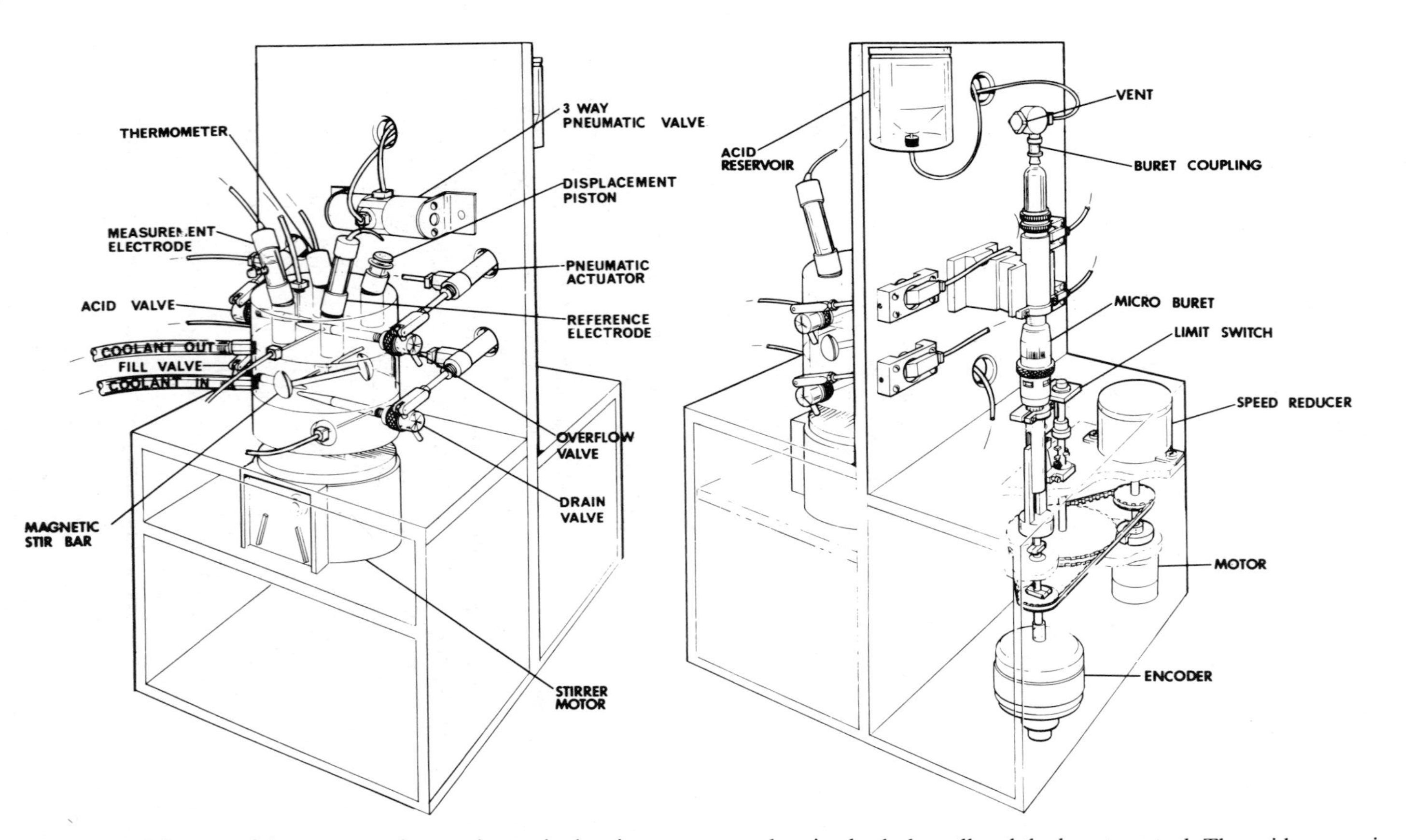

FIGURE 18.1. Diagram of the automated potentiometric titration apparatus showing both the cell and the buret control. The acid reservoir (6-L volume) was larger and mounted somewhat differently than indicated here. Sample inlet was by gravity feed from about 1 m above the apparatus.

The electrode response was checked for stability, and when a stable reading was encountered, the millivolt response of the cell was logged and the next acid increment was automatically added. The criterion for "stability" was a hard-wired function of the microprocessor and cannot be precisely reported here.

Samples, Standards, and Blanks

SAMPLES

Sea-water samples, unfiltered, were drawn from PVC Niskin bottles as soon as possible after recovery of the bottle rosette. Samples (500 mL) were drawn into borosilicate glass bottles, having ground glass stoppers, and were immediately poisoned by the addition of 100 μL of a saturated solution of mercuric chloride to prevent respiration and generation of CO_2 in the sample during storage prior to analysis.

The mercuric chloride addition can potentially change alkalinity and CO_2, both by dilution and hydrolysis of the Hg^{2+} ion., as in

$$Hg^{2+} + H_2O \rightleftarrows HgOH^+ + H^+ \ (\log K = -3.7) \tag{1}$$

and

$$Hg^{2+} + 2H_2O \rightleftarrows Hg(OH)_2 + 2H^+ \ (\log K = -6.3) \tag{2}$$

These effects can be shown to be negligible at the concentrations used.

In comparing the TTO results with those of the GEOSECS expedition, we note that GEOSECS samples were not preserved in any way, in spite of considerable preexpedition debate over this subject (Takahashi et al., 1970; Kroopnick et al. 1970, 1972).

Samples were stored at 25°C in the titration van for several hours prior to analysis.

STANDARDS

The primary standard adopted for the expedition was sodium carbonate. The source was the Fisher alkalimetric standard. The reagent was heated to 285°C for 2 h, cooled in a desiccator, and weighed out in ~0.25-g lots into Pyrex glass ampoules or polypropylene bottles.

The working standard frequently used at sea was sodium borate decahydrate ($Na_2B_4O_7{\bullet}10H_2O$). This was the primary standard used on the GEOSECS expedition. The reagent was recrystallized from borax and stored in a desiccator over a sucrose-sodium chloride solution to maintain the correct humidity. A detailed account of the borate standard preparation is given by Bos and Williams (1982) following the text by Vogel (1951).

The standard solutions of about 2350 $\mu eq \bullet L^{-1}$ were made up from these reagents (at sea) in 2-L volumetric flasks and contained a supporting electrolyte of KCl (Mallinckrodt A. R.) or NaCl (Merck Suprapur) to bring the ionic strength to approximately that of seawater.

Blanks

Nonzero blank values arise from the small quantities of alkaline impurities present in the NaCl or KCl salts used to fortify the standard solutions, and also from any acidic or alkaline impurities in the distilled water used in the preparation of the solutions. The problem is a classic one for potentiometric titrations (Ciavatta 1963), and variations in the blank values were major sources of error in the GEOSECS data set (Bos and Williams 1982).

During the TTO cruise, "blanks" were determined at each standardization by titrating solutions containing only the supporting electrolyte and in the same concentration used to fortify the standard. In each case the salt and the distilled water for both standard and blank solutions were taken from the same batches. Both NaCl and KCl are equally suitable as supporting electrolytes, the choice being dictated by purity and availability. A summary of blank determinations is given in Table 18.1, along with the salt used and the mean and standard deviation of the blank for each leg and for the entire cruise. The individual corrections were applied to the cruise standardizations, as given in the following paragraphs.

Calibration

A major problem is that the cell volume cannot be calibrated gravimetrically at sea. The construction of the cell is complex (Fig. 18.1), and its volume changes with any alteration of electrodes, stir bars, valves, or plunger. Maintenance of the system at sea is necessary, and changes of the kind previously mentioned are a practical reality for expedition work.

The strategy adopted for determining the cell volume is complex and includes the following steps:

1. Assume an approximate cell volume (V_{cell}) from prior knowledge.
2. Fill the cell with standard Na_2CO_3 or $Na_2B_4O_7 \cdot 10H_2O$ solutions.
3. Assume that the acid titrant, which is prepared to be 0.25 *N* by volumetric dilution of Baker Dilutit 0.5 *N* HC1, is exactly 0.25 *N*. The error here is essentially that in the 0.5 *N* value for the source, or about 0.2% or less.
4. Add the measured NaCl or KCl blank alkalinity value to the gravimetric alkalinity of the standard to obtain the true alkalinity of the standard (A_s).
5. Titrate the standard and locate the equivalence point (v_e) on the basis of the assumed volume.
6. Calculate a new cell volume (V'_{cell}) from

$$\frac{0.25\ v_e}{V'_{cell}} = A_s \tag{3}$$

TABLE 18.1. Blank statistics* for the TTO North Atlantic potentiometric titrations.

Leg	0.6 *N* NaCl			0.45 *N* KCl		
	Number	Average value†	Standard deviation	Number	Average value†	Standard deviation
1	1	1.0		0		
2	4	5.6	1.1	6	6.9	1.6
3	3	6.6	3.2	5	8.7	2.5
4	3	2.8	1.3	4	5.5	1.4
5	1	1.2		9	8.5	3.3
6	0			6	10.1	1.6
7				7	7.8	1.5
1-7	12	4.4	2.6	37	8.1	2.5

*Three 0.9 *N* KCl blanks have been omitted. Their average value was 10.2 $\mu eq \cdot L^{-1}$, and their standard deviation was 5.6 $\mu eq \cdot L^{-1}$.
†$\mu eq \cdot L^{-1}$.

7. Iterate the above calculation until the difference between assumed and calculated cell volumes is less than the chosen tolerance (0.01 mL). Normally only one iteration is required.

The "cell volume" thus determined is a rigorously correct estimate of the actual physical volume of the cell only in the case where the acid normality equals exactly 0.25; however, as mentioned above, the deviations from 0.25 are small, and, consequently, the effect of these deviations on total CO_2 and on alkalinity is completely negligible (calculations using titration model data give errors of $<0.005\%$ for deviations up to 5%). The use of this "cell volume" also possesses a feature that the physical volume does not; that is, to a large extent, it absorbs errors in the buret calibration.

If no changes in the apparatus occur, then cell volumes determined in this manner for a cruise leg or segment are averaged, and a mean value is taken. Table 18.2 presents some sample cell volume statistics giving the leg averages and the mean and standard deviation of these values for the entire expedition. Note that of the two titrators used simultaneously during the cruise (designated 1 and 3, the missing integer being held in reserve), it was routine practice to perform the fundamental calibration on one system only. The other was made a slave to the primary one by running sea water replicates on both as a secondary standard.

SENSOR RESPONSE

A critical condition for the analysis is that the response of the electrochemical cell be precisely Nernstian, that is, that the electrode slope be given by the equation

$$E = E_0 + \frac{RT}{F} \ln (H^+) \tag{4}$$

which will yield a slope of 59.16 mV per 10-fold change in (H^+) at the temperature of 25°C maintained in the experiment. The nonideal response

TABLE 18.2. TTO cell volumes (leg averages).

Leg	Volumes (mL)	
	Cell 1	Cell 3
1	111.06*	103.33*
2	110.02	103.10
3	110.13	103.31
4	110.17	103.49
5	110.26	103.10
6	110.03	102.96
7	110.26	103.47
Average (legs 2–7)	110.15	103.24
Standard deviation (legs 2–7)	0.11	0.22

*Both volumes rejected (value for 1 was from standards, value for 3 was from sea-water replicates) and legs 2 to 7 averages used.

of the glass electrode sensor is a frequent concern during potentiometric titrations. Broecker (personal communication) briefly espoused the notion that the nonideal response contributed to the overestimation of CO_2 in the GEOSECS titration data; however, this proved not to be so. The more usual problem occurs from liquid junction effects on the reference electrode. Deterioration of the glass electrode can, however, occur during prolonged use because of leaching from the glass of the Li^+ ions responsible for the conducting path, thus leaving a hydrated and lithium-depleted skin through which diffusion must occur before equilibrium is reached. The manifestation of this would be electrode drift and a response sluggishness. Although electrode life times are considerably longer than the cruise period, it was believed prudent to devise means of checking the electrode slope for overall quality control.

Two techniques of checking the ''practical electrode slope'' [adapted from the work of Johansson and Wedborg (1979)] were devised. The classic approach of calibration versus a standard hydrogen electrode was not attempted at sea. The practical electrode slope (h) was determined from the blank titration data; once the initial addition of acid has protonated alkaline impurities, then the titration is simply that of pure NaCl or KCl, and a linear classic response should result.

From the proton balance equation we have

$$(V_{\text{cell}} + V_{\text{acid}})\{10\,\frac{(E - E_0)}{h} + [HCO_3^-]\} = 0.25\,(V_{\text{acid}} - V_{\text{equiv.point}}) \quad (5)$$

[HCO_3^-, present at < 0.1% of (H^+) in the acid range used, can be neglected. Then

$$(V_{\text{cell}} + V_{\text{acid}})10\,\frac{(E - E_0)}{h} = 0.25\,(V_{\text{acid}} - V_{\text{equiv.point}}) \quad (6)$$

and

$$F(V) = (V_{\text{cell}} + V_{\text{acid}})\,10^{E/h} \propto (V_{\text{acid}} - V_{\text{equiv.point}}) \quad (7)$$

where $F(V)$ is analogous to a Gran (1952) function. The problem then is to find h, which gives the smallest sum of squares of residuals from a linear fit of $F(V)$ to V_{acid}.

An alternate procedure, using the same data source, is to find h by determining the values of h, E_0, and $V_{\text{equiv.point}}$ by a nonlinear least-squares fit of the following equation to E_i and V_{acid}.

$$E_i = E_0 + h\,\log_{10}[\frac{0.25}{(V_{\text{cell}} + V_{\text{acid}})}(V_{\text{acid}} - V_{\text{equiv.point}})] \quad (8)$$

Both procedures were used to check and verify shipboard and shore expedition data from station 151, where 19 stored samples in duplicate were taken for comparison with shipboard results and for study of the stored sample data reproducibility. The calculations were carried out on

TABLE 18.3. Practical electrode slopes (mV) (Station 151 data).

Shipboard fitting techniques		Shore fitting techniques	
Nonlinear	Linear	Nonlinear	Linear
59.44	59.31	59.23	59.38
59.20	59.30	59.26	59.17
59.29	59.65	59.23	59.39
59.50	59.28	59.15	59.19

blank titration data obtained during standardizations that preceded the sample titrations for both ship and shore data. The results are given in Table 18.3. The effects of changes in this property can be illustrated by the observation that a change of -1% in h causes errors of about $+7.5$ $\mu mol \bullet kg^{-1}$ in total CO_2 and -4.5 $\mu eq \bullet kg^{-1}$ in alkalinity. These opposing errors thereby could lead to substantial errors in computed properties such as pCO_2.

Figure 18.2 shows examples of these calculations with the residuals plotted versus V_{acid}.

For TTO shipboard data, the mean "practical" electrode slope determined was 59.36 ± 0.14 mV by the nonlinear fit to the data and 59.39 ± 0.18 by the linear fit technique. The corresponding values for the stored sample data were 59.22 ± 0.05 mV and 59.26 ± 0.10 mV, respectively.

We judge that these small deviations from the theoretical response of 59.16 do not constitute sufficient evidence of nonideal behavior, and,

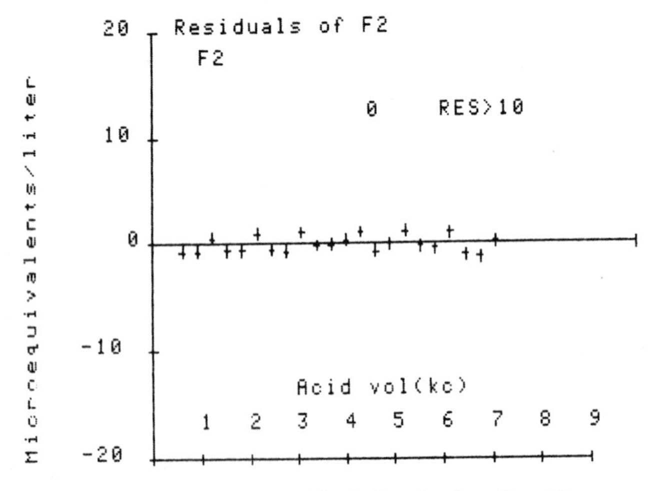

FIGURE 18.2. The plot represents residuals for the function $F2$ versus acid added for a blank titration using a linear fit [Eq. (7)]; these data are used to estimate the "practical" electrode slope.

therefore, for all calculated values from the cruise, the Nernstian slope was assumed.

Location of the Equivalence Points

Although the potential of analyzing the full titration curve by nonlinear techniques exists (Barron et al. 1983), the choice of the linear, modified Gran functions was made. A data window (typically of 8 to 9 points) was taken over roughly a 60-mV range slightly beyond each equivalence point.

The titration curve was analyzed much as described by Bradshaw et al. (1981). The nutrient species phosphate and silicate were determined on each titration sample, and the titration curve was reanalyzed once the analyses were complete and made available.

The thermodynamic constants chosen for the analysis were the acidity constants for carbonic acid and boric acid in sea water from Hansson (1973) as formulated by Almgren et al. (1975). Other values are given in Bradshaw et al. (1981).

Analyses of Stored Samples

Samples (identical to those for shipboard analysis) were taken throughout the cruise for analysis ashore at Woods Hole by a similar titration procedure. For shore-based analyses, an acid normality of 0.05 *N* was used and delivered in 0.2-mL increments by an automatic buret (Dosimat E15). The titration cell was of a different design, principally in its valving, but was again constructed of Lucite and with a volume of ≈105 mL.

The samples were crated at sea and analyzed at Woods Hole several months after sampling. A subset of the results are shown in Table 18.4 together with the shipboard data. A full listing of the analyses will be found in a technical report being prepared by Brewer et al. The essential feature is that the alkalinity determinations agree extraordinarily well—the mean difference between shipboard and shore-based analyses being $0.8 \pm 9.5\ \mu eq \cdot kg^{-1}$. Apparently, the borosilicate glass storage of samples preserved with mercuric chloride has no deleterious effect on sample quality. Equally, the shipboard determinations of alkalinity, in spite of the rigors of expedition work, appear to be both accurate and precise. The error of $0.03 \pm 0.4\%$ is close to the limit achievable for purely volumetric work.

The second noticeable feature is that the total CO_2 analyses do not agree well between shipboard and shore-based determinations, the shore-based samples being high relative to the expedition measurements. The average difference was $21.3 \pm 8.6\ \mu mol \cdot kg^{-1}$ ($1.0 \pm 0.4\%$). This disagreement is disturbingly large; however, there are several possible explanations for the discrepancy:

1. Either shipboard or shore-based analyses suffer from an inaccuracy in calibration.

TABLE 18.4. Potentiometric titration comparison of stored samples with shipboard results for several stations (shore minus ship).

Station 7			Station 113			Station 71			Station 158		
Depth (m)	Alkalinity*	TCO_2†	Depth (m)	Alkalinity*	TCO_2†	Depth (m)	Alkalinity*	TCO_2†	Depth	Alkalinity	TCO_2†
149	−5	22	9	−16	4	8	−10	29	5	4	27
437	7	27	76	7	23	111	−1	21	7	−8	22
780	5	21	195	5	18	214	5	32	19	4	28
878	1	21	314	11	24	313	−5	21	22	−6	28
1181	1	22	460	6	20	413	−9	14	31	4	24
2080	1	18	769	10	18	512	3	19	46	8	8
2705	2	20	986	3	20	611	−8	12	56	−10	31
3718	2	16	1910	11	26	707	3	0	75	−24	21
4713	13	33	2845	8	19	806	−5	13	95	−9	27
						906	−1	20	115	5	23
						1406	−6	24	147	−5	28
						2006	−2	15	231	4	10
						2504	−4	17	282	−5	9
						3498	−9	0	406	8	14
						3989	−16	4	606	−6	18
						4944	−22	12	807	13	14
									999	−3	13
									1398	5	16
									1798	−1	20
									1987	8	21
									2142	−6	16

*$\mu eq \cdot kg^{-1}$
†$\mu mol \cdot kg^{-1}$.

2. Preservation with $HgCl_2$ is inadequate to prevent CO_2 buildup resulting from respiration in the shore-based samples.
3. A contaminant is present.
4. There are procedural differences in sample handling and titration protocol.

The standard Na_2CO_3 reagents used as primary standards were prepared at Woods Hole and used both for expedition and shore-based analyses; electrode slope, blank, and cell volume procedures were all similarly checked. Thus, the calibration is not likely in error.

Preservation of the sample with $HgCl_2$ may not be perfect. One point of note is the markedly covalent character of this salt resulting in decreased ionization in solution. However, identical preservation steps were taken by Keeling (personal communication), and high-precision CO_2 analyses made on his samples by a gas extraction manometric technique do not show such systematic large offsets as those from the shipboard data (see later). It may be that the effect is that of the generation of protolytes masquerading as CO_2 in the titration. Phosphate is such a species (Bradshaw et al. 1981) in which the numerical values of the dissociation constants ($pK_2' = 5.95$; $pK_3' = 8.95$) are closely similar to carbonic acid. However, phosphate itself is present in too low a concentration to cause concern here.

The presence of a contaminant has to be considered. A contaminant in the added $HgCl_2$ preservative (100 uL) would have to be in very high concentrations. Moreover, the same $HgCl_2$ was used for the shorter-term preservation of the shipboard samples. The bottles used for storage of the samples were new as received from the manufacturer. They were rinsed with sea water from the same Niskin bottle immediately before drawing the sample. The sample bottles were not annealed or baked out at high temperature before use; however, an organic carbon residue of any kind would have to metabolize to CO_2 in the presence of the $HgCl_2$ preservative. An invasion of CO_2 gas is not realistic. Deep Atlantic samples, on warming to room temperature, show a pCO_2 far above atmospheric equilibrium, and gas evasion would be more likely.

Evidence that the higher total CO_2 in the stored samples is not due to differences in practical electrode slope values between shore and shipboard titrations has previously been given for station 151 data. For this station, the average differences for shore minus ships were -1.3 ± 4.5 $\mu eq \bullet kg^{-1}$ for alkalinity and 20.2 ± 6.7 $\mu mol \bullet kg^{-1}$ for total CO_2. These values are in good agreement with the results for all stored samples. The replicate results for stored samples from station 151 also show that the reproducibility of the overall effects of sampling, poisoning, storage, and on-shore measurement is good. The average difference between duplicates is 0.0 $\mu eq \bullet kg^{-1}$ for alkalinity and 0.4 $\mu mol \bullet kg^{-1}$ for total CO_2; the corresponding single observation standard deviations are 3.5 $\mu eq \bullet kg^{-1}$ and 3.7 $\mu mol \bullet kg$, respectively.

Sample handling at sea and ashore was very similar. One difference in protocol, however, is in the rapidity of the titration. The time taken at sea for stepping through the titration is approximately 15 min; titration time on the Woods Hole shore-based system is at least twice as long, with the response in the early stages of the titration being the critical feature. The Woods Hole electrode stability criterion was set at a drift of < 0.05 mV per 10 s.

The physical process responsible for the slow response in the higher pH region is that, on acid addition, a localized cloud of acidic CO_2 gas-rich water forms around the buret tip and is swept away by stirring. The sample then reaches equilibrium through hydration of the excess CO_2, which is the classically slow kinetic step. This effect is most certainly real; calculations show that overly rapid titration (e.g., only 95% equilibration) produces a total CO_2 result that is too low. Thus, if we were to invoke this kinetic factor, the agreement between shipboard and shore-based samples would tend to be improved.

Comparisons with the Analyses of Other Investigators

Although the kinetic effect described above would tend to reconcile the shipboard and shore-based titration data sets, it would provide further problems. Checks are available through comparison of calculated values of pCO_2 with those directly measured, or through comparison with analyses of total CO_2, which depend on different physical principles. The latter comparison is available through the work of C.D. Keeling (personal communication), who measured total CO_2 by the gas extraction and manometric technique described in Weiss et al. (1982). This procedure is purported to have an accuracy of 1.0 ± 0.7 μmol • kg^{-1}.

A comparison of the shipboard data with the Keeling total CO_2 analyses is shown in Fig. 18.3. The mean difference between manometric and titrimetric data for samples below a depth of 1000 m is 4.0 ± 2.9 μmol • kg^{-1}. For depths of <1000 m the discrepancy increases to 7.2 ± 5.9 μmol • kg^{-1}. The total data set available consists of 129 comparisons.

The deep water agreement is remarkably good and close to the fundamental limit available because of the inaccuracies in the dissociation constants. For instance, small changes in the boric acid dissociation constant could generate the discrepancy found here.

The large error and uncertainty in the upper ocean samples is perplexing. We previously discussed potential sources of error in the titration procedure. One upper ocean problem could be the presence of unaccounted for protolyte species, for example, the anions of weak organic acids resulting from biological activity. We have, however, no objective evidence of this.

An essential feature of this comparison is nonetheless the overall good agreement. There is therefore no basis for suggesting a priori that the

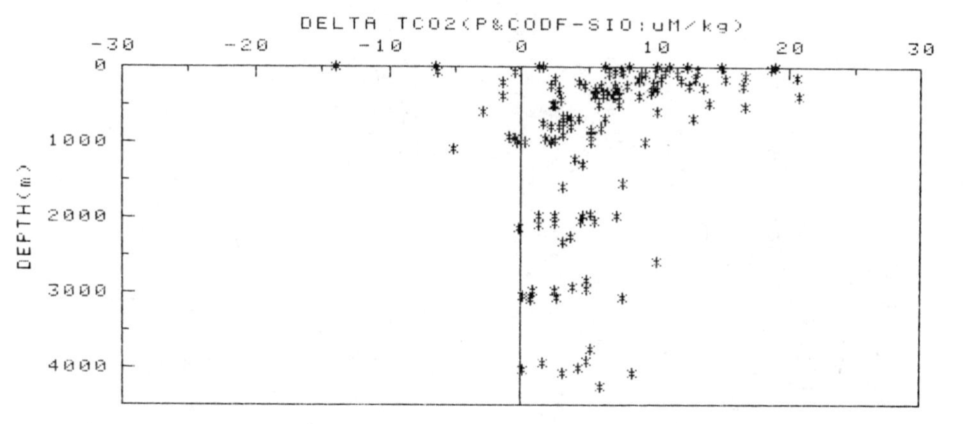

FIGURE 18.3. Comparison of shipboard titration data for total CO_2 made by the Physical and Chemical Oceanographic Data Facility (P & CODF) vs the gasometric total CO_2 analyses of C. D. Keeling of Scripps Institution of Oceanography (SIO) (personal communication). The results are expressed as a difference in μmol $CO_2 \cdot kg^{-1}$. The shipboard data are high relative to the shore-based analyses.

shipboard titration data are too low by the ~20 μmol • kg^1 suggested by comparison with the analyses of stored samples.

On the basis of this comparison, we can say that either a problem exists in the stored samples in that they have become "apparently" enriched in total CO_2; or that overly rapid titration at sea resulted in "low" total CO_2 data, and that an undiscovered substantial inaccuracy exists in the basic theory and practice of the potentiometric titration procedure. We are disinclined to believe the latter hypothesis.

An additional check is available through combination of the directly measured pCO_2 (Takahashi) and total CO_2 (Keeling) properties and calculation of the alkalinity. Here excellent agreement is observed (Takahashi, personal communication) between measured and calculated properties.

Discussion

The principal purpose of this paper is to document the experimental procedures followed so that users of this large data set have an accurate knowledge of its validity and limitations. However, a brief presentation of the oceanic features revealed is also appropriate.

The cruise track followed by the RV "Knorr" is shown in Fig. 18.4. The cruise occupied 200 days of ship time and covered 38,214 km of cruise track. Thus, very wide geographical and seasonal scales were encountered. The details of station work are given in the *Preliminary Hydrographic Data Reports* issued by the Physical and Chemical Oceanographic Data Facility of the Scripps Institution of Oceanography.

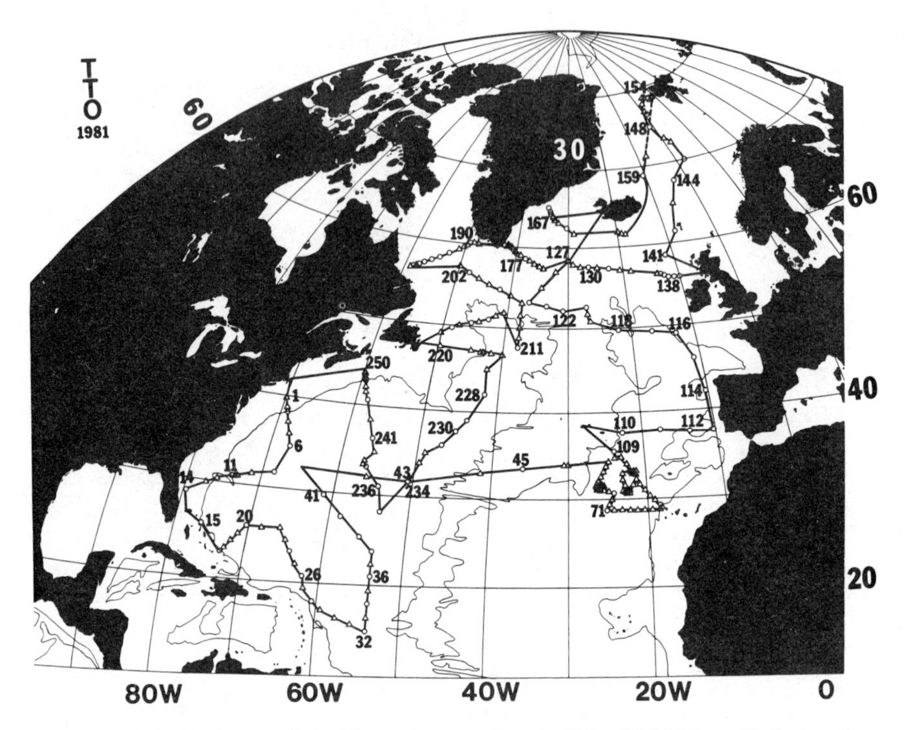

FIGURE 18.4. Cruise track (with station numbers) of the RV "Knorr" during the TTO North Atlantic Cruise, April 1–October 19, 1981.

The surface ocean properties revealed are of particular interest since they define the controls on ocean–atmosphere CO_2 exchange, and also since today's surface waters are those so far most affected by contemporary CO_2 increases.

We earlier showed that of the measured pair, the alkalinity is most accurately defined by the titration data. The correlation of alkalinity with salinity is very strong. In Fig. 18.5 we show the salinity-alkalinity diagram for all TTO surface samples (taken as 0 to 15-m depths). The property plotted is simply titration alkalinity, and it incorporates the protonation of small amounts of phosphate and silicate high latitude encountered in surface waters. The linear correlation is described by

$$\text{Surface alkalinity} = 547.05 + (50.560 \times \text{S } 0/00) \qquad (9)$$

with a correlation coefficient of 0.990 and an estimate of the standard deviation of a single point about the regression line of 9.14 $\mu eq \cdot kg^{-1}$. This is approximately three times the standard deviation of a single measurement. The North Atlantic salinity-alkalinity relationship is apparently very tightly defined in spite of the large seasonal range covered during the expedition. At low salinities, a slightly greater deviation is encountered.

For surface ocean total CO_2 data the variability is much more pro-

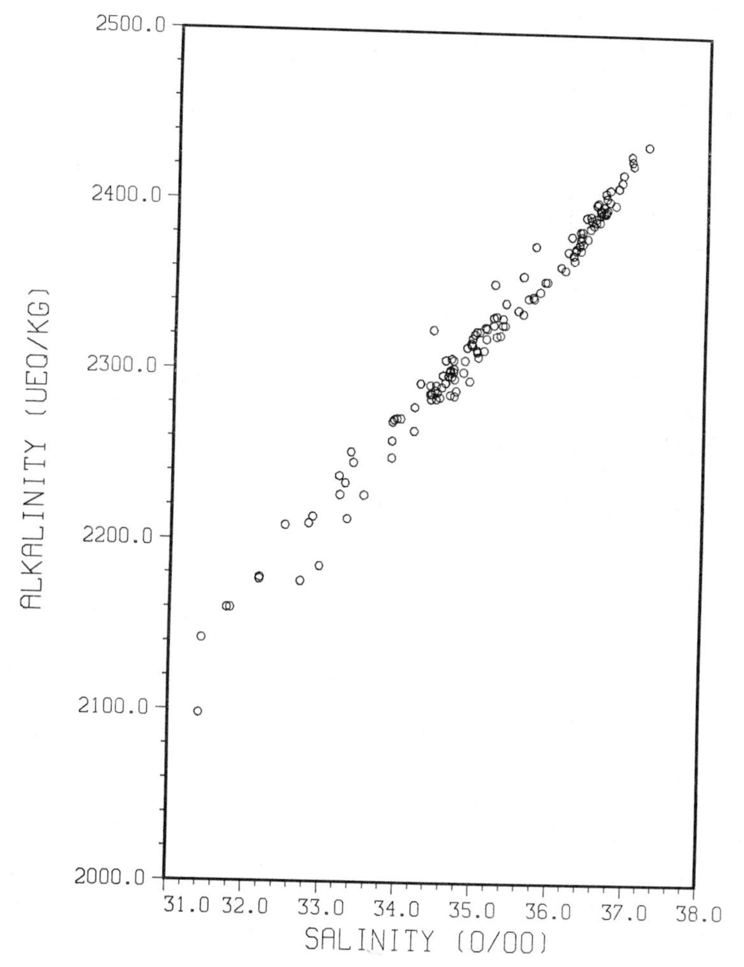

FIGURE 18.5. The salinity-alkalinity relationship for North Atlantic surface waters. The linear relationship is represented by alkalinity = 547.05 + (50.560 × S 0/00.) (Note: μEQ • kg^{-1} = microequivalents per kilogram.)

nounced. There is a tendency to believe that total CO_2 in surface waters is reasonably well correlated with temperature; however, this is not so. The simple correlations (although poor) of surface (0 to 15 m) total CO_2, as defined by the titration data, with temperature and salinity are shown in Figs. 18.6 and 18.7. When total CO_2 is normalized to a constant salinity of 35 0/00, however, the correlation with temperature is reasonably good (Fig. 18.8). Overall, the relationship is defined by

$$\text{Total } CO_2 \text{ (35 0/00)} = 2158 - (8.169 \times t°\text{C}) \qquad (10)$$

with a correlation coefficient of −0.885 and standard deviation of 30 μmol • kg^{-1}.

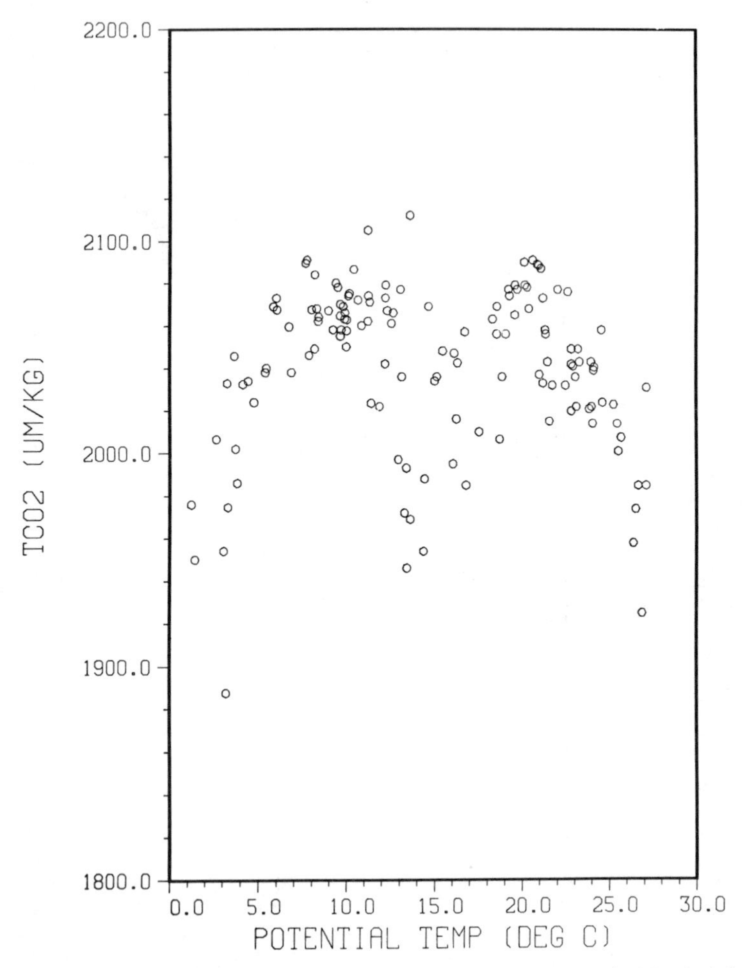

FIGURE 18.6. The correlation between total CO_2 and temperature in North Atlantic surface waters. (Note $\mu M \bullet kg^{-1} = \mu mol \bullet kg^{-1}$.)

This simple linearization ignores the pronounced pattern apparent in the data. Below a temperature of about 13°C, the plot separates into three distinct lobes representing data from different oceanic regions. The lobe of high normalized total CO_2 points extending above the general trend results from the low-salinity waters sampled near Newfoundland and Nova Scotia in late summer. The lobe of low normalized total CO_2 at temperatures roughly from 0° to 10°C represents waters sampled at the extreme northern stations near Spitzbergen. The remaining samples fall on the linear trend established by the warmer lower latitude waters, with a low-temperature end member near Cape Farewell, Greenland (Station 187).

In sampling on such large scales with pronounced seasonality (ranging from 5–100 ($\mu mol \bullet kg^{-1}$), we are of course chasing a moving target. Moreover, the many processes affecting oceanic surface total CO_2 generate

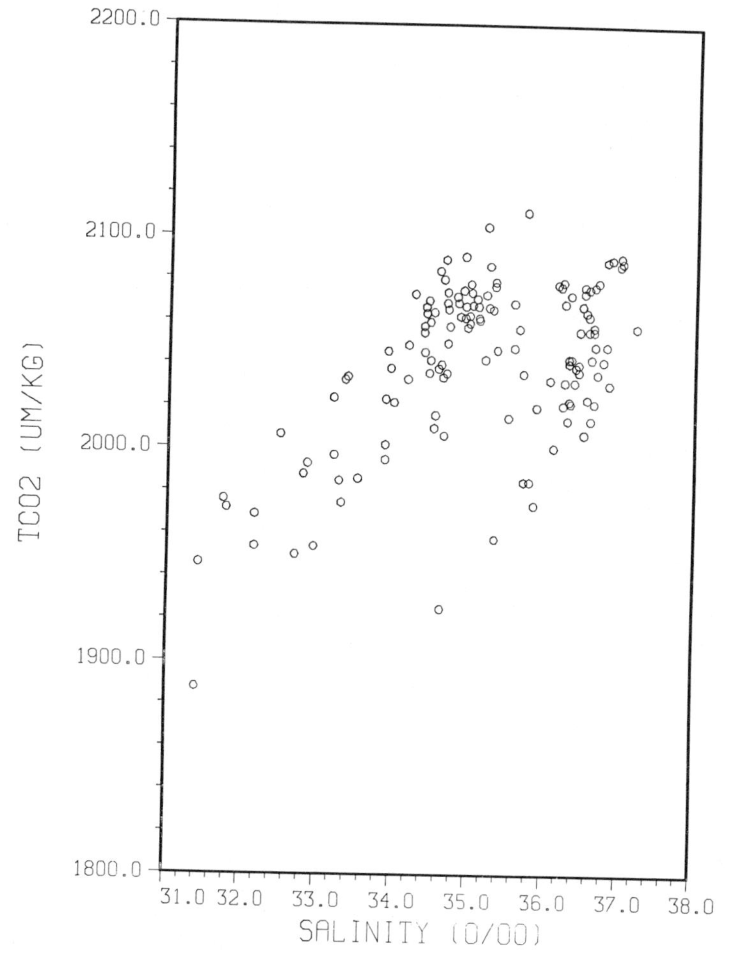

FIGURE 18.7. The correlation between total CO_2 and salinity in North Atlantic surface waters. (Note: $\mu M \cdot kg^{-1} = \mu mol \cdot kg^{-1}$.)

considerable local variability, which directly affects the various schemes to recover the anthropogenic CO_2 signal from ocean waters. The procedure proposed by Brewer (1978) relies on the surface alkalinity-salinity relationship, which we show here to be tightly defined. In contrast, the scheme of Chen and co-workers (Chen 1982) examines deviations from the contemporary salinity-normalized total CO_2-temperature trend, which here is seen to be significantly more variable.

The strong correlations between alkalinity, total CO_2, salinity, and temperature provide important constraints on oceanic processes and on extrapolation to data-poor areas. We are currently exploring several new data presentation and modeling techniques to make use of these relationships.

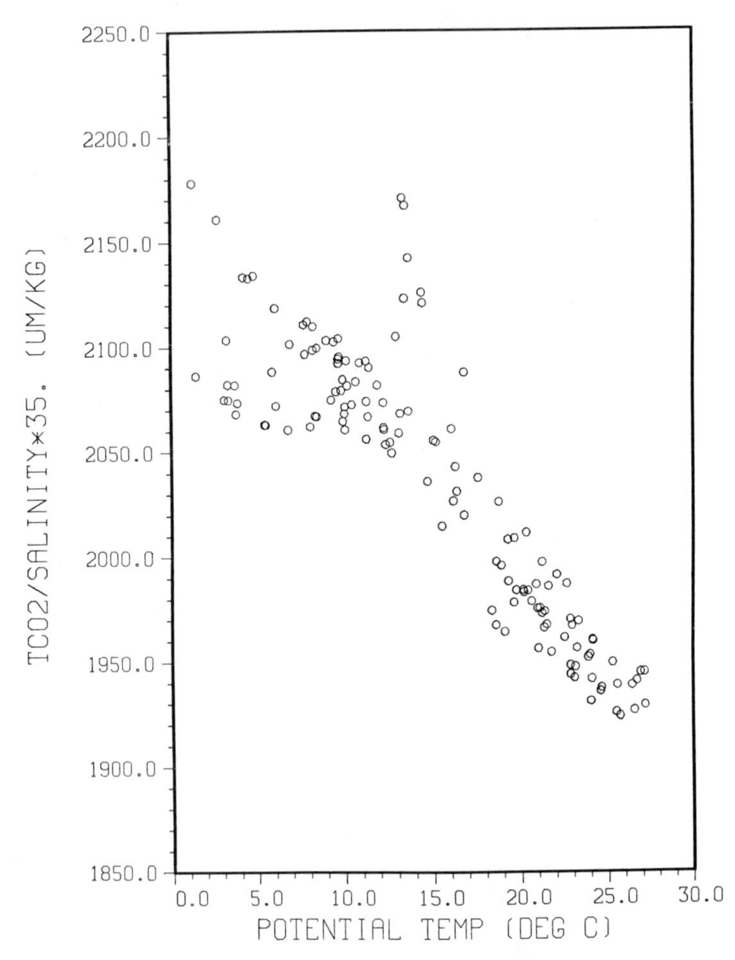

FIGURE 18.8. The correlation between total CO_2, normalized to a constant salinity of 35 0/00, and temperature in North Atlantic surface waters. (Note: $\mu M \bullet kg = \mu mol \bullet kg^{-1}$.)

One important oceanic chemical property is that defined by the alkalinity-total CO_2 relationship. Deffeyes (1965) first advocated the use of alkalinity-CO_2 diagrams as diagnostic tools, and Baes (1982) revived their use. The importance of such diagrams is that isolines of pCO_2, pH, or CO_3^{2-} are defined as linear relationships in this property space. Baes (1982), however, points out that such relationships are valid only under isothermal and isohaline conditions; therefore the use of such relationships on real data is complex. In Fig. 18.9, we show a three-dimensional perspective drawing of the alkalinity-total $CO_2 - pCO_2$ relationship for North Atlantic surface waters. The two-part diagram shows the relationship with a "floor" of 150 ppm pCO_2, so that the entire surface is visible, and also from the

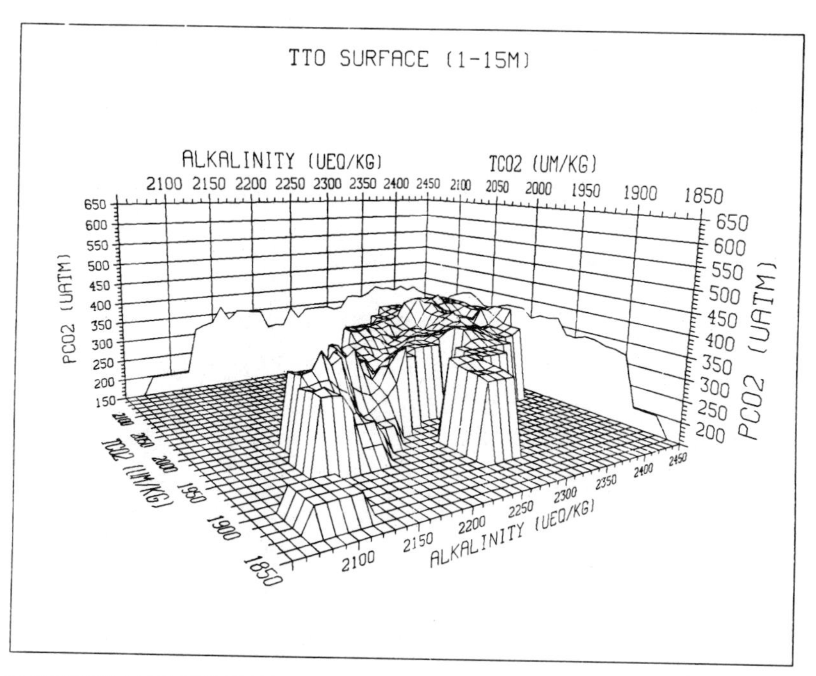

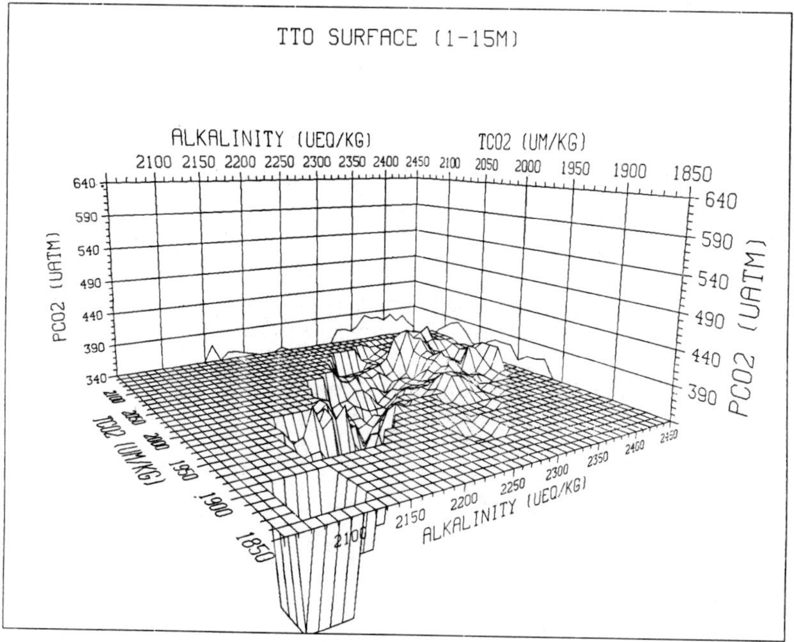

FIGURE 18.9. Three-dimensional perspective drawing of the North Atlantic surface water alkalinity-total CO_2-pCO_2 relationship based on potentiometric titration data. The **upper diagram** represents the relationship with a "floor" of 150 ppm pCO_2. The **lower diagram** represents the same data with a floor of 340 ppm, the atmospheric CO_2 value in 1981. (Note: $\mu eq \bullet kg^{-1}$ = microequivalents per kilogram and $\mu M \bullet kg^{-1}$ $\mu mol \bullet kg^{-1}$.)

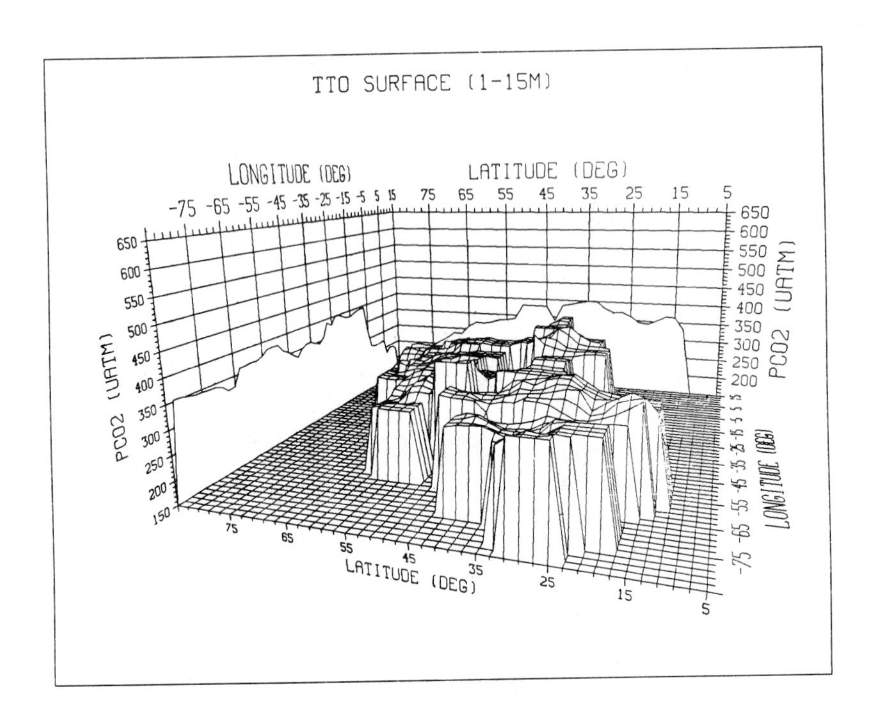

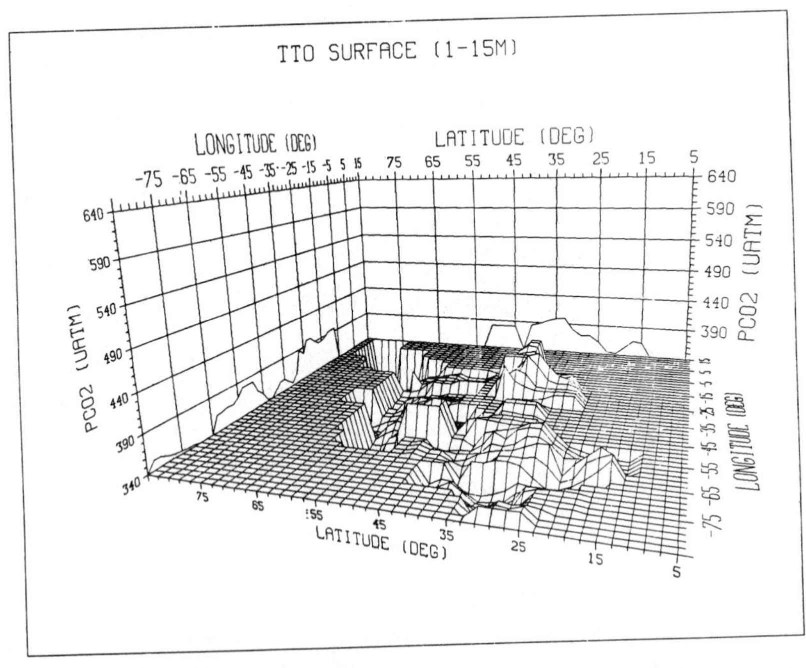

FIGURE 18.10. Three-dimensional perspective drawing of the North Atlantic surface water pCO_2 values, calculated from potentiometric titration data, as viewed from the U.S. East Coast. The **upper diagram** shows the pCO_2 field with a "floor" of 150 ppm; the **lower diagram** shows the field with a floor of 340 ppm, the atmospheric CO_2 value in 1981. Gaseous CO_2 invasion of the ocean takes place in the "holes," and evasion is attributable to the "hills."

perspective of 340 ppm pCO_2, the atmospheric equilibrium value in 1981. Highs represent evasion of CO_2, and lows represent regions of CO_2 uptake in this property space.

Figure 18.10 shows a similar three-dimensional perspective drawing of the geographic variation of pCO_2 (as calculated from potentiometric titration data alone) in North Atlantic surface waters. The view is from the U.S. East Coast. Again, the two-part figure shows pCO_2 both from a floor of 150 ppm and from the atmospheric equilibrium value. The figure graphically illustrates the oceanic loss of CO_2 at the southern and eastern extent of the cruise track and the large CO_2 "hole" in the northern and western regions. The driving forces behind these processes will be explored in a later paper.

Acknowledgments

The authors gratefully acknowledge the assistance of all our TTO colleagues. Particular praise must go to D. Bos for his heroic service in performing several thousand titrations at sea under rigorous expedition conditions. The officers and crew of the RV "Knorr" provided exemplary support. Thanks are due to D. Shafer for data processing and plotting, particularly for Figs. 18.9 and 18.10, to C. D. Keeling and T. Takahashi for access to their unpublished results, and to M. F. Harvey for preparing this manuscript. This research was supported by NSF Grants OCE81-08160 (PGB) and OCE79-25890 (RTW) and the U.S. Department of Energy's CO_2 Research Program.

References

Almgren, T., D. Dyrssen, and M. Strandberg. 1975. Determination of pH on the moles per kg sea water scale (Mw). Deep-Sea Res. 22:635–646.

Baes, C. F. 1982. Ocean chemistry and biology. In W. C. Clark (ed.), Carbon Dioxide Review 1982, pp. 187–211. Clarendon Press, Oxford.

Barron, J. L., D. Dyrssen, E. P. Jones, and M. Wedborg. 1983. A comparison of computer methods for sea water alkalinity titrations. Deep-Sea Res. 30:441–448.

Bos, D. and R. T. Williams. 1982. History and development of the GEOSECS alkalinity titration system. In Workshop on Oceanic CO_2 Standardization. U.S. DOE Report CONF-7911173.

Bradshaw, A. L., P. G. Brewer, D. K. Shafer, and R. T. Williams. 1981. Measurements of total carbon dioxide and alkalinity by potentiometric titration in the GEOSECS program. Earth Planet. Sci. Lett. 55:99–115.

Brewer, P. G. 1978. Direct observation of the oceanic CO_2 increase. Geophys. Res. Lett. 5:997–1000.

Brewer, P. G. 1983*a*. The TTO North Atlantic Study—A progress report. In Proceedings: Carbon Dioxide Research Conference: Carbon Dioxide, Science, and Consensus, pp. II-91–122. U.S. DOE Report CONF-820970.

Brewer, P. G. 1983*b*. Carbon dioxide and the oceans. In Changing Climate. Report of the Carbon Dioxide Assessment Committee, pp. 186–215. National Academy Press, Washington, D.C.

Broecker, W. S., T. Takahashi, H. J. Simpson, and T. H. Peng. 1979. Fate of fossil fuel carbon dioxide and the global carbon budget. Science 206:409–418.

Chen, C. T. 1982. On the distribution of anthropogenic CO_2 in the Atlantic and Southern Oceans. Deep-Sea Res. 563–580.

Ciavatta, L. 1963. Potentiometric purity control of salt media for equilibrium studies with an appendix on the analysis of dilute solutions of strong acids. Ark. Kemi 20:417–435.

Deffeyes, D. S. 1965. Carbonate equilibria: a graphic and algebraic approach. Limnol. Oceanogr. 10:412–426.

Dyrssen, D. and L. G. Sillen. 1967. Alkalinity and total carbonate in sea water: a plea for p-T-independent data. Tellus 19G:113–121.

Edmond, J. M. 1970. High precision determination of titration alkalinity and total carbon dioxide content of sea water by potentiometric titration. Deep-Sea Res. 17:737–750.

Gran, G. 1952. Determination of the equivalence point in potentiometric titrations, II. Analyst 77:661–671.

Hansson, I. 1973. A new set of acidity constants for carbonic acid and boric acid in sea water. Deep-Sea Res. 20:461–478.

Johansson, O. and M. Wedborg. 1979. Stability constants of phosphoric acid in sea water of 5-40 0/00 salinity and temperatures of 5–25°C. Mar. Chem. 8:57–69.

Kroopnick, P., W. G. Deuser, and H. Craig. 1970. Carbon-13 measurements on dissolved inorganic carbon at the North Pacific (1969) GEOSECS station. J. Geophys. Res. 75:7668–7671.

Kroopnick, P., R. F. Weiss, and H. Craig. 1972. Total CO_2, ^{13}C, and dissolved oxygen-^{18}O at GEOSECS II in the North Atlantic. Earth Planet. Sci. Lett. 16:103–110.

National Academy of Sciences. 1983. Changing Climate. Report of the Carbon Dioxide Assessment Committee.

Skirrow, G. 1975. The dissolved gases-carbon dioxide. In J. P. Riley and G. Skirrow (eds.), Chemical Oceanography. Vol. 2. 2nd edit., pp. 1–192. Academic Press, New York.

Stuiver, M., P. D. Quay, and H. G. Ostlund. 1983. Abyssal water carbon-14 distribution and the age of the world oceans. Science 219:849–851.

Takahashi, T. 1977. Consistency of the alkalinity, total CO_2, pCO_2 and pH data obtained in the Pacific and Atlantic GEOSECS expeditions. GEOSECS Internal Report (unpublished document).

Takahashi, T., P. Kaiteris, and W. S. Broecker. 1976. A method for shipboard measurement of CO_2 partial pressure in sea water. Earth Planet. Sci. Lett. 32:451–457.

Takahashi, T., R. F. Weiss, C. H. Culberson, J. M. Edmond, D. E. Hammond, C. S. Wong, Y. H. Li, and A. E. Bainbridge. 1970. A carbonate chemistry profile at the 1969 GEOSECS intercalibration station in the eastern Pacific Ocean. J. Geophys. Res. 75:7648–7666.

Vogel, A. I. 1951. A Textbook of Quantitative Inorganic Analysis, Theory, and Practice. Longmans, Green, and Co.

Weiss, R. F., R. A. Jahnke, and C. D. Keeling. 1982. Seasonal effects of temperature and salinity on the partial pressure of CO_2 in sea water. Nature 300:511–513.

19
Geologic Analogs: Their Value and Limitations in Carbon Dioxide Research

Eric T. Sundquist

The CO_2 research community has recently shown much interest in the use of geologic analogs to verify climate model predictions. This application rests, of course, on the premise that both climate and atmospheric CO_2 have varied in the geologic past. Climate variability is widely documented in the geologic record, and CO_2 changes have recently been documented in ice cores. Although there is no other direct evidence for CO_2 variations in the geologic past, the relatively small size and short residence time of the atmospheric CO_2 reservoir suggest strongly that it must have been sensitive to perturbations in the larger reservoirs with which it exchanges.

However, if the purpose of CO_2 research is to understand the *processes* that control atmospheric CO_2 and climate, then geologic analogs serve this purpose only if they lead to a better understanding of the controlling processes. The use of geologic analogs is not simply a matter of documenting climate and CO_2 fluctuations, because correlations do not necessarily imply cause-and-effect relationships. For example, several recent studies of possible geologic analogs have emphasized the effects of climate on atmospheric CO_2 as well as the influence of CO_2 on climate (e.g., Walker and Hays 1981; Berner et al. 1983). Thus, as geologists continue to provide information about the interaction of geochemical, biological, and climatic processes in the past, the application of geologic analogs to the question of climate model verification becomes more and more complicated.

The time scales considered by geologists are generally much longer than those considered by current CO_2-climate models. Because sediments are deposited episodically and are often stirred after they are deposited, the geologic record is usually inherently averaged or integrated over thousands of years. The time resolution of regional or global reconstructions is further limited by the accuracy of dating and correlating sediment sequences from different locations. Therefore, geologic evidence bears most directly on the long-term geochemical and climatic effects of anthropogenic CO_2.

It is reasonable to question the relevance of these long-term processes to current CO_2 research, which is primarily directed toward resolving significant uncertainties over time scales of decades. This chapter presents two approaches to this question. The first is an examination of the magnitude of plausible natural CO_2 perturbations characteristic of different time scales. This approach leads to the working hypothesis that only long-term processes (i.e., processes acting over time scales of 10^5 years or more) could have caused atmospheric CO_2 to vary in the past by a factor of 2 or greater. The second approach is a model of the effects of anthropogenic CO_2 on ocean chemistry and carbonate sediments. This model suggests that the geochemical effects of anthropogenic CO_2 may persist for 10^4 years or longer.

Review of Recent Geologic Analog Studies

Geologists have long been interested in both the carbon cycle and climate, and the convergence of these interests with CO_2 studies is an exciting and very active development in earth science. Because any attempt to summarize this rapidly evolving field would become outdated in a few months, the following examples are chosen only to illustrate some of the problems under consideration and to exemplify the range of time scales covered.

Interest in geologic CO_2 variations greatly increased after publication of the ice core CO_2 data (Berner et al. 1980, Delmas et al. 1980, Neftel et al. 1982;), which show a correlation between rising CO_2 and the end of the last glacial period. This correlation is tantalizing from the standpoint of climate model verification. However, published efforts to model the geochemical processes that could be responsible for the correlation have unanimously invoked climate change as the primary mechanism driving the deglacial CO_2 increase. For example, Broecker (1981, 1982a,b) and Broecker and Takahashi (1984) pointed out that, although the direct chemical effects of warming the sea surface could explain only a fraction of the observed atmospheric CO_2 increase, climatic effects may have moved CO_2 from the oceans to the atmosphere by altering the oceans' nutrient budgets. Berger (1982) hypothesized similar effects on the oceans' alkalinity budget.

Several mechanisms have been suggested for changing oceanic phosphate and nitrate budgets during deglaciation. Continental shelf and estuarine sediments are significant sites of organic matter sedimentation, and the deglacial rise in sea level must have expanded the shelf area available for organic deposition. Broecker (1981, 1982a,b) calculated that this deposition could have removed approximately 30% of the oceans' dissolved phosphate, thereby reducing algal productivity in the ocean surface and causing dissolved CO_2 in upwelling waters to be released to the atmosphere rather than consumed in the assimilation of organic matter. Although some carbon as well as phosphorus would be removed by shelf sedimentation, the atmospheric effect of organic carbon removal would be negligible compared to the effect of decreased productivity. McElroy (1983) suggested that nitrate rather than phosphate may have been the limiting nutrient removed by sedimentation on the shelves during a rise in sea level. Whatever the nature of the nutrient effect, Shackleton et al. (1983) showed that carbon isotope data from deep sea sediments are consistent with the hypothesized deglacial nutrient depletion and its proposed influence on atmospheric CO_2.

Berger (1982) hypothesized enhanced deposition of shallow carbonate sediments during the deglacial sea level rise. This mechanism would have decreased the alkalinity and increased the CO_2 partial pressure of ocean surface water, causing degassing of CO_2 to the atmosphere. Although this CO_2 would eventually be mixed into the deep ocean and reacted there with carbonate sediments, Berger suggested that the lag time between the release of the excess ocean surface CO_2 and its "neutralization" in the deep sea would be a time of higher atmospheric CO_2 levels.

Another deglacial mechanism proposed by Broecker (1981, 1982a,b) is a decrease in the ratio of carbon to phosphorus fixed by plankton at the ocean surface. This mechanism, which requires fundamental biochemical changes in response to climate-related ecological changes, could have caused dissolved CO_2 in upwelling waters to be released to the atmosphere because algae would have become less efficient assimilators of carbon. A related mechanism, proposed by Broecker and Peng (1984), is a deglacial increase in the ratio of planktonic carbonate to organic carbon production. This change would have affected atmospheric CO_2 in the same way as the shallow carbonate mechanism suggested by Berger.

Broecker and Takahashi (1984) proposed that deglacial changes in ocean mixing may have caused the atmospheric CO_2 increase. They emphasized the importance of deep sea "ventilation," which can be defined as the extent of equilibrium of polar surface waters (viewed as deep water "outcrops") with atmospheric CO_2. This process competes with ocean mixing and the photosynthesis–respiration cycle across the thermocline for control of the distribution of carbon within the ocean–atmosphere system. A deglacial shift in the balance among these processes—such as a shift in the fraction of nitrate and phosphate used by plankton in polar surface waters—might have contributed to the CO_2 increase. This possibility was considered more explicitly by Knox and McElroy (1984), Sarmiento and Toggweiler (1984), and Siegenthaler and Wenk (1984).

While ice core data document CO_2 changes occurring over time scales of perhaps decades to tens of thousands of years, it will be argued in this chapter that the above hypotheses can account only for CO_2 changes that are transient and small compared to changes that are possible over longer time scales. More persistent long-term CO_2 changes have been modeled by Garrels et al. (1976), Budyko and Ronov (1979), and Berner et al. (1983).

Garrels et al. (1976) suggested that, over time scales longer than a few thousand years, the relationship among atmospheric CO_2, oceanic carbon, and carbonate sediments can be approximated by an equilibrium among CO_2, calcite, and the dissolved calcium and bicarbonate in seawater. They constructed a steady-state model based on this equilibrium and on other expressions representing the major weathering, sedimentation, and biological fluxes in the exogenic carbon and oxygen cycles. Perturbations of their steady-state model generated long-term changes in atmospheric CO_2; for example, a tripling of erosion rates caused a 2.5-fold CO_2 increase in 10 million years.

Budyko and Ronov based their model on the observation by Ronov (1976) that Phanerozoic sediments from the Russian platform show a close correlation between carbonate and volcanic rock abundances. They concluded that the rate of carbonate deposition is a response to the rate of CO_2 release by volcanism, and they constructed a simple model in which the rate of carbonate deposition is proportional to the atmospheric CO_2 concentration. This model enabled Budyko and Ronov to estimate Phanerozoic CO_2 levels from Ronov's plot of carbonate rock mass versus time (Fig. 19.1).

Although the model of Berner et al. (1983) was more complex, it retained fundamental similarities to that of Budyko and Ronov. Atmospheric CO_2 concentrations for the past 100 million years were assumed to have been determined primarily by the balance between CO_2 release by tectonism and CO_2 consumption by weathering (and subsequent carbonate deposition). Unlike Budyko and Ronov, however, Berner et al. estimated CO_2 production rates from estimates of sea floor spreading rates, which they assumed to reflect relative levels of volcanic and metamorphic activity. Weathering rates were assumed to respond both to land areas, which were inferred from paleogeographic maps, and to global temperatures, which were calculated by the model from atmospheric CO_2 levels. The resulting nonlinear weathering response to CO_2 was much weaker than the linear feedback assumed by Budyko and Ronov. In spite of these differences, both the results of Berner et al. (Fig. 19.2) and those of Budyko and Ronov suggest mid-Cretaceous (110 up to 85 million years ago) and Eocene (55 up to 40 million years ago) CO_2 peaks. Interestingly, both of these periods were times of relatively warm climate (e.g., see Frakes 1979; Shackleton and Boersma 1981; Barron 1983).

Tectonic processes, chemical weathering, and carbonate sedimentation have also been central to a number of more qualitative proposals for atmospheric CO_2 variations over time scales of tens to hundreds of millions of years. For example, Fischer and Arthur (1977) and Fischer (1981) hypothesized high atmospheric CO_2 concentrations during broad periods of warm climate, high sea level, and high volcanic activity from about 500 up to 350 and about 200 up to 70 million years ago. Other geologists (Pigott and Mackenzie 1979; Mackenzie and Pigott 1981; Sandberg 1983) have documented broad carbonate mineralogical trends that appear to be consistent with these tectonic CO_2 changes (Fig. 19.3). On the other hand, some studies of the last few hundred million years have emphasized bi-

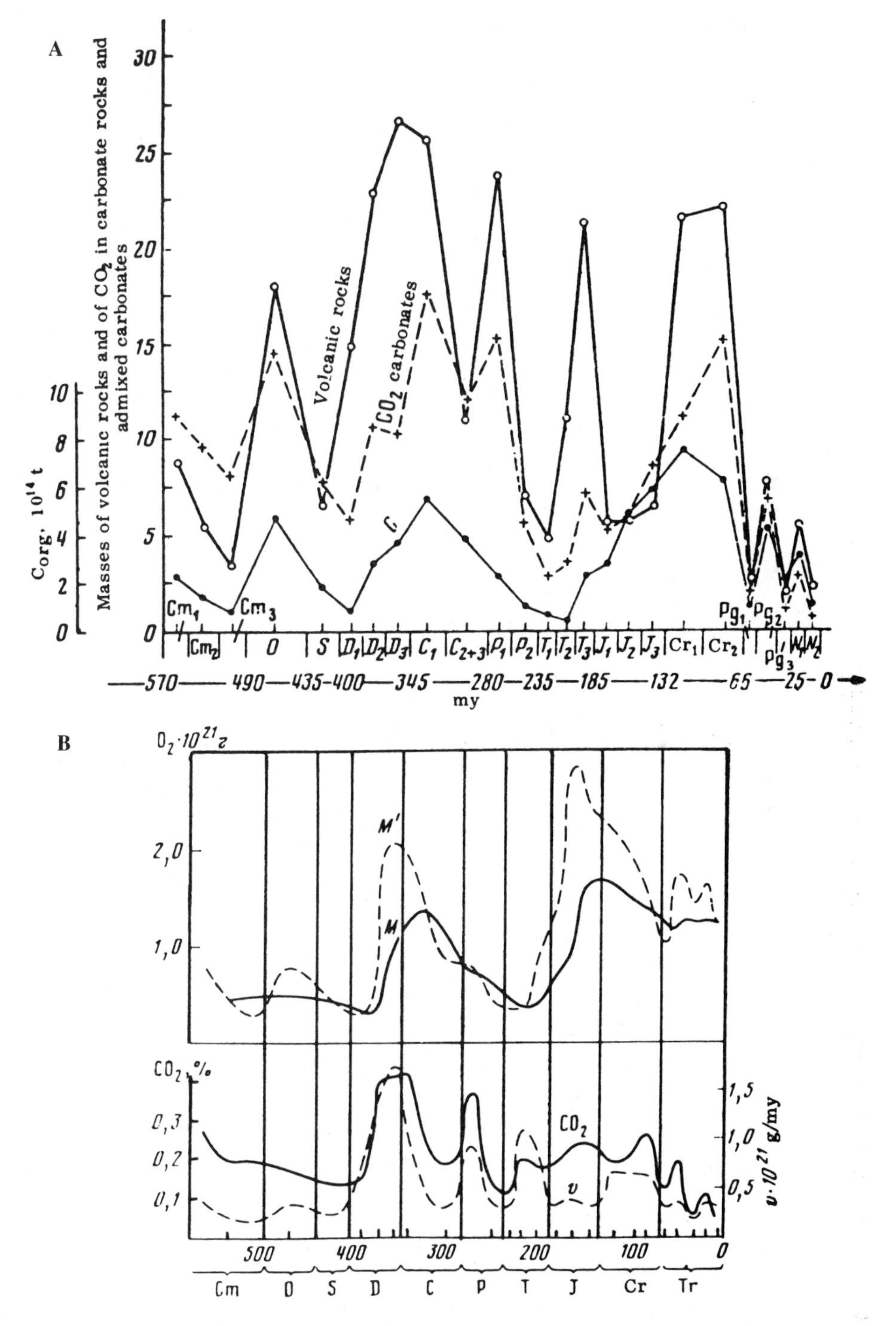

FIGURE 19.1. **A:** Change with time in volumes of volcanic rocks, CO_2 preserved in carbonate rocks and in carbonate impurities in other rocks, and organic carbon in Phanerozoic sediments in continents. (From Ronov 1976.) **B:** Phanerozoic chemical evolution of the atmosphere. (From Budyko and Ronov 1979.)

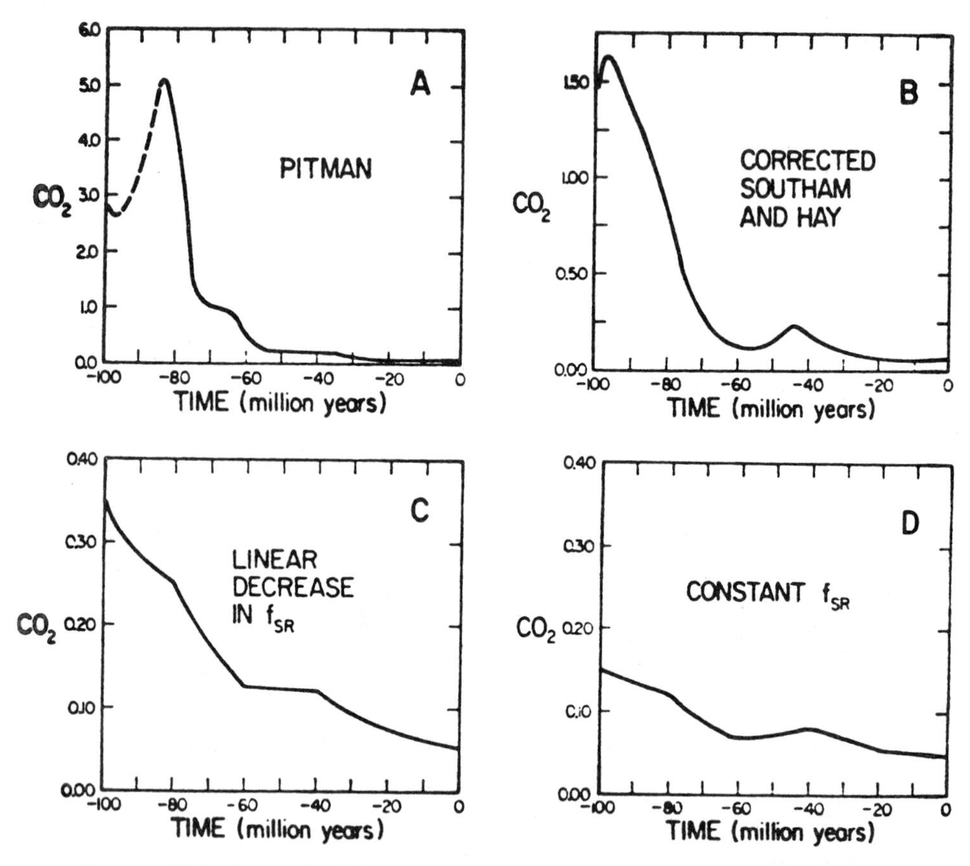

FIGURE 19.2. Atmospheric CO_2 concentrations for the last 100 million years—generated by a geochemical model with four different sea-floor spreading rate scenarios. (From Berner et al. 1983.)

ological rather than inorganic controls on atmospheric CO_2, through changes in both marine (Tappan 1968; Tappan and Loeblich 1971) and terrestrial (McLean 1978) plant productivity.

It has been suggested that the climatic effects of CO_2 may have been essential to the origin of life. According to current understandings of stellar evolution (e.g., Newman and Rood 1977), the luminosity of the sun may have been 25% lower during the Archean period (4 up to 2.5 billion years ago) than at present. Sagan and Mullen (1972) suggested that an atmospheric greenhouse effect would have been necessary to keep the surface of the earth from freezing during this time, when early forms of life were preserved as microfossils in waterborne sediments. Owen et al. (1979) proposed that volcanic CO_2 might have provided the necessary climatic effect, and Walker and Hays (1981) hypothesized that CO_2 and its temperature-dependent reactions with silicate minerals may comprise a long-term negative feedback mechanism that has stabilized the earth's surface temperatures since Archean times.

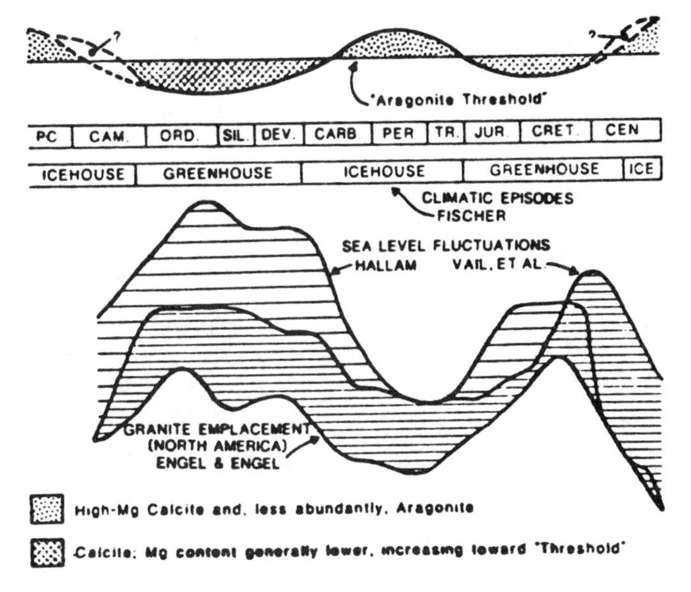

FIGURE 19.3. **Top curve** shows $CaCO_3$ mineralogic trends; *bars* show geologic periods and climatic episodes; **Bottom curves** show sea level fluctuations and granite emplacement activity. (From Sandberg 1983. Reprinted by permission from *Nature* 305 (5929):19–22. ©1983 Macmillan Journals Limited. Adapted from Fischer 1981.)

These geologic analog studies—ranging in time scale from thousands to billions of years—share a preponderant emphasis on circumstantial evidence. Except for the ice core data, which do not extend very far back in geologic time, there is as yet no direct evidence for past atmospheric CO_2 levels. Inferences from carbonate mineralogy and paleoclimate data depend on assumed but not simple chemical and physical relationships. Other inferences, such as relationships between atmospheric CO_2 and tectonism, can be quantified only by using complicated geochemical models. These models must be capable of integrating many separate processes and effects to test hypotheses for consistency with the many lines of evidence available in the geologic record. In developing such models, geochemists must confront some very fundamental questions. To what extent can carbon be redistributed within the ocean–atmosphere–biosphere system? How are the carbon fluxes and reservoirs within this system related to the long-term carbon cycle involving weathering, sedimentation, and tectonism? What are the relationships among atmospheric CO_2, ocean chemistry, and carbonate sedimentation? These questions are among the many that must be answered to understand the processes controlling CO_2 and climate.

Time Scales as Frames of Reference

To cope with the broad range of time scales represented by geologic analogs, it is essential to recognize that the relative importance of various

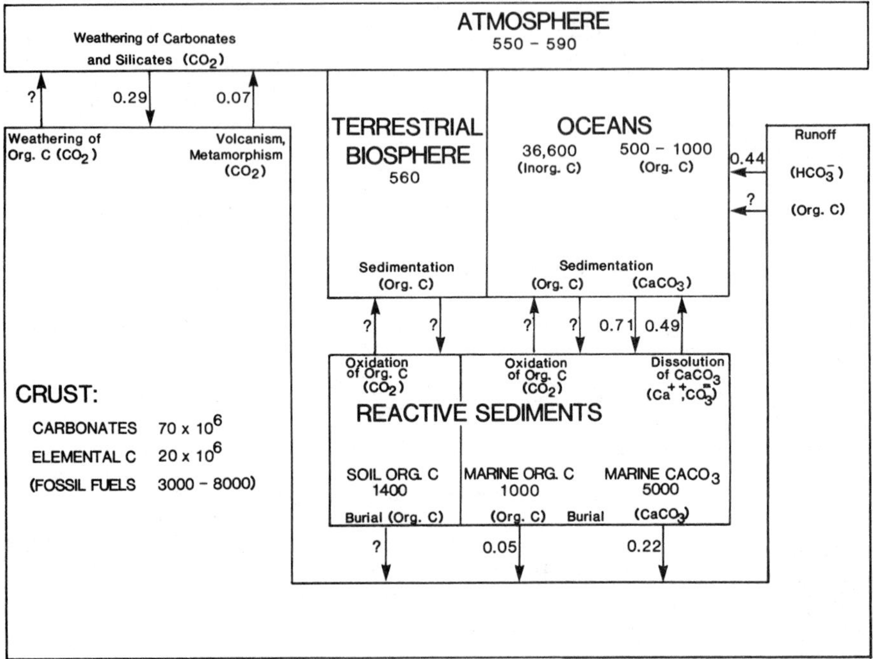

FIGURE 19.4. The carbon cycle over geologic time scales. Geometry corresponds to frames of reference summarized in Table 19.1. Atmospheric reservoir size corresponds to CO_2 partial pressure range of 260 to 280 μatm. Ocean reservoir sizes calculated from GEOSECS data as reported by Takahashi et al. (1981) and organic carbon data of Menzel (1974) and Wangersky (1976). Reactive sediment reservoirs from Broecker and Takahashi (1977) ($CaCO_3$) and Post et al. (1982) (soil organic C). Marine reactive organic C estimate based on assumed 0.25 to 0.5% organic C in sediments contributing to reactive $CaCO_3$ reservoir below shelf break, and 1 to 2% organic C in shelf sediments to depth of 1 m. Fossil fuel estimates correspond to range from identified to ultimately recoverable resources reported by World Energy Conference (1980). Crustal C reservoirs from Holland (1978). Runoff, weathering, volcanism, metamorphism, and carbonate sedimentation fluxes adapted from Berner et al. (1983). Marine organic C burial flux from Berner and Raiswell (1983). Deep ocean $CaCO_3$ dissolution flux from Broecker and Peng (1982). $CaCO_3$ sedimentation flux calculated from sum of burial and dissolution fluxes. Other fluxes, represented by question marks, are more uncertain.

carbon cycle processes and reservoirs often depends on the time scale under consideration. Figure 19.4 shows a useful way of categorizing components of the carbon cycle into time-related frames of reference. The atmosphere, biosphere, and oceans comprise a discrete subsystem that can be considered to be isolated from other components for relatively short time scales. Over somewhat longer time scales, the frame of reference must be expanded to include the "reactive" carbon in soils and in the uppermost layers of marine sediments. Finally, over still longer time scales, the frame of reference must include carbon in the earth's crust.

These frames of reference are summarized in Table 19.1. The time-scale durations specified in the table are extremely uncertain, and the frames of reference and their time scales overlap considerably. For example, the ice core data suggest that the relatively rapid deglacial CO_2 increase should be modeled within the shortest frame of reference; however, deep sea sediments show evidence for deglacial changes in carbonate dissolution that require understanding the 10^3– to 10^5– year frame of reference (Broecker and Broecker 1974; Sundquist et al. 1977). The burning of fossil fuels may be viewed as man's acceleration of normally long-term oxidation processes into a short-term frame of reference. Conversely, although most fossil fuel CO_2 models have been limited to the shortest frame of reference in Table 19.1, the potential magnitude of the fossil fuel perturbation (see Fig. 19.4) implies a need for modeling over longer time scales. The modeling described later in this chapter is a partial effort to meet this need.

Before examining the long-term effects of anthropogenic CO_2, it is useful to consider the range of CO_2 excursions that may have occurred in the past over the time scales summarized in Table 19.1. The approach taken here is subjective and simplistic. It is based on assumptions derived from somewhat arbitrary judgments regarding feasible changes in the natural carbon cycle. These assumptions have also been influenced by a desire to keep the calculations relatively simple. Rather than a systematic consideration of all possible carbon cycle excursions, the following paragraphs present examples. Nevertheless, this approach suggests a working hypothesis that will be very useful in more detailed and sophisticated studies of past CO_2 changes.

It is hypothesized that atmospheric CO_2 perturbations within the 10^2– to 10^3–year and 10^3– to 10^5–year frames of reference are relatively small (less than twofold) because of the limited range of possible perturbations and the existence of effective feedback mechanisms. If there are geologic analogs for atmospheric CO_2 variations as high as those anticipated from burning fossil fuels (fourfold to eightfold), they must be sought in the frame of reference associated with time scales of 10^5 years or longer.* The fol-

TABLE 19.1. Summary of time-dependent frames of reference discussed in text.

Time scale (order of magnitude in years)	Frame of reference
10^2–10^3	Atmosphere, biosphere, and oceans
10^3–10^5	Atmosphere, biosphere, and oceans plus reactive sediments
$>10^5$	Atmosphere, biosphere, and oceans plus reactive sediments and earth's crust

lowing examples illustrate (but do not prove) this hypothesis. They exemplify the kinds of perturbations and feedbacks associated with each time scale, some of the implied effects on marine sediments, and the overlap among perturbations and feedbacks associated with different time scales.

10^2– TO 10^3–YEAR TIME SCALES

Over short† time scales, the atmosphere–ocean–biosphere system can be considered a closed system. This assumption provides a convenient basis for calculating the magnitude and effects of perturbations within the 10^2– to 10^3–year frame of reference. Short-term perturbations are simply redistributions of carbon or alkalinity (or both) within the atmosphere–ocean–biosphere system. For the calculations here, this system is considered to be composed of four simple reservoirs: the atmosphere, the ocean-surface mixed layer, the deep ocean, and the terrestrial biosphere. The ocean-surface reservoir is assumed to be at equilibrium with respect to atmospheric CO_2, and ocean biological productivity is assumed to be manifested in the differences between surface and deep ocean values of total dissolved inorganic carbon (ΣCO_2) and total alkalinity (A_T). Given these assumptions and those of conservation of mass (carbon) and charge (total alkalinity) in a closed system, the effects of simple perturbations can be calculated. These calculations require a computational scheme and choice of appropriate constants for determining the equilibria between the ocean surface and the atmosphere. For longer time scales, the same scheme can be used to calculate equilibria between the deep ocean and carbonate sediments. Throughout this chapter, I have followed the scheme outlined by Takahashi et al. (1980), using the constants of Mehrbach et al. (1973), Lyman (1956), Weiss (1974), Ingle (1975), and Edmond and Gieskes (1970).

Table 19.2 illustrates such a short-term perturbation: a decrease in ocean-surface biological productivity while the terrestrial biosphere is held constant. Here, biological productivity is defined simply by its manifestation in the magnitude of depletion of ΣCO_2 and/or A_T in ocean surface waters. To emphasize the fundamental similarity between the scenarios in Table 19.2 and the deglacial nutrient hypothesis, the scenario calculations are presented for a roughly 30% decrease in productivity. The resulting redistribution of carbon from the deep ocean to the ocean surface and atmo-

---

*An important exception to this hypothesis is the possibility of "catastrophic" carbon cycle perturbations, such as might be caused by meteorite impacts. However attractive this possibility might be as an analog of the pulse-like anthropogenic CO_2 perturbation, there is little if any documentation of such sudden carbon cycle changes in the geologic record.

†To many CO_2 researchers, who are accustomed to "long-term" projections into the next century, 100 to 1000 years will not seem like a short time. All is relative, of course. The principles described here can be applied to time scales of years or decades. For example, for a time scale of 1 to 10 years, the pertinent frame of reference might be only the atmosphere, the ocean-surface mixed layer, and plants with relatively rapid carbon turnover times.

TABLE 19.2. Effects of a productivity decrease on atmospheric pCO_2 and surface and deep ocean A_T and ΣCO_2 values.

	$A_{T_{surf}}$	$A_{T_{deep}}$	$\Sigma CO_{2_{deep}}$	$\Sigma CO_{2_{deep}}$	pCO_2
A:	2320	2400	1981	2263	272
B:	2320	2400	2042	2249	363
C:	2340	2400	2052	2250	352
D:	2318	2378	2041	2239	364
E:	2392	2452	2078	2276	325

Units are 10^{-6} atm, 10^{-6} eq · kg^{-1}, and 10^{-6} mol · kg^{-1}, respectively.
A: Initial conditions. Surface and deep ocean temperatures and salinities are 19°C and 1.5°C, and 35‰, and 34.7‰, respectively. Initial calcite saturation horizon is at 4000 m.
B: Productivity decreased; no change in $CaCO_3$ production.
C: Productivity decreased, including change in $CaCO_3$ production.
D: Calcite saturation horizon returned to 4000 m by calcite loss.
E: Calcite saturation horizon lowered to 5000 m by calcite addition.

sphere causes a CO_2 change very similar to that calculated by Broecker (1982a,b).

Table 19.2 also illustrates possible responses of the ocean alkalinity distribution to ocean productivity changes. If the productivity decrease is accompanied by a comparable decrease in the rate of precipitation of $CaCO_3$ in shallow waters, the resultant atmospheric CO_2 increase is slightly diminished. Other responses may involve interactions with carbonate sediments. The most conspicuous feature in the worldwide distribution of deep-sea sediments is the transition from carbonate-rich sediments at shallow depths to carbonate-depleted sediments in the deep ocean. This transition reflects the difference between shallow and deep seawaters in their state of saturation with respect to the carbonate minerals calcite and aragonite (Li et al. 1969; Broecker and Takahashi 1978; Plummer and Sundquist 1982). The calcite saturation horizon (the depth at which the deep ocean water is saturated with respect to calcite) would be affected by a decrease in productivity. The horizon moves only slightly when the productivity change affects both carbon and alkalinity ($CaCO_3$) removal from surface waters. However, a decrease in $CaCO_3$ precipitation in surface waters over sediments above the saturation horizon would cause an imbalance in the oceans' long-term alkalinity budget, since more river bicarbonate would be delivered to the oceans than could be removed by $CaCO_3$ sedimentation. The response would be an alkalinity increase extending over longer time scales, eventually resulting in a deeper saturation horizon. Such a change is exemplified by the calculations in Table 19.2 for a shift of the calcite saturation horizon from 4000 to 5000 m. Such a shift (although not necessarily of exactly the same magnitude) could restore the ocean-wide $CaCO_3$ sedimentation rate to equality with the river bicarbonate flux.

Table 19.3 shows calculations for a variety of short-term productivity

TABLE 19.3. Cold surface water chemistry calculated from deep water A_T and ΣCO_2 values and various differences between these values and the surface water values. Calcite saturation at 4000 m; units, temperatures, and salinities as in Table 19.2.

ΔA_T	−80	−40	0	−160	−80	−80	−80	−40	−160
$\Delta\Sigma CO_2$	−150	−130	−110	−190	−95	−205	−260	−75	−260
A:				$A_{T_{deep}} = 2400$, $\Sigma CO_{2_{deep}} = 2263$					
A_T	2320	2360	2400	2240	2320	2320	2320	2360	2340
ΣCO_2	2113	2133	2193	2073	2168	2058	2003	2188	2003
Ω_C	3.101	3.388	3.681	2.545	2.408	3.825	4.572	2.679	3.438
pH	8.280	8.317	8.351	8.198	8.153	8.391	8.489	8.197	8.353
pCO_2	272	251	233	326	380	202	154	345	216
B:				$A_{T_{deep}} = 4800$, $\Sigma CO_{2_{deep}} = 4802$					
A_T	4720	4760	4800	4640	4720	4720	4720	4760	4640
ΣCO_2	4652	4672	4692	4612	4707	4597	4542	4527	4542
Ω_C	2.983	3.245	3.521	2.502	2.398	3.675	4.458	2.620	3.304
pH	7.858	7.893	7.927	7.785	7.758	7.955	8.046	7.794	7.913
pCO_2	1635	1514	1405	1919	2084	1289	1027	1925	1404

A: Deep water A_T and ΣCO_2 comparable to present values.
B: Deep water A_T and ΣCO_2 for a hypothetical ocean with values approximately twice as high as present.

changes expressed in terms of various combinations of ΔA_T and $\Delta\Sigma CO_2$ (the total alkalinity and total dissolved inorganic carbon differences between deep water and cold surface water). Although these calculations do not adhere strictly to the closed system mass and charge balance constraints, they do demonstrate the relative magnitude of surface water and atmospheric pCO_2 changes that can be associated with significant productivity changes. Also shown are the ranges of ocean surface pH and Ω_C (the ratio of the $Ca^{2+} \times CO_3^{2-}$ ion product to the solubility of calcite). The range of pCO_2 changes is well within $\pm 50\%$ of the initial value of 272 μatm for A_T and ΣCO_2 values near those presumed to approximate the preindustrial oceans. Also shown are the effects of hypothetical productivity changes on an ocean with A_T and ΣCO_2 values approximately twice the present values. Again, the pCO_2 changes are within $\pm 50\%$ of the initial value of 1635 μatm.

Over time scales of 10^2 to 10^3 years, significant changes in ocean productivity do not appear to be capable of causing CO_2 changes comparable to those that could be generated from burning fossil fuels. Indeed, from Fig. 19.4, it is apparent that only an instantaneous transfer of much of the carbon in the terrestrial biosphere directly to the atmosphere could possibly double the atmospheric CO_2 content. However, in such a case, much of the CO_2 derived from the biosphere would be rapidly dissolved in the oceans. A larger potential CO_2 source is the organic carbon in soils and marine sediments. Natural perturbations from this source require consideration of longer time scales.

10^3– to 10^5–Year Time Scales

Table 19.4 illustrates the interactions with "reactive" sediments characteristic of 10^3– to 10^5–year time scales. The calculations assume that about 35% of the combined soil and marine sediment organic carbon reservoir is oxidized and transferred to the atmosphere and oceans as CO_2. In the absence of any carbonate dissolution feedback, the resulting atmospheric pCO_2 increase is about 40%. However, this increase is cut by more than half if the added CO_2 is allowed to react with carbonate sed-

Table 19.4. Effects of adding oxidized organic carbon to oceans and atmosphere.

	$A_{T_{surf}}$	$A_{T_{deep}}$	$\Sigma CO_{2_{surf}}$	$\Sigma CO_{2_{deep}}$	pCO_2
A:	2320	2400	1981	2263	272
B:	2320	2400	2053	2335	383
C:	2451	2531	2118	2400	313

Units, temperatures, and salinities same as in Table 19.2.
A: Initial conditions; calcite saturation horizon at 4000 m.
B: Oxidized carbon added; no $CaCO_3$ dissolution.
C: Oxidized carbon added; calcite saturation horizon returned to 4000 m by calcite addition.

iments in amounts sufficient to maintain the calcite saturation horizon at its initial depth.

The time scales of reactive sediment reactions are very poorly known. It seems likely that deep-sea carbonate dissolution will respond asymmetrically to changes in deep ocean ΣCO_2 values. If deep ocean ΣCO_2 values increase, the rising saturation horizon will expose previously deposited carbonate-rich sediments to dissolution. If deep ocean ΣCO_2 values decrease, the falling saturation horizon will only open new sea floor areas to contemporaneous carbonate deposition. Thus, the $CaCO_3$ buffering of ΣCO_2 increases is probably more rapid than the $CaCO_3$ buffering of ΣCO_2 decreases. In either case, the effects of $CaCO_3$ buffering must be incorporated into atmospheric CO_2 calculations for time scales of 10^3 to 10^5 years.

Time Scales Longer Than 10^5 Years

Figure 19.4 shows clearly that, given enough time (or enough human activity), the earth's crust contains enough elemental and carbonate carbon to alter drastically the carbon and alkalinity contents of the atmosphere, oceans, biosphere, and reactive sediments. Thus, over time scales greater than 10^5 years, it is clear that large atmospheric CO_2 variations are possible. The geochemist's principal task is to show whether such changes are feasible. One way of doing this is to construct models such as those of Garrels et al. (1976), Budyko and Ronov (1979), and Berner et al. (1983). A different but essential approach was suggested by Bender and Graham (1978) and Bender (1984). Rather than modeling the effects of hypothesized carbon cycle perturbations, they used the range of effects observed in the geologic record to constrain the range of perturbations. Although this approach suffers from limitations in the kind and amount of information available in sediments, it is an absolutely necessary complement to the modeling studies described here.

Bender (1984) concluded that, whereas other factors could probably not have caused large changes in atmospheric pCO_2 during the Cenozoic era (70 to 2 million years ago), ocean ΣCO_2 variations could have been accompanied by large pCO_2 variations. I have hypothesized and attempted to show by example that such changes require relatively long time scales. Another significant constraint on large changes in ΣCO_2 and pCO_2 is imposed by the history of the calcite compensation depth (CCD) throughout the Cenozoic. Many investigators (e.g., Van Andel 1975) have observed that the CCD appears to have remained between depths of 3000 and 5000 m at least since the end of the Cretaceous. Although the detailed relationships between the CCD and the calcite saturation horizon are variable, it seems reasonable to conclude that the Cenozoic saturation horizon has probably remained between depths of about 2000 and 5000 m. If this is true, then changes in deep ocean ΣCO_2 values must have been accompanied by closely related changes in deep ocean A_T values.

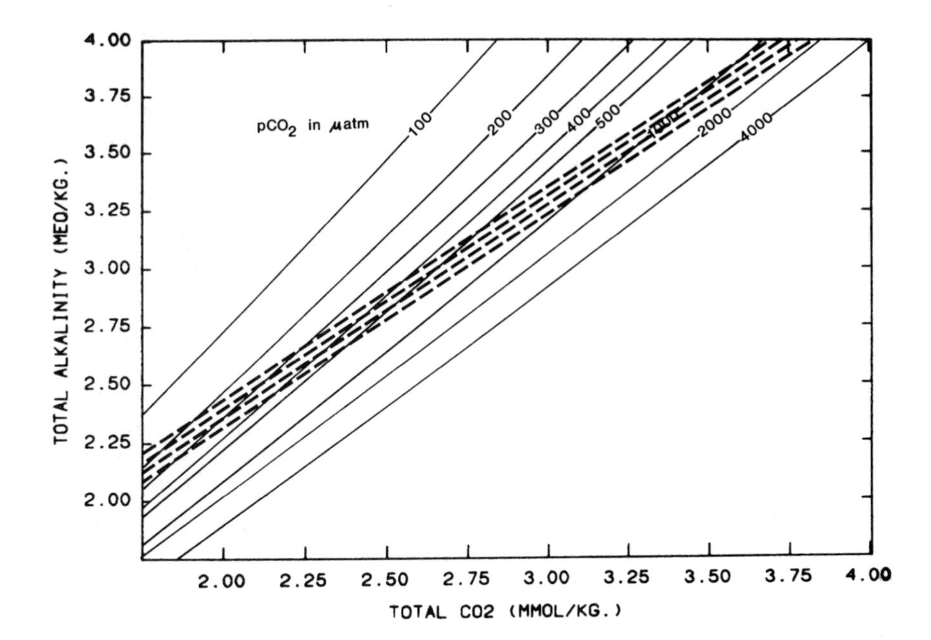

FIGURE 19.5. Deep water A_T and ΣCO_2 values corresponding to calcite saturation horizon depths (*dotted lines*) and atmospheric CO_2 partial pressures (*solid lines*). Saturation horizon contours (from top to bottom, 5000, 4000, 3000, and 2000 m) were calculated for deep water compositions derived from the surface water values shown ($\Delta A_T = -80 \times 10^{-6}$ eq • kg^{-1}; $\Delta\Sigma CO_2 = -300 \times 10^{-6}$ mol • kg^{-1}) assuming a deep ocean salinity of 34.7‰, and temperatures of 2.5°C at 2000 m, 2.0°C at 3000 m, 1.5°C at 4000 m, and 1.0°C at 5000 m. The pCO_2 contours were calculated for warm surface water values shown, assuming a salinity of 35.5‰, and a temperature of 22°C.

Figure 19.5 shows the relationships between deep ocean ΣCO_2 and A_T implied by calcite saturation at depths of 2000, 3000, 4000, and 5000 m. These calcite saturation curves are plotted with surface water pCO_2 contours using axes representing surface rather than deep ΣCO_2 and A_T values. It is very clear that, even given the constraints on the Cenozoic CCD, large differences in surface water and atmospheric pCO_2 are possible. Large pCO_2 variations are feasible even with no changes in productivity, marine $CaCO_3$ precipitation, or the depth of the calcite saturation horizon. However, large pCO_2 variations require drastic and closely interrelated changes in both ocean ΣCO_2 and A_T values. The magnitude of these changes is beyond the capacity of the 10^2– to 10^5–year frames of reference (see Fig. 19.4). If such large changes in oceanic and atmospheric chemistry have occurred, they must have been driven by imbalances among fluxes of carbon through the earth's crust over time scales longer than 10^5 years.

A Model of the Long-Term Effects of Anthropogenic CO_2

One of the most important lessons of geologic analogs for CO_2 research is that short-term perturbations of the carbon cycle can set off long-term geochemical responses. Thus, the burning of fossil fuels raises not only the question of climatic effects during the next century, but also a wide assortment of geochemical questions relating to much longer time scales. For example, how quickly will fossil fuel CO_2 react with carbonate sediments? Will weathering rates increase? Will climatic changes alter ocean circulation?

The study of geologic analogs is essential to answering these questions. Another greatly needed component is the development of predictive models that build on current predictive CO_2-carbon cycle models, but are also appropriate for longer time scales. One such model is depicted in Fig. 19.6. It is basically an ocean–atmosphere box model with an interactive sediment box coupled to each ocean box. In contrast to the two-box ocean assumed in the calculations for Tables 19.2–19.4 and Figure

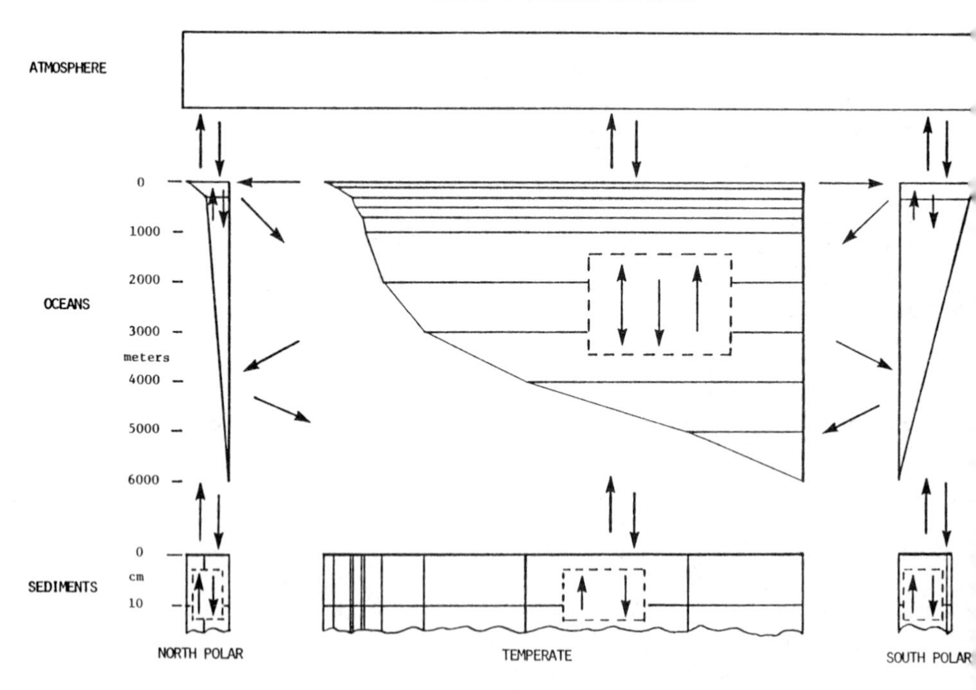

FIGURE 19.6. Atmosphere–ocean–sediment model geometry and transport fluxes. *Double arrow* in temperate ocean represents eddy diffusion; *single arrows* represent generalized exchange fluxes. Figured areas of ocean boxes are proportional to model ocean box volumes, and figured boundaries are proportional to model ocean box areas. See text for further details.

19.4, in this model there are 10 vertically defined "temperate" ocean boxes, representing ocean areas between 50°N and 50°S latitude. There are also four boxes representing north and south polar surface and deep waters. The volumes and areas of each box are calculated from sea-floor hypsometry (for the temperate and south polar boxes, Menard and Smith 1966; for the north polar boxes, Gorshkov 1980). Each sediment box is assumed to be a homogeneous bioturbated layer 10 cm thick.

Whereas most CO_2 models have assumed ocean alkalinities to remain constant, this model requires alkalinities to change through time to simulate the chemical effects of precipitating and dissolving carbonate sediments. Therefore, each ocean box is represented by two differential equations: one for A_T and one for ΣCO_2. The inorganic carbon species are calculated from the salinity, temperature, mean depth, ΣCO_2, and A_T for each box (see Takahashi et al. 1980). Thus, carbonate sediment dissolution rates can be represented by terms that depend on the degree of saturation of the overlying seawater, and gas exchange fluxes can be calculated from surface water equilibrium CO_2 partial pressures with no need for a buffer factor term (Craig 1957, Broecker et al. 1980).

The model incorporates both advective and diffusive ocean mixing terms. The production of cold deep and intermediate waters is parameterized by advective flux terms totaling 40 Sverdrups (1 Sverdrup = 10^6 $m^3 \bullet s^{-1}$) from the deep south polar box to deep temperate boxes 8 through 10*, corresponding to Antarctic bottom water; flux terms totaling 10 Sverdrups from the deep north polar box to deep temperate boxes 6 through 10, corresponding to North Atlantic deep water; and flux terms totaling 20 Sverdrups from the south polar surface box to intermediate temperate boxes, corresponding to Antarctic intermediate water (Gordon and Taylor 1975). These advective fluxes are balanced by upwelling terms in all of the temperate ocean boxes. Vertical diffusive mixing is parameterized in the temperate ocean by exchange coefficients ranging from 1.7 $cm^2 \bullet s^{-1}$ near the surface to 0.6 $cm^2 \bullet s^{-1}$ in the deep ocean (Ku et al. 1980; Li et al. 1984).

The application of this model to relatively long time scales depends primarily on its treatment of marine sediments. As discussed above, the transition from carbonate-rich sediments at shallow depths to carbonate-depleted sediments in the deep ocean reflects the transition from shallow waters that do not dissolve carbonate minerals to deep waters that do. As anthropogenic CO_2 is absorbed by the oceans, the undersaturated regions will become more undersaturated, and some regions that are now supersaturated will become undersaturated. In the model, the state of saturation with respect to calcite is calculated for the depth and composition of each ocean box. By interpolation, the model then estimates the depth

*For reference purposes, the temperature ocean boxes are numbered consecutively from 1, the shallowest, to 10, the deepest.

at which calcite is exactly saturated. The sea-floor area below this calcite saturation horizon is calculated from the sea-floor hypsometric curve. For each box, the model then determines a carbonate dissolution flux that is a function of the sea-floor area exposed to undersaturated water, the degree of undersaturation of the water, and the calcite content of the sediments.

The model dissolution flux calculations are based on laboratory experiments and on a sediment pore water model for calcite dissolution. The exponential rate law derived from dissolution experiments by Keir (1980) has been incorporated into a pore water model [much like that of Keir (1982), but without the assumption of solid carbonate steady-state] to yield the pore water ΣCO_2 and A_T gradients for any degree of undersaturation at the sea–sediment interface. These gradients can be used to calculate the corresponding fluxes from the sediments into the overlying seawater. The model integrates this flux over the total bottom area exposed to undersaturated water for each coupled pair of ocean and sediment boxes. The integration includes an approximation of the effect of the variable degree of undersaturation within each box. The model assumes that the dissolution ΣCO_2 and A_T fluxes entering a particular box are instantaneously mixed throughout the box. Although this approximation obviously minimizes benthic boundary effects, it is consistent with the mixing approximations inherent in all ocean box models.

Several investigators have modeled the effects of carbonate dissolution on a sediment layer that is assumed to be homogeneously mixed by burrowing organisms (e.g., Peng et al. 1977; Sundquist et al. 1977). Peng and Broecker (1978) demonstrated that deep-sea bioturbation is rapid enough so that the bioturbated zone can be treated as well mixed in dissolution models. Thus, the effects of dissolution on the calcite fraction in each sediment box can be modeled using simple equations based on conservation of mass among the flux terms of sedimentation, dissolution, and burial. However, when the ocean–atmosphere–sediment model determines that a box contains a calcite saturation horizon, additional terms and equations are necessary. One term is needed to express the effects of changes in the bottom area exposed to dissolution as the saturation horizon rises or falls. If there is an increase in the bottom area exposed to dissolution, this term must incorporate the effect of the calcite fraction in the sediments newly exposed to dissolution. If there is a decrease in the bottom area exposed to dissolution, there must likewise be a new equation for the calcite fraction in the sediments no longer exposed to dissolution. To keep track of the appropriate calcite fractions, the model divides into two categories the sediments in a box containing a saturation horizon: one category contains all the sediments exposed to dissolution, and the other category contains all the sediments not exposed to dissolution. Each category is characterized by its own values for calcite fraction and bottom area, both of which can vary as the saturation horizon rises or falls. Each sediment box is therefore defined by two mass balance equations, representing sediments that are dissolving and those that are not dissolving. These equations incorporate the effects of transferring sediments from one category to the other as the saturation horizon changes. For each box that does not contain a saturation horizon, the two equations are reduced to one.

In summary, the ocean–atmosphere–sediment model consists of up to 57 differential equations: equations for ΣCO_2 and A_T for each of the 14 ocean boxes, equations for the dissolving and nondissolving calcite fractions in each of the 14 sediment boxes, and an equation representing the CO_2 mass balance in the atmosphere. Before the model can be used to simulate the long-term effects of anthropogenic CO_2, a steady state solution for its system of equations must be found. Even though the real carbon cycle may not have been at steady-state before man's activities, the model must have a steady-state "control" experiment to which "perturbation" experiments can be compared. Because the model's parameterization of sediment dissolution introduces nonlinear constraints with uncertainties that are difficult to estimate, matrix inversion methods have not been used to generate a steady-state solution. Instead, steady-state values for each sediment box are determined by iteratively adjusting the sediment calcite fractions until the value of the sedimentation input is equal to the sum of the values of the burial and dissolution outputs. Then, steady-state constraints for the model as a whole are defined by adjusting the values for parameters representing the "external" CO_2 and alkalinity sources and sinks (i.e., weathering, river input, volcanism, hydrothermal reactions, and sedimentation, see Fig. 19.4). Finally, residual flux terms are calculated for each ocean box so that the residual terms exactly cancel the sums of the previously defined ΣCO_2 and A_T fluxes (i.e., diffusion, advection, precipitation, dissolution, and gas exchange, in addition to the external fluxes). The use of these residual terms means that the steady-state solution for the ocean boxes yields very little information by itself. The ocean part of the model must be viewed essentially as a system of perturbation equations, like most other marine carbon cycle models (for an exception, see Bolin et al. 1983).

This model has many obvious shortcomings. It has no terrestrial biosphere. It ignores the effects of aragonite, high-magnesium calcite, and organic carbon on sediment dissolution. It ignores dissolution of carbonate particles settling through the water column. These omissions are probably relatively insignificant because of the small sizes of the reservoirs relative to the model fossil fuel CO_2 perturbation. Other omissions may be more significant, such as the model's inability to simulate CO_2-induced changes in ocean temperatures, heat transport, or mixing processes. Nevertheless, it is a useful tool for beginning to look at the long-term effects of anthropogenic CO_2 on ocean chemistry and marine sediments. Although the model's many uncertainties preclude all but the most general conclusions, it contributes new perspectives in our evolving understanding of the effects of anthropogenic CO_2.

Model Results

Because there are no reliable preindustrial data for the distribution of dissolved inorganic carbon and alkalinity in the oceans, initial values for these parameters must be estimated using the model. These estimates were determined so that model year 1973 yields the global average ΣCO_2 and A_T distribution calculated by Takahashi et al. (1981) from the GEOSECS data, and then recalculated on a volume-weighted basis to conform to the geometry of the model. The atmospheric CO_2 concentration assumed for the year 1973 was 330 μatm (Bacastow and Keeling 1981). The initial ocean ΣCO_2 and A_T values were estimated by iterative model runs from the year 1700 to the year 1973. The estimated preindustrial atmospheric CO_2 concentration is 295 μatm, which compares favorably to other CO_2 models in which the biosphere is held constant. The model simulated production of CO_2 from fossil fuels and cement by a flux term to the atmosphere box. For the year 1860 to 1980, this term was derived from the annual data of Rotty as compiled by Watts (1982). For extrapolation beyond model year 1980, CO_2 production was assumed to follow the relation

$$\frac{dN}{dt} \quad \alpha \quad N\left[1 - \left(\frac{N}{N'}\right)^n\right]$$

where N is the cumulative CO_2 production, N' is the total amount of recoverable fossil fuels, and n is an exponent that allows the shape of the extrapolation curve to be varied (Perry and Landsberg 1977).

For this study, N' was assumed to be 4.2×10^{17} mol (5000 gigatons) of carbon, a value between the estimates of identified and ultimately recoverable resources reported by the World Energy Conference of 1980 (see Fig. 19.4). The exponent n was set at 0.5, giving an average annual production increase of 2.2% between model years 1980 and 2050. Although this growth rate is higher than many recent projections, the short-term growth rate is not an important factor in the long-term effects considered here. The model CO_2 production data are plotted in Fig. 19.7.

Figure 19.8 shows the model atmospheric CO_2 results for the years 1700 to 11,700. The rapid increase to the year 2200 is typical of models using CO_2 production functions of the type shown in Fig. 19.7. This increase is followed by a rapid decrease during which the effects of sediment interactions, apparent in the lower curves in Fig. 19.8, cause increasingly significant departures from a model run with no sediment interactions, represented by the upper curves. By model year 3500, the model atmosphere stabilizes at a CO_2 partial pressure near 1100 μatm, in the absence of sediment interactions, whereas it falls to about 750 μatm, with buffering by carbonate sediment dissolution. In the model experiment with sediment interactions, the atmospheric CO_2 concentration continues to decrease slowly to the end of the experiment. However, even in model year 11,700, the atmospheric CO_2 concentration is still more than twice its preindustrial value.

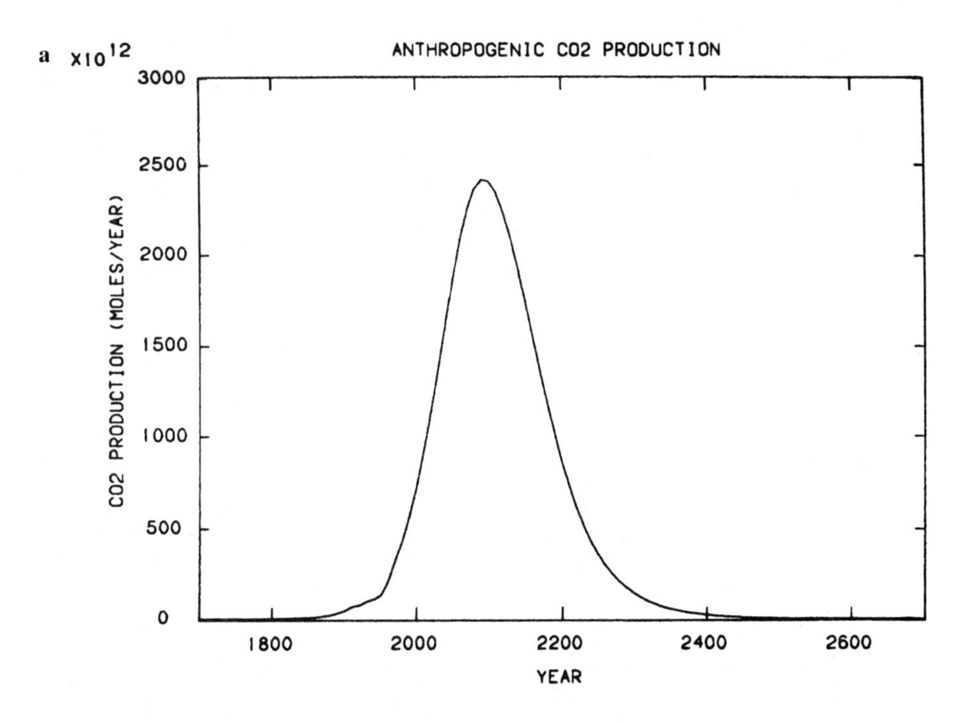

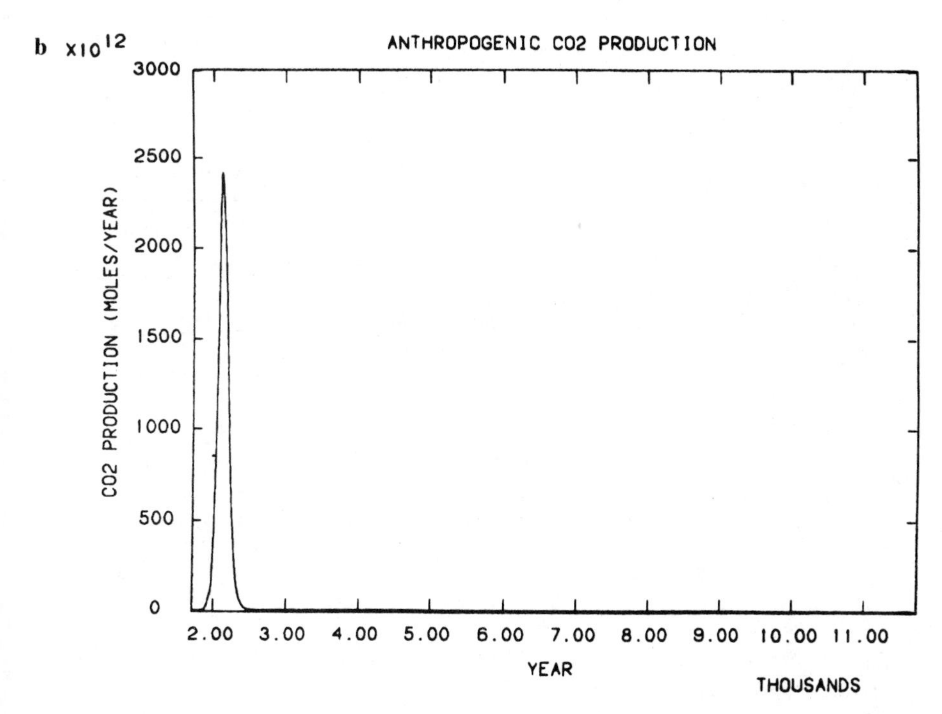

FIGURE 19.7. Model fossil fuel CO_2 production. **a:** 1700–2700. **b:** 1700–11,700.

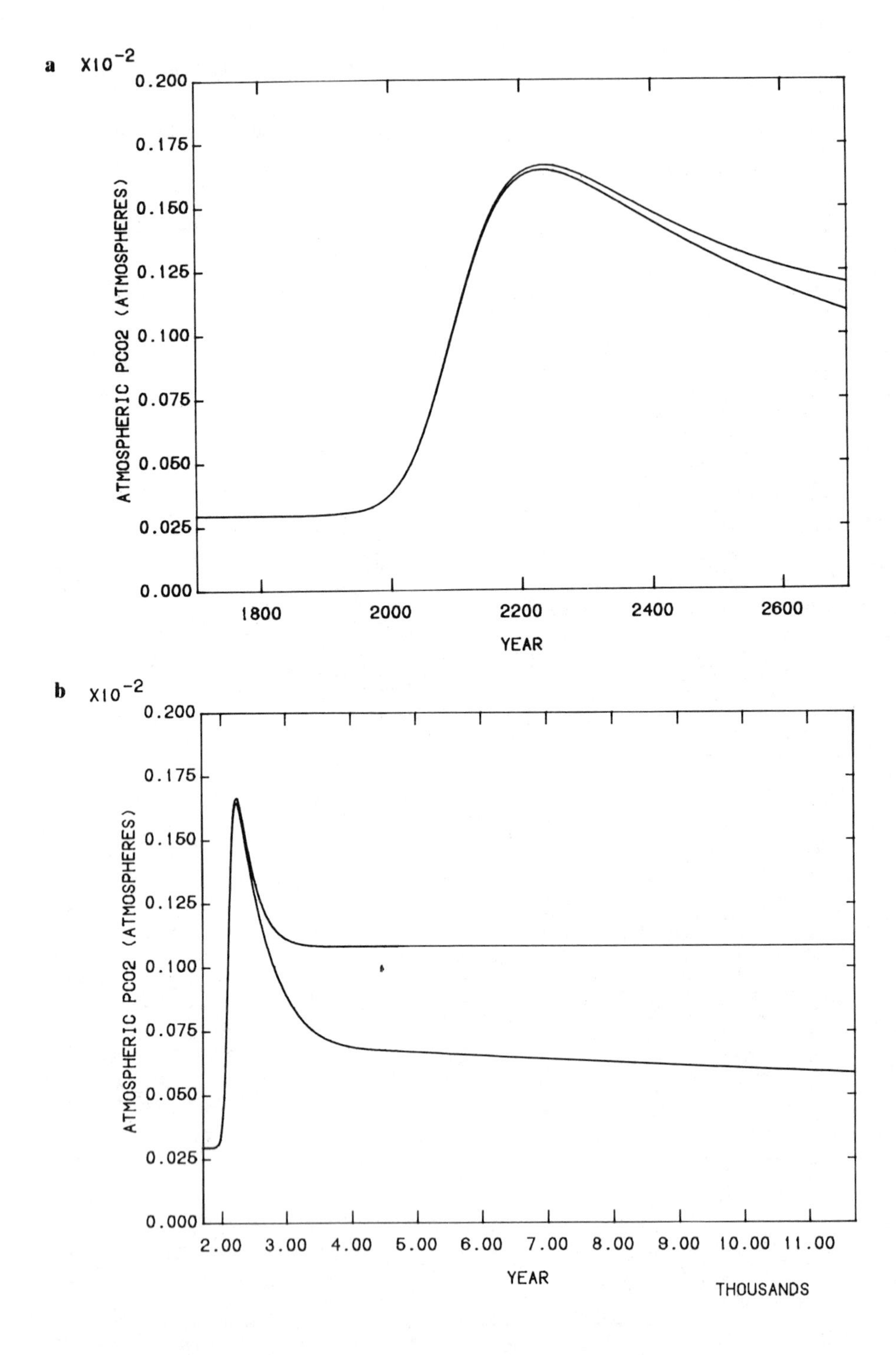

FIGURE 19.8. Model atmospheric CO_2 concentrations. **A:** 1700–2700; **B:** 1700–11,700. **Upper curves** represent simulation with no sediment interactions; **Lower curves** represent simulation with sediment interactions.

Figures 19.9, 19.10, and 19.11 illustrate the significant potential for long-term reactions between anthropogenic CO_2 and carbonate sediments. Model surficial sediments that were comprised mostly of calcite at the beginning of the experiment are left with essentially no calcite after 10,000 model years of extensive dissolution (Fig. 19.9). In the carbonate-rich sediments that were above the initial calcite saturation horizon, the dissolution is so intense that, during part of the experiment, the flux of material removed from the sediments by dissolution actually exceeds the flux of material added by sedimentation. Under these circumstances, the model net accumulation rates become negative (Fig. 19.10). This "negative accumulation" scenario, in which the sediment bioturbated layer dissolves downward into older sediments, may have an analog in the enhancement of dissolution in East Pacific sediments at the end of the last glacial period (Sundquist et al. 1977).

Figure 19.11 shows the extent of calcite undersaturation in the model ocean between the 50° latitude parallels. As expected, the model's large anthropogenic CO_2 input induces a drastic shoaling of the calcite saturation horizon. Transport of CO_2 from the model ocean surface to the oxygen minimum zone, where CO_2 concentrations are already high from oxidizing organic matter, causes these waters to become undersaturated before the waters just below them. Thus, for several decades, the model oceans contain three calcite saturation horizons rather than one.

In the first attempt to quantify the long-term reactions between anthropogenic CO_2 and marine carbonate sediments, Broecker and Takahashi (1977) proposed that the available sediments contain enough carbonate to effectively "neutralize" the CO_2 from all of the world's recoverable fossil fuels. They calculated the amount of carbonate that could be dissolved at water depths below the shelf break before the bioturbated layer becomes depleted of carbonate. They found that this carbonate mass is sufficient to match (on a mole-for-mole basis) the amount of CO_2 that would be produced by burning a reasonable estimate for the world's recoverable fossil fuel resources. Their recoverable resource estimate, sediment carbonate data, sea-floor areas, and sediment-mixing assumptions were virtually identical to those used in this study. Thus, the model results shown in Figs. 19.8 and 19.11 were unexpected. At the end of the experiment, all of the model surficial sediments were essentially calcite-free at depths below the saturation horizon at about 500 m. Yet, the extremely shallow depth of the saturation horizon and the high atmospheric CO_2 concentration at the end of the experiment suggest that the neutralization of fossil fuel CO_2 by carbonate dissolution is less effective than implied by Broecker and Takahashi (1977). From the model alkalinity changes, it can be calculated that the amount of carbonate dissolved is only about half of the "reactive" $CaCO_3$ reservoir estimated by Broecker and Takahashi. This difference can be attributed to different approaches to the fate of "old" carbonate sediments; i.e., sediments that lie below the 10-cm-thick bio-

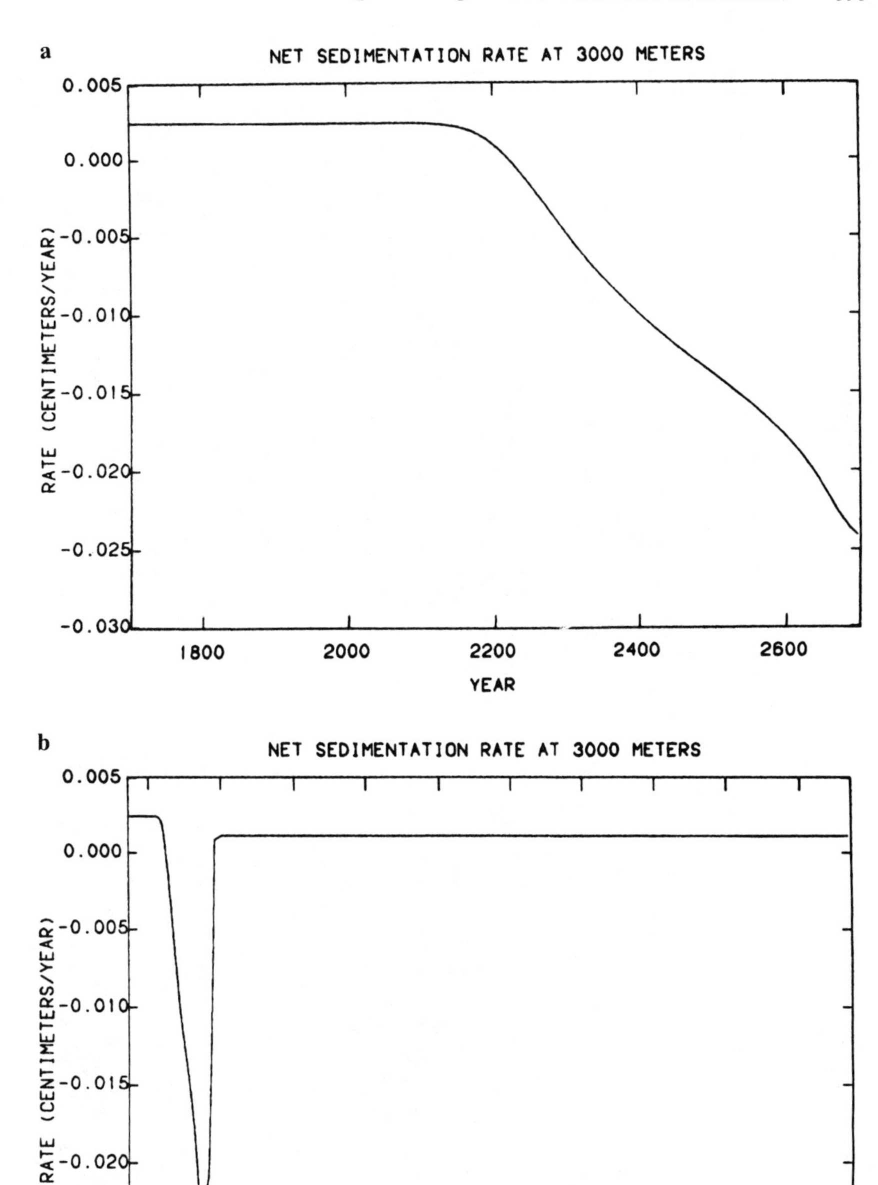

FIGURE 19.10. Model net accumulation rate of sediments at 3000 m. **a:** 1700–2700. **b:** 1700–11,700.

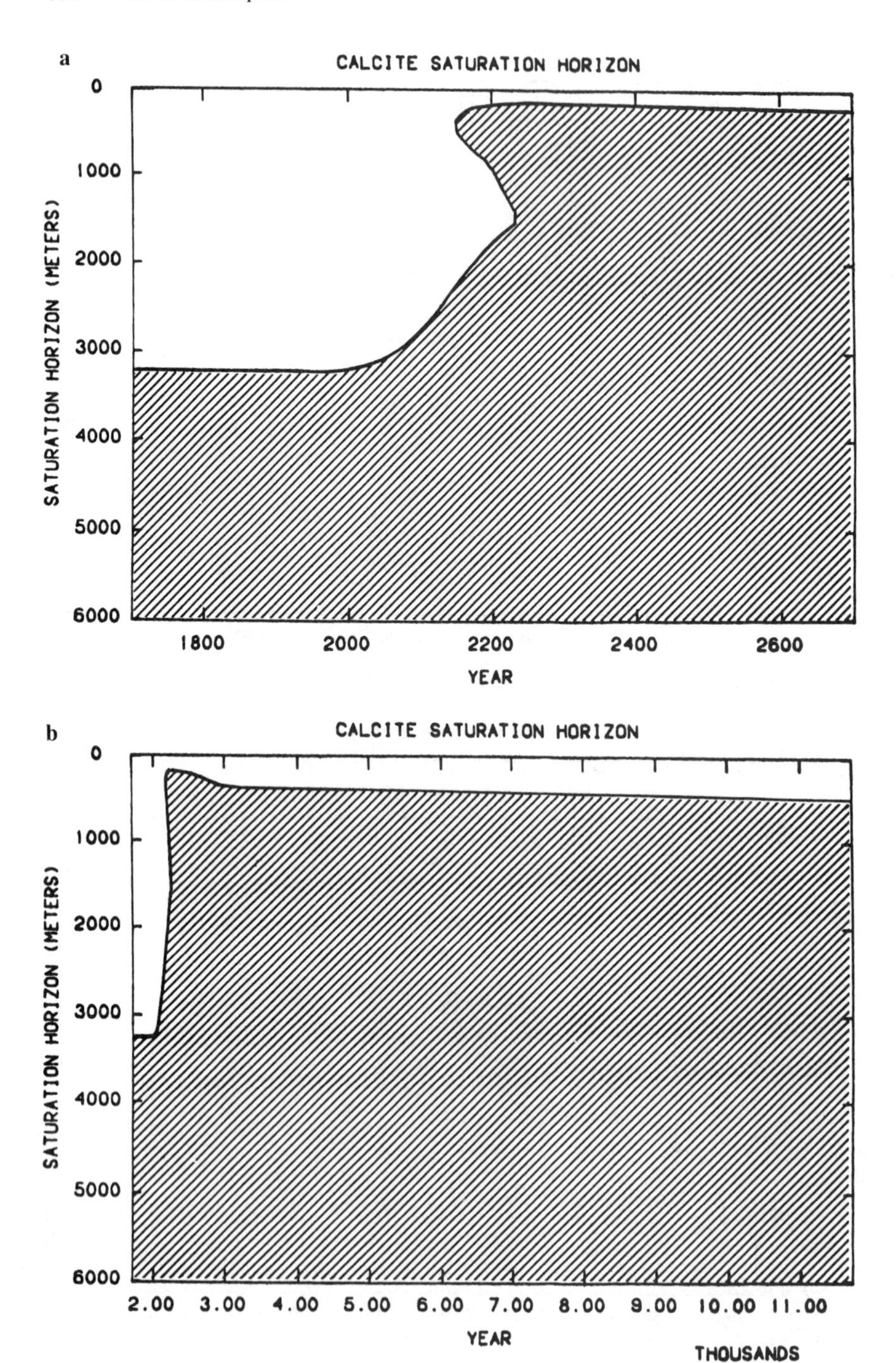

FIGURE 19.11. Model calcite saturation horizon; *shaded area* represents undersaturation with respect to calcite. **a:** 1700–2700. **b:** 1700–11,700.

turbated layer before the addition of fossil fuel CO_2. Broecker and Takahashi in their calculations assumed instantaneous reaction between the fossil fuel CO_2 and "old" sediments to a depth determined by the amount of noncarbonate required to leave a 10-cm-thick carbonate-free layer after the instantaneous dissolution. In the model used here, the dissolution of "old" carbonate sediments is simulated by "negative accumulation" as in Fig. 19.10, but this process is limited by other factors in addition to the fraction of "old" noncarbonate sediments. Ongoing sedimentation of both calcite and noncalcite particles plays a significant role in the model. Preindustrial sediments are buried before they can react with fossil fuel CO_2, and the constant addition of "fresh" calcite slows the reactions with old sediments. Of course, these effects are very sensitive to the dissolution and sedimentation fluxes assumed by the model. Nevertheless, the results presented here suggest that, if man burns all of the world's recoverable fossil fuels, the resulting global geochemical perturbations may persist for longer than 10,000 years.

Eventually, the model will return to a steady-state in which both the atmospheric CO_2 concentration and the calcite saturation horizon are close to their preindustrial values. The time scale of this return can be estimated. At the end of the experiment described in the chapter, the slow decline of atmospheric CO_2 and the deepening of the saturation horizon both result from the fact that calcite sedimentation in the largely undersaturated oceans is not sufficient to remove the bicarbonate added by rivers. The time scale of eliminating this imbalance is comparable to the residence time of river bicarbonate in the oceans—approximately 100,000 years. Thus, the model results fit neatly into the context of time scales as frames of reference. Over time scales of hundreds of years, anthropogenic CO_2 is absorbed and redistributed by the oceans, with little interaction with sediments. Extensive sediment dissolution occurs over time scales of thousands of years. Finally, over time scales of tens of thousands of years, the behavior of the system is determined by the rate of crustal weathering and consequent river transport to the oceans.

Conclusions

This chapter is a beginning attempt to constrain and quantify the relevance of geologic analogs to research on the effects of anthropogenic CO_2. The geologic record is inherently reflective of time scales longer than those usually considered in CO_2 research. However, calculated effects of "natural" perturbations suggest the working hypothesis that, if atmospheric CO_2 has varied by twofold or greater in the geologic past, these variations were most likely caused by processes acting over time scales of 100,000 years or longer. This hypothesis implies that the long-term resolution of the geologic record is not in itself a limitation in the study of geologic analogs of CO_2 changes comparable to those anticipated from burning fossil fuels. Geologic analogs offer not only the possibility of documenting past changes in atmospheric CO_2 but also of understanding the processes that caused them.

An understanding of these long-term processes is also essential to anticipating the long-term effects of anthropogenic CO_2. The model experiment presented here suggests that the geochemical perturbations resulting from anthropogenic CO_2 may persist for tens of thousands of years. Although such long-term projections are subject to many large uncertainties, the overriding conclusion is that significant long-term geochemical effects are possible. The chemical equilibrium relationships in the model are well established, and uncertainties in ocean mixing have diminishing importance over long time scales. Widespread carbonate dissolution in marine sediments, as proposed by Broecker and Takahashi (1977), seems to be geochemically inevitable if man burns all of the world's fossil fuels. This dissolution may be limited in its capacity to "neutralize" such large CO_2 additions.

Implicit in these long-term concerns is the need to look beyond the climatic effects of CO_2 during the next few decades. Although potential climatic effects have justifiably caused great concern and research interest, they must be viewed as part of a continuum of effects that may be manifested over a variety of time scales. The anthropogenic CO_2 perturbation clearly entails important questions relating to time scales longer than the next few decades or even centuries. In a sense, this time frame of responsibility makes the CO_2 problem similar to the problem of storing radioactive waste; that is, the magnitude of the time scales involved demands a clear view of our imprint on future millennia, as well as on present-day events.

References

Bacastow, R. B. and C. D. Keeling. 1981. Atmospheric carbon dioxide concentration and the observed airborne fraction. In B. Bolin (ed.), Carbon Cycle Modelling, Scope 16, pp. 103–112. John Wiley & Sons, Chichester, England.

Barron, E. J. 1983. A warm equable Cretaceous: the nature of the problem. Earth Sci. Rev. 19:305–338.

Bender, M. L. 1984. On the relationship between ocean chemistry and atmospheric pCO_2 during the Cenozoic. In J. E. Hansen and T. Takahashi (eds.), Climate Process and Climate Sensitivity, pp. 352–359. American Geophysical Union, Geophysical Monograph 29, Washington, D.C.

Bender, M. L. and D. W. Graham. 1978. Long term constraints on the global marine carbonate system. J. Marine Res. 36:551–567.

Berger, W. H. 1982. Increase of carbon dioxide in the atmosphere during deglaciation: the Coral Reef hypothesis. Naturwissenschaften 69:S.87.

Berner, R. A., A. C. Lasaga, and R. M. Garrels. 1983. The carbonate-silicate geochemical cycle and its effect on atmospheric carbon dioxide over the past 100 million years. Am. J. Sci. 283:641–683.

Berner, R. A., and R. Raiswell. 1983. Burial of organic carbon and pyrite sulfur in sediments over Phanerozoic time: A new theory. Geochim. Cosmochim. Acta 46:1689–1705.

Berner, W., H. Oeschger, and B. Stauffer. 1980. Information on the CO_2 cycle from ice core studies. Radiocarbon 22:227–235.

Bolin, B., A. Björkström, and K. Holmen. 1983. The simultaneous use of tracers for ocean circulation studies. Tellus 35B:206–236.

Broecker, W. S. 1981. Glacial to interglacial changes in ocean and atmosphere chemistry. In A. Berger (ed.), Climatic Variations and Variability: Facts and Theories, pp. 109–120. D. Reidel Publishing, Boston.

Broecker, W. S. 1982*a*. Glacial to interglacial changes in ocean chemistry. Prog. Oceanography VII:151–157.

Broecker, W. S. 1982*b*. Ocean chemistry during glacial time. Geochim Cosmochim. Acta 46:1689–1705.

Broecker, W. S. and S. Broecker. 1974. Carbonate dissolution on the western flank of the East Pacific Rise. In W. W. Hay (ed.), Studies in Paleo-Oceanography, pp. 44–57. SEPM Special Publishing No. 20.

Broecker, W. S. and T.-H Peng. 1984. The climate chemistry connection. In J. E. Hansen and T. Takahashi (eds.), Climate Processes and Climate Sensitivity, pp. 327–336. American Geophysical Union, Geophysical Monograph 29, Washington, DC.

Broecker, W. S., and T.-H. Peng. 1982. Tracers in the Sea. Lamont-Doherty Geological Observatory, Columbia University, Palisades, New York, 690 p.

Broecker, W. S., T.-H. Peng, and R. Engh. 1980. Modeling the carbon system. Radiocarbon 22:565–598.

Broecker, W. S. and T. Takahashi. 1977. Neutralization of fossil fuel CO_2 by marine calcium carbonate. In N. R. Andersen and A. Malahoff (eds.), The Fate of Fossil Fuel CO_2 in the Oceans, pp. 213–241. Plenum Press, New York.

Broecker, W. S. and T. Takahashi. 1978. The relationship between lysocline depth and in situ carbonate ion concentration. Deep Sea Res. 25:65–95.

Broecker, W. S. and T. Takahashi. 1984. Is there a tie between atmospheric CO_2 content and ocean circulation. In J. E. Hansen and T. Takahashi (eds.), Climate Processes and Climate Sensitivity, Annals of Geophysical Monograph 29, pp. 314–326. American Geophysical Union, Washington, D.C.

Budyko, M. I. and A. B. Ronov. 1979. Chemical evolution of the atmosphere in the Phanerozoic. Geochem. Int. 16:1–9.

Craig, H. 1957. The natural distribution of radiocarbon and the exchange time of carbon dioxide between the atmosphere and sea. Tellus 9:1–17.

Edmond, J. M. and J. M. Gieskes. 1970. On the calculation of the degree of saturation of sea water with respect to calcium carbonate under in situ conditions. Geochim. Cosmochim. Acta 34:1261–1291.

Delmas, R. J., J. M. Ascencio, and M. Legrand. 1980. Polar ice evidence that atmospheric CO_2 20,000 yr BP was 50% of present. Nature 284:155–157.

Fischer, A. G. 1981. Climatic oscillations in the biosphere. In M. H. Nitecki (ed.), Biotic Crises in Ecological and Evolutionary Time, pp. 103–131. Academic Press, San Diego.

Fischer, A. G. and M. A. Arthur. 1977. Secular variations in the pelagic realm. In Deep-Water Carbonate Environments, SEPM Special Publication No. 25, pp. 19–50.

Frakes, L. A. 1979. Climates Throughout Geologic Time. Elsevier, Amsterdam.

Garrels, R. M., A. Lerman, and F. T. Mackenzie. 1976. Controls of atmospheric O_2 and CO_2: past, present, and future. Am. Sci. 64:306–315.

Gordon, A. L. and H. W. Taylor. 1975. Heat and salt balance within the cold waters of the world ocean. In Proceedings of Symposium on Numerical Models of Ocean Circulation, pp. 54–56. National Academy of Sciences, Washington, DC.

Gorshkov, S. G. 1980. Ocean atlas reference tables (in Russian, with map). pp. 156. Department of Navigational Oceanography, Ministry of Defense, USSR.

Holland, H. D. 1978. The Chemistry of the Atmosphere and Oceans. John Wiley & Sons, New York, 351 p.

Ingle, S. E. 1975. Solubility of calcite in the ocean. Marine Chem. 3:301–319.

Keir, R. S. 1980. The dissolution kinetics of biogenic calcium carbonates in seawater. Geochim. Cosmochim. Acta 44:241–252.

Keir, R. S. 1982. Dissolution of calcite in the deep-sea: theoretical prediction for the case of uniform size particles settling into a well-mixed sediment. Am. J. Sci. 282:193–236.

Knox, F. and M. B. McElroy. 1984. Changes in atmospheric CO_2: Influence of the marine biota at high latitude. J. Geophys. Res. 89:4629–4637.

Ku, T. L., C. A. Huh, and P. S. Chen. 1980. Meridional distribution of 226 Ra in the eastern Pacific along GEOSECS cruise tracks. Earth Planet. Sci. Lett. 49:293–308.

Li, Y.-H., T.-H. Peng, W. S. Broecker, and H. G. Ostlund. 1984. The average vertical eddy diffusion coefficient of the ocean. Tellus 36B:212–217.

Li,Y.-H., T. Takahashi, and W. S. Broecker. 1969. Degree of saturation of $CaCO_3$ in the oceans. J. Geophys. Res. 74:5507–5525.

Lyman, J. 1956. Buffer mechanism of sea water. Ph.D. Thesis, University of California, Los Angeles.

Mackenzie, F. T. and J. D. Pigott. 1981. Tectonic controls of Phanerozoic sedimentary rock cycling. J. Geol. Soc. Lond. 138:183–196.

McElroy, M. B. 1983. Marine biological controls on atmospheric CO_2 and climate. Nature 302:328–329.

McLean, D. M. 1978. Land floras: the major Late Phanerozoic atmospheric carbon dioxide/oxygen control. Science 200:1060–1062.

Mehrbach, C., C. H. Culberson, J. E. Hawley, and R. M. Phytkowicz. 1973. Measurement of the apparent dissociation constants of carbonic acid in sea water at atmospheric pressure. Limnol. Oceanogr. 18:897–907.

Menard, H. W. and S. M. Smith. 1966. Hypsometry of ocean basin provinces. J. Geophys. Res. 71:(18)4305–4325.

Menzel, D. W. 1974. Primary productivity, dissolved and particulate organic matter, and the sites of oxidation of organic matter. In E. D. Goldberg (ed.), The Sea. Volume 5: Marine Chemistry, pp. 659–678. John Wiley & Sons, New York.

Neftel, A., H. Oeschger, J. Schwander, B. Stauffer, and R. Zumbrunn. 1982. Ice core sample measurements give atmospheric CO_2 content during the past 40,000 years. Nature 295:220–223.

Newman, M. J. and R. T. Rood. 1977. Implication of solar evolution for the Earth's early atmosphere. Science 198:1035–1037.

Owen, T., R. D. Cess, and V. Ramanathan. 1979. Enhanced CO_2 greenhouse to compensate for reduced solar luminosity on early Earth. Nature 277:640–642.

Peng, T.-H and W. S. Broecker. 1978. Effect of sediment mixing on the rate of calcite dissolution by fossil fuel CO_2. In Sea-Sediment Interface Models, Vol. 5. No. 5, pp. 349–352. Lamont-Doherty Geological Observatory, Palisades, New York.

Peng, T.-H, W. S. Broecker, G. Kipphut, and N. Shackleton. 1977. Benthic mixing in deep sea cores as determined by ^{14}C dating and its implications regarding climate stratigraphy and the fate of fossil fuel CO_2. In N. Andersen and A. Malahoff (eds.), The Fate of Fossil Fuel CO_2 in the Oceans, pp. 355–373. Plenum Press, New York.

Perry, H. and H. H. Landsberg. 1977. Projected world energy consumption. In Energy and Climate, pp. 35–50. National Academy of Sciences, Washington, D.C.

Pigott, J. D. and F. T. Mackenzie. 1979. Phanerozoic ooid diagenesis: a signature of paleo-ocean and atmospheric chemistry. Geol. Soc. Am., Spec. Pap. 11:495–496.

Plummer, L. N. and E. T. Sundquist. 1982. Total individual ion activity coefficients of calcium and carbonate in seawater at 25°C and 35‰ salinity, and implications to the agreement between apparent and thermodynamic constants of calcite and argonite. Geochim. Cosmochim. Acta 46:247–258.

Post, W. M., W. R. Emanuel, P. J. Zinke, and A. G. Strangenberger. 1982. Soil carbon pools and world life zone. Nature 298–156–159.

Ronov, A. B. 1976. Volcanism, carbonate deposition, and life (Patterns of the global geochemistry of carbon). Geochem. Int. 13(4):172–195.

Sagan, C. and G. Mullen. 1972. Earth and Mars: evolution of atmospheres and surface temperatures. Science 177:52–56.

Sandberg, P. A. 1983. An oscillating trend in Phanerozoic non-skeletal carbonate mineralogy. Nature 305:(5929)19–22.

Sarmiento, J. L. and J. R. Toggweiler. 1984. A new model for the role of the oceans in determining atmospheric pCO_2. Nature 308:621–624.

Shackleton, N. J. and A. Boersma. 1981. The climate of the Eocene ocean. J. Geol. Soc. Lond 138:(2)153–157.

Shackleton, N. J., M. A. Hall, J. Line, and C. Shuxi. 1983. Carbon isotope data in core V19-30 confirm reduced carbon dioxide concentration of the ice age atmosphere. Nature 306:319–322.

Siegenthaler, U. and T. Wenk. 1984. Rapid atmospheric CO_2 variations and ocean circulation. Nature 308:624–625.

Sundquist, E. T., D. K. Richardson, W. S. Broecker, and T.-H. Peng. 1977. Sediment mixing and carbonate dissolution in the southeast Pacific Ocean. In N. Andersen and A. Malahoff (eds.), The Fate of Fossil Fuel CO_2 in the Oceans. 429–454. Plenum Press, New York.

Takahashi, T., W. S. Broecker, A. E. Bainbridge, and R. F. Weiss. 1980. Carbonate chemistry of the Atlantic, Pacific, and Indian Oceans. In The Results of the GEOSECS Expeditions, 1972–1978, Technical Report No. 1. CU-1-80. Lamont-Doherty Geological Observatory, Palisades, New York.

Takahashi, T., W. S. Broecker, A. E. Bainbridge, and R. F. Weiss. 1981. Supplement to the alkalinity and total carbon dioxide concentration in the world oceans. In B. Bolin (ed.), Carbon Cycle Modelling, Scope 16, pp. 159–199. John Wiley and Sons, Chichester, England.

Tappan, H. 1968. Primary production, isotopes, extinctions, and the atmosphere. Palaeogeogr. Palaeoclimatol. Palaeoecol. 4:(3)187–210.

Tappan, H. and A. R. Loeblich, Jr. 1971. Geobiologic implications of fossil phytoplankton evolution and time-space distribution. In Symposium on Palynology of the Late Cretaceous and Early Tertiary. Geol. Soc. Am. Spec. Pap. 127:247–340.

Van Andel, T. H. 1975. Mesozoic/Cenozoic calcite compensation depth and the global distribution of calcareous sediments. Earth Planet. Sci. Lett. 26:187–194.

Walker, J. C. G. and P. B. Hays. 1981. A negative feedback mechanism for the long-term stabilization of Earth's surface temperature. J. Geophys. Res. 86:(C10)9776–9782.

Wangersky, P. J. 1976. Particulate organic carbon in the Atlantic and Pacific oceans. Deep Sea Research 23:457–465.

Watts, J. A., Compiler. 1982. The carbon dioxide question: data sampler. In W. C. Clark (ed.), Carbon Dioxide Review: 1982, pp. 456–460. Clarendon Press, Oxford, England.

Weiss, R. F. 1974. Carbon dioxide in water and seawater: the solubility of non-ideal gas. Marine Chem. 2:203–215.

World Energy Conference. 1980. Survey of Energy Resources. Federal Institute for Geosciences and Natural Resources, Hanover, Federal Republic of Germany.

20
Requirements for a Satisfactory Model of the Global Carbon Cycle and Current Status of Modeling Efforts

BERT BOLIN

The concentration of CO_2 in the atmosphere is rising as a result of emissions into the atmosphere by fossil fuel combustion, deforestation, and expanding agriculture. Experiments with numerical models of the earth's climatic system indicate that a further increase of the atmospheric CO_2 concentration might change climate on earth significantly. To understand how rapidly such possible changes might occur, we must be able to project the likely rate of change of atmospheric CO_2 due to future CO_2 emissions. For this purpose a better knowledge of the global carbon cycle is required, particularly of how atmospheric CO_2 is exchanging with the terrestrial ecosystems and the oceans. It has been maintained that about half of the CO_2 emitted into the atmosphere has stayed there; that is, the "airborne fraction" of the emissions has been about 50%. The emissions due to deforestation and changing land use need more careful consideration in this context, and the airborne fraction of the total emissions has not been determined accurately as yet. We need to know more precisely what has happened in the past to be able to validate models of the global carbon cycle.

Before dealing with the problem of modeling the global carbon cycle we shall briefly comment on the data available at present for studies of the carbon cycle, since this data base is decisive for the approach to be chosen in the modeling effort.

Observed Features of the Carbon Cycle Response to Man-Induced CO_2 Emissions

THE AIRBORNE FRACTION OF PAST EMISSION

During the period of accurate observations, 1958 to 1982 (24 years), the atmospheric CO_2 concentration has risen from about 315 ppm to about 341 ppm (Keeling 1983). Preindustrial concentrations, on the other hand, are not accurately known. Recent measurements of the CO_2 concentration in air bubbles trapped in Antarctic glacier ice show values of 270 ± 10 ppm (Oeschger and Stauffer 1985), which may be considered as the most likely concentration at the middle of last century. Thus, there has been an increase of 26 ppm during the last 24 years and of about 75 ppm since 1860, corresponding to 55×10^{15} g C and $150 \pm 20 \times 10^{15}$ g C, respectively. Compared with the preindustrial amount of carbon in the atmosphere, $570 \pm 20 \times 10^{15}$ g C, the increase has been 10% and 27%. It should also be noted that CO_2 measurements in Antarctic ice show that the atmospheric CO_2 concentration has varied between 200 ppm and 340 ppm during the Holocene period, that is, during the last 20,000 years (Oeschger and Stauffer, this volume).

Rotty and Marland (this volume) have estimated the CO_2 emissions to the atmosphere due to fossil fuel burning to have been 96×10^{15} g C for the period 1958 to 1982 (24 years) and to 168×10^{15} g C since 1860, respectively. The net transfer of CO_2 to the atmosphere due to deforestation and expanding agriculture are given by Houghton et al. (1983) to have been 60×10^{15} g C and 180×10^{15} g C during these same two periods. The cited uncertainty of these latter figures is quite large, at least 30%. There are several reasons for this uncertainty. The total amount of carbon stored in the terrestrial ecosystems at the beginning of the industrial period may have been significantly lower than assumed by Houghton et al. (1983) and the emissions from the soils may likewise be too large (see also Schlesinger, this volume). For the following discussion we shall adopt the values $150 \pm 50 \times 10^{15}$ g C and $50 \pm 15 \times 10^{15}$ g C, respectively, for the two periods considered. The ranges of uncertainty include the estimates given by Houghton et al. (1983) but should perhaps even be larger than indicated. On the basis of these estimates we can determine the fraction of the total emissions that have remained airborne. Table 20.1 shows the values of 0.50 ± 0.15 and 0.38 ± 0.04 for the two periods 1860 to1980 and 1958 to 1982. We observe a decrease with time that is barely significant.

It is of interest to compare these figures with the atmospheric CO_2 trend

TABLE 20.1. Estimates of the airborne fraction of the total man-induced emissions of CO_2 into the atmosphere during the periods 1860–1980 and 1958–1982.

	Increase in the atmosphere 10^{15} g C	Fossil fuel emissions 10^{15} g C	Terrestrial biosphere emissions 10^{15} g C	Total emissions 10^{15} g C	Airborne fraction
1860–1980					
maximum	172	180	200	380	0.67
minimum	129	155	100	255	0.34
1958–1982					
maximum	56	103	65	168	0.44
minimum	54	89	35	124	0.32

for the period 1958 to 1980, when more accurate data are available. The fossil fuel emissions increased from 2.30×10^{15} g C • yr^{-1} to 5.26×10^{15}g C • yr^{-1} and the annual values for the years in between are given by Rotty and Marland (this volume). The transfer from the terrestrial ecosystems according to Houghton et al. (1983) remained rather constant at a level of 2.7×10^{15} g C • yr^{-1}. Referring to our previous remarks of a likely overestimate of these emissions, we shall rather adopt a constant value of $2.1 \pm 0.7 \times 10^{15}$g C • yr^{-1} for this same period. Based on these data and the annual increase of atmospheric CO_2 concentrations as given by Keeling (1983) we deduce the annual values of the airborne fraction as shown in Fig. 20.1. Also shown are the corresponding values in the case of negligible emissions due to deforestation and expanding agriculture, as well as the line of regression for the change of the airborne fraction in such a case. We find that in the mean the airborne fraction has increased from 0.33 ± 0.05 in 1958 to 0.42 ± 0.04 in 1980 for the given emissions

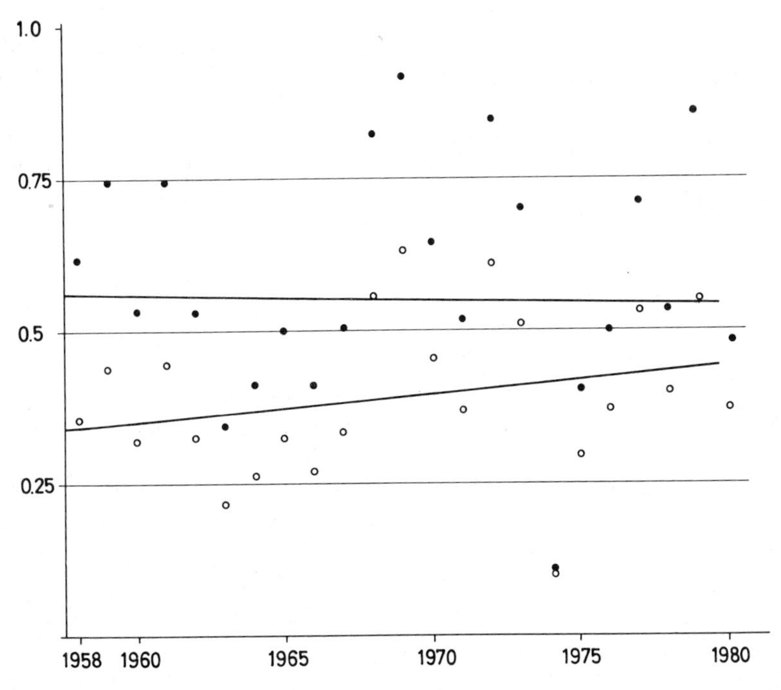

FIGURE 20.1. Airborne fraction of anthropogenic CO_2 emissions to the atmosphere based on the assumption that the annual increase of atmospheric CO_2 concentrations as observed at Mauna Loa (Keeling 1983) is representative for the atmosphere as a whole. The values shown by *filled circles* are obtained (along with the horizontal regression line) if emissions are due only to fossil fuel combustion as given by Rotty and Marland (1985), while *open circles* give values (and the upward sloping line) obtained when a constant net transfer from the terrestrial biosphere of 2.1×10^{15} g C is assumed.

from deforestation and expanding agriculture or has remained almost constant at a value of 0.55 ± 0.04 if no emissions due to man's interference with the terrestrial ecosystems have occurred. It does not, however, seem likely that the latter would have been the case. The airborne fraction may thus possibly have increased during recent decades even though a constant or slightly decreasing value cannot be excluded.

For the later discussion of the response characteristics of the carbon cycle to man-induced emissions, some further facts are of interest. The United Nations statistics on fossil fuel use in the world since 1860 compiled by Rotty and Marland (this volume) show an average rate of increase during this period of 3.3% per year, which corresponds to an *e*-folding time of about 30 years. Until 1914 the rate was about 4% but decreased to <1% during the 30-year period 1914 to 1945. It again rose to about 4% during 1945 to 1973. Between 1973 and 1980 it has been close to 2%, and since then has decreased. The *e*-folding time of external forcing of the carbon cycle by CO_2 input to the atmosphere has thus varied between 25 and 100 years.

Even though the amount of total carbon also must have increased significantly in the oceans due to transfer from the atmosphere, no direct measurements are available that show the magnitude of this change. Brewer (1983) and Chen (1983) by indirect means have deduced increases corresponding to changes of the CO_2 partial pressure between 50 and 80 ppm. There are, however, major uncertainties in attempting such estimations.

CHANGES OF $\Delta^{13}C$ AND $\Delta^{14}C$

Measurements of $\Delta^{13}C$ in the atmosphere (Keeling et al. 1979) and in tree rings (e.g., Freyer, this volume) reveal a downward trend that is about 0.65‰ during the last 20 years and about 1.5‰ since early in the last century.

During the years 1958 to 1962 large amounts of ^{14}C were injected into the atmosphere by nuclear bomb testing, of which a considerable part was directly brought into the stratosphere. The tropospheric ^{14}C concentrations reached their maximum in the Northern Hemisphere in 1963 and in the Southern Hemisphere about 1 year later. Since then a steady decrease has taken place due to a net transfer of ^{14}C into the terrestrial biota and into the sea (cf Nydal and Lövseth 1983). During the period 1964 to 1980 the global average excess ^{14}C in the troposphere decreased by about 6% per year, that is, a characteristic *e*-folding time of 16 years. The decrease in the troposphere might even have been more rapid, implying an *e*-folding time of <16 years, if a replenishment from the stratosphere had not taken place.

There are also observations of the penetration of the excess ^{14}C from the atmosphere into the sea (Tans 1981; Nydal and Lövseth 1983). The increase from a $\Delta^{14}C$ value of −55‰ in 1960 was almost linear with a rate of about 25‰ per year until 1965, after which the increase lessened and at the end of 1969 was at a value of about +105‰ at low latitudes. Since then a very slow decline seems to have occurred.

Finally, the ^{14}C distribution in the sea was first approximately observed for the Atlantic Ocean in 1957 (Broecker et al. 1960), at which time the distribution had hardly been disturbed by bomb testing. Much more detailed observations in all three major oceans were made during the GEOSECS Expedition in 1972 to 1974 (Östlund and Stuiver 1980; Stuiver and Östlund 1980) at which time very significant changes of the $\Delta^{14}C$ distribution were observed. Some of the stations in the northern Atlantic were reoccupied in 1981 to 1982 during the TTO experiment, and these measurements reveal further changes also at greater depths.

A satisfactory global carbon cycle must be able to reproduce the quasi-steady conditions that prevailed before man's interventions were significant as well as the changes that have occurred since then, as briefly described above.

The Principal Components of the Carbon Cycle

We shall be concerned with changes in the carbon cycle on time scales from a few years to centuries, so we shall not consider in any detail the daily and seasonal variations of the carbon system that are primarily induced by the cycling behavior of the biota, nor the long-term processes that involve weathering and sedimentation. We shall assume that the atmosphere is well-mixed and we shall merely carry the annual mean atmospheric CO_2 concentrations as the single variable. This is obviously an important simplification which may possibly introduce significant errors. It is possible, for example, that the seasonal changes of the atmospheric CO_2 concentration in North Polar regions need to be accounted for in determining the net CO_2 flux from the atmosphere to the sea at these latitudes. The development of atmospheric models to deal explicitly with such problems has been made by Fung (this volume), and her findings may well have to be incorporated into global carbon cycle models at a later stage.

Figure 20.2 shows a simplified picture of the global carbon cycle. The terrestrial processes are described by exchange between five major compartments: (1) foliage and grass, where photosynthesis occurs; (2) structural material, particularly wood; (3) surface detritus or litter; (4) soil; and (5) peat. Different ecosystems, however, behave quite differently. It may be necessary to treat them separately in order to describe changes appropriately; attempts in this direction should be pursued. The aggregation depicted in Fig. 20.2 is, however, adequate for our present purposes.

The distinction between the well-mixed surface layer and the remainder of the oceans is an important one but is still insufficient for an accurate

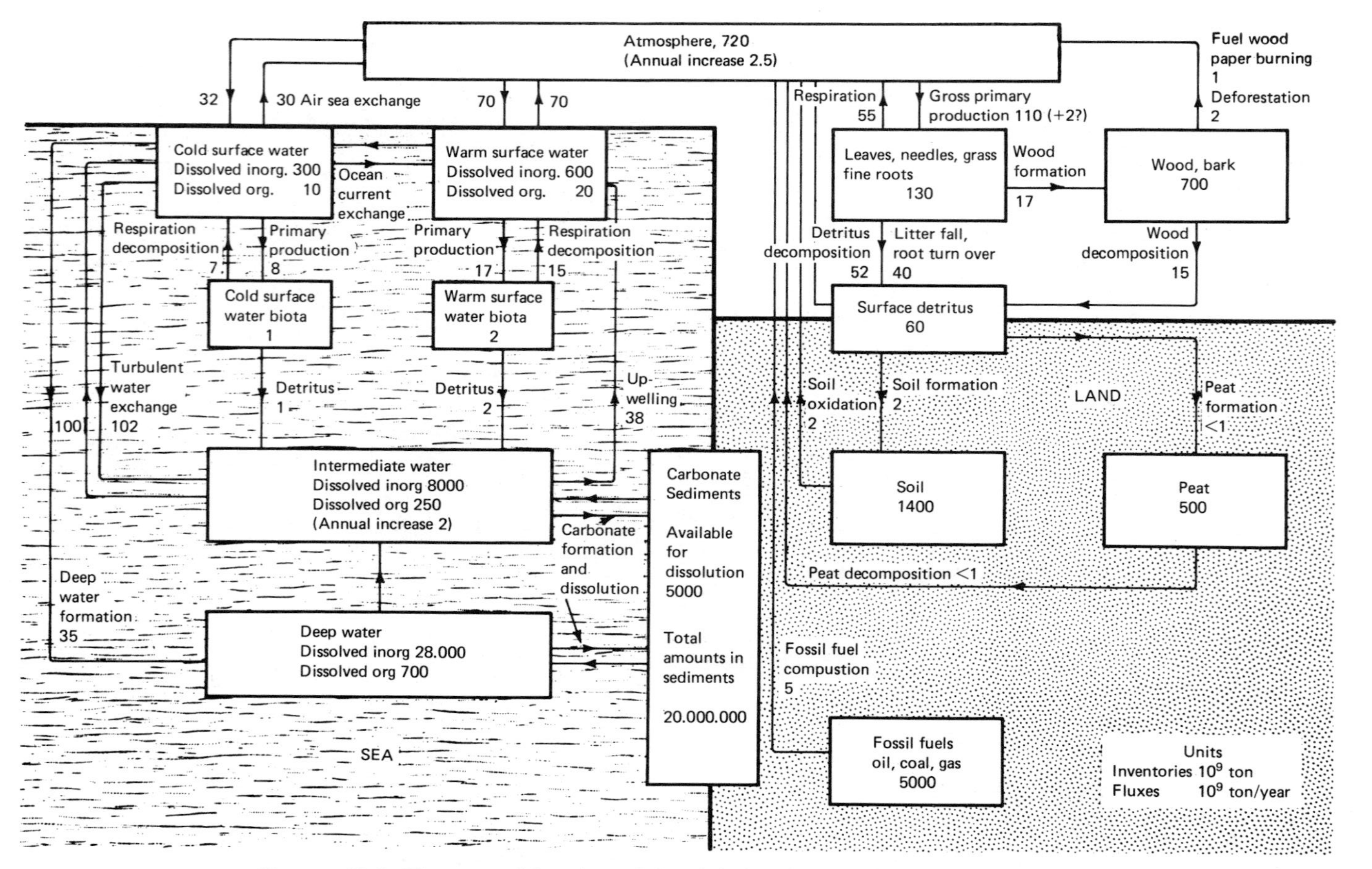

FIGURE 20.2. The general features of the global carbon cycle (From Bolin 1983).

account of the role of the oceans in the carbon cycle. Later discussion will show that a detailed treatment of the surface layer and of the intermediate layer down to at least a depth of 1000 m (at high latitudes to still greater depths) is required. A more accurate description of the oceans also may ultimately require considerations of the seasonal variations. However, there are at present no ocean models that adequately deal with temporal changes of that kind; much development work is required before this will be possible.

We shall in turn consider the major reservoirs of terrestrial biota and soils shown in Fig. 20.2.

Foliage and Grass

This photosynthetically active material has a turnover time from months to a few years. To a first approximation we may describe uptake and release of carbon by $\tau_f^{-1} N_f$, where τ_f is the turnover time and N_f is the total amount of carbon in the form of foliage and grass. To the extent that an increasing atmospheric CO_2 concentration enhances the rate of photosynthesis, the uptake also should be proportional to some power, β, of N_a, where N_a is the atmospheric amount of carbon. It is important to emphasize that even though N_f in this way may increase, it does not necessarily follow that there will be an appreciable accumulation of carbon in other parts of the terrestrial ecosystem (Bolin 1983). This will be discussed further below.

Structural Materials, Particularly Wood

Almost 90% of the carbon in terrestrial ecosystems is found in wood, which thus constitutes a major carbon reservoir in the carbon cycle. The average turnover time varies from little more than a decade in rapidly growing tropical forest systems to a century or more in the forests of cold climates. Two dynamical measures of forest ecosystems, closely related to each other, are relevant:

1. The amount of carbon in wood as a function of time elapsed since its formation. This is of particular importance to determine the role of the terrestrial biota for uptake of bomb-produced ^{14}C.
2. The dependence of the amount of carbon in different terrestrial ecosystems on climate conditions or other environmental factors such as an increasing CO_2 amount in the atmosphere. The carbon storage capacity of an ecosystem will not necessarily increase because of an enhanced rate of photosynthesis (see above) but only if the amount in *storage at climax* will increase in the case of modified environmental conditions. An analysis has not been carried out to determine to what extent this might be the case for an increased atmospheric CO_2 concentration.

In view of the rate of ongoing and expected changes, time integration of models should permit good reproduction of characteristic changes on time scales from 5 to several hundred years.

Litter

Most of the litter decomposes within a few years. This storage pool for carbon can be adequately modeled by an assumption of first-order processes. It is important to consider the variations of turnover rates between different ecosystems. A small fraction, usually <5%, is transferred into soil (Schlesinger 1977).

Soil

The mean turnover time for carbon in soils is of the order of decades, but most of the CO_2 released is from young material (Schlesinger 1977). This implies that the accumulation of soil carbon is slow and its age long, often many hundred years as revealed by ^{14}C analyses. To deal properly with the dynamics of the soil system in a global carbon model, we need to know the transient time distribution function, which is a key to the understanding of its response characteristics.

It is well known that the amount of carbon in soil decreases due to ploughing and other soil management practices whereby the organic matter in the soil becomes more exposed to oxidation. Within a few decades the carbon content may decrease by one-half to one-third of the amount present in undisturbed conditions. This decline varies from one soil type to another (Vitousek 1983) and is most rapid in upper layers (Schlesinger, this volume). When cultivated land is abandoned the net accumulation of carbon in the soil proceeds slowly, since most of the litter formed annually is decomposed quickly, and only a small amount is incorporated. The soil models that so far have been included in global carbon cycle models have seldom accounted for the fundamental features of the soil systems as described above. It should also be recognized that changes of climate will induce changes of the soil system. The rate of decay increases as a function of temperature as long as moisture and oxygen is adequate. A major climatic change, which, for example, might imply a change from a boreal forest to a deciduous forest, would induce a major change of the carbon content of the soil.

Human Impacts and Natural Variations

In modeling terrestrial ecosystems, we must distinguish between (1) the direct impact of human activities, (2) secondary changes as a consequence of (1), and (3) natural internal changes.

The assessment of the first kind of change is basically an accounting problem and requires historical data on land use. Houghton et al. (1983) have attempted detailed analyses of this problem. Secondary changes (2) have also been considered schematically, but our knowledge of ecosystem response to external changes is not sufficiently complete to permit accurate analysis. A considerable uncertainty remains with regard to the impact on the terrestrial ecosystems of the increasing atmospheric CO_2 concentration and also the consequences of air and water pollution in general. We note the rapidly increasing damage to forests in central Europe, presumably due to air pollution and acidification. On the other hand, the deposition of nitrogen compounds may increase the rate of photosynthesis.

The possibility that natural variations of the ecosystems (3) may take place simultaneously further complicates the determination of the present role of the terrestrial ecosystems with regard to the ongoing change of the global carbon cycle. Fundamental research on terrestrial ecology seems most essential.

Modeling the Oceans

The Historical Development

The first attempt to formulate a quantitative model of the oceans was made by Craig (1957) more than 25 years ago. His two-box model with a rather thin, well-mixed surface reservoir and a deep sea has proven to be a very useful first approximation of the real ocean, but the quantitative results obtained were not accurate. Craig (1963) extended his model by considering the cold surface waters separately, thereby permitting the deep sea to be in contact with the atmosphere more directly. This idea was pursued further by Broecker et al. (1971) and more recently by Siegenthaler (1983). Oeschger et al. (1975) introduced the box diffusion model of the oceans, whereby the deep sea is resolved by a series of 43 horizontal layers between which matter is transferred vertically by turbulent diffusion. It is further assumed that the coefficient of eddy diffusivity is constant. The box-diffusion model has been used extensively and has generally been considered as the most adequate first approximation in dealing with the role of the oceans in the carbon cycle. It was calibrated with the aid of the vertical steady-state distribution of ^{14}C. Because of nuclear bomb testing, this profile is not accurately known. We shall later discuss uncertainties inherent in the models for the circulation of carbon in the sea due to this inaccurate data base. Broecker et al. (1980) particularly have emphasized the use of the transient change of the ^{14}C and tritium distributions in the sea to the injections due to bomb testing. Oeschger et al. (1975) and Siegenthaler (1983) have applied this same method for calibrating their models. This certainly has added to the reliability of their results. However, it is becoming obvious that the resolution in present models is insufficient to determine accurately the airborne fraction of injections of carbon or ^{14}C into the atmosphere. Projections into the future also become uncertain. It is also unsatisfactory that the rich data base for the state of the sea, that is, observations of temperature, salinity, current velocity, and chemical tracers, have not been fully exploited in the development of the ocean models for carbon transfer and circulation in the sea.

A Simple Example of Model Sensitivity to Model Resolution

Before discussing the further development of ocean models we will show a simple example of how model sensitivity depends on model resolution.

Consider a two-box model as shown in Fig. 20.3a. The carbon amounts in the two boxes (0) and (1) are the same, and the characteristic exchange of carbon between them is such that the ratio $R = {}^{14}C/{}^{12}C$ relative to some standard at steady state is $R = 1.00$ and $R_1 = 0.99$ for the two boxes, respectively. The steady state is maintained by an appropriate ^{14}C injection (Γ) into box (0). It follows that the first-order exchange coefficients between the two boxes $k_{01} = k_{10} \simeq (80\ yr)^{-1}$.

We next assume that the lower box (1) is divided into two equally large boxes (1) and (2) as shown in Figs. 20.3b or 20.3c. We assume further that the ^{14}C injection rate (Γ) remains unchanged as does $R_0 = 1.00$. For comparison, however, we assign values R_1 and R_2 in the following four alternatives:

$$
\begin{aligned}
(1)\quad R_1 &= 0.993 \\
R_2 &= 0.987 \\
(2)\quad R_1 &= 0.995 \\
R_2 &= 0.985 \\
(3)\quad R_1 &= 0.999 \\
R_2 &= 0.981, \text{ or} \\
(4)\quad R_1 &= 1.000 \\
R_2 &= 0.980
\end{aligned}
$$

The exchange coefficients k_{01}, k_{10}, k_{12}, and k_{21} in case (b) and k_{01}, k_{10}, k_{02}, and k_{20} in case (c) are thereby given and can be determined.

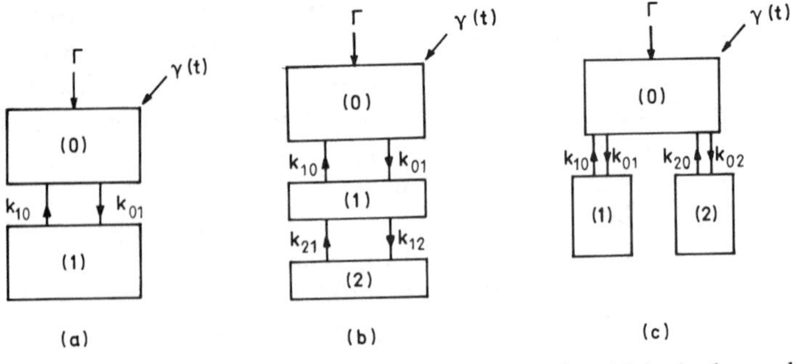

FIGURE 20.3. **a:** Two-box model. **b:** Three-box model, in which the lower box in **(a)** has been divided into two boxes, which are exchanging in series with box (0). **c:** Three-box model, in which the lower box in **(a)** has been divided into two boxes, which are exchanging in parallel with box (0).

We next assume that an injection of CO_2 (γ) into box (0) takes place which is given by

$$\gamma = \gamma_0 \times e^{\alpha t}$$

where we assume that $\alpha = (25 \text{ yr})^{-1}$, to permit some comparison with real conditions. When an adjustment toward an exponential increase in all reservoirs has occurred, we find that the partitioning of the increase between the reservoir (0) and the total injection (which in a sense may be compared with the concept of "airborne fraction" in reality) is given by

	(b)	(c)
(a)	0.808	0.808
(1)	0.869	0.797
(2)	0.828	0.782
(3)	0.705	0.691
(4)	0.638	0.638

This example shows that the response characteristics of this simple system is quite sensitive to the structure of the model and the exact values of R_i that are assigned to the reservoirs chosen. The sensitivity of the results to R_i of course depends upon the slow decay of ^{14}C.

In view of the uncertainties of the prebomb quasi-steady-state distribution of ^{14}C in the upper layers of the oceans, the assumptions of a constant vertical eddy diffusivity in the layers below the well-mixed surface layer and the $\Delta^{14}C$ profile as adopted by Oeschger et al. (1975) are questionable. As a matter of fact, the sensitivity revealed here is already apparent in the computations by Keeling (1973), in which he varied the depth of the well-mixed layer between rather wide limits.

The Use of Transient Tracers for Carbon Cycle Model Calibration

As briefly reviewed earlier, most carbon cycle models have been developed by describing the oceans only by a few well-mixed reservoirs and by using the steady-state distribution of one or several tracers for calibration, particularly ^{14}C. It was inferred that the short time scales are poorly resolved by using such models. Oeschger et al. (1975) and later Broecker et al. (1980) and Siegenthaler (1983) have emphasized the use of transient tracers, for example, bomb-produced ^{14}C, tritium, and chlorofluorcarbon for model calibration to ensure an appropriately rapid response is accounted for by the model. Such tracers reveal features of the carbon cycle on time scales less than the time that has elapsed since their injections began, that is, a few decades. We still lack data that can improve our confidence in how ocean models for the carbon cycle respond on the time scale of a century. This implies that projections of the ocean uptake of CO_2 injections into the atmosphere remain uncertain.

A few further comments on the use of transient tracers in developing ocean models will be made based on the outcrop diffusion model proposed by Siegenthaler (1983). The model can be briefly described as follows.

A well-mixed, warm surface layer that is 75 m deep covers a fraction a_w of the ocean. As in the case of the box-diffusion model, the thermocline region and the deep sea below are described by a series of horizontal layers. They all reach the ocean surface at successively higher latitudes in the cold region of the ocean, which covers a fraction $a_c = 1 - a_w$ of the total ocean surface. Vertical exchange between the successive layers in the thermocline and deep sea is due to turbulence, which is described by a constant eddy diffusivity K. These layers are considered internally well mixed and thus instantly responding to exchange between the atmosphere and the sea in the cold region. The limiting process for this response is thus the rate of air–sea exchange in cold waters, which for CO_2 is described by the flux, F_{as}, between the atmosphere and the sea.

Siegenthaler shows that for $a_c = 0.1$ the average prebomb $\Delta^{14}C$ vertical profile can be reproduced with $K = 2224\ m^2 \bullet yr^{-1} = 0.70\ cm^2 \bullet s^{-1}$ and $F_{as} = 13.3\ mol\ m^{-2} \bullet yr^{-1}$. If we assume an exponentially increasing CO_2 emission into the atmosphere with an *e*-folding time of 22.5 years, the CO_2 uptake by the oceans is 39.5%, that is, the airborne fraction is 60.5%. This latter value is large compared with observed average conditions. Either the ocean model is not adequate or other sinks play a role.

Siegenthaler points out, however, that this model does not account properly for the uptake of bomb-produced ^{14}C during the period 1958 to 1973. To fit the observations optimally within the framework of the model the values $K = 5180\ m^2 \bullet yr^{-1} = 1.65\ cm^2 \bullet s^{-1}$ and $F_{as} = 19.3\ mol\ m^2 \bullet yr^{-1}$ must instead be chosen. The model then attains a steady-state $\Delta^{14}C$ value for the deep sea of $-85‰$, which is much higher than what is observed. In this case, however, the airborne fraction of an exponentially increasing emission becomes 53%.

It seems clear that the difficulty of satisfying all data adequately is due to the generalized description of the surface layers and the thermocline region. More care must be exercised in the formulation of an appropriate ocean model. It is questionable whether a globally averaged model as the one proposed by Siegenthaler is adequate, since the averaging process decreases the uptake capability of the model.

Analyzing Ocean Circulation and Its Role in the Carbon Cycle by Using Inverse Methods*

A thorough development of an ocean model for the transfer of carbon in the sea should, of course, consider more information than that contained

*The following outline of the use of inverse methods for analysis of ocean circulation has been considerably elaborated and a full report will be published elsewhere in 1986.

in the distribution of the three isotopes of carbon and the chemical compounds that they form. We possess a large body of data about the general circulation of the oceans, that is, measurements of temperature, salinity, and currents. Dynamic theory has helped us in interpreting such data. This information is also of great relevance in the development of a global carbon model. So far, only very limited use of such data has been made.

The development of dynamic general circulation models of the oceans has been pursued since the 1960s (Bryan et al. 1975). A fundamental difficulty in this work has been the fact that a considerable part of the kinetic energy of motions in the sea are found on scales of 10 to 100 km. It is obviously not possible to resolve these explicitly in global models for the ocean circulation. In this regard ocean modeling is basically more difficult than atmospheric modeling. We do not know how important these smaller scale motions are for heat transfer and the maintenance of the average temperature and salinity distributions in the sea, nor which role they play for the tracer transfer and thus for the transfer of excess carbon and ^{14}C. Since even the large-scale motions of the sea are not well determined by direct observations, tracer distributions and their changes are important observations for the validation of ocean circulation modeling experiments. Only limited work of this kind has so far been reported, and we do not yet know how well the present general circulation models are able to describe transfer of matter in the sea.

It seems clear that there is still a long way to go before general ocean circulation models will be available for use in the analysis of the global carbon cycle. For this reason and also to carry out more consistent analysis of oceanic data, it is of interest to develop the box-model concept further and, in doing so, to make use of inverse methods. In recent years the latter method has been employed in the analysis of temperature and salinity data in combination of dynamic constraints (Wünsch 1978). Variational calculus has also been proposed with that same objective in mind.

With the study of the carbon cycle in focus and with marine chemistry and biology as a starting point, the *simultaneous* study of tracers in the sea has been a fruitful approach (see, for example, Broecker et al. 1980). A more rigorous application of this methodology is essential and the inverse methods offer an opportunity in this regard. A first attempt has been made by Bolin et al. (1983), in which case the distributions of temperature and salinity were not used. The basic idea is to ask for the patterns of advection and turbulence between the boxes chosen, which are required to maintain steady-state distributions of a number of tracers. Since carbon, ^{14}C, alkalinity, phosphorus, and oxygen were used as tracers, the processes or primary production of organic matter and carbonates and their decomposition and dissolution in the boxes had to be introduced as unknowns and simultaneously solved for.

In a first attempt only 12 boxes were considered, which resulted in 66 unknown and 80 equations. The overdetermined (incompatible) system was solved by minimizing errors in the least-squares sense in satisfying the continuity requirements for the tracers considered. For a more detailed account reference is made to the original publication (Bolin et al. 1983).

The solution derived is qualitatively in good agreement with present knowledge about the general circulation of the oceans. It is clear, however, that the resolution used in this first experiment is inadequate to permit the deduction of quantitatively reliable results. This finding also implies that other carbon cycle models with only a few reservoirs are similarly unsatisfactory for assessing the role of the oceans in the carbon cycle. In a next step a more detailed model of the Atlantic Ocean and adjacent parts of the Arctic and Antarctic seas has been developed to explore the potentials of this methodology further. Using eight layers in the vertical defined by a set of density surfaces (Fig. 20.4) and 12 regions (Fig.20. 5), the total number of boxes considered becomes 84. Using the seven tracers: temperature, salinity, total carbon, ^{14}C, alkalinity, phosphorus, oxygen, and the condition of water continuity, we arrive at a system with 672 conservation requirements. The number of unknowns expressing fluxes of water and turbulent exchange between adjacent boxes and formation, dissolution, and decomposition of detrital matter is less than the number of equations, and the system can again be solved by matrix inversion methods requiring that the errors in satisfying the tracer equations be minimized.

Because we are not dealing with a closed system, boundary conditions must be prescribed in terms of mass and tracer exchange with the atmosphere, the bottom of the sea, and adjacent parts of the world ocean. The few test experiments that have been run show clearly that the system is incompatible (overdetermined) and that the method of least squares yields smooth solutions. We also begin to understand how sensitive the solution is to the boundary conditions chosen, and it is becoming clear that the resolution still may be too coarse to catch features of the ocean circulation that are known to be significant. For example, one would hardly expect to be able to deduce the ocean surface currents well only with the aid of tracer distributions, but they are important because of their role in the intermediate and deep water formation.

Figure 20.6 shows some results from preliminary test computations. (A complete account of systematic testing of the model is forthcoming). The following assumptions have been made in these computations.

Boundary Conditions at the Ocean Surface

The heat flux was prescribed according to estimates by Bunker (1980). The net water flux was derived as the difference between precipitation as given by Baumgartner and Reichel (1975) and evaporation estimated by Ebensen and Kushmir (1981). We prescribe the fluxes of ^{14}C and CO_2 between the atmosphere and the sea in a manner similar to that given by

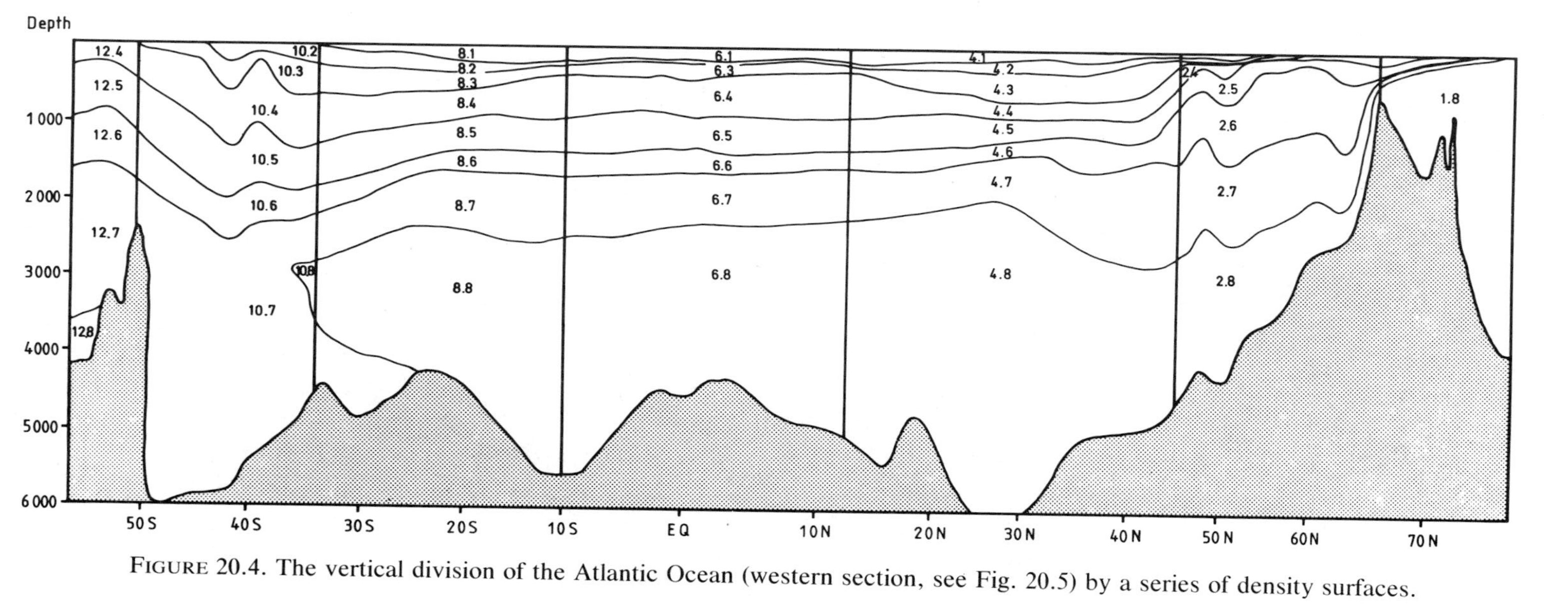

FIGURE 20.4. The vertical division of the Atlantic Ocean (western section, see Fig. 20.5) by a series of density surfaces.

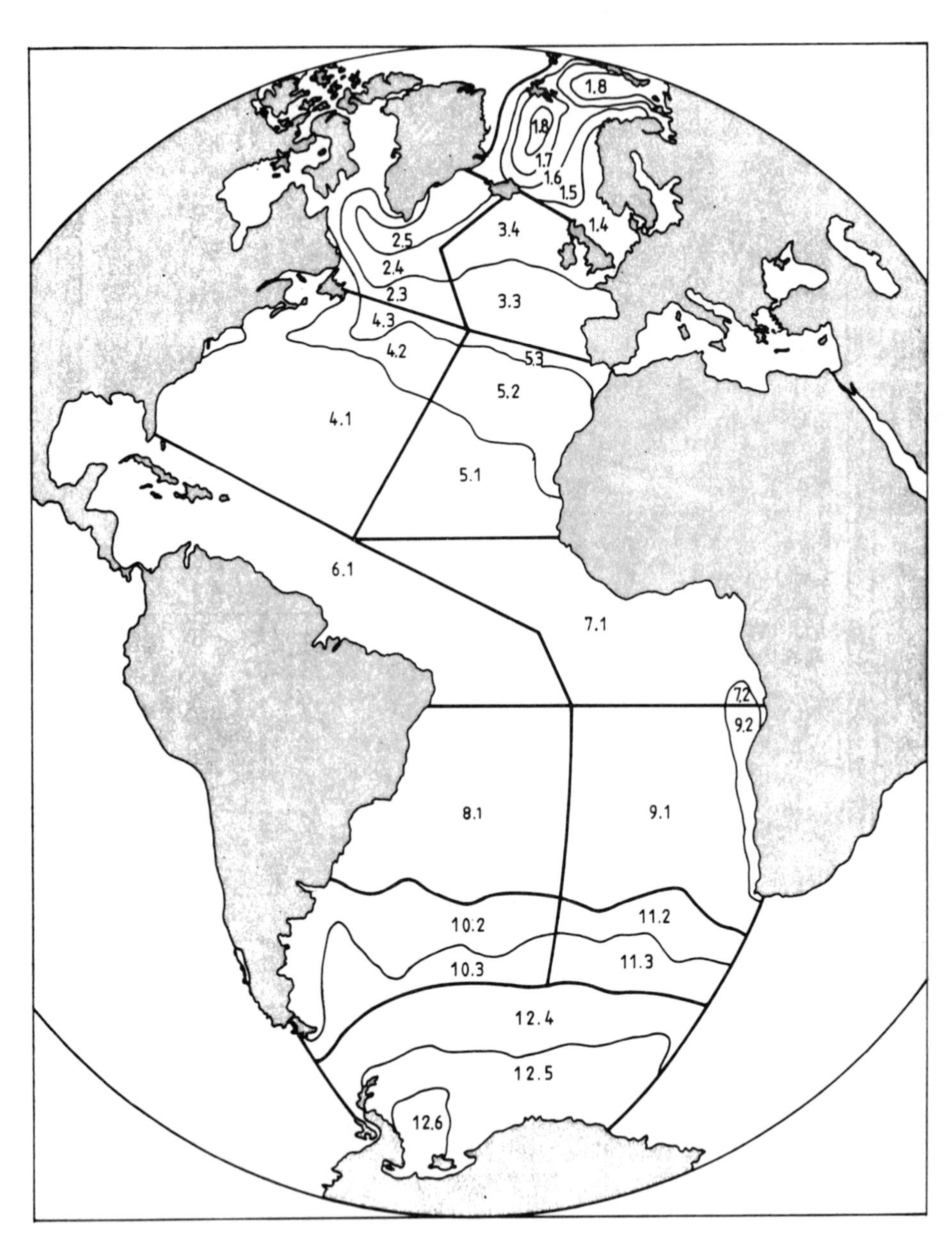

FIGURE 20.5. The horizontal division of the Atlantic Ocean into 12 regions, (*heavy solid lines*), in between which advective and turbulent exchange can take place. *Thin solid lines* show the intersection of the density surfaces, as shown in Fig. 20.4, with the sea surface.

Bolin et al. (1983) except that the average rate of exchange had to be prescribed rather than deduced by assuming a balance between inflow of ^{14}C to the oceans and radioactive decay. No exchange of alkalinity and phosphorus was permitted, and the oxygen concentration in surface waters was kept at saturation pressure with due regard to prevailing temperature.

Inflow by Rivers

The net inflow of water by rivers was considered, but it was assumed that the river transfer of heat and tracers was negligible.

Exchange with Adjacent Seas

The exchange of water with the Mediterranean Sea was accounted for by using the rates given by Sverdrup et al. (1942). Accordingly, a net outflow of water of 0.07 Sv = $21 \times 10^{12}\ m^3 \cdot yr^{-1}$ was adopted with no loss of salt. A water inflow of 2 Sv through the Laborador Sea was assumed having 0°C temperature and 34.74‰ salinity (Stigebrand 1981). To maintain a total balance of water, heat, and tracer material, a net outflow of 0.5 Sv to the Arctic Ocean and 1.5 Sv to adjacent oceans due to the circumpolar current. The exchanges of heat and tracers were prescribed rather arbitrarily to maintain balance. Further testing of the sensitivity of model results to the boundary conditions are being carried out.

Weighting

It has been shown (Bolin et al. 1983) that the results of a matrix inversion is dependent on the relative importance played by the different tracers. The tracer data were first normalized by using the total range of variability. To ascertain that the water and material (detritus) continuity was well handled the appropriate equations were upweighted by a factor of 250. Because the carbon cycle is our prime interest, the continuity equations for dissolved inorganic carbon, ^{14}C, and alkalinity were given the weighting of 25. In this way errors in the minimizing procedure essentially are due to errors in the conservation equations for heat, salt, phosphorus, and oxygen. Extensive testing of how to proceed optimally will be required.

The computations have been carried out on a Prime Computer and one inversion requires about 10 h of computing time. (Later, the use of a CRAY computer has reduced this time to less than 1 min.)

An analysis of the results obtained so far permit the following conclusions (Fig. 20.6).

- Reasonably smooth patterns of advection, turbulence, and detritus formation and dissolution are obtained, but there is considerable noise at small scale.

- There is on the average upwelling at equatorial latitudes and downwelling in the anticyclonic gyres at subtropical latitudes.
- There is generally downwelling (≃10 Sv) in the Arctic Sea and net upwelling (≃7 Sv) in the Antarctic Sea.
- At intermediate levels (layers 2 to 4) we find on the average a northward flow from the Antarctic convergence zone to equatorial latitudes.
- At deeper levels (layers 3 to 7) southward flow takes place from the Arctic Sea, primarily in the western basin.
- Turbulence is confined primarily to vertical exchange between the surface boxes and those immediately below. There is, however, considerable patchiness in the distribution of turbulence elsewhere in the model; it is questionable how reliable these results are.

```
                              WESTERN SECTION ADVECTIONS
  1            2            4            6            8            10           12

                          * 0.086 * 0.023 * 0.937 * 0.010 * 0.261 *
                          *  EAST * >>>>> *  EAST * >>>>> *  WEST *
                            0.113           0.918           0.092   *0.175
                             UP              UP             DOWN    DOWN
                          * 0.765 * 0.560 * 0.403 * 0.251 * 0.032 * 0.483 * 0.250 *
                          *  WEST * >>>>> *  WEST * <<<<< *  WEST * <<<<< *  WEST *
                            0.092           0.295           0.356           0.057
                            DOWN            DOWN            DOWN             UP
             * 0.104 * 0.136 * 0.265 * 0.180 * 0.790 * 0.545 * 0.294 * 0.312 * 0.398 *
             *  WEST * <<<<< *  EAST * <<<<< *  EAST * <<<<< *  WEST * >>>>> *  EAST *
               0.305           0.129           0.129           0.207           0.027   *0.167
               DOWN             UP              UP              UP             DOWN     UP*
*      * 0.000 * 0.011 * 0.019 * 0.058 * 0.267 * 0.050 * 0.074 * 0.118 * 0.104 * 0.205 * 0.314 *      *
*      *       *  WEST * >>>>> *  WEST * >>>>> *  EAST * <<<<< *  WEST * <<<<< *  EAST * <<<<< *      *
  0.159          0.297           0.319           0.163           0.060           0.032           0.152
   UP            DOWN             UP             DOWN             UP             DOWN             UP
*      * 1.075 * 0.440 * 0.180 * 0.009 * 0.065 * 0.157 * 0.011 * 0.003 * 0.043 * 0.486 * 0.363 *      *
*      * <<<<< *  WEST * >>>>> *  EAST * >>>>> *  EAST * >>>>> *  EAST * <<<<< *  EAST * <<<<< *      *
  0.328          0.519           0.213           0.059           0.009           0.134           0.271
  DOWN            UP              UP             DOWN             UP              UP              UP
*      * 0.426 * 0.023 * 0.154 * 0.425 * 0.247 * 0.326 * 0.034 * 0.186 * 0.040 * 0.457 * 0.284 *      *
*      * >>>>> *  WEST * >>>>> *  WEST * >>>>> *  EAST * >>>>> *  EAST * <<<<< *  WEST * >>>>> *      *
  0.271          0.224           0.119           0.053           0.121           0.001           0.090
  DOWN            UP             DOWN             UP              UP              UP              UP
*      * 0.045 * 0.052 * 0.174 * 0.061 * 0.189 * 0.105 * 0.025 * 0.123 * 0.043 * 0.023 * 0.063 *      *
*      * >>>>> *  EAST * >>>>> *  EAST * >>>>> *  EAST * >>>>> *  WEST * >>>>> *  EAST * >>>>> *      *
  0.263          0.405           0.043           0.006           0.015           0.044           0.128
  DOWN            UP             DOWN            DOWN             UP              UP             DOWN
*      * 0.385 * 0.000 * 0.019 * 0.000 * 0.024 * 0.000 * 0.029 * 0.000 * 0.014 * 0.000 * 0.030 *      *
*      * >>>>> *       * <<<<< *       * >>>>> *       * >>>>> *       * >>>>> *       * <<<<< *      *
```

FIGURE 20.6. Deduced fluxes for the western basin of the Atlantic Ocean. Advective water flow (**a**) and turbulent water exchange (**b**) are given in units of 10^{15} m^3 • yr^{-1}; numbers in boxes show the east–west exchange between the corresponding boxes in the western and eastern basins. (**c**) Net organic detritus formation and loss of carbon (−) or decomposition and gain of carbon (+), upper number in 10^{15} mol • yr^{-1}; net carbonate formation and loss of carbon (−) or dissolution and gain of carbon (+), lower number in 10^{15} mol • yr^{-1}.

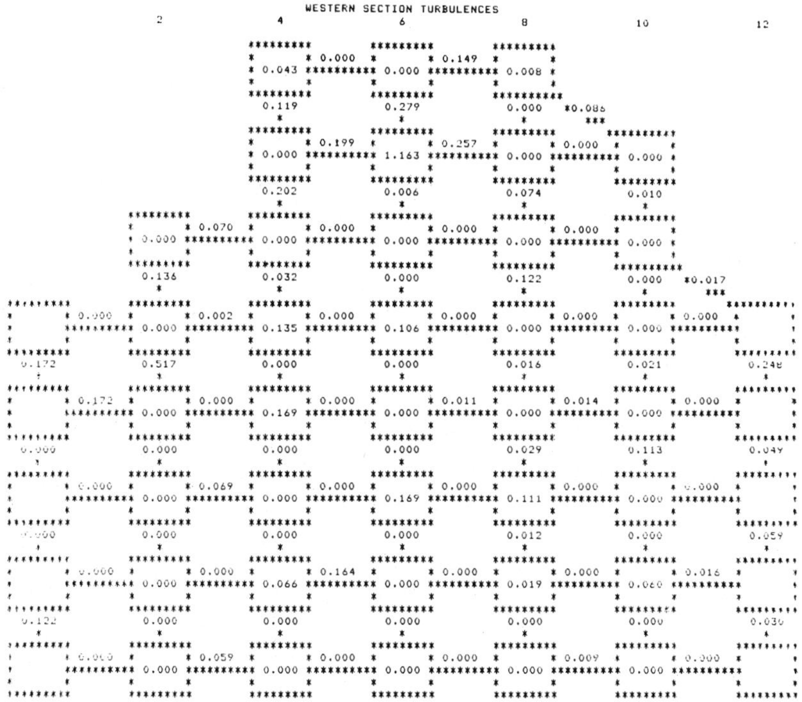

```
                                          WESTERN SECTION TURBULENCES
                     2                4                6                8                10               12

                                  *********        *********        *********
                                  *       * 0.000  *       * 0.149  *       *
                                  * 0.043 ********** 0.000 ********** 0.008 *
                                  *       *        *       *        *       *
                                  *********        *********        **********
                                    0.119            0.279            0.000  *0.086
                                      *                *                *      ***
                                  *********        *********        *********        *********
                                  *       * 0.199  *       * 0.257  *       * 0.000  *       *
                                  * 0.000 ********** 1.163 ********** 0.000 ********** 0.000 *
                                  *       *        *       *        *       *        *       *
                                  *********        *********        *********        *********
                                    0.202            0.006            0.074            0.010
                                      *                *                *                *
                 *********        *********        *********        *********        *********
                 *       * 0.070  *       * 0.000  *       * 0.000  *       * 0.000  *       *
                 * 0.000 ********** 0.000 ********** 0.000 ********** 0.000 ********** 0.000 *
                 *       *        *       *        *       *        *       *        *       *
                 *********        *********        *********        *********        **********
                   0.136            0.032            0.000            0.122            0.000  *0.017
                     *                *                *                *                *      ***
*********        *********        *********        *********        *********        *********        *********
*       * 0.000  *       * 0.002  *       * 0.000  *       * 0.000  *       * 0.000  *       * 0.000  *       *
*       ********** 0.000 ********** 0.135 ********** 0.106 ********** 0.000 ********** 0.000 **********       *
*       *        *       *        *       *        *       *        *       *        *       *        *       *
*********        *********        *********        *********        *********        *********        *********
  0.172            0.517            0.000            0.000            0.016            0.021            0.248
    *                *                *                *                *                *                *
*********        *********        *********        *********        *********        *********        *********
*       * 0.172  *       * 0.000  *       * 0.000  *       * 0.011  *       * 0.014  *       * 0.000  *       *
*       ********** 0.000 ********** 0.169 ********** 0.000 ********** 0.000 ********** 0.000 **********       *
*       *        *       *        *       *        *       *        *       *        *       *        *       *
*********        *********        *********        *********        *********        *********        *********
  0.000            0.000            0.000            0.000            0.029            0.113            0.049
    *                *                *                *                *                *                *
*********        *********        *********        *********        *********        *********        *********
*       * 0.000  *       * 0.069  *       * 0.000  *       * 0.000  *       * 0.000  *       * 0.000  *       *
*       ********** 0.000 ********** 0.000 ********** 0.169 ********** 0.111 ********** 0.000 **********       *
*       *        *       *        *       *        *       *        *       *        *       *        *       *
*********        *********        *********        *********        *********        *********        *********
  0.000            0.000            0.000            0.000            0.012            0.000            0.059
    *                *                *                *                *                *                *
*********        *********        *********        *********        *********        *********        *********
*       * 0.000  *       * 0.000  *       * 0.164  *       * 0.000  *       * 0.000  *       * 0.016  *       *
*       ********** 0.000 ********** 0.066 ********** 0.000 ********** 0.019 ********** 0.060 **********       *
*       *        *       *        *       *        *       *        *       *        *       *        *       *
*********        *********        *********        *********        *********        *********        *********
  0.122            0.000            0.000            0.000            0.000            0.000            0.030
    *                *                *                *                *                *                *
*********        *********        *********        *********        *********        *********        *********
*       * 0.000  *       * 0.059  *       * 0.000  *       * 0.000  *       * 0.009  *       * 0.000  *       *
*       ********** 0.000 ********** 0.000 ********** 0.000 ********** 0.000 ********** 0.000 **********       *
*       *        *       *        *       *        *       *        *       *        *       *        *       *
*********        *********        *********        *********        *********        *********        *********
```

FIGURE 20.6b.

```
                                                                 ORG.
                                     WESTERN SECTION DETRITUS FLUXES   INORG.
    1                2                4                6                8                10               12

                                  *********        *********        *********
                                  * 0.000 *        * -.104 *        * 0.000 *
                                  *       *        *       *        *       *
                                  * -.004 *        * 0.000 *        * 0.000 *
                                  *********        *********        *********

                                  *********        *********        *********        *********
                                  * -.001 *        * 0.085 *        * 0.000 *        * 0.000 *
                                  *       *        *       *        *       *        *       *
                                  * 0.000 *        * 0.000 *        * 0.000 *        * 0.000 *
                                  *********        *********        *********        *********

                 *********        *********        *********        *********        *********
                 * -.020 *        * -.003 *        * 0.016 *        * 0.000 *        * -.010 *
                 *       *        *       *        *       *        *       *        *       *
                 * -.003 *        * 0.001 *        * 0.000 *        * 0.000 *        * -.000 *
                 *********        *********        *********        *********        *********

*********        *********        *********        *********        *********        *********        *********
* -.005 *        * -.006 *        * 0.000 *        * 0.004 *        * 0.000 *        * 0.004 *        * -.012 *
*       *        *       *        *       *        *       *        *       *        *       *        *       *
* 0.000 *        * -.000 *        * 0.000 *        * 0.000 *        * 0.000 *        * 0.000 *        * -.003 *
*********        *********        *********        *********        *********        *********        *********

*********        *********        *********        *********        *********        *********        *********
* -.003 *        * 0.016 *        * 0.003 *        * 0.000 *        * 0.000 *        * 0.005 *        * 0.010 *
*       *        *       *        *       *        *       *        *       *        *       *        *       *
* -.000 *        * 0.003 *        * 0.001 *        * 0.000 *        * 0.000 *        * 0.000 *        * 0.002 *
*********        *********        *********        *********        *********        *********        *********

*********        *********        *********        *********        *********        *********        *********
* -.001 *        * 0.002 *        * 0.000 *        * 0.000 *        * 0.000 *        * 0.000 *        * -.000 *
*       *        *       *        *       *        *       *        *       *        *       *        *       *
* 0.001 *        * 0.000 *        * 0.000 *        * 0.000 *        * 0.000 *        * 0.000 *        * -.001 *
*********        *********        *********        *********        *********        *********        *********

*********        *********        *********        *********        *********        *********        *********
* -.001 *        * 0.000 *        * 0.000 *        * 0.000 *        * 0.000 *        * 0.000 *        * 0.002 *
*       *        *       *        *       *        *       *        *       *        *       *        *       *
* -.000 *        * 0.001 *        * 0.000 *        * 0.000 *        * 0.000 *        * 0.000 *        * 0.001 *
*********        *********        *********        *********        *********        *********        *********

*********        *********        *********        *********        *********        *********        *********
* 0.009 *        * 0.007 *        * 0.000 *        * 0.000 *        * 0.000 *        * 0.000 *        * 0.000 *
*       *        *       *        *       *        *       *        *       *        *       *        *       *
* 0.000 *        * 0.000 *        * 0.001 *        * 0.000 *        * 0.000 *        * 0.000 *        * [illegible] *
*********        *********        *********        *********        *********        *********        *********
```

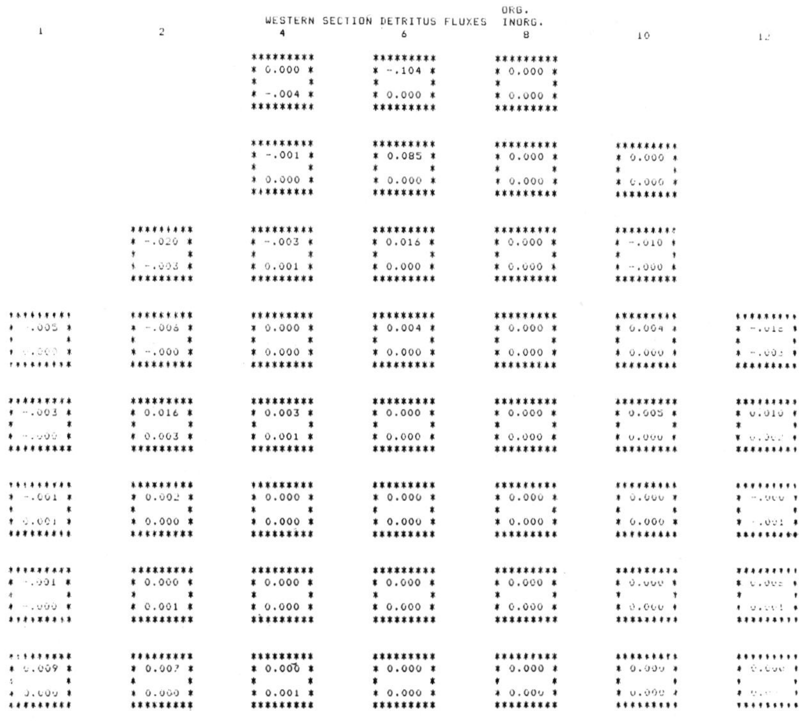

FIGURE 20.6c.

- There is considerable noise in the distribution of detritus formation, decomposition, and dissolution. We note, however, that the total amount of organic detritus leaving the surface reservoirs is about twice the amount deduced with the 12-box model (Bolin et al. 1983). This also implies a doubling of the decomposition rate at deeper levels and corresponding increases of the fluxes of dissolved inorganic carbon, phosphorus, and oxygen. Obviously, this model of the Atlantic Ocean is more accessible for carbon uptake than the one depicted as part of the 12-box model.

It is premature to discuss these preliminary results in any further details. Even though many inadequacies are apparent, the results are promising; further improvements are possible. The approach permits the inclusion of dynamical constraints. We may thus take advantage of the fact that the wind stress determines a pattern of horizontal divergence and convergence in the Ekman layer. If the top layer of the ocean as defined in the model approximately coincides with the Ekman layer we may identify the vertical motions, which are induced by the wind stress, as vertical motions at the interphase between the two uppermost layers of the model. It may also be possible to prescribe a pattern of ocean currents in the surface layer of the model based on dynamical considerations of a well-mixed surface layer subject to a surface wind stress.

We may also impose a quasigeostrophic constraint. The density distribution as defined by the temperature and salinity fields can be translated into a vertical change of the isopycnic advective velocity between adjacent boxes. This adds a good number of equations. As a matter of fact we would in this way combine traditional methods used in diagnostic studies of the density field in the oceans with a powerful method of interpreting steady-state tracer distributions.

However, we are not limited to the assumption of a steady-state tracer distribution. The GEOSECS and TTO data for ^{14}C, tritium, and freons clearly reveal the transient behavior of these tracers, which have been introduced by man during the last several decades. For the northern Atlantic Ocean we can determine reasonably well how the ^{14}C and tritium fields have changed during the 10-year period 1972 to 1982. Thus we are able to incorporate the term $\partial q^m/\partial t$ (q^m being the tracer concentration) into the known vector b which constitutes the inhomogeneous term of our set of equations (see Eqs. [2.1] and [2.9] in Bolin et al. 1983). We of course implicitly assume that the general circulation and biological activity, which are our unknowns (vector x), do not change with time. Experiments of this kind are under way.

Conclusions

Work during the last two decades has given us a general understanding of the global carbon cycle, and we are also able to estimate its approximate response to anthropogenic emissions of CO_2 into the atmosphere. Present models, however, are inadequate for more precise projections into the future. Even though the role of the oceans is better understood than that of the terrestrial biota, considerable improvements in the modeling of all parts of the carbon cycle are essential. It is not until the response characteristics are well described on all time scales for a few years to one or two centuries that we will be able to analyze more subtle processes, which may be quite important when projecting until the time when atmospheric CO_2 concentrations will have doubled. Detailed models are needed for comparison of what is implied by a synthesis of our present knowledge in terms of a model and the real behavior of the carbon cycle as revealed by observations.

Acknowledgments

Important contributions to the results described in the present paper have been made by Dr. B. Moore, Complex Systems Research Center, University of New Hampshire (supported by the National Science Foundation under Contract No. DEB-8110477; by Dr. A. Björkström and Mr. K. Holmen, Department of Meterology, University of Stockholm (with support from the Swedish Natural Science Research Council [NFR] under contract E-EG 223-111); and Mr. U. Cederlöf, Department of Physical Oceanography, University of Gothenburg. A full account of the work with the multiple-box ocean model will be presented as a joint paper by the group as a whole.

References

Baumgartner, A. and E. Reichel. 1975. The World Water Balance. Elsevier/North Holland, Amsterdam.

Bolin, B. 1983. Changing global biogeochemistry. In P. Brewer (ed.), Oceanography. The Present and the Future, pp. 305–326. Springer Verlag, New York.

Bolin, B., A. Björkström, K. Holmen, and B. Moore. 1983. The simultaneous use of tracers for ocean circulation studies. Tellus 35B, 206–236.

Brewer, P. 1983. Carbon dioxide and the oceans. In Changing Climate, pp. 188–215. Report of the Carbon Dioxide Assessment Committee. National Academy Press, Washington, D.C.

Broecker, W. S., R. Gerard, M. Ewing, and B. C. Heezen. 1960. Natural radiocarbon in the Atlantic Ocean. J. Geophys. Res. 65:2903–2931.

Broecker, W. S., I.-H. Li, and T.-H. Peng. 1971. Man's unseen artifact. In D. W. Hood (ed.), Impingement of Man on the Oceans, pp. 287–324. Wiley-Interscience, New York.

Broecker, W. S., T.-H. Peng, and R. Engh. 1980. Modeling the carbon system. Radiocarbon 22:565–598.

Bryan, K., S. Manabe, and R. C. Pacanowski. 1975. A global ocean-atmosphere-climate model. Part II. The oceanic circulation. J. Phys. Oceanogr. 5:30–46.

Bunker, A. F. 1980. Trends of variables and energy fluxes over the Atlantic Ocean from 1948 to 1972. Mon. Weather Rev., 108:720–732.

Chen, C.-T. A. 1983. The distribution of anthropogenic CO_2 in the Atlantic and Southern Oceans. (in press).

Craig, H. 1957. The natural distribution of radiocarbon and the exchange time of carbon dioxide between atmosphere and sea. Tellus 9:1–17.

Craig, H. 1963. The natural distribution of radiocarbon: mixing rates in the sea and residence times of water and carbon. In J. Geiss and E. D. Goldberg (eds.), Earth Sciences and Meteorites, pp. 103–114. North Holland Publishing Co., Amsterdam.

Ebensen, S. K. and Y. Kushmir. 1981. The heat budget of the global ocean: an atlas based on estimates from surface marine observations. Report No. 29, Climate Res. Inst., Oregon State University, Corvallis.

Houghton, R. A., J. E. Hobbie, J. M. Melillo, B. Moore, B. J. Peterson, G. R. Shaver, and G. Woodwell. 1983. Changes in the carbon content of terrestrial biota and soils between 1860 and 1980: a net release to the atmosphere. Ecol. Monogr. 53(3):235–262.

Keeling, C. D. 1983. The global carbon cycle: what we know and could know from atmospheric, biospheric, and oceanic observations. In Carbon Dioxide, Science and Consensus, U.S. Department of Energy, CO_2 Conf. 820970, pp. II.3–II.62. National Technical Information Service, Springfield, Virginia.

Keeling, C. D., W. G. Mook, and P. P. Tans. 1979. Recent trends in the $^{13}C/^{12}C$ ratio of atmospheric carbon dioxide. Nature 277:121–123.

Nydal, R. and K. Lövseth. 1983. Tracing bomb ^{14}C in the atmosphere 1962–1980. J. Geophys. Res. 88, C6:3621–3642.

Oeschger, H., U. Siegenthaler, U. Schotterer, and A. Gugelmann. 1975. A box diffusion model to study the carbon dioxide exchange in nature. Tellus 27:168–192.

Östlund, H. G. and M. Stuiver. 1980. GEOSECS Pacific radiocarbon. Radiocarbon 22:25–53.

Schlesinger, W. H. 1977. Carbon balance in terrestrial detritus. Ann. Ecol. Syst. 8:51–81.

Siegenthaler, U. 1983. Uptake of excess CO_2 by an outcrop-diffusion model of the ocean. J. Geophys. Res., 88, C6, 99:3599–3608.

Stigebrand, A. 1981. A model for the thickness and salinity of the upper layer in the Arctic Ocean and the relationship between ice thickness and some external parameters. J. Phys. Oceanogr. 11:1407–1422.

Stuiver, M. and H. G. Östlund. 1980. GEOSECS Atlantic radiocarbon. Radiocarbon 22:1–24.

Sverdrup, H. U., M. W. Johnson, and R. H. Fleming. 1942. Oceanography, Prentice Hall, Englewood Cliffs, New Jersey.

Tans, P. 1981. A compilation of bomb ^{14}C data for use in global carbon model calculations. In B. Bolin (ed.), Carbon Cycle Modeling, pp. 131–157. John Wiley & Sons, New York.

Vitousek, P. M. 1983. The effects of deforestation on air, soil, and water. In B. Bolin (ed.), The Major Biogeochemical Cycles and Their Interactions. pp. 223–245. John Wiley & Sons, New York.

Wunsch, C. 1978. The general circulation of the North Atlantic west of 50° W determined from inverse methods, Rev. Geophys. Space Phys., 16:583–620.

21
The Use of Observations in Calibrating and Validating Carbon Cycle Models

IAN G. ENTING AND GRAEME I. PEARMAN

To determine the most appropriate data for calibrating and validating carbon cycle models, it is first necessary to determine the aims of the modeling study. Among the main uses of such studies, we can identify in particular:

1. Prediction of future atmospheric CO_2 concentrations (to assess climatic and biological impacts)
2. Reconstruction of past atmospheric CO_2 concentrations (to determine the driving force when attempting to analyze past climatic records in the search for a response to changes in CO_2)
3. Interpretation of current measurements involving the carbon cycle (to determine the major carbon fluxes)
4. Cross-comparisons of models with different degrees of resolution, including different dimensionality (to determine which models are most appropriate for particular studies).

A very general requirement is that the model calibration should be as precise as possible, and that it should be performed in a way that reduces the possibility of biased results being obtained from the use of inappropriate data. There are two main aspects to the approach that we have used in attempting to satisfy the requirements of model calibration. The first is our choice of calibration techniques based on the principles of constrained inversion, and the second is an analysis of the way in which each item of information that is used influences the final calibration and predictions of the model. Such an analysis of the influence of data is most commonly associated with constrained inversion formalisms [see for example the discussion by Jackson (1972) of the marginal utility of data]. This association is traditional rather than necessary, and an analysis of the influence of particular data items could (and we suggest should) be undertaken even if some other approach to the model calibration is used.

The organization of the remainder of this chapter is as follows. First we describe the general characteristics that are desirable in the data used for calibrating carbon cycle models. The next two sections describe how these characteristics relate to the techniques of constrained inversion that we used in our calibrations and the extension of this approach into the sensitivity analysis. We then include a section presenting results from a one-dimensional model and discusses the relative importance of the various model parameters and the various items of data used in the calibration. The following section extends this analysis to consider some of the types of data that we have not used either because of doubts about their applicability or because the observational programs are still in progress. By performing model calibrations using various possible data values, we can assess the extent to which such data could reduce the uncertainties in our knowledge of the global carbon cycle. Finally, we discuss some of the issues involved in the choice of data for modeling the carbon cycle with two- or three-dimensional models.

Criteria for Data Selection

The main qualities required of the data are consistency with the model, relevance to the modeling study, lack of circularity in the definition, and sufficient precision to add to our understanding of the carbon cycle.

The requirement for consistency means that the data must refer to quantities that the model is capable of representing in a manner that does not result in biased values. In particular, the quantity must not be influenced by processes that are not included in the model, and must be defined for space and time scales consistent with those being modeled. This will require that, in the averaging process used to produce low-resolution lumped or box models, the quantities are not biased by covariance effects, or else that such effects can be absorbed by constructing a model that uses "effective parameters" whose values can be adjusted to allow for the covariance effects.

The question of the relevance of data is, to some extent, determined only by using the data to calibrate the model and assessing the extent to which particular data items contribute to reducing the uncertainties in the calibration. This is the study of what Jackson (1972) has called "the marginal utility of data"; Twomey (1977) has called it the "information content of the data."

The principal way in which circularity can arise in the use of carbon cycle data is when in trying to bring the data into a form consistent with the model, we find it (the data) subjected to some transformation (e.g., averaging, filtering, selection, deconvolution, etc.) that involves the use of a carbon cycle model in the specification of the transformation. Circularity can arise if the transformation is defined by a carbon cycle similar to the one being calibrated, and, of course, if the models are rather different, then such transformations may result in the data being inconsistent with the model result to which it is compared.

The precision of the data principally determines its utility, and the question of whether the data are sufficiently precise to contribute to the calibration must be assessed by attempting to use it. We believe the precision of the data is particularly important, to the extent that the same criteria of consistency, relevance, and lack of circularity must be applied to both the values of the data to be fitted and to the estimates of the uncertainties in this data. For example, in using data constructed from averages over spatially varying quantities, we have in most cases followed the suggestion of Bolin et al. (1981) that the appropriate measure of the uncertainty is given by the range of values across the whole of the region.

One of the major uncertainties in our current understanding of the carbon cycle concerns the so-called "missing carbon problem" (see for example Broecker et al. 1979). In its simplest terms, the problem is that when carbon cycle models are calibrated using ^{14}C data, predictions are that the amount of carbon remaining in the atmosphere after fossil fuel release is greater than that implied by direct observations of increasing CO_2 concentrations.

There are three main solutions to the problem that have been suggested:

1. That current ocean models cannot be adequately calibrated using ^{14}C
2. That the biosphere is currently acting as a sink for carbon
3. That the biosphere has in the past been a net source of carbon.

The analysis by Peng and Broecker (1984) suggests that alternative solutions involving changes in the organic carbon cycle in the oceans are unacceptable. The question as to which of these solutions is correct is vital for the choice of data in carbon cycle model calibrations. The first solution is that ocean mixing processes are so complex that current low-resolution models are unable to describe consistently the uptake of both total carbon and ^{14}C. If this was true, it would then mean that the use of ^{14}C data to calibrate carbon cycle models would be inappropriate, and that the best way to make predictions of future CO_2 concentrations would be to assume a constant airborne fraction, the value of which can be estimated from Mauna Loa data as suggested by Laurmann and Spreiter (1983).

There have been a number of studies that considered more complex ocean models to determine whether these could be consistent with the observed ocean uptake for both CO_2 and ^{14}C. Siegenthaler (1983) analyzed an outcrop-diffusion model that included direct ventilation of deep ocean layers as an additional feature in the basic box-diffusion model (Oeschger et al. 1975). He found that, assuming no biospheric changes, the airborne fraction was 0.67 with no outcrop (i.e., the basic box-diffusion model) and calibration using natural ^{14}C, 0.60 with 10% of the ocean area as outcrop and calibrated using natural ^{14}C, and 0.53 with 10% outcrop but calibrated using anthropogenic ^{14}C. (The Mauna Loa observations give an airborne fraction of 0.59.)

Bolin et al. (1983) analyzed a 12-reservoir ocean model that was calibrated by applying constrained inversion techniques to data for five tracers

(^{14}C, C, O, P, and alkalinity). They found an airborne fraction of 0.78, a large value that they attributed in part to the poor vertical resolution of their model.

Viecelli et al. (1981) constructed a three-zone ocean model, with advective mixing and calibrated using anthropogenic ^{14}C. They pointed out that the inclusion of advection emphasizes the differences between the long-term deep-ocean processes that determine the natural ^{14}C distribution and the short-time-scale processes that have been involved in the uptake of anthropogenic CO_2 and ^{14}C. They found that they obtained good agreement with the Mauna Loa data and Suess effect data if they included a net release of carbon from the biosphere occurring around the end of the 19th century.

A comprehensive review of the applicability of the box-diffusion model has been given by Broecker et al. (1980). They used bomb-produced ^{14}C as the principal data for calibrating the ocean uptake and mixing rates, and also showed that the resulting rates were consistent with the distribution of bomb-produced tritium and with other carbon cycle data. They also constructed a more detailed regional ocean model with a more realistic circulation than that used by Viecelli et al. (1981). They found that the rates of carbon uptake did not differ greatly from the rates given by the box-diffusion model. They concluded that given the disagreements between direct estimates of biomass changes and the changes required to give agreement with the Mauna Loa record, "the error lies in the biomass estimates rather than in existing ocean models."

The possible resolution of the missing carbon problem in terms of biospheric changes (explanations [2] and [3] above) and the implications for the calibration of carbon cycle models are most easily discussed using the notation introduced by Oeschger and Heimann (1983), who derived the following equation for linear atmosphere/ocean systems with the biosphere acting as an external source:

$$r^*_{af} = r_{af} - (\Delta a + B_p - B_h)/\Delta Q_f \tag{1}$$

The various quantities are:

r_{af} = the observed airborne fraction of CO_2;
r^*_{af} = the airborne fraction that would hold if no biospheric perturbations (and no natural noise) were present;
t_1, t_2 = the beginning and end, respectively, of the period for which r_{af} is observed;
Δa = the change in atmospheric CO_2 concentration over the period (t_1,t_2) because of natural variability;
B_P = the change in CO_2 concentrations over the period (t_1,t_2) because of biospheric releases during that period;
B_h = the change in CO_2 concentrations over the period (t_1,t_2) because of biospheric releases prior to t_1;

ΔQ_f = the amount by which the fossil carbon release over the period (t_1,t_2) would have changed the atmospheric CO_2 concentration if all such carbon had remained in the atmosphere.

Oeschger and Heimann related the biospheric contributions B_p and B_h to the net biospheric release rate $b(t)$ by

$$B_p = \int_{t_1}^{t_2} b(\tau)\, R(t_2 - \tau) d\tau \quad (2)$$

$$B_h = \int_{-\infty}^{t_1} b(\tau)\, [R(t_1 - \tau) - R(t_2 - \tau] d\tau \quad (3)$$

where $R(t)$ is the response function that describes the rate of decay of a pulse of atmospheric CO_2.

The missing carbon problem arises if it is assumed that the unperturbed airborne fraction r^*_{af} can be estimated from carbon cycle models. Values given by Oeschger and Heimann (1983) for the period 1959 to 1978 are

$$(r^*_{af} - r_{af})\, \Delta Q_f + \Delta a = B_h - B_p = 3.9 \pm 2.2 \text{ ppmv} \quad (4)$$

This calculation is inconsistent with the absence of a biospheric change, although the inconsistency disappears if the value of r^*_{af} (taken as 0.67) is reduced to 0.60 (the box-diffusion model as calibrated by Broecker et al. 1980) or 0.55 (the regional model of Broecker et al.).

However, ecosystem modeling (see for example Moore et al. 1981) suggests that there has been a net release of carbon from the biosphere over the last century with a current release of about 2.5 Gt • yr^{-1} (see also Houghton et al. 1983 and Houghton, *this volume*). For such a release, Oeschger and Heimann calculate B_p = 17.6 ppmv and B_h = 5.3 ppmv, so that even for models with $r^*_{af} \sim r_{af}$ there is still a large disagreement with Eq. (4).

Alternatively, other estimates (Olson 1982) suggest a net loss of carbon from the biosphere of 0.5 to 2 Gt • yr^{-1}. A release of 1 Gt • yr^{-1} gives a contribution B_p = 8 ppmv if the release rate was maintained over the entire period of the Mauna Loa observations. This release is thus inconsistent with any of the ocean models unless it is balanced by additional releases before 1958. These releases must be very large since a constant release of 1 Gt • yr^{-1} over the period 1860 to 1958 gives B_h = 4 ppmv.

As an alternative to large net releases of biospheric carbon in the past (or very large past releases if there is at present a net release), a number of studies have proposed that the missing carbon problem can be resolved by having a net biospheric uptake of carbon, that is, $b(t) < 0$ in the notation used above. This uptake has been attributed to enhanced growth of vegetation due to the increasing atmospheric CO_2 concentrations (Bacastow and Keeling 1973; Oeschger et al. 1975). Such a behavior would give B_h, $B_p < 0$ with $B_h - B_p > 0$. A comparison of the calculations quoted above with the discrepancy indicated by Eq. (4) indicates that the discrepancy could be resolved by an uptake whose magnitude was one-sixth to one-half of the release proposed by Moore et al. (1981). If the estimates of r^*_{af} given by Broecker et al. (1980) were correct, the discrepancies would then be smaller (and possibly vanish) and the implied biospheric uptake rates would be correspondingly smaller.

The question of which of these explanations is correct is vitally important for the way in which carbon cycle models are calibrated to predict future CO_2 concentrations. If the biosphere is currently a sink of carbon due to enhancement of growth by increasing CO_2 levels, then this effect would be expected to continue for some time, and the discrepancy between the observed and calculated airborne fractions will persist. The most accurate predictions would then be based primarily on direct CO_2 observations, thus ignoring the ^{14}C data. Similarly, if the ocean mixing processes do act to produce a genuine difference between total carbon, which increases on time scales of decades, and ^{14}C with longer time scales (i.e., the slower equilibration with the ocean due to buffering [Broecker et al. 1980] and the natural 5730-year half-life), then again the ^{14}C data are inappropriate for model calibration, and the most reliable predictions for the future will be made on the basis of the Mauna Loa data.

If, however, the discrepancy in the airborne fractions is due to the effect of past biospheric changes, then these effects will be expected to decrease with time so that the observed airborne fraction will increase to equal the ^{14}C-derived airborne fraction. Longer-term predictions will be most accurate if they are based on the latter value.

Thus, the choice of data can be critically important, particularly when there is an uncertainty about the processes that should be modeled. The calculations presented in the later section describing model results are designed to take into account all three of the possibilities previously described. Because all of the possibilities are considered, there is less danger of the data set being inconsistent with the model. In more specialized model calculations that consider only one of the possible explanations (whose validity is therefore conditional on the correctness of the assumptions), the dangers of using data that are in some way inconsistent with the model are somewhat greater.

The Constrained Inversion Formalism

Constrained inversion techniques have been used in a variety of geophysical problems, usually in the analysis of systems that are in some way not completely understood. As described by Enting (1985a), there are a number of considerations that suggest that such techniques might be appropriate in carbon cycle modeling:

1. In the carbon cycle model described by Enting and Pearman (1982), it was found that the best-fit used in the calibration was poorly determined.
2. As described by Wunsch (1978), indirect methods (i.e., the inversion

of tracer data) are not adequate for determining the general circulation of the oceans. To the extent that the carbon cycle depends on this circulation, the incomplete description of ocean circulation implies an incomplete description of the carbon cycle.
3. The uncertainties in the present and past changes in the size of the biosphere are very large.
4. The fact that fossil fuel use has increased in a roughly exponential manner means that, to the extent that the system is linear, this past behavior gives no information about the natural response rates of the system.

Of these aspects, the third is the most important. It must be noted that Eq. (4) gives one number (or more precisely one number for each time period for which an independent estimate of r_{af} can be obtained) from which to estimate the function $b(t)$. It is only in studies such as those of Bacastow and Keeling (1973) and Oeschger et al. (1975), where it is assumed that $b(t)$ is determined entirely by the atmospheric CO_2 concentration, that there is indeed only one number, the enhancement factor, to be estimated from Eq. (4) to determine the uptake rate.

Rodgers (1977) has pointed out that constrained inversion techniques involve the use of additional information to supplement data sets that do not adequately specify unique, numerically stable solutions to geophysical problems. Such additional information can consist of arbitrary assumptions, analogies with related problems, or information from outside the domain of the original problem. Even the use of arbitrary assumptions can be an acceptable procedure if it can be shown that the particular choice does not affect those aspects of the final result that are of primary interest. Specifically, such arbitrary assumptions can be computationally useful if the parameters depend strongly on the assumptions and the dependence is such that the predictions are virtually independent of these assumptions. The most common such assumptions are those of smoothness of various functions in circumstances where there may well be rapid variations, but where such variations do not affect the long-term behavior. In carbon cycle studies, smoothness assumptions are sometimes required in calculations such as the deconvolution of ^{13}C in tree rings (Peng et al. 1983). The main form of constrained inversion we are considering here (described by Enting 1983, 1984) supplements conventional carbon cycle data (e.g., atmospheric CO_2 levels and natural and post-bomb ^{14}C levels) with additional geophysical information that acts to constrain the range of possible parameter values in the model. The formalism makes use of J measurements of carbon cycle data, m_j, with standard deviations v_j. The model predictions for the m_j are denoted y_j. The y_j are functions of K parameters x_k, and the functional relation is expressed as $y_j(\mathbf{x})$. The constrained inversion requires some independent prior estimates of the x_k. These prior estimates are denoted q_k, and the standard deviations of these prior values are u_k.

The formalism obtains estimates of the parameters by choosing, as estimates of x_k, the values $\hat{x}_k$ that minimize:

$$\Theta = \sum_{j=1}^{J} [m_j - y_j(\mathbf{x})]^2/v_j^2 + \gamma \sum_{k=1}^{K} (x_k - q_k^2/u_k^2 \qquad (5)$$

In constrained inversion methods the quantity γ determines the relative contribution of the two types of information. The desirable situation is to have the final estimates relatively insensitive to the value of γ that is chosen. If the results depend strongly on γ, it can be an indication that the observations m_j are inconsistent with the prior parameter estimates q_k, at least within the context of the given model. Such inconsistency would require a reassessment of either the model, the way in which it is parameterized, or the type of data that are used to fit it.

The case $\gamma = 1$ is the most direct example of Rodgers' view that constrained inversion corresponds to including extra information (either explicitly or implicitly), since $\gamma = 1$ gives an optimally weighted least-squares fit to the two sets of information. Alternatively, Enting (1983, 1985a) has pointed out that the inversion formalism can be interpreted as Bayesian inference, with the carbon cycle observations m_i being used to obtain the refined (posterior) parameter estimates from prior estimates q_k.

The formalism has been applied to the box-diffusion model (Enting 1985a) and indicates an inconsistency with independent estimates of the natural ^{14}C production rate (O'Brien 1979), apparently because the absence of any detrital transport leads to an underestimate of the amount of total carbon and ^{14}C in the deep oceans. The formalism has also been applied to a more complex two-region model (Enting and Pearman 1983) as described below.

It is of interest to compare this formalism to two other applications of constrained inversion to problems related to the carbon cycle.

In the ocean model considered by Bolin et al. (1983) (see also Moore and Björkström, *this volume*), the parameters x_α are the transport parameters describing advective fluxes and diffusion coefficients, source/sink strengths, and detrital flux. The observations are m_i, $i = 1{,}60$, where i ranges over five tracers in 12 reservoirs. For concentrations y_i, the rates of change can be written in the general form:

$$\begin{aligned} \frac{\partial y_i}{\partial t} &= \sum_{\alpha i k} \mathrm{R}_{ik\alpha} y_k x_\alpha + \sum_{\alpha} \mathrm{S}_{i\alpha} x_\alpha + \sum_{j} \mathrm{T}_{ij} y_j + \mathrm{U}_i \qquad (6) \\ &= \sum_{\alpha} \mathrm{A}_{i\alpha}(\mathbf{y}) x_\alpha - \mathrm{b}_i(\mathbf{y}) = \mathrm{g}_i \\ &= 0 \text{ for equilibrium} \end{aligned}$$

In determining the parameters x_α, Bolin et al. considered two cases.

The "indeterminate" case is solved by taking the solutions that satisfy

$$g_i = 0, i = 1,60 \tag{7a}$$

$$y_i - m_i = e_i = 0, i = 1,60 \tag{7b}$$

and minimize Σx_α^2 with respect to the x_α subject to these constraints. The "incompatible" case is solved by requiring

$$y_i - m_i = e_i = 0 \tag{8}$$

and minimizing Σg_i^2 subject to this constraint. In this case, additional constraints of the form $\sum_\alpha V_\alpha x_\alpha > 0$ were included, mainly to require positive diffusion coefficients. These inequalities are handled with linear programming techniques.

Bolin et al. (1983) recognized several conceptual problems. First, there is the distinction between the two cases that has no obvious physical meaning. Then there is the lack of justification for minimizing the sum Σx_α^2, and finally there is the arbitrariness in the choice of the relative scales for the terms in the sums that are minimized.

On the other hand, the constrained inversion formalism from Eq. (5) avoids all these problems and, for the ocean model, reduces to

$$g_i = 0 \text{ (thus defining the } y_k \text{ as a function of the } x_\alpha) \tag{9a}$$

and, subject to Eq (9a), minimizing Θ with respect to the x_α where

$$\Theta = \sum_i (e_i/v_i)^2 + \sum_a (x_\alpha - q_\alpha)^2/u^2_\alpha \tag{9b}$$

In this case all the scaling is involved in the v_i, which are known from the observations, and the u_α, which described the plausible ranges of the x_α. Any information concerning the values of the x_α is incorporated in the q_α, although, in some aspects of ocean transport, the prior information is so scanty that zero is indeed an appropriate prior estimate.

However, these conceptual advantages are associated with the practical disadvantage that Eq. (9a) and (9b) give a nonlinear set of equations. A more detailed comparison between constrained inversion formalisms is given by Enting (1985a).

Another recent study that should be considered in terms of constrained inversion is the deconvolution of a $\delta^{13}C$ record from tree rings by Peng et al. (1983). In this study the carbon cycle model (a modified box-diffusion model) served to define a ^{13}C response function that was to deconvolve the tree-ring data. The main uncertainties considered were those associated with uncertainties in the tree-ring data. Apart from the influence of the air–sea fractionation factor, the implications of uncertainties in the model were not considered. As emphasized in the following section, a complete assessment of the uncertainties in any quantity, such as the reconstructed $b(t)$, requires assessment of all uncertainties simultaneously in the search for a "worst-likely-case." An appropriate formalism for incorporating the

^{13}C data would be to include it directly into the calibration—possibly with weaker constraints on the release function.

Sensitivity Analysis

The aim of a sensitivity analysis is to determine the extent to which uncertainties in the parameter estimates affect the predictions of the model. This is a particularly important point: The sensitivity has to be assessed in the context of predicting a particular quantity (such as the CO_2 concentration) at a particular time in the future. In the following analysis the symbol Z will be used to denote the prediction for which the uncertainty is being investigated.

The full assessment of uncertainties in Z has to include the uncertainties in all the parameters, including any parameters that are not estimated from the carbon cycle data. If there are some parameters that are not estimated but still subject to significant uncertainties, then the total uncertainty is a combination of the uncertainties in the estimation and the uncertainties from the parameters that are not estimated. The combined uncertainty should be assessed by repeating the estimation procedure and assessing the uncertainties for all possible combinations of parameters that are not estimated.

Conventional estimation based on carbon cycle data can normally only handle a small number of parameters and, as indicated above, the full assessment of the uncertainties is a complicated procedure.

On the other hand, the formalism based on Eq. (5) uses the extended data set that formally allows the estimation of many more parameters, and the formalism is simply a conventional least-squares fit to the extended data set.

The form of sensitivity analysis described below concentrates on the extreme case—the greatest probable deviations in the prediction Z. This is achieved by taking the most extreme Z values subject to fixed deviations in Θ, the sum of squares of deviations. Extremes subject to constraints are determined by introducing Lagrange multipliers—the quantity α in Eq. (10).

The quantity to be minimized is

$$\varphi = \Theta - 2\alpha Z \tag{10}$$

More specifically, Enting (1983, 1985a) has shown that if $Z(\alpha)$ is the value of Z for the parameter set that minimizes φ for a particular α value, then the variance of the estimated prediction $\hat{Z}$ is, in a linear approximation,

$$\mathrm{Var}(\hat{Z}) = [Z(\alpha) - \hat{Z}(O)]/\alpha \tag{11}$$

This form of sensitivity analysis is also used as part of the study of the utility of the data by considering the extent to which particular items of information contribute to reducing the uncertainties in the predictions made by the model.

Some One-Dimensional Results

The specific results described in the next two sections were obtained using the one-dimensional carbon cycle model developed in the Division of Atmospheric Research (CSIRO, Australia); a number of comparisons are made with other models when appropriate. The one-dimensional model was first described by Pearman (1980) and in more detail by Enting and Pearman (1982). Details of a number of minor refinements to the model and a description of a calibration using the techniques described above have been given by Enting and Pearman (1983).

In brief, the model is a box model with one atmospheric reservoir, two biospheric reservoirs, and ocean reservoirs divided into two regions (warm and cold), with each region having one surface reservoir and, for the present calculations, six subsurface reservoirs.

Transport of carbon in the oceans is parameterized in a manner that should, in principle, be able to describe a wide range of uptake rates. The advection is described as a sum of two components, one that injects equal amounts of cold surface water into each subsurface layer, and one that injects cold surface water into the lowest layer. (Each of these fluxes is parameterized by its contribution to the upwelling velocity at the bottom of the surface layer). Two parameters are used to describe an eddy diffusion coefficient that varies linearly with depth, and a final parameter describes a detailed carbon flux from the surface to subsurface layers. (There is also an additional horizontal diffusion in the surface layers that is found to have little effect on the system but was originally included to study its significance in the steady-state cycle of carbon between the atmosphere and the oceans).

The ocean model is in some ways comparable with that described by Viecelli et al. (1981), the main difference being that there is less spatial resolution but more detailed descriptions of the various processes. Table 21.1 shows the full set of parameters. The precise way in which the parameters are used (and the sources for the prior estimates q_k) are described by Enting and Pearman (1983).

In view of the ambiguities concerning the biospheric history, the parameterization of this history is described here in more detail.

The release $b(t)$ is expressed as

$$b(t) = \sum_{i=1}^{7} a_i b_i(t) \tag{12a}$$

where

$$\int_{-\infty}^{\infty} b_i(t)dt = 1 \tag{12b}$$

and b_i is a quadratic B-spline, with nodes at 25-year intervals and being nonzero only for the 75 years between $1790 + 25i$ and $1865 + 25i$ for $i =$

TABLE 21.1. Estimates and ranking of parameters used in the one-dimensional model.

No. (k)	Parameter	Units	Prior (g_k)	SD (u_k)	Best-fit	Parameter ranking: Best-fit	Parameter ranking: Perturbation
1	Preindustrial CO_2	ppmv	270	50	264.32	18	5
2	^{14}C from weapons	10^{26} atom · megaton^{-1}	2.0	0.5	1.401	3	18
3	Preindustrial $\delta^{13}C$	‰	−6.0	3.0	−5.896	24	21
4	a_1	Gt	0.0	10.0	0.07	27	23
5	a_2	Gt	15.0	20.0	15.14	28	14
6	a_3, Biospheric	Gt	32.5	40.0	31.38	25	4
7	a_4, Release	Gt	37.5	60.0	15.40	20	2
8	a_5, Coefficient	Gt	57.5	80.0	27.08	14	15
9	a_6	Gt	102.5	100.0	−35.08	1	16
10	a_7	Gt	87.5	100.0	−20.37	4	24
11	Cold ocean temperature	°C	10.3	5.0	10.80	21	3
12	Warm ocean temperature	°C	25.2	2.0	26.47	9	13
13	Alkalinity	mol · m^{-3}	2.369	0.021	2.351	8	11
14	Uniform upwelling	m · yr^{-1}	1.8	1.35	0.210	2	17
15	Varying upwelling	m · yr^{-1}	2.5	2.0	0.526	7	10
16	K_v, vertical diffusion	m^2 · yr^{-1}	3500	2000	5283	6	1

17	$\frac{d}{dz}K_v$	$m \cdot yr^{-1}$	0	1.0	0.242	15	22
18	Horizontal diffusion	$m^2 \cdot s^{-1}$	2.5×10^4	2×10^4	23,807	22	29
19	^{14}C from cosmic rays	$mol \cdot s^{-1}$	1.484×10^{-5}	3.7×10^{-6}	1.331×10^{-5}	13	26
20	u_*, cold	$m \cdot s^{-1}$	1.7	1.2	1.569	19	24
21	u_*, warm	$m \cdot s^{-1}$	0.8	0.5	1.308	5	8
22	Detrital fallout	$Gt \cdot yr^{-1}$	2.0	3.0	3.537	10	7
23	Young biosphere	Gt	140	70	169	12	20
24	Old biosphere (after release)	Gt	1400	700	1040	11	9
25	NPP	$Gt \cdot yr^{-1}$	100	50	103.2	23	31
26	Old biosphere turnover time	year	60	30	67.0	16	19
27	$\delta^{13}C$ Depletion of detritus	‰	−20	10	−20.07	26	28
28	Mixed layer depth	m	100	60	100.1	29	6
29	f_{as}		0.99795	0.00025	0.99795	30	27
30	f_{sa}		0.99005	0.00025	0.99005	31	30
31	f_{ab}		0.982	0.004	0.9828	17	12

Details of the model, the parameter definitions, and the determination of prior estimates $q_k \pm u_k$ are given by Enting and Pearman (1982, 1983). The best-fit estimates are obtained using the constrained inversion formalism described in the text. The two parameter rankings (1 = most important) define the importance of parameters for determining the value and precision, respectively, of future CO_2 concentrations.

1 to 7. Quadratic splines with zero values before 1815 can be represented in this way using the 7 a_i values which, by virtue of Eq. (12b), give the total carbon release for the B-spline contribution over the 75-year periods. The prior estimates in Table 21.1 (parameters 4 to 10) give an approximation to the release given by Moore et al. (1981). The uncertainties in the a_i are given magnitudes close to the release estimates following the argument that larger changes will be associated with larger uncertainties, and the current uncertainties are comparable to the estimated release rates (see for example Olson 1982). It would be preferable to relate the uncertainties more directly to the underlying ecosystem model, with the ultimate approach being the simultaneous estimation of all parameters in a combined carbon cycle/ecosystem model. Not only would this give more objective estimates of the uncertainties, but it could also give a realistic indication of the correlation between uncertainties at different times.

Figure 21.1 shows the reconstructed release $b(t)$ (solid curve) obtained by minimizing Eq. (5) compared to the constraint embodied in the prior estimate based on Moore et al. (1981) (dashed curve). Probably the best that can be expected from such reconstructions is that they act as a guide to discrepancies and anomalies, and in this case the discrepancies are clearly very large. In particular, the discrepancy is so large that it overshadows any effects that might arise from the fact that our prior estimate is only an approximation to the $b(t)$ given by Moore et al, i.e., quadratic

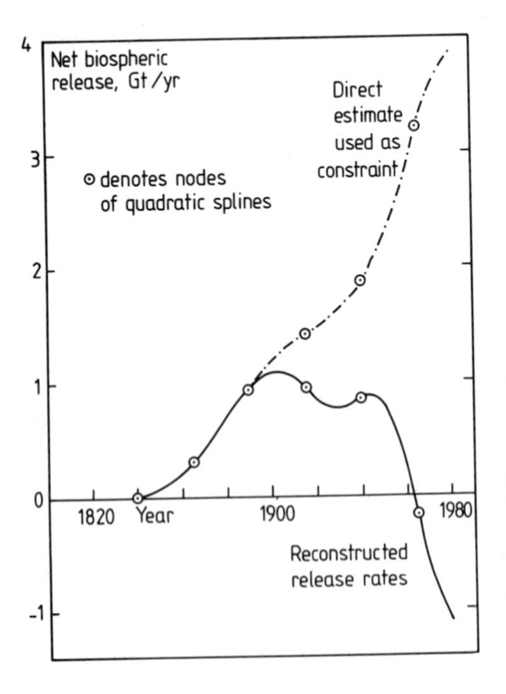

FIGURE 21.1. Reconstructed net release rate of carbon from the biosphere. The *solid curve* is the reconstruction that is consistent with the carbon cycle data, and yet is as close as possible (in a least-squares sense) to the direct estimates (*dashed curve*) from ecosystem modeling. (From Moore et al. 1981.)

splines with 25-year nodes cannot provide an exact fit to the latter function. In interpreting the reconstruction it must be recalled that, subject to the requirements that it be consistent with the carbon cycle model and the data used in the calibration, the reconstructed release is as close as possible (albeit in terms of a somewhat arbitrary definition of "close") to the prior estimate. Other release functions will be consistent with the carbon cycle model but will differ even more from the prior estimates.

To investigate the relative importance of the various parameters and data items, we investigate the size of the residuals that are summed to give Θ in Eq. (5). This means that the importance is being compared on a scale that is proportional to the uncertainties in the information that is fitted. This is a natural scale for comparison: If an observation is repeated with higher precision, then this will lead to such an observation becoming a more important part of the calibration, and conversely, the least precise observations will be of lesser importance.

There are two types of studies of residuals that are relevant. First, we can consider the residuals at the best-fit point; the relative sizes will be a measure of the relative importance that the various pieces of information play in determining the parameters. Second, we can consider the changes in the residuals associated with perturbations about the best-fit. In particular, we can undertake the sensitivity analysis described earlier, looking for the largest possible variations in some prediction of the model. The quantities whose residuals change most with these perturbations will be the quantities that are most important in constraining the predictions that are being investigated.

The two types of residuals that lead to the two different rankings of both data and parameters in Tables 21.1 and 21.2 can be regarded as ranking the importance of the information for determining the accuracy and precision of the predictions. The "best-fit" column of each table gives the importance of the information in determining the actual predicted (most likely) value of the CO_2 concentration in 2050. The final column in each table ranks the information according to its importance in reducing the statistical uncertainties that are assigned to the prediction.

To present a specific example we consider the model's predictions of the atmospheric CO_2 concentration in the year 2050, calculated assuming a fossil carbon release that increases at 2.25% per annum from 1979 to 2000 and is constant thereafter. This type of flattening is more extreme than is likely to be achievable. We consider it as an extreme case and one in which the uncertainties in the model predictions are expected to be large (Laurmann and Rotty 1983, Laurmann and Spreiter 1983). To standardize the comparisons, the biospheric releases are set to zero after 1985.

Table 21.2 lists the various data items, m_j, that were used in calibrating the model together with the standard deviations (SD) assigned and a rank-

TABLE 21.2. Data ranking.

Quantity	Reservoir	Date	Value	SD	Units	Importance	
						Best-fit	Predict
$\Delta^{14}C$	Biosphere	1890	0	5	‰	25	14
$\Delta^{14}C$	Biosphere	1910	−5	5	‰	29	19
$\Delta^{14}C$	Biosphere	1930	−12	5	‰	26	23
$\Delta^{14}C$	Biosphere	1950	−22	5	‰	11	9
$\Delta^{14}C$	Warm surface	1955.5	−58	10	‰	6	11
$\Delta^{14}C$	Cold surface	1955.5	−58	10	‰	18	26
$\delta^{13}C$	Atmosphere	1956.2	−6.69	0.13	‰	14	12
C_a	Atmosphere	1959.5	313.76	0.3	ppmv	15	3
C_a	Atmosphere	1963.5	316.81	0.3	ppmv	5	6
$\Delta^{14}C$	Atmosphere	1967	650	50	‰	19	16
$\Delta^{14}C$	Warm surface	1968.5	150	70	‰	9	27
$\Delta^{14}C$	Atmosphere	1969	580	50	‰	21	18
C_a	Atmosphere	1970.5	323.61	0.3	ppmv	13	5
$\Delta^{14}C$	Atmosphere	1971	500	50	‰	10	13
$\Delta^{14}C$	Atmosphere	1973	450	30	‰	12	28

C_a	Atmosphere	1973.5	327.92	0.3	ppmv	1	1
$\Delta^{14}C$	Warm surface	1973.5	120	50	‰	20	15
$\Delta^{14}C$	Cold surface	1973.5	100	50	‰	8	7
$\Delta^{14}C$	Cold-deep	1973.5	−150	50	‰	24	22
$\Delta^{14}C$	Warm-deep	1973.5	−150	50	‰	7	8
ΣCO_2	Cold surface	1973.5	2.132	0.102	$mol \cdot m^{-3}$	16	17
ΣCO_2	Warm surface	1973.5	2.009	0.041	$mol \cdot m^{-3}$	2	4
ΣCO_2	Cold-deep	1973.5	2.337	0.041	$mol \cdot m^{-3}$	22	25
ΣCO_2	Warm-deep	1973.5	2.358	0.062	$mol \cdot m^{-3}$	30	30
$\Delta^{14}C$	Atmosphere	1975	420	30	‰	4	10
$\Delta^{14}C$	Atmosphere	1977	350	30	‰	28	29
C_a	Atmosphere	1977.5	331.73	0.3	ppmv	3	20
$\Delta^{14}C$	Warm surface	1978.5	100	50	‰	27	24
$\delta^{13}C$	Atmosphere	1978.5	−7.24	0.05	‰	23	21
C_a	Atmosphere	1980.5	336.46	0.3	ppmv	17	2

Shown are data used in calibrating the one-dimensional model by combining it with the $q_k \pm u_k$ from Table 21.1 and using the formalism of constrained inversion described in the text. The rankings are as for Table 21.1. Reference for the data values and the reasons for choosing the given standard deviation are given by Enting and Pearman (1983).

ing from 1 (largest residuals) to 30 (smallest residuals), indicating the relative extent to which the various items influenced the model calibration. The two cases are for the "best-fit" residuals and for the residuals for the ±1 SD in the model predictions of the concentration in 2050, that is, the residuals that constrain the prediction in the most sensitive (least constrained) direction in parameter space. Because the best-fit is not an exact fit, there will be an asymmetry between the residual constraining the +1 SD and the −1 SD. The data items are ranked by using the larger of the two values in each case.

The precise order of the ranking is not particularly significant, but a number of broad features deserve comment. The Mauna Loa data play an important part in the calibration by revealing the main discrepancy in the system; that is, that the missing carbon problem occurs and that the biospheric releases proposed by Moore et al. (1981) and Houghton et al. (1983) do not resolve it. Beyond that, we observe that the Seuss effect data (biospheric $\Delta^{14}C$) is of relatively low importance, presumably because the same aspects of the model are calibrated with greater accuracy by other data such as the atmospheric ^{14}C data. The large influence of the warm ocean surface carbon concentration may be indicating an inadequacy in the model structure in that, in the model, about half of all the advective flux passes through this reservoir; therefore, the mixing of water with higher carbon concentrations raises the warm surface concentration in the model above what is actually observed. The overall distribution of residuals is roughly as expected by chance; however, it may still be appropriate to examine the advective flux formulation in the light of more detailed oceanographic knowledge.

A similar analysis of the importance of parameters is shown in Table 21.1, which gives a description of the parameter, its prior estimate, and the standard deviation of the prior estimate. Some of the parameters that are poorly known or that have large regional variations have been assigned large standard deviations in order to avoid biasing the results. In terms of the Bayesian interpretation of constrained inversion (see Enting 1985), these parameters have been given "noninformative" prior distributions. A full description of the way in which the parameters are used and the justification for prior estimates are given by Enting and Pearman (1982, 1983). Also shown are the estimates of the parameters obtained by using the constrained inversion formalism. The ranking of "parameter importance" is more accurately described as a rank of the potential importance of additional information about the parameters. The ranking compares, for example, the relative importance that a 50% reduction in uncertainties would have in the ability to predict CO_2 concentrations. The column "best-fit" ranks the parameters according to the residuals contributing to Eq. (5), and the perturbation column ranks the parameters according to changes associated with the extreme predictions of the 2050 concentration.

The Utility of Some Current Programs

The analysis in the previous section gave an indication of the relative importance of the various pieces of information that are available at present. In this section we consider the utility of some results that may arise from various observational programs that are currently in progress. The way in which we do this is simply to repeat the calibration process using one of a set of hypothetical values for the results in question and then look at the extent to which the uncertainties are reduced. There are two possibilities that must be considered here, depending on the value of the hypothetical observation. If the observed value (actual or hypothetical) is in agreement with what the model would predict, then the use of the additional observation will reduce the uncertainties, and the extent of this reduction will depend on the precision of the observation. If, however, the observation is inconsistent with the previously calibrated model, we must enlarge our original estimates of the uncertainties to include all of the augmented data set. (In the most extreme case we may have to abandon the model and try some more general structure). In either case, after making and using the additional observation, we will know more than we did before; however, finding an inconsistent observation indicates that we previously knew less than we thought we knew.

To illustrate this approach, we consider a number of specific types of measurements and investigate the extent to which they can determine our ability to calculate future CO_2 concentrations. All of these calculations use the same CO_2 release scenario that was used in the previous section.

The examples we present here are some of the simpler cases. In particular, we will not discuss the use of ^{13}C in tree rings as a means of deducing biospheric changes, partly because such a study is beyond the scope of this presentation, partly because it would be necessary to include data on ^{13}C throughout the system to constrain the possible interpretation of a ^{13}C signal in tree rings, but mainly because of doubts about the extent to which any representative signal can be extracted (see for example Francey and Farquhar 1982). However, some of these matters will be discussed by Francey (this volume—see also Freyer, this volume, Peng and Freyer, this volume, and Stuiver, this volume), and the results of a generalized deconvolution calculation (as suggested earlier) will be presented elsewhere.

The Air–Sea Exchange Rate

We have based our prior estimates of the rate of air–sea exchange of CO_2 mainly on the wind-tunnel studies described by Liss (1983), although our formal expressions are given in terms of the formalism described by Deacon (1977). (This means that we work with effective wind stresses that are weighted to take into account the observed nonlinearities in the air-sea exchange). There is, however, a considerable degree of uncertainty in the wind-speed dependence of air–sea gas exchange; in particular, the open-sea observations using the radon deficit method (Peng et al. 1979) appear to indicate that the wind-speed dependence is absent or at least quite small (Deacon 1981). In modeling terms this means that the exchange rates must be determined primarily from ^{14}C data, but in this subsection, we examine the extent to which the calibration of the model might be made more precise if a global value of the air–sea exchange rate could be obtained by direct measurements.

This study has been confined to redefining the prior estimates of the effective u_* values (parameters 20 and 21 in Table 21.1), replacing them by the best-fit values with 10% SD. The result is that the uncertainties in the predictions are not significantly reduced, essentially because whatever the uncertainties, the air–sea exchange of CO_2 is sufficiently fast for the atmosphere-mixed-layer system to be always near equilibrium. (In our model it is a dynamic equilibrium because of the temperature difference between the ocean regions).

The Preindustrial CO_2 Concentration

While there are as yet no definitive data concerning the preindustrial (e.g., early 19th century) level of CO_2 in the atmosphere, a recent study of CO_2 concentrations in air bubbles from polar ice cores has produced some interesting results. It now appears likely that at least for some gases, including CO_2, the composition of the air remains relatively unchanged for thousands of years. French and Swiss studies (see Barnola et al. 1983, WMO 1983; see also Oeschger and Stauffer, this volume) are obtained similar concentrations in the range 258 to 270 ppmv. Later work (Neftel et al. 1985) suggests that these values may have been too low. Although there still remains the possibility of systematic error in these results, the method appears most promising for the future determination of preindustrial CO_2 concentrations.

To study the extent that a more precise knowledge of the preindustrial concentration would improve our ability to predict the future, we have repeated the calibration using a number of different values (for the preindustrial concentration) ranging from 240 ± 10 ppmv to 300 ± 10 ppmv. (These were the values for $q_i \pm u_i$). Figure 21.2 shows some of the corresponding release rates obtained from the constrained inversion. These curves show that the "best-fits" account for the different amounts of concentration change over the period 1900 to 1968 by changing the amount of carbon released from the biosphere and adjusting the timing of the release so that the current airborne fraction is given correctly. Only small adjustments are made to the ocean uptake rates, and as a consequence, the predictions of future concentrations depend only slightly on the hypothetical data, the prediction varying by 15 ppmv as the preindustrial concentration is varied from 240 to 300 ppmv. The standard deviations

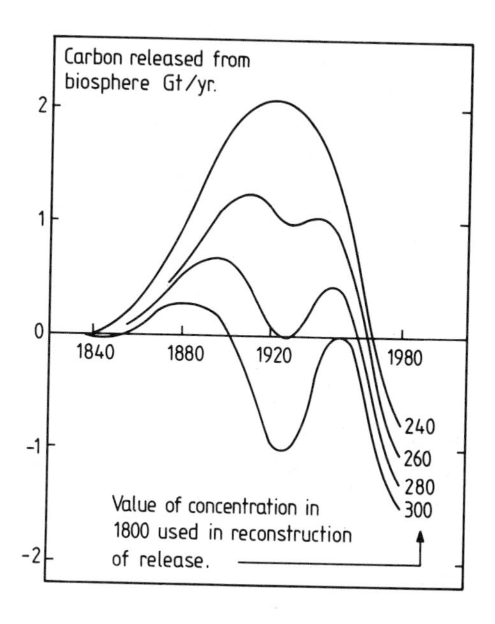

FIGURE 21.2. Reconstructed releases constructed as in Fig. 21.1 but also required to be consistent with hypothetical additional information giving the preindustrial (i.e., 1800) atmospheric CO_2 concentration with an uncertainty of ± 10 ppmv.

of these predictions [as determined by Eq. (11)] are practically independent of the hypothetical data, and as discussed in the following subsection, the importance of such data would lie in a resolution of the missing carbon problem.

Of particular interest in Figure 21.2 is the curve corresponding to a preindustrial concentration of 280 ppmv, ie the value obtained from ice-cores by Neftel et al. (1985). The release plotted in Figure 21.2 shows two peaks with the earlier peak being larger. A deconvolution of the full atmospheric CO_2 history given by Neftel et al. also shows some aspects of the two peak form although in this case the peaks tend to merge leading to the trough between them being relatively shallow.

THE 1900 CO_2 CONCENTRATION

Toward the end of the 19th century there were a number of attempts to directly measure atmospheric CO_2 concentrations. These data have been reviewed by Callender (1958), who rejected as unsatisfactory two extensive data sets that have recently been given further consideration. One made at Montsouris Observatory, Paris, was recently reviewed by Stanhill (1982), and the other made by French scientists included data of special interest from Cape Horn (Muntz and Aubin 1886). Both of these data sets show characteristics such as day-to-day, seasonal, and meridional variability, which are known from modern observations (e.g., Pearman et al. 1983; Fraser et al. 1983*b*) to be unrealistic, thus there are serious questions about the accuracy of the methods used and the possibility of local contamination (especially during nocturnal inversion conditions). For further discussion, see Fraser (this volume).

Another possibility for measuring atmospheric CO_2 concentrations in the late 19th century is based on the use of archived solar spectra (Stokes 1982; see also Stokes and Barnard, this volume). Even for this period, the ice cores would seem to be the most promising source of reliable atmospheric CO_2 concentrations. Although bubbles in any particular ice layer will have been trapped over a range of times, the sharply peaked form of the trapping time distribution should allow a good reconstruction of an atmospheric CO_2 history. In his theoretical analysis of measurements of trapped bubble volume by Schwander and Stauffer (1984), Enting (1985b) pointed out that the necessary deconvolution should be more stable than most other geochemical inversion problems.

Again because of the current importance, we have performed calculations using a number of possible values. Figure 21.3 shows some of the biospheric release reconstructions that are obtained. Qualitatively, the results are similar to the previous case, although when the value 310 ± 5 ppmv is used, the reconstructed release shows rather extreme oscillations. Again the predictions are insensitive to the value fitted, and the inclusion of hypothetical data for the 1900 CO_2 concentration does not reduce the uncertainties in the predictions.

In this and the previous case, the prime importance of such data will be to help to resolve the missing carbon problem rather than to improve the accuracy of existing models. The ultimate result will either be that the models are found to be inadequate or that our confidence in the models is increased. These results correspond to either finding that no possible biospheric release is compatible with existing models or that the discrepancies with ecosystem models are resolved and a possible release is found.

The studies in these last two subsections have considered the utility of single data items. Because of the stability of the deconvolution of CO_2 concentrations measured in polar ice (Enting, 1985b), ice-core studies should be able to give a reconstruction of the entire recent atmospheric CO_2 history. Such a data set has been presented by Neftel et al. (1985). The analysis of the trapping time distribution indicates that it may be possible to reduce the uncertainties in the effective ages assigned to their concentrations. A full utilisation of such data must await a more complete study of the role of impermeable layers in bubble trapping and a resolution of discrepancies between the results of Neftel et al. (1985) and those of Barnola et al. (1983). For the present purpose of discussing the utility of data, it is not yet appropriate to go beyond the use of single data items since there are so many possible functions for the atmospheric CO_2 history that a systematic investigation is beyond the scope of this paper.

However, as an indication of the relation between biospheric release histories and atmospheric CO_2 concentrations, Fig. 21.4 plots the CO_2 concentrations arising from some of the reconstructed releases shown in Fig. 21.1 through 21.3.

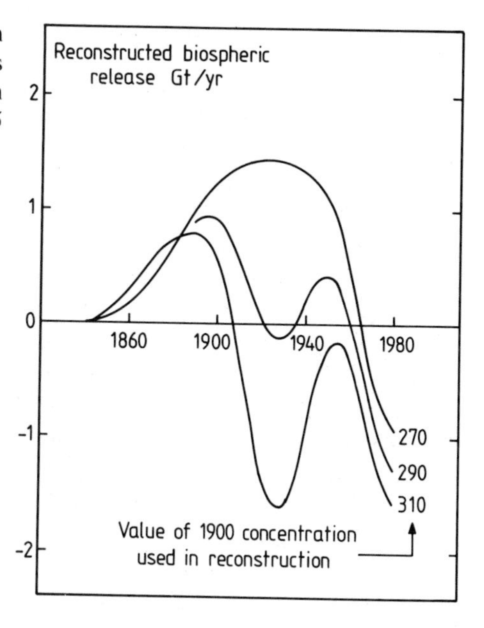

FIGURE 21.3. Reconstructions as in Fig. 21.2 using hypothetical values of the atmospheric concentration in 1900 with an uncertainty of ±5 ppmv.

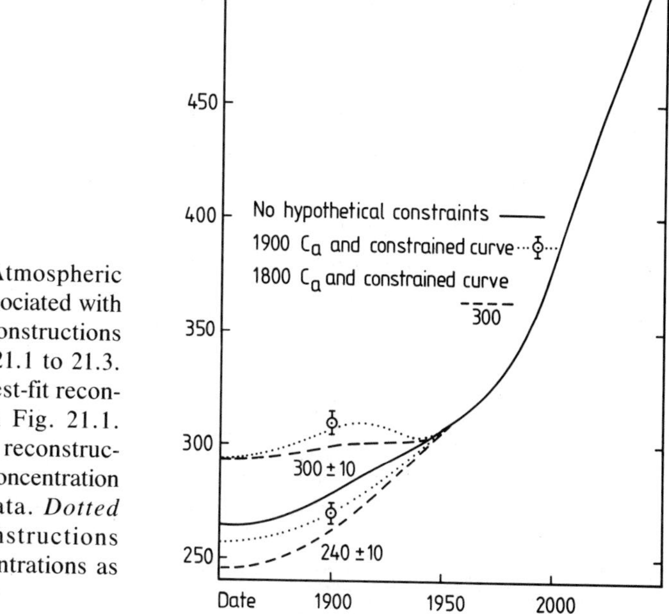

FIGURE 21.4. Atmospheric CO_2 histories associated with some of the reconstructions shown in Figs. 21.1 to 21.3. *Solid curve* = best-fit reconstructions from Fig. 21.1. *Dashed curve* = reconstructions with 1800 concentration as additional data. *Dotted curves* = reconstructions with 1900 concentrations as additional data.

The Current Rate of Biospheric Change

In view of the importance of biospheric changes, we have considered the significance of information concerning the current rate of change. Olson (1982) estimates a current release rate in the range 0.5 to 2.0 Gt • /yr^{-1}—significantly less than the estimates of Moore et al. (1981) and Houghton et al. (1983).

We recalibrated the model using values of $b(1980) \pm$ SD of b (1980) of -1.5 ± 0.5 Gt • yr^{-1} to 2.5 ± 0.5 Gt • yr^{-1} as an additional data item, and then reconstructed the function $b(t)$ as before. As with all other pieces of information, we did not require an exact fit between the value that was used as input to the calibration and the value of the reconstruction obtained as a result of the calibration. Figure 21.5 shows the discrepancies, plotting the 1980 release rate on both axes—the reconstruction on the vertical axes and the value that was to be fitted on the horizontal axes. The departure from the diagonal bands, defined by ±1 SD and ±2 SD, is an

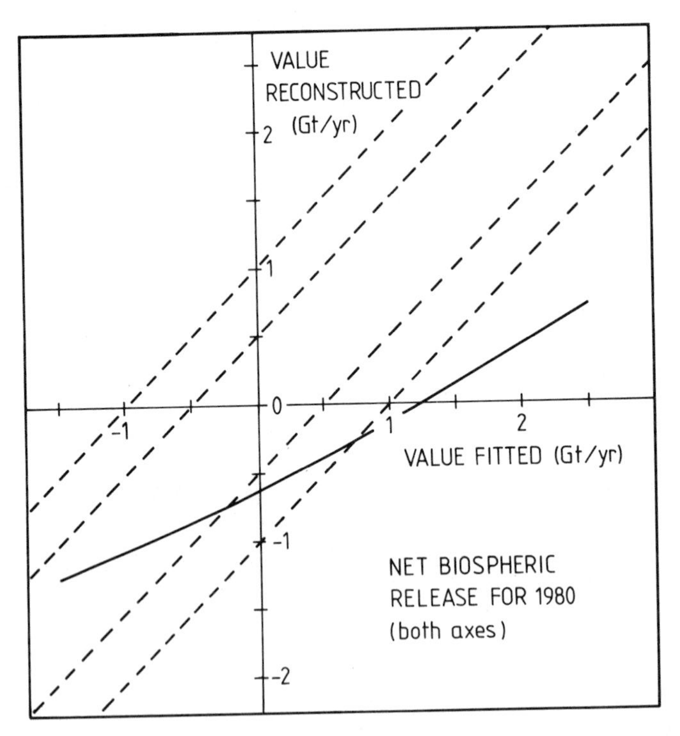

Figure 21.5. Release rates in 1980 from reconstructions that use this rate as a constraint. The horizontal axis shows the hypothetical value used as additional data (with uncertainty ±0.5 Gt • yr^{-1}). The *vertical axis* shows the rate reconstructed on the basis of the best compromise between this value and all the other data. The *diagonal lines* define bands of ±1-SD and ±2-SD discrepancies between the reconstruction and the hypothetical data.

indication of the extent to which the bulk of information is inconsistent with a significant current net release of carbon from the biosphere.

THE MAUNA LOA CO_2 RECORD

This subsection differs from the others in that it considers the possibility that the uncertainties we have assigned to the values are too small. There are important reasons for carrying out such a study. First, since we are using the Mauna Loa record (with a constant correction) to represent the atmospheric mean, we need to explore the effects of possible inadequacies in this assumption. Second, the Mauna Loa record shows the effects of a number of variations on time scales of several years (see for example Rust et al. 1979). Some of these variations have been correlated with specific geophysical phenomena. In calibrating our model we assumed that such variations will influence the calibration only as one of the "noise" terms that prevent us from obtaining an exact fit to the data. The fact that we can fit the observations to within the specified precision confirms this assumption, but recalibrating the model with a lower precision specified for the Mauna Loa data enables us to check the consequences of having inadvertently chosen a data set that is in some way biased.

In addition, the Mauna Loa data has to some extent an anomalous role in the calibration. To within a linear approximation, the Mauna Loa data has almost no influence on the predictions of the model but serves rather to define the magnitude of the missing carbon problem. There is, however, a strong nonlinear dependence of the predictions on the Mauna Loa data; therefore, the linearized theory described in earlier sections may not be adequate to determine the extent to which the Mauna Loa data set determines the uncertainties in the predictions. The direct tests described in this subsection are desirable to check whether the precision assigned to the Mauna Loa data also has an influence on the actual value predicted.

In the light of the previously described criteria for data selection, we use the Mauna Loa data in the following ways:

1. We use values based on a single site (Mauna Loa) to ensure consistency in the record.
2. We fit the model to a set of individual data points.
3. We omit data from the period 1964 to 1969 because of the problems described by Keeling et al. (1982).
4. We correct the data to produce an atmospheric average based on the studies of Fraser et al. (1983*b*) and Pearman et al. (1983). Enting and Pearman (1983) showed how calibrating a model with a single-reservoir atmosphere (by fitting the airborne fraction) leads to an effective uptake parameter that gives the same uptake rate as a more precisely calibrated model with a two-reservoir atmosphere.
5. Because we are interested in changes in time, in this case, we have not used the global range of variation or CO_2 concentrations as a measure of the uncertainty of the values.

The C_a data given in Table 21.2 represent annual means from Mauna Loa, with 1.9 ppmv subtracted to give an atmospheric average. The standard deviation is 0.3 ppmv.

The alternative calibration replaces this uncertainty with 0.6 ppmv. In consequence, the sum of squares Θ drops from 28 to 21, but the other changes are minor. The biospheric release reconstruction moves slightly closer to the prior estimate, but the future predictions are essentially unchanged. A possible interpretation of this is that the 0.6 ppmv is a more realistic estimate of the size of the "noise" terms that are not modeled [i.e., the contribution to Δa in Eq. (1)]. With the lower value of Θ, the two sets of information make roughly equal contributions to Θ, which would be expected since the same numbers of points are involved, whereas if the Mauna Loa data is assigned a precision of ± 0.3 ppmv, the carbon cycle data accounts for about two-thirds of Θ.

The Rate of Change of Atmospheric $\delta^{13}C$

One of the original motivations for the development of our carbon cycle model was for the interpretation of our Division's program of atmospheric ^{13}C observations. (For a discussion of a number of aspects of this program, see Francey, this volume).

The present calibration makes relatively little use of ^{13}C, which is one reason why the fractionation factors f_{as}, f_{sa}, and f_{ab} appear as parameters of low importance in Table 21.1. The data used are from Keeling et al. (1979) and give atmospheric $\delta^{13}C = -6.69 \pm 0.13‰$ (1956) and $= -7.24 \pm 0.05‰$ (1978). The 1956 value is an indirect estimate. There are a number of observational programs involved in determining the rate of change of $\delta^{13}C$. We have incorporated hypothetical data of this form by using the 1982 value as an additional point. In practice, before any such real data can be included, it is necessary for the data to span a 4-year period or else be shown as measured on a $\delta^{13}C$ scale equivalent to that used by Keeling et al. (1979).

Figure 21.6 is analogous to Fig. 21.5 except that the 1982 value of $\delta^{13}C$ is plotted on each axis. The horizontal axis shows the value that the model tries to fit, and the vertical axis shows the value that is reconstructed. It also shows the reconstructed value of the 1978 $\delta^{13}C$. Rates of change larger than 0.03‰/year seem inconsistent with the other data.

The reconstruction acts by fitting the 1978 and 1982 $\delta^{13}C$ values as closely as possible while maintaining a rate of change $\delta^{13}C$ of 0.03‰/year.

One interesting aspect of these tests is that these changes are not associated with any significant change in the current biospheric release rates. The changes are rather accounted for by changes in the biospheric fractionation factor and changes in the earlier parts of the biospheric release.

An additional point is that not only are the changes in the $\delta^{13}C$ fit not associated with changes in the biospheric release rate, but that when the biospheric release rate is changed the changes in the $\delta^{13}C$ gradient are very small. In the calculations on which Fig. 21.5 is based where the bio-

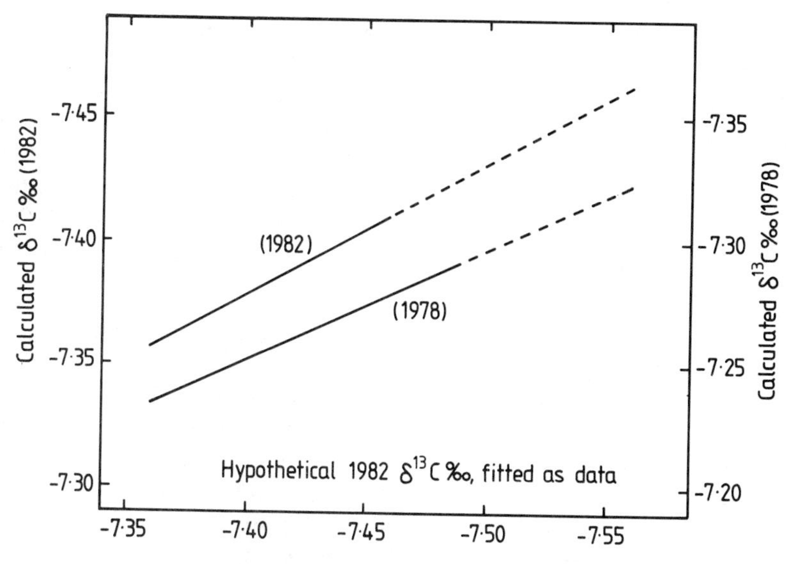

FIGURE 21.6. Results from attempting to fit atmospheric $\delta^{13}C$ values of -7.24‰ for 1978 (Keeling et al 1979), and for 1982, a range of hypothetical values as shown on the *horizontal axis*. The *dashed sections of the curve* correspond to points when the deviations between the calculations and the data fitted exceed 1 SD (taken as 0.05‰). In spite of the flexibility in the biospheric release and the possibility of adjusting the fractionation factors, the model appears to be inconsistent with the rates of change exceeding 0.03‰/year.

spheric release rate was changed from -1.26 Gt • yr^{-1} to 0.42 Gt • yr^{-1} (by fitting values from -1.5 Gt • yr^{-1} to 2.5 Gt • yr^{-1}), the $\delta^{13}C$ gradient only changed from 0.030‰/year to 0.036‰/year.

It is this sort of indirect effect that gives us an important justification of the calibration procedure that we have been using. The parameter estimation procedure is so highly interdependent, and the parameter estimates are so highly correlated, that it is generally impossible to make a one-to-one correspondence between a single carbon cycle observation and a single aspect of the model.

Data for Calibrating and Validating Two- and Three-Dimensional Atmospheric Models

Carbon cycle models of two and three dimensions are used to examine the carbon cycle with greater spatial and temporal resolution than is possible with global "box" models. The difference between such objectives and the objectives identified in the first section will be reflected in the differences in the type of data that is used. The high-resolution studies can be used for checking that the detailed description of the carbon cycle is in fact consistent with the "lumped" description used in global models; and when the global models indicate discrepancies, the high-resolution models can supply additional constraints on the possible solutions for the difficulties.

Bolin and Keeling (1963) and Junge and Czeplak (1968) have studied seasonal variations using atmospheric models with latitude as the single dimension; however, in this section we will concentrate on the use of such data in two- and three-dimensional models. The two main types of smaller-scale variations are the seasonal changes and the latitudinal gradient. (For further discussion of some aspects of these variations, see Heimann et al. this volume).

1. The seasonal variation of CO_2 is due principally to seasonal variations in the net exchange of carbon between the atmosphere and the biosphere. Carbon cycle models can be used to deduce these net sources from the observations, as in the two-dimensional study by Pearman and Hyson (1980), which deduced a seasonal net exchange of 6 Gt of carbon. Alternatively, the models can be used to check that the atmospheric observations are consistent with independent estimates of the release (see also Fung, this volume). Fung et al. (1983) used a three-dimensional model and found that the observations were consistent with biospheric net exchanges of 13 Gt of carbon each season. The disagreement in biospheric releases is disappointing, since the seasonal variations give some of the most accurate observations of smaller-scale variation because they are unaffected by problems of interstation comparability. Possible resolutions of the discrepancy are discussed below in connection with the requirements for validating high-resolution models.
2. Data concerning the latitudinal variations of CO_2 concentration is subject to difficulties of intercomparison between stations, except in the GMCC "flask" program (Deluisi 1981), which analyzes all flask samples in a single laboratory. Pearman et al. (1983) have used a two-dimensional model to study the relationship between gradients and fluxes. Their general result that gradients of $\sim$1 ppmv/30° are associated with meridional transports of $\sim$3 Gt • yr^{1} determines the precision with which gradients must be measured if fluxes are to be deduced. They concluded that observed gradients were consistent with an $\sim$5-Gt • yr^{-1} fossil fuel source in the northern hemisphere, 2.3 Gt • yr^{-1} interhemispheric flux, and 4.4 Gt • yr^{-1} high-latitude ocean uptake. They also concluded that the observed gradients were inconsistent with net biospheric releases of 2 Gt • yr^{-1} or more in the tropics.

The discrepancies between the seasonal biospheric releases considered by Pearman and Hyson (1980) and Fung et al. (1983) appear to be because the three-dimensional model predicts large differences in amplitude of the seasonal variations between oceanic and continental regions (see Fig. 6 of Fung et al. 1983).

Furthermore, in the three-dimensional model, the large-amplitude differences at the surface persist with little attenuation throughout the full depth of the troposphere, as indicated by mass balance calculations in general, and, in particular, by the vertical CO_2 transport at 500 mb (Fig. 11 of Fung et al. 1983). If this distribution were correct, then the amplitudes at the near-ocean stations would be much lower than the zonal average amplitudes. The biospheric fluxes deduced by Pearman and Hyson (1980), using the former amplitudes, would be correspondingly low.

On meteorological grounds, the persistence of the land-sea differences throughout the full depth of the troposphere seems unrealistic (J. R. Garratt, private communication). In addition, the amplitude of the annual cycle is observed to be 10 ppmv and 7 ppmv at Point Six Mountain (47°N, 114°W) and at Niwot Ridge (40°N, 105°W), respectively (Bodhaine and Harris 1982). The three-dimensional model predicts ~20 ppmv at the surface. Since similar amplitudes are implied throughout the troposphere, this particular comparison avoids the problem of deciding whether the observations, which are from mountain sites, should be compared to the model at the model's surface or at the actual height. Further discussion of such difficulties is presented by Fraser et al. (1983*b*).

Essentially, the modeling studies with both the two- and three-dimensional models must be regarded as preliminary. Additional studies (which are in progress for both models—see in particular Fung, *this volume*) are needed to establish the range of validity of the models (particularly the two-dimensional model) and to consider the problems of calibration and validation.

The two-dimensional model developed in our laboratory (Hyson et al. 1980) uses observed zonal mean wind-fields and diffusion coefficients that were tuned on the basis of the horizontal and vertical distribution of CCl_3F (Freon-11). A degree of validation is provided by (unpublished) studies of the distribution of ^{14}C from nuclear weapons (Pittock 1983). The three-dimensional model used wind fields generated by a general circulation model, and the validation of the model is based on comparing the model results with real climatic data. Although this comparison is a very powerful test, it is not a direct test of the model's performance in passive tracer studies, and more direct tests would be of considerable interest. In particular, comparison of the vertical gradient of CO_2 (as given by Fung et al. 1983, Fig. 7) against observations for the North Atlantic (Bolin and Bischof 1970) and Hawaii (Bodhaine and Harris 1982) tend to confirm the suggestion that, in the three-dimensional model, the vertical mixing associated with small-scale processes is too strong. More recently, Prather (1984) has used the same three-dimensional model to study halocarbon distributions. He found it necessary to tune the horizontal transport due to small scale processes in order to match known interhemispheric gradients.

The tuning performed by Prather is an adequate way of dealing with a well-identified defect in the model, but in more complex situations this approach would become unmanageable due to the large number of degrees of freedom involved in the transport fields. The difficulties in tuning the transport coefficients are exacerbated by the fact that there are few if any tracers for which there are adequate observations of concentrations together with adequate knowledge of the sources. Even for the halocarbons, which have been used to tune two- and three-dimensional models, it has been suggested that the release estimates may have significant errors (Fraser et al. 1983*a*). Enting (1984) briefly indicated the possible relevance of constrained inversion techniques to this type of severely underdetermined calibration problem and stressed the importance of having a good first guess for the transport fields.

In general terms, our comments on the requirements for the calibration and validation apply equally well to both high-resolution and low-resolution models, and the refinement of the models must include refinement of the calibration techniques.

Discussion

The analysis in preceding sections has to a large part been concerned with the missing carbon problem. The attempts to calibrate the model have revealed striking discrepancies between direct estimates of releases of biospheric carbon and reconstructions of such release constraints by a requirement of consistency with the other carbon cycle data. The formal statistical analysis described above would indicate that these uncertainties have little effect on our ability to predict future CO_2 concentrations with the present models. The analysis effectively reproduces the result of Broecker et al (1980) that the bulk of the carbon cycle data is consistent with existing models; consequently, the most plausible explanation of the discrepancies is that "the error lies in the biomass estimates...."

Nevertheless, the existence of discrepancies must continue to cast doubt on the validity of the models.

The conventional approach to model validation is to use only part of the data for calibrating the model and the remaining data for validation (Frenkiel and Goodall 1978). More particularly, the data used for validation should include those quantities that are to be predicted by the model. For carbon cycle studies, this would mean that the data on atmospheric CO_2 concentrations would have to be used for validation and not for calibration. This is effectively what has been done in a number of studies (see for example Oeschger et al. 1975); the models are calibrated using mainly ^{14}C data, and the results are then compared to the Mauna Loa record as a check. What happened of course is that the check failed, and the "missing carbon problem" was revealed. To go beyond this point requires the use of the atmospheric CO_2 concentration data in the calibration, and indeed the use of all the available data. (In practice, later stages of the calibration can omit data that are found to be of low utility, but such data will also be of little use for model validation). A common check used in geophysical studies is to check the consistency of the two data sets, the prior information, and the carbon cycle observations. Part of such a comparison can be performed by varying the parameter γ that appears in the calibration formalism so that the results vary from those based only on prior information to those based only on the carbon cycle data. Enting and Pearman (1982) have pointed out that the fact that the carbon cycle observations can be reproduced by a model that uses parameter values that are in reasonable agreement with independent estimates is in fact an important test of the validity of the model structure. They also pointed out that such agreement also acts to validate the use of the same independent geophysical information in models of higher dimensionality where, because of the larger number of degrees of freedom, there is even less hope of estimating all the parameters from carbon cycle observations than there is in one dimension. In addition to checking for inconsistencies between the two data sets taken as a whole, the final calibration set can be checked for "outliers," that is, individual items of data (from either set) that are poorly fitted. Since it is these items that have the most influence on the final results, the role of these data must be carefully checked to determine the origin of the discrepancies. Disagreements could come because of an inadequacy in the way a particular quantity is modeled or because of inadequacies in the model structure.

Mosteller and Tukey (1977) have described a number of techniques for data analysis that could be used to extend the studies described above. In particular, they describe a cross-validation technique that uses each of N available data items in turn as a validation of calibrations based on the other $N - 1$ items. Such an analysis has not yet been performed on our model; however, it might be an appropriate way of searching for discrepancies in the ocean model to see if these indicate possible resolutions of the missing carbon problem. As mentioned earlier the investigation of discrepancies in the biospheric release is probably most appropriately investigated by simultaneously modeling the carbon cycle and the ecosystem responses.

At present the framework described above with systematic calibration techniques, quantifiable discrepancies, and suggestions of possible refinements to the techniques of data analysis exists only for one-dimensional models. For multidimensional models there is not yet a comparable breadth of modeling experience, and the questions of calibration and validation have not yet been approached systematically. For multidimensional models, the greater number of degrees of freedom make such a systematic approach even more important than for one-dimensional models. It is to be hoped that, with the experience of one-dimensional modeling as a guide, the development of appropriate techniques for calibration and validation can progress concurrently with the development of multidimensional models.

References

Bacastow, R. B. and C. D. Keeling. 1973. Atmospheric carbon dioxide and radiocarbon in the natural carbon cycle. II. Changes from AD 1700 to 2100 as deduced from a geochemical model. In G. M. Woodwell and E. V. Pecan (eds.), Carbon and the Biosphere, CONF-729519, pp. 86–135. U.S. Atomic Energy Commission, Washington, D.C.

Barnola, J. M., D. Raynaud, A. Neftel, and H. Oeschger. 1983. Comparison of CO_2 measurements by two laboratories on air bubbles in polar ice. Nature 303:410–413.

Bodhaine, B. A. and J. M. Harris (eds.). 1982. Geophysical Monitoring for Climatic Change, No. 10, Summary Report 1981. U.S. Department of Commerce, NOAA/ERL/GMCC, Boulder, Colorado.

Bolin, B., and W. Bischof. 1970. Variations of the carbon dioxide content of the atmosphere in the northern hemisphere. Tellus 22:431–442.

Bolin, B., A. Björkström, K. Holmen, and B. Moore. 1983. The simultaneous use of tracers for ocean circulation studies. Tellus 35B:206–236.

Bolin, B., A. Björkström, C. D. Keeling, R. B. Bacastow, and U. Siegenthaler. 1981. In B. Bolin (ed.), Carbon Cycle Modelling, Scope 16, pp. 1–28. John Wiley & Sons, New York and Chichester, England.

Bolin, B. and C. D. Keeling. 1963. Large-scale atmospheric mixing as deduced from the seasonal and meridional variations of carbon dioxide. J. Geophys. Res. 68:3899—3920.

Broecker, W. S., T.-H. Peng, and R. Engh. 1980. Modeling the carbon system. Radiocarbon 22:565–598.

Broecker, W. S., T. Takahashi, H. J. Simpson, and T.-H. Peng. 1979. Fate of fossil fuel carbon dioxide and the global carbon budget. Science 206:409–418.

Callender, G. S. 1958. On the amount of carbon dioxide in the atmosphere. Tellus 10:243–248.

Deacon, E. L. 1977. Gas transfer to and across an air–water interface. Tellus 29:363–374.

Deacon, E. L. 1981. Sea-air gas transfer: The wind-speed dependence. Boundary-Layer Meteorol. 21:31–37.

Deluisi, J. J. (ed.). 1981. Geophysical Monitoring for Climatic Change, No. 9, Summary Report 1980. U.S. Department of Commerce, NOAA/ERL/GMCC, Boulder, Colorado.

Enting, I. G. 1983. Error analysis for parameter estimates from constrained inversion. CSIRO Division of Atmospheric Research Technical Paper No. 2. Commonwealth Scientific and Industrial Research Organization, Australia.

Enting, I. G. 1984. Preliminary studies with a two-dimensional model using transport fields derived from a GCM. Paper presented at the CSIRO-ABM Meeting on the Scientific Application of Baseline Observations of Atmospheric Composition, November 7–9, 1984, Aspendale, Australia.

Enting, I. G. 1985a. Principles of constrained inversion in the calibration of carbon cycle models. Tellus (37B:7–27).

Enting, I. G. 1985b. A lattice statistics model for the age distribution of air bubbles in polar ice. Nature 315:654–655.

Enting, I. G. and G. I. Pearman. 1982. Description of a one-dimensional global carbon cycle model. CSIRO Division of Atmospheric Physics Technical Paper No. 42. Commonwealth Scientific and Industrial Research Organization, Australia.

Enting, I. G. and G. I. Pearman. 1983. Refinements to a one-dimensional carbon cycle model. CSIRO Division of Atmospheric Research Technical Paper No. 3. Commonwealth Scientific and Industrial Research Organization, Australia.

Francey, R. J. and G. D. Farquhar. 1982. An explanation of $^{13}C/^{12}C$ variations in tree rings. Nature 297:28–31.

Fraser, P. J., P. Hyson, I. G. Enting, and G. I. Pearman. 1983*a*. Global distribution and southern hemisphere trends of atmospheric CCl_3F. Nature 302:692–695.

Fraser, P. J., G. I. Pearman, and P. Hyson. 1983*b*. The global distribution of atmospheric carbon dioxide 2. A review of provisional background observations, 1978–1980. J. Geophys. Res. 88C:3591–3598.

Frenkiel, F. N., and D. W. Goodall (eds.). 1978. Simulation Modelling of Environmental Problems, Scope 9. John Wiley & Sons, New York and Chichester, England.

Fung, I., K. Prentice, E. Matthews, J. Lerner, and G. Russell. 1983. Three-dimensional tracer model study of atmospheric CO_2: response to seasonal exchanges with the terrestrial biosphere. J. Geophys. Res. 88C:1281–1294.

Houghton, R. A., J. E. Hobbie, J. M. Melillo, B. Moore, B. J. Peterson, G. R. Shaver, and G. M. Woodwell. 1983. Changes in the carbon content of terrestrial biota and soils between 1860 and 1980: a net release of CO_2 to the atmosphere. Ecol. Monogr. 53:235–262.

Hyson, P., P. J. Fraser, and G. I. Pearman. 1980. A two-dimensional transport simulation model for trace atmospheric constituents. J. Geophys. Res. 85C:4443–4455.

Jackson, D. D. 1972. Interpretation of inaccurate, insufficient, and inconsistent data. Geophys. J. R. Astron. Soc. 28:97–109.

Junge, C. E. and G. Czeplak. 1968. Some aspects of the seasonal variation of carbon dioxide and ozone. Tellus 20:422–434.

Keeling, C. D., R. B. Bacastow, and T. P. Whorf. 1982. Measurements of the concentration of carbon dioxide at Mauna Loa observatory, Hawaii. In W. C. Clark (ed.), Carbon Dioxide Review 1982, pp. 377–385. Clarendon Press, Oxford, England.

Keeling, C. D., W. G. Mook, and P. P. Tans. 1979. Recent trends in the $^{13}C/^{12}C$ ratio of atmospheric carbon dioxide. Nature 277:121–123.

Laurmann, J. A. and R. M. Rotty. 1983. Exponential growth and atmospheric carbon dioxide. J. Geophys. Res. 88C:1295–1299.

Laurmann, J. A. and J. R. Spreiter. 1983. The effects of carbon cycle model error in calculating future atmospheric carbon dioxide levels. Climatic Change 5:145–181.

Liss, P. 1983. Gas transfer: experiments and geochemical implications. In P. Liss and W. G. N. Slinn (eds.), Air-Sea Exchange of Gases and Particles. 241–298 D. Reidel, Dordrecht.

Moore, B., R. D. Boone, J. E. Hobbie, R. A. Houghton, J. M. Melillo, B. J. Peterson, G. R. Shaver, C. J. Vörösmarty, and G. M. Woodwell. 1981. A simple model for analysis of the role of terrestrial ecosystems in the global carbon budget. In B. Bolin (ed.), Carbon Cycle Modelling, Scope 16, pp. 365–385. John Wiley & Sons, New York and Chichester, England.

Mosteller, F. and J. W. Tukey. 1977. Data Analysis and Regression: A Second Course in Statistics. Addison-Wesley, Reading, Massachusetts.

Muntz, A. and E. Aubin. 1886. Recherches sur l'acide carbonique de l'air. Du Cap horn et de l'ocean Atlantique. Recherches sur la constitution chimique de l'atmosphere. Tome 3. Gaunthier-Villars, Imprimeur-Libraire, Paris.

Neftel, A., E. Moor, H. Oeschger and B. Stauffer. 1985. Evidence from polar ice cores for the increase in atmospheric CO_2 in the past two centuries. Nature 315:45–47.

O'Brien, K. 1979. Secular variations in the production of cosmogenic isotopes in the earth's atmosphere. J. Geophys. Res. 84:423–431.

Oeschger, H. and M. Heimann. 1983. Uncertainties of predictions of future atmospheric CO_2 concentrations. J. Geophys. Res. 88C:1258–1262.

Oeschger, H., U. Siegenthaler, U. Schotterer, and A. Gugelmann. 1975. A box-diffusion model to study the carbon dioxide exchange in nature. Tellus 27:168–192.
Olson, J. S. 1982. Earth's vegetation and atmospheric carbon dioxide. In W. C. Clark (ed.), Carbon Dioxide Review 1982, pp. 388–398. Clarendon Press, Oxford, England.
Pearman, G. I. 1980. Preliminary studies with a new global carbon cycle model. In Carbon Dioxide and Climate: Australian Research, pp. 79–91. Australian Academy of Science, Canberra.
Pearman, G. I. and P. Hyson. 1980. Activities of the global biosphere as reflected in atmospheric CO_2 records. J. Geophys. Res. 85C:4468–4474.
Pearman, G. I., P. Hyson, and P. J. Fraser. 1983. The global distribution of atmospheric carbon dioxide: 1. Aspects of observations and modelling. J. Geophys. Res. 88C:3581–3590.
Peng, T. -H. and W. S. Broecker. 1984. Ocean life cycles and the atmospheric CO_2 content. J. Geophys. Res. 89C:8170–8180.
Peng, T.-H., W. S. Broecker, H. D. Freyer, and S. Trumbore. 1983. A deconvolution of the tree-ring-based δ^{13} C record. J. Geophys. Res. 88C:3609–3620.
Peng, T.-H., W. S. Broecker, G. G. Mathieu, and Y.-H. Li. 1979. Radon evasion rates in the Atlantic and Pacific Oceans as determined during the GEOSECS programs. J. Geophys. Res. 84C:2471–2486.
Pittock, A. B. 1983. The atmospheric effects of nuclear war. In M. Denborough (ed.), Australia and Nuclear War, pp. 136–160. Croom Helm, Fyshwick, ACT.
Prather, M. 1984. Simulations of chlorofluorocarbons with a three-dimensional model. Paper presented at the CSIRO-ABM Meeting on the Scientific Application of Baseline Observations of Atmospheric Composition, November 7–9, 1984, Aspendale, Australia.
Rodgers, C. D. 1977. Statistical principles of inversion theory. In A. Deepak (ed.), Inversion Methods in Atmospheric Remote Sounding, pp. 117–134. Academic Press, New York.
Rust, B. W., R. M. Rotty and G. Marland. 1979 Inferences drawn from atmospheric CO_2 data. J. Geophys Res. 84C:3115–3122.
Schwander, J. and Stauffer, B. 1984. Age difference between polar ice and air trapped in its bubbles. Nature 311:45–47.
Siegenthaler, U. 1983. Uptake of excess CO_2 by an outcrop-diffusion model of the ocean. J. Geophys. Res. 88C:3599–3608.
Stanhill, G. 1982. The Montsouris series of carbon dioxide abundance: An archival study of spectroscopic data. In W. C. Clark (ed.), Carbon Dioxide Review: 1982, pp. 385–388. Clarendon Press, Oxford, England.
Stokes, C. M. 1982. Atmospheric carbon dioxide abundance: An archival study of spectroscopic data. In W. C. Clark (ed.), Carbon Dioxide Review 1982, pp. 385–388. Clarendon Press, Oxford, England.
Twomey, S. 1977. Introduction to the Mathematics of Inversion in Remote Sensing and Indirect Measurements. Elsevier, Amsterdam.
Viecelli, J. A., H. W. Ellsaesser, and J. E. Burt. 1981. A carbon cycle model with latitude dependence. Climatic Change 3:281–302.
World Meterological Organization. 1983. Report of the WMO (CAS) meeting of experts on the CO_2 concentrations from pre-industrial times to IGY. Boulder, Colorado, June 22–25, 1983. WMO, Geneva.
Wunsch, C. 1978. The North Atlantic general circulation west of 50°W determined by inverse methods. Rev. Geophys. Space Phys. 16:583–620.

22

Analysis of the Seasonal and Geographical Patterns of Atmospheric CO_2 Distributions with a Three-Dimensional Tracer Model

INEZ Y. FUNG

A great amount of information on the sources and sinks of atmospheric CO_2 is contained in the geographical, seasonal, and interannual variations of the global atmospheric CO_2 distribution. The measured concentrations of CO_2 at several locations illustrate large variations in the amplitude and phase of the seasonal cycle superimposed on an increasing long-term trend (see e.g., Gammon, this volume; Keeling 1983). Recent analysis of CO_2 records by Keeling and his collaborators (1983) reveals that the amplitude of seasonal cycle has detectable interannual variations and may be increasing with time.

We have initiated a modeling effort to study the prospects of extracting some of the potential information of CO_2 sources and sinks from the observed CO_2 variations. The approach is to use a three-dimensional (3-D) global transport model, based on winds from a 3-D general circulation model (GCM), to advect CO_2 noninteractively, that is, as a tracer, with specified sources and sinks of CO_2 at the surface. If the model can reproduce the general character of observed CO_2 variations on the basis of physically justified sources and sinks, it may then be used for experiments to determine the sensitivity of the global CO_2 distribution to various assumptions about CO_2 sources and sinks. It is anticipated that this approach may lead to useful quantitative limits on some CO_2 sources and sinks.

In this chapter, we attempt to identify what we know and do not know about the sources and sinks that contribute to the geographical and seasonal variations of atmospheric CO_2. The 3-D tracer transport model used in this study is discussed first. The ensuing sections will describe model experiments with seasonal exchange of CO_2 with the terrestrial biota, the dominant contributor to the seasonal cycle, the role of the oceans in the seasonal oscillation of atmospheric CO_2, and the atmospheric CO_2 distribution resulting from a perturbation in atmospheric circulation.

Tracer Model

The tracer model uses winds generated from the Goddard Institute of Space Studies general circulation model (GCM) (Hansen et al. 1983) to advect

CO_2 as an inert trace constituent. In the GCM, the large-scale processes operating in the atmosphere are represented, and the wind, temperature, and moisture fields are calculated from first principles. The simulated atmospheric circulation shows reasonable agreement with the large body of atmospheric observations. Although the GCM is able to reproduce the main features of atmospheric circulation, there are still uncertainties in the model that may affect the transport of CO_2. The model resolution usually employed (~8° latitude by 10° longitude) is sufficient to produce fairly realistic atmospheric long waves and large-scale eddies, such as those that occur at middle and high latitudes; it cannot, however, resolve the smaller-scale tropical disturbances, and this may affect transports at low latitudes. Also, our understanding and modeling ability of and for many physical processes in the atmosphere are rudimentary. In particular, the present lack of global observations about the intensity of convective activity hampers efforts to obtain accurate parameterization of moist convection, which may be especially important for simulating subgrid-scale vertical transports. This process is at present treated in very simple ways in GCMs.

It is important to understand the limitations of present 3-D modeling (discussed by Hansen et al. 1983), since these affect the ability to interpret the characteristics of the global CO_2 distribution.

Seasonal Exchange with Terrestrial Biota

As has been noted by many investigators, terrestrial biota is the major contributor to the seasonal oscillations of CO_2 in the atmosphere. CO_2 is absorbed via photosynthesis and is released via respiration and decay. Fung et al. (1983) used a global vegetation map (Matthews 1983) and simple ad hoc definitions of CO_2 exchange between the biosphere and the atmosphere as input to the tracer model to study geographical variations of the seasonal cycle of atmospheric CO_2. The study demonstrates that large longitudinal variations exist in the atmospheric CO_2 distribution, and a 3-D approach is necessary for accurate analysis of the global carbon cycle.

We present here the results of three experiments with different biospheric CO_2 exchange functions. These functions are based on Machta (1972), Pearman and Hyson (1980), and a global net primary productivity (NPP) map indicative of the vegetation type to which Azevedo's (1982) seasonal CO_2 uptake and release curves were applied.

Experiments 1 and 2

Machta (1972) and Pearman and Hyson (1980) have obtained tables of carbon exchange by month and latitude belts. For each month, we perform an area-weighted interpolation for their latitude zone to our model's ~8° latitude zone. The monthly carbon flux within each model zone is then uniformly distributed over the land areas in the zone.

EXPERIMENT 3

Using about 90 sources, Matthews (1983) constructed a global, 1° × 1° resolution vegetation file containing about 200 vegetation types. Annual NPP values were assigned to each vegetation type in each 1° × 1° cell and then grouped to obtain the resultant NPP map at the model resolution. The calculated global NPP is 45×10^{15}g C • yr^{-1}, within the range from 38×10^{15} g C • yr^{-1} (Lieth 1973) to 78×10^{15} g C • yr^{-1} (Bazilevich et al. 1971) found in the literature.

Azevedo (1982) constructed simple curves of CO_2 uptake and release by the biosphere. These curves are based loosely on the assumption that uptake and release are governed by air and soil temperatures and that these processes are switched off when temperatures fall below critical values. From 40° to 70°N, biospheric uptake of CO_2 is concentrated in the months May through August, whereas release occurs throughout the year but with a maximum in the same months. From 10° to 40°N, release occurs uniformly throughout the year, whereas maximum uptake occurs from May to August. Like Machta (1972), Azevedo shifts the curves by 6 months for the southern hemisphere.

In this experiment we combined the NPP map with Azevedo's carbon uptake and release curves to form the monthly flux of carbon atmosphere:

$$\text{FLUX}(\lambda,\theta,t) = \text{NPP}(\lambda,\theta) \times [\text{RELEASE}(\theta,t) - \text{UPTAKE}(\theta,t)]$$

where λ is longitude, Θ is latitude, and t is time.

The UPTAKE and RELEASE curves are normalized so that no net annual exchange of carbon occurs. Note that this exchange, like the NPP, possesses both latitudinal and longitudinal variations, even with these extremely simple seasonal distributions.

The tracer model, with these biospheric sources and sinks, was run for 25 months until the annual cycles repeated themselves. The CO_2 concentrations are defined relative to a globally uniform (and arbitrary) background concentration, so that a positive (or negative) CO_2 mol fraction means that the simulated concentration is greater (or less) than the background.

Seasonal oscillations of CO_2 concentrations simulated at grid boxes corresponding to five observation sites are shown in Fig. 22.1 together with the observed annual cycles. As one would expect from the magnitudes of the source functions in the three experiments, the peak-to-peak amplitudes of the CO_2 oscillations at these locations increase from experiment 1 to 3. Machta's biospheric exchange function underestimates the amplitudes, as has been noted by other investigators (Pearman and Hyson 1980; Azevedo 1982). In experiment 1, the underestimation is about 50% at Mauna Loa and 63% at Point Barrow. Pearman and Hyson's source functions, constructed to duplicate the observations with their model, produce a peak-to-peak amplitude of 6.0 ppm at Mauna Loa, close to that observed.

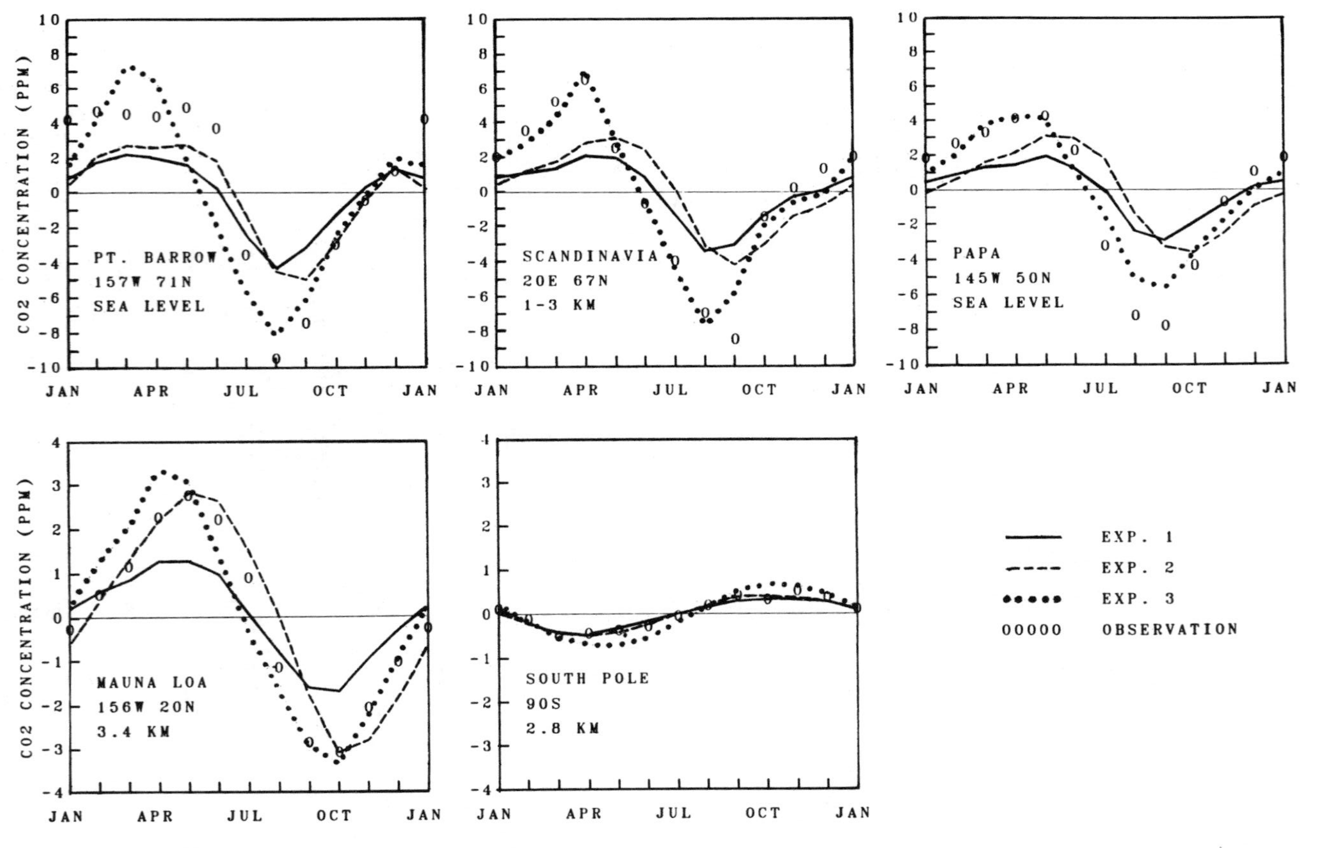

FIGURE 22.1. Simulated and observed annual cycles of CO_2 at five locations. The simulated cycles are responses to seasonal exchanges with terrestrial biota. The experiments (1,2,3) are described in the text. (From Fung et al. 1983.)

However, the amplitudes at Ocean Weather Station Papa and Pt. Barrow are underestimated by about 45%. At Mauna Loa and the South Pole, the amplitudes are within 10% of those observed. The amplitude of CO_2 at Ocean Weather Station Papa is underestimated in all three experiments, even though the amplitudes at the land stations are reasonably simulated in experiment 3. The simulated amplitudes at the National Oceanic and Atmospheric Administration monitoring stations are also found to compare reasonably well with those observed (Gammon et al. this volume).

The phase of the CO_2 annual cycles at the five station locations can be seen in Fig. 22.1. In all three experiments the months of predicted maxima and minima at the station locations are within 2 months of those observed. However, the simulated annual cycles at the northern stations, especially at Point Barrow, lack the asymmetry seen in the observed cycles. This may have occurred because we assumed, even at 70°N, that the growing season starts in May instead of in early June when snow melts at this latitude.

The azonal nature of CO_2 distributions is apparent in Fig. 22.2, which shows the seasonal amplitudes at the surface as simulated in experiment 3. Except in the northern hemisphere tropics, which is relatively well mixed

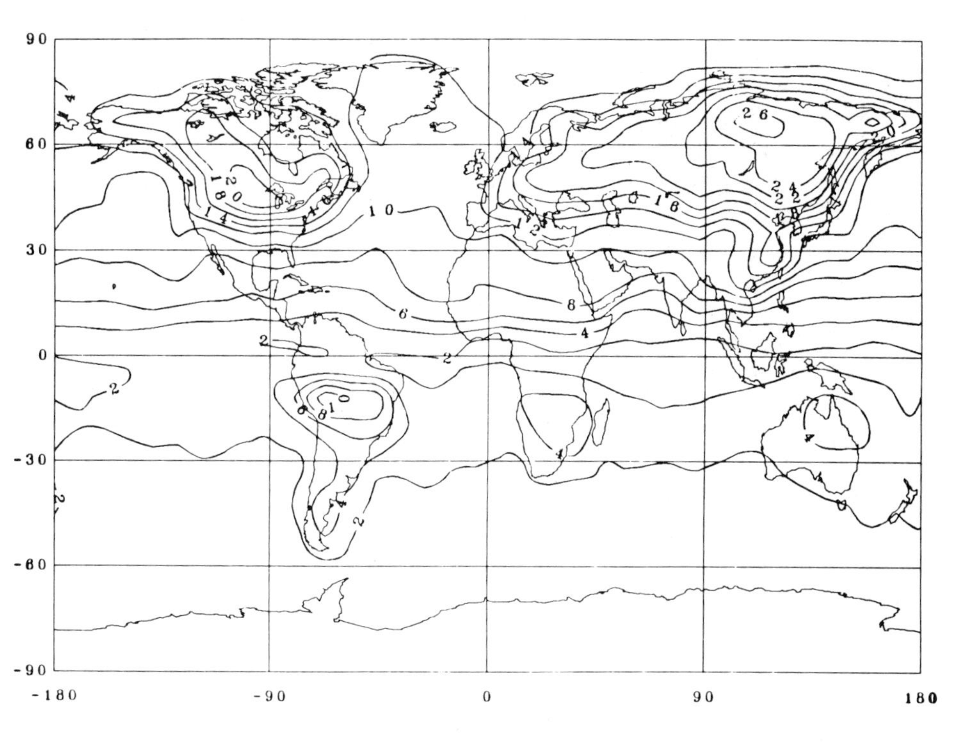

FIGURE 22.2. Simulated peak–peak amplitudes (ppm) of surface CO_2 oscillations induced by seasonal exchanges with terrestrial biota. (From Fung et al. 1983.)

zonally, i.e., across a latitude band, the isopleths of amplitude closely parallel the coastlines. Large contrasts in CO_2 concentration between land and sea are created by the highly productive and seasonal land vegetation and, in this experiment, the absence of oceanic sources and sinks. These contrasts are not smoothed effectively by the wind. The amplitudes simulated over ocean are only half the maximum amplitudes simulated over land at similar latitudes. The isopleths tighten and wrap the coastlines more closely at higher latitudes as land biomes become more strongly seasonal. Amplitudes >20 ppm are seen in the boreal forests of North America and Siberia. In South America, Azevedo's seasonality applied to the rain forests south of 10°S results in an amplitude of 10 ppm. Validation of model results in these regions is at present impossible because of the limited observations of CO_2.

Of the three biospheric exchange functions investigated, that of experiment 3, constructed from a global NPP map and Azevedo's (1982) seasonal exchange curves, produces CO_2 annual cycles at monitoring station locations most similar to those observed. In this particular exchange function, the net flux of carbon to the biosphere during the growing season (the GSNF) is 5.4×10^{15} g in the region north of 45°N. This flux is larger than Pearman and Hyson's estimate of 2.5×10^{12} g and Bolin and Keeling's (1963) estimate of 4.1×10^{12} g for the same region. These biospheric exchange functions, when input to the respective models, all reproduce reasonably well the amplitudes of the observed CO_2 cycles. However, only one tracer transport model and one biospheric exchange function can be correct. The differences in the GSNF thus underlie the need for an ecological model of CO_2 exchange rather than one based on model requirements.

Seasonal Exchange with the Upper Ocean

The magnitude and geographical distribution of the exchange of CO_2 between the atmosphere and the ocean is proportional to the gradient of pCO_2 across the air–sea interface. The concentration of CO_2 in surface waters is determined by total carbon concentration, total alkalinity, sea surface temperature (SST), salinity, and concentrations of borates, silicates, and other minor constituents. Apart from SST, there are few observations of the seasonal behavior of quantities that determine CO_2. Observations in the Sargasso Sea (Takahashi et al. 1982) and in the north and south tropical Pacific (Weiss et al. 1982) reveal that there are large seasonal oscillations of pCO_2 and that these oscillations are correlated with those in SST. Recent data from TTO (Weiss, 1983) show that pCO_2 at ~70°N in the North Atlantic is ~85 ppm higher in the winter than in the summer. This pCO_2 cycle suggests that the effect of the winter increase in total carbon concentration in the surface waters overwhelms the effect of the decrease in SST. The observed cycle has a timing that is opposite that expected from SST effects alone. Whatever the cause, these seasonal variations of pCO_2 have amplitudes of order 30 ppm or greater, and may contribute to the oscillations of atmospheric CO_2. Indeed, these oceanic contributions have been found in the $^{13}C/^{12}C$ data at a few land stations (e.g., Mook et al., 1984).

The magnitude of the air–sea flux of CO_2 depends also on the rate of air–sea exchange, or the "piston velocity," as well as on the pCO_2 gradient between the atmosphere and the surface waters. Laboratory experiments show clearly that the piston velocity increases with wind speed and decreases with temperature (e.g., Broecker et al., 1978; Broecker and Siems, 1984; Ledwell, 1984; Kerman, 1984). However, the exact form of this dependence, especially when applied to the ocean, is under debate (e.g., Deacon, 1977; Peng et al., 1979).

The discussion above highlights the difficulty in assessing the effect of seasonal oscillations in the upper ocean on the oscillations of atmospheric CO_2. To gain an estimate of the magnitude of this effect, we have performed a tracer model experiment to study the atmospheric CO_2 response to the effect of sea surface temperature variations alone. Monthly sea surface temperatures are prescribed by the climatology of Alexander and Mobley (1974), and temperature-dependent carbon chemistry (Takahashi 1976) is included in the mixed layer. This results in oceanic pCO_2 oscillations with peak-to-peak amplitudes >100 ppm in mid-latitudes in the northern hemisphere, where the amplitudes of SST oscillations exceed 10°C. A globally uniform piston velocity of 1700 $m \cdot yr^{-1}$ is used in this model experiment. This corresponds to a global mean exchange time constant of ~10 years, consistent with that used by Oeschger et al. (1975) for a globally averaged model.

The peak-to-peak amplitude of CO_2 oscillations in surface air induced by this seasonal cycle of oceanic pCO_2 is shown in Fig. 22.3. In the northern hemisphere, this seasonal effect on atmospheric CO_2 is small compared to that induced by exchange with the terrestrial biosphere. Nevertheless, the simulated amplitudes at observation sites are about 10% or more of those observed; the amplitude at Ocean Weather Station Papa is ~3 ppm and ~1 ppm at Mauna Loa and at the South Pole. In the southern hemisphere, the SST-induced oscillations have amplitudes comparable to those caused by biospheric exchange, consistent with the findings of Mook et al. (1984). We note that, at middle to high latitudes, although the pCO_2 cycles may not have amplitudes as large as those modelled here, the strong winds there would yield a piston velocity greater than the globally averaged value. The effect of the piston velocity increase would partially compensate for the possible reduction in the amplitude of the pCO_2 cycle, so that the oceanic contribution to the atmospheric CO_2 cycle may be of a magnitude comparable to those simulated here.

Figure 22.4 shows (for the five observation sites) the annual cycles of atmospheric CO_2 induced by SST and biospheric exchange and the combined annual cycles. At these sites, the SST-induced cycles are out of phase with the biosphere-induced cycles. Thus, when compared to the biosphere-induced cycles, the combined cycles have reduced amplitudes and little phase shift at the northern stations, where the amplitudes of the SST-induced cycles are much less than the amplitudes of the biosphere-induced cycles. At the South Pole, however, where the amplitudes are of similar magnitudes, the combined cycle shows significant phase shift and little amplitude change.

In this treatment of seasonal ocean exchange, only the SST effect has been included. Other oceanic processes, such as mixed-layer depth variations and changes in total alkalinity and total carbon concentration resulting from oceanic transports and activities of the marine biota, may have impact on pCO_2 variations in the ocean and consequently in the atmosphere. At present, except for a few observations, we do not know whether these processes act to enhance or to oppose the SST-induced oscillations of pCO_2 of the global ocean. The experiment does illustrate, however, that as long as the oceanic pCO_2 exhibits seasonal oscillations with amplitudes ~30 ppm or greater, regardless of the phasing, there will be small but significant atmospheric CO_2 responses to the seasonality of the upper ocean. Hence, to gain information about the activities of the

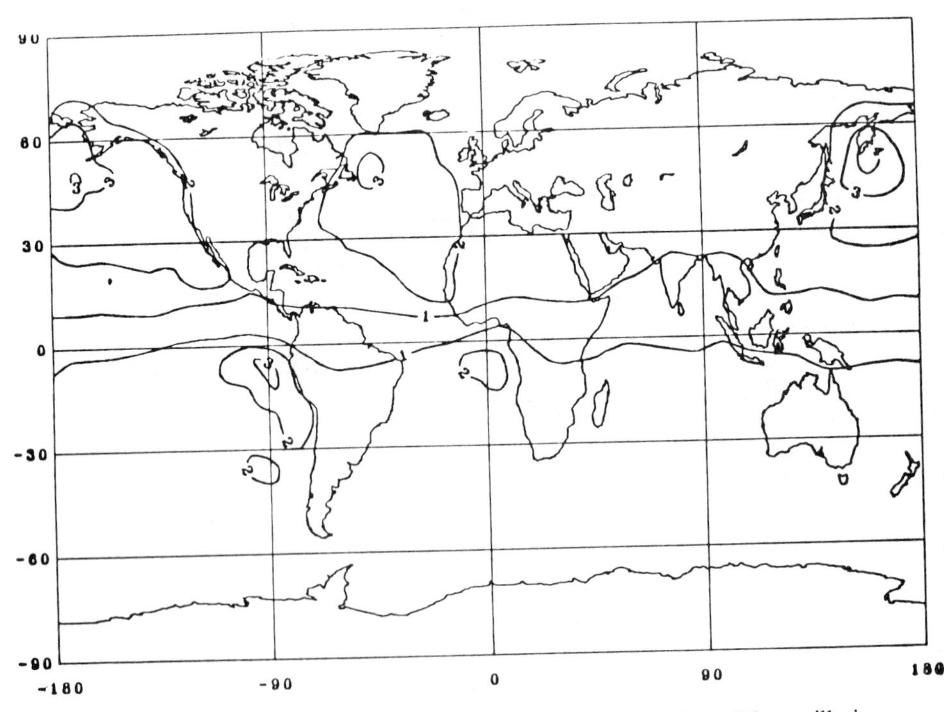

FIGURE 22.3. Simulated peak–peak amplitudes (ppm) of surface CO_2 oscillations induced by seasonal cycle of sea surface temperatures.

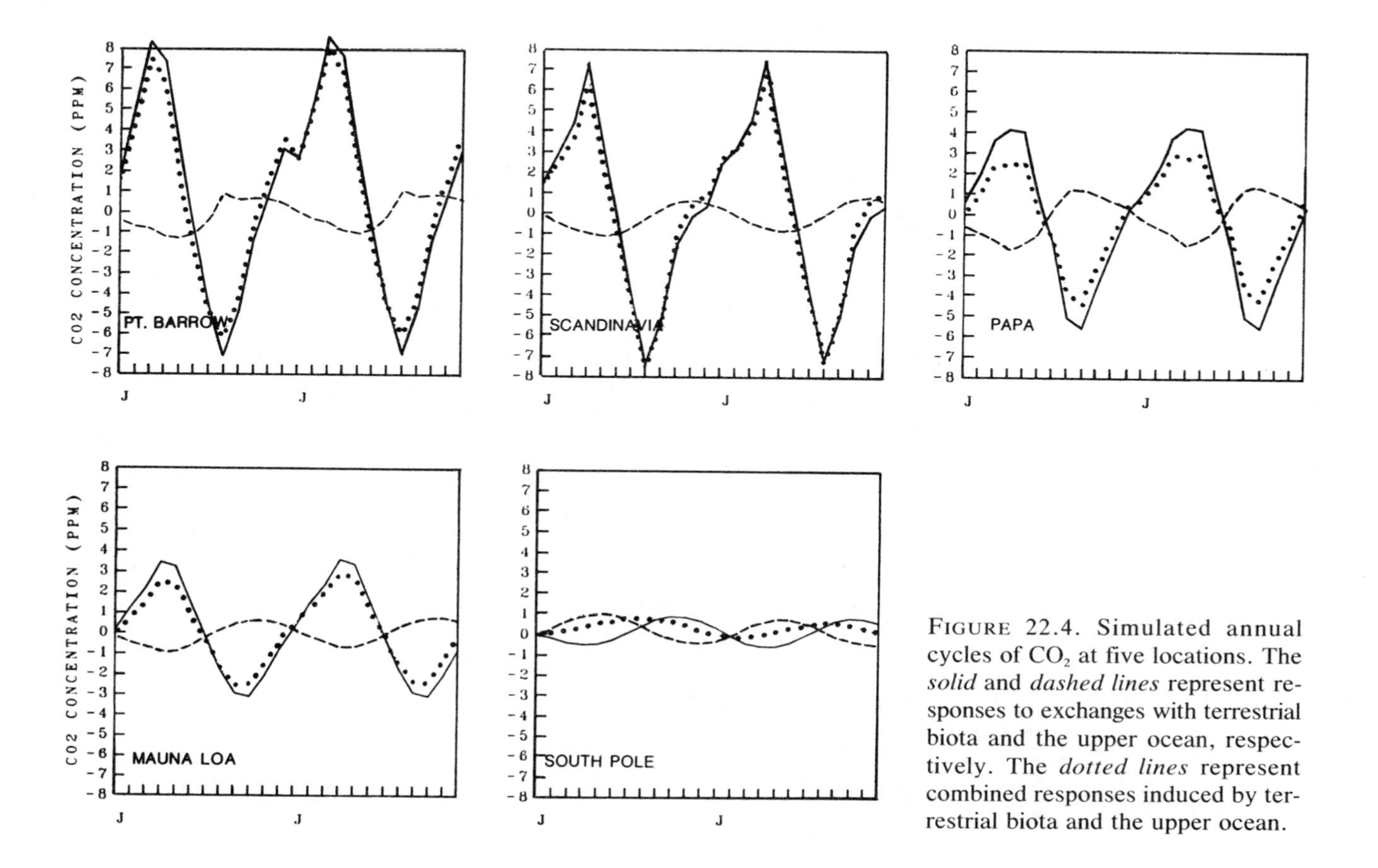

FIGURE 22.4. Simulated annual cycles of CO_2 at five locations. The *solid* and *dashed lines* represent responses to exchanges with terrestrial biota and the upper ocean, respectively. The *dotted lines* represent combined responses induced by terrestrial biota and the upper ocean.

terrestrial biota and the changes in these activities, the oceanic contribution to the seasonal cycle should not be neglected.

Sensitivity to Atmospheric Circulation Anomalies

The circulation pattern of the atmosphere exhibits significant and persistent interannual variation, as exemplified by the El Nino-Southern Oscillation (ENSO) events. ENSO events are characterized by persistent, positive sea–surface temperature anomalies in the equatorial Pacific accompanied by global shifts in wind, temperature, and precipitation patterns. These changes could affect the sources and sinks of carbon and the transport of CO_2 in the atmosphere.

We have begun some tracer model experiments to investigate the factors that contribute to the interannual variation of CO_2 in the atmosphere. The GCM has been run for a year to simulate the atmospheric circulation during an anomalous year, such as an ENSO event. The sea surface temperature for this GCM run was prescribed by the addition of monthly sea surface temperature anomalies [compiled by Rasmusson and Carpenter (1982)] to the monthly SST climatology of Alexander and Mobley (1974). The winds and convective mixing frequencies thus generated were saved on tape for use in the tracer model.

To isolate the effect of circulation changes on CO_2 distribution in the atmosphere, we used vegetation and oceanic sources and sinks described earlier as inputs to the tracer model. In this way, we neglected possible contributions from biotic and oceanic changes to interannual variations found in atmospheric CO_2 records.

Figure 22.5 shows the annual cycles of CO_2 concentrations simulated at five monitoring station locations for mean atmospheric conditions and the anomalous year. The simulated cycles for the normal and anomalous years are similar. However, amplitude and phase changes are detectable. At Point Barrow and at Mauna Loa, the peak–peak amplitudes decrease by ~15% during the anomalous year, and the maximum concentrations are delayed by about a month. At Ocean Weather Station Papa, however, the amplitude increases by ~1 ppm during the anomalous year, and the maximum concentration occurs about 1 month earlier. At Scandinavia the amplitude decreased by ~10%, and no change in phasing was noted. The effects of circulation changes on the annual cycle at the South Pole are small.

During the anomalous year, changes in the simulated annual cycle of CO_2 are global. The departure of anomalous-year amplitudes from normal-year amplitudes is shown in Fig. 22.6. The magnitudes of the amplitude change are largest at middle to high latitudes in the northern hemisphere (where the amplitudes of the CO_2 cycles are large) and decrease southwards. In general, these amplitude differences represent about 10 to 20% of the amplitudes shown in Fig. 22.2. The sign of the change does not

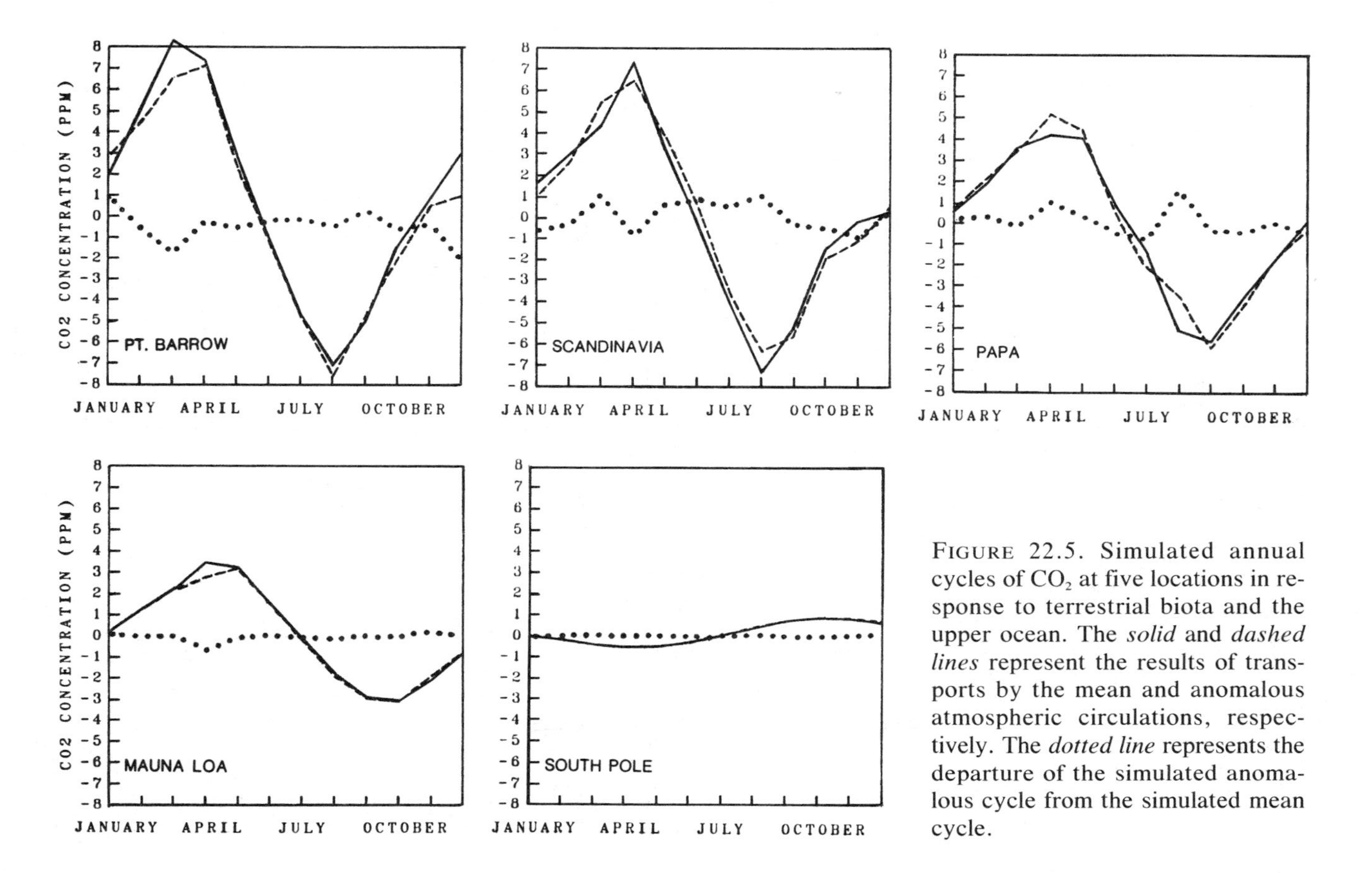

FIGURE 22.5. Simulated annual cycles of CO_2 at five locations in response to terrestrial biota and the upper ocean. The *solid* and *dashed lines* represent the results of transports by the mean and anomalous atmospheric circulations, respectively. The *dotted line* represents the departure of the simulated anomalous cycle from the simulated mean cycle.

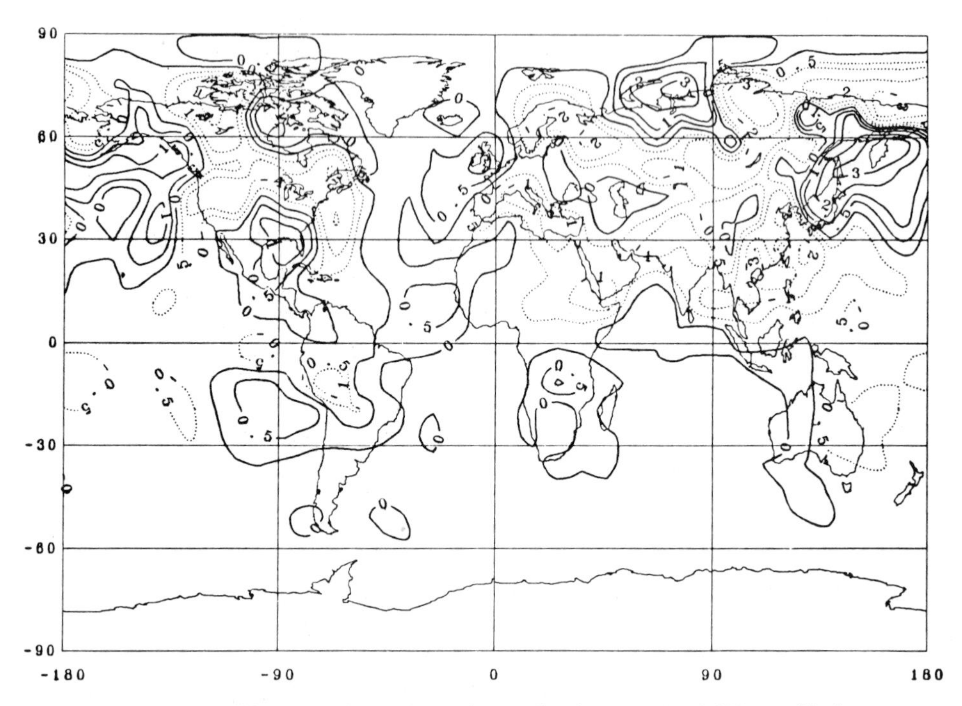

FIGURE 22.6. Differences in peak–peak amplitudes (ppm) of CO_2 oscillations at the surface between the simulated anomalous and mean years. Positive values indicate that amplitudes are higher during the anomalous year.

show any smooth latitudinal progression; however, it is consistent with the anomalous circulation pattern. Amplitude decreases or increases where the anomalous winds advect CO_2 away from or into the region. These changes could not be ascribed to changes in the sources and sinks, as the modelled sources and sinks were prescribed to be the same in these experiments. The changes are the results of global redistribution by the anomalous circulation alone.

Keeling (1983) has shown that there are interannual variations in the CO_2 cycles at Mauna Loa and at Ocean Weather Station Papa. The amplitude variations are about 10 to 20% of the mean amplitude, similar to that simulated here. Of course, variations in the behavior of the terrestrial biosphere and in air–sea exchange may contribute to interannual variation. This study demonstrates, however, that variations in the annual cycle of CO_2 at a single monitoring station may not necessarily reflect global variations in CO_2 sources and sinks alone. The variations at a single station could have resulted from changes in the atmospheric redistribution processes. It is only when global variations are found over a long period of time, and when these variations are shown to be greater than the natural variations at the individual stations, that we can confidently use the CO_2 data to extract information about long-term changes in biotic activity and atmosphere–ocean exchange of CO_2.

Discussion

Initial attempts to model atmospheric CO_2 distribution (with the ocean and biosphere as seasonal sources and sinks) encourage the notion that this approach will lead to useful quantitative constraints on CO_2 fluxes and changes in these fluxes. Realization of this objective will require:

1. Continued improvement in the realism of 3-D global transport modeling.
2. Information about the geographical distribution and seasonal variations of marine biota, total alkalinity, and total carbon concentrations in the surface waters.
3. Realistic models of the seasonal behavior of terrestrial biota that are based on knowledge of the behavior of the biota rather than derived by simple transport models.
4. Extended timelines of atmospheric CO_2 monitoring, with improved precision and definition of the uncertainties in the measured CO_2 amounts. Many of the potential applications depend on measurement of perturbations of the CO_2 distributions and changes in time or in geographical distribution. Consequently, accurate calibration and intercalibration among different observing stations is important.

Acknowledgments

It is a pleasure to acknowledge discussions with Taro Takahashi, who also provided the carbonate chemistry program. The author also thanks Jean Lerner, Elaine Matthews, Katie Prentice, and Bob Souzzo for discussions and assistance during various phases of this study. This work is supported by Oak Ridge National Laboratory Subcontract UCC-19X 22292C with the Lamont-Doherty Geological Observatory of Columbia University.

References

Alexander, R. C. and R. L. Mobley. 1974. Monthly average sea surface temperature and ice-pack limits on a 1° global grid. Report 4-1310-ARPA. Rand Corporation, Santa Monica, California.

Azevedo, A. E. 1982. Atmospheric distribution of CO_2 and its exchange with the biosphere and the oceans. Ph.D. Thesis. Columbia University, New York, New York.

Bazilevich, N. I., L. E. Rodin, and N. N. Rozov. 1971. Geographical aspects of biological productivity. Sov. Geogr. Rev. Transi. 12:293–371.

Bolin, B. and C. D. Keeling. 1963. Large-scale atmospheric mixing as deduced from the seasonal and meridional variations of carbon dioxide. J. Geoph. Res. 68:3899–3920.

Broecker, H. C., J. Peterman, and W. Siems. 1978. The influence of wind on CO_2-exchange in a wind-wave tunnel, including the effects of monolayers. J. Mar. Res. 36:595–610.

Broecker, H. C. and W. Siems. 1984. The role of bubbles for gas transfer from water to air at high wind speeds. Experiments in the wind-wave facility in Hamburg. In W. Brutsaert and G. H. Jirka (eds.), Gas Transfer at Water Surfaces, pp. 229–236. D. Reidel, Dordrecht, Netherlands.

Deacon, E. L. 1977. Gas transfer to and across an air-water interface. Tellus 19:363–374.

Fung, I., K. Prentice, E. Matthews, J. Lerner, and G. Russell. 1983. Three-dimensional tracer model study of atmospheric CO_2: response to seasonal exchanges with the terrestrial biosphere. J. Geophys. Res. 88:1281–1294.

Hansen, J., G. Russell, D. Rind, P. Stone, A. Lacis, S. Lebedeff, R. Ruedy, and L. Travis. 1983. Efficient three-dimensional global models for climate studies: model I and II. Mon. Weather Rev. 111:609–662.

Keeling, C. D. 1983. The global carbon cycle: what we know and could know from atmospheric, biospheric, and oceanic observations. In Proceedings: Carbon Dioxide Research Conference; Carbon Dioxide, Science, and Consensus, Sept. 19–23, 1982, Berkeley Springs, West Virginia. U.S. Department of Energy CONF-820970.

Kerman, B. R. 1984. A model of interfacial gas transfer for a well-roughened sea. J. Geophys. Res. 89:1439–1446.

Ledwell, J. R. 1984. The variation of the gas transfer coefficient with molecular diffusivity. In W. Brutsawert and G. H. Jirka (eds.), Gas Transfer at Water Surfaces, pp. 293–302. D. Reidel, Dordrecht, Netherlands.

Lieth, H. 1973. Primary production: terrestrial ecosystems. Human Ecology 1:303–332.

Machta, L. 1972. Mauna Loa and global trends in air quality. Bull. Am. Meteorol. Soc. 53:402–420.

Matthews, E. 1983. Global vegetation and land use: new high-resolution data bases for climate studies. J. Appl. Meteorol. 22:474–487.

Mook, W. G., M. Koopmans, A. F. Carter, and C. D. Keeling. 1984. Seasonal, latitudinal, and secular variations in the abundance and isotopic ratios of atmospheric carbon dioxide. 1. Results from land stations. J. Geophys. Res. 88:10915–10933.

Oeschger, H., U. Siegenthaler, U. Schotterer, and A. Gugelmann. 1975. A box diffusion model to study the carbon dioxide exchange in nature. Tellus 27:168–192.

Pearman, G. I. and P. Hyson. 1980. Activities of the global biosphere as reflected in atmospheric CO_2 records. J. Geophys. Res. 85:4457–4467.

Peng, T.-H., W. S. Broecker, G. G. Mathieu, and Y. H. Li. 1979. Radon evasion rates in the Atlantic and Pacific Oceans as determined during GEOSECS program. J. Geophys. Res. 84:2471–2486.

Rasmusson, E. and T. Carpenter. 1982. Variations in tropical sea surface temperature and surface wind fields associated with the Southern Oscillation/El-Nino. Mon. Weather Rev. 110:354–384.

Takahashi, T. 1976. Carbonate chemistry of the Atlantic, Pacific, and Indian Oceans: the results of the GEOSECS expedition 1972–1978. Lamont-Doherty Geological Observatory Technical Report No. 1, CV1-80.

Takahashi, T., D. Chipman, and T. Volk. 1982. Geographical, seasonal, and secular variations of the partial pressure of CO_2 on surface waters of the North Atlantic Ocean: the results of the North Atlantic TTO Program. In Proceedings, Carbon

Dioxide Research Conference: Carbon Dioxide, Science, and Consensus, Sept. 19–23, 1983, Berkeley Springs, West Virginia. U.S. Department of Energy CONF-820970.

Weiss, R. H. 1983. Paper presented at the XVIII General Assembly of the International Union of Geodesy and Geophysics, Hamburg, Germany, August 18, 1983.

Weiss, R. F., R. A. Jahnke, and C. D. Keeling. 1982. Seasonal effects of temperature and salinity on the partial pressure of CO_2 in seawater. Nature 300:511–513.

23 Fossil Fuel Combustion: Recent Amounts, Patterns, and Trends of CO_2

RALPH M. ROTTY AND GREGG MARLAND

Several types of human activity have introduced perturbations that impinge on the natural global carbon cycle. During the past century or so, one of the major perturbations has been the release of carbon from long-term storage through the combustion of fossil fuels. During the next 100 years fossil fuel use will almost certainly be *the* major source causing increased levels of CO_2 in the atmosphere.

Fossil fuel combustion now releases CO_2 at the rate of 5 gigatons of carbon per year. The annual emissions from fossil fuel combustion during the period following World War II grew exponentially at a near constant value of 4.5% per year until 1973. Since then the global growth rate has been <2% per year. The annual rate of emission of CO_2 from fossil fuel combustion is generally conceded to be one of the better known facts about the CO_2 problem. Only the increase in the amount of CO_2 in the atmosphere during the past 25 years can be documented with greater confidence.

Because the fossil fuel recoverable resources are so vast, it is very difficult to foresee any world scenario that denies continuing use of very large amounts of fossil fuels. As a consequence, the annual emissions are likely to continue to grow and become even more central to any developing CO_2 problem.

Calculating CO_2 Releases from Fossil Fuel Use

The amount of CO_2 released by fossil fuel combustion is fundamental data in analyses of the global carbon cycle. Therefore, it is appropriate to document how the estimates of these emissions are made. For global-scale annual CO_2 emissions, we assemble fuel production information for each of the three types of fossil fuel, that is, gases, liquids, and solids. The CO_2 estimation procedure is expressed as

$$CO_{2_i} = (P_i)\,(FO_i)\,(C_i) \qquad (1)$$

where the subscript i refers to each of the three fuel types, P is the fuel production, FO is the fraction of the fuel that is oxidized, and C is the carbon content per unit of fuel production.

The fuel data used must not only have a high level of accuracy and reliability but must also be complete for all parts of the world. The data must be consistent from year to year and form a historical time series for as long a period as possible. Further, the units used must be homogeneous and consistent from country to country, and the unit used for each fuel type should acknowledge the quality of the fuel. We have accepted the energy data series compiled by the United Nations (UN) Statistical Office (United Nations 1983) as best meeting these needs.

In defining fuel types, care must be taken to count each fuel unit only once. For gases we have used the amount of natural gas reported after extraction of natural gas liquids, and care has been taken not to include gas that is flared. CO_2 from flared gas is handled separately. In the liquids category, we start with the crude petroleum produced and add the mass of natural gas liquids. In determining a total liquid fuels category, the UN multiplies the mass of natural gas liquids by a factor 1.06 before adding the data to the crude petroleum production. This is appropriate when considering energy because the natural gas liquids have a higher hydrogen-to-carbon ratio than crude petroleum and thus a greater energy content. Because we seek the amount of carbon present, we treat crude petroleum and natural gas liquids as having the same carbon content and then sum the masses without the energy normalizing factor. Solid fuels, as reported by the UN Statistical Office (United Nations 1983), include brown coal, lignite, and peat as well as hard coal, all on a ton-of-coal equivalent basis. When we use quantities for fuel production in units as given in the UN tabulations (i.e., natural gases in terajoules, liquids in tons of crude petroleum, and solids in tons of coal equivalent), we also fix the units for the term C in the CO_2 emission equation. The use of other units for the P_i's would require modification of the C_i's.

The fraction oxidized, FO_i, requires an estimate of the fraction diverted from fuel uses as well as an estimation of the fraction not oxidized in the combustion process. Data required to obtain the nonfuel use of natural gas globally are not available; therefore we have based our estimates on U.S. data where a 7-year average (1970 to 1976) of 3.16% of gas production is diverted to nonfuel uses. Because world total natural gas production is heavily dominated by the United States, the United Kingdom, Canada, and the Netherlands, all of which use gas largely as a fuel, we estimate that about 3% of world production goes to nonfuel uses. A large fraction of the 3% nonfuel use goes to ammonia production during which the carbon in the gas is mostly oxidized; the remainder goes to uses in which the rate of carbon oxidized varies over a period of several years. We assumed that an equivalent of about two-thirds of the carbon in the nonfuel-use gas is oxidized each year. Thus, 1% of the gas produced each year is assumed to remain unoxidized for long periods. In addition, a small amount of the carbon in natural gas burned as fuel will remain unoxidized (as soot). This amount is impossible to determine on a global scale, but in modern combustion systems it is small. We assign 1% of the carbon in the gas produced as the amount remaining unoxidized during combustion processes. Thus the total fraction of gas remaining unoxidized is 0.02, and the fraction of gas oxidized equals 0.98.

Some of the crude oil and natural gas liquids produced each year end up as fibers, lubricants, paving materials, or in other uses in which the carbon is oxidized only over a longer time interval. The UN statistics separate refinery output into energy products and nonenergy products. In 1979 the nonenergy products made up 10.2% of the total products. Liquified petroleum gas (LPG) and ethane from gas liquids plants provide feedstocks for the petrochemical industry in addition to their use as fuels. The other natural gas liquids are used as fuels, as is most of the LPG from refineries. We estimate the LPG and ethane used in the petrochemical industry as a fraction of that produced in the natural gas liquids plants alone. (We do not mean that all LPG used in the petrochemical industry originated in gas processing plants, only that the global amount can be estimated as a fraction of LPG produced there.) We estimate that, worldwide, about 40% of the LPG and ethane produced in natural gas processing plants ends up in materials that are not soon oxidized. Naphthas are largely used as chemical feedstocks, especially in Europe. We estimate that about 80% of the naphthas produced globally remain unoxidized for long times in plastics, tires, and fabrics. Asphalt and much of the lubricants produced remain unoxidized, and although this discussion omits some partially nonoxidizing uses (e.g., portions of waxes, paint solvents, and dry-cleaning materials), the overestimations and the underestimations are approximately balanced by using the results of Table 23.1. [A somewhat more detailed discussion of these arguments is given in Marland and Rotty (1983).] Using 6.7% of production for the unoxidized petroleum liquids in the nonfuel path seems satisfactory. An adjustment for incomplete combustion of petroleum-derived fuels is also required. Relying largely on Environmental Protection Agency data for emissions from several types of combustion processes in the United States, we estimated global quantities for some

TABLE 23.1. Estimate of 1980 world petroleum liquids produced but not oxidized (10^3 tons).

40% of LPG + ethane (from gas plants)	25,754
80% of naphtha	58,541
Asphalt	97,941
50% of lubricants	19,649
Total	201,885
Liquids produced, %	6.53
10-Year average, %	6.67

From Marland and Rotty (1983).

TABLE 23.2. Nonoxidized nonfuel solids.

	Light oil and tar produced in U.S. coke plants*		
Year	Crude light oil and crude tar (millions of tons of coal equivalent)	Coke to coke plants	Coal converted to light oil and tar %
1980	3.6	60.5	6.00
1979	4.0	69.9	5.74
1978	3.7	64.8	5.68
1977	4.0	70.5	5.72
1976	4.4	76.8	5.68
1975	4.4	75.8	5.80
1974	4.7	81.8	5.71
1973	5.0	85.4	5.87
1972	5.0	79.6	6.33
1971	4.6	75.5	6.11
1970	5.2	87.5	6.00
1969	5.4	84.7	6.33
		12-Year mean	5.91

From Marland and Rotty (1983).

*Globally: coal in nonoxidation uses = (0.591)(0.75)(0.18) = 0.008 (see text).

of the major processes and concluded that an equivalent fraction of 1.5% of the carbon in the produced crude passes through burners without being oxidized. Thus $FO_l = 0.918$.

For solid fuels, we must also consider the nonfuel use of coal and incomplete combustion of coal used as fuel. Most recent available data for the U.S. light oil and tar produced in coke plants is given in Table 23.2. About 75% of the light oil and tar from coke plants is used in applications where oxidation is long delayed. We deduce from Table 23.2 that (0.75)(0.0591) of the coal supplied to coke plants remains unoxidized as a result of nonfuel use. Globally, about 18% of the coal produced is supplied to coke plants; consequently, about 0.8% of the coal produced remains unoxidized in nonfuel uses. From studies by the United Nations Environment Programme (UNEP 1979) and data from the Tennessee Valley Authority, we have concluded that the unoxidized fraction of carbon in the combustion of coal is about 1%. [Greater details on the arguments supporting these estimates are also given in Marland and Rotty (1983).] Thus, our estimate of $FO_s = 1 - (0.01) - (0.008) = 0.982$.

The final term in the CO_2 emission equation denotes the carbon content of the fuel. For several average gases, we have found that the linear regression,

$$C_g = 57.357 + 1.459 \times 10^{-3} (\Delta H_H - 8898) \tag{2}$$

where C_g is in grams carbon per 1000 kcal and ΔH_H is the higher heating value in kcal • m^{-3}, closely defines the carbon content per unit of energy of the gas. The value for C_g is not very sensitive to changes in heating value of the gas, and this fact, to a large extent, justifies using a constant world value for ΔH_H of 8900 kcal • m^{-3}, even though UN data (United Nations 1982) show slowly decreasing values for the annual average during the 1970s. Thus our estimated value for C_g = 57.36 g carbon per 1000 kcal = 0.0137 × 10^6 tons carbon per 1000 terajoules. In Marland and Rotty (1983), we also argue that the carbon content for flared gas is probably close to that of dry marketed gas in the United States, and C_f = 0.525 tons C • 1000 m^{-3}.

The carbon content of crude (petroleum) for liquid fuels is well defined by a simple relation to density of the crude. Figure 23.1 shows the regression line, and, using UN single-country values for specific gravity (United Nations 1982), the weighted average specific gravity for the 10 largest crude-producing countries in 1978 was 33.6°API. Other estimates (e.g., Carter 1979), show slightly lower values for the 1970 to 1979 period. Although the mean carbon content of crude will vary slightly from year to year as the distribution of sources changes, we conclude that C_l = 0.85 adequately describes the mean composition of world crude oil for our period of interest.

In the case of coal, the carbon content correlates closely with heating value. Using a higher heating value of coal of 7000 cal • g^{-1}, which is a reasonable standard for coal equivalent, we found the mean carbon content to be 0.707 tons carbon per ton of coal equivalent. The UN now tabulates

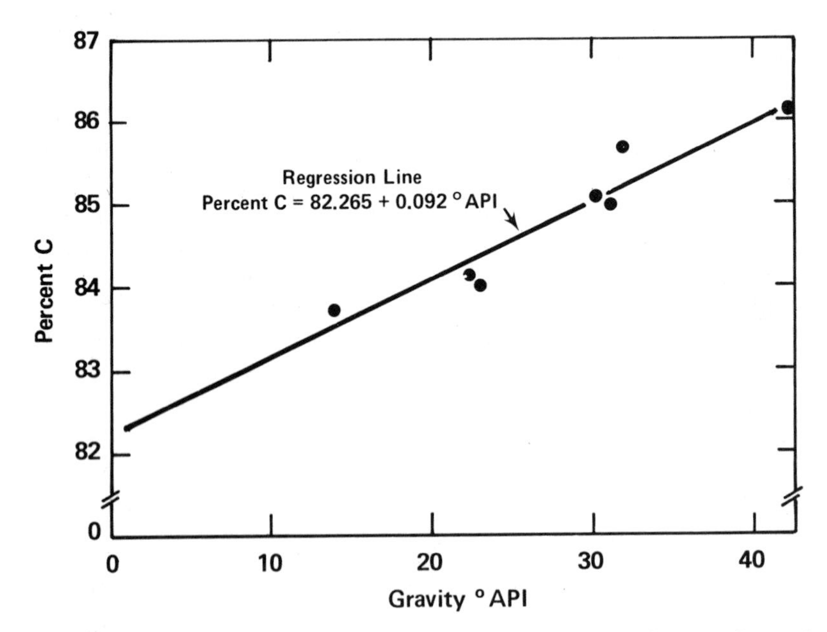

FIGURE 23.1. Percent carbon as a function of specific gravity for samples typical of North American crude petroleum.

solid fuels in coal equivalents on a 7000 cal • g^{-1} "net heat value" (lower heating value) basis. Comparing UN and U.S. Department of Energy published vales of U.S. coal production for 1970 to 1979, we deduced that the UN used a constant factor to adjust tons of coal equivalent from the higher to lower heating value basis. This factor was found to be 1.055; hence, we estimate C_s to be (0.707)(1.055) = 0.746 (to be used with the UN coal statistics reported on a "net heating value" basis).

The factors and units used for calculating annual CO_2 emissions from global production fuel data are summarized in Table 23.3.

Resulting Global CO_2 Emissions from Fossil Fuel Production Data

Using the procedures and factors developed above and the current UN data set found in the *Yearbook of World Energy Statistics* (United Nations

TABLE 23.3. Factors and units for calculating annual CO_2 emissions from global fuel production data.*

$$[CO_{2_i} = (P_i)(FO_i)(C_i)]$$

From natural gas production

CO_{2_g} = CO_2 emissions in 10^6 tons carbon

P_g = Annual production in thousands of 10^{12} J ($\pm \approx 10\%$)
FO_g = Effective fraction oxidized in year of production = $0.98 \pm 1\%$
C_g = Carbon content in 10^6 tons per thousand 10^{12} J = $0.0137 \pm 2\%$

From crude oil and natural gas liquids production

CO_{2_l} = CO_2 emissions in 10^6 tons carbon

P_l = Annual production in 10^6 tons ($\pm \approx 8\%$)
FO_l = Effective fraction oxidized in year of production = $0.918 \pm 3\%$
C_l = Carbon content in tons carbon per ton crude oil = $0.85 \pm 1\%$

From coal production

CO_{2_s} = CO_2 emissions in 10^6 tons carbon

P_s = Annual production in 10^6 tons coal equivalent ($\pm \approx 11.2\%$)
FO_s = Effective fraction oxidized in year of production = $0.982 \pm 2\%$
C_s = Carbon content in tons carbon per ton coal equivalent = 0.746† $\pm 2\%$

From natural gas flaring

CO_{2_f} = CO emissions in 10^6 tons carbon

P_f = Annual gas flaring in 10^9 m^3 ($\pm \approx 20\%$)
FO_f = Effective fraction oxidized in year of flaring = $1.00 \pm 1\%$
C_f = Carbon content in tons per thousand cubic meters = $0.525 \pm 3\%$

*All masses are in metric tons (10^3 kg).

†The 0.746 value includes a heating value adjustment to recognize that the carbon content, developed on a higher heating value basis, must be increased when used with UN production data (United Nations 1982) based on "net" or lower heating values.

TABLE 23.4. CO_2 emissions from fossil fuels, 1950–1981.

Year	Gas fuel production (10^{12} J) (thousands)	CO_2 from gas fuel (10^6 tons C)	Liquid fuel production (10^6 tons oil)	CO_2 from liquid fuel (10^6 tons C)	Solid fuel production (10^6 tons coal equiv.)	CO_2 from solid fuel (10 tons C)	Gas flared and vented (10^9 m^3)	CO_2 from flared gas (10^6 tons C)	Total CO_2 from fossil fuels (10^6 tons C)
1950	7193	97	542	423	1468	1075	44	23	1618
1951	8562	115	613	478	1549	1135	46	24	1752
1952	9241	124	646	504	1534	1124	50	26	1778
1953	9772	131	684	534	1542	1130	51	27	1822
1954	10,250	137	714	557	1528	1119	52	27	1840
1955	11,140	150	801	625	1654	1212	58	30	2017
1956	12,027	161	870	679	1743	1277	61	32	2149
1957	13,247	178	915	714	1793	1314	67	35	2241
1958	14,296	192	938	732	1829	1340	66	35	2299
1959	15,976	214	1012	790	1877	1375	69	36	2415
1960	17,501	235	1089	850	1946	1426	75	39	2550
1961	18,865	253	1161	906	1825	1337	79	41	2537
1962	20,592	276	1258	982	1868	1368	84	44	2670
1963	22,364	300	1351	1054	1947	1426	90	47	2827
1964	24,447	328	1458	1138	2018	1478	97	51	2995

1965	26,172	351	1563	1220	2048	1500	104	55	3126
1966	28,313	380	1696	1323	2077	1522	115	60	3285
1967	30,500	409	1822	1422	1998	1464	126	66	3361
1968	33,130	445	1988	1551	2040	1494	139	73	3563
1969	36,268	489	2139	1669	2077	1522	152	80	4013
1970	38,440	516	2349	1833	2152	1576	167	88	4013
1971	41,204	553	2492	1945	2146	1572	171	90	4160
1972	43,391	583	2632	2054	2163	1585	180	95	4317
1973	45,284	608	2868	2238	2190	1604	213	112	4562
1974	45,897	616	2877	2245	2202	1613	204	107	4581
1975	46,232	621	2730	2130	2297	1683	183	96	4530
1976	48,043	645	2958	2308	2357	1727	210	110	4790
1977	49,275	662	3077	2401	2410	1765	206	108	4936
1978	51,742	695	3106	2424	2450	1795	201	106	5020
1979	54,376	730	3238	2527	2583	1892	200	105	5254
1980	53,443	718	3091	2412	2641	1935	194	102	5167
1981	54,115	727	2909	2270	2658	1947	178	93	5037
1982	54,171	727	2771	2162	2729	1999	175	92	4980

1983) and the supporting data tapes, we have estimated the CO_2 emissions for each fossil fuel type for each year from 1950 through 1982. Our series starts in 1950 because that is the year selected by the UN staff for the beginning of their master energy data file. When methods or definitions are changed, the implied data changes are made in the master file back to 1950. Our results are given in Table 23.4 and shown graphically in Fig. 23.2.

The 1973 change in the pattern of steady exponential growth of global CO_2 emissions is clear. That this was a result of changes in the oil and gas segments of the total is also clear, and neither the oil nor the gas data suggest any early return to the pre-1973 pattern. Not quite so obvious in the figure is the fact that emissions for 1980 through 1982 were each lower than emissions for the previous year. This is the only time that global CO_2 emissions have decreased for 3 years in a row. This observation supports the hypothesis that CO_2 emissions are heavily influenced by economic

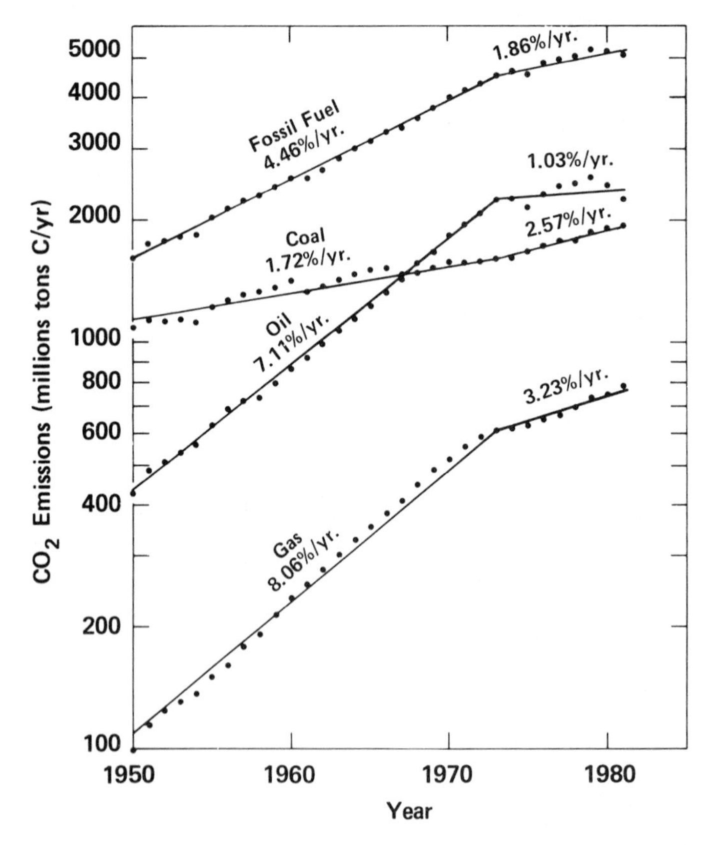

FIGURE 23.2. Annual CO_2 production from each fossil fuel group and total fossil fuels.

factors, and early indications are that emissions for 1983 will show an end to the downward trend.

Whether CO_2 in the atmosphere becomes a serious problem for humanity in the future will depend on the rate at which fossil fuels are used in the next century. Questions such as "When will the next basic change in growth in annual CO_2 emissions occur?" and "What will be the nature of that change?" are central to the issue. Many scenarios have been and will continue to be employed in the analysis of the carbon cycle, and we do not intend to argue for or against any of them. We believe strongly, however, that to be taken seriously, such scenarios must be consistent with present happenings and trends. Detailed examination of some of our data on CO_2 emissions can provide additional insight to ensure this consistency.

Calculating CO_2 Emissions Using Fuel Consumption Data

To learn more about the patterns of CO_2 emissions, we have disaggregated the world into six more-or-less homogeneous segments. The disaggregation required a basic conceptual change in the calculation of CO_2 emission rates. To calculate CO_2 emissions for individual nations or regions, we had to use fuel consumption data rather than production data. For example, the demand (consumption) for oil in many nations determines the production of oil in the Mideast (and vice versa). Changes in demand will result in changes in production, and reasons for the changes (and the effects of the changes) may be very different in various parts of the world.

The quality of UN fuel data has improved considerably during the past decade, and frequent revisions of data back to 1950 have resulted in consistent data sets for most nations, for both fuel production and "apparent" fuel consumption. Although the production data are the primary data set and are more reliable than the consumption data, to determine what emissions have occurred in each area of the world, consumption data are necessary. However, to ensure consistency with the global CO_2 emissions calculated from fuel production data, we developed our own "apparent" consumption data set. Using these data we were able to estimate CO_2 emissions disaggregated by world region and have determined the difference in growth rates from one world section to another. We emphasize that the sum of all regional CO_2 emissions, based on our consumption data, will not equal the global annual emissions calculated with fuel production data. We have been able to reconcile the two to within about 2% when we include CO_2 from flared gas and from fuel used in bunkers and thus not assigned to any individual country.

Not only has the annual rate of global CO_2 emissions from fossil fuels more than tripled during the interval 1950 to 1980 (from 1.6 million tons carbon to nearly 5.2 million tons carbon), but the fraction of the total that was released from North America and Western Europe has dropped markedly. Growth rates in poorer parts of the world have been much greater than those in more industrialized nations. Within some regions, individual countries show emission growth patterns much different from those of nearby neighbors. Reasons for these differences among nations have not yet been examined in sufficient detail to offer explanations; however, this will be necessary information in projections for the future.

Using fuel consumption data requires some modification of the procedure developed for CO_2 calculations from fuel production data. Using UN data, we developed an intermediate data set for consumption in each fuel category. These data were put in a format conducive to easy computation of annual CO_2 emissions rates for each country, using the basic equation

$$CO_{2_i} = (FC_i)\,(FO_i)\,(C_i) \tag{3}$$

where FC_i is fuel consumption, FO_i is the fraction oxidized, and C_i is the carbon content of the fuel.

For gases consumed as fuel in a given country, we started with the gross production of natural gas, subtracted gas flared or reinjected, corrected for extraction losses, added the excess of imports over exports, and subtracted any increase in stocks. Secondary gases such as coke oven gas and refinery gas were included in the computation only to the extent that there was a net trade or a change in stocks. Refinery gas, for example, will appear as liquids consumption unless it is stored or traded between nations, at which point the primary liquid fuel has been consumed and gas fuel is in commerce. All of the gas data used were given in terajoules and taken from the UN computer tape of world fuel data (also published as *1980 Yearbook of World Energy Statistics*, United Nations 1982). Quantities of gas given in terajoules were converted to tons of carbon emitted as CO_2 through the above equation, with factors FO_g and C_g having the same values as with fuel production data.

For solid fuel consumption the computation was analogous to that for gases. United Nations solid fuel production data (including peat and oil shale) in tons of coal equivalent were increased by the amount imports exceeded exports less the net increase in stocks. Secondary solids such as coke and briquettes entered the computation only when there was a net trade or change in stocks. The factors used in the CO_2 equation, FO_s and C_g, were taken to be the same as those used with fuel production data.

The treatment of liquid fuels was more complicated. Statistics for liquid fuels include many secondary products (e.g., gasoline, kerosene, refined fuel oil, jet fuel, waxes), some of which are for nonfuel applications. The crude petroleum produced plus the natural gas liquids (which are added to the crude petroleum to balance the liquids accounts) must supply the raw material for all the liquid fuel products, plus the petrochemicals that go into nonfuel products, plus most of the energy used in the refining process. We treated oil refineries like any other industry using fuel and producing goods. Production of nonfuel products was subtracted from primary fuel production, and nonfuel products were assumed not to result in CO_2 emissions even though they might be oxidized during use or disposal. In effect, we calculated CO_2 emissions from energy uses of primary fuels. An exception was fuel used in international bunkers. Global CO_2 emissions exceed the sum of the individual country totals by the amount of CO_2 resulting from fuels in bunkers, nonfuel uses of liquids that oxidize carbon, and flaring of natural gas (see Fig. 23.3). The consumption of liquid fuels (in tons) within each country was determined from the production of crude oil and natural gas liquids by adding imports minus exports, subtracting amounts used in bunkers and any increase in stocks, and subtracting production of secondary nonfuel products, net exports of products, and the increase in stocks of secondary liquid fuel products. All of these data are on the UN data tape. The CO_2 emissions associated with this fuel consumption were calculated using factors FO_l and C_l developed for use with fuel production data, except that FO_l was modified to acknowledge that nonfuel uses of liquid hydrocarbons are no longer in the accounts.

CO_2 Emissions by World Region

Using the United Nations fuel data, we estimated the CO_2 emission rates for six major divisions of the world for each year during 1950 to 1980. The regions were selected for some geographical, political, and economic homogeneity, and also so that each had emission rates of similar magnitude (the six regional emission rates fell within a factor of 4 in 1980). The sum of the emission rates for the six regions was reconciled with the global total emission rates (computed above) on the basis of fuel production data. This reconciliation required consideration of worldwide flaring of natural gas, fuel used in bunkers for international commerce, and estimated oxidation of nonfuel petroleum products. Figure 23.3 shows the changes over the 30-year period of record—an increase by greater than a factor of 3 in global emissions and an increasing fraction of the total attributed to the industrially developing parts of the world and the Asian countries with centrally planned economies (e.g., Peoples Republic of China). Reduction in the fraction of total emissions from North America is marked.

Figure 23.4 is a plot of the same data for each region on a time-dependent basis. This figure shows clearly the effect of changing fuel price structure after 1973 on fuel use (and consequently on CO_2 emission rates) in North America, Western Europe, Japan, and Australia. The numbers indicated on the straight line segments in Fig. 23.4 are the exponential growth rates determined by a least-squares best fit of an exponential curve for the period indicated. (No such numbers are given for the USSR and Eastern Europe or for the Asian-planned economies, because these growth rates are con-

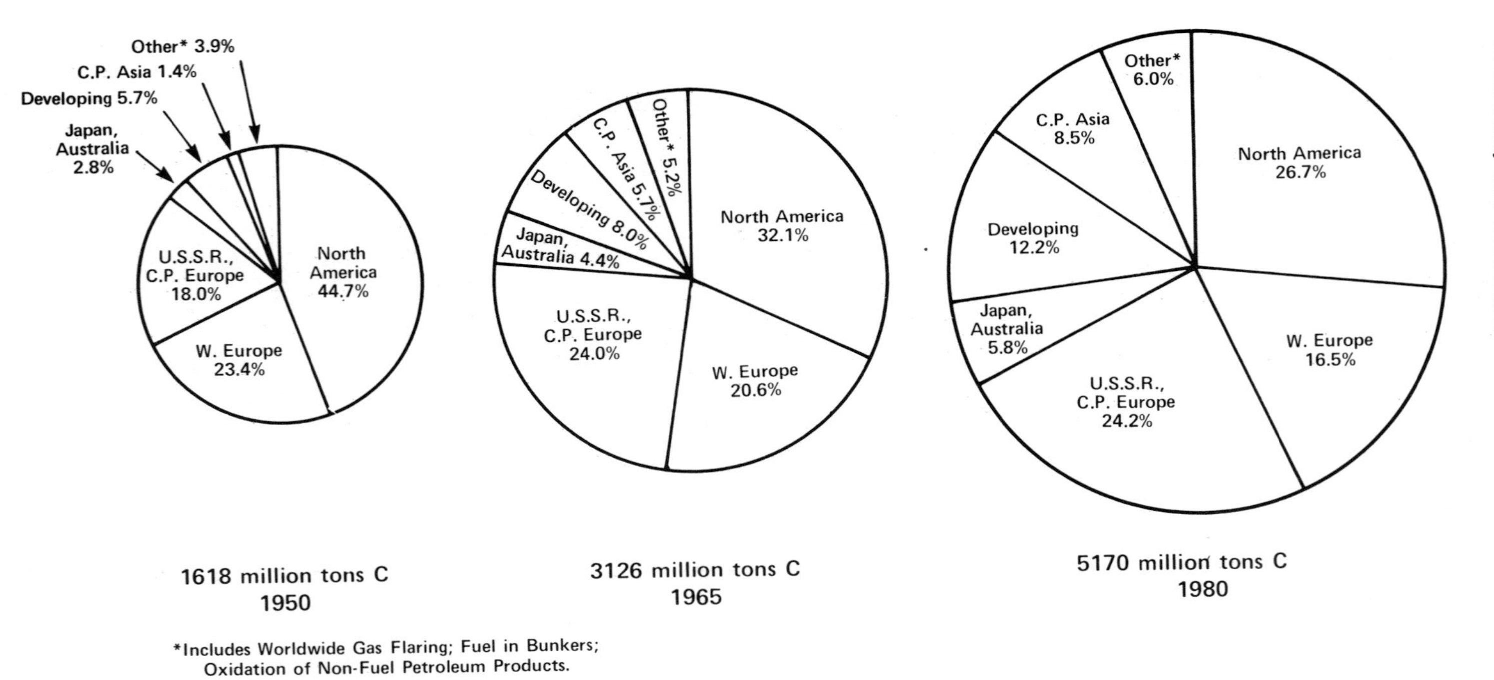

FIGURE 23.3. The changing pattern of global CO_2 emissions.

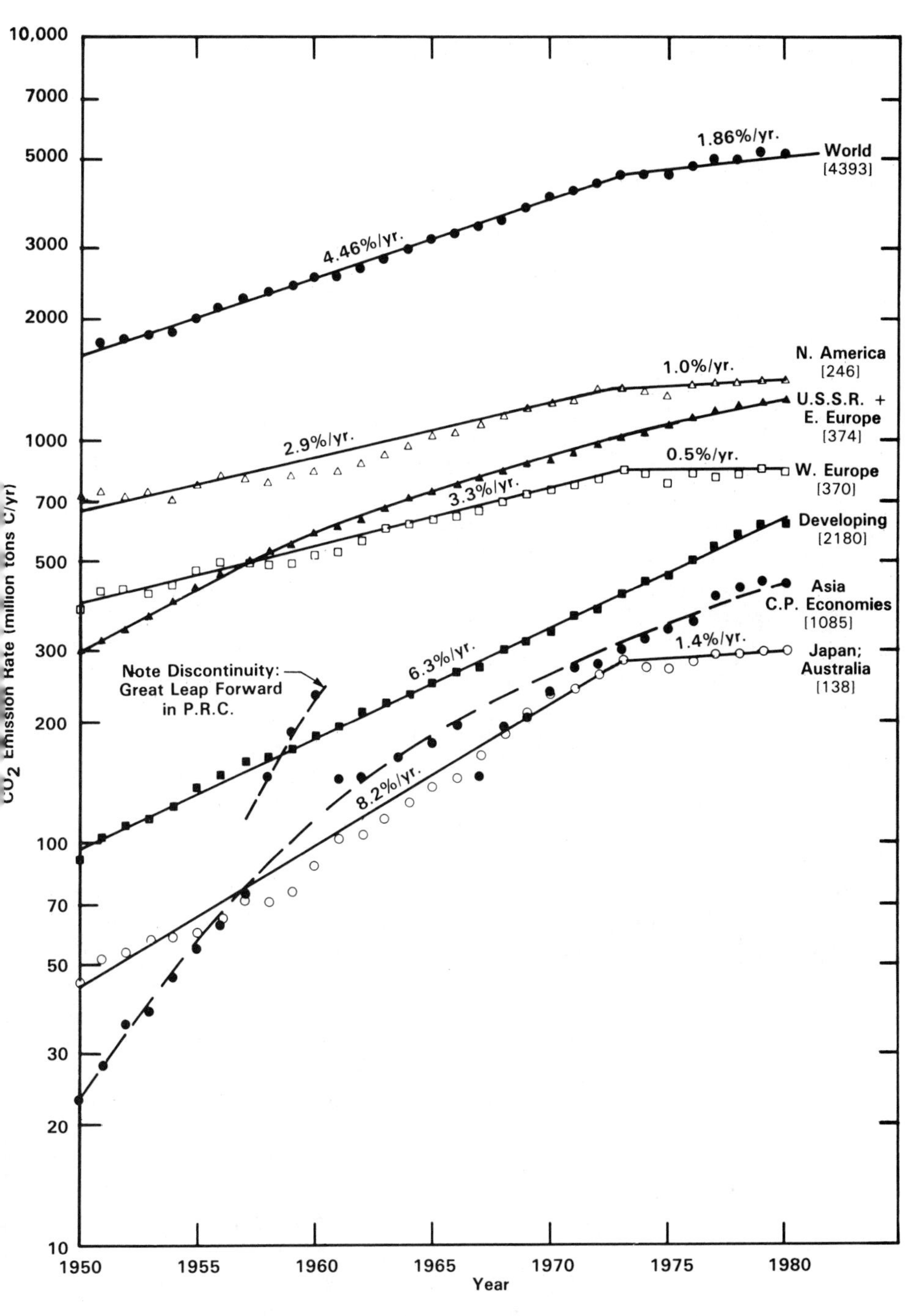

FIGURE 23.4. Global fossil fuel CO_2 emissions by world segments.

stantly changing.) The numbers in brackets under each regional label indicate the approximate 1980 population for the region.

Regions with centrally planned economies (both in Europe and in Asia) show a different pattern. There are no sustained periods of constant exponential growth and no abrupt change in the growth rate attributable to the changes in oil prices. Rather, this semilog plot shows smooth curves with continually diminishing but still positive slope, indicating growth each year but at a steadily reducing rate. In the USSR and Eastern Europe, there are no abrupt changes in growth rate (as observed in the data for the industrialized Western countries circa 1973), and the same tendency appears present in the data for the planned economies of Asia. Data from the Asian region are dominated by those from the Peoples Republic of China, and one must concede that the fuel data for China are erratic. Data for the period 1958 to 1960 are associated with the "Great Leap Forward," and those for 1967 to 1970 are associated with the "Cultural Revolution."

The industrially developing world shows continued, steady, exponential growth at 6.3% per year with no indication of a discontinuity at the time of changing oil prices—even though this group is most dependent on oil as its main energy supply. During the 1970s and into the early 1980s, this world segment showed the largest growth rate (by far) in CO_2 emissions. These contrasting growth rates for different parts of the world must be considered in projecting emissions into both the short- and long-term future.

The data for CO_2 emissions can also be expressed in terms of carbon emissions per capita. The most industrialized nations use the greatest amounts of energy and, because most of the energy comes from fossil fuels, produce the largest amounts of CO_2. Many of the less industrialized nations have very large populations, and therefore even with very large fossil fuel consumption and associated CO_2 releases, the emissions per capita are quite low (e.g., China). Table 23.5 summarizes the carbon emissions per person for the various regions of the world. These numbers offer a rough correlation to the standard of living the average person has in each part of the world and reveal those areas where increases in emissions are most likely. Table 23.5 also suggests the importance of fossil fuels in improving living standards. Until a suitable substitute is widely employed on a global scale, energy derived from fossil fuels will be required as a basic ingredient in efforts to improve living standards among the poorer areas.

The concept that world carbon emissions average just over 1 ton per year for each person is worth special note. Assuming that the world population doubles in the next 50 years or so, as seems likely, and that alternative (nonfossil) energy sources do not replace most fossil fuels, and that no relative improvement in living standards is possible without an increase in CO_2 emissions per person above the present 1.15 tons carbon per year, the global CO_2 emissions should exceed 10 billion tons carbon per year.

The unusually high per capita CO_2 emissions from the German Democratic Republic (East Germany) deserve more complete analysis. At this time it seems that a large amount of traditional heavy industry coupled with a population that has been severely reduced by emigration during the past two or three decades is largely responsible for the greater per capita emissions. In Western Europe, the Federal Republic of Germany (West Germany) has the highest emissions rate per capita, probably reflecting the emphasis on certain types of industry. However, the problems

TABLE 23.5. Carbon emissions per person (1980).

	Average tons carbon per person per year		
	World	By region	By country
World	1.15		
North America		5.6	
United States			5.7
Canada			4.7
USSR and Eastern Europe		3.4	
German Democratic Republic			4.9
USSR			3.4
Poland			3.4
Other Eastern Europe			2.9
Western Europe		2.3	
Federal Republic of Germany			3.3
United Kingdom			2.9
France			2.5
Italy			1.8
Other Western Europe			1.8
Japan, Australia, etc.		2.2	
Oceania (including Australia)			2.7
Japan			2.1
Asia, centrally planned economies		0.41	
Vietnam, North Korea, Mongolia			0.56
Peoples Republic of China			0.40
Developing world		0.29	
Mideast		0.65	
Latin America		0.60	
Venezuela			1.5
Argentina			1.0
Mexico			1.0
Chile			0.5
Brazil			0.4
Bolivia			0.2
Africa		0.28	
South and Southeast Asia		0.16	

for the future that are most evident from Table 23.5 are the very low carbon emissions from the Asian countries with centrally planned economies, and from most parts of the developing world. Even small improvements in per capita energy use will mean very large increases in CO_2 emissions because of the huge populations in these regions—75% of the world total in 1980.

References

Carter, C. O. 1979. What world crude analyses imply. Hydrocarbon Processing. 103–108.

Marland, G., and R. M. Rotty. 1983. Carbon dioxide emissions from fossil fuels: a procedure for estimation and results for 1950–1981. TR-003. Carbon Dioxide Research Division, Office of Energy Research, U.S. Department of Energy, DOE/NBB-0036.

United Nations. 1982. 1980 Yearbook of World Energy Statistics. Department of International Economic and Social Affairs, Statistical Office.

United Nations. 1983. 1981 Yearbook of World Energy Statistics. Department of International Economic and Social Affairs, Statistical Office.

United Nations Environment Programme. 1979. The environmental impacts of production and use of energy, Part I, Fossil Fuels. ERS-1-79, Nairobi.

24 Distribution and Quantitative Assessment of World Crude-Oil Reserves and Resources*

CHARLES D. MASTERS, DAVID H. ROOT, AND WILLIAM D. DIETZMAN

The enumeration of world crude-oil reserves and resources is a difficult undertaking. One must rely on data gathered by others and, through various spot-check and cross-check mechanisms, distinguish the reasonable from the unreasonable. Estimating techniques vary; standards of measurement vary; classification of the reported elements differ; and, in some cases, even the recognition of what is being reported is obfuscated by language and nomenclatural ambiguities. Therefore, to improve understanding of world petroleum resources, it is prudent to describe the methodologies and measurement standards explicitly, to disaggregate the data as much as possible to gain insight into the source of the differences, and to standardize the classification and nomenclature scheme to establish unambiguous communication.

The purpose of this chapter is to give an estimate of world crude-oil resources, their distribution, and their API gravity (American Petroleum Institute). The resource classification diagram of the U.S. Geological Survey (USGS) and the U.S. Bureau of Mines (USGS 1980) allows for the reporting of both reserves and resources with the degree of geologic certainty of the data expressed on the horizontal axis and the degree of economic potential expressed on the vertical axis. Utilizing this classification format, Fig. 24.1 shows the estimate for world resources of conventional crude oil. By conventional oil, we mean the oil >10° API gravity (excluding natural gas liquids) which is broadly considered to be economically recoverable by conventional primary, water-flood, and pressure-maintenance techniques; the extra-heavy oil (<10° API and >10,000 centipoise) of the Orinoco Petroleum Belt of Venezuela and the bitumen of the Athabasca tar sands of Canada, for example, are not included with conventional crude

---

*Paper originally presented at the Sixth ORNL Life Sciences Symposium in October 1983 and has since also been published in proceedings of 11th World Petroleum Congress using metric tons as the unit of oil measure. Paper was also published as U.S. Geol. Survey Open File Report 83–728 using barrels as the unit of oil measure and with an expanded Table 1 to show additional numerical detail by counting.

Area: World Units: Billions barrels

Cumulative production 445	Identified resources		Undiscovered resources		
	Demonstrated (Measured, Indicated)	Inferred	Probability range 95%	Mode	5%
Economic	Reserves 723	Inferred reserves	321	550	1417
Marginally economic	Marginal reserves each 1% world average recovery increase equals 34	NOT ASSESSED Inferred marginal reserves			
Sub-economic	Demonstrated subeconomic resources 2267	Inferred subeconomic resources	623	1068	2750
Other occurrences	Extra-heavy oil and bitumen				

FIGURE 24.1. Classification of world ultimate crude oil resources. Subeconomic quantity based on world average conventional recovery of 34%.

oil but are listed separately. The resources of crude oil are classified as being identified or undiscovered.

The left side of Fig. 24.1 shows identified (or "discovered") resources. Economically recoverable identified resources are called reserves and are classified in three subdivisions: Measured reserves (or "proved," of the U.S. Energy Information Administration, 1982) are defined as the expected recovery, by conventional techniques, considering the full extent of each recognized reservoir; indicated reserves [or "indicated additional," as defined by American Petroleum Institute (API 1980)] are defined as the additional amount of oil that could be recovered given installation of conventional water-flood facilities; and inferred reserves are defined as the amount of oil derived from the expected growth in the reported size of fields, given the addition of new pools or reservoir extensions not envisioned in the measured reserve calculation. The combination of the measured and indicated reserve is a demonstrated reserve. Industry custom prior to 1970 was to report measured (proved) reserves in the restricted sense of only what had actually received sufficient development drilling to permit reliable flow testing. Since then, however, it has become customary to report an estimate of the maximum expectable conventional recovery from the recognized reservoirs, whether or not they have been completely developed by drilling and whether or not they have been completely fitted for optimum water-flood recovery. Our definitions and estimates are consistent with current custom and are reported as demonstrated reserves. The remainder of the in-place crude oil may be defined as subeconomic resources, or, if some portion of the remainder is judged to be economic under a specified set of conditions, it can be reported as marginal reserves. In effect, when we gain the capability to assess oil deriving from enhanced oil recovery (EOR) (i.e., recovery techniques other than those herein defined as conventional), the quantities likely will be identified initially as marginal reserves; even though there is limited production from EOR, its low level of 81 million barrels of oil per year in the United States and 18 million barrels per year in the remainder of the world (Meyer et al., 1984) suggests that the activity is still in the pilot-plant stage; consequently, any estimates of potential reserves are premature.

The right side of the classification diagram (Fig. 24.1) shows subdivisions of the undiscovered resources based on the same principles as those considered in classifying identified resources. The estimate of the economically recoverable undiscovered resources is based broadly on existing price-cost relationships, and their quantity may be expressed graphically through a full range of probabilities; commonly, we report two points on the range encompassing a 90% probability of occurrence, as well as the mode or most likely occurrence.

Other occurrences of oil, which are defined as including both extra-heavy oil and tar sand (bitumen), are referred to at the bottom of Fig. 24.1. These resources are set apart from conventional oil at least in part because of their vastly different, technically achievable reserve-to-annual production ratios (R/P). The estimate for "other occurrences" of oil is discussed in a later section.

Methodology for Assessment of Crude-Oil Reserves and Their API Gravities

Reserves were estimated by field and aggregated by country and by region (Table 24.1). Prime sources of data were the well and field files of Petroconsultants S.A., scientific and trade literature, and government publications. Approximately 80% of world reserves (field by field) have been specifically examined by W. D. Dietzman and his colleagues with the EIA; they have estimated for each field the quantities of demonstrated oil in place and the original reserves. (For an example of one of several publications, see Dietzman et al. 1981.) The weighted-average recovery factor from these studies is 34%; this recovery factor includes only the conventional (primary plus water-flood and pressure-maintenance) oil recovery.

Approximately 20% of the world's known oil is in countries for which adequate field data were not available. In those countries, reserves are estimated from available production data and from published national estimates. It is possible that the published estimates of national reserves correspond to a definition of reserves more strict than that used here. We

TABLE 24.1. World estimate of original recoverable resources of conventional crude oil (in billions of barrels).

Area	Production*	Reserves*		Undiscovered recoverable resources† Probability range			Ultimate recoverable resources (mode)
	Cumulative	Demonstrated	Original	95%	Mode	5%	
North America	142.2	62.7	205.6	104	163	322	369
U.S.A.	124.0	29.8	153.8	64	80	105	
Canada	10.1	6.4	16.5	19	26	48	
Mexico	8.1	26.5	34.6	26	50	170	
Other			0.7	1	2	8	
Cuba							
Guatemala							
Greenland				1	2	8	
Percent original reserves by average	10–20°	20–25°	25–35°	>35°			
API gravity	11	16	37	36			
South American	47.2	34.2	81.5	20	33	69	115
Venezuela	36.1	25.5	61.6	12	17	38	
Other	11.2	8.7	19.9	10	14	28	
Argentina	3.5	2.5	5.9				
Bolivia	0.2	<0.1	0.3				
Brazil	1.2	1.6	2.8				
Chile	0.3	0.1	0.4	8	12	26	
Columbia	2.2	0.8	2.9				
Ecuador	0.7	2.3	3.0				
Peru	1.2	0.9	2.1				
Trinidad	2.0	0.5	2.5	1	2	4	

Percent original reserves by average API gravity	10–20°	20–25°	25–35°	>35°			
	8	15	63	14			
Europe							
(less U.S.S.R)	11.2	26.5	37.7	13	20	49	58
Western	5.7	24.9	30.6	12	17	40	
U.K.	1.9	14.0	15.9	9	15	34	
Norway	0.8	8.8	9.6				
Other (including Mediterranean)	3.0	2.1	5.2	1	2	10	
Austria	0.6	0.1	0.8				
Denmark	<0.1	0.3	0.3				
Ireland		0.1	0.1				
France	0.4	<0.1	0.5				
W. Germany	1.3	0.4	1.7				
Greece		0.2	0.2				
Italy	0.3	0.4	0.6				
Netherlands	0.4	0.3	0.7				
Spain	0.1	0.1	0.2				
Eastern	5.5	1.6	7.1	1	2	4	
Romania	3.9	0.9	4.8				
Other	1.6	0.7	2.3				
Albania	0.2	0.2	0.4				
Bulgaria	<0.1	<0.1	0.1				
Czechoslovakia	<0.1	<0.1	0.1				
E. Germany	<0.1	<0.1	0.1				
Hungary	0.4	0.2	0.6				
Poland	0.4	<0.1	0.4				
Yugoslavia	0.5	0.3	0.7				
Percent original reserves by average API gravity	10–20°	20–25°	25–35°	>35°			
	2	7	18	73			

TABLE 24.1. *(Continued)*

Area	Production*	Reserves*		Undiscovered recoverable resources† Probability range			Ultimate recoverable resources (mode)
	Cumulative	Demonstrated	Original	95%	Mode	5%	
U.S.S.R.	67.8	69.8	137.6	59	107	343	245
Percent original reserves by average API gravity	10–20°	20–25°	25–35°	>35°			
Africa	32.1	52.9	85.0	28	46	105	131
Libya	12.8	24.3	37.1	4	7	25	
Algeria	6.4	11.7	18.0	3	5	17	
Egypt	2.4	4.0	6.3	1	2	12	
Tunisia	0.4	0.7	1.1	1	2	9	
Nigeria	8.4	7.9	16.3	2	6	23	
Other	1.7	4.3	6.1	10	21	45	
Angola	0.6	0.7	1.3				
Cameroon	<0.1	0.4	0.4				
Gabon	0.9	0.6	1.5	1	3	11	
Ghana		<0.1	<0.1				
Ivory Coast		0.8	0.9				
Congo	0.1	0.4	0.5				
West Sahara				0	0		
Morocco	0.04	0.02	0.1	0.1	0.2	2	
Benin		<0.1	<0.1				
Chad		1.0	1.0	0.8	3	8	
Sudan		2.0	0.4	1.6	6	14	
Zaire	<0.1	0.1	0.1				
Niger				0	0.1	1	

Mali				0.0	0.1	0.5	
Mauritania				0.0	0.1	0.5	
Ethiopia				0.1	0.5	2	
Somalia				0.1	2.5	6	
Percent original reserves by average	10–20°	20–25°	25–35°	>35°			
API gravity	2	3	29	66			
Middle East	123.6	441.7	565.3	72	125	337	690
Saudi Arabia	40.8	170.5	211.3	23	40	109	
Kuwait	20.3	88.6	108.9	1	2	7	
Neutral Zone	3.3	12.6	15.9	1	2	4	
Iran	30.0	63.8	93.8	11	19	51	
Iraq	15.8	50.8	66.6	32	56	150	
Abu Dhabi }							
Dubai }	7.1	46.5	53.6	3	5	13	
Other	6.3	8.9	15.2	<1	1	4	
Bahrain	0.7	0.3	1.0	0	0		
Oman	1.5	3.5	5.0	<1	1	4	
Qatar	3.2	3.6	6.8	0	0		
Syria	0.6	1.2	1.8				
Israel	<0.1	<0.1	<0.1				
Turkey	0.4	0.2	0.6				
Percent original reserves by average	10–20°	20–25°	25–35°	>35°			
API gravity	5	3	68	24			
Asia/Oceania	21.0	34.6	55.6	33	58	176	114
China	6.1	16.3	22.4	14	34	90	
Indonesia	9.4	10.5	19.9	5	9	35	
Other	5.6	7.7	13.3	12	21	34	
Australia }				4	6	11	
New Zealand }	1.6	2.1	3.7	0.05	0.15	0.5	

TABLE 24.1. *(Continued)*

Area	Production*	Reserves*		Undiscovered recoverable resources† Probability range			Ultimate recoverable resources (mode)
	Cumulative	Demonstrated	Original	95%	Mode	5%	
Malaysia / Brunei	2.2	2.4	4.6	3.1	5.5	15.4	
Thailand		<0.1	<0.1	trace	1		
Vietnam				1	3	8	
Philippines		0.2	0.2				
Papua N.G.							
Afghanistan							
Pakistan	0.1	0.5	0.6				
India	1.0	2.4	3.4				
Bangladesh				3	5	9	
Burma	0.4	0.1	0.6				
Japan	0.2	<0.1	0.3				
Percent original reserves by average	10–20°	20–25°	25–35°	>35°			
API gravity	2	12	45	41			
Antarctica	0	0	0	0	0	19	0
World Total	445.1	723	1168	321	550		1718
Percent original reserves by average	10–20°	20–25°	25–35°	>35°			
API gravity	5	6	57	32			

[a]From Masters, C. D., D. H. Root, and W. D. Dietzman.
*Reserves and production as of January 1, 1981.
†Resources as of March 1983.

have not compensated for this by adding an extra quantity of oil to these countries, but remain mindful of its possible existence.

Table 24.1 also shows the distribution of crude oil by API gravity. Data are derived from Meyer (1984), from Petroconsultants S.A., and from personal communication with many interested persons. The aggregate data base is almost complete for most countries, and at the regional level, the data are entirely credible. As shown in Fig. 24.2 and Table 24.1, 89% of conventional world oil is lighter than 25° API gravity, 57% lies between 25° and 35° API, and 32% is lighter than 35° API. Actually, we have no basis for the specific subdivisions shown; a technical rationale is needed for classification, for which the data apparently are available.

Methodology for Assessment of Undiscovered Resources of Crude Oil

Undiscovered economically recoverable resources of crude oil (Table 24.1) were assessed for this study by utilizing the techniques employed by the USGS for domestic petroleum assessment (Dolton et al. 1981). In general, the intent is to estimate economically recoverable resources, but in the case of very deep water and of Arctic/Antarctic ice conditions, we have assumed eventual economic recovery for large-field occurrence, even though technical capability may not yet be demonstrated.

The assessment technique requires study of a given area, paying particular attention to the geologic factors controlling the occurrence, quality, and quantity of the petroleum resource. Standardization of critical elements of the investigations is achieved by the preparation of data forms for each basin that call for specific volumetric, areal, and rock-quality measure-

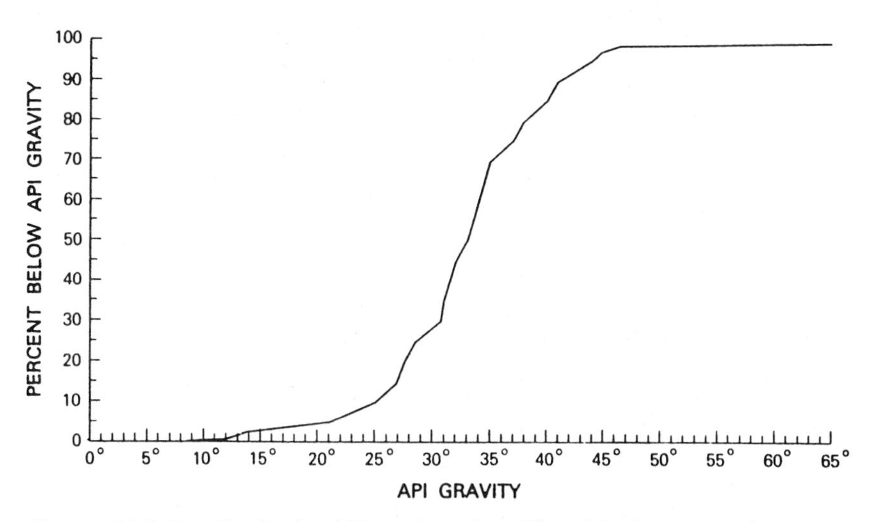

FIGURE 24.2. Distribution by API gravity of world's original reserves of crude oil. (API gravity = 141.5 specific gravity − 131.5.)

ments, as well as the determination of basin analogues for comparison. In addition, finding-rate histories and projections are constructed whenever possible. The absence in some areas of reliable dry-hole data is a problem that unquestionably weakens the assessment process and may result in a high-side bias to the assessment. From these data and analyses, a set of resource numbers is calculated by making volumetric comparisons with different analogue basins, by projecting discovery and production curves, and/or by play analysis.

The assessment process itself is subjective; the results of the geologic investigation and of the resource calculations are presented to a team of USGS assessment specialists (in this case, a team of 10 to 12 scientists, each having an average of about 20 years petroleum studies experience), each of whom makes his personal estimates conditional on economically recoverable resources being present. Initial assessments are made for each of the assessed provinces as follows:

1. A low-resource estimate corresponding to a 95% probability of more than that amount
2. A high-resource estimate corresponding to a 5% probability of more than that amount
3. A modal (most likely) estimate of the quantity of the resource associated with the greatest likelihood of occurrence.

Individual estimates are then posted and averaged, and the results are debated from the perspectives of the personal experiences of the individual assessors. A second and third iteration of the procedure may follow, depending on consensus. If no commercial oil has yet been discovered in the basin, a marginal probability is subjectively assessed to record the probability of any commercial oil being present. The results of the final estimates are averaged, and the numbers are computer-processed using probabilistic methodology (Crovelli 1983) to show graphically the range of the economically recoverable resource values.

Approximately 90% of the world's undiscovered resources of crude oil were assessed in this way over a period of about 3 years utilizing ~18 person-years. (For an example of one of several publications, see Masters et al. 1982.) Many small basins, or those considered after preliminary examination to have limited petroleum potential, were assessed on the basis of exploration or production projections or by cursory volumetric and analogue determinations. In our judgment, no area having significant crude-oil potential remains unassessed at this stage of data development and availability.

Analysis of Estimates of Reserves of Conventional Crude Oil and Their Discovery Rates

Because reserves by definition are economic and because economics varies with the perspective of the operator, unavoidable inconsistencies exist in reserves reporting. Some world reserves estimates, however, appear to be clearly too low and others arguably too high. That the series of numbers reported, for example, in the API Basic Petroleum Data Book (API 1981) is likely too low is evidenced by the following analysis: (1) The amounts of oil discovered over the past 10 years, as reported for this study (Fig. 24.3) or by Exxon (1980), are considerably less than the amounts of oil produced over the same period. We show the differential to be about 80 BBO over the 10-year period, whereas Exxon shows it to be about 60 BBO. Whichever data set proves to be most nearly correct, there would seem to have been a deficit for the 10-year period of about 70 BBO. (2) During this same period, the reserves reported in the API Basic Petroleum Data Book did not decline but rather increased by about 41 BBO. (3) Reserves, therefore, must have been larger than stated, or they grew, through additions from the inferred reserves, by at least 111 BBO. The conundrum is how much longer we will be able to reevaluate upwards because of improved recovery practices or because of new pools and unpredicted field extensions (inferred reserves) enlarging the demonstrated reserves. On a smaller scale, in the United States during the last 10 years,

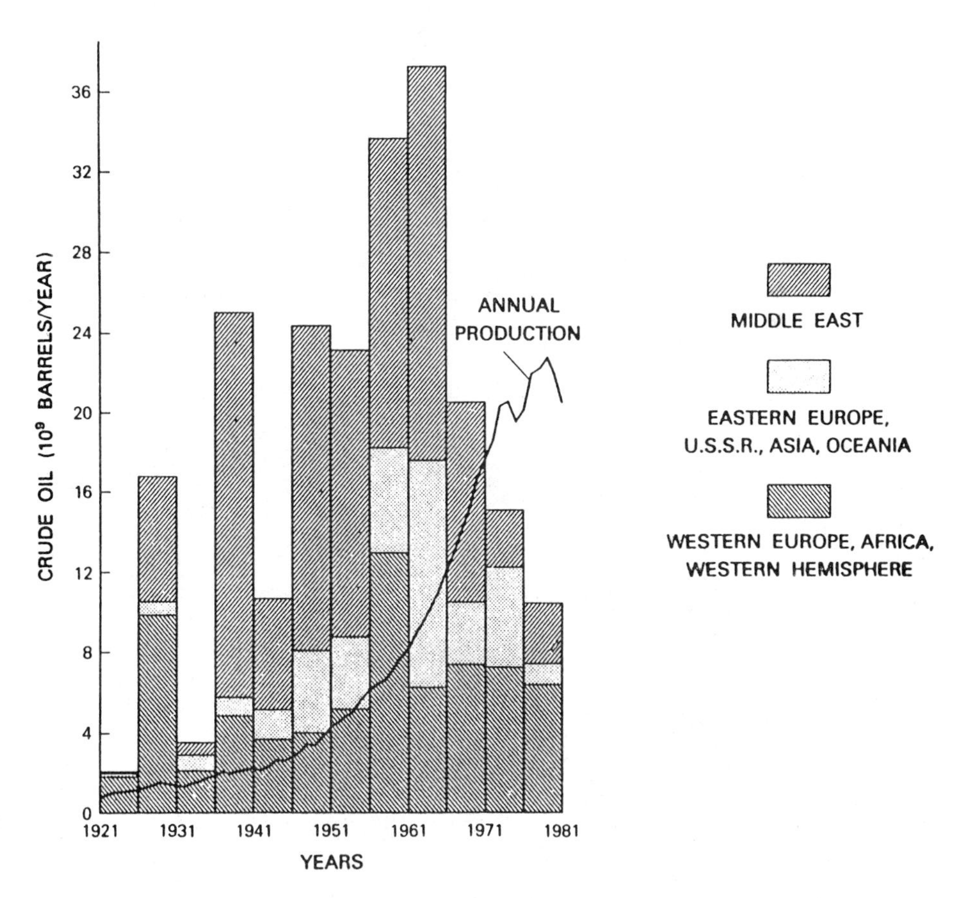

FIGURE 24.3. World crude-oil discovery rate averaged over 5-year periods.

production exceeded discoveries by more than 22 BBO, yet reported reserves declined by only about 12 BBO. A possible explanation for this lies in the USGS estimate of 23 BBO of inferred reserves (Dolton et al. 1981). The United States clearly is the most extensively explored and developed region in the world, yet the data show the need for revisions of the reserves reporting, year by year, on the order of 1 BBO, or one-third of our annual production. It is important for us to learn where that oil is coming from because some day that cushion is going to be exhausted; it is equally important, for the same reasons, that we establish, on the world scale, a known relationship between discovery, production, and reserves.

To the contrary, some reserve numbers appear to us, definitionally, to be too high, because the authors assumed 40% average conventional recovery (Halbouty and Moody 1979). When the higher assumed recovery factors are considered, however, the estimates of these particular authors are approximately consistent overall with those in Table 24.1.

Four years ago, Halbouty and Moody (1979) reported a modest decline in reserves over the previous decade; we concur with their findings and with the continuation of that trend (Fig. 24.4). We also concur with their

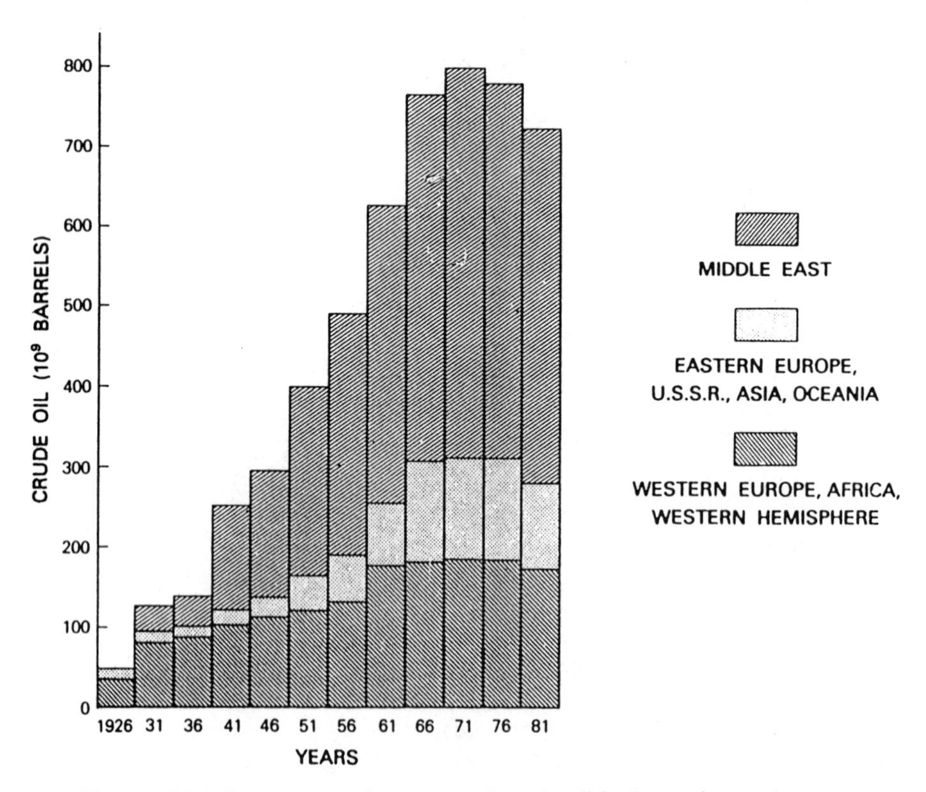

FIGURE 24.4. Demonstrated reserves of crude oil in 5-year intervals.

findings and those of Exxon (1980) of a declining discovery rate now extending over two decades. Our data, however, show an even sharper decline (Fig. 24.3); we are conscious of the possibility that additions from inferred reserves might enlarge the reserves estimates of the discoveries of the past 5 or 10 years but would suggest that, for reasons described above, those additions may be modest.

The elements of accurate reserves reporting appear to be within our grasp. In a few countries, understanding of definitions and availability of public data for independent analysis are still lacking, but overall, we are confident that these estimates are proportionally and geographically satisfactory. Specifically unaccounted for is an unassessed quantity of indicated reserves, as described above, along with some inferred reserves in the form of new pools and unpredicted extensions of known reservoirs; we believe the quantities unaccounted for can be no more than a few percent of the total and likely could be assessed statistically, given greater certainty in production and discovery data.

Analysis of the Undiscovered Resources of Conventionally Recoverable Crude Oil

Our assessment of the world potential for undiscovered resources of conventionally recoverable crude oil indicates a 90% probability that the value lies between 321 and 1417 BBO, with a most likely (modal) value being 550 BBO (Table 24.1). Our assessment (95 to 5% probability range) falls inside the range (90 to 10% probability range) assessed by Halbouty and Moody (1979) for the Tenth World Petroleum Congress of 180 to 2415 BBO, and our modal value is significantly less than their expected value of 987 BBO. Whereas comparison of the estimates by region (Fig. 24.5 of this paper and Table 4 of Halbouty and Moody 1979) discloses remarkable similarities in the modal assessments of the United States and of the Middle East, the other world subdivisions show considerable differences. Because the areas corresponding to the assessments are not the same, however, it is not possible to be precise about where all the differences lie. Nehring (1982) has recently published a disaggregated reserve and resource analysis with which we find we are in substantial agreement, both with respect to total numbers and to regional distribution of crude oil. Because of reporting differences, however, a dialogue is nonetheless required to ascertain the conceptual differences between oil additions deriving from discovery as opposed to oil additions deriving from increased recovery.

A further difference between the assessment provided to the Tenth World Petroleum Congress and this assessment lies in the high-side values which differ by a factor of almost 2. Our view at this point is that world data are fully adequate to support a relatively narrower range of resource estimates. To achieve the higher of the two assessments, a discovery of

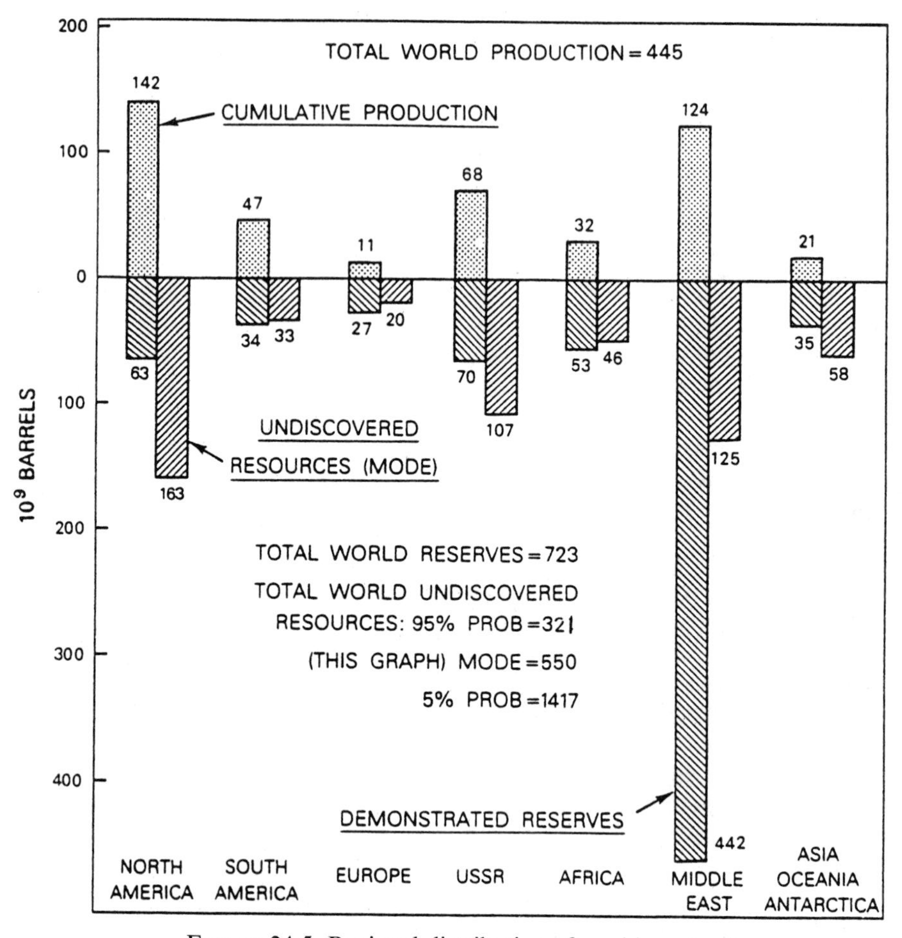

FIGURE 24.5. Regional distribution of world crude oil.

the dimensions of another Middle East province clearly must be considered, and our studies indicate no possibility for such an occurrence.

The low-side assessments, on the other hand, remain essentially identical. They, of course, reflect the potential of relatively well known provinces, and the assessment process has, in this case, the advantage of more complete information.

For our modal assessment of 550 BBO of undiscovered resources to be sustained, a massive amount of exploration is still required. We have not completed the studies of trends of finding-rate efficiencies, but our general perception is that numbers of exploratory wells are increasing yet discoveries are declining; hence exploration efficiency is also declining. Needed to reverse that decline is the discovery of new major provinces (ultimate resources in excess of 20 BBO or 1 year of production for the world) or significant discovery increases in the giant Middle East province or in other identified petroleum-favorable regions of the world. Our analysis suggests the likelihood that most of the new oil will come from the established provinces, but the unknown of the Arctic, deep water, and other frontier areas must always remain a consideration.

In attempting to address the issue of contributions of frontier basins to new oil discovery, we tried to analyze the often repeated notion that there is a certain significant number of still untested basins. We have not yet completed the work to permit a comprehensive statement, but we can note that most of these basins are untested for good geologic or economic reasons. In particular, we examined the Arctic region for clues to the prediction of undiscovered major provinces. The region is so big that it is awkward to downgrade almost any part for lack of favorable oil characteristics. However, we do not think that the evidence is encouraging, for example, for any significant extension under the Barents Sea of the highly favorable Devonian source rock of the Volga Urals basin or the mature Jurassic source rock of the North Sea basin; in all other areas of the Arctic, likewise, there does not appear to be that combination of geologic factors necessary for truly large oil occurrence. On the basis of area and volume alone, a significant amount of oil must be assessed for the Arctic, but we see little possibility there for the discovery of a major province. In addition, the Arctic poses logistic and environmental problems that necessarily will deter commercial development.

Little is known about oil reserves in deep-water areas. Although an attempt was made to extend geologic understanding of coastal regions out to a water depth of 2500 m, we must acknowledge inconsistency in this effort; certainly, data became sparse. However, in most areas, we also judge that the merit of the petroleum geology declines as well.

Our still limited understanding of China reflects, in part, only the time constraints for preparation of this paper. There is much published literature to be digested and vast, minimally drilled areas to be tested. Nonetheless, it has been possible to develop a broad, general understanding of China's petroleum geology sufficient to render a credible assessment (Table 24.1). Published regional analyses are excellent, and sufficient quantitative data are available to begin to get an idea of their reserves, the dimension of their basins, and various other specific petroleum geology parameters. The most critical data still missing are drilling results to permit exploration-maturity analysis.

Other Occurrences of Oil Resources—Extra-Heavy Crude Oil and Bitumen

Although large quantities of extra-heavy crude oil and bitumen are known in various parts of the world (Meyer et al., 1984), these commodities are not at present serious economic contenders in the market. We suggest, therefore, that the issue is not how much is in the ground but rather how much can be produced. Considering that production facilities have a finite service life, we define reserves as the amount of oil that can be produced over a 25-year period given installed production facilities; the otherwise recoverable resources are classified as marginal reserves; the remainder of the in-place resources, considering approximately a 10% recovery factor, is shown as subeconomic resources (Fig. 24.6). Although considerable quantities of bitumen are available in different parts of the world, only in Canada, from the Athabasca tar sands, is a major extraction effort under way—now producing approximately 60 million bbl per year. The overwhelmingly dominant resource of extra-heavy oil is in the Orinoco heavy-oil belt of Venezuela; on a pilot-plant basis, some 3 million bbl are being produced. Elsewhere in the world, annual production of 3 million bbl is also being achieved in Italy and 1 million bbl per year is being produced in both the United States and Mexico, for a total extra-heavy-oil production of 8 million bbl per year. Extra-heavy oil and bitumen cannot contend economically with conventional crude oil, but their presence in large quantities in various parts of the world assures mankind of the essentially limitless availability of naturally occurring oil.

Conclusions

There is an immense quantity of conventional crude oil in the world; our capacity for consumption, however, is perfectly capable of challenging that immensity. Although there is much yet to be discovered and produced, there is no room for complacency. The slack is coming out of reserve estimates, and to the extent that discoveries remain less than production, world demonstrated reserves, as defined herein, will continue to decline. Even if enhanced oil recovery and the production of extra-heavy oil and bitumen become a ubiquitous reality, they are unlikely to achieve rates

Area: World Units: Billions Barrels

Cumulative production	Extra-heavy oil	0.227
	Bitumen	0.427
Reserves	Extra-heavy oil	0.2
	Bitumen	1.5
Marginal reserves	Extra-heavy oil	59
	Bitumen	172
Subeconomic resources	Extra-heavy oil	500
	Bitumen	1,500

FIGURE 24.6. Original measured resources of extra-heavy oil and bitumen.

of production even close to matching those of primary and secondary recovery in the past; the resulting patterns of consumption of crude oil in the decades ahead, then, will be quite different.

The distribution of crude oil remains restricted, and there is little chance that condition will change (Fig. 24.5). The Middle East region dominates known world petroleum occurrence, and we judge it still to be the richest hunting ground. Even though most of the world's oil will continue to be supplied from traditional areas, exciting discoveries of great local significance will be made in many other areas of the world. However, one should not anticipate riches from so-called untested areas just because exploration activity has not been obvious.

A further concern about estimates of world petroleum availability is the present limited discovery rate relative to production. Without evaluating in detail the geology of all exploration activity, the general level of exploration, as measured by numbers of wells, has increased over time, yet annual discoveries are declining, suggesting the possible reality of low-side assessments. If the low side be reality, the need for alternate energy sources becomes increasingly critical for most of the world's countries.

References

American Petroleum Institute. American Gas Association, and Canadian Petroleum Association, Vol. 34. Washington, D.C.

American Petroleum Institute. 1980. Reserves of crude oil, natural gas liquids, and natural gas in U.S. and Canada as of December 31, 1979.

American Petroleum Institute. 1981. Basic petroleum data book. Vol. 1, No. 3, Sect. 11, Table 1. American Petroleum Institute, Washington, D.C.

Crovelli, R. A. 1983. Procedures for petroleum resource assessment used by the U.S. Geological Survey—statistical and probabilistic methodology. U.S. Geol. Sur. Open-File Rep. 83–402.

Dietzman, W. D. et al. 1981. Middle East—crude oil potential from known deposits. DOE/EIA-0298. U.S. Government Printing Office, Washington, D.C.

Dolton, G. L. et al. 1981. Estimates of undiscovered recoverable conventional resources of oil and gas in the United States. U.S. Geol. Sur. Circ. 860, 87 pp.

Exxon Corporation. 1980. World energy outlook. Public Affairs Department, New York.

Halbouty, M. T. and J. D. Moody. 1979. 10th World Petroleum Congress. Vol. 2, pp. 291–301, London.

Masters, C. D. et al. 1982. Assessment of undiscovered recoverable resources of the Arabian-Iranian basin. U.S. Geol. Sur. Circ. 881, pp. 1–12.

Meyer, R., J. Wynn, and J. Olsen (eds.). 1984. The Future of Heavy Crude and Tar Sands, 1388 p. McGraw-Hill, New York.

Nehring, R. 1982. Annual Review of Energy, Vol. 7. pp. 175–200. Annual Reviews Inc. Palo Alto, Ca.

U.S. Energy Information Administration. 1982. Annual report 1981. DOE/EIA-0216(81). U.S. Government Printing Office.

U.S. Geological Survey. 1980. Principles of a resource/reserve classification for minerals. U.S. Bureau of Mines and U.S. Geol. Sur. Circ. 831.

25
Long-Term Energy Projections and Novel Energy Systems

HANS-HOLGER ROGNER

Energy demand and supply projections are bound inextricably with external realities. Consequently, they reflect the dominant views on the development of those factors that primarily determine energy consumption. Some of these factors are highly uncertain or so difficult to quantify that it is impossible to model their effects. Other factors, such as economic activity or population development, can serve as variables that, to a large extent, determine energy consumption. Hence, parallel to the prospect of global economic activity slowly shifting from exponential growth (until the 1970s considered the norm) to a prolonged period of low growth, there have been similar downward revisions in the corresponding energy consumption. But, there are other reasons for the recent downward revisions of energy projections. In contrast to energy forecasters of the 1970s, today's analysts can base their projections on empirical evidence about the response of industries and households to two unprecedented oil price increases of the last decade. Adjustments to the realities of high energy costs have resulted in energy efficiency improvements along the entire energy chain from resource extraction to consumption, as well as interfuel substitution, structural economic changes, and changes in individual attitudes.

During the 1970s, reactions to the changed energy supply environment focused on identifying strategies for speedy transition from the current energy system, based mainly on depletable fossil fuels, to one that would be sustainable. Quick technical "fixes" were sought to supply enough energy to satisfy the anticipated continuation of the exponential growth trend. The major increase in energy prices resulted not only in an accelerated search for additional fossil resources and the development of alternative energy technologies, but it also enlarged the global base of economically recoverable reserves. Further, the downward revisions of energy demand projections had the effect of increasing the lifetime of known resources beyond the time horizon originally calculated on the basis of "business as usual."

Nevertheless, many facets of global energy development explored during the 1970s, such as the findings of the IIASA study reported in *Energy in a Finite World* (Häfele et al. 1981), have remained unchanged. One of the study's main concerns was understanding the path from today's energy systems a system of sustainable energy sources. Another part of the study explored the question of whether fossil fuels would run out and when; during the 1970s it was generally expected that the exhaustion of fossil fuel resources would seriously threaten world energy supplies within the next 50 years. Analyses have shown, however, that neither of these two major issues of long-term energy development is likely to take place, at least not in the simple way envisaged. It proved impossible to verify that the supply of fossil fuels would indeed run out in the decades ahead. Also, it became evident that the transition to a sustainable energy system exploiting nonfossil energy technologies could not be completed until the end of the 21st century. The IIASA study concluded that in the interim there will be a transition of a different kind; the world will utilize fossil fuels at an increasing rate, albeit fossil fuels of an increasingly lower quality measured in terms of the hydrogen-to-carbon ratio.

The recent surplus of oil production capacity worldwide and the decrease in the world market price for oil point to a longer reliance on fossil fuels. One reason for this condition is that as oil prices declined, a number of major alternative resource development projects were hastily cancelled. Hence the present energy outlook suggests that the global-scale introduction of new and advanced technologies will be shifted further into the future. The cumulative effect of these developments has reinforced the tendency of decision makers to avoid becoming actively involved in tackling long-term energy issues. This attitude may well change when the inevitable impacts on the ecosphere associated with increased use of "dirty" fuels force themselves into consideration. The enhanced use of fossil fuels with lower hydrogen-to-carbon ratios in conventional technologies implies a corresponding higher production of carbon-based wastes, in the form of CO_2 releases, not to mention other problem emissions of fossil fuel combustion, such as SO_2 and NO. The atmospheric concentration of CO_2 could reach levels that might have serious adverse impacts on the world's climatic conditions. This chapter does not purport to elaborate on the likely development of atmospheric CO_2 concentration or the consequences thereof, but rather to review recent long-term global energy projections and to identify the relative share of fossil fuels in total energy supply. It presents energy projections and perspectives that conform to the realities of the early 1980s, a period characterized by the second oil price hike, a sluggish world economy, and a depressed oil market. It is therefore important to bear in mind the most up-to-date global energy projections, paying special attention to the future contribution of fossil fuels to global energy supply. This is done in the sections that follow.

The International Energy Workshop

The International Energy Workshop (IEW) is a joint venture begun in 1981 between Stanford University and IIASA. Its aim is to compare recently published energy projections and to explore the reasons for divergent views on future developments and thus on current energy policies. The workshop process entails iterative polling of projections of economic growth, crude oil prices, primary energy production and consumption, and energy trade, along with subsequent meetings to assess the implications. The second IEW was held at IIASA in June 1983. Energy projections for all the regions of the world were examined. Seventy groups and individuals contributed to the 1983 IEW poll, generally providing a reference scenario and, in some instances, alternative growth scenarios. The period covered was 1980 to 2010. The IIASA contribution was the IIASA '83 scenario, which considered global energy development for the period 1980 to 2030.

When tackling the CO_2 question, time horizons longer than 20 to 30 years are required. Thus, the energy projections provided by IEW participants serve as a point of departure, indicating the probable trend in fossil fuel consumption over the next 30 years. The IIASA '83 scenario extends the analysis to the year 2030; projections for the later period up to the year 2100 are based on an extrapolation of this scenario.

Let us first briefly consider the IEW results. Figure 25.1 depicts the

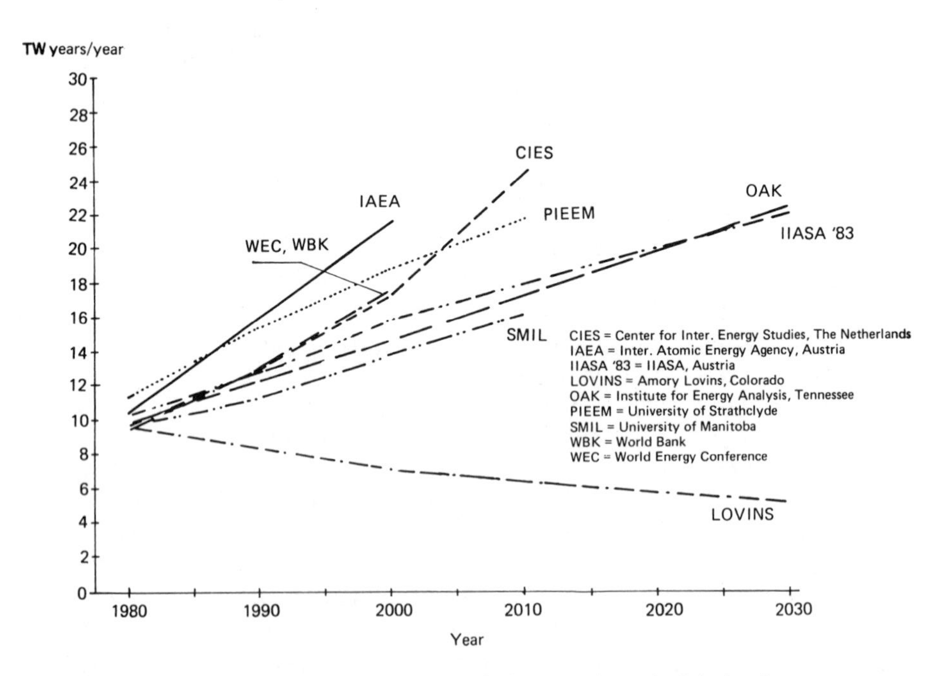

FIGURE 25.1. International Energy Workshop: projected global primary energy consumption. (After Manne and Schrattenholzer 1983.)

general range of projected global primary energy consumption as presented at the IEW in 1983. The differences in primary energy consumption for the reference year, 1980, to a certain extent, reflect problems of definition. This was particularly the case for noncommercial energy. Most of the projections in Fig. 25.1 suggest a continuous growth in global energy consumption. Although the differences in the energy consumption trajectories cannot be attributed to a single factor, there is a significant correlation between primary energy consumption and the assumptions concerning one of the principal factors determining future energy use, namely gross domestic product (GDP). In other words, the specific economic outlook of various individuals has markedly influenced some of the projections for the development of energy consumption. Table 25.1 compares the real GDP and primary energy consumption projections for the period 1980 to 2000 of selected contributors to the IEW. The three highest GDP projections correspond to the three highest energy consumption trajectories.

Even more revealing than the growth indices are the corresponding income elasticities, also given in Table 25.1. Virtually all projections indicate income elasticities of significantly <1, which points to a shift away from the causal relationship previously assumed to hold between economic growth and energy consumption. The low elasticity (0.58) resulting from the World Bank projection is particularly interesting, since this scenario foresees the highest economic activity of any contribution to the IEW. The higher economic growth rate causes an acceleration in the replacement of capital assets on both the producers' and consumers' sides. This gives producers and consumers greater flexibility and speed in adapting or replacing their energy-consuming equipment in response to energy price trends. All other factors being equal, that is, in the absence of any other incentives to conserve energy, the autonomous technical progress embodied in new plant and equipment results in a more energy-efficient economy. In the low-growth scenarios such improvements are not found,

TABLE 25.1. International Energy Workshop: Global gross domestic product (GDP), primary energy consumption for the year 2000 (1980 = 100), and income elasticity of energy demand.

Contributors*	Gross domestic product	Primary energy consumption	Primary energy-GDP elasticity
WBK	248	170	0.58
CIES	234	178	0.68
WEC	220	180	0.75
OAK	188	151	0.65
IIASA '83	178	152	0.73
SMIL	160	138	0.69
LOVINS	183	78	−0.41

*Acronyms are defined in the legend to Fig. 25.1.
After: Manne and Schrattenholzer (1983).

and the main factor behind reduced energy consumption is simply the quantitative reduction in economic activity.

In contrast to all the other scenarios, the extremely low-energy intensities underlying the scenario of Amory Lovins result in a sharp divergence between economic activity and future energy consumption trends, as shown by the negative income elasticity value in Table 25.1.

The range of projected global primary energy consumption varies considerably by the year 2000 (see Fig. 25.1). The highest and lowest projections differ by a factor of almost 3. Not surprisingly, both extreme projections stem from parties that represent two extremes—the International Atomic Energy Agency (IAEA) and Amory Lovins. Whereas the IAEA envisages a more than twofold increase in global energy use between 1980 and 2000, Lovins advocates a reduction of >20%. The remaining projections more or less cluster around a median increase in primary energy consumption of 50% between 1980 and 2000.

Another observation can be derived from the energy consumption projections presented at the 1983 IEW. There is a tendency that the longer the time horizon, the lower the resulting energy consumption projections. There may be several explanations for this. First of all, long-term scenarios require a more imaginative view of the future than a simple "business as usual" extrapolation. Such views necessarily encompass measures concerning energy conservation, interfuel substitution, and technical progress in general. Further, the analysis of long-term and global scenarios highlights the increasing rate at which inexpensive fossil fuels are being depleted and, at the same time, increases awareness of the unavoidable environmental burden of steadily growing energy consumption and the switch to "dirtier" fuels.

This leads to the question: How does the long-term and global energy supply picture look, particularly regarding the contribution of fossil fuels? According to the IEW, the period up to the turn of the century will be characterized by relatively stable oil consumption with little or no increase in international oil trade. The major substitutes for oil are coal and nuclear energy and, to a lesser extent, natural gas. Taking the median of the various projections,* the majority of the IEW poll participants projected significant increases in coal consumption, resulting in a level close to that of oil by the year 2000. Although other fuels and technologies are projected to grow more rapidly than coal (e.g., nuclear and solar, renewable forms of energy), their low initial contributions in 1980 will not allow them to be globally significant contributors to the world's energy supply by the year 2000.

Figure 25.2 depicts the projected developments of fossil fuel consumption worldwide according to the IEW. The majority of the projections

*Note that this is a somewhat doubtful procedure, and any conclusions derived from such medians must be taken with a "pinch of salt," because the procedure neglects the principal differences in the exogenously determined assumptions.

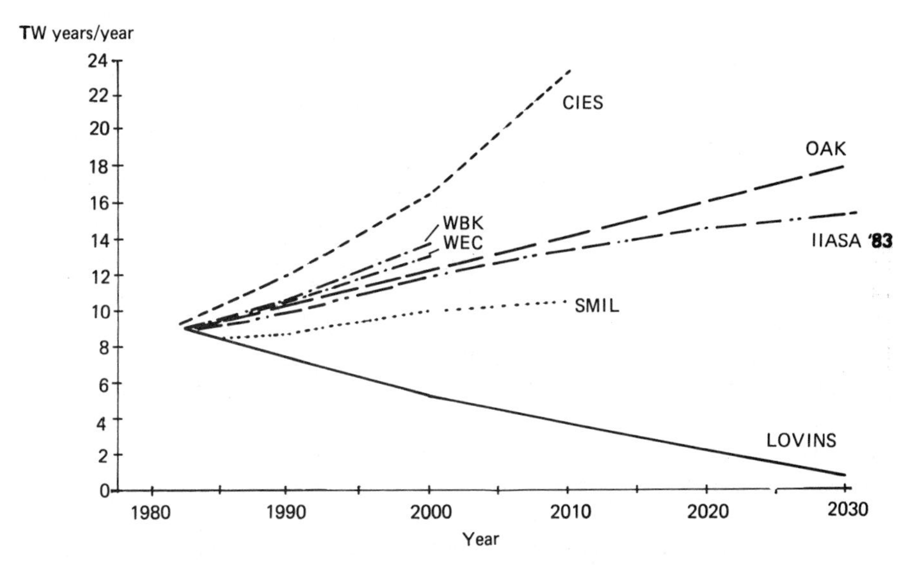

FIGURE 25.2. International Energy Workshop: projected global fossil fuel consumption. (After Manne and Schrattenholzer 1983.)

foresee a 40 to 60% increase in fossil fuel consumption between 1980 and the year 2000. Over the next two decades, the relative share of fossil fuels in total energy consumption will not be reduced drastically. Thus, the CO_2 emissions will—roughly speaking and for the moment neglecting fossil interfuel substitution and thus the differences in levels of CO_2 generation—increase in parallel with global primary energy consumption.

But as already mentioned, any questions related to CO_2 emissions require a look further into the future than just two decades. Thus the remainder of this chapter will concentrate on the IIASA '83 Global Scenario of Energy Development (Rogner 1983), which covers the period 1980 to 2030. The scenario development up to the year 2030 then serves as input to a market penetration analysis by which the trends of the period 1980 to 2030 will be extended up to the end of the 21st century. We will attempt to identify the contributions of oil, gas, coal, and nonfossil fuels to the global energy supply, and then assess the CO_2 output these contributions imply.

The IIASA '83 Scenario of Energy Development

Altogether there were a number of reasons for repeating (to a certain extent) the procedure of scenario writing which, at the end of the 1970s, led to the IIASA high and low scenarios reported in *Energy in a Finite World*. First of all, external realities have changed over the past decade. In particular, coal, which in both the high and low scenarios is the largest single primary energy carrier by the year 2030, has recently lost much of its popularity. For example, in the United States, an officially released paper has linked coal combustion with the acid rain problem (U.S. Federal Government 1983). Although, as with the CO_2 problem, some of the major physical questions are not totally understood (e.g., those relating to the chemistry, dispersion processes, and other interdependences), the problem has certainly become more vivid and more crucial.

Second, the role of natural gas required reevaluation. During the 1970s, gas was considered a form of energy mainly for local use, and intercontinental trade in natural gas was not really foreseen. Technological advances in many gas-related technologies, however, have opened new avenues. Deep gas drilling has been successfully extended to depths of 7000 m and more. This increases the natural gas potential by several orders of magnitude. But apart from the straightforward increase in the availability of natural gas, such as that from geopressure zones, "tight" formations, and Devonian shelves, the prospects for transporting gas over large distances have also improved significantly.

In addition to gas in gaseous form, we must now also take into account liquefied gas. Liquefied gas would not only fit into the existing long-distance transport infrastructure developed for oil—thus deriving similar economic advantages—but also would be more consistent with energy end-use requirements. Converting gas to meet liquid fuel demand would be one step in transforming gas into a global fuel. There are a number of possibilities: liquefied natural gas (LNG) is one route, while the chemical conversion of natural gas into methanol or other alcohols are other possible approaches. The LNG route implies the application of cryogenic engineering and the construction of special tankers and port facilities—technologies that are neither easy nor cheap—accompanied with considerable conversion losses; the methanol approach requires special conversion plants, whose costs would require proportionally large distances to be bridged in order to become economically attractive, given today's crude oil prices. But a planning horizon of 50 years calls for the inclusion of technologies not yet fully competitive under present market conditions.

Last but not least, advances in methodology now permit a more comprehensive analysis of the global energy system. The IIASA set of energy models, and in particular the energy supply model MESSAGE II, has been advanced, and now encompasses the entire energy chain from resource extraction to end-use conversion (Messner 1984). The energy demand analysis was restricted to the determination of energy service requirements such as passenger kilometers, square meters heated in private homes and commercial buildings, process heat requirements of various industrial production processes, etc. The selection of the appropriate technology and final energy source to satisfy these demands is now part of the overall optimization process of the MESSAGE II model.

Although most of the exogenous assumptions and the methodology involved differ from the earlier scenarios, the principal approach underlying *Energy in a Finite World* was maintained. In particular, the process of iteration between scenario design, model application, and scenario revision was retained. That is, the scenario assumptions—the most important of which will be presented later in this chapter—are not assumptions per se, but should be viewed as the results of this iteration.

The geographical disaggregation of the globe shown in Fig. 25.3 was also retained. Admittedly, the disaggregation may appear somewhat arbitrary, but it is the result of a necessary compromise. Analyzing each country in the world separately would have been an impossible undertaking. On the other hand, neglecting international differences in energy demand densities and energy resource endowments would have meant omitting the principal causes of interregional dependence, trade, and

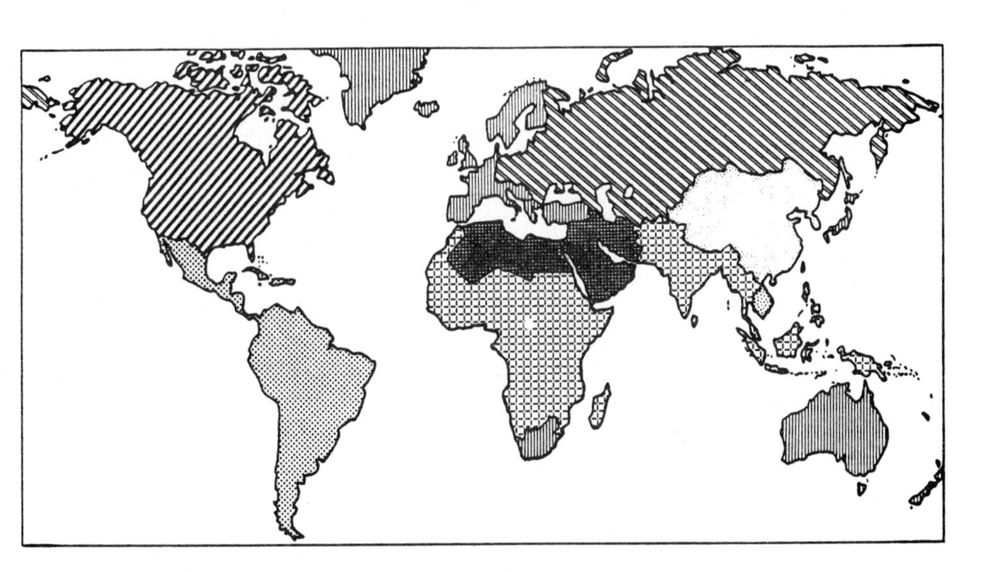

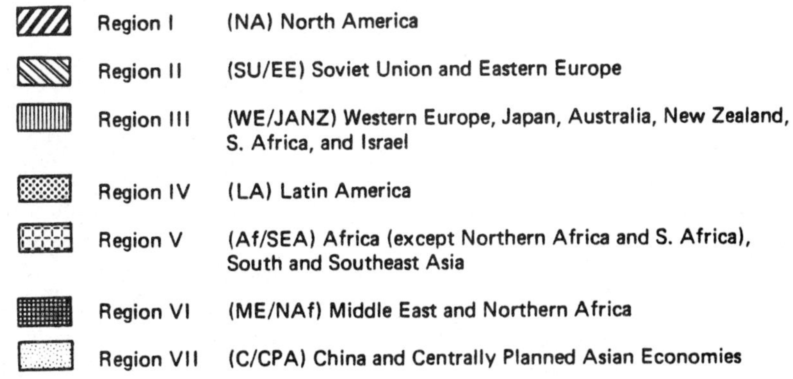

FIGURE 25.3. IIASA world regions. (Häfele et al. ENERGY IN A FINITE WORLD: A GLOBAL SYSTEMS ANALYSIS, Copyright 1981 by International Institute for Applied Systems Analysis. Reprinted with permission from Ballinger Publishing Company.)

competition. Thus, although the nations of the world were grouped into seven distinct regions, the composition of each region was not determined solely by geographical proximity but rather according to similarities in economic infrastructure, energy resource endowment, and demographic development. For example, Region I (North America) is industrialized, is endowed with sufficient energy resources, and has a relatively low rate of population growth. In marked contrast to Region I is Region V (Africa/ Southeast Asia), which is a developing region with few energy resources but high population growth rates.

Major Scenario Assumptions

The major aggregate scenario assumptions concern those variables that, to a large extent, determine future energy consumption, such as population growth, economic activity, and technical progress. The projected future development of regional population was taken from Keyfitz (1982) and is summarized in Table 25.2. The assumptions with respect to population development were the only ones not adjusted and revised by the iterative scenario-writing approach, since these projections are usually considered subject to the least uncertainty.

The economic activity projections are naturally more difficult to obtain. Here the IIASA '83 scenario began with the general philosophy behind the IIASA low scenario projections of *Energy in a Finite World*. Again, it was assumed that the interrelations between regional economies result in the interdependency of the economic well-being of the different world regions. Thus, the economic growth of the developing regions was considered to be linked to the economic performance of the industralized regions (Hicks et al. 1976). The per capita gross domestic product (GDP) growth rates per region are also shown in Table 25.2. The per capita income for all regions is increasing but at different rates. In particular, the developing Regions V and VII (China and other centrally planned Asian economies) achieve significantly higher per capita growth rates than the Organisation for Economic Cooperation and Development (OECD) countries of Regions I and III (Western Europe, Japan, Australia, Israel, and New Zealand). This development may be attributed to their strong economic performance over the last decade, as well as to the reduction in population growth rates, especially in Region VII, and the changes in their prospective energy resource base. The general economic development foreseen in this scenario follows a quasi-S-shaped path. During the 1980s only modest economic growth was assumed worldwide. The two decades prior to and immediately after the turn of the century are characterized by accelerated economic activity in all regions, whereas the last two decades of the study horizon are marked by declining absolute GDP growth rates. The per capita growth rates, however, distort the prospects of the developing countries relative to those of the developed regions. The ab-

TABLE 25.2. IIASA '83 scenario: population and gross domestic product (GDP) per capita (1980 prices and exchange rates)*.

Region†	1980		2000		2030	
	Population (in 10^6)	GDP/cap (in 10^3\$)	Population (in 10^6)	GDP/cap (in 10^3\$)	Population (in 10^6)	GDP/cap (in 10^3\$)
I (NA)	251.3	11.2	284	15.2	315	20.6
II (SU/EE)	379.4	4.4	437	7.8	482	13.8
III (WE/JANZ)	561.0	8.6	680	11.2	767	14.9
IV (LA)	357.0	1.9	570	2.5	797	4.3
V (Af/SEA)	1613.0	0.4	2518	0.7	3550	1.2
VI (ME/NAf)	148.2	2.1	247	2.9	353	4.4
VII (C/CPA)	983.0	0.5	1127	1.0	1185	2.0
World	4292.9	2.7	5863	3.45	7449	4.8

*Population projections based on data from N. Keyfitz (1982).
†Regions are defined in legend to Fig. 25.3.
After: Rogner (1983).

solute values certainly do not show any significant narrowing of the gap between the developing and industralized parts of the world.

The assumptions concerning technical progress are hidden in the numerous detailed technical parameters describing energy conversion, transportation, and distribution technologies. A detailed presentation of these parameters is, however, beyond the scope of this chapter. To indicate the order of magnitude of the technical progress assumed in this analysis, we examined the aggregate energy intensity of the various world regions, that is, the primary energy input per dollar of output produced (Fig. 25.4). Whereas the industralized regions are characterized by a continuous decline in energy requirements per unit of GDP, this is not the case for the developing regions where energy efficiency remains relatively stable at its 1980 value for two to three decades. This points to the fact that building up an industrialized infrastructure is more energy intensive (even when one assumes the installation of only the most modern plant and equipment) than the operation of one that already exists. By the end of the projection horizon the contribution of energy efficiency to GDP in the industrialized regions doubles, while the developing regions achieve an improvement of >25%. It is interesting to note that by the year 2030, the developing regions' energy intensity in kilowatt hours per dollar GDP (at 1980 prices and exchange rates) is approaching the 1980 value for the industrialized regions.

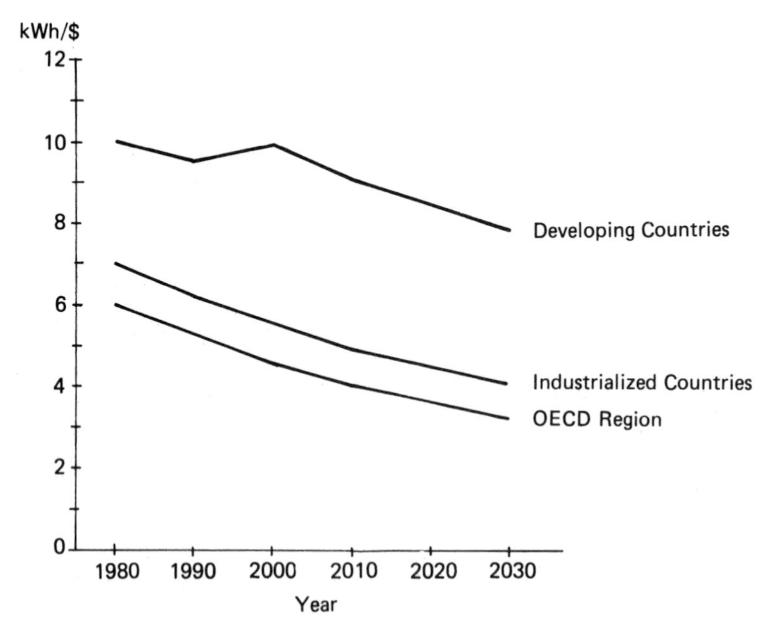

FIGURE 25.4. IIASA '83 scenario: energy intensity in kilowatt hours per U.S. dollars, gross domestic product (1980 prices and exchange rates), 1980–2030. (From Rogner 1983.)

Primary Energy Consumption

The primary energy consumption per region as calculated within the IIASA '83 scenario is given in absolute terms in Table 25.3 and Fig. 25.5, which seems to suggest that over the next 50 years the present fundamental differences in the magnitude of regional primary energy consumption will diminish. In particular, the share of the developing Regions IV, V, VI, and VII in total global energy consumption increases from 25% in 1980 to >47% by the year 2030.

As already mentioned, improvements in the energy efficiencies of the industrialized—and to a lesser extent the developing—regions enable the economies to perform according to the scenario assumptions while still showing only modest increases in primary energy consumption over the entire projection period. Total world product is assumed to grow by a factor of slightly more than 3, whereas total primary energy consumption roughly doubles from 10.0 TW years per year* in 1980 to 21.9 TW years per year by 2030.

The global primary energy picture changes drastically if one looks at the various regional per capita values for primary energy consumption

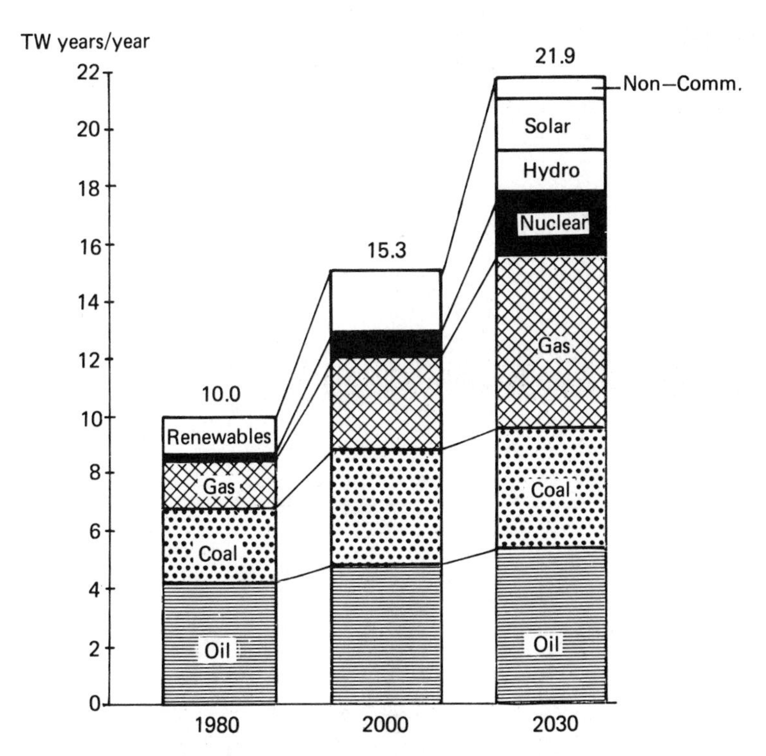

FIGURE 25.5. IIASA '83 scenario: global primary energy consumption by source, 1980–2030. (From Rogner 1983). One terawatt (TW) year = 10^9 kW year = 1.076 x 10^9 tons of coal equivalent (tce).

TABLE 25.3. IIASA '83 scenario: primary energy consumption (in TW years/year) and primary energy consumption per capita (in kW years/year/cap 1980–2030.

Region*	1980		2000		2030	
	PEC	PEC/cap	PEC	PEC/cap	PEC	PEC/cap
I (NA)	2.8	11.2	3.1	10.8	3.4	10.7
II (SU/EE)	2.2	5.8	3.5	8.0	4.6	9.5
III (WE/JANZ)	2.5	4.5	3.2	4.7	3.5	4.6
IV (LA)	0.6	1.6	1.2	2.2	2.7	3.3
V (Af/SEA)	0.8	0.5	1.7	0.7	3.6	1.0
VI (ME/NAf)	0.2	1.2	0.6	2.2	1.2	3.2
VII (C/CPA)	0.9	1.0	2.1	1.8	2.9	2.5
World	10.0	2.3	15.4	2.6	21.9	2.9

*Regions are defined in legend to Fig. 25.3.
After: Rogner (1983).

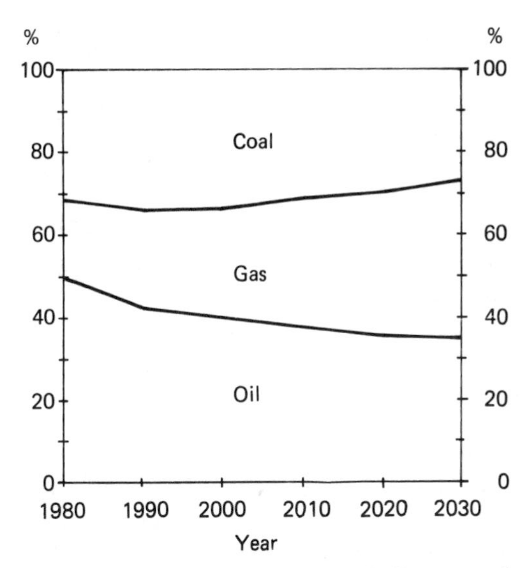

FIGURE 25.6. IIASA '83 scenario: contribution of oil, gas, and coal to fossil fuel supply, 1980–2030. (From Rogner 1983.)

shown in Table 25.3. Showing behavior similar to economic development, the per capita values continue to differ considerably between the industrialized and developing regions. But there are some interesting differences. For example, the total values for Region I (North America) and Region III remain at their 1980 levels, while the developing regions accomplish at least a doubling of their per capita energy consumption. The faster population growth in the developing regions simply offsets the rise in total regional primary energy consumption.

In summary, the overall global pattern can be characterized by noting that the doubling time for primary energy consumption has increased significantly compared to the period 1950 to 1975. During that period the growth of global energy use corresponded to a doubling time of 15 years; the IIASA '83 scenario implies a doubling period of little less than 50 years. This certainly has some major implications for future emissions of CO_2 and its accumulation in the atmosphere. However, the actual CO_2 releases are determined by the contributions of the various forms of primary energy source to global consumption. In general, one expects a decline in the relative contributions of fossil fuels to primary energy supply as new nonfossil fuel technologies penetrate the energy markets. Figure 25.6 illustrates the relative contributions of fossil fuels to overall energy consumption. In 1980 fossil fuels accounted for 81% of global energy use, but this figure declines to 71% by the end of the study period. The principal nonfossil technologies and energy carriers that substitute for fossil fuels are various forms of renewable energy (labeled "solar" in Fig. 25.5) and, to a lesser extent, nuclear power.

Two major features are revealed in Fig. 25.6. First of all, the composition of fossil fuels between 1980 and 2030 displays some recognizable dynamics. This is important if long-term CO_2 releases are a matter of concern. Of all the fossil fuels, natural gas is the least hazardous to the environment (not only in terms of CO_2 emissions), followed by oil products, coal, and oil shales. According to Fig. 25.6, liquid fuels (crude oil and derivatives thereof) and coal lose market shares to natural gas. In 1980 natural gas held a market share of 20% in fossil fuel consumption, and this expands to 38% by 2030. In fact, natural gas turns into the largest single fuel contributor to global primary energy consumption, with a market share of 27% by 2030.

The second aspect relates to the utilization of renewable forms of energy. The consumption of noncommercial energy is steadily supplanted by commercial renewable energy technologies. The utilization of noncommercial energy was assumed to increase within certain limits, namely, by roughly 30% above the 1980 level. This assumption is intended to reflect the serious environmental impacts, such as deforestation and soil erosion, of recent noncommercial energy consumption in the developing regions. Therefore, it became necessary to introduce commercial forms of renewable energy at an increasing rate to keep the exploitation of biomass at a level consistent with regeneration.

Given the relative contributions of primary energy sources to global energy consumption, one important question remains open: How fast can nonfossil resources be tapped and utilized on a scale that allows them not only to keep pace with the overall growth of energy demand, but at the same time permit substantial reductions in the use of fossil fuels? According to Table 25.3, the absolute contributions of all primary energy sources are increasing between 1980 and 2030. The fields of application of each type of fuel vary considerably. Coal, for example, penetrates primarily the electricity generation and heat markets, the latter via cogenerated district heat. The direct use of coal in the residential and commercial sectors is decreasing, while in some regions the industrial use of coal increases slightly. Coal gasification and liquefaction do not enter the picture in this scenario; this is definitely a consequence of the increasing introduction of natural gas.

Crude oil and natural gas liquids together supply the demand for liquid fuels. As natural gas extraction is increasing worldwide, there is a corresponding increase in the amount of associated liquids available; after the turn of the century these account for an increasing share in liquid fuel supply. Overall, the consumption of liquid fuels grows proportionately more slowly than total primary energy consumption. Interfuel substitution, primarily in the residential heating sector and in electricity generation, is the main explanatory factor here. In other words, liquids are increasingly restricted to fields of application such as transportation, where substitutability is difficult.

Relative to the IIASA low scenario of *Energy in a Finite World*, Table 25.3 displays some important differences. The role of nuclear power has changed significantly. While in the IIASA low scenario this technology contributed 23% or 5.17 TW years per year to the global primary energy supply (by 2030), this share declines to 10% or 2.19 TW years per year in the IIASA '83 scenario. Major factors responsible for this reduction include economic disincentives associated with safety and environmental concerns in many industrialized countries, and the downward revisions of expected electricity demand attributed to the sluggish performances of many economies, which has resulted in cancellations of a number of plants already ordered. The postponement of investment decisions related to advanced technologies shifts the gestation times even further into the future. For example, in Region I (North America), the 1980 position with respect to planned nuclear power plants, in conjunction with the lead times experienced in the construction of such plants, results in a practically constant nuclear power capacity throughout the 1980s and 1990s. Only after the year 2000, when the depletion of inexpensive coal and natural gas necessitates the extraction of more costly resources, does nuclear power gain momentum and expand rapidly along the upper boundary of the market penetration constraint.

The reductions in the contributions of nuclear power worldwide are only partly offset by increases for renewable (solar and noncommercial) energy technologies. As hydropower is already operating at its upper economically feasible limit in all the world regions, fossil fuels have to fill the gap caused by the setbacks experienced by nuclear energy. Compared to the IIASA low scenario, in IIASA '83, the share of fossil fuels increases from 67 to 71%, while its composition (as already mentioned) differs considerably from that of the earlier scenario.

Extension of IIASA '83 to the Year 2100

The IIASA '83 scenario projections for fossil fuels development differ significantly from most projections: the generally accepted annual growth rate of fossil fuel consumption underlying many studies of the buildup of CO_2 in the atmosphere was 4.3%(Ausubel and Nordhaus 1983), whereas in the IIASA '83 scenario for the period 1980 to 2030 it is 1.3%. However, this does not mean IIASA is assuming that the CO_2 problem will disappear automatically. Although reduced growth rates for fossil fuel consumption may shift the eventual doubling of the preindustrial atmospheric concentration further into the future,* they do not solve the underlying problem of CO_2 emissions. To arrive at a clearer picture of how the IIASA '83

*A doubling of the atmospheric CO_2 concentration may cause global mean temperature changes on the order of $3 \pm 1.5°C$ (Smagorinsky 1983).

scenario could be translated into CO_2 emission paths, it was necessary to extend the time frame of the scenario up to the year 2100. The market penetration model developed by Marchetti and Nakicenovic (1979) was applied, based on the dynamics of the primary energy development calculated in the IIASA '83 scenario for the period 1980 to 2030. The basic assumption underlying this approach is that every technology competing in a market passes through three distinct phases: growth, saturation, and decline. The dynamics of each phase are assumed to follow logistic substitution curves. Thus, the substitutions between the various fossil and nonfossil fuels identified in the IIASA '83 scenario proceed to completion in these three distinct phases.

Figure 25.7 shows the contributions of the various primary energy sources as fractions of the total market up to the year 2100, as well as the development in absolute values. Coal and oil are steadily forced out of the market and eventually supply about 10% each of total energy consumption by the end of the time horizon. Natural gas slightly increases its market share until the year 2040, when it reaches the saturation level. According to this analysis the maximum share of natural gas in total primary energy supply will not exceed the 30% ceiling. The decades after 2040 are characterized by a gradual decline of natural gas to little more than 10% by the year 2100. The "young" technologies at the beginning of the 21st century—nuclear and others (solar, renewables)—enjoy a phase of rapid growth and expand their market shares throughout the projection period. The continuous increase in their contributions is due to the fact that no new competitors are introduced after the year 2030. The market penetration model does not predict the introduction of new technologies. Only the introduction of new competitors would eventually limit the growth and initiate the saturation phase of nuclear and other energy sources. But this is of no immediate concern here, since we are primarily interested in the future development of fossil fuels. By the year 2100 the total fossil fuel contribution to global primary energy consumption amounts to 30% or 13.2 TW-years per year.

The calculation of CO_2 emissions associated with the extended IIASA '83 scenario required the application of appropriate carbon coefficients as well as the determination of total primary energy consumption up to the year 2100. The carbon coefficients actually applied, that is, liquids 19.2, solids 23.8, and gases 13.7 Tg of carbon per exajoule of consumption, respectively, were adapted from Rotty and Marland (1980). Global primary energy demand was assumed to grow at 0.9% annually, reaching 44 TW by the year 2100—a simple trend extrapolation of the period 2020 to 2030 from the IIASA '83 scenario.

Figure 25.8 depicts the global CO_2 emissions up to the year 2100. During the period 1980 to 2040 carbon releases appear to grow at an average rate of 1% per year. By the year 2050, carbon output reaches its maximum with an annual emission of 10 gigatons of carbon, compared to 5.4 gigatons

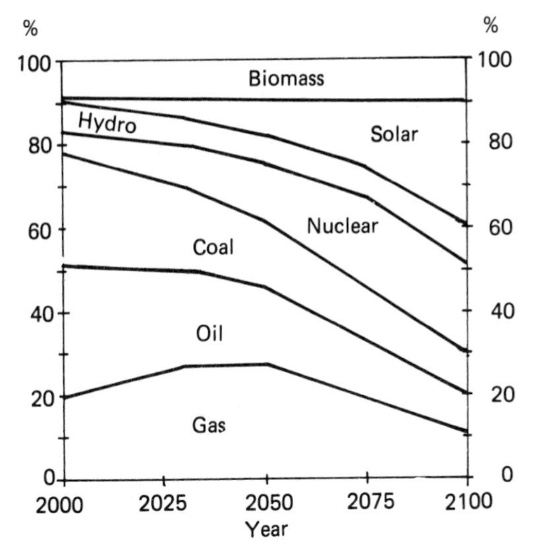

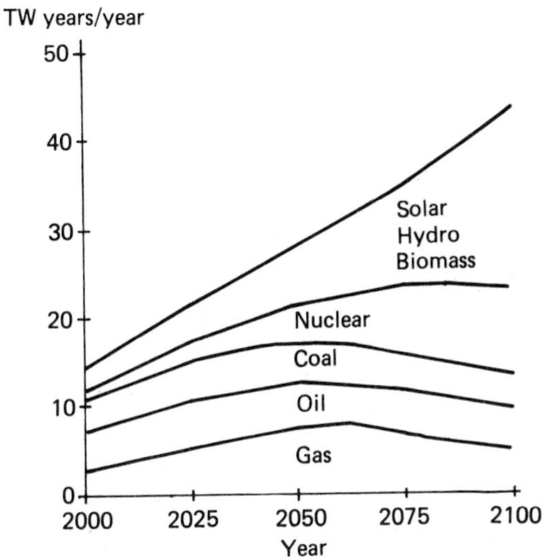

FIGURE 25.7. IIASA '83 scenario: global primary energy consumption, 2000–2100. (Based on Rogner 1983 and Marchetti and Nakicenovic 1979.)

in 1980. After the middle of the 21st century the continuous reduction in fossil fuel consumption results in declining carbon releases (i.e., 7.8 gigatons by 2100).

The translation of carbon emissions into atmospheric CO_2 concentrations and subsequently into possible consequences such as global average temperature changes is still quite a complex undertaking, and it has not been done in the context of the IIASA '83 scenario. However, a quick cross reference to the earlier IIASA low scenario provided some helpful insights.

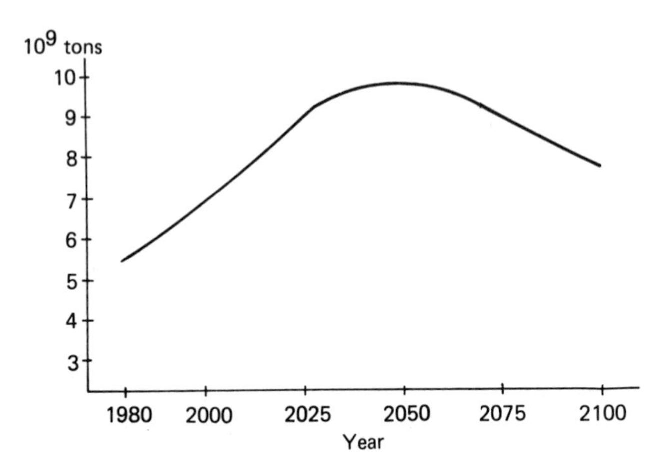

FIGURE 25.8. IIASA '83 scenario: global CO_2 emissions, 1980–2100. (Based on Rogner 1983 and Marchetti and Nakicenovic 1979.)

For example, carbon emissions for the year 2030 are almost identical in the IIASA low and in the IIASA '83 scenarios. The higher *total* fossil fuel consumption in IIASA '83 is more than offset by the differences in the contributions of oil, gas, and coal. Thus, conclusions with respect to the accumulation of CO_2 in the atmosphere and the accompanying global temperature changes may be judiciously drawn from *Energy in a Finite World*. This report estimated that the doubling of carbon emissions between 1980 and 2030 would push the atmospheric CO_2 concentration up to 430 ppm (compared to 335 ppm in 1978), causing a temperature change of 0.8°C.

A Novel Energy System

Present knowledge about the consequences of increased atmospheric CO_2 concentrations is inconclusive, in particular when one attempts to arrive at quantitative statements. Broadly speaking, the evidence suggests that some increase in the global average temperature should be expected, and that this may cause other climatic changes, such as shifts in precipitation patterns. However, CO_2-related problems are not the only adverse environmental impacts associated with fossil fuel consumption. Concern is also growing about the effects of sulfur- and nitrogen-based emissions. But the damage apparently caused to date by acid rain seems to be on a different and shorter time frame. Whereas the emission of CO_2 appears to be a problem with a longer time horizon (and with impacts that may be difficult to reverse), the effects of SO_2 are being felt right now. It is not the purpose of this chapter to elaborate in any detail on the status of research concerning the adverse effects of fossil fuel consumption, but rather to explore the implications of such adverse environmental impacts on the development of energy systems.

As mentioned above, a major conclusion of *Energy in a Finite World* was the recognition that the world will not be constrained by the simple exhaustion of energy resources. There are plentiful fossil resources, albeit of increasingly lower quality. The enhanced use of even dirtier fuels than those in use today would add to the amount of hazardous emissions. Therefore, there is a pressing need for energy production and conversion systems characterized by quasi-zero emissions. The concept of energy systems with zero emissions has been advanced in Häfele, Barnert, and Sassin (1982), Sassin (1982), Rogner (1982), Barnet (1983), Häfele and Nakicenovic (1984), and Häfele et al. (1984). Simply stated, zero emission systems involve shifting the processes of purification, primary conversion, waste handling, and quality and pollution control from the consumer end to the producer of secondary energy; consequently, the consumer is supplied with a high quality and clean energy. Examples of such systems in use are electricity, district heat, and town gas. These systems have two common features: they are grid-dependent and subject to significant economies of scale. However, these examples do not sound too convincing as prototypes for future zero-emission systems, since the conversion stations for these forms of secondary energy are at present the main sources of CO_2, SO_2, and NO emissions. Figure 25.9 shows the present energy system with its vertically integrated subsectors and the locations of the major points of pollution emissions. The current system depicted can be summarized as one that externalizes the environmental costs of consuming fossil fuels.

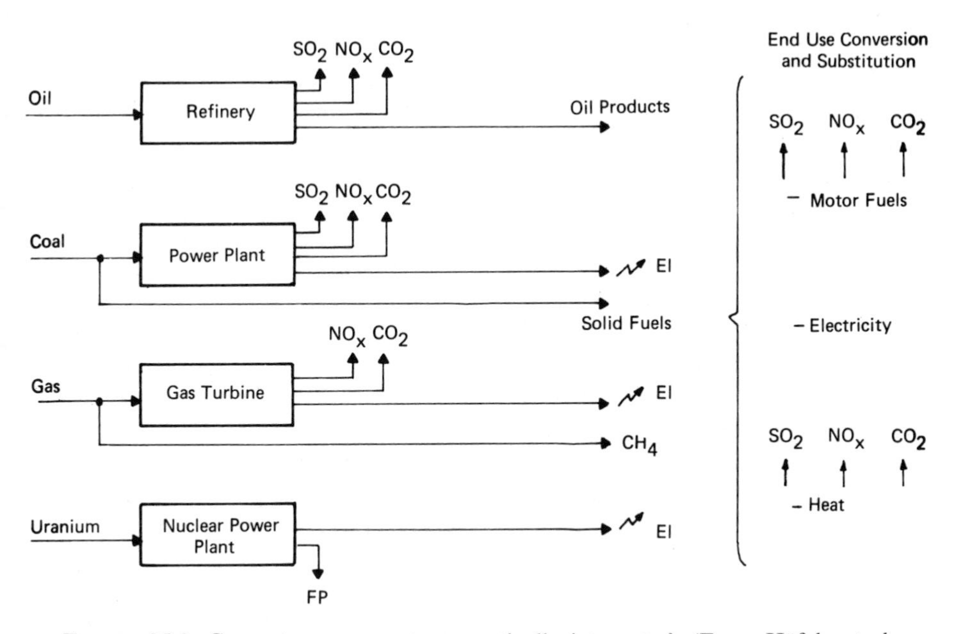

FIGURE 25.9. Current energy system, vertically integrated. (From Häfele et al. 1984.)

The principal idea behind the novel, zero-emission energy system proposed here and illustrated in Figure 25.10 is to avoid closing the fossil fuel cycle through the atmosphere. Unlike the vertically structured sectors of the existing energy system, the new system would consist of a horizontally integrated system consisting of four distinct stages. Previously independent sectors such as the coal, oil, or gas industries are merged into a interconnected system.

The principal idea of this proposed integrated energy system (IES) is to decompose and purify all fossil fuels at the front end of the energy chain. The net result of this process would be a variety of clean chemical products. These products then would be stoichiometrically supplemented for conversion or synthesis into energy forms needed to meet final energy requirements.

Specifically, the system starts with a number of energy and nonenergy inputs. According to Fig. 25.10 the system's inputs include coal (solids or oil residues), natural gas, fissionable material, water, and air. Outputs of this decomposition and cleansing stage are the clean intermediate energy carriers carbon monoxide (CO), hydrogen (H_2), and oxygen (O_2). The technological processes deployed at this stage would comprise partial oxidation of solids using pure oxygen from air separation plants to form synthesis gas (a mixture of CO and H_2). High-temperature combustion would help avoid the generation of complex hydrocarbon compounds, sulfur emissions, etc. The molten iron bath process is one suitable technology for this purpose.

Rather than deploy the inefficient operation of combusting part of the CO to split water (the net effect of all gasification, liquefaction, and cracking schemes), strategic choice in the IES would be different. Natural gas or methane would supply the stoichiometrically missing amounts of hydrogen for the synthesis of liquid fuels by means of methane reforming. Another possibility for the provision of additional hydrogen (and oxygen) foreseen in this configuration is off-peak electrolysis.

Chemical reforming of natural gas with steam yields synthesis gas plus one additional molecule of hydrogen. Steam reforming is an established technology that operates with clean substances and is thermodynamically attractive: it provides an inlet for high-temperature heat in the range of 700 to 900°C, which can split the water molecule at the expense of decomposing methane. Conceivably, rather than apply natural gas process heat for the shift reaction (which is both resource consumptive and environmentally polluting) high-temperature reactors or solar tower plants can be the source of process heat. In this way, methane reforming can serve as an interface to the nonfossil part of integrated energy systems.

The second state of such systems consists of the stoichiometric complementation/supplementation of the decomposed and purified intermediate energy products. This step controls the mass flows of CO, H_2, and O_2, and directs the flows to the different conversion and synthesis pro-

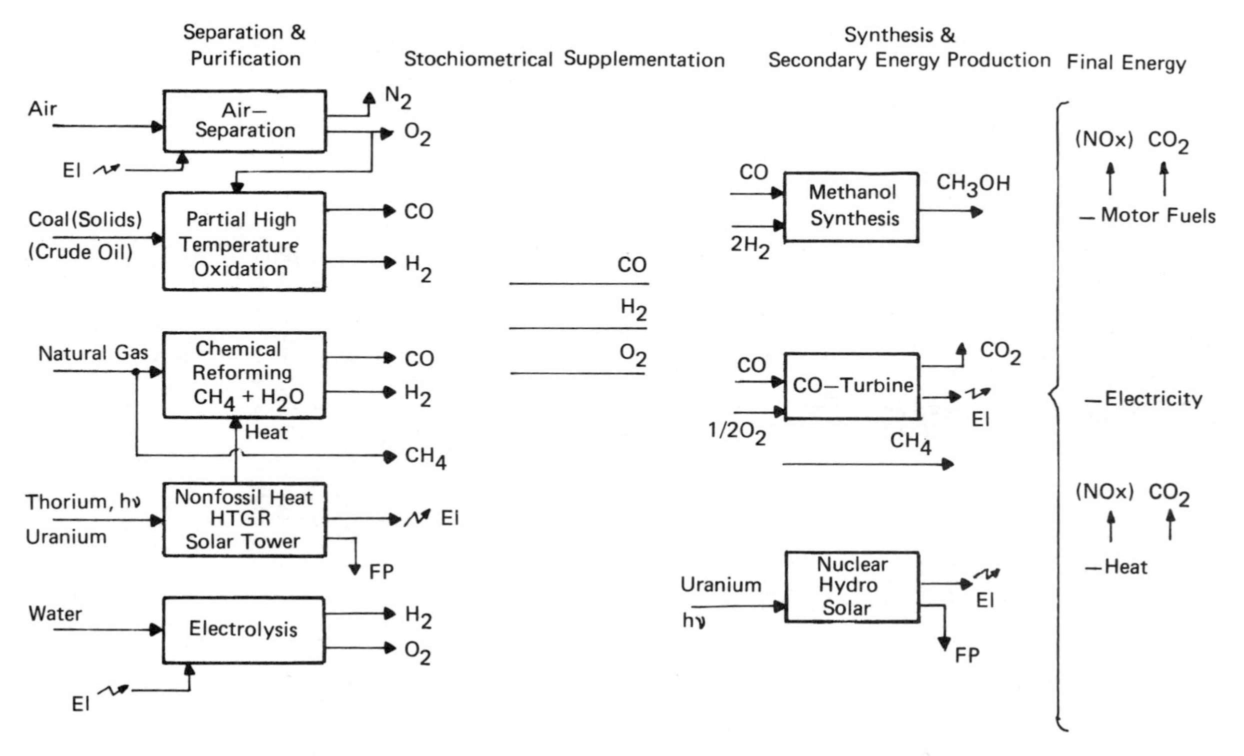

FIGURE 25.10. Novel energy systems, horizontally integrated. (From Häfele et al. 1984.)

cesses (the third stage). The various forms and quantities of final energy demand determine the mode and operation of the technologies in stage three, the outputs of which are electricity, a synthetic liquid fuel (here considered to be methanol), and natural gas. Note that stage three includes also "conventional" electricity production by hydropower, nuclear, and solar sources which traditionally are not associated with CO_2, SO_2, and NO_x emissions. Similarly, natural gas is piped through the system to the sites of end-use conversion in a conventional fashion.

A newly introduced technology to this configuration is the CO turbine. In combination with pure oxygen, this turbine is ideally suited for peak-load electricity production. The isothermal expansion of both CO and O_2 yields high efficiencies on the order of 60% as well as a significant reduction in generated CO_2. This scheme could operate without emitting any stack gases. The combustion product, CO_2, would be contained and used, say, for enhanced recovery in crude oil production, for other chemical needs, or in liquid form as a convenient solvent to absorb other waste products for storage in depleted gas fields.

The conversion of synthesis gas to a liquid fuel is another strategic technology of the third stage. Currently, this well-established process is associated with emissions resulting from incomplete chemical processes. Work is underway to confine these emissions by means of recycling of purge gases, etc. Methanol appears to be a prime output candidate in this system, at least as an intermediate form of energy for long-distance transportation and as a substance for blending classic hydrocarbons.

Furthermore, the configuration of the integrated energy system as shown in Fig. 25.10 eliminates the need for intermediate electricity production: the use of off-peak nuclear or other nonfossil electricity for specially designed low capital cost electrolysis contributes sufficient hydrogen and oxygen to the overall system. The present competition between coal and nuclear power in the electricity generation subsystem would be gradually resolved and coal would become competitive where it matters: as a substitute for oil. Further, one should note that all major conversion facilities produce certain amounts of waste heat that could easily be supplied to district heating systems and hence further improve the overall system performance.

The fourth stage of the integrated energy system is the end-use conversion part of the energy chain. Given the configuration of Fig. 25.10, the conversion of the various forms of final energy into energy services is accompanied essentially only by CO_2 emissions (and by some 50% less NO_x emissions than in conventional end-use systems).

The full advantage of this integrated energy system would be realized by providing a few interconnecting pipelines (e.g., for H_2, O_2, CO, and methanol (CH_3OH)) and storage capacity. These so called "grids" would allow for flexibility and a certain robustness in the energy system and would help exploit otherwise almost inaccessible energy opportunities.

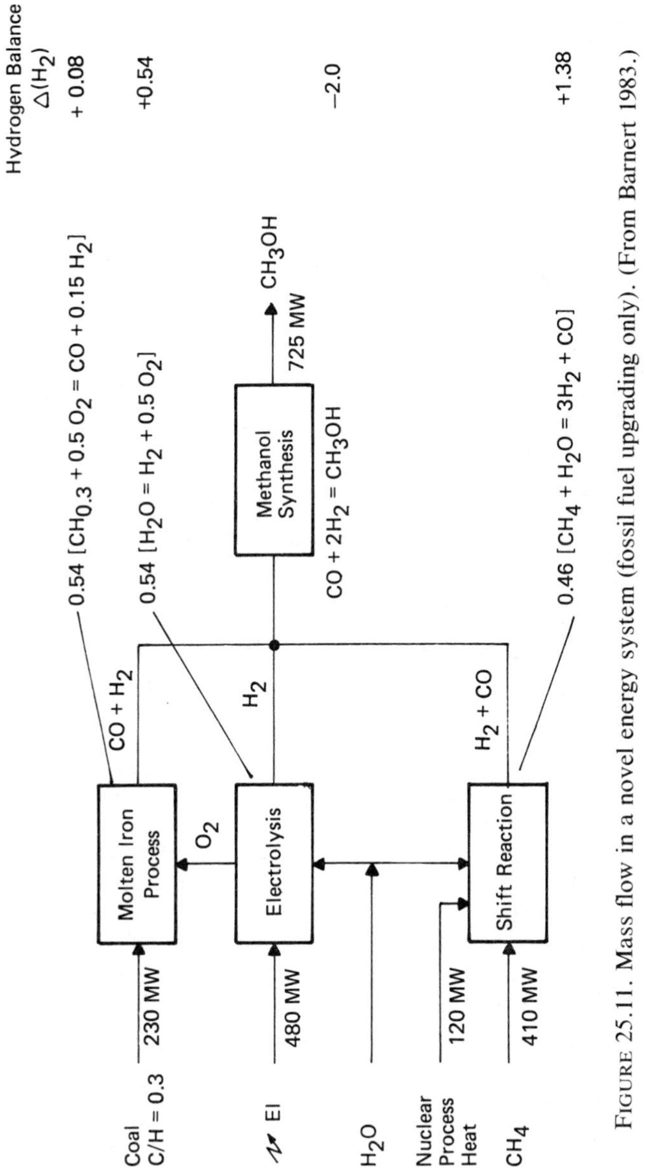

FIGURE 25.11. Mass flow in a novel energy system (fossil fuel upgrading only). (From Barnert 1983.)

The most valuable characteristic of the IES is the decoupling of the production of liquid fuels from any particular fossil fuel resource. By varying the input ratios of methane, nuclear or other nonfossil energy, and carbon-containing components, it would be possible to compensate for any changes in the availability or market price of carbon-based fuels. Also, the system, in principle, would result in unprecedented levels of overall reduction in pollution emissions without giving up any opportunities regarding energy resources, technologies, or economies.

Figure 25.11 depicts the mass flows connected with the configuration of the IES presented in Fig. 25.10. This is, of course, only one of many possibilities. Here we see that 410 MW* of methane upgrade 230 MW of coal (with a hydrogen-to-carbon ratio of 0.3) to 725 MW of methanol, yielding a carbon efficiency of 113%. Other inputs to this process comprise electricity for electrolysis (480 MW) and high-temperature heat for the methane shift reaction (120 MW). The overall efficiency of this system amounts to 59%. Note that in this system configuration there are no CO_2 releases.

The concept of integrated energy systems is currently being explored by a consortium of research institutes, including the Nuclear Research Center Jülich (Kernforschungsanlage Jülich, F.R.G.), Massachusetts Institute of Technology, and the Institute for Hydrogen Systems, Canada. Efforts are underway to better understand the actual chemistry involved and to determine the most appropriate technological system components, their costs and investment requirements. Comprehensive analyses of the potential of such integrated energy systems will then follow.

References

Ausubel, J. and W. D. Nordhaus. 1983. A review of estimates of future carbon dioxide emissions. In Changing Climate, pp. 153–185. National Academy of Sciences, Washington, D.C.

Barnert, H. 1983. Kohleveredelung mit dem HUMBOLDT-Kohleveredelungsverfahren. Internal publication. Nuclear Research Center Jülich, Jülich, FRG.

Häfele, W. et al. 1981. Energy in a finite world: a global systems analysis. Report by the Energy Systems Program Group of IIASA. Ballinger, Cambridge, Massachusetts.

Häfele, W., H. Barnert, and W. Sassin. 1982. Künftige fossile Brennstoffe: Ihre Nutzung und Einbettung in moderne Energiesysteme. Nuclear Research Center, Jülich, F.R.G.

Häfele, W., H. Barnert, S. Messner, M. Strubegger, and J. Anderer. 1984. The concept of novel horizontally integrated energy systems: the case of zero emissions. Nuclear Research Center, Jülich, F.R.G.

*One megawatt (MW) = 10^3 kW.

Häfele, W., and N. Nakicenovic. 1984. The contribution of oil and gas for the transition of long range novel energy systems, proceedings of the Eleventh World Petroleum Congress. John Wiley & Sons, Chichester-New York.

Hicks, N. L. et al. 1976. A model of trade and growth for the developing world. Eur. Econ. Rev. 7:239–255.

Keyfitz, N. 1982. Global prospects for population growth and distribution. WP-82-36. International Institute for Applied Systems Analysis, Laxenburg, Austria.

Manne, S. A. and L. Schrattenholzer. 1983. International energy workshop: individual poll responses. International Institute for Applied Systems Analysis, Laxenburg, Austria.

Marchetti, C. and N. Nakicenovic. 1979. The dynamics of energy systems and the logistic substitution model. RR-79-13. International Institute for Applied Systems Analysis, Laxenburg, Austria.

Messner, S. 1984. User's guide for the matrix generator of MESSAGE II, part 1: model description and implementation guide; part 2: appendices. WP-84-71a and WP-84-71b. International Institute for Applied Systems Analysis, Laxenburg, Austria.

Rogner, H-H. 1982. Substitution of coal and gas for oil—some global considerations. Presentation at the First US–China Conference on Energy, Environment, and Resources, Beijing, Nov. 7–12, 1982.

Rogner, H-H. 1983. IIASA '83 scenario of energy development: summary. International Institute for Applied Systems Analysis, Laxenburg, Austria.

Rotty, R. M. and G. Marland. 1980. Constraints on carbon dioxide production from fossil fuel use. ORAU/EA-80-9(M). Institute for Energy Analysis, Oak Ridge, Tennessee.

Sassin, W. 1982. Fossil energy and its alternatives—a problem beyond costs and prices. Presentation at the International Economic Association Conference on Economics of Alternative Sources of Energy, Tokyo, Sept. 7–Oct. 1, 1982.

Smagorinsky, J. 1983. Effects of carbon dioxide. In Changing Climate, pp. 266–284. National Academy of Sciences, Washington, D.C.

U.S. Federal Government. 1983. Report of the inter-agency task force on acid precipitation. Washington, D.C.

26
Atmospheric CO_2 Projections with Globally Averaged Carbon Cycle Models

JOHN R. TRABALKA, JAMES A. EDMONDS,
JOHN M. REILLY, ROBERT H. GARDNER,
AND DAVID E. REICHLE

The principal objective of this analysis is to attempt to project the level of CO_2 in the atmosphere over the next century (to the year 2075). Knowledge of future atmospheric CO_2 concentrations is necessary to establish whether there are likely to be significant climatic and biological consequences resulting from continued fossil fuel combustion. To estimate the effects of projected fossil fuel emissions on future atmospheric CO_2 levels, it is necessary to know how this anthropogenic source impacts the natural biogeochemical cycle of carbon—a complex and dynamic set of processes linking the atmosphere, the oceans, and the terrestrial environment. Because of the complicated and changing nature of the anthropogenic source terms and the carbon cycle, simulation models that deal with relationships between economic patterns, energy use, and CO_2 emissions and with the major components of the global biogeochemical cycle are needed. These models appear to offer the only practical means to integrate the vast array of detailed quantitative data needed to provide projections of future atmospheric change. Such modeling efforts should be internally consistent with the natural carbon fluxes among atmospheric, oceanic, and terrestrial pools; they should accurately reflect the impact of fossil and biospheric releases over past centuries; and, of course, they should satisfactorily account for empirical measurements of recent decades (e.g., 316 ppmv in 1959 rising to 345 ppmv at present).

Approach

The most important potential future source of CO_2 emissions is the burning of fossil fuels. There are many factors which shape the future production and use of fossil energy resources. These include economic, technological, social, demographic, environmental, and political factors such as global population growth, gross national product (GNP) growth, changes in energy use efficiencies, emergence of new energy supply technologies, energy resource constraints, and the relationship between energy, income, and price. The analyses presented here are the result of a continuing effort,

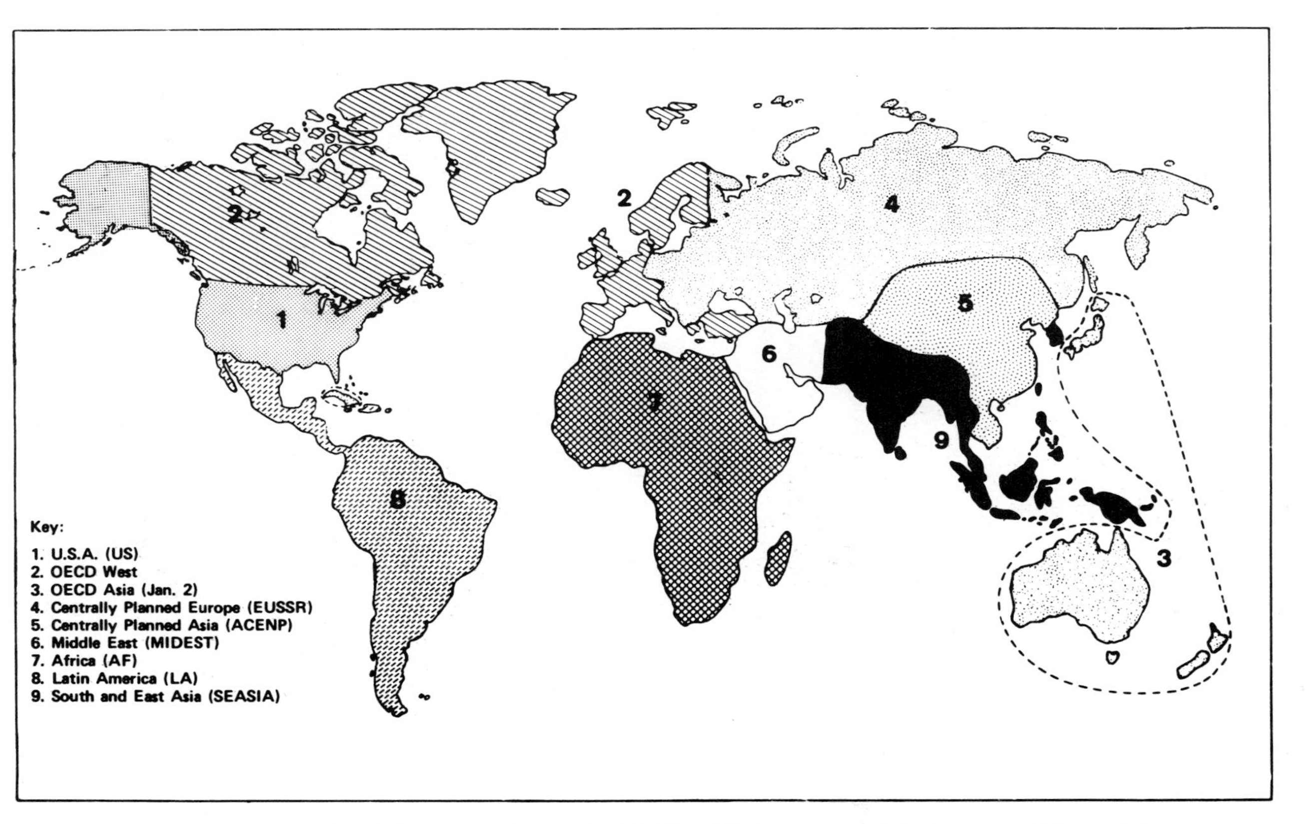

FIGURE 26.1. Regional disaggregation of the IEA/ORAU long-term, global, energy-CO_2 model.

including model development, documentation, data development, and scenario analysis, over the past 4 years and are documented in several publications (e.g., Edmonds and Reilly 1983*a,b,c*; Edmonds et al. 1984; Edmonds and Reilly 1985).

Because emissions are of concern over periods ranging from decades to centuries, a fair amount of uncertainty exists in the key determinants of future emissions. To accommodate this reality, a parametric model of global energy use and CO_2 emissions developed for the U.S. Department of Energy (DOE) by the Institute for Energy Analysis/Oak Ridge Associated Universities (IEA/ORAU) (Edmonds and Reilly 1983*a*) was used. Although the model was developed for DOE and the DOE remains its principal user, it also has been used to support work by the Massachusetts Institute of Technology (MIT)-Energy Laboratory, the Gas Research Institute, the Electric Power Research Institute, and by the U.S. Environmental Protection Agency. The model and its associated documentation are available through the Carbon Dioxide Information Center at Oak Ridge National Laboratory (ORNL).

The model is disaggregated into nine geopolitical regions (Fig. 26.1). It is long term, based to 1975 using United Nations (1982) fossil emissions data, with benchmark years 2000, 2025, 2050, and 2075. Nine types of primary energy technologies are currently considered: conventional and unconventional oil, conventional and unconventional gas, coal, biomass, and hydro, solar, and nuclear electric power (Table 26.1). In addition, fossil fuels and biomass may be transformed into electricity. Final consumption consists of four secondary energy categories: solids, liquids, gases, and electricity. The model estimates the supply and demand for energy by region and forecast period as well as by a set of world and regional energy prices consistent with global energy balance. Assumptions about the CO_2 content of fuels at the point of combustion were taken from Marland and Rotty (1983). A schematic diagram of the model structure is given in Fig. 26.2.

TABLE 26.1. Distribution of IEA/ORAU global model energy supply technologies across supply categories.

	Supply categories		
	Resource-constrained conventional energy	Resource-constrained renewable energy	Unconstrained energy resources
Energy technologies	Conventional oil Conventional gas	Hydro	Unconventional oil Unconventional gas Solids Solar Nuclear

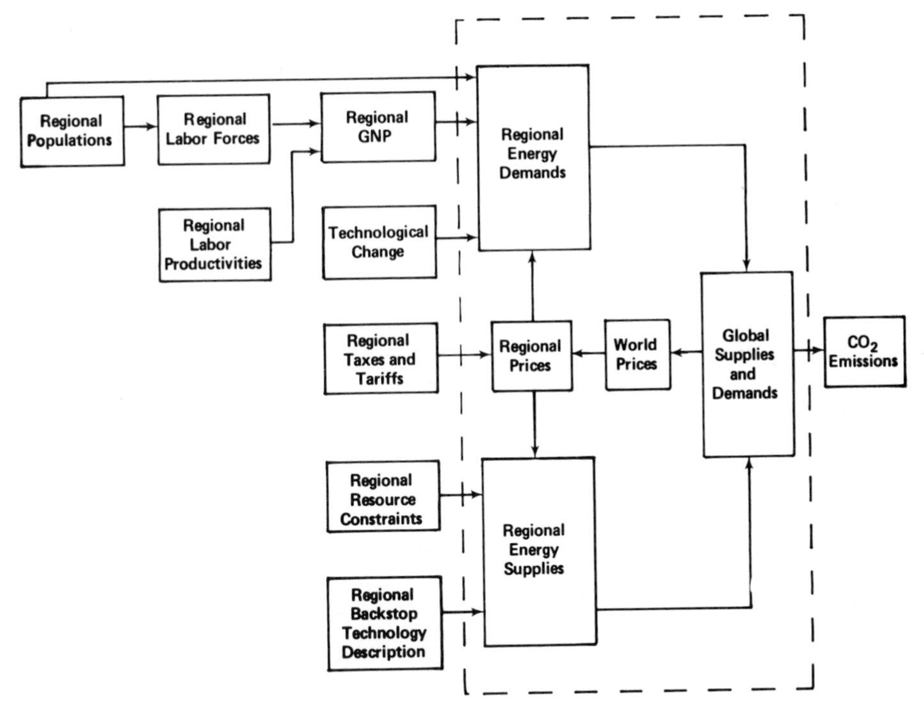

FIGURE 26.2. Structure of the IEA/ORAU long-term, global, energy-CO_2 model.

The IEA/ORAU global CO_2 emissions model was used to project CO_2 emissions from fossil fuel combustion, in conjunction with three sets of alternative assumptions about the future, to create three future scenarios denoted Cases A, B, and C. This exercise is documented in Edmonds et al. (1984). This approach permits disaggregation of resource, economic, and technological factors governing future energy demand so that scenario assumptions can be analytically examined and evaluated. These energy production and use scenarios are translated into equivalent CO_2 emissions that serve as input terms to a second suite of dynamic simulation models of the global carbon cycle.

Two state-of-the-art globally averaged ocean carbon cycle models utilized the three projections of fossil fuel CO_2 emissions and three terrestrial biospheric release scenarios to construct a spectrum of 18 atmospheric CO_2 projections from the year 1800 until year 2075. An additional exercise was carried out to assess the contribution of carbon cycle model parameter uncertainty to the projection exercise. These projections represent a range of possible future atmospheric CO_2 levels that can be used to evaluate plausible climatic consequences resulting from increased atmospheric CO_2. These atmospheric CO_2 projections should be interpreted as the output of a set of models that, through careful and thoughtful use, can improve the understanding of the global carbon cycle, past and future, rather than as precise forecasts of energy-demand or as recommendations for world energy-use policies.

Future Fossil Fuel CO_2 Emissions

The construction of future emissions scenarios has focused on those parameters most important in determining global CO_2 emissions and that have received attention in the literature. A standard case (B) fossil emissions scenario was based on median estimates of model parameters (see Edmonds et al. 1984). Given that uncertainty is a significant factor in long-term projections, a single carbon release scenario is of little use unless some quantification of the degree of confidence one can place in the estimate is possible. The motivation for developing two extreme cases, in addition to a case based on the standard assumptions, is to describe a range. The two additional scenarios were developed by varying the assumptions concerning solar/nuclear electric power generation costs and factors demonstrated by model sensitivity analysis to be important determinants of future CO_2 emissions levels: energy efficiency improvement in end use, population, Gross National Product (GNP), and coal and shale oil supplies. The assumptions were varied jointly within the bounds of what are currently considered to be likely future values (cf. Edmonds and Reilly 1983*c*) to explore extreme scenarios of carbon emissions as opposed to extremes in energy use or energy prices. While scenarios A and C were constructed to explore extreme CO_2 emissions trajectories, neither case can be taken as an absolute bound on CO_2 emissions.

There are several reasons for this caveat. First, only a small subset of variables, albeit a subset containing variables to whose values the model displays greatest sensitivity, were varied. Second, neither the uncertainty surrounding the entire data set nor the relationship between inputs and outputs was explored using state-of-the-art probabilistic numerical methods. The relationships between input uncertainties can be highly nonlinear. Thus, even though median input assumptions were used to generate the standard Case B, there is no guarantee that the standard Case B outputs are not significantly different from median outputs in a direction that we cannot know a priori. We must therefore emphasize the exploratory nature of the exercise. As a follow up to this exercise, a formal uncertainty analysis is being conducted jointly by IEA/ORAU and ORNL.

A comparison of the sets of assumptions varied between the three cases is presented in Table 26.2. The resulting model outputs based on these three sets of assumptions generate a high carbon emissions case (A), a median emissions case (B), and low emissions case (C) (Table 26.3). For all three cases developed in this report, the rate of CO_2 emissions is higher in 2075 than in 1975. However, in all three cases the rate of growth of emissions declines over the period:

	1975–2075 (%/year)	2050–2075 (%/year)
Case A	3.0	2.7
Case B	1.4	1.0
Case C	0.3	0.0

Based on the choice of parameters, it appears highly likely that the carbon emissions scenarios represented by Cases A and C are reasonable, though not absolute, bounds on future emissions rates. That is, we think it likely (greater than a 50–50 chance) that future emissions will fall within those bounds. Fossil fuel CO_2 emissions are projected to reach 91 $\times$ 10^{15}g C in Case A by the year 2075.

Fuel Types

The role of various fuels changes both over time and across scenarios (Fig. 26.3). The percentage contribution of conventional fuels (oil and gas) declines in all three scenarios over time. By 2075, conventional oil and gas, combined, account for $<10\%$ in all three cases and $<2\%$ in Case A. As a consequence, the existence of high rates of carbon emissions is possible only if coal and shale oil are mined in quantity. More than half of the carbon emissions from Case A in 2075 are produced from unconventional oil (principally shale oil), with coal and synthetic gas accounting for the bulk of the remainder (Fig. 26.3).

The role of shale oil in determining CO_2 emissions is minor in all cases except Case A. The heavy use of shale oil in Case A depends on optimistic assumptions about both the resource base and the technology employed to extract the fuels (Edmonds et al. 1984). Such high rates are, however, not out of the question, and it is the heavy dependence on this carbon-intensive fuel to supply liquids to a rapidly growing world economy that allows carbon emissions to continue to grow rapidly in this scenario. Thus, carbon emissions continue to grow more rapidly than overall energy demand because the fuel mix is shifting from one that relies on fuels with low carbon intensity (liquids, gases, hydro, solar, and nuclear) to one that employs high-carbon fuels (coal and shale oil). Carbon emissions in the median case (B) are dominated by coal, with approximately equal contributions from the direct use of coal and coal-derived synthetic oil; projected emissions in 2075 are 18.8 $\times$ 10^{15} g C. Coal is also the principal source of carbon emissions in the low case (C). Emissions in Case C reach 6.7 $\times$ 10^{15} g C in 2025, and then remain nearly constant through 2075.

TABLE 26.2. Model parameter values varied to produce global carbon emissions cases A, B, and C (1975–2075)

Year	Case	Region 1	2	3	4	5	6	7	8	9	Global
		Base GNP assumptions (1975 US$ x 10^{12})									
1975	A, B, C	1.5	1.8	0.6	1.0	0.3	0.1	0.2	0.3	0.2	6.1
2000	A	3.5	4.4	1.9	2.1	1.0	0.6	0.6	1.3	0.9	16.3
	B	2.9	3.6	1.6	1.8	0.9	0.5	0.5	1.1	0.7	13.4
	C	2.5	3.2	1.4	1.6	0.8	0.4	0.4	1.0	0.6	11.9
2025	A	6.8	9.5	4.7	4.2	2.6	1.8	1.7	4.2	2.5	37.8
	B	4.6	6.4	3.2	2.8	1.8	1.2	1.1	2.8	1.7	25.7
	C	3.6	5.0	2.5	2.2	1.4	1.0	0.9	2.2	1.3	20.2
2050	A	12.3	18.6	9.6	7.7	6.0	4.4	3.8	9.2	5.6	77.2
	B	6.8	10.4	5.3	4.3	3.3	2.5	2.1	5.2	3.2	43.2
	C	4.7	7.2	3.7	3.0	2.3	1.7	1.5	3.6	2.2	30.0
2075	A	22.2	33.7	16.9	14.0	12.2	9.1	8.0	19.3	11.8	147.0
	B	10.1	15.4	7.8	6.4	5.6	4.2	3.7	8.9	5.4	67.6
	C	6.2	9.4	4.8	3.9	3.8	2.6	2.3	5.5	3.3	41.5

Energy efficiency improvement (%/year, 1975–2075)

Regions	Sector	Case A	B	C
1, 2, 3	Residential/commercial, and transport	0.0	0.0	0.5
1, 2, 3	Industrial	0.0	1.0	1.5
4, 5, 6, 7, 8, 9	Aggregate	0.0	0.5	1.0

Base rates of expansion and breakthrough prices of unconstrained supply technologies

Year	Case	Unconventional oil	Coal	Global base GNP
		(%/year)		
1975–2000	A	2.1	3.3	4.0
	B	1.2	2.3	3.2
	C	0.5	1.2	2.7
2000–2025	A	8.4	3.8	3.4
	B	6.2	2.8	2.6
	C	4.4	1.6	2.1
2025–2050	A	5.7	2.9	2.9
	B	4.4	2.3	2.1
	C	3.3	1.2	1.6
2050–2075	A	3.9	1.7	2.6
	B	2.9	1.2	1.8
	C	2.1	0.9	1.3
		(1975 US$/GJ)		
1975 break through price		13.3	0.26	
2010 break through price	A	3.85	0.26	
	B	7.35	0.26	
	C	19.25	0.26	

Nuclear and solar costs (1975 US$/GJ)

Case	Cost
A	100
B	18
C	9

TABLE 26.3. CO_2 emissions from fossil fuel burning 1975–2075 for three cases (10^{15} g C).

	1975	2000	2025	2050	2075
Case A	4.7	8.4	19.0	47.4	91.1
Case B	4.7	7.2	10.3	14.5	18.8
Case C	4.7	6.2	6.7	6.8	6.8

PRIMARY ENERGY

Primary energy production and use for scenarios A, B, and C are displayed in Fig. 26.4. (Total global primary energy production and total global primary energy use are equal in all scenarios because the model enforces global energy balance. Individual regions will generally be either net importers or exporters, however.) In all three cases, energy production and use increase over time, though the primary energy consumption of Case A is > 600% higher than that of Case C. Similarly, the rate of growth of primary energy use decreases in all three cases as global population and

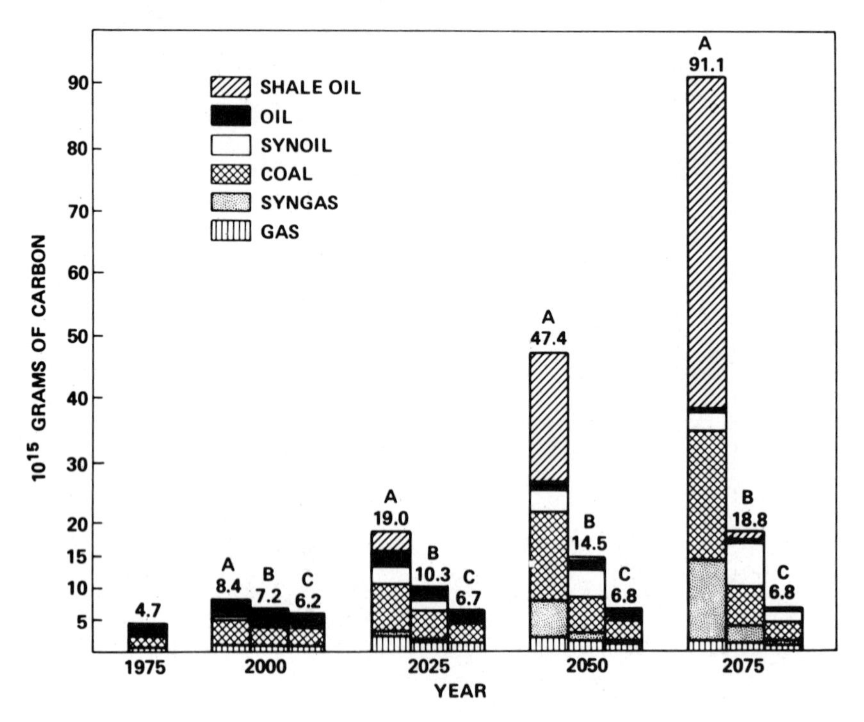

FIGURE 26.3. Annual CO_2 emissions by fuel source for cases, A, B, and C: 1975–2075.

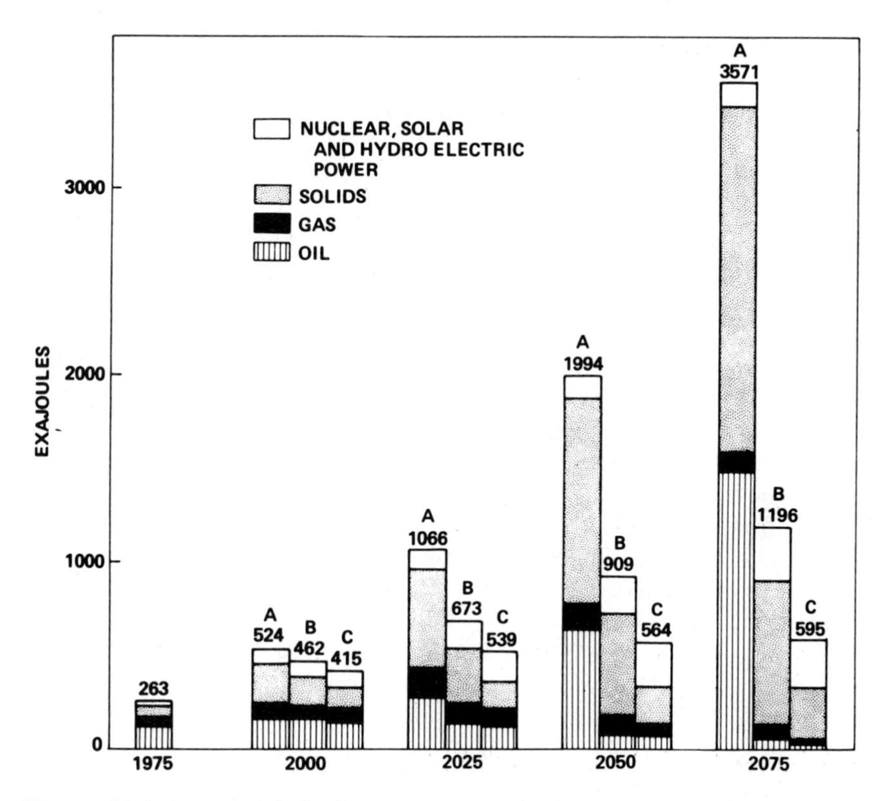

FIGURE 26.4. Annual global primary energy use by fuel and case: 1975–2075 (exajoules).

GNP growth slow in time in all cases. The differences in rates of growth of primary energy are very great indeed:

	1975–2075 (%/year)	2050–2075 (%/year)
Case A	2.6	2.4
Case B	1.5	1.1
Case C	0.8	0.2

In all cases the fuel mix changes dramatically over the next 100 years (Fig. 26.4). Conventional resources of oil and gas are all but exhausted. Conventional oil production peaks around the year 2025. The model is then forced to decide what form of energy to choose to meet growing demands. In all cases solids are used heavily. They are relatively cheap and even in Case C production more than triples from 1975 to 2075, accounting for almost one-half of total primary energy use by 2075. Case B shows the heaviest proportional dependence on solids (64% by 2075). Case

A shows the heaviest absolute use of solids 1852 EJ (52% of the 2075 total and more than total primary production in Case B). Most of the solids production comes in the form of coal but about 10% of this total takes the form of biomass in Cases B and C, with this share rising to 17% in Case A in response to the high price of solids generated in this scenario.

With coal holding a relatively similar market share in all three scenarios, its overall contribution to CO_2 emissions is heavily influenced by the level of overall energy use, which in turn is heavily dependent on GNP. The remaining supply of energy takes one of two forms, either unconventional oil and gas sources or direct electric power from nuclear, solar, and hydro sources. In Case A, unconventional fossil fuels dominate the scenario, supplying > 40% of primary energy. The lion's share of this production (95%) is unconventional oil. Less than 5% of primary energy comes from CO_2-benign direct electricity production. Case B provides a balance between these two primary energy sources. By 2075, 25% of primary energy comes from direct electric production, while <10% comes from unconventional oil and gas. In contrast, Case C uses direct electricity to generate 45% of primary energy, with only 1% coming from unconventional fossil fuels in the form of gas.

The heavy dependence on coal and unconventional oil raises the question of how cumulative consumption compares to the resource base. Coal reserves and resources are (see Edmonds and Reilly 1985):

Proved recoverable reserves	20,000 EJ
Total recoverable resources	165,000 EJ

Cumulative production (using a linear extrapolation between benchmark years) over the period 1975–2075 for the three cases is:

Case A	40,000 EJ
Case B	19,000 EJ
Case C	12,000 EJ

Only in Case A are known recoverable reserves exhausted in the time horizon of the model. Even in Case A only 24% of total recoverable resources are exhausted.

Shale oil is the dominant form of unconventional oil. Taking oil yields greater than 38×10^{-3} $m^3 \cdot Mg^{-1}$ (10 gallons per ton) of shale as the economic resource grade at prices greater than \$4/GJ, the resource base can be disaggregated into three categories according to uncertainty:

Known resources	20,000 EJ
Possible extensions of known resources	65,000 EJ
Undiscovered and unappraised resources	2,000,000 EJ
Total resource base	2,000,000 EJ

Cumulative production in the three cases over the period 1975 to 2075 is:

Case A	36,000 EJ
Case B	500 EJ
Case C	0 EJ

Review of Emissions Scenarios

Figure 26.5 contrasts these scenarios with other selected energy and CO_2 scenarios. The assemblage of scenarios is not meant to represent a comprehensive review; in selecting scenarios, fairly complete coverage of extreme scenarios was sought. As a result, there are many other studies that would fall within the broad limits described by Cases A and C.

In the figure, one observes that Case C is lower than all but the Niehaus-Williams scenario in their 30-Terawatt (TW)-solar/nuclear case [NW(L)] and Lovins et al. (1981) low CO_2 case (L). Case B, on the other hand, is remarkably close to the IIASA low scenario as well as the Nordhaus and Yohe 50th percentile or median case [NY(M)]. It is also very close to the Reister and Rotty (1983) scenario; Reister and Rotty (1983) and Seidel and Keyes (1983) (neither shown in the figure) developed scenarios independently of the U.S. DOE effort but used the IEA/ORAU computer model employed here. The IEA/ORAU model was also used by Rose et al. (1983) to derive a series of 13 scenarios for the 1975–2050 period based on an

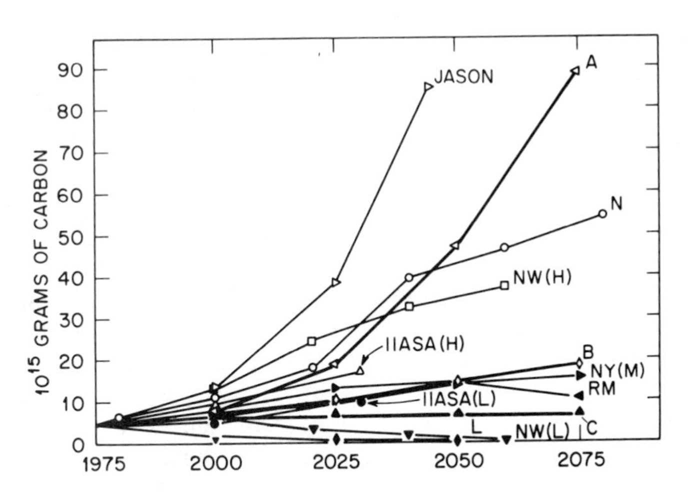

FIGURE 26.5. Comparison of selected long-term, global CO_2 emissions cases. A, Case A; B, Case B; C, Case C; N, Nordhaus 1979; NW(H), NW(L), Niehaus and Williams 1979. H, high case (50-TW-fossil fuel scenario); L, low case (30-TW-solar-nuclear scenario); NY(M), Nordhaus and Yohe 1983. M, median case; IIASA(H), Häfele 1981; RM, Rotty and Marland 1980; JASON, JASON 1979; L, Lovins et al. 1981.

alternative set of assumptions about end-use efficiencies, solar/nuclear contributions, fuel prices, and availability of Mideast oil. Emissions in the year 2050 ranged from 2.7 to 15.2 $\times$ 10^{15} g C • yr^{-1}.

Case A is a high CO_2 emissions scenario with over 90 $\times$ 10^{15} g of carbon emitted in 2075. While Case A is high relative to most studies, it is lower, in all projection years, than a simple extrapolation of the pre-1970s experience as in the JASON (1979) study.

In many cases, the comparison of scenarios involves different assumptions, different model structures, and different study objectives. Thus, comparisons must be made with caution. For example, the JASON projection is a simple extrapolation of historical rates of increase. The Niehaus-Williams and Lovins scenarios both use the IIASA scenarios as a base line and then make various alterations to demonstrate certain points, Niehaus-Williams to show the potential impact of fuel mix on the dimensions of the CO_2 issue, and Lovins to suggest that energy strategies consistent with the IIASA low assumptions but that yield no CO_2 problem are possible. Nordhaus and Yohe, on the other hand, attempt to make specific predictive statements, e.g., the probability of future emissions falling between two points is X. The International Institute for Applied Systems Analysis (IIASA) developed two energy scenarios saying that "the real future could certainly be 'higher' than the high scenario or 'lower' than the low scenario, but the range was selected such that the expected future value is more likely within the range than outside it." (Häfele 1981, p. 424). Cases A and C were constructed to explore potential future extremes in carbon emissions rather than in energy use. Carbon emissions from Cases A and C in the year 2075 would fall at the 99th and 10th percentile levels, respectively, in the projection set developed by Nordhaus and Yohe (cf. Fig. 2.1 in Nordhaus and Yohe 1983). However, the corresponding levels of energy use for the same cases (Fig. 26.4) would fall at the 95th and 15th percentiles of the corresponding Nordhaus and Yohe distributions, respectively. Fuel mix (Fig. 26.3) is an important determinant of future CO_2 emissions.

To the extent that scenarios are developed within a common modeling framework, differences in results can in principal be directly attributed to differences in model input assumptions. The 11 scenarios developed by Rose et al. (1983) can be directly compared to the "base case" presented in Edmonds and Reilly (1983c) and in turn to cases A, B, and C since all employ the same IEA/ORAU model. In fact, the differences between CO_2 emissions trajectories for all scenarios derived from a common model are directly comparable.

The IEA/ORAU model tended to be most sensitive to variations in income elasticity of demand for energy, coal supply, rate of improvement in energy end-use efficiency, the breakthrough price of unconventional oil, GNP, and the unconventional oil supply (Edmonds et al. 1984). This ranking was not unique, however. The price of unconventional oil, for example, was important (elasticity >0.5) only in Case A.

Carbon Cycle Modeling

The three CO_2 emissions scenarios (A, B, and C) served as inputs to two globally averaged ocean carbon cycle models. These models have been documented in Björkström (1979) and Emanuel et al. (1984*b*); the former was a slightly modified version of the original formulation (see Edmonds et al. 1984). The Björkström ocean model is an advective/diffusive model, while the Emanuel et al. model is diffusive in character (although diffusivity is effectively varied with depth). At present, it is difficult to discriminate among models of the global carbon cycle that are applicable to projecting future atmospheric CO_2 concentrations for different fossil fuel use scenarios. Given the available data for calibration and testing, a large number of alternative models can be derived that appear satisfactory (Edmonds et al. 1984). These two formulations are representative of globally averaged models of the atmosphere-ocean carbon system currently under consideration. Both models are applied in two stages. The first is a calibration step in which parameter values and initial conditions are calculated consistent with an assumed preindustrial steady-state condition. In the second stage, the solution of the model equations to fossil fuel CO_2 release and other perturbations is generated.

Because of current uncertainty about the historical role of the biosphere in the global carbon balance, three alternative biospheric release scenarios were used. The simplest, though not very realistic, scenario is that there has been no net historical flux from terrestrial ecosystems to the atmosphere. Thus, the first set of paired model projections was run with a neutral biosphere (zero net carbon exchange with the atmosphere), beginning with a CO_2 concentration in the atmosphere of 292 ppmv for the year 1800. Next, the initial atmospheric CO_2 value was changed to incorporate reconstructions of terrestrial biospheric carbon exchange based on the $^{13}C/^{12}C$ tree-ring record (Peng et al. 1983) and on independent forest history data from land-use clearing (Houghton et al. 1983). For both biospheric release scenarios it was necessary to utilize an initial 1800 atmospheric CO_2 concentration of 245 ppmv to fulfill the condition that the model projection approach the trajectory of the Mauna Loa record. The initial atmospheric concentrations for 1800 of 245 and 292 ppmv bracket the approximate range of uncertainty for this value (Machta 1983). Total estimated biospheric carbon releases were 210×10^{15} and 260×10^{15} g C over the pre-1980 period, respectively, for the scenarios based on land-use change and on the tree-ring record. These scenarios were projected into the future by making the assumption that the annual release rates in

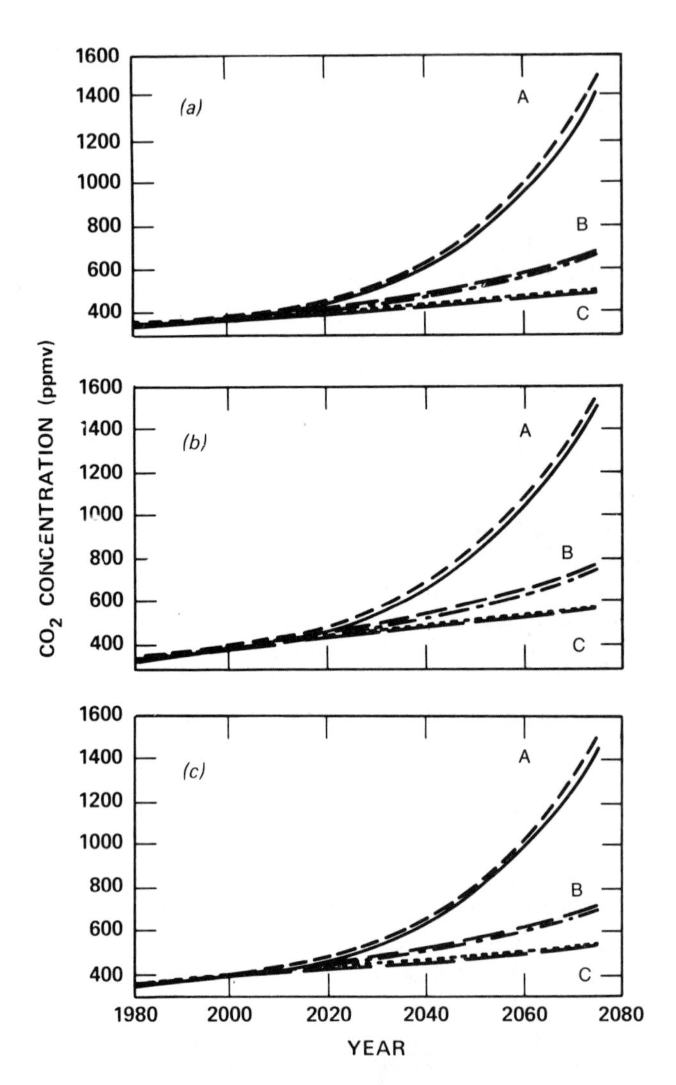

FIGURE 26.6. Simulated atmospheric CO_2 concentration. **a**: No terrestrial source (neutral case). **b**: Terrestrial source based on ^{13}C data. **c**: Terrestrial source based on land-use and demographic data. (A) high fossil fuel CO_2 emissions case; (B) standard fossil fuel CO_2 emissions case; (C) low fossil fuel CO_2 emissions case. *Upper curve* represents the response of the modified Björkström (1979) ocean model and the *lower curve* the response of the Emanuel et al. (1984*b*) ocean model in each case.

TABLE 26.4. Model-calculated atmospheric CO_2 concentrations based on global carbon emissions cases A, B, and C for fossil fuels combined with various biospheric inputs, as described in the text.

	Atmospheric CO_2 concentration (ppmv) Year			
Case	2000	2025	2050	2075
A	380–400	480–530	760–840	1400–1550
B	380–400	440–480	540–600	670–760
C	370–400	420–460	470–520	510–580

1980, 2.7×10^{15} and 1.2×10^{15} g C, respectively, remained constant until 2075 (Edmonds et al. 1984).

Future Atmospheric CO_2 Concentrations

The results from the carbon cycle modeling exercise are shown in Figure 26.6 and Table 26.4 as projected atmospheric CO_2 concentrations for the period from 1975 to 2075. The data in Table 26.4 are presented as a range of values for each date (2000, 2025, 2050, and 2075) and each fossil fuel emissions case, because six combinations of two globally averaged carbon cycle models and three potential terrestrial biospheric release histories were run for each emissions case. The range of values serves to provide the reader with some sense of the uncertainty contributed by carbon cycle modeling to the overall projection exercise. It is readily apparent from Fig. 26.6 and Table 26.4 that, given the model and terrestrial history combinations used, variations in the fossil fuel emissions cases dominate the atmospheric CO_2 response by the year 2050. A relatively tight range of projected atmospheric CO_2 concentrations is produced for the year 2000 (370 to 400 ppmv range will accommodate results from all cases), but significant divergence occurs after the year 2025. Median atmospheric CO_2 concentrations are projected to reach 540, 710, and 1480 ppmv by 2075 using emissions cases C, B, and A, respectively.

Given the uncertainties about the preanthropogenic atmospheric CO_2 concentration, the estimated date for a doubled atmospheric concentration falls between 2025 to 2050 in results from Case A, near 2050 using Case B, and in, or beyond, 2075 for Case C. Thus, this modeling exercise indicates that doubling of atmospheric CO_2 is unlikely to occur before the year 2025 and may not occur within the next century.

Carbon Cycle Model Uncertainty Analysis

A variety of techniques have been developed to determine the dependence of predictions of simulation models on inputs, initial conditions, and parameters. One approach has already been described (permutations and combinations of emissions cases, terrestrial flux reconstructions, and carbon cycle models). Monte Carlo methods are also appropriate for questions involving carbon cycle simulations because costs are very low and recent developments have drastically improved both the efficiency and the precision of this analytical tool (Gardner 1985). However, limits must be placed on the use of Monte Carlo methods for carbon cycle models because of the interdependence of model parameters and contemporary patterns of atmospheric CO_2 increases. In practical terms, this means that one must either relax the condition that the model output satisfactorily approaches the trajectory of the modern atmospheric CO_2 record on its path from an assumed nineteenth century "steady-state" towards future time, or apply a "filter" to the model outputs to reject simulations which do not satisfy this constraint. The latter is more desirable since we make better use of existing information in setting bounds on atmospheric futures, but this requires a large number of model runs to obtain a sample of "filtered" outputs large enough for statistical analyses.

Monte Carlo methods with Latin hypercube sampling were used to randomly generate parameters (Gardner et al. 1983) over a range of specified values (Gardner 1985). These parameters were used to produce a series of simulations of one globally averaged carbon cycle model (Emanuel et al. 1984*b*) from which the sources of uncertainty associated with atmospheric CO_2 predictions could be estimated. The ocean component of the model was identical to one used in the projection exercise just described. The terrestrial component has been fully documented by Emanuel et al. (1984*a*,*b*). A schematic diagram of the global carbon cycle model is shown in Fig. 26.7.

Inspection of the relationship among model parameters led to the selection of 32 independent parameters and model inputs that are subject to uncertainty (Gardner et al., 1985). All other values in the model can be calculated from the assumed behavior of the system (i.e., at steady state) or calculated directly from this set of 32 parameters. Actual ranges, or uncertainties, in parameters were approximated for 24 parameters by setting the a priori probability distribution function as a uniform distribution with a range equal to ± 20% of the nominal value. In most cases, this appears to be a satisfactory approximation of actual uncertainty (e.g., total carbon in vegetation and soils, see Post et al. 1982; Olson et al. 1983), and it will be shown that the results are insensitive to the ranges chosen for these 26 parameters. Critical parameters or model inputs were identified by sensitivity analysis and were assigned specific ranges based on a review of the literature (Gardner et al. 1985).

A compromise "baseline" terrestrial biospheric release scenario was used for the uncertainty analysis. The scenario allows for a linear increase in the rate of forest clearing until the year 1920; the level of forest clearing is then held constant from 1920 to 2075. This results in a curvilinear net carbon flux from the biosphere to the atmosphere rising from 0 in 1800

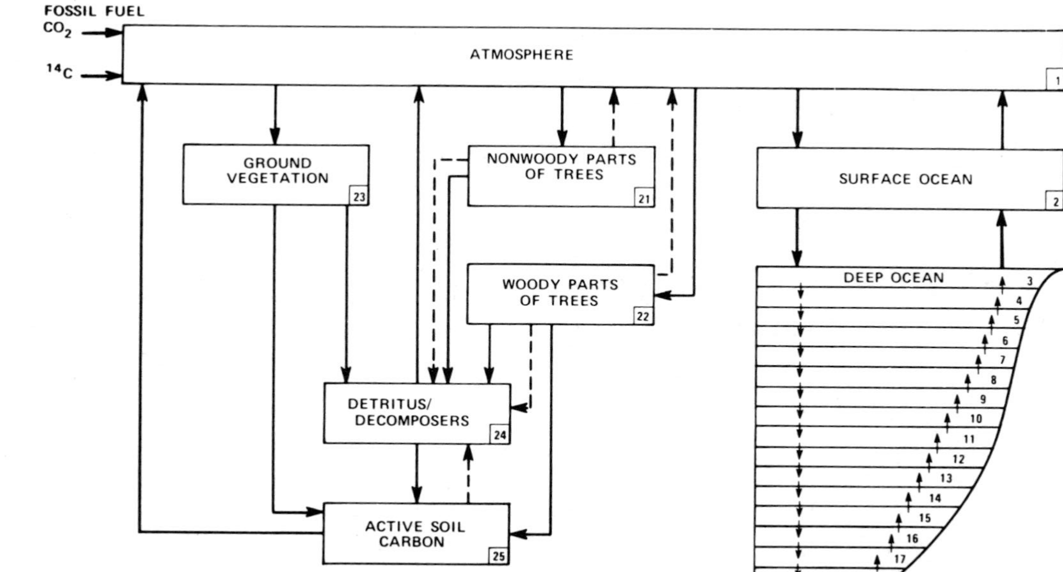

FIGURE 26.7. Compartment diagram for a globally averaged model of the world carbon cycle. The atmosphere, which is treated as a single well-mixed reservoir, exchanges carbon with the inorganic pool in the oceans and with vegetation and soils on land. The "surface ocean" corresponds to waters above 75 m. Carbon turnover in the "deep ocean" is by exchanges between adjacent layers. Carbon in plants is divided between "trees" and "ground vegetation," and assimilation from the atmosphere is associated with net primary production. Carbon in dead organic matter is divided between "detritus/decomposers" corresponding to litter at the soil surface and "active soil carbon." *Solid arrows* indicate naturally occurring carbon fluxes, while *dashed arrows* are associated with forest clearing. There are two inputs of carbon to the system, release of CO_2 by fossil fuel combustion and natural or weapons-produced ^{14}C, indicated by *double arrows*. Indexes in each compartment indicate the variable assignment in model equations.

to 1.5×10^{15} g C • yr^{-1} in 1920 (Fig. 26.8). Although the rate of forest clearing is held constant after 1920, the average flux rate decays slowly to a value of 1.3×10^{15} g C • yr^{-1} in 1980 and to 1.2×10^{15} g C • yr^{-1} in 2075. The allowable Monte Carlo deviations from the baseline clearing scenario (±80%) were applied equally to each year from 1800 to 2075, i.e., varying the elevation of the curve, but maintaining the basic shape. The basic shape and range of variation were based on data in Peng et al. (1983), Woodwell et al. (1983*a,b*), and Stuiver et al. (1984). Coupled with the uncertainties in other terrestrial model parameters, the dynamic range in net fluxes allowed from the biosphere to the atmosphere in 1980 was 0.3 to 2.7×10^{15}g C • yr^{-1}, essentially spanning the range of the three biospheric release scenarios described earlier.

Historical fossil fuel emissions rates were assigned an estimated weighted mean variability of ±12% (Keeling 1973; Marland and Rotty 1983). Future emissions were based on Case B and were assumed to have zero uncertainty for this particular exercise because the focus was on carbon cycle modeling uncertainties.

Six hundred Monte Carlo iterations provided the necessary samples of possible model behavior for statistical analysis. Only 47 solutions obeyed the constraint that model solution not deviate by more than ±2 ppmv from the average measured atmospheric CO_2 concentration between 1958 and 1980 (Mauna Loa Record, Keeling et al. 1982). This result serves to remind us that our current understanding of the carbon cycle is unsatis-

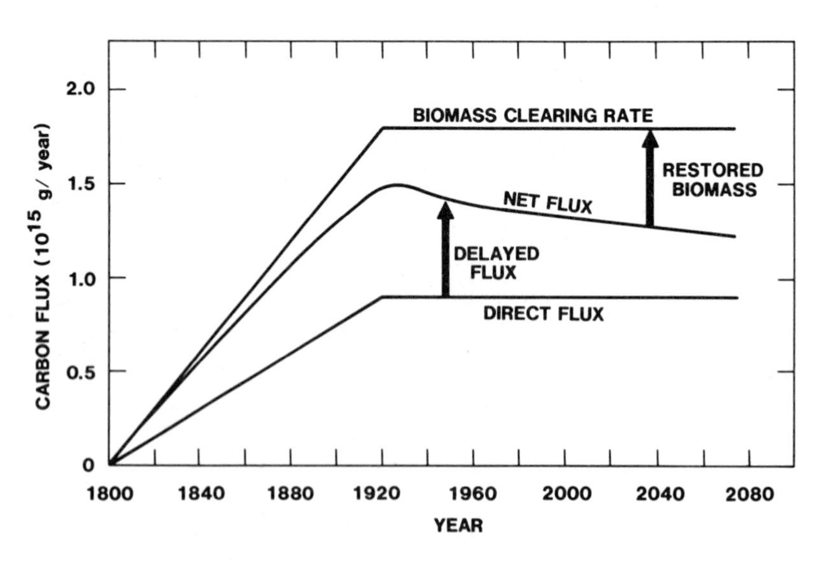

FIGURE 26.8. Baseline scenario of forest clearing rate and atmospheric CO_2 fluxes from direct deforestation, delayed decomposition processes, and forest biomass regrowth for the period 1800 to 2075.

TABLE 26.5. Percentage contributions from carbon cycle model parameter variations to variance in atmospheric CO_2 concentrations with future fossil emissions case B*

Parameter (Range of allowable variation)	Year 1959—1980† Mean	Year 1959—1980† Trajectory‡	2025	2075
Forest clearing rate (±80%)	48.3	22.8	35.1	21.6
Ocean tracer profile§	16.5	46.1	41.9	61.1
Initial atmospheric CO_2 concentration in 1800 (240–280 ppm)	27.2	0.9	12.8	6.7
Pre-1980 fossil emissions (±12%)	1.6	22.8	0.9	0.3
2 x CO_2 temperature increase in ocean surface layer (1.5–4.5°C)	0.7	1.8	1.6	2.1
All others (±20%)¶	4.4‖	3.6‖	6.0‖	5.9‖
Total variance explained	98.7	98.0	98.3	97.7

*Analysis of 600 Monte Carlo iterations (discussion of parameter ranges in Gardner et al. 1985).
†Mauna Loa Record (Keeling et al. 1982).
‡Deviation of model-calculated atmospheric CO_2 trajectory from observed trajectory represented by annual mean values.
§Range of equivalent vertical diffusion coefficient calculated by the method of Oeschger (1975) was 0.95 to 5.2 cm^2/s^{-1}.
¶27 other parameters varied by ±20% to approximate uncertainty; one exception, initial ocean surface temperature in 1800, was allowed to vary only ±2°C (Gardner et al. 1985).
‖No individual contribution >2.3% at any time.

factory. As a consequence, high variability exists for some model parameters, particularly those which define the historical role of the biosphere.

The principal contributors to deviations from the observed level of atmospheric CO_2 was the uncertainty in the forest clearing rate and ocean tracer data used to parameterize vertical mixing (Table 26.5). Although forest clearing, past fossil emissions, and the estimate of the preanthropogenic atmospheric CO_2 content are initially significant sources of uncertainty, their importance declines with time. Error contributions from other parameters are either negligible or insignificant by comparison. By the year 2075, uncertainty in the ocean model dominates the uncertainty in atmospheric CO_2 because fossil emissions (15×10^{15}g C • yr^{-1} in Case B) are the prominent atmospheric inputs. Thus, ocean-atmosphere exchange, and the ocean tracer data, the surrogate for uptake by the deep ocean CO_2 sink, correspondingly dominate uncertainty in predictions of atmospheric CO_2.

Within-model variation, calculated by a Monte Carlo analysis of parameter and input uncertainties, was somewhat greater than, but still quite comparable to, variation introduced by using two carbon cycle models and by varying terrestrial history. For example, the 95% confidence bands for the atmospheric CO_2 concentration were 376 to 394 ppmv (2000), 438 to 478 ppmv (2025), 526 to 598 ppmv (2050), and 651 to 759 ppmv (2075); cf. Case B data in Table 26.4. Thus, the apparent carbon cycle contribution to overall uncertainty in atmospheric CO_2 in the year 2075 is relatively small (~15% of the variation). This is so even though uncertainty estimates for some model parameters would be considered conservative (large) by some authorities (Gardner et al. 1985). These estimates of relative parameter-related contributions to projection uncertainties are similar to those obtained by others using different carbon cycle models (Laurmann and Spreiter 1983, Oeschger and Heimann 1983). It should be recognized, however, that our results are unique to the carbon cycle model and release scenario combinations employed. Thus the impact of future changes in model structure could be greater than that produced by parameter and input uncertainties and cannot be evaluated as part of this exercise.

Conclusions

Although a number of well-posed recent efforts have been made to project the concentration of atmospheric CO_2 into the future (Keeling and Bacastow 1977; Revelle and Munk 1977; Siegenthaler and Oeschger 1978; Niehaus and Williams 1979; Pearman 1980; Bach 1983; Laurmann and Rotty 1983; Laurmann and Spreiter 1983; Markley and Hurley 1983; Nordhaus and Yohe 1983; Oeschger and Heimann 1983; Seidel and Keyes 1983; Perry 1984), it is difficult to make exacting comparisons between the results of this effort and those of others because of major differences in assumptions in models to represent energy/economic forecasts (including exponential growth as a simple model), and in models to represent the carbon cycle (including airborne fraction as a simple model). In economic models, the role of resources, prices, income, technologies, and reaction lags and penetration rates for new technologies are debated. In carbon cycle models, key issues include the efficiency of ocean CO_2 uptake, initial atmospheric CO_2 concentration, and terrestrial biospheric CO_2 release history. Potential variations in initial atmospheric CO_2 concentration and the time series of CO_2 concentration during the industrial era due to effects of terrestrial biospheric history generally have not been rigorously considered as part of a future CO_2 projection effort. In only three of these previous efforts (Bach 1983, Nordhaus and Yohe 1983, Seidel and Keyes 1983) were the outputs of a rigorous energy-economic model of CO_2 release from fossil fuel emissions incorporated into the projections. In only one instance (Nordhaus and Yohe 1983) was an attempt made to estimate the statistical confidence level of the projections of future CO_2 emissions. In the latter exercise, a modified airborne fraction concept rather than a global carbon cycle model was used to make projections of atmospheric CO_2 levels. The assumptions used in this airborne fraction model result in lower predicted retention of atmospheric CO_2 than state-of-the-art carbon cycle models and make it more difficult to compare with other recent studies. It should be noted that our results do fall well within the envelope of results reported in these other exercises, and despite differences in methodologies, our basic conclusions about the future growth of atmospheric CO_2 are virtually identical to those of Nordhaus and Yohe (1983).

Because of the obvious difficulty associated with forecasts of energy/economic/CO_2 emissions futures, our ability to accurately project future atmospheric CO_2 levels is limited. Our exercise indicates that doubling of atmospheric CO_2 is unlikely to occur before the year 2025 and may not occur within the next century. Thus, our results agree quantitatively with the conclusions of Nordhaus and Yohe (1983) even though totally different energy-economic-CO_2 emissions and carbon cycle modeling approaches were used in the two studies. It seems noteworthy that the median estimates from three recent projection exercises (i.e., Nordhaus and Yohe 1983, Seidel and Keyes 1983; Edmonds et al. 1984) indicate that a doubling of the preanthropogenic atmospheric CO_2 level, and most probably the contemporary level as well, could occur by the year 2075. This should provide added impetus for more accurate assessments of climatic and indirect effects of CO_2.

One must be very cautious about making projections, particularly at high rates of fossil fuel emissions, because of the potential for future expression of feedbacks or nonlinear responses in the global carbon system. Existing carbon cycle models may not be valid at higher atmospheric CO_2 concentrations, or at higher release rates, or at later dates used in this CO_2 projection exercise. However, these concerns do not mean that we cannot attempt predictions with existing models, but, rather, that we are aware of their limitations. We are particularly concerned about potential feedbacks that could increase the rates of growth of atmospheric CO_2 above the levels of our projections. These concerns are currently overshadowed by the magnitude of the potential uncertainty in future fossil fuel emissions.

Although projections of global fossil fuel emissions are the greatest apparent source of error in predicting future atmospheric CO_2, it may not be reasonable to assume that a significant reduction in this uncertainty is possible until we better understand the forcing functions of the global energy-economic system. There exist obvious limits to our ability to predict economic/energy-use futures. There is no reason to believe that CO_2 emissions cannot grow as fast as 3% per year. Neither resources nor technology provide absolute constraints on CO_2 emissions of that magnitude. Similarly, there is no guarantee that emissions rates will increase at all. Uncertainties in emissions sources may always remain dominant in CO_2 projections. They are at once a forecasting problem and an expression of man's potential to control his own destiny. The real challenge is to forge policy decisions intelligently in the face of uncertainty.

Yet the study of CO_2 emissions from fossil fuel use has resulted in significant changes in thinking about the problem of CO_2 emissions on both the qualitative and quantitative levels. The mode of analysis has shifted dramatically from simple time trend analysis to energy-economic analysis. The vehicles used for projecting future CO_2 emissions have changed from simple log-linear extrapolation models to behavioral energy-economic models designed specifically to investigate long-term global energy-economic relationships. Thus the focus of research has shifted from CO_2 emissions to long-term global energy systems, and to the demographic, socioeconomic, and technological factors that will define and limit the energy choices that are available to society over the next century. The analytical perspective has broadened. The single "best estimate" scenario has been supplanted by an explicit recognition of uncertainty, e.g., Nordhaus and Yohe (1983) constructed CO_2 emissions projections using a Monte Carlo simulation and Edmonds et al. (1984) developed subjective scenario bounds, and by alternative scenario development, e.g. Rose et al. (1983), Häfele (1981), Seidel and Keyes (1983). As a consequence, our understanding of likely future CO_2 emissions rates has changed dramatically as well, and despite the broad uncertainty that remains, the current consensus view of CO_2 emissions from fossil fuels finds a continuation of the postwar time trend extremely unlikely. Emissions of CO_2 from fossil fuel use are likely to increase, but at a lower rate than in the past. The "best estimate" scenarios show that rate to be not only lower, but significantly lower than the postwar time trend.

Variability in the fossil fuel forcing has such a dramatic impact on atmospheric CO_2 prediction uncertainty that one can be misled to conclude that further research on the global carbon cycle has a very low priority, or, alternatively, that carbon cycle research ought to be focused exclusively on the oceanic component. However, our view of the uncertainty analysis results is quite different if the atmospheric concentration must not exceed 500 ppmv or, alternatively, must be reduced to 350 ppmv as quickly as possible to minimize deleterious environmental and human impacts. All major systems in the global carbon cycle then contribute significantly to uncertainty in atmospheric CO_2 content (see results for year 2025 in Tables 26.4 and 26.5). Since we would then require much more accurate projections of atmospheric levels, uncertainty in *all* systems of the global geochemical cycle, as well as the fossil fuel CO_2 emissions component would have to be reduced. Further, mitigation strategies or lesser CO_2-climate feedbacks involving the natural carbon cycle, which would not be considered at very high levels of atmospheric CO_2, e.g., in the context of limits to atmospheric content, now could play an important role in controlling atmospheric concentrations. It should also be recognized that immediate implementation of mitigation strategies might not be reflected in stabilized or reduced atmospheric CO_2 levels for 50 to 100 years because of the inertia inherent in responses of human political, economic, and technological systems and of the global carbon cycle (Perry et al. 1982; Bach 1983; Laurmann 1983). Finally, by improving our understanding of the sources of uncertainty we may be better able to formulate approaches that reflect that underlying reality. For example, CO_2 emissions may remain low unless the rate of end-use energy efficiency is low. This information provides a bellwether upon which to focus attention and a narrowed range over which intervention might be considered.

The presence of other atmospheric trace constituents (e.g., methane, nitrous oxide, Freons) could augment the predicted climatic response due to CO_2. This might be of particular significance if trace gas concentrations continue to increase at rates observed over the past several decades and if fossil fuel CO_2 releases continue at the reduced rates observed over the past several years (Marland and Rotty 1983; Wuebbles et al. 1984; low emissions cases suggested by others and Case C, this work). Readers of our work should realize, therefore, that we have presented only a portion of the atmospheric gas concentration data needed to attempt modeling of future climatic responses, and that the carbon cycle is only one of several element cycles that must be understood to address climate-related issues.

Acknowledgments

We thank William Emanuel and Berrien Moore for performing atmospheric CO_2 projections for us. We also acknowledge the contributions of many reviewers. This research was supported by the U.S. Department of Energy, Office of Energy Research, Carbon Dioxide Research Division, under contract No. DE-AC05-84-OR21400 with Martin Marietta Energy Systems, Inc., and contract No. DE-AC05-760R0033 with the Institute for Energy Analysis, Oak Ridge Associated Universities.

References

Bach, W. 1983. Carbon dioxide/climate threat: fate or forbearance? In W. Bach et al. (eds.), Carbon Dioxide: Current Views Developments in Energy/Climate Research, pp. 461–509. D. Reidel, Dordrecht.

Björkström, A. 1979. A model of CO_2 interaction between atmosphere, oceans, and land biota. In B. Bolin, E. T. Degens, S. Kempe, and P. Ketner (eds.), The Global Carbon Cycle, Scope 13, pp. 403–457. John Wiley & Sons, New York.

Edmonds, J. A. and J. Reilly. 1983*a*. A long-term global energy-economic model of carbon dioxide release from fossil fuel use. Energy Econ. 5(2):74–88.

Edmonds, J. A. and J. Reilly. 1983*b*. Global energy production and use to the year 2050. Energy 8:419–432.

Edmonds, J. A. and J. Reilly. 1983*c*. Global energy and CO_2 to the year 2050. Energy J. 4(3):21–47.

Edmonds, J. A. and J. Reilly. 1985. Global Energy: Assessing the Future. Oxford University Press, New York.

Edmonds, J. A., J. Reilly, J. R. Trabalka, and D. E. Reichle. 1984. An Analysis of Possible Future Atmospheric Retention of Fossil Fuel CO_2, DOE TR-013. U. S. Department of Energy Technical Report Series. National Technical Information Services, Springfield, Virginia.

Emanuel, W. R., G. G. Killough, W. P. Post, and H. H. Shugart. 1984*a*. Modeling terrestrial ecosystems in the global carbon cycle with shifts in storage capacity by land-use change. Ecology 65:970–983.

Emanuel, W. R., G. G. Killough, M. P. Stevenson, W. M. Post, and H. H. Shugart. 1984*b*. Computer implementation of a globally averaged model of the world carbon cycle, DOE TR-010. U.S. Department of Energy Technical Report Series. National Technical Information Services, Springfield, Virginia.

Gardner, R. H. (1985) Error analysis and sensitivity analysis in ecology. In Macan Singh (ed.) Encyclopedia of Systems and Control. Pergamon Press, London.

Gardner, R. H., B. Rojder, and U. Bergstrom. 1983. PRISM: A systematic method for determining the effect of parameter uncertainties on model predictions, AB Report/NW-83/555. Studsvik Energiteknik, Nykoping, Sweden.

Gardner, R. H., J. R. Trabalka, and W. R. Emanuel. (1985) Methods of uncertainty analysis for a global carbon dioxide model, DOE TR-024. U. S. Department of Energy Technical Report Series. National Technical Information Services, Springfield, Virginia.

Häfele, W. 1981. Energy In a Finite World. Ballinger, Cambridge, Massachusetts.

Houghton, R. A., J. E. Hobbie, J. M. Melillo, B. Moore, B. J. Peterson, G. R. Shaver, and G. M. Woodwell. 1983. Changes in the carbon content of terrestrial biota and soils between 1860 and 1980: a net release of CO_2 to the atmosphere. Ecol. Monogr. 53:235–262.

JASON. 1979. The Long-Term Impact of Atmospheric Carbon Dioxide on Climate. Technical Report JSR-78-07. SRI International, Arlington, Virginia.

Keeling, C. D. 1973. Industrial production of carbon dioxide from fossil fuels and limestone. Tellus 25:174–198.

Keeling, C. D. and R. B. Bacastow. 1977. Impact of industrial gases on climate. In Energy and Climate, pp. 72–95. National Academy of Sciences, Washington, D. C.

Keeling, C. D., R. B. Bacastow, and T. P. Whorf. 1982. Measurements of the concentration of carbon dioxide at Mauna Loa Observatory, Hawaii. In W. C. Clark (ed.), Carbon Dioxide Review: 1982, pp. 377–385. Oxford University Press, New York.

Laurmann, J. A. 1983. Strategic issues and the CO_2 environmental problem. In W. Bach et al. (eds.), Carbon Dioxide: Current Views and Developments in Energy/Climate Research, pp. 415–460. D. Reidel, Dordrecht.

Laurmann, J. A. and R. M. Rotty. 1983. Exponential growth and atmospheric carbon dioxide. J. Geophys. Res. 88:1295–1299.

Laurmann, J. A. and J. R. Spreiter. 1983. The effects of carbon cycle model error in calculating future atmospheric carbon dioxide levels. Clim. Change 5:145–181.

Lovins, A. B., L. H. Lovins, F. Krause, and W. Bach. 1981. Energy Strategies for Low Climate Risks. Prepared for the German Federal Environmental Agency, San Francisco International Project for Soft-Energy Paths, San Francisco, California.

Machta, L. 1983. The atmosphere. In Changing Climate, pp. 242–251. National Academy Press, Washington, D.C.

Markley, O. W. and T. J. Hurley III. 1983. A brief technology assessment of the carbon dioxide effect. Technol. Forecast. Soc. Change 23:185–202.

Marland, G. and R. M. Rotty. 1983. Carbon dioxide emissions from fossil fuels: a procedure for estimation and results for 1950–1981, DOE TR-0003, U. S. Department of Energy Technical Report Series. National Technical Information Services, Springfield, Virginia.

Niehaus, F. and J. Williams. 1979. Studies of different energy strategies in terms of their effects on the atmospheric CO_2 concentration. J. Geophys. Res. 84(c6):3123–3129.

Nordhaus, W. D. 1979. The Efficient Use of Energy Resources. Yale University Press, New Haven, Connecticut.

Nordhaus, W. D. and G. Yohe. 1983. Future carbon dioxide emissions from fossil fuels. In Changing Climate, pp. 87–153. National Academy Press, Washington, D.C.

Oeschger, H. and M. Heimann. 1983. Uncertainties of predictions of future atmospheric CO_2 concentrations. J. Geophys. Res. 88:1258–1262.

Olson, J.S., J. A. Watts, and L. J. Allison. 1983. Carbon in Live Vegetation of Major World Ecosystems, ORNL-5862. Oak Ridge National Laboratory, Oak Ridge, Tennessee.

Pearman, G. I. 1980. The global carbon cycle and increasing levels of atmospheric carbon dioxide. In Carbon Dioxide and Climate: Australian Research, pp. 11–20. Australian Academy of Sciences, Canberra, Australia.

Peng, T.-H., W. S. Broecker, H. D. Freyer, and S. Trumbore. 1983. A deconvolution of the tree-ring-based ^{13}C record. J. Geophys. Res. 88:3609–3620.

Perry, A. M. 1984. Atmospheric retention of anthropogenic CO_2: Scenario dependence of the airborne fraction. EPRI EA-3466. Oak Ridge National Laboratory, Oak Ridge.

Perry, A. M., K. J. Araj, W. Fulkerson, D. J. Rose, M. M. Miller, and R. M. Rotty. 1982. Energy supply and demand implications of CO_2. Energy 7:991–1004.

Post, W. M., W. R. Emanuel, P. J. Zinke, and A. G. Stangenberger. 1982. Soil carbon pools and world life zones. Nature 298:156–159.

Reister, D. B. and R. M. Rotty. 1983. Scenario analysis of future global fossil fuel consumption. Energy 8(4):283–289.

Revelle, R. and W. Munk. 1977. The carbon dioxide cycle and the biosphere. In Energy and Climate, pp. 140–158. National Academy of Sciences, Washington, D.C.

Rose, D., M. Miller, and C. Agnew. 1983. Global Energy Futures and CO_2-Induced Climate Change, MITEL 83-015, MIT Energy Laboratory. Massachusetts Insitute of Technology, Cambridge, Massachusetts.

Rotty, R. M. and G. Marland. 1980. Constraints on Carbon Dioxide Production From Fossil Fuel Use. ORAU/IEA-80-9(m). Institute for Energy Analysis, Oak Ridge, Tennessee.

Seidel, S. and D. Keyes. 1983. Can We Delay a Greenhouse Warming? U.S. Environmental Protection Agency, Washington, D.C.

Siegenthaler, U. and H. Oeschger. 1978. Predicting future atmospheric carbon dioxide levels. Science 199:388–395.

Stuiver, M., R. L. Burk, and P. D. Quay. 1984. $^{13}C/^{12}C$ Ratios and the transfer of biospheric carbon to the atmosphere. J. Geophys. Res. 89:11731–11748.

Woodwell, G. M., J. E. Hobbie, R. A. Houghton, J. M. Melillo, B. J. Peterson, G. R. Shaver, and T. A. Stone. 1983*a*. Deforestation Measured by LANDSAT: Steps Toward a Method, DOE TR-005. U. S. Department of Energy, Technical Report Series. National Technical Information Services, Springfield, Virginia.

Woodwell, G. M., J. E. Hobbie, R. A. Houghton, J. M. Melillo, B. Moore, B. J. Peterson, and G. R. Shaver. 1983*b*. Global deforestation: contribution to atmospheric carbon dioxide. Science 222:1082–1086.

Wuebbles, D. J., M. C. MacCracken, and F. M. Luther. 1984. A Proposed Reference Set of Scenarios for Radiatively Active Atmospheric Constituents, TR-015. U. S. Department of Energy Technical Report Series. National Technical Information Services, Springfield, Virginia.

27
Possible Changes in Future Use of Fossil Fuels to Limit Environmental Effects

ALFRED M. PERRY

Full use of the world's recoverable fossil fuels could lead to atmospheric CO_2 concentrations up to 10 times the preindustrial concentration (Keeling and Bacastow 1977; Siegenthaler and Oeschger 1978; Perry et al. 1982) (See Fig. 27.1). It is not possible at present to specify a maximum allowable CO_2 concentration, i.e., a level at which the incremental losses would begin to exceed the incremental benefits, in some aggregate sense. Nevertheless, it is often assumed that a concentration of 600 parts per million by volume (ppm) would induce significant climatic and other changes that should perhaps be avoided if possible. Calculations of future CO_2 concentrations are usually based on estimates for future worldwide consumption of fossil fuels that do not incorporate any restrictions on CO_2 emissions. Some of us at the Oak Ridge National Laboratory, along with colleagues at the Institute for Energy Analysis and at the Massachusetts Institute of Technology, have asked how difficult it might be (and when it might be necessary to begin) to modify the future use of fossil fuels so that the atmospheric CO_2 concentration would never exceed some specified level, such as 500, 600, or 800 ppm (Perry et al. 1982).

Approach

The choice of CO_2 limits between 500 and 800 ppm was not arbitrary, but was related to perceived uncertainties in (1) the magnitude of climate changes that might be considered "acceptable," (2) the sensitivity of climate change to increasing atmospheric CO_2, and (3) the role of other atmospheric trace gases that may also influence climate and whose concentrations may also be influenced by human activities. Climate change is represented by a range of allowable changes, ΔT, in the average global surface temperature, $2K \leq \Delta T < 5K$ (Perry et al. 1982). These very tentative limits were suggested by consideration of past climatic changes and by concern for the large changes that might accompany an ice-free Arctic Ocean (Flohn 1982). The doubling ΔT of CO_2, i.e., the temperature rise, ΔT_2, that would result from a doubling of the CO_2 concentration, was

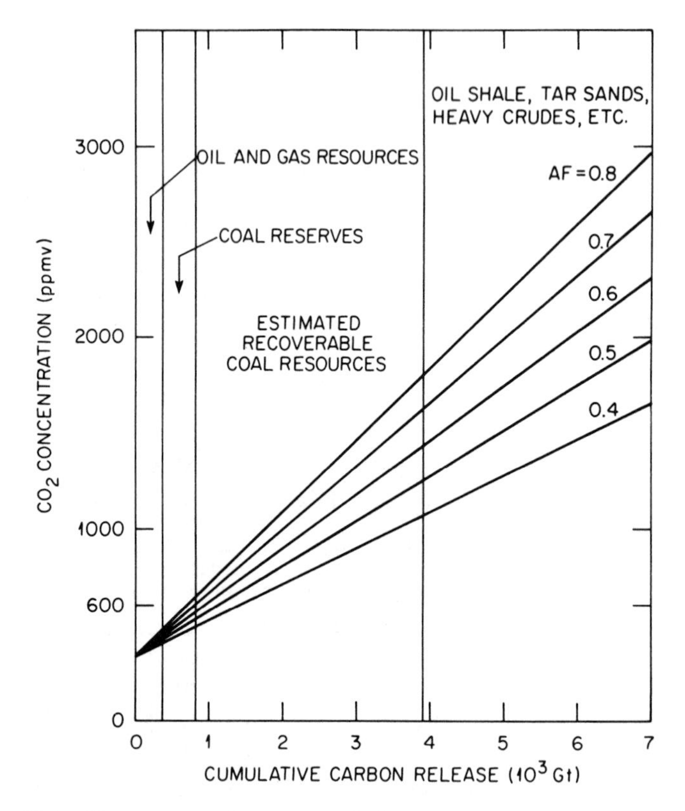

FIGURE 27.1. CO_2 concentration vs cumulative post-1980 carbon release. AF, Airborne fraction.

assumed to lie between 1.5 and 4.5K, though higher or lower values cannot be excluded. The effect of other trace gases that also absorb infrared radiation (e.g., methane, nitrous oxide, chlorofluorocarbons) was incorporated by multiplying ΔT_2 by a factor, n, assumed to be between 1.0 and 1.6, although, once again, higher values cannot be excluded. In any case, n (the total temperature rise divided by the rise due to CO_2) would not be a constant, but would depend on the sources and sinks for each trace gas. To the extent that uncertainties in the climatic effect of increasing trace-gas concentrations stem primarily from uncertainties in the climatic response to changes in radiative forcing, rather than from uncertainties in the infrared (IR) absorption properties of the gases, the uncertainties in the respective contributions of CO_2 and of other trace gases to global climate change will be correlated. Thus the augmentation factor n will depend mainly on the relative emission rates of the various gases and on their chemical behavior in the atmosphere.

A feeling for the maximum allowable CO_2 concentration may be obtained from the expression

$$C = C_0 \exp(\tau \ln 2) \tag{1}$$

where C is the allowable CO_2 concentration, C_0 is the initial concentration (e.g., 300 ppm), and $\tau = \Delta T/(n\Delta T_2)$ is the allowable ΔT expressed in terms of the generalized sensitivity parameter, $n\Delta T_2$ (see Table 27.1). Thus, a CO_2 limit of 500 ppm ($\tau = 0.74$) or less corresponds, for example, to the combination of a low allowable global temperature increase (such as 2°K) with an intermediate to high sensitivity parameter, $n\Delta T_2 \geqslant 3$. Indeed, a value $n\Delta T_2 = 5$, together with a ΔT limit of 2°K, suggests a CO_2 limit of 400 ppm, which we believe will almost inevitably be exceeded. A CO_2 limit of 800 ppm ($\tau = 1.42$) or more corresponds, for example, to the combination of a high allowable global temperature increase (such as 4 to 5°K) with an intermediate to low sensitivity parameter, $n\Delta T_2 \leqslant 3.5$.

Action Initiation Time

It is our impression that the collection and sequestering of CO_2 from combustion gases is probably not a viable approach to restricting CO_2 emissions to the atmosphere (Baes et al. 1980; but see also Steinberg 1984). We have assumed that restricting CO_2 emissions would require corresponding restrictions in the use of fossil fuels. Since we have postulated increasing worldwide demand for energy (see below), restricting CO_2 emissions implies an accelerated transition from fossil to nonfossil sources of energy. The maximum CO_2 concentration would depend on the date when policies designed to curtail the use of fossil fuels would begin to take effect [which we have called the Action Initiation Time (AIT)], and on how rapidly the transition from fossil to nonfossil energy sources could take place. We were particularly interested in the required AIT as a function of the specified maximum CO_2 concentration and as a function of parameters characterizing the rapidity of the transition. How urgent is the problem?

Table 27.1. Maximum allowable CO_2 concentration (ppm) as a function of allowable temperature rise and the generalized sensitivity parameter, $n\,\Delta\,T_2$

Sensitivity	T, Allowable temperature rise			
$n\,\Delta\,T_2$	2K	3K	4K	5K
2	600	849	1200	1697
2.5	522	689	909	1200
3	476	600	756	952
3.5	446	543	662	808
4	424	505	600	714
5	396	455	522	600
6	378	424	476	534
7	366	404	446	492

Do we have time to learn more about the carbon cycle, climate effects, and so forth before moving to limit the use of fossil fuels?

In our analysis, total energy demand was based on the two scenarios presented by the International Institute for Applied Systems Analysis (IIASA 1981). Although the CO_2 concentrations for these scenarios (based on the distribution of energy sources given by IIASA) would probably not exceed 500 ppm before 2030 (depending on what the airborne fraction would actually prove to be for these scenarios), it was clear that preventing the maximum CO_2 concentration from ever exceeding 500 ppm would require a departure from the fossil fuel consumption of the IIASA scenarios beginning substantially before 2030.

We described the transition from fossil to nonfossil energy sources, following the AIT, in two different ways, which proved to give very similar shapes for the time profiles of fossil and nonfossil energy use. In one method, we specified the maximum CO_2 concentration and linked the concentration at the AIT to the specified maximum value by a logistic function

$$C(t) = \lambda + \beta(1 + \gamma e^{-\alpha t})^{-1} \qquad (2)$$

where the parameters α, β, γ, and λ are determined by the CO_2 concentrations at the AIT and asymptotically, and by requiring continuity of the first and second time derivatives of C at the AIT.

A constant airborne fraction of 0.53 was assumed, as well as a constant CO_2 emission rate per unit of fossil energy consumption (0.634 gigatons of carbon per TW year*). The continuity conditions correspond to continuity of fossil energy consumption and its first time derivative at the AIT. The use of a constant cumulative airborne fraction is not strictly correct, and certainly not for projections far into the future when annual CO_2 emissions will have diminished to low levels. However, comparison with results of calculations with detailed carbon cycle models indicates that the asymptotic CO_2 concentration obtained with a constant airborne fraction is a reasonably good measure of the maximum concentration to be expected for scenarios, such as those considered here, in which the use of fossil fuels decreases fairly rapidly after the time of peak use. The error incurred by using a constant airborne fraction and by using the asymptotic CO_2 concentration as an indication of the transient maximum concentration is generally smaller than the uncertainty in the true magnitude of the airborne fraction.

In the second method for describing the fossil-to-non-fossil transition, we assumed for each base scenario (i.e., the IIASA high and low scenarios) a fixed projection for the production of oil and gas, corresponding to total post-1980 production of 2000 billion barrels (273×10^9 metric tons) of oil

*TW-year = 8.76×10^{12} kilowatt-hours.

and 9000 trillion cubic feet ($255 \times 10^{12}m^3$) of natural gas. Together, these quantities of oil and gas contain about 370 gigatons of carbon and would yield an increase of only about 90 ppm in the CO_2 concentration. We assumed that use of these premium fuels would be little affected by concern over the greenhouse effect and their profiles were not altered in the CO_2-limiting scenarios.

Growth rates of coal consumption in the IIASA scenarios averaged approximately 2%/year and 3%/year for the low and high scenarios, respectively. For this study, we assumed initial coal growth rates of 3%/year and 4%/year for the IIASA high scenario and 2%/year and 3%/year for the low scenario; the higher values in each case correspond to an assumption that much of IIASA's nuclear energy contribution would be replaced by coal. These scenarios result in average growth rates of CO_2 emissions over the period 1980 to 2000 of approximately 1.5, 1.9, 2.4, and 2.8%/year.

Starting at the AIT, the rate of growth of coal consumption (held constant for each scenario from 1980 until the AIT) was assumed to decrease at either 0.1%/year per year or 0.2%/year per year until a negative slope of −1, −2, or −4%/year was reached. At this point, the coal consumption profile was smoothly joined to a negative exponential "tail." Typically, a "tail" of −1%/year contributed nearly three-fourths of the total post-1980 coal consumption; a tail of −2%/year contributed above half, and a tail of −4%/year, one-fourth or less of the total post-1980 coal use. Thus, it hardly seemed worthwhile to consider a wider range of possibilities.

For a deceleration rate of −0.1%/year per year, the time after the AIT during which the consumption of coal remains above its value at the AIT is very long, i.e., 80 years or more for an initial growth rate of 4%/year and 40 years or more for an initial growth rate of 2%/year. For a deceleration of −0.2%/year per year, these times are cut in half. On the basis of the above observations, we considered a deceleration rate of −0.1%/year per year, combined with a tail of −1%/year or −2%/year to be a relatively slow and easy transition, whereas a deceleration rate of −0.2%/year per year, combined with a tail of −2%/year or −4%/year was considered a relatively rapid and difficult transition.

Other measures of difficulty were also considered. The first and second time-derivatives of E_{NF}, the nonfossil component of energy supply, represent respectively the required rate of introduction of nonfossil energy facilities and the required rate of expansion of manufacturing and construction capabilities for providing these facilities. These were viewed as major constraints on the fossil-to-nonfossil transition and hence on the AIT. Although it is difficult to justify specific limiting values of these parameters, the analysis showed clearly that the second derivative, especially, is a sensitive function of the AIT (Araj 1982; Perry et al. 1982). For scenarios judged to be feasible in terms of deceleration rate and exponential tail (e.g., −0.2%/year per year deceleration, −2%/year tail), the rate of expansion of manufacturing capability for nonfossil energy facilities was typically less than 30 Gw (thermal)/ year per year, which is comparable to adding each year a capability equal to roughly one-third of the present worldwide capacity for constructing nuclear power plants.

We considered also the implied market penetration rates for nonfossil sources as described by Marchetti at IIASA (IIASA 1981). Marchetti concluded that the time required for a new energy source to increase its market share from 10% to 50% is typically about 50 years. That is, the market share, f, is given by $f = x(1 + x)^{-1}$ where $x = (1/9)\exp[(\ln 9)(t/T)]$ and T is the market penetration time defined above. Scenarios with "feasible" combinations of deceleration rate and exponential tail typically had market penetration times on the order of 40 to 50 years.

Results

Detailed results of these analyses are presented in Perry et al. (1982). Here, I present a simplified summary for generalized scenarios defined solely in terms of the initial rate of growth of CO_2 emissions (starting at 5.3 gigatons of carbon per year in 1980), and in terms of the AIT and the transition parameters (deceleration and exponential tail). Figure 27.2 shows the required AIT for CO_2 limits of 500 and 800 ppm. (For these computations, a slightly higher airborne fraction of 0.55 was assumed.)

The AIT, as defined here, implies sustained and effective policies to moderate the future growth of fossil-fuel use, requiring some measure of international agreement and cooperation. It would certainly be preceded by some years of discussion and negotiation. It seems unlikely that the AIT, in this sense, could occur much before the turn of the century.

We concluded (Perry et al. 1982; Perry 1982) that it would be very difficult to prevent atmospheric CO_2 from reaching 500 ppm, even if strong international efforts to do so (by shifting away from fossil fuels) were begun very soon. Of course, the ease or difficulty of limiting CO_2 depends in part on how rapidly annual carbon emissions would continue to grow before policies to slow down that growth would start to be effective. The much lower growth rates (e.g., 1 to 2% per year), which now seem likely anyway in comparison with those expected a few years ago (4 to 5% per year), certainly eases matters considerably. We found that a limit of 600 ppm should be significantly easier to satisfy, even if the slowing down of growth did not begin for a few years. A limit of 800 ppm or more, we concluded, would require no near-term actions specifically to limit growth in fossil fuel use. Even with growth in annual carbon emissions of 3% per year, there appeared to be a grace period of two or three decades before curtailment policies would have to take effect. Growth rates of 1 or 2% per year, which now seem more likely, could continue well into the next century without jeopardizing the 800-ppm limit.

Role of Other Greenhouse Gases

In reaching these conclusions, as noted above, we did not consider in detail other anthropogenic greenhouse gases. We did make a proportional allowance for them, however, assigning them a fraction of the allowable temperature rise, which was absorbed in the CO_2-temperature sensitivity parameter and contributed to the perceived uncertainty in that parameter (Perry et al. 1982). Thus, in effect, we assumed that if climate change proves to be enough of a problem to provoke major changes in fossil fuel consumption, it would probably also require efforts to control the other greenhouse gases. However, we did not look at what such efforts might entail.

The U.S. Environmental Protection Agency (EPA) has reported the results of an analysis (Seidel and Keyes 1983) that underscores the importance of the other greenhouse gases. The central conclusion of this analysis is that even very radical restrictions on the future use of fossil

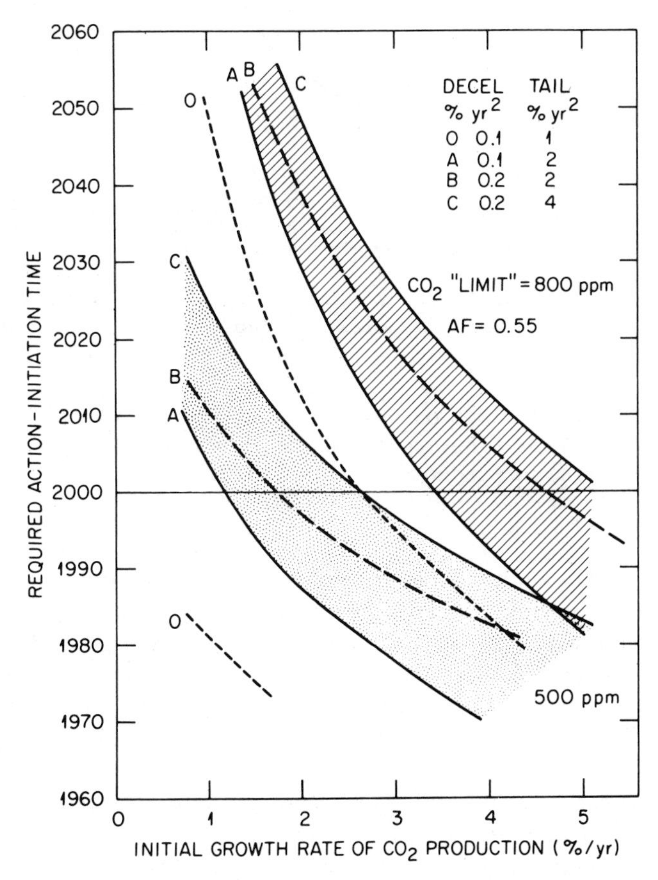

FIGURE 27.2. Required action initiation time (AIT) as a function of initial growth rate of annual carbon release, limiting CO_2 concentration and rapidity of transition away from fossil fuels (starting at the AIT).

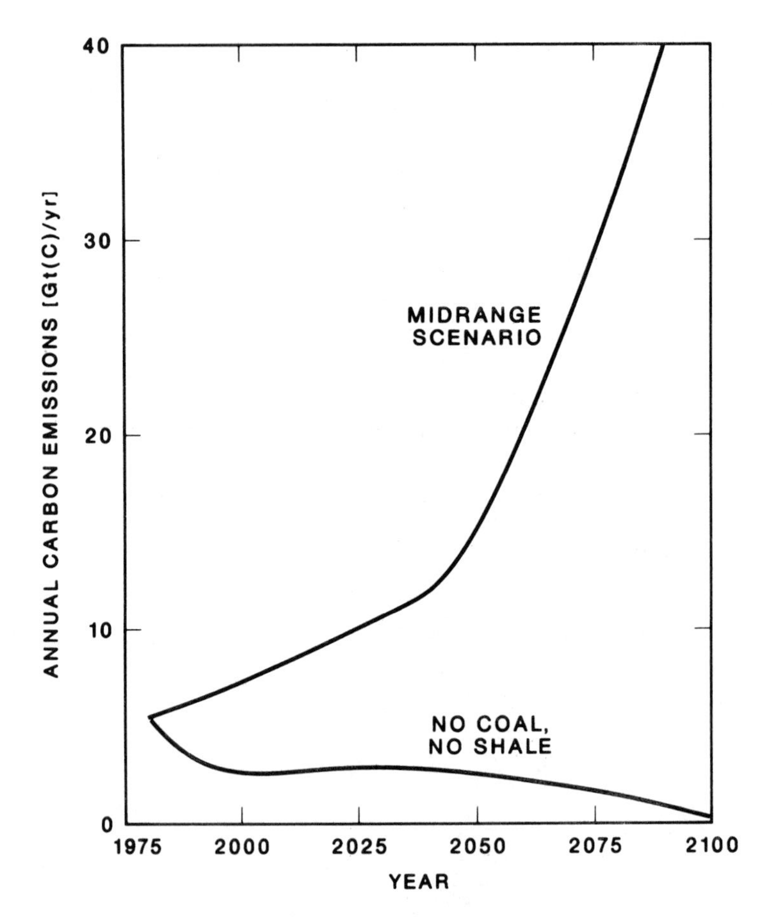

FIGURE 27.3. Annual carbon release rate (Gt per year) for two EPA scenarios (Seidel and Keyes 1983) showing the mid-range scenario and the extreme low-carbon scenario (no coal and no shale oil).

fuels could delay an inevitable greenhouse warming by only a few years. The basis for this conclusion can be understood by reference to Figs. 27.3 and 27.4, redrawn here from Fig. 4-14 of the EPA report. Fig. 27.3 shows annual carbon emissions (as CO_2) for EPA's most likely scenario and for a scenario with very severe restrictions on future fossil fuel use. Figure 27.4 shows the resulting global temperature rise for these scenarios. To calculate the temperature rise, the EPA has combined models for determining how much of the emitted CO_2 remains in the atmosphere, how the radiation balance is affected by CO_2 and other atmospheric trace gases, and how the temperature rise may be delayed by the thermal capacity of the oceans.

Figure 27.4 indicates that even for such radically different carbon-emission scenarios, the time when the global temperature rise reaches 2°C is delayed only 25 years, from 2040 to 2065. This time of reaching a 2°C ΔT is taken by EPA as the primary test of the effectiveness of the bans on fossil fuels, although the report also points out the substantial reduction in the calculated temperature rise at the year 2100.

This selection for the primary test is a bit unfortunate, since it diverts attention from other important differences between the scenarios. To illustrate this point, Fig. 27.5 shows two extremely different scenarios (not from the EPA report) for future carbon emissions. The high one is the upper limit case of Siegenthaler and Oeschger 1978. All that matters here, however, is that one is very high and the other one markedly lower.

The upper portion of Fig. 27.6 shows the CO_2 concentrations for these two scenarios as calculated with Killough and Emanuel's (1981) model 3b (also used in the EPA calculations). The high scenario reaches almost 3000 ppm, whereas the low one tops out at 600 ppm. The temperature

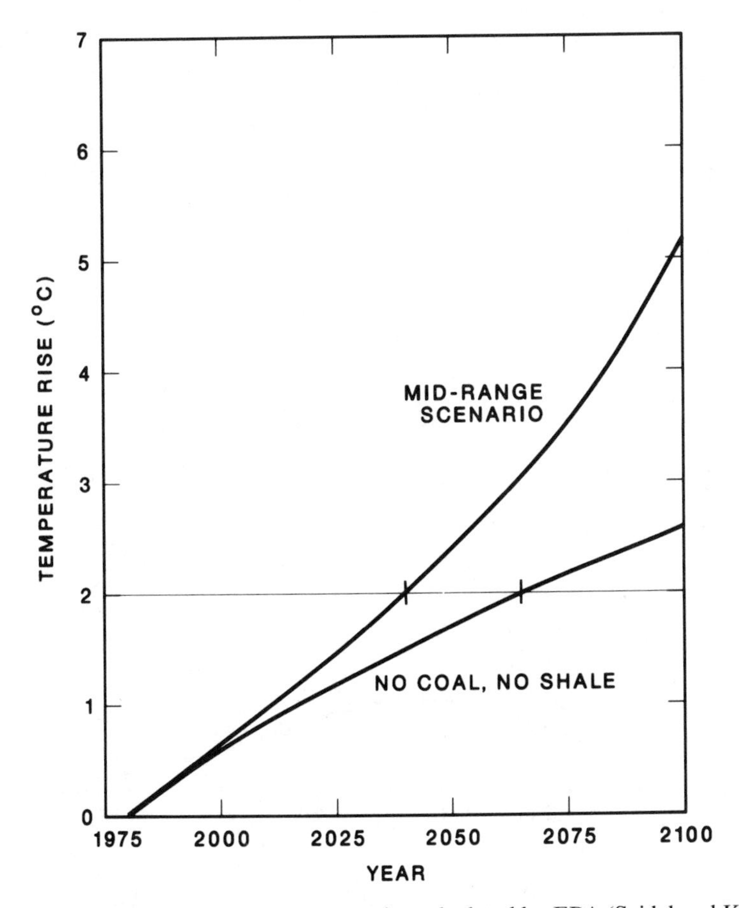

FIGURE 27.4. Average global temperature rise calculated by EPA (Seidel and Keyes 1983) for the two scenarios shown in Fig. 27.3.

rises for these two cases, calculated with a nominal temperature sensitivity of 3°C per CO_2 doubling (as in the EPA report) and taking only the CO_2 into account, are shown in the lower portion of Fig. 27.5. These, too, are very different—about 9°C for the high scenario and only 2.5°C for the low one. However, the time of reaching 2°C differs only by 30 years. We note that these are "equilibrium" temperatures, that is, not accounting for ocean thermal lag. These differences in temperature serve to illustrate the point that this primary test does not really differentiate adequately between two such radically different cases.

Why is it that the nearly complete elimination of CO_2, as illustrated in Figs. 27.3 and 27.4, makes so little difference? The reason, of course, is that there must be a large contribution from something else; that is, the

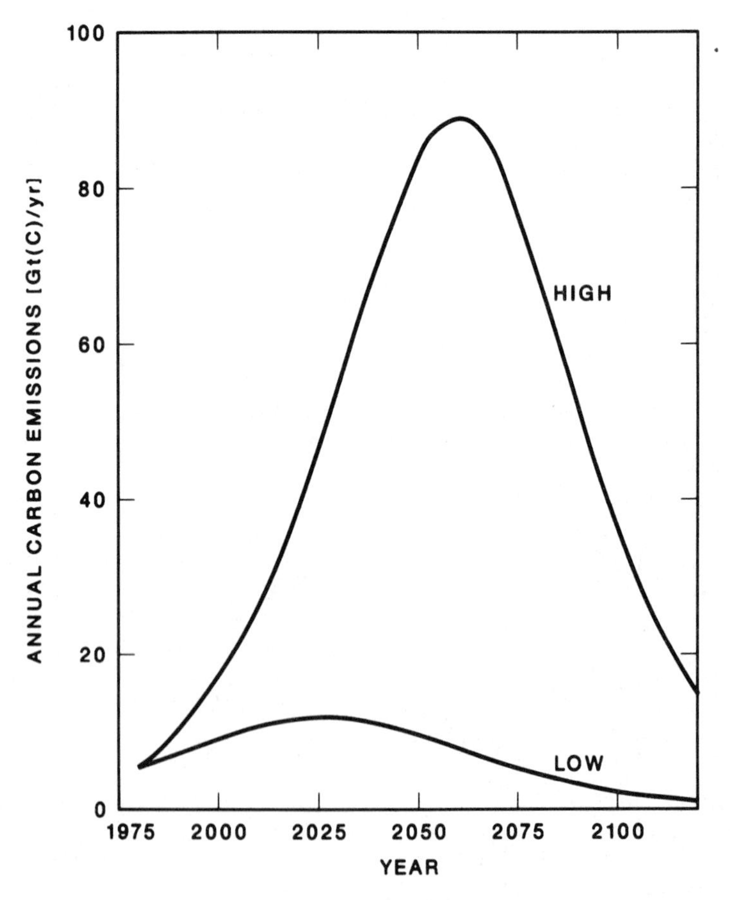

FIGURE 27.5. Two hypothetical carbon-release scenarios, one low and one very high.

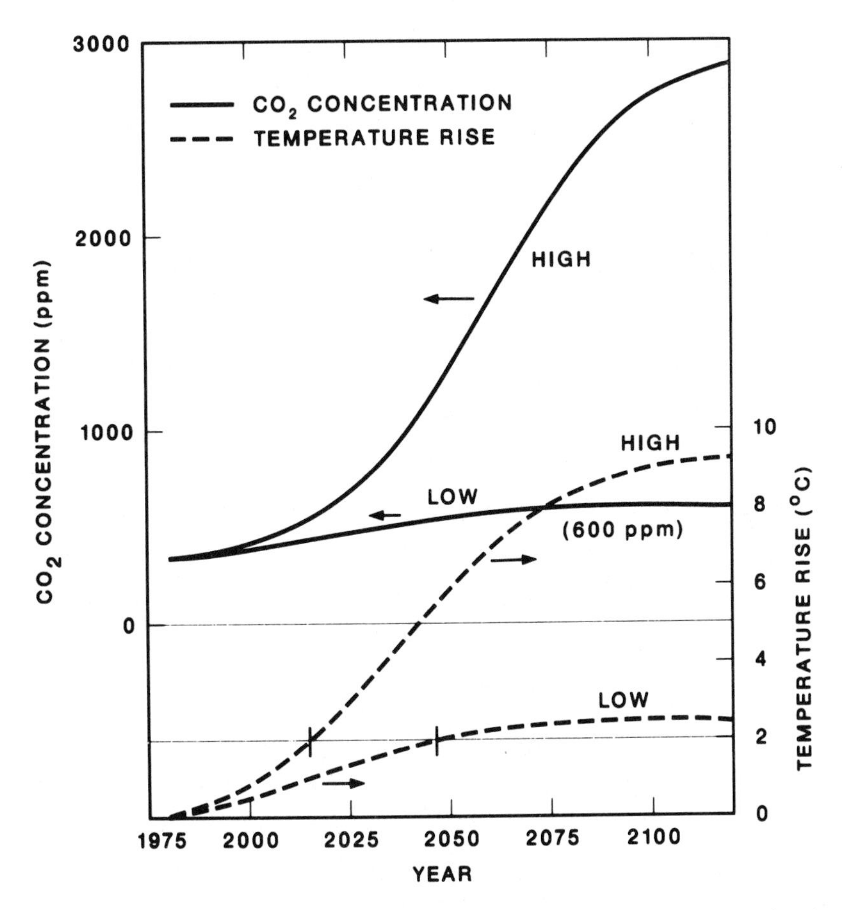

FIGURE 27.6. CO_2 concentrations and average equilibrium temperature rise (no ocean thermal lag) for the scenarios shown in Fig. 27.5.

other greenhouse gases, and notably (according to the EPA analysis) methane, nitrous oxide, and the chlorofluoromethanes. Figure 27.7 shows approximately the respective contributions of CO_2 and the other greenhouse gases to the total calculated temperature rise in EPA's mid-range reference scenario. Up to about 2060, less than half of the temperature rise can be attributed to CO_2.

Figure 27.8 shows a similar allocation for the no coal-no shale scenario. In this case, only about one-fifth of the temperature rise is caused by increasing CO_2, and one begins to see why reducing the CO_2 emissions does not appear to be very effective. The contributions of the other greenhouse gases are also shown in Fig. 27.8. The major contributor is methane, whose concentration in the atmosphere was assumed to increase at 2%/year over the entire period from 1980 to 2100. It is important to note that

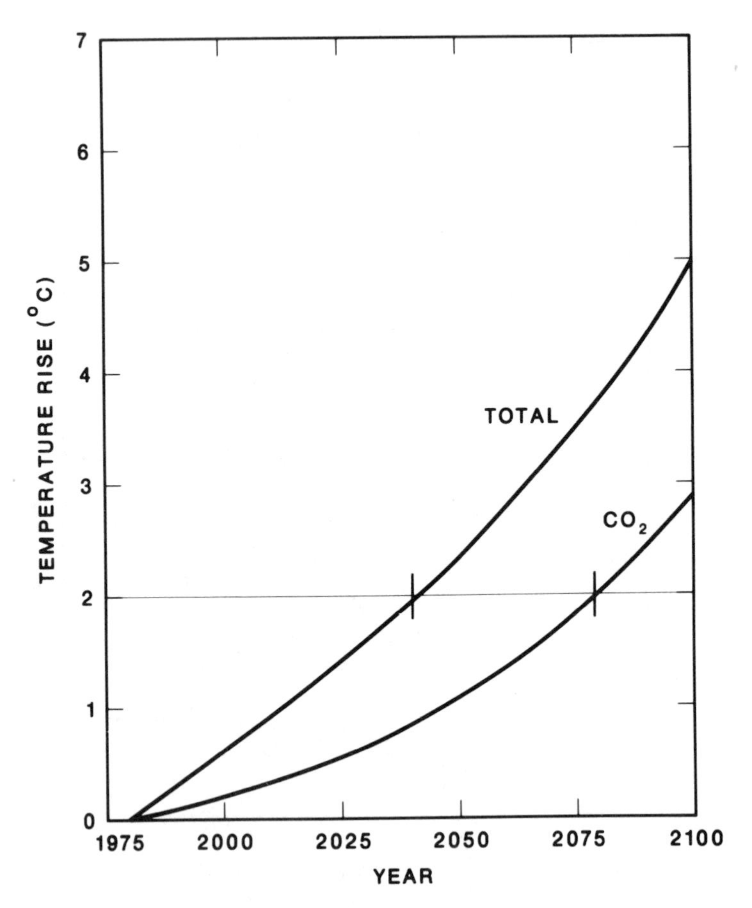

FIGURE 27.7. Average global temperature rise (with ocean thermal lag) for the EPA mid-range scenario, showing the CO_2 contribution.

the allocation of temperature rise among the various gases was based on an expression given in Appendix D of the EPA report (Seidel and Keyes 1983). This expression probably overestimates the effects of CH_4 and N_2O for a few decades after 1980 because the concentrations are expressed in terms of the excess over the 1980 concentrations, which are then raised to fractional powers to allow for saturation of the IR absorption bands. Also, the assumed 2%/year growth rate for the CH_4 concentration may be too high. A review by Wuebbles et al. (1984) suggests that the growth rate at present appears to be about 1 ± 1%/year. On the other hand, the estimated concentrations of the chlorofluoromethanes F11 and F12, which were based on constant emissions at present rates, may prove to be too low.

Conclusions

The role of the other greenhouse gases in altering future climate, and especially their future concentrations in the atmosphere, are at this stage subject to very large uncertainties. The conclusions that a greenhouse warming of a few degrees Celsius is inevitable and that cutting back on CO_2 emissions from fossil fuels would not help much, therefore, seem to be somewhat premature. However, the analysis undertaken in the EPA report does point up dramatically the need to get a much better understanding of the potential effect of the other greenhouse gases as well as that of CO_2. It may well be that restrictions on fossil fuel use would not be very effective in limiting future climate change unless simultaneous

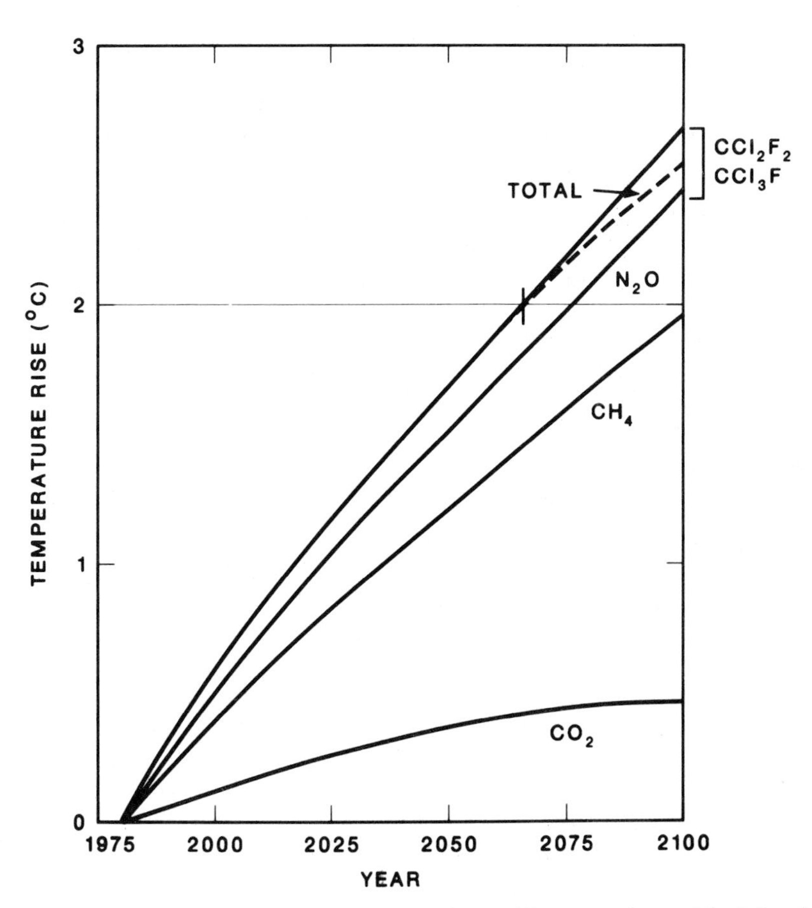

FIGURE 27.8. Average global temperature rise (with ocean thermal lag) for the EPA low-carbon scenario (no coal, no shale oil), showing the contributions of individual greenhouse gases.

measures were taken to restrict emissions of other anthropogenic greenhouse gases.

References

Araj, K. 1982. World energy strategies and the buildup of carbon dioxide: an assessment. Doctoral dissertation, Massachusetts Institute of Technology, Cambridge, Massachusetts.

Baes, C. F. Jr., S. E. Beall, D. W. Lee, and G. Marland. 1980. Options for the collection and disposal of carbon dioxide. Oak Ridge National Laboratory Report ORNL-5657. Oak Ridge, Tennessee.

Flohn, H. 1982. Climate change and an ice-free Arctic Ocean. In W. C. Clark (ed.), Carbon Dioxide Review: 1982, pp. 143–179. Oxford University Press, New York.

IIASA. 1981. Energy in a finite world. International Institute for Applied Systems Analysis, Energy System Program, W. Häfele, Program Leader. Ballinger, Cambridge, Massachusetts.

Keeling, C. D. and R. B. Bacastow. 1977. Impact of industrial gases on climate. In Energy and Climate, pp. 72–95. National Academy of Sciences, Washington, D.C.

Killough, G. G. and W. R. Emanuel. 1981. A comparison of several models of carbon turnover in the ocean. Tellus 33:274–290.

Perry, A. M. 1982. Carbon dioxide production scenarios. In W. C. Clark (ed.), Carbon Dioxide Review: 1982, pp. 335–363. Oxford University Press, New York.

Perry, A. M., K. Araj, W. Fulkerson, D. J. Rose, M. M. Miller, and R. M. Rotty. 1982. Energy supply and demand implications of CO_2. Energy 7:991–1004.

Seidel, S. and D. Keyes. 1983. Can we delay a greenhouse warming? U.S. Environmental Protection Agency report, Washington, D.C.

Siegenthaler, U. and H. Oeschger. 1978. Predicting future atmospheric carbon dioxide levels. Science 199:338–395.

Steinberg, M. 1984. Analysis of concepts for controlling atmospheric carbon dioxide. U.S. Department of Energy Technical Report (TR 007). National Technical Information Service, Springfield, Virginia.

Wuebbles, D. J., M. C. MacCracken, and F. M. Luther. 1984. A proposed reference set of scenarios for radiatively active atmospheric constituents. U.S. Department of Energy Technical Report Series (TRO15). National Technical Information Service, Springfield, Virginia.

Index

Page numbers referring to tables or figures are followed by (T) or (F) respectively.